# The Geology of Stratigraphic Sequences

AF393360

Andrew D. Miall

# The Geology of Stratigraphic Sequences

## Second Edition

Prof. Andrew D. Miall
University of Toronto
Dept. Geology
Toronto ON M5S 1A1
Canada
miall@geology.utoronto.ca

ISBN 978-3-642-42306-2        ISBN 978-3-642-05027-5 (eBook)
DOI 10.1007/978-3-642-05027-5
Springer Heidelberg Dordrecht London New York

© Springer-Verlag Berlin Heidelberg 2010
Softcover re-print of the Hardcover 2nd edition 2010
This work is subject to copyright. All rights are reserved, whether the whole or part of the material is concerned, specifically the rights of translation, reprinting, reuse of illustrations, recitation, broadcasting, reproduction on microfilm or in any other way, and storage in data banks. Duplication of this publication or parts thereof is permitted only under the provisions of the German Copyright Law of September 9, 1965, in its current version, and permission for use must always be obtained from Springer. Violations are liable to prosecution under the German Copyright Law.
The use of general descriptive names, registered names, trademarks, etc. in this publication does not imply, even in the absence of a specific statement, that such names are exempt from the relevant protective laws and regulations and therefore free for general use.

*Cover design*: deblik

Printed on acid-free paper

Springer is part of Springer Science+Business Media (www.springer.com)

# Preface to Second Edition

It has been more than a decade since the appearance of the First Edition of this book. Much progress has been made, but some controversies remain.

The idea that the stratigraphic record could be subdivided into sequences and that these sequences store essential information about basin-forming and subsidence processes remains as powerful an idea as when it was first formulated. L. L. Sloss and P. R. Vail are to be credited with the establishment of the modern era of sequence stratigraphy. The definition and mapping of sequences have become a standard part of the basin-analysis process. Subsurface methods make use of advanced seismic-reflection analysis, with three-dimensional seismic methods, and seismic geomorphology adding important new dimensions to the analysis. Several advanced textbooks have now appeared that deal with the recognition and definition of sequences and their interpretation in terms of the evolution of depositional systems, the recognition and correlation of bounding surfaces, and the interpretation of sequences in terms of changing accommodation and supply. This is not one of these books.

The main purpose of this book remains the same as it was for the first edition, that is, to situate sequences within the broader context of geological processes, and to answer the question: why do sequences form? Geoscientists might thereby be better equipped to extract the maximum information from the record of sequences in a given basin or region.

Central to the concept of the sequence is the deductive model that sequences carry messages about the "pulse of the earth". In the early modern period of sequence stratigraphy (the late 1970s and 1980s) the model of global eustasy was predominant, and it was to offer a critique of that model that provided the impetus for developing the first edition of this book. Model-building has been central to the science of geology from the beginning; it was certainly a preoccupation of such early masters of the science as Lyell, Chamberlin, Barrell, Ulrich and Grabau. A historical evaluation of the contrasting deductive and inductive approaches to geology has been added to this edition of the book, in order to provide a background in methodology and a historical context.

Standard sequence models have become very well described and understood for most depositional settings, and are the subject of several recent texts. Two chapters are provided in this edition in order to outline modern ideas, and to provide a framework of terminology and illustration for the remainder of the book.

A major component of the first edition was devoted to a documentation and illustration of the main types of sequence in the geological record, ranging from those representing hundreds of millions of years to the high-frequency sequences formed by rapid cyclic processes lasting a few tens of thousands of years. Such documentation remains a major component of the book, and has been updated with new examples.

The central core of the first edition was composed of a detailed description and evaluation of the major processes by which sequences are formed. This remains the central focus of the book and has been updated.

Perhaps partly in response to this book, many geoscientists have recognized the complexity of the geological record, have adopted a rigorous inductive approach to their analyses and remain committed to a multi-process interpretation of their rocks. Such an approach can provide a rich array of ideas regarding regional tectonics and basin analysis. However, the original Vail model of global eustasy remains convincing to many, and a powerful guide to interpretation. The practical, theoretical and methodological issues surrounding this still controversial area are the focus of a concluding section of the book. The philosophy and methodology that are the basis for the ongoing work to construct the geological time scale constitute an essential background to this discussion.

Toronto, ON                                                          Andrew D. Miall
September 2009

# Acknowledgements

My own interest in sequence stratigraphy began slowly, as my work on regional basin analysis for the Geological Survey of Canada matured in the late 1970s, and I am grateful to this organization for introducing me to the scope and sweep of large-scale regional analysis. My developing knowledge of basin analysis provided me with a practical view of the subject that induced skepticism. In particular, my work in the Canadian Arctic included attempts to adjudicate debates between various biostratigraphic specialists who could not agree on the dating of certain subsurface sections that I was trying to correlate. My critique of sequence stratigraphy as a chronostratigraphic tool developed from this starting point. A few individuals in GSC discussed the early concepts with me and helped me to realize that something important was going on. Among these Ashton Embry stands out. Later, Jim Dixon's work provided food for thought.

Discussions with the main protagonists of sequence stratigraphy have met with mixed success. I would, however, like to acknowledge these colleagues for contributing to the development of my ideas: Phil Allen, Bert Bally, Chris Barnes, Octavian Catuneanu, Sierd Cloetingh, Jim Coleman, Bill Galloway, Jake Hancock, Makoto Ito, David James, Alan Kendall, Chris Kendall, Dale Leckie, Peter McCabe; Dag Nummedal, Tobi Payenberg, Guy Plint, Henry Posamentier, Brian Pratt, Larry Sloss, John Suter, Peter Vail, John Van Wagoner, Roger Walker, Tony Watts and Yongtai Yang.

This book began life as an in-house report prepared for the exclusive use of the Japan National Petroleum Corporation in 1993. I am grateful to the Corporation for permission to publish their report, and to my employer, the University of Toronto, for providing the time for me both to write the original report and to prepare the material for the revisions incorporated into the final book.

Much of the material in Sect. 14.5 appeared in a contribution to the PaleoScene series in Geoscience Canada in 1994. I am grateful to Darrel Long for stimulating the writing of the paper, and to series editor Godfrey Nowlan and critical reviewers John Armentrout and Terry Poulton for their invaluable comments.

The treatment of geological methods in Chap. 1 and the discussion of stratigraphic paradigms in Chap. 12 could not have been written without the essential role played by my partner and co-investigator, Charlene Miall. Her deep knowledge of scientific methods and her exploration of the sociology of science led us to prepare three papers between 2001 and 2004, from which these parts of the book have been adapted. I thank Bill Berggren and John Van Couvering for giving us the opportunity to publish the very first two articles in the new journal *Stratigraphy*.

The entire manuscript of the first edition was critically read by Brian Pratt and Phil Allen. The second edition was reviewed in detail by Ray Ingersoll. Octavian Catuneanu advised on sequence concepts and models (Chap. 2). Uli Wortmann assisted with my understanding of chemostratigraphy. I am most grateful to these individuals for undertaking these onerous tasks, for their painstaking efforts in completing them, and for their numerous thoughtful and helpful comments. Remaining errors and omissions are, of course, my responsibility.

And once again, I must thank my wife Charlene and my children Christopher and Sarah for their encouragement, love and support.

# Contents

# Structure of the Book

## Part I: The Emergence of Modern Concepts

The first chapter of the book provides some essential historical background to the modern story of sequence stratigraphy. The history of the study of stratigraphy includes two parallel but largely independent strands of research that have been underway since at least the early twentieth century. They are characterized by some profound differences in underlying principles, references and research methods, one research method being essentially empirical and inductive in approach, while another groups of researchers has attempted to develop deductive, theoretical models for understanding Earth history. Chapter 1 is based largely on four papers which explored this history (Miall and Miall, 2001, 2002, 2004; Miall, 2004). It is to be hoped that readers will not skip this chapter, because experience suggests that students of geology do not learn enough about the history, philosophy, or methodology of their science.

The present book is not a treatise on the recognition, mapping and classification of sequences. Modern work on this subject has been provided in such texts as Emery and Myers (1996), Posamentier and Allen (1999, focusing on clastic sequences), Coe (2003), Schlager (2006, focusing on carbonate sequences) and Catuneanu (2006), the last being the most authoritative. The reader is referred to Catuneanu et al. (2009) for a compilation of widely-held concepts concerning sequence classification and nomenclature. Two chapters dealing with sequence models are provided in the present book as a basis for the subsequence discussion. Chapter 2 sets out the main framework of our current ideas about the sedimentology and architecture of sequences, including definitions and explanations of terminology. Chapter 3 describes some of the lesser-known, mostly older, techniques, that have been used to explore the importance concepts of accommodation and sea-level change.

## Part II: The Stratigraphic Framework

The last dozen years of research have revealed that there are five broad types of sequence, in terms of their origins or driving mechanisms. These types were summarized in a review article by Miall (1995), and little has occurred to change this basic summation since that time. The classification is set out and illustrated in Chap. 4, and this is followed by three chapters documenting the evidence in greater detail.

One of the points made in 1995 was that the original "order" hierarchy of sequences, based on their duration, that had been proposed by Vail et al. (1977) is not supportable. The "orders" overlap in time, and the classification does not discriminate between sequences having entirely different driving mechanisms.

The worldwide collection of case studies that provide the basis for the rest of Part II range in scale from the hundreds-of-millions-of-years-long supercontinent cycle, to the climate cycles driven by orbital forcing on a time scale of tens of thousands of years, and the regional cycles of comparable duration that reflect tectonic effects on accommodation and sediment supply. Chapter 5, 6, and 7 subdivide this material on the basis of the temporal scale of the sequences, tens to hundreds of millions of years (Chap. 5), millions of years (Chap. 6) and less than a million years (Chap. 7). The term "episodicity" is used, rather than "periodicity" for this temporal scale because, of the various driving mechanisms that have been recognized as the causes of sequence generation, only orbital forcing may be said to be truly cyclic, in the sense of a predictable, mathematical regularity. Nor should the chapters be taken to indicate a subdivision or classification in terms of driving mechanisms, because these overlap considerably in temporal scale.

## Part III: Mechanisms

Eustatic sea-level change is but one of a suite of mechanisms that govern accommodation, and which act over widely varying time scales, some regional, some global in scope, and commonly, in the geological past, acting simultaneously. These are the subject of Part III of the book. Controversies arising from Vail's original model of global eustasy (Vail et al., 1977) have triggered a vigorous program of research to explore, verify, or challenge this concept, and much new work is reported in this part of the book. A short introductory chapter (Chap. 8) summarizes the various mechanisms, with an emphasis on rates and scales.

The most important forward strides that have been made since the first edition of this book was published in 1997 are in three main areas: Firstly, we are reaching a better understanding of the importance of dynamic topography—the effects of mantle thermal processes on the surface elevation of the Earth's crust. This is the most significant new element in our understanding of $10^7$-year cycles—what have come to be called *Sloss sequences*, after their founder, the "grandfather" of modern sequence stratigraphy (Chap. 9). Secondly, much new detailed regional mapping, incorporating refined methods of local and regional correlation, has provided many new case studies of regional sequence successions and their relationship to local tectonism. This has provided new insights into $10^4$–$10^6$-year-scale tectonism and sedimentation (Chap. 10). Thirdly, an increasing number of case studies of high-frequency sequence stratigraphy of the early Cenozoic and Cretaceous record, is casting new light on the importance of orbital forcing as a sequence-generating mechanism (Chap. 11). Intriguing new arguments from the stable-isotope record are being used to argue that even during the Cretaceous, supposedly a lengthy period when the earth was characterized by a greenhouse climate, there were small, short-lived ice caps on Antarctica, and these may have been the driving mechanism for high-frequency accommodation changes that may therefore have been glacioeustatic in origin (Chap. 11).

Chapter 11 concludes with a short section that explores the ways by which apparently similar high-frequency sequences may be identified as either tectonic in origin or caused by orbital forcing.

## Part IV: Chronostratigraphy and Correlation: An Assessment of the Current Status of "Global Eustasy"

To set the stage for this final part of the book, Chap. 12 presents much of the methodological discussion developed by A. D. Miall and C. E. Miall (2001). In this paper we explored how research methods and the development of technical language set the Vail/Exxon "school" of sequence stratigraphy apart (to understand the social framework within which this happened the reader is referred to C. E. Miall and A. D. Miall, 2002). As suggested by the heading to Sect. 12.5, the global-eustasy paradigm could be said to be a "revolution in trouble."

Global correlation is a key criterion (necessary but insufficient) for the testing of a model of global eustasy and the validity of any global cycle chart. It is argued in this part of the book that the case for global correlation has not yet been made, except in some very specific cases, which are enumerated in Chap. 14. Concepts of deep time, the hierarchy of time, and its expression in the geological record, are discussed in Chap. 13.

Standard methods of stratigraphic correlation and dating are described in some detail in the first part of Chap. 14. These have dramatically improved in recent years, with the use of multiple correlation criteria, the establishment of international working groups to explore specific intervals of geologic time, and, even more importantly, the launching of an authoritative website www.stratigraphy.org, managed by the International Commission on Stratigraphy, which is continuously updated with new facts and references. An updated time scale sets out the methodology and latest results (Gradstein et al., 2004), some of which are already superceded by information posted online.

With these new concepts and methods in hand, the second half of Chap. 14 then examines modern tests of the global eustasy paradigm, focusing on a few key areas, notably New Jersey and New Zealand, where modern data sets provide the basis for new interpretations. This author concludes that a case can still not be made for global eustasy as a primary sedimentary control prior to the Neogene, when the well-documented series of glacioeustatic fluctuations commenced. Overall, however, the new style of detailed stratigraphic research, supported by meticulous chronostratigraphic documentation, is encouraging. A eustatic signal seems to be emerging for the Neogene record—perhaps not surprisingly, given the importance of glacioeustasy during this most recent period of Earth history.

The methodological discussions of Chaps. 1 and 12 are then brought to bear on another emerging paradigm: cyclostratigraphy. While valuable work on the construction of a highly accurate time scale is underway, some similarities to the Vail/Exxon pitfalls are pointed out, and some cautions expressed. This author is skeptical that a reliable time scale can be developed for the pre-Neogene, which must largely rely on "hanging" sections.

The book concludes with a brief discussion of modern research methods and remaining issues and problem (Chap. 15).

# The Emergence of Modern Concepts

Modern sequence stratigraphy began with the work of L. L. Sloss, although it was founded on observations and ideas that developed during the nineteenth century. The subject did not move to the centre of the stratigraphic stage until modern developments in seismic stratigraphy were published by Peter Vail and his Exxon colleagues in the mid-1970s. The purpose of this first part of the book is to set out the historical and theoretical background, and to present current basic sequence concepts, building from the original Exxon models with the incorporation of recent work on sequence architecture and nomenclature. Supporting and parallel work by other authors is referred to, some of the problems with the methods and concepts are briefly touched upon, and some of the major current areas of research are listed. However, the main critical analysis of the methods and results is contained in Parts III and IV of the report.

- It was intimated in the introduction to the symposium on the classification and nomenclature of geologic time divisions published in the last number of this magazine [Journal of Geology] that the ulterior basis of classification and nomenclature must be dependent on the existence or absence of natural divisions resulting from simultaneous phases of action of world-wide extent (Chamberlin, 1898, p. 449).

- Nature vibrates with rhythms, climatic and diastrophic, those finding expression ranging in period from the rapid oscillation of surface waters, recorded in ripple mark, to those long-defended stirrings of the deep titans which have divided earth history into periods and eras (Barrell, 1917).

- Psychologists, anthropologists, and philosophers of science have long recognized the fact that there is a fundamental need in man to explain the nature of his surroundings and to attempt to make order out of randomness . . .. The Western mind does not willingly accept the concept of a truly random universe even though there may be much evidence to support this view. . . .. Science, to an extent matched by no other human endeavor, places a premium upon the ability of the individual to make order out of what appears disordered (Zeller, 1964, p. 631).

- In the late decades of the eighteenth century, geologists were striving toward a stratigraphic taxonomy within which their observations could find organization and structure. Some of the early schemes of classification were largely descriptive and relatively free of the taint of genetic implication . . . By the middle of the 19th century, the gross elements of geochronology and chronostratigraphy, the periods and corresponding systems (Cambrian, Cretaceous, and so on) were widely recognized and accepted . . . transplanting classical chronostratigraphic units to the New World, whether defined by unconformities and other physical changes or by paleontologic changes, was not a simple or wholly satisfying operation. . . . As the twentieth century advanced, stratigraphers were made increasingly aware of the necessity of distinguishing between what are now termed "lithostratigraphic" and "chronostratigraphic" units. . . . In the same decades, the three-dimensional view of stratified rocks provided by subsurface exploration and the practical requirements of subsurface stratigraphic nomenclature in the service of industry and government produced an environment within which nonclassical approaches were fostered and developed (Sloss, 1988a, pp. 1661–1662).

- The interpretation of the stratigraphic record has been greatly stimulated over the past few years by rapid conceptual advances in "sequence stratigraphy," i.e., the attempt to analyze stratigraphic successions in terms of genetically related packages of strata. The value of the concept of a "depositional sequence" lies both in the recognition of a consistent three-dimensional arrangement of facies within the sequence, the facies architecture, and the regional (and inter-regional) correlation of the sequence boundaries. It has also been argued that many sequence boundaries are correlatable globally, and that they reflect periods of sea-level lowstand, i.e. "sequence-boundaries are subaerial erosion surfaces." (Nummedal, 1987, p. iii)

# Chapter 1

# Historical and Methodological Background

## Contents



## 1.1 Introduction

Many observations and hypotheses regarding regional and global changes in sea level, and the ordering of stratigraphic successions into predictable packages, were made early in the twentieth century. The word *eustatic* was first proposed for global changes of sea level by Suess (1885; English translation: 1906). He recognized that sea-level change could be determined by plotting the extent of marine transgression over continental areas, and by studying the changes in water depths indicated by successions of sediments and faunas. Observations, models and hypotheses regarding regional stratigraphy and the processes driving subsidence and sedimentation were made by such scientists as Lyell, Grabau, Chamberlin, Blackwelder, Ulrich, and Barrell. The North-American mid-continent cyclothems are a particularly interesting case of cyclic sedimentation. Workers such as Wanless, Weller, Moore and Shepard began studying these in the 1930s. The reader is referred to Dott (1992a) and the memoir of which this paper is a part, for fascinating descriptions of the early controversies, many of them having a very modern flavour.

Modern work in the area of sequence stratigraphy evolved from the research of L. L. Sloss, W. C. Krumbein, and E. C. Dapples in the 1940s and 1950s, beginning with an important address they made to a symposium on *Sedimentary facies in geologic history* in 1949 (Sloss et al., 1949). H. E. Wheeler also made notable contributions during this period, particularly in the study of time as preserved in stratigraphic sequences.

Ross (1991) pointed out that all the essential ideas that form the basis for modern sequence stratigraphy were in place by the 1960s. The concept of repetitive episodes of deposition separated by regional unconformities was developed by Wheeler and Sloss in the 1940s and 1950s. The concept of the "ideal" or "model" sequence had been developed for the mid-continent cyclothems in the 1930s. The hypothesis of glacioeustasy was also widely discussed at that time. Van Siclen (1958) provided a diagram of the stratigraphic response of a continental margin to sea-level change and variations in sediment supply that is very similar to present-day sequence models. An important symposium on cyclic sedimentation convened by the Kansas Geological Survey marks a major milestone in the progress of research in this area (Merriam, 1964); yet the subject did not "catch on." There are probably two main reasons for this. Firstly, during the 1960s and 1970s sedimentologists were preoccupied mainly by


A.D. Miall, *The Geology of Stratigraphic Sequences,* 2nd ed.,
DOI 10.1007/978-3-642-05027-5_1, © Springer-Verlag Berlin Heidelberg 2010

autogenic processes and the process-response model, and by the implications of plate tectonics for large-scale basin architecture. Secondly, geologists lacked the right kind of data. It was not until the advent of high-quality seismic-reflection data in the 1970s, and the development of the interpretive skills required to exploit these data, that the value and importance of sequence concepts became widely appreciated. Shell and Exxon were both actively developing these skills in their research and development laboratories in the 1970s. Peter Vail, working with Exxon, was the first to present his ideas in public, at the 1975 and 1976 annual meetings of the American Association of Petroleum Geologists (Vail, 1975; Vail and Wilbur, 1966). This was the beginning of a major revolution in the science of stratigraphy.

As the discussion in this chapter demonstrates, there have been two strong but sometimes conflicting methodological approaches to the science of geology: (1) inductive science, or empiricism, and (2) deductive science, the building and testing of models. Peter Vail was a model builder, but his approaches to stratigraphy have very deep roots in the earth sciences. Some of his concepts were anticipated by others in the early 1970s, the 1950s, and in fact as far back as the first two decades of the twentieth century. It is instructive to follow this history. But first, it is necessary to set out the nature of the scientific method.

## 1.2 Methods in Geology

Geology is historically an empirical science, firmly based on field data. Hallam (1989, p. 221) has claimed that "Geologists tend to be staunchly empirical in their approach, to respect careful observation and distrust broad generalization; they are too well aware of nature's complexity." However, interpretive models, including the modern trend towards numerical modeling, have become increasingly important in recent years. As discussed by Miall and Miall (2001), specifically with reference to sequence stratigraphy, there are two broad approaches that can be taken to geological research. Most of this section is based on that study.

The empirical approach to geology, including the building of models, is *inductive* science, whereas the use of a model to guide further research is to employ the *deductive* approach. This methodological difference was clearly spelled out for geologists by

Johnson (1933). Frodeman (1995) recently reviewed the work of the German philosopher Heidegger, arguing that the practice of the science of geology illustrates a process termed the *hermeneutic circle*, in which induction and deduction supposedly follow each other in an iterative process of observation, generalization and theorizing (*induction*), followed by the construction of hypotheses and the seeking of new observations in order to test and abandon or refine the theory (*deduction*). Ideally, this is a continuous and circular process (Fig. 1.1), but it has been argued elsewhere (Miall and Miall, 2001; and see below) that at present there are separate groups of stratigraphic researchers that are following these two different methodological approaches in partial isolation from each other. One of the purposes of this chapter is to argue that this dichotomy has deep historical roots; that from the mid-nineteenth century to the present, the inductive and deductive approaches to the science of stratigraphy have largely been followed by different groups of researchers having different objectives, and that through much of the history of the science, the groups have had little to do with each other.

Since modern stratigraphic studies began in the late eighteenth century a central theme of stratigraphic research has been the empirical construction of a vast data base of descriptive stratigraphy, focusing on the occurrence and relative arrangements of formations and their contained fossils. This data base now constitutes what has come to be called the chronostratigraphic time scale. In recent years, methods of determining the ages of beds by other means, such as by radiometric dating, magnetostratigraphy and chemostratigraphy have added depth to this data base.

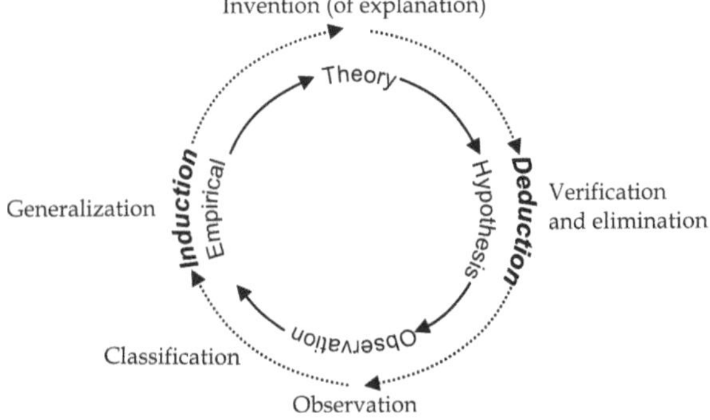


**Fig. 1.1** The hermeneutic circle, based on Wallace (1969). The terms around the periphery are those of the five "stages of analysis" of Johnson (1933)

As we show here, research into the preserved record of deep geological time has grown into an enormously complex, largely inductive science carried out mainly in the academic realm. From this, a descriptive (inductive) classification of Earth history has been built, consisting of the standard eras, periods, epochs, etc.

At various times, deductive models of Earth history have been proposed that have had varying levels of success in contributing to our understanding of Earth's evolution. There have also been many attempts to develop deductive models of stratigraphic processes, including the *cyclothem* model of the 1930s, and modern *facies models* and *sequence stratigraphy*. Ideas about the tectonic setting of sedimentary basins have also included several bold attempts at model building, including the pre-plate tectonic *geosyncline theory* of Kay (1951), the modern *petrotectonic assemblage* concepts of Dickinson (1980, 1981), the various geophysically-based basin models of McKenzie (1978), Beaumont (1981) and many later workers, and the *actualistic plate-tectonic models* of Dickinson (1974), Miall (1984) and Ingersoll (1988).

The concepts of sequence stratigraphy that evolved from seismic-stratigraphy in the 1970s constitute the basis for the most recent and most elaborate attempts to develop deductive stratigraphic models. These included the Exxon global cycle chart (Vail et al., 1977; Haq et al., 1987, 1988a), which, if it had proved to be a successful explanation of the stratigraphic record, could potentially have become the dominant paradigm, entirely replacing the old inductive classification of geologic time, and largely supplanting the complex method with which it was being constructed. However, the two distinct intellectual approaches have resulted in the development of two conflicting and competing paradigms which are currently vying for the attention of practicing earth scientists (Miall and Miall, 2001, 2002). It is argued later in this book (Sect. 14.7) that current research in the field of cyclostratigraphy may be following a similar pattern of development.

The history of stratigraphy since the end of the eighteenth century has encompassed the following broad themes:

(1) Recognition of the concept of stratigraphic order and its relationship to Earth history, and the growth from this of an empirical, descriptive stratigraphy based on sedimentary rocks and their contained fossils.

(2) The emergence of the concept of "facies" based on the recognition that rocks may vary in character from place to place depending on depositional processes and environments. This was one of the first deductive models developed to facilitate geological interpretation.

(3) Recognition that rocks are not necessarily an accurate or complete record of geologic time, because of facies changes and missing section, and the erection of separate units for "time" and for "rocks".

(4) Development of a multidisciplinary, empirical approach to the measurement and documentation of geologic time, an unfinished science still actively being pursued.

(5) Attempts at different times to recognize patterns and themes in the stratigraphic succession and to interpret Earth processes from such patterns. Facies models and sequence stratigraphy are amongst the main products of this effort.

(6) Attempts to extract regional or global signals from the stratigraphic record and to use them to build an alternative measure of geologic time, based on an assumed "pulse of the Earth."

These themes may be further generalized into a descriptive, inductive approach to the science (themes 1–4), which may be categorized as the *empirical paradigm* of stratigraphy, and a distinctly different, interpretive approach to the subject (themes 5 and 6), that may be termed the *model-building paradigm*. The model of global eustasy as applied to sequence stratigraphy is discussed below. The use of cyclostratigraphy as a potential basis for a refined geologic time scale is discussed in Sect. 14.7.

The purpose of the present chapter is to summarize the parallel development of these two broad approaches to stratigraphy, The discussion is not intended to be historically complete; there is a substantial literature that addresses the history of stratigraphy. By establishing the history of stratigraphy, the body of ideas from which the modern controversy over sequence stratigraphy may be better understood. The mainstream of stratigraphic research has been the development of an ever more precise and comprehensive chronostratigraphic time scale based on the accumulation and integration of all types of chronostratigraphic data. This paradigm is essentially one of meticulous empiricism which makes no presuppositions about the history of the Earth or the evolution of life or of geological events in general. The Vail/Exxon sequence models exemplify a quite different paradigm, but one that reflects an equally long intellectual history. They are the latest manifestation of the idea that there is some kind of "pulse of the earth" that is amenable to elegant classification and broad generalization.

These two strands of research correspond to two of the modes or "cognitive styles" of research in geology that were described by Rudwick (1982). Descriptive stratigraphy and the development of the geological time scale corresponds to the "concrete" style of Rudwick (1982, p. 224), who cites several of the great nineteenth-century founders of stratigraphic geology as examples (e.g., William Smith, Sedgwick, Murchison). These individuals were primarily concerned with documenting and classifying the detail of stratigraphic order. In contrast, the "abstract" style of research includes individuals such as Hutton, Lyell, Phillips and Darwin, who sought underlying principles in order to understand Earth history.

### 1.2.1 *The Significance of Sequence Stratigraphy*

The business of research is new ideas. Few things, however, are more unpopular with researchers than truly new ideas. (Vail, 1992, p. 83)

Emerging in the 1970s, *seismic stratigraphy,* as developed by Peter Vail and his coworkers at Exxon Corporation (Vail, 1975; Vail et al., 1977), brought about a revolution in the study of stratigraphy (Miall and Miall, 2001). It re-energized a discipline, 200 years in the making, offering new theoretical possibilities for knowledge of Earth history and fundamental Earth processes. It also provided a potentially powerful new analytical and correlation tool for use by practicing basin analysts, especially those in the petroleum industry. Amongst the new concepts and methods constituting seismic stratigraphy were the following:

(1) The use of a new form of data—reflection-seismic records—for the generation of stratigraphic information;

(2) Demonstration that the new data could provide images of large swaths of a basin at once;

(3) Demonstration of the complex internal architecture of basin fills;

(4) Demonstration that stratigraphic successions consist of "sequences," which are packages of conformable strata bounded by regional unconformities;

(5) The proposal that the bounding unconformities are mostly global in extent and were generated by repeated eustatic changes in sea level. We refer here to this hypothesis as the *global eustasy model.*

(6) The proposal that Earth's stratigraphic record, consisting of a global record of sequences, could be characterized by a *global cycle chart*, which could be used as a universal correlation template.

Within the framework of scientific revolutions established by Kuhn (1962, 1996), seismic stratigraphy could be said to constitute a new paradigm. Kuhn (1962, 1996) explained "paradigms" as universally recognized scientific achievements that for a time provide model problems and solutions to a community of practitioners. These new developments have proved to be of profound importance to the science of stratigraphy. Building on points 1–4, stratigraphers developed an entirely new way of practicing their craft, including application of the concepts to outcrop and subsurface well data, in a new science termed *sequence stratigraphy* (Posamentier et al., 1988; Posamentier and Vail, 1988; Van Wagoner et al., 1990). In this science, the architecture and predictability of sequences are amongst its most valuable components.

In the early days of seismic stratigraphy, in the late 1970s and early 1980s, the global cycle chart was considered an inseparable part of the new method. The 1970s were a period when globalization was a theme in many facets of scientific and societal development. Marshall McLuhan (1962) was teaching us about a "global village" of humans, linked by the mass media; economists were beginning to argue for increasing globalization of trade and commerce; and, in the earth sciences, the paradigm of plate tectonics was having an enormous impact on our understanding of earth history (e.g., Dewey, 1980; Dickinson 1974). Indeed, the global reach of seismic stratigraphy was one of its most persuasive features.

Miall and Miall (2002) examined social factors shaping the development, dissemination, and initial validation of seismic stratigraphy and the social organizations in which these occurred. The main objective of this chapter (based on Miall and Miall, 2001) is to examine the methods that underlie the global-eustasy model, and its precursors in stratigraphic modeling, as far back as the work of Charles Lyell in the mid-nineteenth century. Later sections deal with the period of the initial enthusiastic reception of seismic stratigraphy by the scientific community in the late 1970s, to a period of increasing doubt about the global-eustasy model that extends to the present. Miall and Miall (2002) argued that beyond the undoubted new facts and invigorating new ideas that seismic stratigraphy brought to geology, the popularity of the new paradigm within the geological community owed much to human, "social" factors. The analysis draws extensively on Kuhn's work regarding the development and acceptance of new *paradigms* in science. The paradigm model has been usefully applied to the study of several areas of research in the earth sciences, including the development of turbidite concepts (Walker, 1973), methods of grain-size analysis in sedimentology (Law, 1980), the acceptance of plate-tectonic theories about the earth's crust (Stewart, 1986), and the controversy surrounding the abiogenic origins of oil (Cole, 1996).

It is also argued here that because of controversies surrounding ideas about the origins of sequences, sequence stratigraphy has now evolved into two distinct, competing paradigms. During the 1980s, a series of anomalies and conceptual problems about the global-eustasy model emerged. It is argued that these have not been fully addressed by proponents of the global-eustasy model, many of whom continue to use this model as a central theme in their stratigraphic work, despite a growing controversy that surrounds it. In subsequent chapters an examination of the nature of stratigraphic data and its importance in the process of validating Vail's new ideas is carried out (Part IV).

The chapter is offered as a contribution to the study of Dott's (1998) question: "What is unique about geological reasoning?" As an outcome of the analysis it is to be hoped that geologists will be alerted to the bias that preconceptions and group processes can bring to observations and interpretations in the geological record.

### 1.2.2 Data and Argument in Geology

The traditional view of science, as described by the tenets of analytical philosophy (Frodeman, 1995),

is that science proceeds from careful, objective observation and replication of data, to hypotheses and theory through the workings of the scientific method. According to Popper (1959), scientists engage in the practice of proposing and falsifying testable hypotheses, based on dispassionate experimentation. There is some basis for arguing that this is what actually happens in the so-called "hard" sciences, where ideas may be tested by carefully designed experiments. Even here, however, human factors can intervene, as the controversy surrounding "cold fusion" a few years ago attests. The proponents of a new model of cold fusion were able to convince themselves that their experiments had provided evidence for it, but attempts at replication by other observers demonstrated the falsity of the original claims, and the new hypothesis was quickly discarded (Peat, 1989).

It is not so simple in geology, where we are attempting to understand a past that cannot be replicated by experiment. As Dott (1998) pointed out, there is much that we can replicate, such as the chemistry of mineral formation, or flume experiments to model bedform generation, but we cannot recreate the past in its complex entirety. The new science of numerical modeling is providing us with powerful new techniques for simulating complex processes, such as stratigraphic accumulation under conditions of varying rates of basin subsidence and climate change, but these are just simple models, not complete replication experiments, so geologists have to search for what Frodeman (1995) called "explanations that work." Much of geological practice constitutes what Dott (1998) referred to as *synthetic science*.

Our views of "what works" have changed dramatically as the science has evolved (Fig. 1.2). Consider the science of stratigraphy, for example. Until the 1950s, stratigraphic practice consisted primarily of what we now term *lithostratigraphy*, the mapping, correlating and naming of formations based on their lithologic similarity and their fossil content. In the 1960s, the revolution in process sedimentology led to the emergence of a new science, *facies analysis*, and a focus on what came to be termed "autogenic" processes, such as the meandering of a river channel or the progradation of a delta. Most stratigraphic complexity was interpreted in facies terms, and the science witnessed an explosion of research on process-response models, otherwise termed *facies models*. The revolution in seismic stratigraphy in the late 1970s changed the

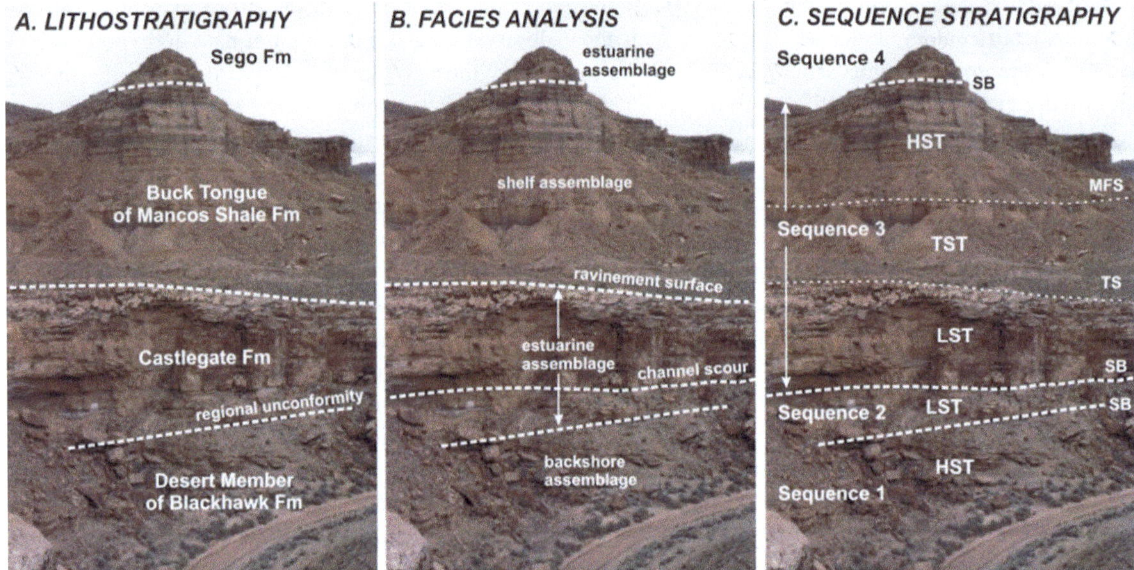



**Fig. 1.2** An outcrop in the Book Cliffs of Utah (Tusher Canyon, near Green River) interpreted using three successive deductive models of stratigraphic interpretation. (**a**) based on Fouch et al. (1983); (**b**), (**c**) based on van Wagoner et al. (1990) and Yoshida (2000). Sequence-stratigraphic terminology: SB = sequence boundary, TS = transgressive surface, MFS = maximum flooding surface, LST, TST, HST = lowstand, transgressive and highstand systems tracts

face of stratigraphy yet again, with a new focus on large-scale basin architecture and regional and global basinal controls.

In the course of 20 years, therefore, the kinds of data geologists looked for in the rocks, and the "explanations that worked" in explaining them, underwent two wholesale changes. The rocks did not change, but the "objective" facts that geologists extracted from them did. This attests to our improved understanding of our own subject, but the details of the evolution of this science, as with any other, are also influenced by human factors. The fact that scientific journals include "Commentary" or "Discussion" sections attests to the fact that apparently dispassionate observation can, nonetheless, lead to different interpretations and to controversy. Scientists accept this, while they remain reluctant to accept that human factors play an important role in scientific development. How important is the reputation of the scientist in furthering a new idea? How important is "fashion"? Those of us who were active in the 1970s may remember a time when *turbidite* models were first popularized, and all thinly bedded lithic arenites tended to be reinterpreted as turbidites, and later, when the new *sabkha* model became the explanation for all evaporites.

As argued by Miall and Miall (2001), social scientists have helped to show that science, like other human endeavors, is a human activity, subject to the same social influences as non-scientific endeavors (Kuhn, 1962, 1996; Cole, 1992; Barnes et al., 1996). Mulkay (1979) for example, has concluded that physical reality constrains, but does not uniquely determine, the conclusions of scientists. Much depends on the preconceptions of the investigator. Further, the evolution of a body of ideas depends on the social and work conditions in which the practitioners find themselves (e.g., Barnes et al., 1996). Particular data sets and models are not necessarily tidbits of universal truth to be teased from the ether by dispassionate scientists, but are very much a product of the social conditions within which the scientific work is carried out. Even primary scientific observations are not value-free, but have a context. This context may include an array of hypotheses; these may be made using special observation methods, and these methods may reflect assumptions or simplifications of their own (Barnes et al., 1996, p. 2). Barnes et al. (1996) have made a distinction between a simple "observation" and that which is made and reported as part of an hypothesis-testing exercise. The latter they have termed an "observation report." They have suggested that "our thoughts could influence our perceptions as well as our perceptions influence our thoughts." Indeed, it has been argued that there really is no such thing as a "pure" observation. There may be many assumption sets in a given field. For example, as noted above, the same stratigraphic section may have been described in at least three different ways since the 1950s. As Kuhn (1962. p. 129) has noted, "one and the same operation, when it attaches to nature through a different paradigm, can become an index to a quite different aspect of nature's regularity." This is not intended as a criticism or negation of the scientific method, but as an attempt to throw light on the very human processes involved in the action of "doing science", so that we might understand it better.

Such an analysis of the very foundations of the scientific method, which points to and incorporates the various contextual characteristics of the work, is a dramatically different way of viewing science than the analytical approach. Based on the philosophical ideas of Heidegger, this approach is termed *hermeneutics*, and the back-and-forth thought processes between observation and interpretation are conceptualized as the *hermeneutic circle* (Heidegger, 1927, 1962; Frodeman, 1995; Barnes et al., 1996). Frodeman (1995), in a discussion of the philosophy of science, noted that Heidegger identified three types of "prejudgement" or "forestructures" that scientists bring to each situation: preconceptions; ideas or presumed goals, and a sense of "what will count as an answer" (Frodeman, 1995, p. 964); and the particular tools and methods of the research. Hermeneutics argues that our original goals and assumptions result in certain facts being discovered rather than others, which in turn lead to new avenues of research and sets of facts. Rudwick (1996) argued that, beyond the basic data directly measurable by technological means, such as the density of pyrite, or the drilling depth of the Toronto Formation, "there are no theory-free facts in geology." While extreme, this viewpoint provides a useful counterweight to the power of a newly popular theory.

One of the themes followed through this book is an attempt to show how "human factors" influenced, and still influence, the progress of a single geological model, that of global eustasy and its implications for sequence stratigraphy.

### 1.2.3  *The Hermeneutic Circle and the Emergence of Sequence Stratigraphy*

Sequence concepts, as first developed by Sloss et al. (1949) and Sloss (1963), constituted a resurrection of the deductive approach to the documentation and interpretation of the stratigraphic record (one with a long prior tradition, as discussed below), but sequence stratigraphy had not become the prevailing stratigraphic method when approaches based on *seismic stratigraphy* were first introduced in the mid 1970s. The new work that emerged from Exxon in the 1970s brought about what Kuhn would term a "crisis" in sedimentary geology; that is, the introduction of a new method and a body of ideas that could not be reconciled with existing paradigms. As Kuhn (1996, p. 181) has stated:

> Crises need not be generated by the work of the community that experiences them and that sometimes undergoes revolution as a result. New instruments like the electron microscope or new laws like Maxwell's may develop in one specialty and their assimilation creates crises in another.

The development of high-quality reflection-seismic data and the new methods of stratigraphic interpretation based on these data (e.g. the focus on stratigraphic surfaces and terminations, and on three-dimensional architecture) clearly constituted a new "instrument" in the specialty of petroleum geophysics. These data then affected the field of conventional, largely university- and state-survey-based stratigraphy. To this extent, Vail's work created "crisis", which his work was, in turn, designed to resolve (Miall and Miall, 2001). Indeed, the emergence of sequence stratigraphy resulted in two additional challenges to what Kuhn (1962) would have referred to as the "normal science" of stratigraphic interpretation:

(1) The proposal that the global cycle chart was a superior standard of geologic time to that based on conventional chronostratigraphy. For example, Vail et al. (1977, p. 96) stated: "One of the greatest potential applications of the global cycle chart is its use as an instrument of geochronology." Vail and Todd (1981, p. 217) stated, with regard to correlations in the North Sea Basin: "several unconformities cannot be dated precisely; in these cases

their ages are based on our global cycle chart, with age assignments based on the basis of a best fit with the data." They proceeded to revise biostratigraphic ages based on the correlations suggested by their chart.

(2) The attribution of all changes in sea level to two favored eustatic mechanisms—eustasy (typically glacioeustasy for high-frequency sequences), and changes in ocean-basin volumes (Vail et al., 1977, pp. 92–94)—and the assertion that other regional processes, including tectonism and changes in sediment supply, affected only the amplitude but not the timing of sea-level changes (Vail et al., 1991, p. 619).

Unlike that which occurred during the development and acceptance of plate tectonics (Stewart, 1986), there was no "crisis" in normal science, in the sense originally intended by Kuhn (1962). Other geologists were not dissatisfied with the body of "normal" stratigraphic science. Conventional chronostratigraphic methods, based on biostratigraphy, radiometric dating, chemostratigraphy, etc., were, and remain, the primary means for determining geologic age (e.g., Harland et al., 1990; Holland, 1998; Gradstein et al., 2004) and, to the extent that there had been no widespread search for global mechanisms for stratigraphic processes, most were comfortable with the prevailing views regarding the complexity of geological processes. Nonetheless, it could have been expected that a geological community, alerted to the power and influence of the new global tectonics (plate tectonics) in the mid- to late 1970s, would view a new global model with interest and excitement.

Vail himself attempted to describe his own scientific procedures. In his memoir on the evolution of sequence stratigraphy (Vail, 1992), he discussed how research creativity could be optimized through the application of a set of procedures which seemingly reflected the hermeneutic approach. These included what we could term *forestructures*, such as the establishment of a thematic research program with clearly defined goals for the overall research, and the definition of concepts that would drive the thematic research—*driving concepts*. However, Vail argued that it was important to define driving concepts because it would then be possible "to acknowledge and nurture ideas that challenge, and may prove superior to the existing driving concepts" (Vail, 1992, p. 84). He also argued (on the same page) that,

truly new, worthwhile ideas based on competing driving concepts may not be accepted within the framework of thematic research. These competing driving concepts are commonly ignored or put aside because of the priority of other work.

Despite this awareness that driving concepts in thematic research can direct attention away from anomalous observations and hypotheses, and based on his own recollections of his time at Exxon, Vail appears to have established, within his own seismic research group, a structure likely to yield the kind of hermeneutic circle which did not allow for competing driving concepts, as argued in the remainder of this section. According to Vail (1992, p. 89), two organizing principles appeared to guide the research. First, a working environment was fostered in which problems were accurately defined and interrelated through the establishment of a thematic research program informed by the driving concepts mentioned above. These concepts directed the group's attention to the problems to be solved, the methods to be used in obtaining solutions, and the types of phenomena to be studied (from Vail, 1992, p. 87):

What this driving concept showed was that seismic sections are a high-resolution tool for determining chronostratigraphy-the time lines in rocks. This was a "eureka" at that time. We had found the tool and developed the methodology to make regional chronostratigraphic correlations and to put stratigraphy into a geologic time framework for mapping and the understanding of paleogeography. ... The fact that seismic reflections follow time lines is the second basic driving concept

As Sloss (1988a, p. 1661) put it,

the sequence concept was alive and well in a research facility of ... Exxon. here, Peter Vail and a cohort of preconditioned colleagues seized upon the stratigraphic imagery made available by multichannel, digitally recorded, and computer-massaged reflection seismography to establish the discipline of seismic stratigraphy.

Second, an environment was created where both teamwork and individual responsibility co-existed. As Vail (1992, p. 89) has observed, in his article on the evolution of seismic stratigraphy and the global sea-level curve,

As a group, we developed an overall plan. We would then try to identify the person who was most interested and knowledgeable for each task, and then endeavor to give each person a maximum amount of responsibility for his or her project area... We tried to develop a situation wherein each researcher had a clear-cut area of responsibility, but we made sure it overlapped with as many other areas as possible. This ensured good communication, because each person was vitally interested in what the others were doing.

What this also ensured, as Law and Lodge (1984) might argue, was that each member of the group had an interest and an investment in the success of seismic stratigraphy and the global eustasy model. The work of the Exxon team, as described by Vail, is a good example of the "socially constructed" nature of group scientific work. As Clarke and Gerson (1990) have argued, scientific theories, findings and facts are socially constructed (although, of course, based on observation and measurement). They note that to solve research problems, scientists will make commitments to specific theories and methods, to other scientists, to research sponsors, and to various organizations. Clark and Gerson (1990, p. 184) further directed attention to the importance of "... structural conditions of work, and the concrete processes of actually collating different lines of evidence." These, they argued, are critical to the emergence and maintenance of belief in the results of a particular line of research in the face of what they term the "buried uncertainties" of the research.

It must be emphasized that the Exxon work is being used here to illustrate common themes in scientific research, not in order to criticize the method, but in order to help focus in on the importance of group dynamics as the research evolved and came under scrutiny by the wider geological community.

It is instructive to situate Vail's methodological approach in a historical context. In the next sections the development of two parallel traditions in stratigraphy is followed from their origins in the nineteenth century.

### 1.2.4 Paradigms and Exemplars

As Kuhn (1962, p. 121) has noted, "given a paradigm, interpretation of data is central to the enterprise that explores it, but that interpretive enterprise ... can only articulate a paradigm, not correct it." Kuhn (1962, pp. 23–24) has further argued that,

The success of a paradigm ... is at the start largely a promise of success discoverable in selected and still incomplete examples. Normal science consists in the actualization of the promise, an actualization achieved by extending the knowledge of those facts that the paradigm displays as particularly revealing, by increasing the extent of the match between those facts and the paradigm's predictions, and by further articulation of the paradigm itself.

The fundamental building blocks of new science, he goes on to state, are "solved problems." These he referred to as exemplars, because they are used as examples for teaching and in order to guide further research (See also, Barnes et al., 1996, pp. 101–109).

In an historical science like geology, it is difficult to arrive at "truth" or "proof" by the normal process of experimentation and replication (though not always impossible, as demonstrated by Dott, 1998). It could be argued that exemplars are "explanations that work;" what Frodeman (1995) has referred to as a type of understanding having "narrative logic." Typical examples of exemplars in this case are: "In the North Sea, 15 of 17 unconformities identified on the basis of well results and seismic data matched Exxon's global sea-level curve." (P. R. Vail at a meeting in Woods Hole Massachusetts in April as reported by Kerr, 1980, p. 484). And again: "We can correlate ten of them [unconformities] perfectly with the Vail curve, five correlate pretty well, and the others still have a few problems." (Tom Loutit and James Kennett commenting on their work in New Zealand as reported by Kerr, 1980, p. 485). In a follow-up report on the Vail method 4 years later, Kerr (1984) reported several more of these "X out of Y" comparisons, where Y is slightly greater than but never equal to X.

As we documented elsewhere (Miall and Miall, 2002), the publication of exemplars of this type helped to rapidly convince the geological fraternity of the power and importance of the new global eustasy model. This, despite the lack of supporting documentation in Vail et al. (1977), including, for example, interpreted seismic lines or biostratigraphically documented subsurface sections. Missing, therefore, are the "observations" that are supposedly so critical to the scientific method (Barnes et al., 1996). At best, we have what Barnes et al. (1996) might have called "a report that such observations exist". Given the lack of published documentation, there was no opportunity for outsiders to influence the workings of the hermeneutic circle on the development of the global eustasy model.

In their discussion of the development of scientific knowledge in general, Barnes et al. (1996) discussed how an important experiment in physics (determination of the charge of the electron) was gradually refined by repeated experimentation, documented by extensive laboratory note-taking, and how the results of a rival set of experiments that generated different results were eventually discarded because the results were gradually shown to be inconsistent with the results of new work. Similarly, at some point, Vail and his coworkers decided that a given set of sea-level events represented global eustatic signals, and the first global cycle chart (Vail et al., 1977) was the result. The global eustasy hypothesis seems to have begun to work a powerful influence on the selection and collation of additional data. But how were these events selected? How did Vail know that these were the "right" ones? How were other events discarded as the result of "local tectonics", or as having been "offset" by biostratigraphic imprecision? When consideration is given then to the "solved problems" or exemplars of the global cycle chart, the method of documentation and cross-checking and virtually all of the primary data are missing. For example, the reader's attention is directed to the only published diagram in AAPG Memoir 26 (Vail et al., 1977, Fig. 5 on p. 90) that illustrates the synthesis of individual cycle charts into the global average. There are several events appearing in the average curve that are not well represented in the individual charts, and vice versa (see also discussion of this subject in Chap. 12). These points have not been explained by Vail or his coworkers.

In his discussion of the consensus model of science, Kuhn (1962) has observed that the real function of experiment is not the testing of theories. He showed that commonly, theories are accepted before there is significant empirical evidence to support them. Results which confirm already accepted theories are paid attention to, while disconfirming results are ignored. Knowing what results should be expected from research, scientists may be able to devise techniques that obtain them (Kuhn, 1977; Cole, 1992, pp. 7–8). The exercise of correlating new stratigraphic sections to the global cycle chart entails the dangers of self-fulfilling prophecy. As noted by Kuhn (1962, pp. 80, 84):

> The bulk of scientific practice is thus a complex and consuming mopping-up operation that consolidates the ground made available by the most recent theoretical breakthrough and that provides essential preparation for the breakthrough to follow. In such mopping-up operations, measurement has its overwhelmingly most common scientific function. ... Often scientists cannot get numbers that compare well with theory until they know what numbers they should be making nature yield.

The lack of published experimental documentation and detailed analysis in support of the global-eustasy model makes it difficult for scientists in general to evaluate the importance of these human processes that Kuhn described.

Why did geologists eagerly embrace the global-eustasy model despite the lack of published data that they could see for themselves? We suggest that it was (1) because of the current willingness to accept global explanations of earth processes in light of the new plate-tectonics paradigm; (2) the ideas offered a utilitarian application—the global correlation template—of apparently considerable potential use in the exploration business; (3) because of the assumed authority of corporate geophysics, or what we term "the Exxon Factor" (Miall and Miall, 2002); and (4) because working petroleum geologists had no investment in the complex, confusing and "academic" science of conventional chronostratigraphy. Dott (1992b) suggested an additional factor, the "innate psychological appeal of order and simplicity" of a pattern of cyclicity based on Milankovitch periodicity.

Use of the global cycle chart appeared to offer a simple global solution to the problem of stratigraphic correlation so, perhaps, it was not surprising that working geologists eagerly adopted it. Indeed, the non-availability of a data base, with all the messiness, incompleteness and inconsistency such as normally characterizes stratigraphic successions from diverse areas characterized by different tectonic histories, undoubtedly made it easier to accept the global cycle chart as is. To cite Law's (1980, p. 13) summation of Mannheim's philosophy, the application of the global cycle chart displayed

> the attributes of Mannheim's <u>natural law</u> style of thought – they are atomistic, generally quantitative, emphasize the routine nature of scientific practice, seek or utilize general laws, stress continuity, and are in general reductionist.

The techniques for identifying sea-level events in seismic records and (later) stratigraphic sections, and equating them to existing events in the global cycle chart, became a routine operation ideally suited to petroleum-exploration work. The alternative, holistic approach to stratigraphy is the traditional one of correlation by biostratigraphy and the erection of stratotypes with no built-in assumptions of global events. It was this approach that Vail's chart seemed destined to replace. While Vail and his colleagues have repeatedly reaffirmed the importance of biostratigraphic dating and correlation in the testing of the global cycle chart, the emphasis in much of the work of this group is on the superiority of the global cycle chart as a method of correlation and stratigraphic standardization (Vail and Todd, 1981, p. 230; Vail et al., 1984, p. 143; Baum and Vail, 1988, p. 322; Vail et al., 1991, pp. 622, 659; see Chap. 12).

In the early 1980s, however, problems with the global eustasy model began to emerge as alternative ideas about sequence generation began to appear, and doubts emerged about the accuracy and precision of chronostratigraphic methods available to test global correlations. These are discussed in Chap. 12, where the global cycle chart is examined in detail.

## 1.3 The Development of Descriptive Stratigraphy

### 1.3.1 The Growth of Modern Concepts

The development of the science of descriptive stratigraphy is described by Hancock (1977), Conkin and Conkin (1984), and Berry (1987). Earlier discussions that include much valuable historical detail include those by Teichert (1958) and Monty (1968). The following summary is intended only to emphasize the evolution in methodologies that took place from the latest eighteenth century until they stabilized during the mid-twentieth century.

Stratigraphy is founded on the ideas of Richard Hooke and Nicolaus Steno, physician to the Grand Duke of Tuscany, although Vai (2007) suggested that several key points were anticipated by Leonardo da Vinci in about 1500. The *Law of Superposition* was enunciated by Nicolaus Steno in his work *Prodromus*, published in 1669. This law, simply stated, is that in any succession of strata, the oldest and first formed must be at the bottom, with successively younger strata arranged in order above. As described by Berry

(1987), several rock successions were described during the eighteenth century, primarily because of their importance to mining operations, but no fundamental principles emerged from this work until the four-fold subdivision of the Earth's crust was proposed by Abraham Gottlob Werner.

The foundation of modern stratigraphy is attributed to William Smith, a surveyor for contemporary canal builders, who became interested in the rocks that were being dug into as a series of canals were constructed across southern England (Hancock, 1977, pp. 3–4; Berry, 1987, pp. 56–57). His knowledge of geology was self-taught, owing nothing to such illustrious predecessors as James Hutton. As Hancock (1977, p. 5) noted, Smith's work was entirely empirical, free of any attempt at grand theory, and free of any influence of theology—an important point considering the powerful influence of biblical teachings at the time. Smith's work began in the Jurassic strata around Bath, in southwest England. He recognized that the stratigraphic succession was the same wherever he encountered it, and that particular strata could also be characterized by particular suites of fossils. From this inductive base, Smith evolved the deductive principle that he could identify the stratigraphic position of any outcrop by its distinctive rock types and fossil content. He committed his observations to maps that showed the outcrop patterns of his succession, and over a period of about 25 years he gradually compiled a complete geological map of England and Wales, the first such map of its kind ever constructed (Smith, 1815). Because Smith was not a member of the landed or aristocratic class in England, his work was largely ignored until late in his life, when he was appropriately honoured by the Geological Society of London. The story is told in detail by Winchester (2001) and Torrens (2001). Others were describing local successions of strata during the late eighteenth and early nineteenth centuries, such as Cuvier and Brongniart who documented the Tertiary strata of the Paris Basin in the first two decades of the nineteenth century (Conkin and Conkin, 1984; Berry, 1987, p. 66), but Smith was the first to show that rocks, with their contained fossils, constituted mappable successions. Cuvier was more concerned with the history of life on earth (Hancock, 1977, p. 6). Brongniart was amongst the first to appreciate the importance of Smith's contribution in creating the possibility of long-distance correlation based on fossils alone, independent of rock type (Hancock, 1977, p. 7).

As knowledge of regional stratigraphy evolved in various parts of Europe, the fourfold primary subdivision of Werner was broken down locally into various "series", and these, in turn, were commonly subdivided into local "formations." This was an entirely piecemeal operation, reflecting local interests, but from this gradually evolved a body of descriptive knowledge of rock successions and their contained fossils. As noted by Berry (1987, p. 63):

> Many of the widely used descriptive units did bear fossils that, when analyzed using the principle of faunal succession, proved to be a fossil aggregate diagnostic of a time unit in an interpretive scale; thus many descriptive units became interpretive ones, and today bear the same names. Among the major units of the interpretive time scale that were originally descriptive rock units are the Cambrian, Carboniferous, Jurassic, Cretaceous and Tertiary. Units that were based on interpretation of fossils from their inception are the Ordovician, Devonian, Permian, and the Tertiary Epochs.

During the 1820s and 1830s such workers as Young and Bird in Yorkshire, and Eaton in New York, recognized that some formations changed in character as they were traced laterally (Hancock, 1977, pp. 7–8). Amand Gressly (1838; see translation in Conkin and Conkin, 1984, pp. 137–139) was the first to systematize the observation of such changes, with the introduction of the term and concept of *facies*, based on his work on the Jurassic rocks of the Jura region of southern France. For example, he was aware of the differences between the limestones with contained fossils of coral-bank environments, oolitic deposits (which we now recognize to be beach deposits), and the "oozy" deposits of deeper-water environments, all of which may form at the same time, and may also form one above the other as environments shift over time. This type of change can be documented by careful observation of rocks in outcrop, by studying the vertical succession of rock types or by tracing an individual set of beds laterally, perhaps for many kilometres. Gressly proposed two new laws: the first that in different places formations ("terrains" in the French terminology) may consist of rocks of different petrologic and paleontological character (the original meaning of the term *facies*, which we still retain), and, secondly, that a similar succession of facies may occur in both vertical and lateral arrangement, relative to the bedding. As Hancock (1977, p. 9) pointed out, this predated Walther's proposal of the law of the correlation of facies by some 56 years. The study of facies became a central activity of stratigraphic work in the 1960s, with the establishment of the *process-response facies model*, as noted above.

With Gressly's concept of facies in place, the stage was set for the next important development, that of the introduction of the concept of the *stage*, by another French geologist, d'Orbigny (1842–1852). He recognized the vertical variability in fossil assemblages within individual series, and realized that stratigraphic successions could be subdivided into smaller units based on careful categorization of these succeeding fossil assemblages. These he called *stages*. D'Orbigny also used the term *zone*, but nowhere clearly defined it (Monty, 1968). Teichert (1958) argued that d'Orbigny was inconsistent in his usage, sometimes using the term *zone* as a synonym for *stage*, and sometimes as a subdivision of a stage. He established, if informally, the idea of the ideal "type" of succession, a locality where the stage was well represented, and from this has grown the concept of the *type section* or *stratotype*, to which formal importance has now been assigned as the first point of reference for establishing the character of a stratigraphic unit. Many of the stage names d'Orbigny erected are still those used worldwide. He was aware of the concept of facies change and of the variability in the nature of stage boundaries, from conformable to unconformable. As Conkin and Conkin (1984, p. 83) noted, the importance of d'Orbigny's work is his consolidation and adaptation of ideas that already existed in embryonic form, and the scope of his stage classification, which included the erection of some 27 stages for the Paleozoic and Mesozoic.

Hancock (1977, p. 12) suggested that the true foundation of biostratigraphy came with the work of the German stratigrapher Albert Oppel (1856–1858), whose work also concentrated on the Jurassic succession of western Europe. Oppel extended the ideas of d'Orbigny about the subdivision of successions based on their contained fossils to a more refined level. He recognized that careful study of the contained fossils would permit a much more detailed breakdown of the rocks, into what he called *zones*. He investigated "the vertical distribution of each individual species at many different places ignoring the mineralogic character of the beds" (Oppel, as quoted in Berry, 1987, p. 127). Some species were discovered to have short vertical ranges, others long ranges. Each zone could be characterized by several or many fossil species, although commonly one species would be chosen to be used as the name of the zone. Oppel built up stages from groups of zones. Stages were referred to as *zonegruppen*, or groups of zones (Teichert, 1958). These would usually fit into the stages already defined by d'Orbigny, but as Hancock (1977, p. 13) noted, in some places his zones spanned already-defined stage boundaries. This was the beginning of a practical problem that has still to be fully resolved; but in many cases, such as at the base and top of the Jurassic, Oppel's zone boundaries coincided with the System boundaries. In practice, zones became the foundation upon which the whole framework of biostratigraphy, the zone, stage, series and system, was gradually built. Teichert (1958, p. 109) emphasized the importance of Oppel's original description of zones as "paleontologically identifiable complexes of strata," not as subdivisions of time. The original concept was therefore clearly inductive—the recognition of a zone depended on the field geologist finding specific fossils in the rocks.

In their summation of the work of d'Orbigny and Oppel, Teichert (1958, p. 110) and Hancock (1977, p. 11) were at pains to emphasize the empirical, descriptive nature of the stage and zone concepts. They suggested that they were later distorted by the introduction of concepts about time that, they claimed, served to confuse the science of stratigraphy for some years. Teichert (1958) attributed these misconceptions to individuals such as H. Hedberg and O. H. Schindewolf. Hancock (1977) argued that Hedberg's influence was detrimental to the development of clear stratigraphic concepts. These problems are addressed below. However, careful examination and translation of d'Orbigny's original statements by Monty (1968) and Aubry et al. (1999, Appendix A) cast a different light on this historical work. Aubry et al. (1999, p. 137) pointed out that in the mid-nineteenth century no clear distinction between rocks and time had been made, and that paleontology was the only means of long-distance correlation. Selected translations from d'Orbigny's work clearly indicate that he envisioned stages as having a time connotation (Aubry et al., 1999, pp. 137–138).

Shortly after Oppel's work was completed, Charles Darwin's *The origin of species* was published (1859), and provided the explanation for the gradual change in the assemblage of species that Oppel had observed. However, resistance to the concepts of the stage and the zone remained strong through the remainder of the nineteenth century in Britain and the United States,

where the concept of facies had also still not taken hold, and reliance for correlation tended to still be placed on the lithology of formations. (Hancock, 1977, pp. 14–15).

International agreement on the definition and usage of most stratigraphic terms was attempted at the first International Geological Congress in Paris in 1878. A commission was established, that subsequently met in Paris and in Bologna. At the latter meeting, in September–October 1880, the commission:

> Decided on definitions of stratigraphic words like series and stage, and listed their synonyms in several languages .... Rocks, considered from the point of view of their origin, were formations; the term was not part of stratigraphic nomenclature at all, but concerned how the rock had been formed (e.g., marine formations, chemical formations). Stratigraphic divisions were placed in an order of hierarchy, with examples, thus: group (Secondary Group) [what we would now term the Mesozoic], system (Jurassic System), series (Lower Oolite Series), stage (Bajocian Stage), substage, assise (Assise à *A. Humphriensianus*), stratum. A distinction was made between stratigraphic and chronologic divisions. The duration of time corresponding to a group was an era, to a system a period, to a series an epoch, and to a stage an age (Hancock, 1977, p. 15).

Definitions of the terms "zone" and "horizon" were added to the published record of this meeting by the secretary of the commission, based on national reports submitted by the delegates, but full international agreement was slow in coming (Hancock, 1977, p. 16). Nonetheless, by the early twentieth century, the following basic descriptive terms and the concepts on which they were based had become firmly established, if not universally used:

Law of superposition of strata
Stratigraphic outcrop maps based on the succession of
  sedimentary rocks with its contained fossils
Facies
Stage
Type section or stratotype
Zone

At the 8th International Geological Congress in Paris in 1900 the stratigraphic hierarchy era, system/period, epoch/series, age/stage, phase/zone was accepted (Vai, 2007).

According to Teichert (1958), the term *biostratigraphy* was introduced by the Belgian paleontologist Dollo in 1904, for the "entire research field in which paleontology exercises a significant influence on historical geology."

Codification of these principles by the mid-twentieth century is illustrated by the work of Schenk and Muller (1941). Various national and international stratigraphic codes and guides have been developed that standardize and formalize the definitions of stratigraphic terms and set out the procedures by which they should be used. An international guide was published in 1976 (Hedberg, 1976), with a major revision appearing in 1994 (Salvador, 1994). A code for North America, based on the international guide, appeared in 1983 (NACSN, 1983).

### 1.3.2  Do Stratigraphic Units Have "Time" Significance?

As long ago as 1862 Huxley wrote: "neither physical geology nor paleontology possesses any method by which the absolute synchronism of two strata can be demonstrated. All that geology can prove is local order of succession." (quoted in Hancock, 1977, p. 17). As an example, Huxley suggested that there was no way to prove or disprove that "Devonian fauna and flora in the British Isles may have been contemporaneous with Silurian life in North America, and with a Carboniferous fauna and flora in Africa." The variations in fauna and flora could simply be due to the time it took for the organisms to migrate. There is therefore a need for a distinction between "'homotaxis' or 'similarity of arrangement' and 'synchrony' or 'identity of date'" (Hancock, 1977, p. 17). Conkin and Conkin (1964) suggested that this concept was first enunciated by DeLapparant (1885; as cited and translated by Conkin and Conkin, 1984, p. 243), although Callomon (2001, p. 240) stated that the Principle of Biosynchroneity", whereby beds with similar fossils are assumed to be the same age, is "usually ascribed to William Smith". As geologists accumulated a very detailed knowledge of the succession of fossils, the assumption that the same succession of fossil assemblages indicated synchroneity assumed the status of a truism. However, until the development of radiometric dating and the growth of modern chronostratigraphy (see below) the true "time" value of fossils remained a problem, because biostratigraphy provides only relative ages. The basic assumption about the temporal value of fossils was first made most clearly by Lyell

(1830–1833) and, according to Conkin and Conkin (1984; see in particular their Table 1, p. 2), subsequent developments by Bronn (1858), Phillips (1860), Lapworth (1879), DeLapparant (1885) and Buckman (1893) established the main framework upon which this part of modern stratigraphy is built.

One of Lyell's (1830–1833) most important contributions was his detailed study and classification of Tertiary deposits, based on their contained fossils. Berggren (1998) provided a succinct summary of this important contribution. Lyell's subdivisions of the Tertiary were based on the idea that, through the course of time, contemporary faunas become more and more like those found at present. Under a heading "The distinctness of periods may indicate our imperfect information" he stated:

> In regard to distinct zoological periods, the reader will understand . . .. That we consider the wide lines of demarcation that sometimes separate different tertiary epochs, as quite unconnected with extraordinary revolutions of the surface of the globe, as arising, partly, like chasms in the history of nations, out of the present imperfect state of our information, and partly from the irregular manner in which geological memorials are preserved, as already explained. We have little doubt that it will be necessary hereafter to intercalate other periods, and that many of the deposits, now referred to a single era, will be found to have been formed at very distinct periods of time, so that, notwithstanding our separation of tertiary strata into four groups, we shall continue to use the term *contemporaneous* with a great deal of latitude (Lyell, 1833, vol. 3, pp. 56–57).

This quote contains most of the modern concept that units defined on the basis of their fossil content may have global significance with regard to contemporaneity, but that the preserved record may be imperfect. Lyell's "lines of demarcation" are what we would now define as chronostratigraphic boundaries. These were commonly drawn at unconformities until the introduction of modern practices, as described below. Vai (2007, p. 87) pointed out that Lyell's original units were clearly described as physical rock bodies. Therefore, the separation of "rock" and "time" is not as clear in this early work as has sometimes been reported (e.g., Berggren, 1998).

Phillips (1860, p. xxxii), in comparing fossil successions between localities in different parts of the world (he mentioned several Paleozoic successions that had been described from Europe and North America), suggested that "the affinity of the fossils is accepted as evidence of the approximate contemporaneity of the rocks." Phillips (1860, p. xxxvi) referred to the work of Charles Darwin to explain the succession of forms, replacing the theologically-based assumptions about the catastrophic destruction and remaking of life that had dominated earlier interpretations of the geological record.

Towards the end of the nineteenth century geologists developed some terms to distinguish between rock subdivisions and implied time; for example, the 1880 Bologna congress recommended the use of the term "age" for the rock equivalent of the time term "stage." The need for a "dual system" of nomenclature for time and for rocks was emphasized by Williams (1894), and led to the dual hierarchy accepted at the 1900 IGC, as noted in the previous section (see Vai, 2007).

S. S. Buckman, in a series of papers on the biostratigraphy of the English Jurassic strata, proposed a new concept and a term to encompass it. This was the *hemera*, defined by Buckman (1893, p. 481) as "the chronological indicator of the faunal sequence." Buckman intended the hemera to be "the smallest consecutive divisions which the sequence of different species enables us to separate in the maximum development of strata." This unit of time was intended to correspond to the *acme zone*, the rock unit representing the maximum occurrence of a particular zone species. If the record is complete, the span of time of a given hemera should be present in the rocks even beyond the facies changes that limit the extent of the original zone fossils. There has always been the potential for confusion between a reference to a time span and the rocks that were deposited during that time span. Buckman (1898, p. 442) noted that, for example, terms such as Bajocian and Jurassic had been used to refer to rocks of that age and also to a specific span of time.

Most of the work in the nineteenth and early twentieth centuries that addressed the issue of how geologic time is represented in the rocks approached the subject from the point of view of the fossil record. We have touched on some of the key developments in the preceding paragraphs. For example, Buckman (1893, p. 518) said

> Species may occur in the rocks, but such occurrence is no proof that they were contemporaneous …their joint occurrence in the same bed [may] only show that the deposit in which they accumulated are embedded very slowly.

And in a later paper:

> The amount of deposit can be no indication of the amount of time … the deposits of one place correspond to the gaps of another (Buckman, 1910, p. 90).

A very different approach was taken by Barrell (1917), in what became a classic paper, exploring the origins of stratigraphic units in terms of depositional processes. Aided by the new knowledge of the Earth's radiogenic heat engine and a growing understanding of sedimentary processes, Barrell worked through detailed arguments about the rates of sedimentation and the rates of tectonism and of climate change.

Combining many of these ideas together, Barrell (1917, Fig. 5) constructed a diagram showing the "Sedimentary Record made by Harmonic Oscillation in Baselevel" (Fig. 1.3). This is remarkably similar to diagrams that have appeared in some of the Exxon sequence model publications since the 1980s, and represents a thoroughly modern deductive model of the way in which "time" is stored in the rock record. Curve A–A simulates the record of long-term subsidence and the corresponding rise of the sea. Curve B–B simulates an oscillation of sea levels brought about by other causes—Barrell discussed diastrophic and climatic causes, including glacial causes, and applied these ideas to the rhythmic stratigraphic record of the "upper Paleozoic formation of the Appalachian geosyncline" in a discussion that would appear to have provided the foundation for the interpretations of "cyclothems" that appeared in the 1930s (see below). Barrell showed that when the long-term and short-term curves of sea-level change are combined, the oscillations of base level provide only limited time periods when sea-level is rising and sediments can accumulate. "Only one-sixth of time is recorded" by sediments (Barrell, 1917, p. 797). This remarkable diagram anticipates (1) Jervey's (1988) ideas about sedimentary "accommodation" that became fundamental to models of sequence stratigraphy ("accommodation" is defined as the space made available for sediment to accumulate as a result of a rise of base level above the basin floor), and (2) Ager's (1973) point that the sedimentary record is "more gap than record." This important paper did not appear to influence thinking about the nature of the stratigraphic record as much as it should, as demonstrated by the fact that the rediscovery of the ideas by Jervey, Ager and others is largely attributed to the rediscoverers, not to Barrell (Wheeler, 1958,

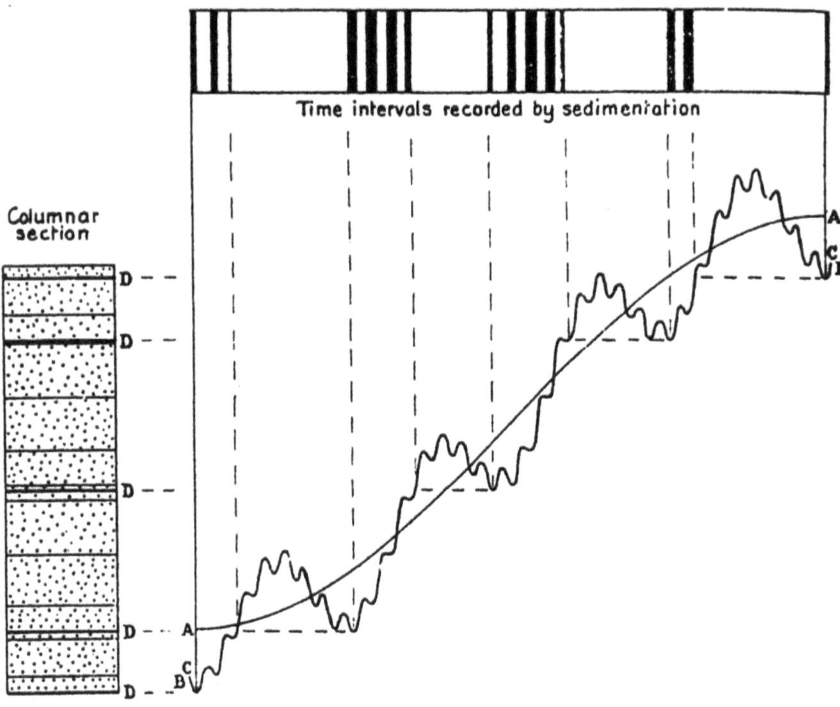



**Fig. 1.3** Barrell's (1917) explanation of how oscillatory variations in base level control the timing of deposition. Sedimentation can only occur when base level is actively rising. These short intervals are indicated by the black bars in the top diagram. The resulting stratigraphic column, shown at the left, is full of disconformities, but appears to be the result of continuous sedimentation

in the first of an important series of papers to which we return below, comments favourably on Barrell's "frequently neglected base-level concept"). The point relevant to the discussion here is that Barrell demonstrated how fragmentary the stratigraphic record is, and how incomplete and unreliable it is as a record of the passage of the continuum of geologic time.

Other workers of this period who were cognizant of the significance of gaps in the stratigraphic record were Grabau (1913), who first defined the term *disconformity* as a major time break between units that nevertheless remained structurally parallel—conformable—to one another, and Blackwelder (1909) who wrote an essay on unconformities. Barrell (1917, p. 794) added the new term *diastem*, for minor sedimentary breaks.

A much-needed updating in stratigraphic concepts and terminology was undertaken by Schenk and Muller (1941). They "tried to clarify the distinction between the interpretive nature of 'time' and 'time-stratigraphic' units in contrast with the purely descriptive rock or stratigraphic term 'formation.'" (Berry, 1987, p. 7). They formalized the system of nomenclature that had been proposed at the Paris IGC in 1900 (see Vai, 2007), and is in use today:

| Time division (for abstract concept of time) | Time-stratigraphic division (for rock classification) |
| --- | --- |
| Era | Erathem |
| Period | System |
| Epoch | Series |
| Age | Stage |
| Phase | Zone |

Arkell (1946), the specialist in the Jurassic System, said: "A stage is an artificial concept transferable to all countries and continents, but a zone is an empirical unit" (cited in Hancock, 1977, p. 18). By this statement he was essentially adopting the rock-time concepts of Buckman's hemera for units of the rank of the stage, suggesting that stages had some universal time significance. This statement represents a deductive interpretation of the meaning of the fossil record, and perpetuated the confusion between "rocks" and "time." Hancock (1977) blamed Arkell for bringing into the modern era the controversy over the relative meaning of chronostratigraphic and biostratigraphic (zone) concepts. A biozone is an empirical unit based on the rock record, and can only be erected and used for correlation if the fossils on which it is based are present in the rocks. Stages are simply groupings of zones.

Confusion between the meaning of "rock" units and "time" units appeared to be widespread during the early part of the twentieth century. Teichert (1958) summarized the various approaches taken by French, German, British and North American stratigraphers and paleontologists up to that time. To him it was clear that there were three distinct concepts (summarized here from Teichert, 1958, p. 117):

*Biostratigraphic units*: the zone is the fundamental unit of biostratigraphy, consisting of a set of beds characterized by one or more fossil species.

*Biochronologic units*: the unit of time during which sedimentation of a biostratigraphic unit took place. These are true relative time units.

*Time-stratigraphic units*: units of rock which have been deposited during a defined unit of time. "Arrangement of rock-stratigraphic units within a time-stratigraphic unit and assignment to such a unit are generally made on a variety of lines of evidence both physical and paleontologic, including extrapolation, conclusion by analogy, and sometime merely for reasons of expedience."

The distinguished American stratigrapher Hollis Hedberg picked up on Arkell's idea about the meaning of the stage. He stated (Hedberg, 1948, p. 456) "The time value of stratigraphical units based on fossils will fluctuate from place to place in much the same manner as the time value of a lithologic formation may vary." He proposed defining a separate set of chronostratigraphic units that corresponded to, and could be used to define, units of time. In 1952 he became the Chairman of a newly established International Commission on Stratigraphy, and one of his achievements was to establish this new set of units. These ideas became formalized and codified after many years work in a new international stratigraphic guide (Hedberg, 1976). For example, he defined a chronozone as a zonal unit embracing all rocks anywhere formed during the range of a specified geological feature, such as a local biozone. In theory, a chronozone is present in the rocks beyond the point at which the fossil components of the biozone cease to be present as a result of lateral facies changes. Hedberg (1976) used a biostratigraphic example to illustrate the chronozone concept but, clearly, if the fossil components are not present, a chronozone cannot be recognized on biostratigraphic grounds, and its usefulness as a stratigraphic concept may be rather hypothetical (Johnson, 1992a). However, other means may be available to extend the chronozone, including local marker beds or magnetostratigraphic data. This is discussed this in the next section.

Hedberg (1976) suggested that the stage be regarded as the basic working unit of chronostratigraphy because of its practical use in interregional correlation. As Hancock (1977, p. 19) and Watson (1983) pointed out, Hedberg omitted to mention that all Phanerozoic stages were first defined on the basis of groups of biozones, and they are therefore, historically, biostratigraphic entities. In practice, therefore, so long as biostratigraphy formed the main basis of chronostratigraphy, no useful purpose was served by treating them as theoretically different subjects. Stages could be defined by more than one system of biozones, which extended their range and reduced their facies dependence, but, although this improved their chronostratigraphic usefulness, it did not change them into a different sort of unit. We return to this point in the next section (the modern definition of the term "stage" is quite different, as discussed below).

Harry E. Wheeler of the University of Washington pointed out the problems in Hedberg's concepts of time-stratigraphic units (Wheeler, 1958, p. 1050) shortly after they appeared in the first American stratigraphic guide (American Commission on Stratigraphic Nomenclature, 1952). He argued that a time-rock unit could not be both a "material rock unit," as described in the guide, and one whose boundaries could be extended from the type section as isochronous surfaces, because such isochronous surfaces would in many localities be represented by an unconformity. Wheeler developed the concept of the chronostratigraphic cross-section, in which the vertical dimension in a stratigraphic cross-section is drawn with a time scale instead of a thickness scale (Fig. 1.4). In this way, time gaps (unconformities) become readily apparent, and the nature of time correlation may be accurately indicated. Such diagrams have come to be termed "Wheeler plots." Wheeler cited with approval the early work of Sloss and his colleagues, referred to in more detail below:

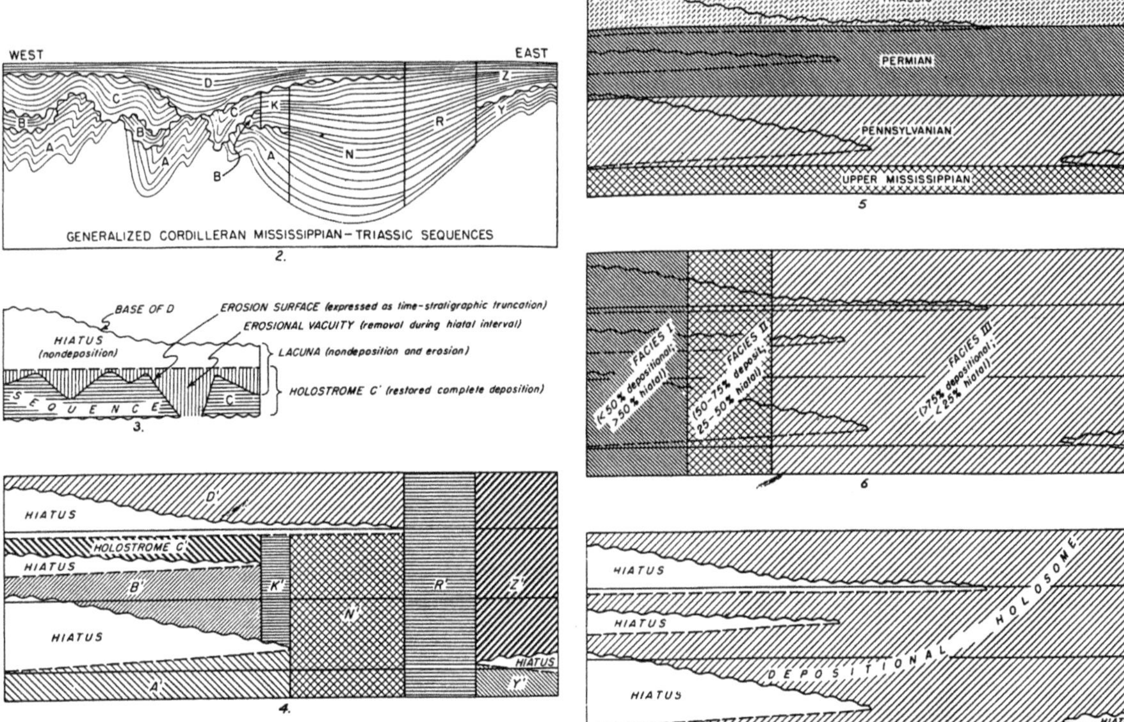

**Fig. 1.4** Wheeler's development of chronostratigraphic diagrams, showing a stratigraphic cross section plotted with a vertical time axis in order to portray accurately the duration of stratigraphic units and unconformities from place to place. In the first diagram a generalized cross-section is provided. A time cross-section of sequence C is shown below, providing the definition of some of the terms used by Wheeler. Remaining diagrams show conversion of these data into chronostratigraphic charts (Wheeler, 1958)

As a tangible framework on which to hang pertinent faunal and lithic data, the *sequence* of Sloss, Krumbein and Dapples (1949, pp. 110–11) generally fulfills these requirements. Paraphrasing these authors' discussion, a *sequence comprises an assemblage of strata exhibiting similar responses to similar tectonic environments over wide areas, separated by objective horizons without specific time significance* (Wheeler, 1958, p. 1050; italics as in original).

Sequences came later to be called simply "unconformity-bounded units," whereas Wheeler's description of them is a logically inconsistent mixture of empirical description and tectonic interpretation. He proposed a new term for time-rock units, the *holostrome*, which consists of a sequence (in the Sloss et al. sense) together with the "erosional vacuity" representing the part of the sequence lost to erosion. Such a vacuity would not be obvious on a conventional stratigraphic cross-section, appearing simply as the line corresponding to an unconformity. However, an erosional vacuity might constitute a significant area of a Wheeler time plot and would require some knowledge of the lateral variations in the age of the units immediately above and below the unconformity in order for it to be drawn in accurately. Although Wheeler's concepts and plots are now commonplace in geology (they are cited in Vail's early work), the term *holostrome* has not become an accepted term.

The recognition of the importance of suites of stratigraphic units bounded by unconformities, based on the work of Levorsen, Sloss, Wheeler and others, led to suggestions for the formal recognition of such units in stratigraphic guides and codes. The proposal for a type of unit called a "synthem" by Chang (1975) represents the first formal proposal of this type. Chang (1975, pp. 1546–1547) considered the importance of tectonic cycles in the generation of such deposits, quoting Grabau's (1940) concept of "pulsations", and he was aware of Vail's early work on eustatic cycles (citing presentations by Vail's group at a Geological Society of America meeting in 1974). However, he concluded that "In recognizing the unity or individuality of a synthem, the involvement of as little subjective judgment as possible is desirable." In other words, Chang was at pains to adhere to a descriptive, empirical concept in his definition. Synthems were included in the International Stratigraphic Guide (Hedberg, 1976, p. 92) as an additional class of time-significant unit, although, because of the variable age of the bounding unconformities, they were not categorized as strictly chronostratigraphic in character. Unconformity-bounded units were accorded their own chapter in the revised version of the International Stratigraphic Guide (Salvador, 1994, Chap. 6), and also formed the basis for a new type of stratigraphy, termed *allostratigraphy*, in the North American guide (NACSN, 1983). But not everybody was content with these developments. Murphy (1988, p. 155) pointed out that "The statement that they [synthems] are objective and non-interpretive …. assumes that particular unconformities have tangible qualities by which they can be distinguished and identified; this assumption is false." He went on to argue that many unconformities can only be recognized on the basis of knowledge of the units above and below, and interpretation of the nature of the unconformity surface itself, and that therefore unconformity-bounded units are not a class of objective stratigraphic unit. This objection has been almost totally ignored in the rush to adopt sequence stratigraphic methods, although attempts to incorporate sequence concepts into formal stratigraphic guides and codes have yet to achieve agreement.

### 1.3.3   The Development of Modern Chronostratigraphy

One of the central themes of the development of stratigraphy has been the work to establish an accurate geological times scale. Why? McLaren (1978) attempted to answer this question. Here are his nine reasons:

> Some of these geological problems and questions include: (1) rates of tectonic processes; (2) rates of sedimentation and accurate basin history; (3) correlation of geophysical and geological events; (4) correlation of tectonic and eustatic events; (5) are epeirogenic movements worldwide … (6) have there been simultaneous extinctions of unrelated animal and plant groups; (7) what happened at era boundaries; (8) have there been catastrophes in earth history which have left a simultaneous record over a wide region or worldwide; and (9) are there different kinds of boundaries in the geologic succession (That is, "natural" boundaries marked by a worldwide simultaneous event versus "quiet" boundaries, man-made by definition).

It is, in fact, fundamental to the understanding of the history of Earth that events be meticulously correlated in time. For example, current work to investigate the history of climate change on Earth during the last few tens to hundreds of thousands of years has demonstrated how important this is, because of the rapidity of climate change and because different geographical regions and climatic belts may have had histories of climate change that were not in phase. If we are to understand Earth's climate system thoroughly enough to determine what we might expect from human influences, such as the burning of fossil fuels, a detailed record of past climate change will be of fundamental importance. That we do not now have such a record is in part because of the difficulty in establishing a time scale precise enough and practical enough to be applicable in deposits formed everywhere on Earth in every possible environmental setting.

Until the early twentieth century, the geologic time scale in use by geologists was a relative time scale dependent entirely on biostratigraphy. The standard systems had nearly all been named., based on European data, by about 1840 (Berry, 1987; Callomon, 2001). Estimates about the duration of geologic events, including that of chronostratigraphic units, varied widely, because they depended on diverse estimation methods, such as attempts to quantify rates of erosion and sedimentation (Hallam, 1989). The discovery of the principle of radioactivity was fundamental, providing a universal clock for direct dating of certain rock types, and the calibration of the results of other dating methods, especially the relative scale of biostratigraphy. Radiometric dating methods may be used directly on rocks containing the appropriate radioactive materials. For example, volcanic ash beds intercalated with a sedimentary succession provide an ideal basis for precise dating and correlation. Volcanic ash contains several minerals that include radioactive isotopes of elements such as potassium and rubidium. Modern methods can date such beds to an accuracy typically in the $\pm 2\%$ range, that is, $\pm 2$ million years at an age of 100 Ma (Harland et al., 1990), although locally, under ideal conditions, accuracy and precision are now considerably better than this ($\pm 10^4$–$10^5$ years; see Fig. 14.22 and discussion thereof). Where a sedimentary unit of interest (such as a unit with a biostratigraphically significant fauna or flora) is overlain and underlain by ash beds it is a simple matter to estimate the age of the sedimentary unit. The difference in age between the ash beds corresponds to the elapsed time represented by the succssion of strata between the ash beds. Assuming the sediments accumulated at a constant rate, the rate of sedimentation can be determined by dividing the thickness of the section between the ash beds by the elapsed time. The amount by which the sediment bed of interest is younger than the lowest ash bed is then equal to its stratigraphic height above the lowest ash bed divided by the rate of sedimentation, thereby yielding an "absolute" age, in years, for that bed. This procedure is typical of the methods used to provide the relative biostratigraphic age scale with a quantitative basis. The method is, of course, not that simple, because sedimentation rates tend not to be constant, and most stratigraphic successions contain numerous sedimentary breaks that result in underestimation of sedimentation rates. Numerous calibration exercises are required in order to stabilize the assigned ages of any particular biostratigraphic unit of importance. I return to this issue later in Chap. 14.

Initially, the use of radiometric dating methods was relatively haphazard, but gradually geologists developed the technique of systematically working to cross-calibrate the results of different dating methods, reconciling radiometric and relative biostratigraphic ages in different geological sections and using different fossils groups. In the 1960s the discovery of preserved ("remanent") magnetism in the rock record led to the development of an independent time scale based on the recognition of the repeated reversals in magnetic polarity over geologic time. Cross-calibration of radiometric and biostratigraphic data with the magnetostratigraphic record provided a further means of refinement and improvement of precision. The techniques are described in all standard textbooks of stratigraphy (e.g., Miall, 1999; Nichols, 1999).

These modern developments rendered irrelevant the debate about the value and meaning of Hedberg's (1976) hypothetical chronostratigraphic units. The new techniques of radiometric dating and

magnetostratigraphy, where they are precise enough to challenge the supremacy of biostratigraphy, could have led to the case being made for a separate set of chronostratigraphic units, as Hedberg proposed. However, instead of a new set of chronostratigraphic units, this correlation research is being used to refine the definitions of the existing, biostratigraphically based stages. Different assemblages of zones generated from different types of organism may be used to define the stages in different ecological settings (e.g., marine versus nonmarine) and in different biogeographic provinces, and the entire data base is cross-correlated and refined with the use of radiometric, magnetostratigraphic and other types of data. The stage has now effectively evolved into a chronostratigraphic entity of the type visualized by Hedberg (1976). This is the essence of the procedure recommended by Charles Holland (1986, Fig. 10), one of the leading spokespersons of the time for British stratigraphic practitioners. For most of Mesozoic and Cenozoic time the standard stages, and in many cases, biozones, are now calibrated using many different data sets, and the global time scale, based on correlations among the three main dating methods, is attaining a high degree of accuracy. The Geological Society of London time scale (GSL, 1964) was an important milestone, representing the first attempt to develop a comprehensive record of these calibration and cross-correlation exercises. Wheeler's (1958) formal methods of accounting for "time in stratigraphy" (the title of his first important paper), including the use of "Wheeler plots" for showing the time relationships of stratigraphic units, provided much needed clarity in the progress of this work. Time scales for the Cenozoic (Berggren, 1972) and the Jurassic and Cretaceous (Van Hinte, 1976a, b) are particularly noteworthy for their comprehensive data syntheses, although all have now been superseded. More recent detailed summation and reconciliation of the global data base were provided by Harland et al. (1990) and Berggren et al. (1995). Gradstein et al. (2004) provided a comprehensive treatment of the subject, and nowadays, the global data base and updates of definitions are maintained at an official website: www.stratigraphy.org.

In the 1960s, several different kinds of problems with stratigraphic methods and practice had begun to be generally recognized (e.g., Newell, 1962). There are two main problems. Firstly, stratigraphic boundaries had traditionally been drawn at horizons of sudden change, such as the facies change between marine Silurian strata and the overlying nonmarine Devonian succession in Britain. Changes such as this are obvious in outcrop, and would seem to be logical places to define boundaries. Commonly such boundaries are unconformities. However, it had long been recognized that unconformities pass laterally into conformable contacts (for example, this was described by Whewell, 1872). This raised the question of how to classify the rocks that formed during the interval represented by the unconformity. Should they be assigned to the overlying or underlying unit, or used to define a completely new unit? When it was determined that rocks being classified as Cambrian and Silurian overlapped in time, Lapworth (1879) defined a new chronostratigraphic unit—the Ordovician, as a compromise unit straddling the Cambrian-Silurian interval. The same solution could be used to define a new unit corresponding to the unconformable interval between the Silurian and the Devonian. In fact, rocks of this age began to be described in central Europe after WWII, and this was one reason why the Silurian-Devonian boundary became an issue requiring resolution. A new unit could be erected, but it seemed likely that with additional detailed work around the world many such chronostratigraphic problems would arise, and at some point it might be deemed desirable to stabilize the suite of chronostratigraphic units. For this reason, the development of some standardized procedure seemed to be desirable.

A second problem is that to draw a significant stratigraphic boundary at an unconformity or at some other significant stratigraphic change is to imply the hypothesis that the change or break has a significance relative to the stratigraphic classification, that is, that unconformities have precise temporal significance. This was specifically hypothesized by Chamberlin (1898, 1909) who, as discussed below, was one of many individuals who generated ideas about a supposed "pulse of the earth." In the case of lithostratigraphic units, which are descriptive, and are defined by the occurrence and mappability of a lithologically distinctive succession, a boundary of such a unit coinciding with an unconformity is of no consequence. However, in the case of an interpretive classification, in which a boundary is assigned time significance (such as a stage boundary), the use of an unconformity as the boundary is to make the assumption that the unconformity has time significance; that is, it is of the same age everywhere. This

places primary importance on the model of unconformity formation, be this diastrophism, eustatic sea-level change or some other cause. From the methodological point of view this is most undesirable, because it negates the empirical or inductive nature of the classification. It is for this reason that it is inappropriate to use sequence boundaries as if they are chronostratigraphic markers.

How to avoid this problem? A time scale is concerned with the continuum of time. Given our ability to assign "absolute" ages to stratigraphic units, albeit not always with much accuracy and precision, one solution would be to assign numerical ages to all stratigraphic units and events. However, this would commonly be misleading or clumsy. In many instances stratigraphic units cannot be dated more precisely than, say, "late Cenomanian" based on a limited record of a few types of organisms (e.g., microfossils in subsurface well cuttings). Named units are not only traditional, but also highly convenient, just as it is convenient to categorize human history using such terms as the "Elizabethan" or the "Napoleonic" or the "Civil War" period. What is needed is a categorization of geological time that is empirical and all encompassing. The familiar terms for periods (e.g., Cretaceous) and for ages/stages (e.g., Aptian) offer such a subdivision and categorization, provided that they can be made precise enough and designed to encompass all of time's continuum. A group of British stratigraphers (e.g., Ager, 1964) is credited with the idea that seems to have resolved the twin problems described here. McLaren (1970, p. 802) explained the solution in this way:

There is another approach to boundaries, however, which maintains that they should be defined wherever possible in an area where "nothing happened." The International Subcommission on Stratigraphic Classification, of which Hollis Hedberg is Chairman, has recommended in its Circular No. 25 of July, 1969, that "Boundary-stratotypes should always be chosen within sequences of continuous sedimentation. The boundary of a chronostratigraphic unit should never be placed at an unconformity. Abrupt and drastic changes in lithology or fossil content should be looked at with suspicion as possibly indicating gaps in the sequence which would impair the value of the boundary as a chronostratigraphic marker and should be used only if there is adequate evidence of essential continuity of deposition. The marker for a boundary-stratotype may often best be placed *within* a certain bed to minimize the possibility that it may fall at a time gap." This marker is becoming known as "the Golden Spike."

By "nothing happens" is meant a stratigraphic succession that is apparently continuous. The choice of boundary is then purely arbitrary, and depends simply on our ability to select a horizon that can be the most efficiently and most completely documented and defined (just as there is nothing about time itself that distinguishes between, say, February and March, but to define a boundary between them is useful for purposes of communication and record). This is the epitome of an empirical approach to stratigraphy. Choosing to place a boundary where "nothing happened" is to deliberately avoid having to deal with some "event" that would require interpretation. This recommendation was accepted in the first International Stratigraphic Guide (Hedberg, 1976, pp. 84–85), although Hedberg (1976, p. 84) also noted the desirability of selecting boundary stratotypes "at or near markers favorable for long-distance time-correlation", by which he meant prominent biomarkers, radiometrically-datable horizons, or magnetic reversal events. Boundary-stratotypes were to be established to define the base and top of each chronostratigraphic units, with a formal marker (a "golden spike") driven into a specific point in a specific outcrop to mark the designated stratigraphic horizon. Hedberg (1976, p. 85) recommended that such boundary-stratotypes be used to define both the top of one unit and the base of the next overlying unit. However, further consideration indicates an additional problem, which was noted in the North American Stratigraphic Code of 1983 (NACSN, p. 868):

> Designation of point boundaries for both base and top of chronostratigraphic units is not recommended, because subsequent information on relations between successive units may identify overlaps or gaps. One means of minimizing or eliminating problems of duplication or gaps in chronostratigraphic successions is to define formally as a point-boundary stratotype only the base of the unit. Thus, a chronostratigraphic unit with its base defined at one locality will have its top defined by the base of an overlying unit at the same, but more commonly, another locality.

Even beds selected for their apparently continuous nature may be discovered at a later date to hide a significant break in time. Detailed work on the British Jurassic section using what is probably the most refined biostratigraphic classification scheme available for any pre-Neogene section has demonstrated how common such breaks are (Callomon, 1995; see Miall and Miall, 2002). The procedure recommended by NACSN (1983) is that, if it is discovered that a

boundary stratotype does encompass a gap in the temporal record, the rocks (and the time they represent) are assigned to the unit below the stratotype. In this way, a time scale can be constructed that can readily accommodate all of time's continuum, as our description and definition of it continue to be perfected by additional field work. This procedure means that, once designated, boundary stratotypes do not have to be revised or changed. This has come to be termed the concept of the "topless stage."

The modern definition of the term "stage" (e.g., in the online version of the International Stratigraphic Guide by Michael A. Murphy and Amos Salvador at www.stratigraphy.org) indicates how the concept of the stage has evolved since d'Orbigny. The Guide states that "The stage has been called the basic working unit of chronostratigraphy. ... The stage includes all rocks formed during an age. A stage is normally the lowest ranking unit in the chronostratigraphic hierarchy that can be recognized on a global scale. ... A stage is defined by its boundary stratotypes, sections that contain a designated point in a stratigraphic sequence of essentially continuous deposition, preferably marine, chosen for its correlation potential."

The first application of the new concepts for defining chronostratigraphic units was to the Silurian-Devonian boundary, the definition of which had begun to cause major stratigraphic problems as international correlation work became routine in post-WWII years. A boundary stratotype was selected at a location called Klonk, in what is now the Czech Republic, following extensive work by an international Silurian-Devonian Boundary Committee on the fossil assemblages in numerous well-exposed sections in Europe and elsewhere. The results are presented in summary form by McLaren (1973), and, more extensively, by Chlupáč (1972) and McLaren (1977) (see also Miall, 1999, pp. 116–117). As reported by Bassett (1985) and Cowie (1986), the establishment of the new procedures led to a flood of new work to standardize and formalize the geological time scale, one boundary at a time. This is extremely labour-intensive work, requiring the collation of data of all types (biostratigraphic, radiometric and, where appropriate, chemostratigraphic and magnetostratigraphic) for well-exposed sections around the world, and working to reach international agreement amongst ad-hoc international working groups set up for the purpose. In many instances, once such detailed correlation work is undertaken, it is discovered that definitions for particular boundaries being used in different parts of the world, or definitions established by different workers using different criteria, do not in fact define contemporaneous horizons (e.g., Hancock, 1993a). This may be because the original definitions were inadequate or incomplete, and have been subject to interpretation as practical correlation work has spread out across the globe. Resolution of such issues should simply require international agreement; the important point being that there is nothing significant about, say, the Aptian-Albian boundary, just that we should all be able to agree on where it is. Following McLaren's idea that boundaries be places where "nothing happens", the sole criterion for boundary definition is that such definitions be as practical as possible. The first "golden spike" location (the Silurian-Devonian boundary at Klonk: Chlupác, 1972) was chosen because it represents an area where deep-water graptolite-bearing beds are interbedded with shallow-water brachiopod-trilobite beds, permitting detailed cross-correlation among the faunas, thereby permitting the application of the boundary criteria to a wide array of different facies. In other cases, the presence of radiometrically datable units or a well-defined magnetostratigraphic record may be helpful. In all cases, accessibility and stability of the location are considered desirable features of a boundary stratotype, because the intent is that it serves as a standard. Perfect correlation with such a standard can never be achieved, but careful selection of the appropriate stratotype is intended to facilitate future refinement in the form of additional data collection.

Despite the apparent inductive simplicity of this approach to the refinement of the time scale, further work has been slow, in part because of the inability of some working groups to arrive at agreement (Vai, 2001). In addition, two contrasting approaches to the definition of chronostratigraphic units and unit boundaries have now evolved, each emphasizing different characteristics of the rock record and the accumulated data that describe it. Castradori (2002) provided an excellent summary of what became a lively controversy within the International Commission on Stratigraphy. The first approach, which Castradori described as the *historical and conceptual approach*, emphasizes the historical continuity of the erection and definition of units and their boundaries, the data base for which has continued to grow since the nineteenth century by a process of inductive accretion. Aubry

et al. (1999, 2000) expanded upon and defended this approach. The alternative method, which Castradori terms the *hyper-pragmatic approach*, focuses on the search for and recognition of significant "events" as providing the most suitable basis for rock-time markers, from which correlation and unit definition can then proceed. The choice of the term "pragmatic" is unfortunate in this context, because the suggested method is certainly not empirical. The followers of this method (see response by J. Ramane, 2000, to the discussion by Aubry et al., 2000) suggest that in some instances historical definitions of units and their boundaries should be modified or set aside in favour of globally recognizable event markers, such as a prominent biomarker, a magnetic reversal event, an isotopic excursion, or, eventually, events based on cyclostratigraphy. This approach explicitly sets aside McLaren's recommendation (cited above) that boundaries be defined in places where "nothing happened," although it is in accord with suggestions in the first stratigraphic guide that "natural breaks" in the stratigraphy could be used or boundaries defined "at or near markers favorable for long-distance time-correlation" (Hedberg, 1976, pp. 71, 84). The virtue of this method is that where appropriately applied it may make boundary definition easier to recognize. The potential disadvantage is that is places prime emphasis on a single criterion for definition. From the perspective of this book, which has attempted to clarify methodological differences, it is important to note that the hyper-pragmatic approach relies on assumptions about the superior time-significance of the selected boundary event. The deductive flavour of hypothesis is therefore added to the method. In this sense the method is not strictly empirical. As has been demonstrated elsewhere, assumptions about global synchroneity of stratigraphic events may in some cases be misguided (see Miall and Miall, 2001, 2002).

The hyper-pragmatic approach builds assumptions into what has otherwise been an inductive method free of all but the most basic of hypotheses about the time-significance of the rock record. The strength of the historical and conceptual approach is that it emphasizes multiple criteria, and makes use of long-established practices for reconciling different data bases, and for carrying correlations into areas where any given criterion may not be recognizable. For this reason, this writer is not in favour of the proposal by Zalasiewicz et al. (2004), supported by Carter (2007), to eliminate the distinction between time-rock units (chronostratigraphy) and the measurement of geologic time (geochronology). Their proposal hinges on the supposed supremacy of the global stratotype boundary points. History has repeatedly demonstrated the difficulties that have arisen from the reliance on single criteria for stratigraphic definitions, and the incompleteness of the rock record, which is why "time" and the "rocks" are so rarely synonymous in practice (see also Sect. 14.5.6 and Aubry, 2007, on this point).

Ongoing work on boundary stratotypes is periodically recorded in the IUGS journal *Episodes*, and is summarized in web pages at www.stratigraphy.org. The reader is referred to Aubry et al. (2000, including the discussion by Remane which follows) and to Castradori (2002) for additional details about this controversy. The latter article provides several case studies of how each approach has worked in practice.

For our purposes, the importance of this history of stratigraphy is that the work of building and refining the geological time scale has been largely an empirical, inductive process (with the exception of the hyper-pragmatic approach discussed above). Note that each step in the development of chronostratigraphic techniques, including the multidisciplinary cross-correlation method, the golden spike concept, and the concept of the topless unit, are designed to enhance the empirical nature of the process. Techniques of data collection, calibration and cross-comparison evolved gradually and, with that development came many decisions about the nature of the time scale and how it should be measured, documented, and codified. These decisions typically were taken at international geological congresses by large multinational committees established for such purposes (See Vai, 2007, for the early history). For example, the International Stratigraphic Guide, first published by Hedberg (1976), was an official product of the International Subcommission on Stratigraphic Classification of the International Union of Geological Sciences' Commission on Stratigraphy. For our purposes, the incremental nature of this method of work is significant because it is completely different from the basing of stratigraphic history on the broad, sweeping models of pulsation or cyclicity that have so frequently arisen during the evolution of the science of geology, a topic to which we now turn.

## 1.4 The Continual Search for a "Pulse of the Earth"

The self-appointed task of geologists is to explain the Earth. Given that Earth is a complex object affected by multiple processes, there is a natural drive to attempt to systematize and simplify these processes in our hypotheses of how Earth works. Numerical modeling, which has become popular in many fields with the advent of small but powerful and cheap computers, is but the most recent manifestation of this tendency, and is now widely used by earth scientists. The purpose of this section is to show how the idea of a worldwide stratigraphic pattern, as exemplified by the Exxon sequence model of the 1970s, is but the latest example of a theme that runs through the entire course of modern Geology.

Two themes that recur throughout the evolution of geological thought are *pattern recognition* and *cyclicity*. Zeller (1964) demonstrated the ability of geologists to recognize patterns in data where none exists. In a famous psychological experiment he constructed simulated stratigraphic sections from lists of random numbers (in fact, digits from lists of phone numbers in a city phone directory), using the numbers to determine rock types and bed thicknesses. Professional geologists were then asked to "correlate" the sections, that is, to identify "beds" that extended from one "section" to the next. All were able to do so and, moreover, were able to develop comparisons with actual patterns of repetitive vertical order of rock types (sedimentary cyclicity) that had been well documented in the local outcrop geology and were well known to the professional geological community. Zeller explained these results thus:

> Psychologists, anthropologists, and philosophers of science have long recognized the fact that there is a fundamental need in man to explain the nature of his surroundings and to attempt to make order out of randomness …. The Western mind does not willingly accept the concept of a truly random universe even though there may be much evidence to support this view. ….. Science, to an extent matched by no other human endeavor, places a premium upon the ability of the individual to make order out of what appears disordered (Zeller, 1964, p. 631).

Dott (1992a) compiled studies of a particular recurring obsession of geologists, that of the idea of repeated changes in global sea-level. The idea that the formation and subsequent melting of continental ice caps would affect sea levels by first locking up on land, and then releasing back to the oceans, large volumes of water, is attributed to a newspaper publisher, Charles Maclaren in 1842, and appeared in his review of Louis Agassiz' glacial theory. The Scotsman James Croll was the first to develop these ideas into a hypothesis of orbital forcing in 1864, but the idea received no serious attention until the 1920s. The word "eustatic," as applied to sea levels, and meaning sea-level changes of global scope, was proposed by Suess (1888) (these historical developments are summarized by Dott, in his introduction to the volume). But, as Dott's book demonstrates, ideas about the repetitiveness or periodicity of earth history have existed since at least the eighteenth century.

> Why should periodicity be such a powerful opiate for geologists? Obviously, periodicity comes naturally through the universal human experience of diurnal, tidal, and seasonal cycles. And it has ancient roots in the Aristotelian Greek world view of everything in nature being cyclic. The answer must lie more directly, however, in the innate psychological appeal of order and simplicity, both of which are provided by rhythmically repetitive patterns. For geologists the instinctive appeal to periodicity constitutes a subtle extension of the uniformity principle, which is in turn a special geological case of simplicity or parsimony (Dott, 1992b, p. 13).

To Charles Lyell, the founder of modern Geology, uniformitarianism included the concept that the Earth had not fundamentally changed throughout its history, and would not do so in the future. Earth history was not only directionless, but might also be cyclic (Rudwick, 1998; Hallam, 1998a). Lyell did not accept Darwin's ideas about organic evolution until late in his career, but held the opinion that most life forms had always been present on Earth, and if any were absent from the fossil record it was because of local environmental reasons or because the record had been destroyed by post-depositional processes, such as metamorphism. Lyell believed that in the future:

> Then might those genera of animals return, of which the memorials are preserved in the ancient rocks of our continents. The huge iguanodon might reappear in the woods, and the ichthyosaur in the sea, while the pterodactyl might flit again through the umbrageous groves of tree ferns (Lyell, 1830; cited in Hallam, 1998a, p. 134).

Lyell's ideas about the circularity of earth history were quickly discredited and discarded. However, his combination of inductive and deductive science and the attempt at building a grand, all-encompassing model that ultimately failed is uncannily similar to the

modern story of sequence stratigraphy set out by Miall and Miall (2001), as discussed in Chap. 12.

Through the latter part of the nineteenth century and, in fact, until the modern era of plate tectonics, most theories of Earth processes included some element of repetition or cyclicity. These theories were developed in the absence of knowledge of the Earth's interior, an absence that was not to be fully corrected until development of the techniques of seismic tomography in the 1970s (Anderson, 1989), which revealed for the first time how Earth's mantle really works. The impetus for the development of theories of cyclicity presumably arose from the tendency to seek natural order, as described by Zeller, Dott, and others. The more well-known of such theories were proposed by some of the more prominent geologists of their times, and typically seemed to represent attempts to reconcile and explain their knowledge of Earth's complex history accumulated over a lifetime's work.

Among the more important such theories was the model of worldwide diastrophism proposed by Chamberlin (1898, 1909; useful summaries and interpretations of Chamberlin's ideas are given by Conkin and Conkin, 1984 and Dott, 1992c) and elaborated by Ulrich (1911). In some fundamental ways this model contains the basis of modern concepts in sequence stratigraphy, although the papers are not cited by the main founder and "grandfather" of modern sequence stratigraphy, L. L. Sloss, in his first major paper (Sloss, 1963), or in his later work.

Chamberlin opened his paper with this remark:

> It was intimated in the introduction to the symposium on the classification and nomenclature of geologic time divisions published in the last number of this magazine [Journal of Geology] that the ulterior basis of classification and nomenclature must be dependent on the existence or absence of natural divisions resulting from simultaneous phases of action of world-wide extent (Chamberlin, 1898, p. 449).

Chamberlin made note of the widespread transgressions and regressions that could be interpreted from the stratigraphic record, and he understood the importance of regional uplift and erosion as the cause of widespread unconformities, which he termed "base-leveling." He suggested that "correlation by base-levels is one of the triumphs of American geology." (Chamberlin, 1909, p. 690) and emphasized that *the base-leveling process implies a homologous series of deposits the world over*" (emphasis by italics as in the original).

The concept of widespread unconformities, which was later to form the basis for the sequence stratigraphy of Sloss and Vail (see below) appears to have been an inevitable, inductive product of the mapping and data collection that was gradually being carried out at this time to document the North American continent. As noted by Carter (2007, p. 191), "North American geologists have a long history of recognizing and naming these large regional sediment packages, which they first described as dynasties or terranes. For example, Williams (1893, p. 284, 290) referred to the Green Mountain, Appalachian, Rocky Mountain and Glacial terranes as the main unconformity-bounded Phanerozoic sediment packages of North America. Each such unit was said to represent:"

> periods of continuity of deposition for the regions in which they were formed, separated from one another by grand revolutions interrupting the regularity of deposition, disturbed by faulting, folding and sometimes metamorphosing the older strata upon which the following strata rest unconformably and form the beginnings of a new system.

Blackwelder (1909), in an essay on unconformities, published a diagram (Fig. 1.5) that contains, in embryo, the sequences eventually documented and named in detail by Sloss (1963). Knowledge of these broad stratigraphic relationships seems to have formed the basis for much subsequent theorizing, although few references to this specific paper can be found in later work. Barrell (1917) referred to a different study by Blackwelder. Wheeler refers to it in his 1958 paper.

Chamberlin suggested that the base-levelings were caused by "diastrophism," that is, regional uplift and subsidence of the Earth's crust. He suggested that the movements were periodic.

> Reasons are growing yearly in cogency why we should regard the earth as essentially a solid spheroid and not a liquid globe with a thin sensitive crust. I think we must soon come to see that the great deformations are deep-seated body adjustments, actuated by energies, and involving masses, compared to which the elements of denudation and deposition are essentially trivial. Denudation and deposition seem to me clearly incompetent to perpetuate their own cycles. It seems clear that diastrophism is fundamental to deposition, and is a condition prerequisite to epicontinental and circum-continental stratigraphy (Chamberlin, 1909, p. 693).

According to Chamberlin the worldwide episodes of diastrophism would have four important outcomes:

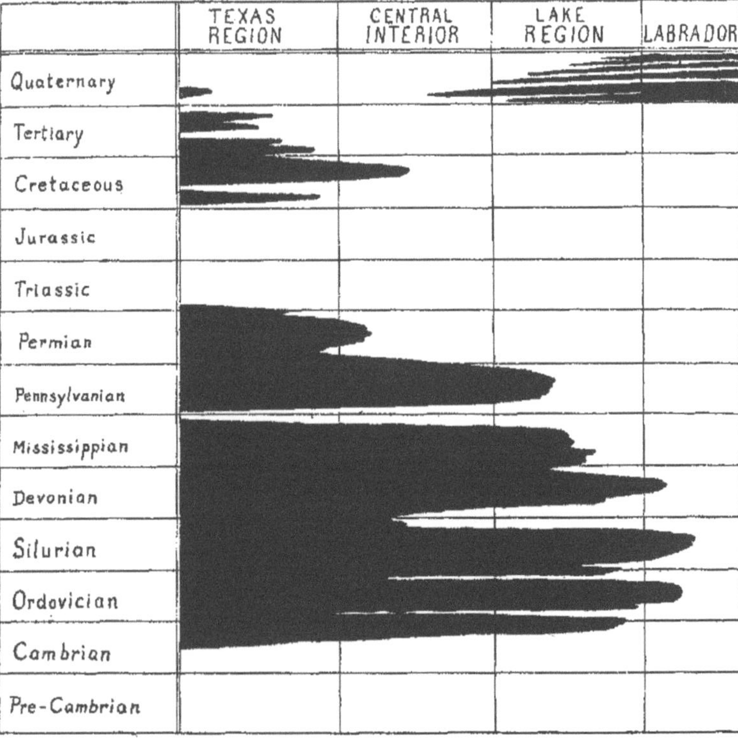


**Fig. 1.5** The broad stratigraphic patterns of the North American continent, as they were known in the early twentieth century. From Blackwelder (1909). This diagram was drawn to illustrate the concept of the widespread unconformity, and the data it illustrated formed the basis for the concepts of rhythmic diastrophism of Ulrich (1911), the regional petroleum evaluation studies of Levorsen (1943), and the sequences of Sloss (1963)

(1) diastrophic uplift and subsidence of the Earth' surface would cause the development of worldwide unconformities; (2) such episodes of uplift and subsidence would affect global sea levels (Chamberlin did not have a term for this. The word "eustasy" emerged later, from the work of Suess, as noted above); (3) the rise and fall of the ocean, in alternately expanding and contracting the area and depth of the seas, would affect the living space and ecology of life forms, and would therefore be a major cause of organic evolution, which would explain the worldwide synchroneity of successive faunas; and (4) uplift and subsidence would also affect the area of the Earth undergoing erosion, which would, in turn, control the level of carbon dioxide in the atmosphere. Chamberlin was one of the first to realize the importance of $CO_2$ as a greenhouse gas (he did not use this term, either), and attributed Earth's changes in climate through the geologic past to this process. As Dott (1992c, p. 40) noted, with this theory Chamberlin provided much of the foundation of modern sequence stratigraphy and of modern ideas about climate change. He illustrated the formation of continental-margin sediment wedges by progradation, the sediment being derived by uplift and "base-leveling." These process, because of their effect on the stratigraphic record, provided the "ultimate basis of correlation," for Chamberlin.

In his paper, Ulrich (1911) developed Chamberlin's ideas further. He complained (p. 289) about the "Paleontological autocrat," a symbolic representation of the authority of biostratigraphic correlation which was, by its massing of detail, making it difficult to perpetuate the broad, sweeping generalizations about stratigraphic correlation that he preferred. He was also dubious about the supposed diachroneity of rock units, regarding such a process as insignificant relative to the regional correlatability of geological formations (Ulrich, 1911, p. 295). Here is an excellent example of the model-building paradigm at work—in assessing the stratigraphic record Ulrich placed higher value on his interpreted generalizations than on the actual empirical evidence from the rocks. Ulrich opposed the idea of "dual nomenclatures" for rocks and for time, preferring to see his natural stratigraphic subdivisions as a sufficient basis for stratigraphic classification. The following quotations from this paper provide a remarkable foretelling of many of the principles of sequence stratigraphy:

In my opinion a rhythmic relationship connects nearly all diastrophic movements. For a few the meter is very long, for others shorter, and for still others much shorter. The last may be arranged into cycles and these again into grand cycles, the whole arrangement probably corresponding in units to the divisions of an ideal classification of stratified rocks and, so far as these go, of geologic time. .... As I shall endeavor to show .... Diastrophism affords a true basis for intercontinental correlation of not only the grander cycles by also of their subordinate stages. .... The principle of rhythmic periodicity being recognized, it seems to me merely a matter of time and close comparative study of sedimentary rocks and faunal associations to determine the time relations of interruptions in sedimentation in any one section to similar interruptions in another (Ulrich, 1911, p. 399).

Displacement of strandline chiefly relied on in proving periodicity of deformative movements.—The only thing that moves ... and which, therefore, offers the most reliable criteria in determining the periodicity and contemporaneity of diastrophic events, is the level of the sea. .... Whatever the qualifications, there yet remains the fact that the strandline is contemporaneously and universally displaced (Ulrich, 1911, pp. 401–402).

Accuracy in correlation, whether narrow or intercontinental in scope, depends solely on the uniform application of the criteria and principles adopted, and that if our practice is thoroughly consistent we shall finally succeed in discovering physical boundaries what will separate the systems so that none will include beds of ages elsewhere referred to either the preceding or succeeding period (Ulrich, 1911, p. 403).

In these three paragraphs we see in embryo the concepts of a cycle hierarchy, the idea of sedimentary accommodation, and the idea of the preeminent importance of the sequence boundary as a time marker. His model of diastrophic periodicity is illustrated in Fig. 1.6.

Ulrich is referring here to the idea of "natural" subdivisions of geologic time into what we would now call sequences, as a practice to be preferred to the use of the European-based stage and series nomenclature. In the North American successions with which Ulrich was familiar, most of the boundaries between the series and stages occurred within conformable stratigraphic successions, and this was reason for him to question their validity and usefulness. In his paper Ulrich provided diagrams that illustrate sedimentary overlap, and discussed the implications of these structural arrangements for documenting marine regression and transgression.

The Chamberlin-Ulrich model was very influential on later generations of geologists. It undoubtedly influenced Joseph Barrell of Yale University, whose classic 1917 paper begins in this way:

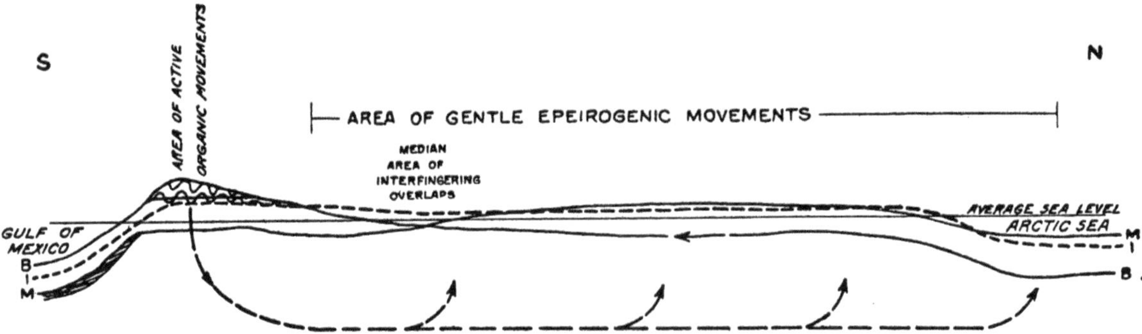


**Fig. 1.6**  Ulrich's (1911) model of diastrophic periodicity

Nature vibrates with rhythms, climatic and diastrophic, those finding expression ranging in period from the rapid oscillation of surface waters, recorded in ripple mark, to those long-defended stirrings of the deep titans which have divided earth history into periods and eras.

Barrell (1917, p. 750) was aware of climatic cycles and discussed what we would now call orbital-forcing mechanisms (e.g., the "precession cycle of 21,000 years"). He discussed the major North American orogenic episodes and their influence on the broad patterns of stratigraphy, referring (p. 775) to the rise and ebb of sea level, which "pulsates with the close of eras, falling and then slowly rising again," as "the most far-reaching rhythm of geologic time." However, the main focus of this important paper is on attempts to establish the rates of geological processes and the measurement of the length of geologic time, given the new impetus to the study of this problem provided by the discovery of radioactivity. He refers to "diastrophic oscillation," but only from the understanding such a process may provide for the interpretation of the stratigraphic record, not as a fundamental mechanism to be used as a basis for the definition of geologic time.

Having compiled a great deal of information about the nature and rates of Earth processes, and having assessed the ages of the major eras in earth history, including that of the major diastrophic episodes, Barrell (1917, p. 888) suggested that "There appears to run through geologic time a recurrence of greater crescendos which in their average period approach in round numbers to 200,000,000 years." But then, after some discussion of this periodicity, he warned that "There is a human tendency, however, to seek for over-much regularity in nature and it is doubtful if much weight should be attached to this cycle of

approximately 200,000,000 years. Although extremely suggestive of a new perspective, there are not enough terms, nor are they sharply enough defined, above those of lesser magnitude to give this indication more than such suggestive value" (Barrell, 1917, pp. 889–890). On the basis of this discussion Barrell does not appear to have been one of those who regarded some "pulse of the earth" as central to geologic history.

A succession of late Paleozoic deposits that is widely exposed in the continental interior of the United States has had an exceptionally important influence on the development of ideas about cyclicity in the geological record. Johan August Udden is credited with being the first to recognize (in 1912) that a coal-bearing succession of Pennsylvanian age in Illinois contained a repetition of the same succession of rock types, which he attributed to repeated inundations of the sea during basinal subsidence (Langenheim and Nelson, 1992; Buchanan and Maples, 1992). In 1926, the Illinois Geological Survey began a stratigraphic mapping study of these deposits, under the direction of J. Marvin Weller. "As this study proceeded he [Weller] was impressed by the remarkable similarity of the stratigraphic succession associated with every coal bed. . . .. Their studies showed that the Pennsylvanian system in the Eastern Interior basin consists of repeated series of beds or cyclothems, each of which is composed of a similar series of members" (Wanless and Weller, 1932; they did not formally acknowledge Udden in this work). This paper contained the definition of the term *cyclothem*, for a particular type of cyclic or repeated pattern of sedimentation. Cyclothems are typically no more than a few tens of metres in thickness and, we now know, each represents a few tens to hundreds of thousands of years of geologic time. They are particularly characteristic of upper Paleozoic successions, for reasons that Shepard and Wanless (1935) were to suggest. Mapping by Weller and his colleagues was the first to demonstrate that these cyclothems underlie much of the continental interior of the United States. Weller suggested diastrophism as the cause of the cycles (Weller, 1930), but a different mechanism was proposed a few years later. "It happens that there is abundant evidence of the existence of huge glaciers in the southern hemisphere during the very times when these curious alternations of deposits were being formed. A relation between these continental glaciers and the sedimentary cycles has been proposed recently by the writers" (Shepard and Wanless, 1935). The authors proceeded to provide a sedimentological interpretation of how climatic and eustatic oscillations associated with the formation and melting of continental ice caps could have generated the succession of deposits that characterize the cyclothems. They relegated diastrophic causes to a secondary role in cyclothem generation, suggesting that tectonic movements would have been too slow. Thus was borne a very important hypothesis about the relationship between cycles of glaciation, sea-level change and sedimentation, although nobody seems to have made the connection between the Wanless-Shepard cyclothem hypothesis and the MacLaren-Croll orbital forcing concept until relatively recently (Crowell, 1978). The Wanless and Weller paper also clearly established the cyclothem as a stratigraphic concept, in the sense that the cyclothem constitutes a distinct type of mappable unit, distinct from the formation, which they described merely as "a group of beds having some [lithologic] character in common" (Wanless and Weller, 1932, p. 1003).

Through the first half of the twentieth century theories of Earth processes tended to include ideas about periodicity or rhythmicity. In part these ideas were fueled by the new knowledge of the driving force of radiogenic heat in the Earth's interior (e.g., Joly, 1930). As noted by Dott (1992b, p. 12) "by the 1940s the enthusiasm for global rhythms was overwhelming." This can be seen in the title of some of the major books of this period: Grabau's (1940) *The rhythm of the ages*, which contained his "pulsation theory" (Johnson, 1992b), and *The pulse of the Earth* (Umbgrove, 1947) and *Symphony of the Earth* (Umbgrove, 1950). Other "pulsation" theories of the period are noted by Hallam (1992a, b).

Of particular importance to our theme is the work of Grabau (1940), who developed a comprehensive theory of eustatic sea-level change based on the ideas of cyclic crustal expansion of the ocean basins (Johnson, 1992b). Grabau compiled a eustatic sea-level curve for the Paleozoic, based on his own wide-ranging stratigraphic compilation, which showed episodes of continental transgression interspersed with episodes of tectonic uplift and regression (Fig. 1.7). Grabau based his documentation of sea-level events on offlap-onlap relationships, just as did Vail some 30 years later (Fig. 1.8). Although Grabau was noted for his massive data compilation, he did very little field work of his own after 1920, shortly after he moved from the United States to China (Johnson, 1992b). As to his method, M. E. Johnson (1992b, p. 50) quoted Grabau as follows:

> It is not a question of coining a plausible theory of world evolution and then attempting to apply it superficially to the history of all continents. The theory is rather a summation of the critical study of stratigraphic and pale-ontological facts from all parts of the works assembled by me during a period of more than 30 years. (Grabau, 1936a, p. 48).

In this statement Grabau was in effect claiming to be carrying out inductive science—the building of a hypothesis from dispassionately collected data. He was answering a criticism by Hans Becker (cited in Grabau, 1936a) in which "Becker doubted the wisdom of applying an untested theory to the whole world." Becker argued that the proper approach would be to "begin such an attempt in one continent and check the results with the facts gathered in other parts of the earth." Becker clearly suspected that Grabau was being model-driven, and he argued for the classic empirical observation and replication approach. M. E. Johnson's (1992b, p. 50) conclusion about all this is that Grabau's ideas were "not a theory in search of data, but rather a set of data somewhat reluctantly entrusted to a theory of murky crustal mechanisms." Johnson argues that Grabau came late in his career to his model of eustasy and that it therefore represents an empirical construction.

Another influential model was that of Hans Stille (1924) who postulated an alternation of epeirogenic and orogenic episodes affecting all the continents. He named some thirty orogenic episodes which were believed to have global significance (his work is summarized by Hallam, 1992b). As recently as the late 1970s, Fischer and Arthur (1977) plotted graphs

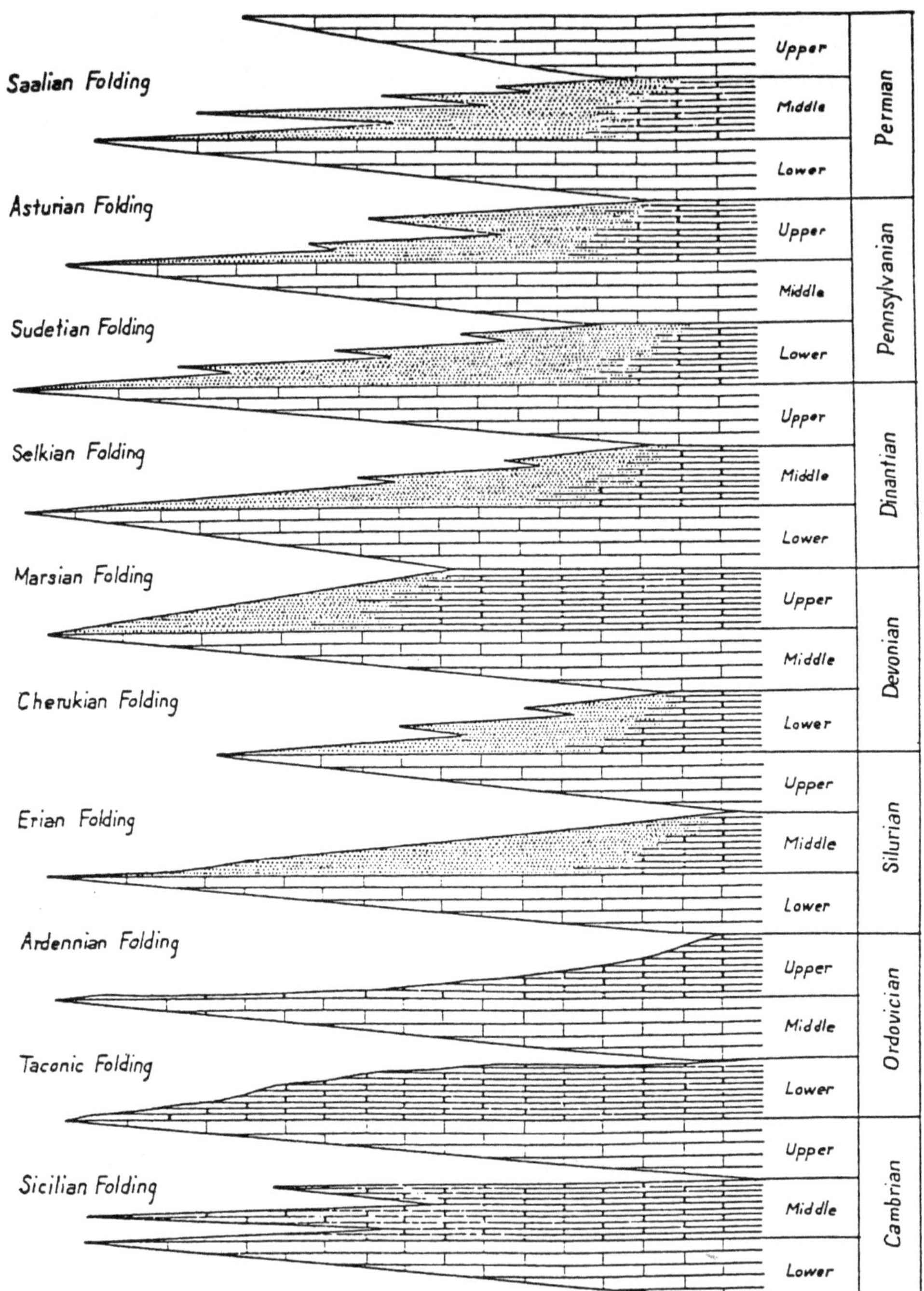



**Fig. 1.7** Grabau's curve of eustatic sea-level events, illustrating his "Pulsation Theory" (Grabau, 1936b)

of organic diversity through the Mesozoic-Cenozoic, which they compared with Grabau's eustatic cycles, and believed demonstrated a 32-million-year cyclicity.

By the late 1920s the ideas of Chamberlin and Ulrich about the periodicity of earth processes had become very popular, but were strongly opposed by some skeptics. For example, Dott (1992c, p. 40) offered the following quote from this period:

So much nonsense has been written on various so-called ultimate criteria for correlation that many have the faith or the wish to believe that the interior soul of our earth governs its surface history with a periodicity like the

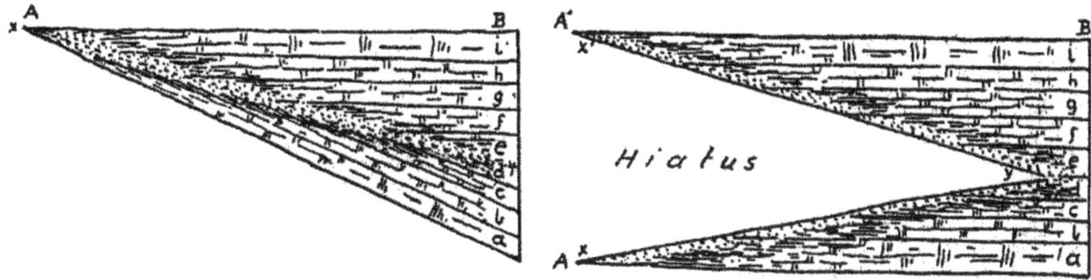

**Fig. 1.8** Grabau's model of offlap-onlap relationships, shown on the *right* in the form of a chronostratigraphic diagram (Grabau, 1906)

clock of doom, and that when the fated hour strikes strata are folded and raised into mountains, epicontinental seas retreat, and the continents slide about, the denizens of the land and sea become dead and buried, and a new era is inaugurated. This picture has an epic quality which is very alluring and it makes historical geology *so very understandable*, but is it a true picture? (Berry, 1929, p. 2; italics as in original).

The work of Chamberlin, Ulrich, and Grabau, and the development of the cyclothem concept were essentially academic and theoretical, and did not appear to directly affect the practice of stratigraphy, particularly as it was carried out by petroleum geologists. The distinguished petroleum geologist A. I. Levorsen was one of the first to describe in detail some examples of the "natural groupings of strata on the North American craton:"

A second principle of geology which has a wide application to petroleum geology is the concept of successive layers of geology in the earth, each separated by an unconformity. They are present in most of the sedimentary regions of the United States and will probably be found to prevail the world over (Levorsen, 1943, p. 907).

This principle appears to have been arrived at on the basis of practical experience in the field rather than on the basis of theoretical model building (Fig. 1.9).

These unconformity-bounded successions, which are now commonly called "Sloss sequences," for reasons which we mention below, are tens to hundreds of metres thick and, we now know, represent tens to hundreds of millions of years of geologic time. They are therefore of a larger order of magnitude than the cyclothems. Levorsen did not directly credit Grabau, Ulrich, or any of the other theorists, cited above, who were at work during this period, nor did he cite the description of unconformity-bounded "rock systems" by Blackwelder (1909). Knowledge of these seems to have been simply taken for granted. It presumably was based on long practical experience carrying out regional petroleum exploration across the continent.

A general model for the architecture of continental-margin sedimentation was proposed by Rich (1951; see Fig. 1.10) and was used by Van Siclen (1958) in his cyclothem model (Fig. 1.11). This model was developed to explain the cyclothemic deposits of late Paleozoic age as they draped across the margin of the craton in central Texas (e.g., see Fig. 7.41). It anticipated modern sequence models by some 20 years. The *undaform* environment corresponds to the continental margin, a region up to several hundreds of kilometers wide and under water depths of less than about 200 m,

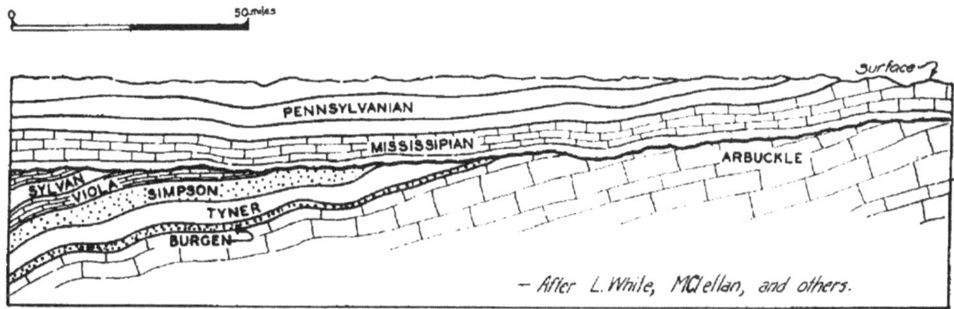


**Fig. 1.9** Examples of Levorsen's "Layers of geology" (Levorsen, 1943). AAPG © 1943. Reprinted by permission of the AAPG whose permission is required for further use

**Fig. 1.10** The three environments of Rich (1951)

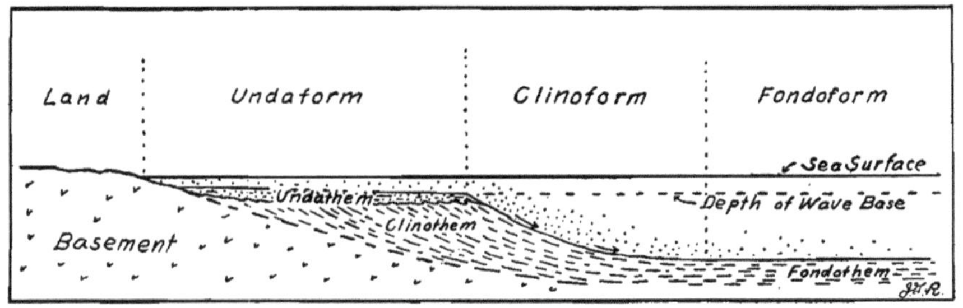


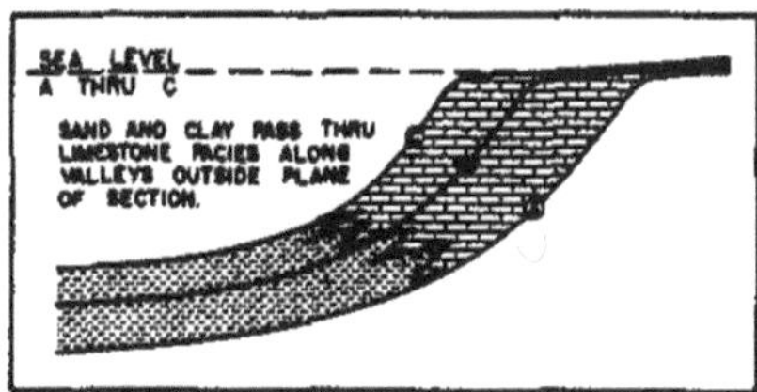


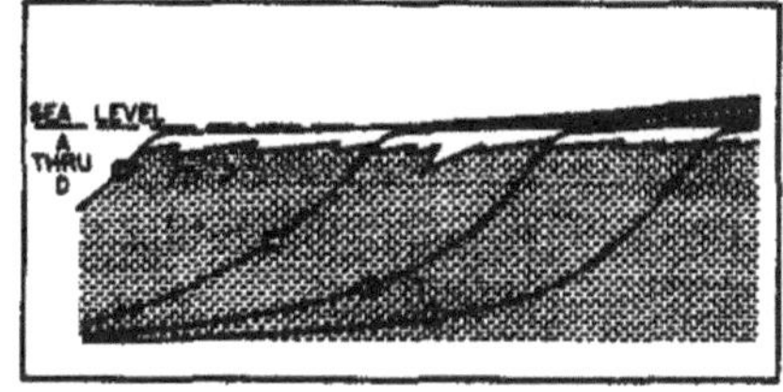


A. STABLE SEA LEVEL

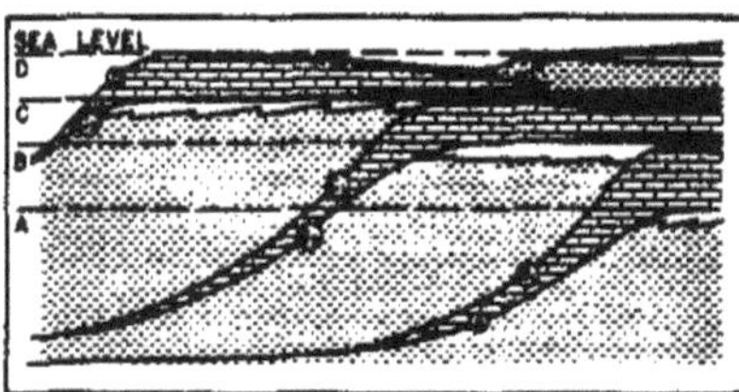


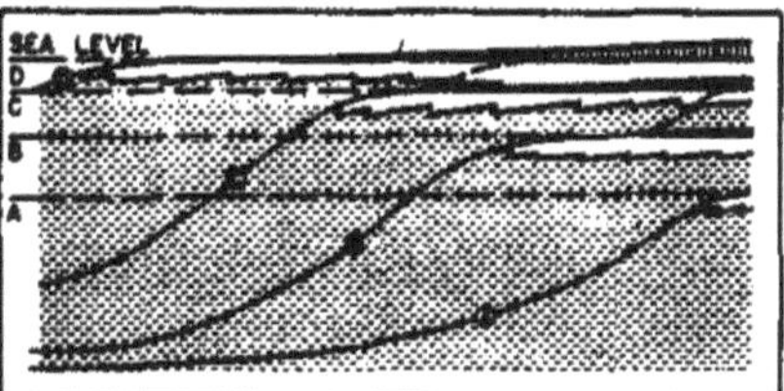


B. INTERMITTENTLY RISING SEA

DASHED LINES SHOW SOME INTERMEDIATE SURFACES OF DEPOSITION

**Fig. 1.11** Subsurface exploration of the Upper Paleozoic section along the shelf margin in central Texas after WW2 generated shelf-to-basin cross-sections that displayed a strong cyclothemic cyclicity. This is the set of models developed by Van Siclen (1958) to explain the stratigraphic architecture in terms of different patterns of sea-level change. AAPG © 1958. Reprinted by permission of the AAPG whose permission is required for further use

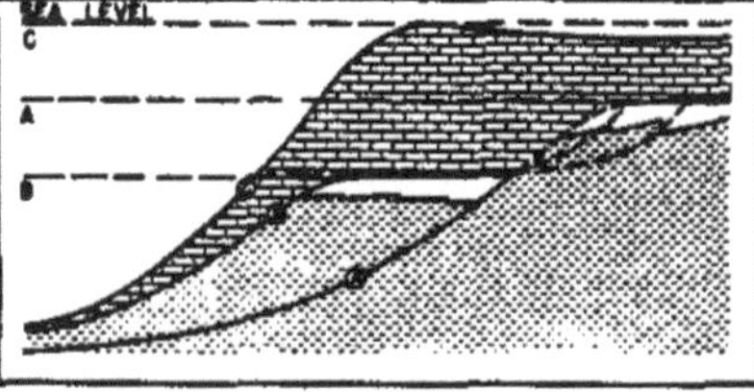


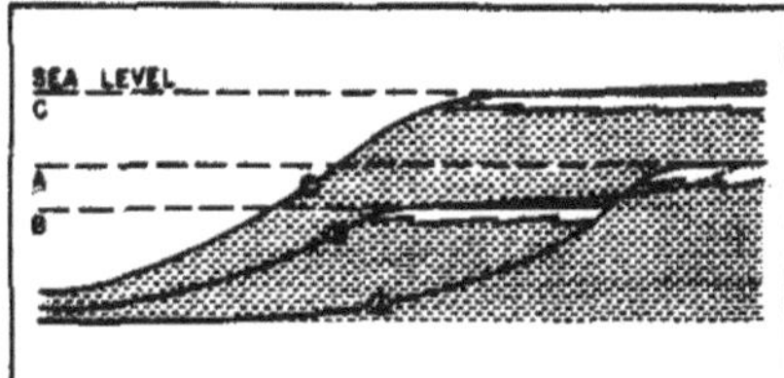


C. DROP FOLLOWED BY RISE ABOVE ORIGINAL SEA LEVEL

DASHED LINES INDICATE EROSION

commonly only a few tens of metres. The *undathem* is the body of sediments developed within the undaform, and consists of shelf and platform deposits, such as carbonate reef and backreef sediments, and shelf sands deposited by storms and tides. The continental slope is the clinoform. The dipping packages of strata that constitute the *clinothem* are a very important component of continental-margin sequences, and can readily be seen on reflection-seismic data, as shown later. The *fondoform* is a deep-water environment, where sediment supply and sedimentation rates are usually slow.

Although we now know that most continental margins, especially those on extensional-margin settings, have this basic architecture, the terminology that was proposed for the three subenvironments and their deposits has not survived. The term *clinoform* is the only one of the original terms that is in regular use. It is applied not to the environment but to the sediments that develop on the continental slope, as characterized by their distinctive depositional dips.

Larry Sloss of Northwestern University is commonly regarded as the "grandfather" of sequence stratigraphy, for two reasons. Firstly, his classic 1963 paper (an elaboration of the original Sloss et al., 1949 article) provided the foundation for the modern science, with its detailed documentation of the six fundamental sequences into which the North American cratonic Phanerozoic record could be subdivided (Fig. 1.12). Secondly, Sloss was the doctoral supervisor of Peter Vail, who showed how sequences could be recognized from modern seismic-reflection data and thereby provided a critical practical tool for petroleum geologists (Vail et al., 1977). Sloss (1963, p. 111) cited Levorsen's 1943 paper, and referred to the ideas as Levorsen's "layer cake" geology. Sloss (1963) suggested that the sequence concept "was already old when it was enunciated by the writer and his colleagues in 1948" and that "many other workers of wide experience have informally applied the sequence concept since at least the 1920s," although he did not cite any of the earlier work of Chamberlin, Ulrich or Barrell. He would undoubtedly have been aware of (but did not cite) Blackwelder's (1909) essay on unconformities, which includes a diagram of "the principle periods and areas of sedimentation" within North America, a diagram which contains Sloss's sequences in embryonic form (Fig. 1.5). Sloss may also have been thinking of cyclothems as representing a type of sequence,

although these units are of a smaller order of thickness than Sloss's six major sequences, and are not mentioned in his paper. Building on the work of Rich (1951), Van Siclen (1958) had already developed a sedimentological model for cycles such as the cyclothems (Fig. 1.11), which Sloss also did not cite, but which was to be re-invented by sequence stratigraphers Henry Posamentier and John Van Wagoner as part of their adaptation of seismic stratigraphy for use on outcrop and drill data (Van Wagoner et al., 1990).

A useful history of the development of modern sequence stratigraphy has been provided by the main protagonist, L. L. Sloss (1988a), who described the evolution of the field based on the studies by himself, his colleagues and his students, of the cratonic sedimentary cover of North America. His most important early paper (Sloss, 1963) established the existence of six unconformity-bounded sequences (Fig. 1.12) which he named using Indian tribal names in order to distinguish them from the conventional litho- and chronostratigraphic subdivisions, the latter having been mainly imported from Europe. This paper built on earlier work by Sloss et al. (1949) and was paralleled independently by Wheeler (1958, 1959a, b, 1963).

During this period prior to the appearance of seismic stratigraphy (1960s and early to mid 1970s) a few other workers were interested in the subject of global stratigraphic correlations, the possibility of eustatic sea-level change, and the geometries of stratigraphic units formed under conditions of fluctuating sea level. For example, Hallam (1963) reviewed the evidence for global stratigraphic events and was amongst the first to discuss the idea that global changes in sea-level may have occurred in response to changes in the volumes of oceanic spreading centres. Many of the ideas incorporated into current models of continental-margin sequence architecture were developed by workers analysing the Cenozoic record of the Gulf and Atlantic coasts of the United States. Curray (1964) was among the first to recognize the relationships between sea-level and sediment supply. He noted that fluvial and strandplain aggradation and shoreface retreat predominate under conditions of rising sea level and low sediment supply, whereas river entrenchment and deltaic progradation predominate under conditions of falling sea level and high sediment supply (Morton and Price, 1987). Curtis (1970) carried these ideas further, illustrating the effects of variations in the balance between subsidence and sediment supply

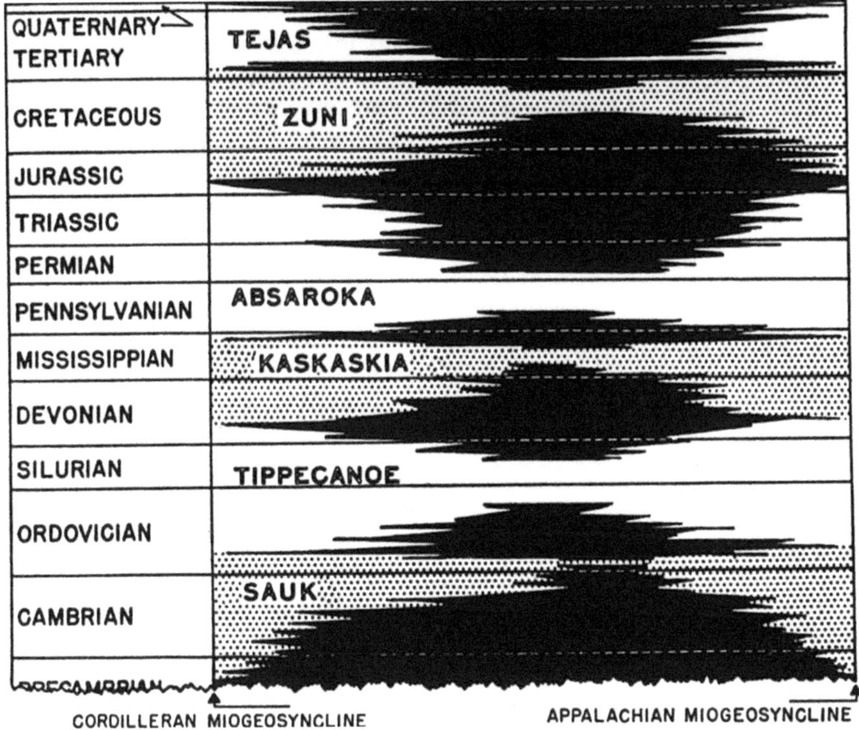



**Fig. 1.12**  The six classic North American sequences of Sloss (1963)

as controls on the stacking patterns of deltas, concepts that are now encapsulated by the terms progradation, aggradation and retrogradation (Fig. 1.13). Frazier (1974) subdivided the Mississippi deltaic successions into transgressive, progradational, and aggradational phases (Fig. 1.14), and discussed autogenic (delta switching) and glacioeustatic sedimentary controls. Brown and Fisher (1977) in a paper that actually deals with seismic data and appears in AAPG Memoir 26, summarized the ideas of an important group of stratigraphers at the Bureau of Economic Geology, University of Texas, that later became an integral part of the Exxon sequence-stratigraphy interpretive framework—the use of regional facies concepts to define *depositional systems* and *systems tracts*. Soares et al. (1978) attempted to correlate Phanerozoic cycles in Brazil with those of Sloss, and Hallam continued his detailed facies studies of the Jurassic sedimentary record, leading to successive refinements of a sea-level curve for that period (Hallam, 1978, 1981).

Sloss (1963) defined stratigraphic sequences as "rock-stratigraphic units of higher rank than group, megagroup, or supergroup, traceable over major areas of a continent and bounded by unconformities of interregional scope." With the advent of seismic stratigraphic research sequences much smaller than group in equivalent rank were recognized. This raised a nomenclature problem, as discussed below.

Wheeler (1958) is credited with the introduction of chronostratigraphic charts, in which stratigraphic cross sections are plotted with a vertical time axis rather than a thickness axis (Fig. 1.4). These diagrams are now referred to as *Wheeler diagrams* (Sloss, 1984). They are a useful way of indicating the actual range in age from place to place of stratigraphic units and unconformities. Vail and his team made use of these concepts in their analysis of seismic stratigraphic data (Fig. 1.15).

Sequence stratigraphy remained a subject of relatively minor, academic interest throughout the 1960s and 1970s (Ross, 1991), until the publication of a major memoir by the Exxon group in 1977, which revealed the practical utility of the concepts for basin studies and regional, even global, correlation (Payton, 1977). The work was led by Peter R. Vail, the

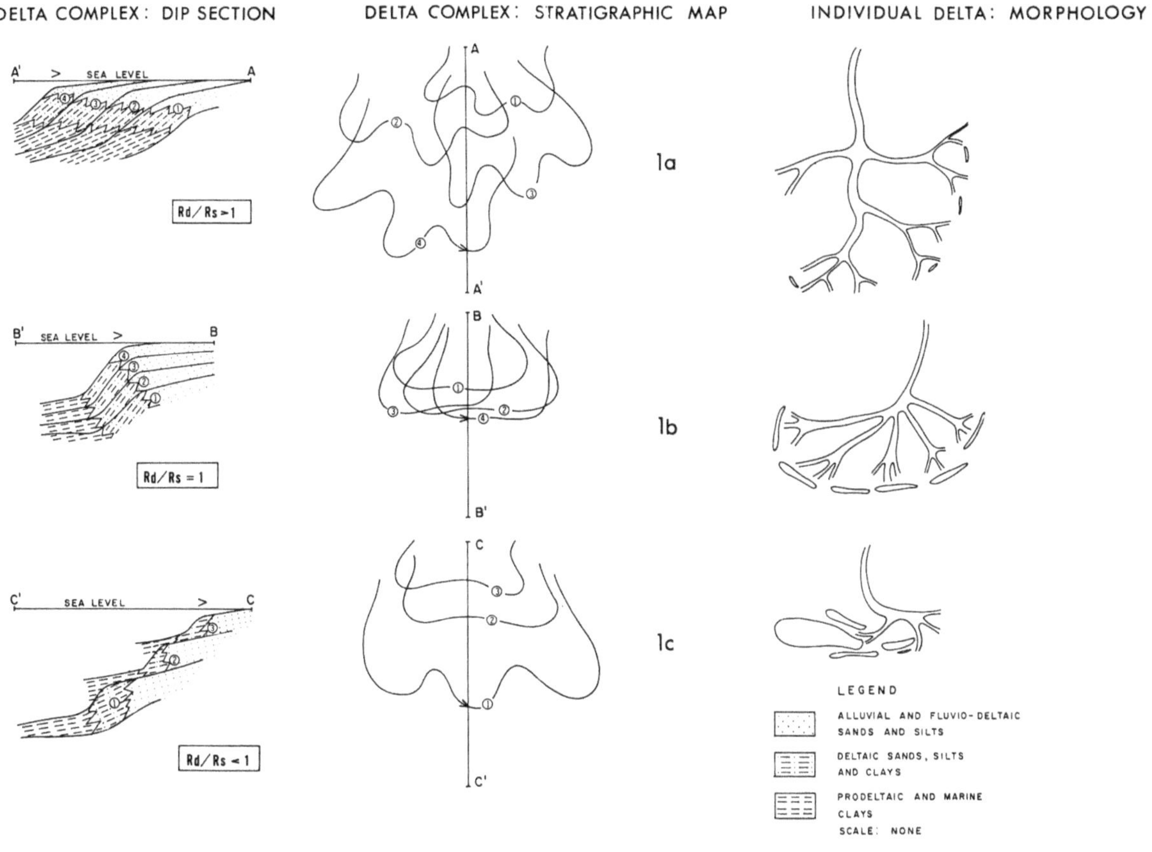



**Fig. 1.13** Relationship between rate of deposition (Rd) and rate of subsidence (Rs) in a delta complex. (**a**) Progradational, (**b**) aggradational, (**c**) retrogradational. Maps at *right* show successive positions of delta fronts (Curtis, 1970)

former graduate student of Sloss, and the research team also included several other former students of Sloss. Over a period of more than a decade in the 1960s and 1970s Vail and his coworkers studied seismic-reflection data, and the methods and results gradually evolved that eventually appeared in AAPG Memoir 26 (Vail, 1992). Seismic stratigraphy makes use of the concept that seismic reflections parallel bedding surfaces and are therefore of chronostratigraphic significance, enabling widespread correlation to be readily accomplished, although there are limitations to this general rule, where complex facies relationships may not be completely resolved by the seismic method (e.g., the pseudo-unconformities of Schlager, 2005, pp. 126–129; see also Christie-Blick et al., 1990, Cartwright et al., 1993). This contrasts with the correlation methods used in conventional outcrop basin analysis, in which lithofacies contacts are known to be typically diachronous. The resulting lithostratigraphic classification may have limited local applicability, and can only, with considerable effort, be integrated into a reliable chronostratigraphic framework (see discussion of mapping methods in Miall, 1999, Chap. 5).

Seismic stratigraphy permitted two major practical developments in basin analysis, the ability to define complex basin architectures in considerable detail, and the ability to recognize, map and correlate unconformities over great distances. Architectural work led to the development of a special terminology for defining the shape and character of stratigraphic surfaces. A particular emphasis came to be placed on the nature of bedding terminations because of the significance these carry with regard to the processes of progradation,

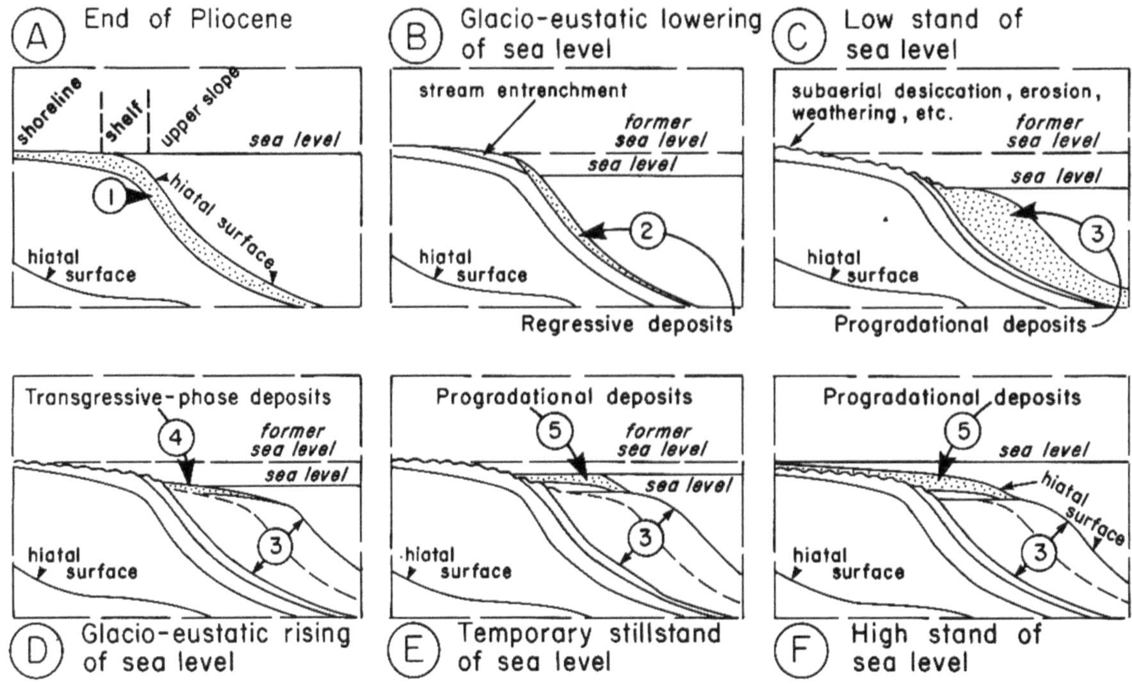


**Fig. 1.14**  The sequence-stratigraphic concepts of Frazier (1974)

aggradation and erosion (Fig. 1.16). Seismic work by Vail and his team eventually led to recognition of the sequence-stratigraphic model for the interpretation of seismic records and the building of regional and global stratigraphic syntheses. The basic components of this model are illustrated in Figs. 1.15 and 1.16, and a complete summary of modern Exxon models is contained in Chap. 2.

One of the original objectives of Vail's work with Exxon, which presumably reflected the early influence of Sloss, was to attempt to correlate seismic sequences from basin to basin in order to test the idea of regional and global cyclicity. As reported by Sloss (1988a), a large data base consisting of seismic sections, well records, biostratigraphic interpretations and related data was assembled within the Exxon group of companies worldwide. From this emerged the famous global cycle chart, which was first published in AAPG Memoir 26, and has subsequently gone through several revisions and refinements. The chart is introduced in Chap. 12 and discussed at some length in Chap. 14.

Vail's work is the latest manifestation of the "cyclicity" theme in the evolution of geologic thought,

and is dependent on the acceptance of the reality of "patterns" in the rock record. In contrast to the work on descriptive stratigraphy and chronostratigraphy described in the previous sections of this chapter, Vail's science is clearly deductive in nature, and constitutes a distinct paradigm that has, since the 1970s, coexisted with the paradigm of empirical stratigraphy.

## 1.5 Problems and Research Trends: The Current Status

Vail's work, beginning with AAPG Memoir 26 (Payton, 1977) has, of course, revolutionized the science of stratigraphy, and Vail himself has been much honoured as a result. This is as it should be. However, as with the development of any major new paradigm, many problems, some critical, have developed in the application of sequence concepts, and much research remains to be carried out. Considerable controversy remains regarding the existence of a worldwide sequence framework, and in the

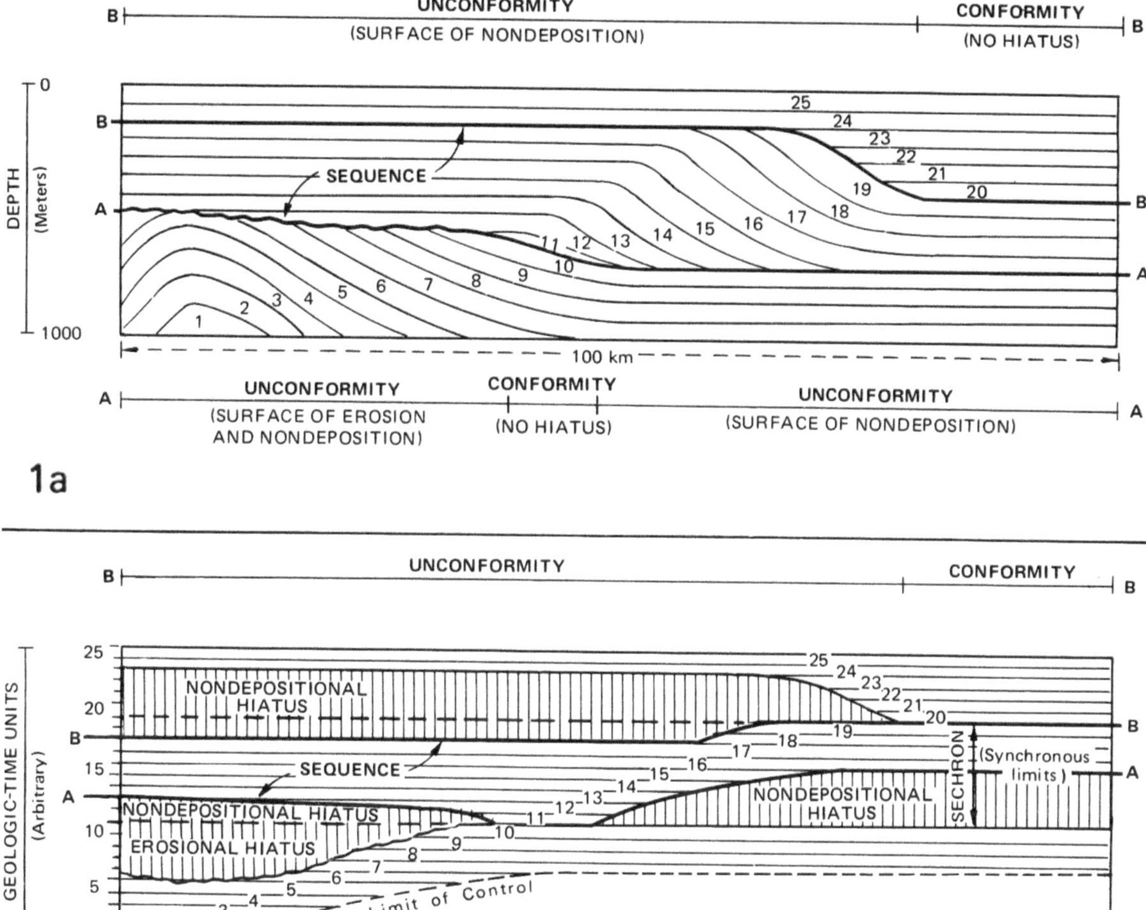



**Fig. 1.15** Basic concepts of the depositional sequence. (a) Stratigraphic geometry. Three sequences are shown, separated by unconformities (a) and (b). (b) Chronostratigraphic chart (Wheeler diagram) of the same succession as in (a), emphasizing the time breaks in the succession (Mitchum et al., 1977b). AAPG © 1977. Reprinted by permission of the AAPG whose permission is required for further use

interpretation of the origins of several types of sequence. There are several separate but related types of problem:

(1) The basic sequence models erected by the Exxon group were intended for a specific type of tectonic setting, that of extensional continental margins, and should be used with caution in other types of setting. The original models were also very simplistic with regard to the nature of the balance among the three major controls of basin architecture: subsidence, sea-level change, and sediment supply, and were developed primarily for siliciclastic sediments. Carbonate and evaporite sequence models required major revisions of the Exxon models (Sect. 2.3.3).

(2) The problem of causality is a critical one. Much work has been done and remains to be done to

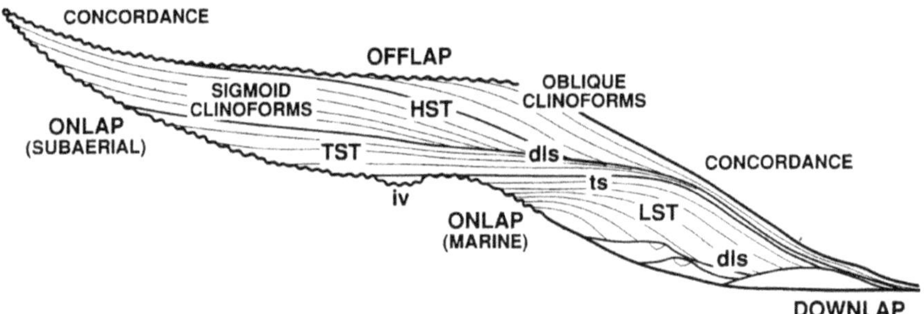


**Fig. 1.16** The basic sequence-stratigraphic model, illustrating strata geometries and terminology. Clinoforms are dipping stratal surfaces that indicate lateral progradation or accretion of the section. Abreviations: *dls*, downlap surface, *ts*, transgressive surface, *iv*, incised valley, *LST*, lowstand systems tract, *TST*, transgressive systems tract, *HST*, highstand systems tract (Christie-Blick, 1991)

investigate the processes that generate unconformities. They are the result of erosion following changes in sea level or the elevation of the continents. Some causal mechanisms are regional in scope, others global. Some are rapid in their effects, others slow, even by geological standards of time (Part III). Distinguishing between all these mechanisms involves a consideration of the third problem.

(3) The problem of correlation. Are sequences regional or global in scope? Answering this problem is one of the first important tasks in attempts to distinguish between the effects of global processes, such as eustatic changes in sea level, and the results of regional processes, including various types of tectonism. Correlation among sequences and other stratigraphic and tectonic events is critical in answering this question. However, correlation is a problem, because of imprecisions in the geological time scale and difficulties in generating sufficiently accurate dates for any given stratigraphic successions. Serious questions remain about the construction, meaning, and utility of the Exxon global cycle chart (Part IV).

Approaches to these various problems have evolved into several categories of research (see also Miall, 1995a):

(1) Theoretical geophysical studies of crust and mantle processes in a search for mechanisms of continent and ocean elevation change (Chaps 9 and 10).

(2) Modelling of basins, employing numerical manipulation and graphical computer simulation to integrate the effects of subsidence, sea-level change and sediment supply in various tectonic settings. Such studies are commonly tested against data from real basins in the process known as forward modeling, using the research described in the next two paragraphs.

(3) Detailed stratigraphic studies, employing refined chronostratigraphic methods to assess the significance of unconformities, sequence boundaries etc., and to relate these to regional tectonic events and the global cycle chart (so-called "tests" of the Vail curve). These studies are focusing primarily on the Mesozoic and Cenozoic, for which the stratigraphic record is reasonably complete (Part IV).

(4) Detailed stratigraphic studies of the Cenozoic record have become extremely sophisticated, especially the late Cenozoic, for which the marine stratigraphic record provides excellent undisturbed sections to which many separate techniques can be applied (magnetostratigraphy, strontium- and oxygen-isotope stratigraphy, refined biostratigraphy), and correlations can be carried out with the record of glacioeustasy (facies and paleoecological changes in the sedimentary record) and the Milankovitch astronomical periodicities. This has evolved into a new subdiscipline, *cyclostratigraphy* (Chap. 11, Sect. 14.7).

(5) Investigation of modern and very recent sea-level change around the world, and the effects of glacioeustasy, carried out as a means of providing a base-line of well-constrained studies of what is happening now, largely as a way to document the possible effects on oceanic volumes of melting continental ice caps. This specialized work carries the subject to levels of measurement detail and chronostratigraphic refinement that are beyond the scope of most geological research, and the work is not discussed in this book.

## 1.6  Current Literature

Good overviews of the subject of sequence stratigraphy, from various perspectives, can be obtained from several recent books, special journal issues, and review articles. These include the following:

1. *Books*: Miall's (1999) textbook on basin analysis includes a fairly succinct review of sequence stratigraphy and global stratigraphic cycles written at the graduate level. It carries the development of techniques up to the late 1990s. Hallam (1992a) discussed sea-level changes throughout the Phanerozoic, and evaluated the various mechanisms that have been proposed for sea-level change. Walker and James (1992) provided a thorough, updated version of the Geological Association of Canada "Facies Models" book, written at the advanced undergraduate level. It contains extensive discussions of the stratigraphic effects of sea-level change, but there is little discussion of mechanisms in this book beyond a useful introductory chapter. Exxon models and terminology are deliberately avoided in this book.

Detailed treatment of sequence-stratigraphic models, including the recognition and documentation of sequences in outcrop, drill core, well logs and seismic data, have been provided by several recent books, including Emery and Myers (1996), Posamentier and Allen (1999, focusing on clastic sequences), Coe (2003), Schlager (2005, focusing on carbonate sequences) and Catuneanu (2006). These books all focus on the recognition and definition of sequences, and do not examine the mechanisms for generating sequences, or the problems and controversies surrounding the issues of sequence correlation and its bearing on the testing of eustatic models of sequence

generation. That by Catuneanu (2006) is comprehensively referenced, contains examples and illustrations from around the world, and is characterized by a consistent emphasis on the temporal significance of sequences, systems tracts and bounding surfaces; it seems likely to achieve recognition as the standard work on the subject.

2. *Research syntheses* (books and special journal issues): Essential reading is the first major publication on sequence stratigraphy, the AAPG Memoir which established seismic stratigraphy as a major new technique (Payton, 1977). A second AAPG collection of articles on seismic stratigraphy appeared in 1984, and constituted an attempt to examine some of the major premises of the Exxon work, such as the significance of condensed successions in the sedimentary record, and the ability to correlate unconformities using the limited well and seismic data available from specific continental margins (Schlee, 1984). Bally (1987) compiled three volumes of seismic-stratigraphic studies in a large, atlas format. These include important introductory papers dealing with Exxon methods, and numerous case studies from many types of basin around the world. Most of these are superbly illustrated with long seismic sections, many in colour. Berger et al. (1984) edited a major research compilation dealing with Milankovitch processes, including astronomical and climatic studies and many papers describing the geological evidence.

Two research collections published by the Society of Economic Paleontologists and Mineralogists, one edited by Nummedal et al. (1987), and the other by Wilgus et al. (1988), contain many important papers. The first focuses on studies of Quaternary sea-level change, and provides comparisons of coastal stratigraphy with the ancient record; the second contains a major suite of papers by the Exxon group, dealing with their sequence models. These papers provided contemporary expositions of the Exxon sequence-stratigraphy research, although much has changed since these book appeared. The papers by Jervey, Posamentier and Vail in Wilgus et al. (1988) constitute the next most important exposition of the Exxon models following Payton (1977), and contain the first detailed treatment of the sedimentology of sequences. Practical examples of this work are illustrated by Van Wagoner et al. (1990) in a well-illustrated review of outcrop and subsurface examples of Exxon-type sequences. In my opinion this is the best of the Exxon products because it contains numerous actual examples, and is less "model-driven" than the other papers by this group, although there are problems with some concepts and terminology (see Chap. 2).

Other collections of case studies include that edited by James and Leckie (1988), which draws particularly on the wealth of subsurface detail available for the Western Canada Sedimentary Basin. Collinson (1989) edited a compilation of studies of relevance to the petroleum industry, including techniques of correlation, with examples from the Arctic and North Sea regions. Cross (1990) put together a unique compilation of research articles describing quantitative approaches to basin modelling. Many important details of the sequence-stratigraphy story emerged in this book. Ginsburg and Beaudoin (1990) collected research on the Cretaceous system—a synthesis of the first major project of the Global Sedimentary Geology Program. Fischer and Bottjer (1991) provided an introduction to a special issue of Journal of Sedimentary Petrology dealing with Milankovitch rhythms. Macdonald (1991) compiled case studies of stratigraphic architecture in convergent and collisional plate settings. One of the major focuses of this book is to examine sequence stratigraphies in a tectonic context. A wealth of stratigraphic detail is contained in this book, but there are no synthesis or overview articles, except that by Carter et al. (1991), who tested the sequence-stratigraphic model and the global-cycle-chart model of Exxon against data from New Zealand. Revelle (1990) edited an important collection of review articles on sea-level change and its causes, including several referenced separately, below. Swift et al. (1991a) published a collection of research articles on shelf sedimentation. This includes a major set of theoretical papers that attempted to establish a quantitative framework for shelf sedimentation within a sequence framework. Cloetingh (1991) edited a special issue of Journal of Geophysical Research consisting of a collection of papers that examined the measurement, causes and consequences of long-term sea-level change. Advanced computer modelling and the use of refined stratigraphic data sets are two of the features of this collection. Another collection of papers in this area was edited by Biddle and Schlager (1991). Einsele et al. (1991a) edited a multi-authored compilation of chapters dealing with many aspects of cyclic and event stratification, including several useful overview chapters. An updated version of the Exxon approach to

sequence-stratigraphic analysis, including their first realistic appraisal of the importance of tectonism, is included in a chapter by Vail et al. (1991). Franseen et al. (1991) edited a collection of articles on the subject of sedimentary modeling, focusing on high-frequency cycles. Another useful collection of papers on tectonics and seismic sequence stratigraphy is that edited by Williams and Dobb (1993).

Weimer and Posamentier (1993) and Loucks and Sarg (1993) focus on the broader principles of siliciclastic and carbonate sequence stratigraphy, respectively. Many useful books consisting largely of case studies have been published during the last decade or so. Eschard and Doligez (1993) edited a suite of papers demonstrating how detailed outcrop studies, including documentation of high-resolution sequence stratigraphy, could be of use in the development of our understanding about petroleum reservoirs. A book edited by Posamentier et al. (1993) contains papers dealing with concepts and principles, methods and applications, and case studies. Van Wagoner and Bertram (1995) edited a collection of papers dealing with the sequence stratigraphy of foreland basins. The book contains an introductory article by Van Wagoner that provides revised definitions of many of the terms used in sequence stratigraphy. Norwegian work in the North Sea and Svalbard led to two important collections (Steel et al., 1995; Gradstein et al., 1998). The very useful book edited by Gradstein et al. (1998) contains a lengthy historical article by J. P. Nystuen, and several theoretical studies. Other sets of case studies have been compiled by Armentrout and Perkins (1991) and Hailwood and Kidd (1993).

One of the issues dealt with in Nystuen's (1998) article is the existence of several competing models for sequence documentation and classification. This has led to extensive debate regarding the relative importance of various surfaces that may be recognized in stratigraphic successions (see Sect. 1.7, Chap. 2). Hunt and Gawthorpe (2000) compiled a book that deals with one aspect of this debate, the significance of the process termed "forced regression," which is the term used to refer to the seaward migration of the shoreline, shallow-shelf wave ravinement erosion and shoreface sedimentation under conditions of falling sea level. Processes of sea-level change and its stratigraphic record are discussed in detail in a special issue of Basin Research edited by Fulthorpe et al. (2008).

Milankovitch cycles are discussed in books by Schwarzacher (1993) and Weedon (2003) that provide detailed treatments of the nature of orbital perturbations and spectral analysis of orbital and cyclic frequencies. A collection of research papers was published by de Boer and Smith (1994a). House and Gale (1995) focus on orbital forcing and cyclic sequences. More recent work includes the book edited by D'Argenio et al. (2004a). Shackleton et al. (1999) edited a collection of research papers dealing with developments in a cyclostratigraphic time scale.

A special issue of the journal *Stratigraphy* (vol. 4, pp. 2–3) published in 2007 contains many useful articles discussing the history and development of chronostratigraphy, including articles on current debates about principles and methods. Christie-Blick et al. (2007) contributed a very thoughtful article about sequence stratigraphy to this collection, an article that is cited at several places in this book.

3. *Review articles*: Useful reviews of Milankovitch-type cycles were given by Fischer (1986) and Weedon (1993). Burton et al. (1987) reviewed the lack of reference frames in attempts to quantify sea-level change, and the fact that it is very difficult to quantify and isolate the effects of the three main depositional controls, subsidence, sea-level change, and sediment supply. Cross and Lessenger (1988) reviewed the methods of seismic stratigraphy, and discussed some of the constraints on the use of seismic data in stratigraphic studies. Three useful companion studies by Christie-Blick et al. (1990), Christie-Blick (1991) and Christie-Blick and Driscoll (1995) reviewed current work on mechanisms of sea-level change and numerical modelling techniques. Schlager (1992a) provided a brief but well illustrated discussion of the sequence stratigraphy of carbonate depositional systems. This is an important contribution, containing many original ideas making the case that the standard Exxon model for clastic sequences cannot readily be applied to carbonate systems. Miall (1995a) outlined recent developments in research in the field of stratigraphy, including sequence stratigraphy. Wilson (1998) provided a set of personal reflections on the sequence stratigraphic "revolution" expressing some skepticism regarding the model of global eustasy.

Catuneanu et al. (2009) published a landmark study concerning the documentation, definition and classification of sequences, the objective of which is to reconcile the many debates that have raged in the lecture theatre, in panel debates, and in the published literature, for more than two decades. This paper is discussed below.

4. *Web sites*: There are numerous Internet resources that now deal with sequence stratigraphy. Many are teaching sites established by universities and consist primarily of repetitions of standard textbook fare. A few are exceptional and worth highlighting. That maintained by C. G. St. C. (Chris) Kendall at http://strata.geol.sc.edu/ is undoubtedly the most useful. At this site there are numerous examples, teaching modules, movies, galleries of photographs, and summaries and online debate concerning sequence definition and classification. Octavian Catuneanu has posted most of the illustrations from his 2006 book on a website from which they may be copied: http://research.eas.ualberta.ca/catuneanu/.

The site at the University of Georgia http://www.uga.edu/~strata/sequence/index.html provides a useful source of definitions. Basic concepts are defined by J. W. Mulholland at http://www.aseg.org.au/publications/articles/mul/ss1/htm. There are many other websites, including an article in Wikipedia, but none particularly recommends itself as a primary information source.

A different type of website is the official website of the International Commission on Stratigraphy at http://www.stratigraphy.org, which contains the online version of the International Stratigraphic Guide and maintains an up to date reference source for global stratotypes and the definitions and ages of global stage boundaries.

## 1.7 Stratigraphic Terminology

Sloss (1963) used the term "sequence" for his packages of strata. The term has had a varied usage since that time, having been employed informally, in a non-genetic sense, as a synonym for "succession" in some literature, and for cyclic or unconformity-bounded units of varying dimensions and time-spans. Sloss's original sequences represent sea-level cycles that were tens of millions of years in duration. They were termed second-order cycles by Vail et al. (1977; see Chap. 3), whereas the term "sequence" was used for much smaller packages of strata, representing sea-level cycles of a few million years duration (third-order cycles of Vail et al., 1977) in the first seismic work, that of Vail and his co-workers in AAPG Memoir 26 (Payton, 1977). Attempts to systematize the terminology have met with mixed success. Chang (1975) proposed the term synthem for unconformity-bounded sequences, the intent being that units would be defined and named using the word synthem in the same way as the term "formation" is used. This proposal was not formally adopted by the North American Commission on Stratigraphic Nomenclature (NACSN, 1983), but the International Subcommission on Stratigraphic Classification (ISSC, 1987) later approved the suggestion, and added the additional variants supersynthem, subsynthem and miosynthem for groups of synthems, and for two scales of subdivided synthem. Other proposed terms, such as interthem and mesothem were also discussed in this document (ISSC, 1987) but were not formally accepted. These terms are provided in the *International Stratigraphic Guide* (Salvador, 1994; but not in the online version). However, few, if any, geologists, have made use of them, and some (e.g., Murphy, 1988) were actively opposed to their adoption.

An alternative approach, contained in the North American code (NACSN, 1983), has met with more interest, perhaps only because of the more euphonious nature of the terminology. This is the method of allostratigraphy. "An allostratigraphic unit is a mappable stratiform body of sedimentary rock that is defined and identified on the basis of its bounding discontinuities." (NACSN, 1983, p. 865) A hierarchy of units, including the allogroup, alloformation and allomember, was proposed, and rules were established for defining and naming these various types of unit. Allostratigraphic methods enable the erection of a sequence framework that avoids the cumbersome nature of lithostratigraphy; for example, lateral changes in facies within a unit of comparable age, may involve a change in name, and similar units of similar age separated by facies change are typically assigned to different units. In the past, because of the localised nature of much stratigraphic research, different lithostratigraphic frameworks have commonly been erected for similar successions in separate geographic areas, and this has led to much confusion. An example of the allostratigraphic method is illustrated in Fig. 1.17, in which it can be seen that using an allostratigraphic approach to the subdivision of a fluvial-lacustrine assemblage, the natural subdivision of the succession into four sequences, bounded by breaks in sedimentation, can readily be formalised.

Several groups of workers are now explicitly employing allostratigraphic methods. For example, Autin (1992) subdivided the terraces and associated sediments in a Holocene fluvial floodplain succession into alloformations. R. G. Walker, A. G. Plint and

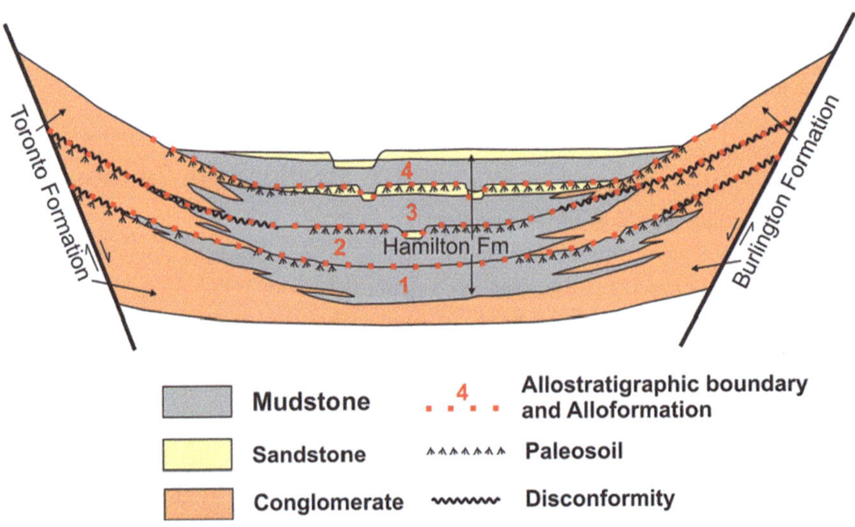



**Fig. 1.17** Example of the allostratigraphic classification of a fluvial-lacustrine assemblage in a graben. Numbers 1–4 correspond to alloformations, which are defined by breaks in sedimentation and cut across facies boundaries. Using lithostratigraphic methods each gravel, sand and clay unit would typically be given separate names (adapted from North American Commission on Stratigraphic Nomenclature, 1983)

their students and coworkers have employed allostrati-graphic terminology in their study of the sequence stratigraphy of part of the Alberta Basin, Canada. Their first definition of unconformity-bounded units is described in Plint et al. (1986), where the defining concepts are referred to as event stratigraphy, following the developments of ideas in this area by Einsele and Seilacher (1982). Explicit use of allostratigraphic terms appears in their later papers (e.g., Plint, 1990) and has become the standard method for the work of this research group (e.g., Varban and Plint, 2008). The standard Canadian text on facies analysis that builds extensively on the work of this group recommends the use of allostratigraphic methods and terminology as a general approach to the study of stratigraphic sequences (Walker, 1992). Martinsen et al. (1993) compared lithostratigraphic, allostratigraphic and sequence concepts as applied to a stratigraphic succession in Wyoming. As they were able to demonstrate, each method has its local advantages and disadvantages.

One of the achievements of seismic stratigraphy has been development of the ability to trace unconformity-bounded units into areas where the unconformable bounding surfaces are no longer recognizable. Thus Mitchum et al. (1977b, p. 53) defined a depositional sequence as "a stratigraphic unit composed of a relatively conformable succession of genetically related strata and bounded at its top and base by unconformities or their correlative conformities." Recognizing the bounding contacts in a conformable succession might, in practice, be difficult. The concept of a correlative conformity does not appear in the NACSN code, although some (e.g., Walker, 1992, p. 9) have recommended that it should (see Sect. 2.2). Great care needs to be taken in assessing the chronostratigraphic significance of unconformities and correlative conformities. Considerable disagreement exists regarding the correlation of sequence-bounding unconformities from basin margins, where they are commonly of subaerial origin and may be accompanied by major facies shifts, into deeper parts of the basin (Sect. 2.4). These concerns are echoed by Christie-Blick et al. (2007) who remarked (p. 222) that "some level of diachroneity [is] unavoidable" and that "at some scale, unconformities pass laterally not into correlative conformities but into correlative intervals."

Sequence classifications and allostratigraphic units are based on concepts of sequence scale and duration that are hierarchical in character (first- to sixth-order sequences; synthem and its variants; alloformations and allomembers). However, it is increasingly clear that sequences occur over a wide range of time scales and physical scales (e.g., thicknesses) that show no significant natural breaks, as would justify hierarchical classification (Sect. 4.2). In this book, sequences are described with reference to their duration (Chaps 5, 6 and 7), but formal hierarchical classifications are largely avoided, except where it is necessary to make reference to earlier literature. At the time of writing this book, Working Groups of the IUGS International Subcommission on Stratigraphic Classification and the North American Commission on Stratigraphy were wrestling with the problem of how to define sequences and how to incorporate sequence concepts into formal systems of stratigraphic nomenclature. A review paper by Catuneanu et al. (2009) represents the efforts of a large and diverse group of stratigraphers worldwide to develop recommendations for the formalization of sequence terminology. The current state of the debate is touched on in Chap. 2, but this book is not primarily about sequence terminology and models (for which the reader is referred to Catuneanu, 2006, and Catuneanu et al., 2009), but about the origins of sequences.

# Chapter 2

# The Basic Sequence Model

## Contents



## 2.1 Introduction

The purpose of this chapter is to present a succinct summary of sequence concepts, focusing on what has become the standard, or most typical, sequence model, and making only brief references to exceptions and complexities. These are described more completely elsewhere, most notably by Catuneanu (2006).

The nomenclature for sequence architecture and systems tracts was established initially by the Exxon group, led by Peter Vail, and included much new terminology, new concepts and new interpretive methods. It is important to be familiar with these, as the terms and concepts came to be used by virtually all stratigraphers. Application of the methods to different types of data (outcrop, well-log, 2-D seismic, 3-D seismic) from around the world permitted a quantum leap in understanding of basin architectures, but also revealed some problems with the Exxon concepts and terminology that have stimulated vigorous debate over the last two decades. The purpose of this chapter is to bring the debate up to date by presenting current concepts, with enough historical background that readers familiar with the debates will be able to comprehend the need for revisions and new approaches.

The history of sequence concepts is presented in Chap. 1. Current concepts and definitions are presented by Catuneanu (2006) and Catuneanu et al. (2009), from which much of this chapter is summarized. The reader is referred to these two sources which, at the time of publication of this book, could be considered the most up to date and definitive sources. The introduction to Chap. 5 in Catuneanu (2006, pp. 165–171) is particularly useful in providing a history of the controversies, the main players and publications, and the concepts and terminologies about which there have been dispute.

Sequence stratigraphic concepts are, to a considerable extent, independent of scale (Schlager, 2004; Catuneanu, 2006, p. 9). Posamentier and Vail (1988) and Posamentier et al. (1992) gave examples of sequence architecture evolving from base-level changes in small natural systems, and much experimental work has successfully simulated sequence development in small laboratory tanks (e.g., Wood et al., 1993; Paola, 2000). Interpretations of the temporal significance of sequences can, therefore, be difficult. However, as discussed later in this book, some aspects of the driving process, notably tectonism and climate change, generate facies and architecture that are sufficiently distinctive to yield reliable interpretations.


A.D. Miall, *The Geology of Stratigraphic Sequences,* 2nd ed.,
DOI 10.1007/978-3-642-05027-5_2, © Springer-Verlag Berlin Heidelberg 2010

## 2.2 Elements of the Model

A practical, working geologist faces two successive questions: firstly, is his/her stratigraphy subdivisible into stratigraphic sequences? And, secondly, what generated these sequences: regional tectonism, global eustasy, orbital forcing, or some other cause? The first part of this chapter deals with the methods for analyzing the sequence record. These include the following:

- The mapping of unconformities as a first step in identifying unconformity-bounded sequences.
- Clarifying the relationship between regional structural geology and the large-scale configuration of sequences.
- The mapping of onlap, offlap and other stratigraphic terminations in order to provide information about the internal architectural development of each sequence.
- The mapping of cyclic vertical facies changes in outcrop or well records in order to subdivide a stratigraphic succession into its component sequences and depositional-systems tracts, and as an indicator of changes in accommodation, including changes in relative sea-level.

The first three steps may be based on seismic-reflection data, well records or outcrops; the last step cannot be accomplished using seismic data alone. The fourth step may make use of facies-cycle and other data, as described in Chap. 3.

Sequence stratigraphy is based on the recognition of *unconformity-bounded units*, which may be formally defined and named using the methods of *allostratigraphy*, as noted in Sect. 1.7. Mitchum et al. (1977b, p. 53) defined a depositional sequence as "a stratigraphic unit composed of a relatively conformable succession of genetically related strata and bounded at its top and base by unconformities or their correlative conformities." As noted in Sect. 1.7, an unconformity may be traced laterally into the deposits of deep marine environments, where it may be represented by a *correlative conformity*. However, recognizing the bounding contacts in a conformable succession might, in practice, be difficult. Documentation of the two- and three-dimensional architecture of sequences was one of the most important breakthroughs of the seismic method, as explained in the Sect. 2.2.2.

Sequences reflect the sedimentary response to *base-level cycles*—the rise and fall in sea level relative to the shoreline, and changes in the sediment supply. Change in sea-level relative to the shoreline may result from *eustasy* (absolute changes in sea-level elevation relative to the centre of Earth) or from vertical movements of the basin floor as a result of tectonism. Because of the difficulty in distinguishing between these two different processes, the term *relative sea-level change* is normally used in order to encompass the uncertainty. These basic controls are explained in Sect. 2.2.1. In nonmarine settings, upstream controls (tectonism and climate change) are the major determinant of sequence architecture (Sect. 2.3.2).

The cycle of rise and fall of base level generates predictable responses in a sedimentary system, such as the transgressions that occur during rising relative sea level, and the widespread subaerial erosion and delivery of clastic detritus to the continental shelf, slope, and deep basin during a fall in relative sea level. The depositional systems that result, and their vertical and lateral relationships, provide the basis for subdividing sequences into systems tracts. These are described and explained in Sect. 2.2.2.

Unconformities provided the basis for the first definitions of sequences, by Blackwelder, Levorsen, Sloss and Vail (see Chap. 1). The unconformities that are the key to sequence recognition are those that develop as a result of subaerial exposure. Unconformities may develop below sea level as a result of submarine erosion, but are not used as the basis for sequence definition. Where subaerial unconformities are present, as in nonmarine and coastal settings, sequence definition is relatively straightforward. Carrying a correlation into the offshore, including recognition of a correlative conformity, is not necessarily simple; in fact this has been the cause of considerable debate and controversy, as discussed in Sect. 2.2.3. Several methods of defining sequences evolved from these controversies, and this has inhibited further progress in the establishment of sequence terminology and classification as a formal part of what might be called the official language of stratigraphy. Catuneanu (2006) and Catuneanu et al. (2009) have gone a long way towards resolving these problems, and their proposals are summarized in Sect. 2.3.

## 2.2.1 Accommodation and Supply

Sequence stratigraphy is all about *accommodation* (Fig. 2.1). Accommodation is defined as *the space available for sedimentation*. Jervey (1988) explained the concept this way:

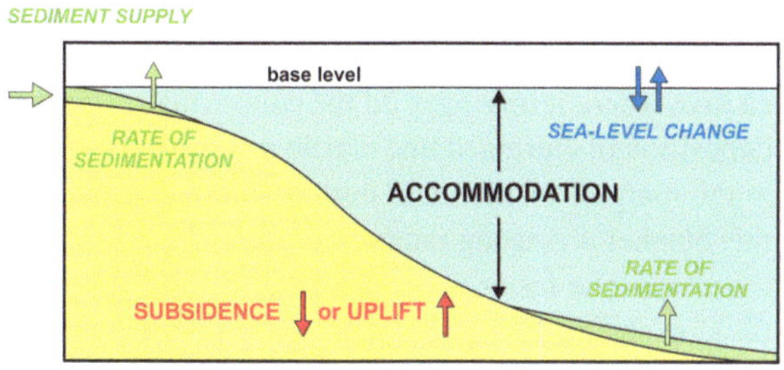

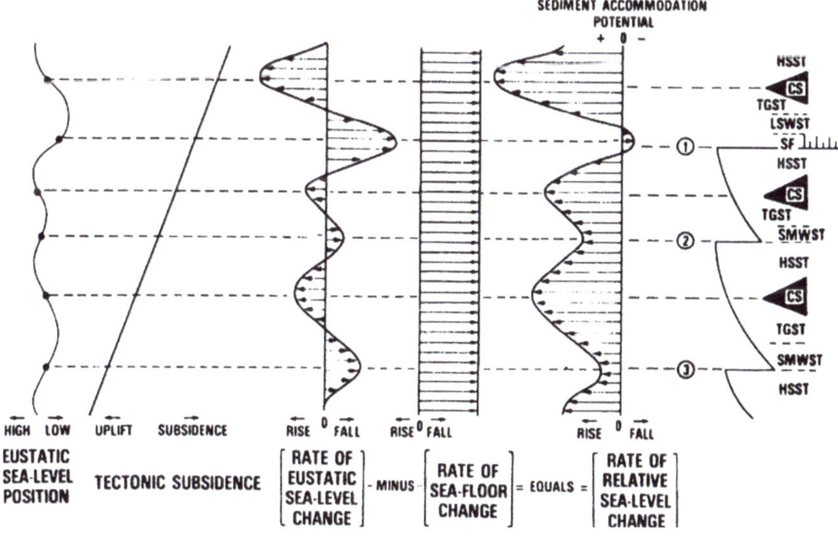



**Fig. 2.1** Accommodation, and the major allogenic sedimentary controls. Total accommodation in a basin is that generated by the subsidence of the basin floor (measured by backstripping methods). At any moment in time, the remaining accommodation in the basin represents that not yet filled by sediment, and is measured by water depth (the space between sea level and the sediment-water interface). Changes in accommodation (eustasy + tectonism) almost never correlate with bathymetric changes (eustasy + tectonism + sedimentation). Water depth reflects the balance between simultaneous creation (eustasy + tectonics) and consumption (sedimentation) of accommodation

In order for sediments to accumulate, there must be space available below base level (the level above which erosion will occur). On the continental margin, base level is controlled by sea level and, at first approximation, is equivalent to sea level. ... This space made available for potential sediment accumulation is referred to as accommodation.

In marine basins this is equivalent to the space between sea level and the sea floor. In nonmarine basins, a river's graded profile functions as sedimentary base level (Holbrook et al., 2006). Sequences are a record of the balance between accommodation change and sediment supply. As accommodation is filled by sediment, the remaining space is measured by the depth of water from the sea surface to the sediment-water interface at the bottom of the sea. Total accommodation increases when the basin floor subsides or sea level rises faster than the supply of sediment to fill the available space. Barrell (1917) understood this decades before geologists were in a position to appreciate its significance (Fig. 1.3). Where supply > accommodation, progradation results. Where supply < accommodation, retrogradation results. These contrasting scenarios were recognized many years ago, and are illustrated in Fig. 1.13 with reference to the stacking patterns of deltas on a continental margin. Figure 2.2 illustrates the initial Exxon concept



**Fig. 2.2** The standard Exxon diagram illustrating the relationship between eustasy and tectonism and the creation of "sediment accommodation potential". Integrating the two curves produces a curve of relative change of sea level, from which the timing of sequence boundaries can be derived (events 1–3 in right-hand column). However, changing the shape of the tectonic subsidence curve will change the shape and position of highs and lows in the relative sea-level curve, a point not acknowledged in the Exxon work. This version of the diagram is from Loutit et al. (1988). CS=condensed section, HSST=highstand systems tract, LSWST=lowstand wedge systems tract, SF=submarine fan, SMWST=shelf-margin wedge systems tract, TGST=transgressive systems tract

of how "sediment accommodation potential" is created and modified by the integration of a curve of sea-level change with subsidence. A diagram very similar to that of Barrell's was provided by Van Wagoner et al. (1990; see Fig. 2.3) and used to illustrate the deposition of shoaling-upward successions (*parasequences*).

The three major controls on basin architecture, subsidence/uplift (tectonism), sea-level change and sediment supply are themselves affected by a range of ultimate causes. Crustal extension, crustal loading, and other regional tectonic processes (summarized in Chaps. 9 and 10), provide the ultimate control on the size and architecture of sedimentary basins. Sea level change is driven by a range of low- and high-frequency processes, as discussed at length in Chaps. 9, 10 and 11. Sediment supply is affected by the tectonic elevation of the source area, which controls rates of erosion, and by climate, which affects such factors as rates of erosion, the calibre, volume and type of erosional detrital product, and the rates of subaqueous biogenic carbonate production (see Sect. 5.1).

### 2.2.2 Stratigraphic Architecture

The recognition of *unconformities* and *stratigraphic terminations* is a key part of the sequence method (Vail et al., 1977). This helps to explain why sequence stratigraphy developed initially from the study of seismic data, because conventional basin analysis based on outcrop and well data provides little direct information on stratigraphic terminations, whereas these are readily, and sometimes spectacularly, displayed on seismic-reflection cross-sections.

Unconformities may be used to define stratigraphic sequences because of two key, interrelated characteristics: (1) The break in sedimentation that they represent has a constant maximum time range, although parts of that time range may be represented by sedimentation within parts of the areal range of the unconformity; (2) The sediments lying above an unconformity are everywhere entirely younger than those lying below the unconformity.

There are a few exceptions to the second rule, that of the age relationships that characterize unconformities. There are at least two situations in which diachronous unconformities may develop, such that beds below the unconformity are locally younger than certain beds lying above the unconformity. The first case is that where the unconformity is generated by marine erosion caused by deep ocean currents (Christie-Blick et al., 1990, 2007). These can shift in position across the sea floor as a result of changes in topography brought about by tectonism or sedimentation. Christie-Blick et al. (1990) cited the case of the Western Boundary Undercurrent that flows along the continental slope of the Atlantic Ocean off the United States. This current is erosive where it impinges on the continental slope, but deposition of entrained fine clastic material takes place at the margins of the main current, and the growth of this blanket is causing the current to gradually shift up the slope. The result is onlapping of the deposits onto the slope below the current, and erosional truncation of the upslope deposits.

The second type of diachronous unconformity is that which develops at basin margins as a result of syndepositional tectonism. Continuous deformation during sedimentation may lead to migration of a surface of erosion, and subsequent rapid onlap of the erosion surface by alluvial sediments (Riba, 1976; Anadón et al., 1986). Typically these unconformities are associated with coarse conglomeratic sediments and die out rapidly into the basin. There is, therefore, little danger of their presence leading to the development of erroneous sequence stratigraphies.

As discussed in Sect. 2.3.3, breaks in sedimentation, called *drowning unconformities*, occur in carbonate sedimentary environments, as a result of environmental change, having nothing to do with changes in relative sea level.

A *correlative conformity* (defined by Mitchum et al., 1977) is the deep-offshore equivalent of a subaerial unconformity. In practice, as noted later in this section, in many cases this surface is conceptual or hypothetical, occurring within a continuous section bearing no indication of the key stratigraphic processes and events taking place contemporaneously in shallow-marine and nonmarine environments. It may be possible to determine an approximate position of the correlative conformity by tracing seismic reflections, but this may be quite inadequate for the purpose of formal sequence documentation and classification. As pointed out by Christie-Blick et al. (2007, p. 222), given that an unconformity represents a span of time, not an instant, "at some scale, unconformities pass laterally not into correlative conformities but into correlative intervals.

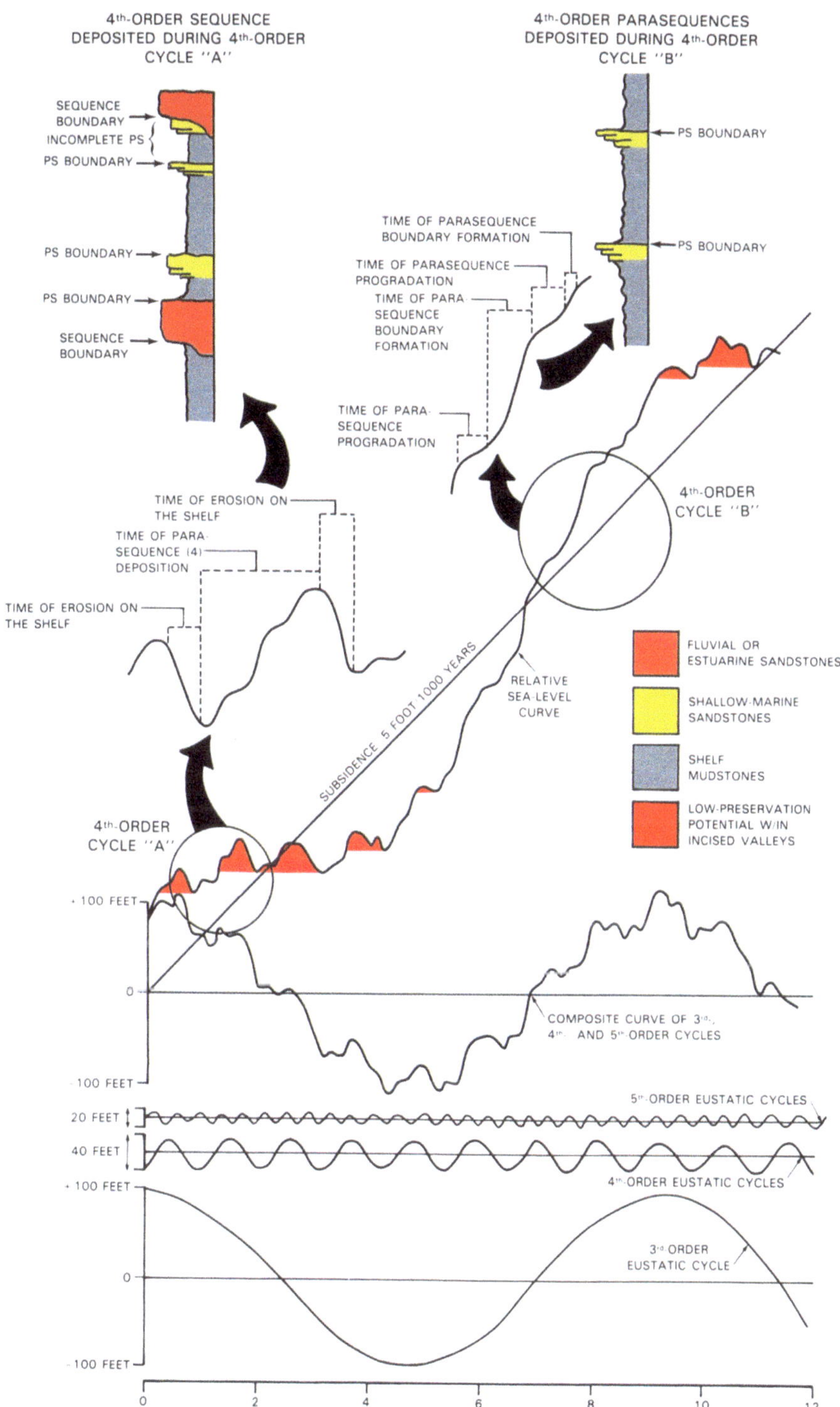


**Fig. 2.3** A modern version of Barrell's diagram (Fig. 1.3), showing the relationship between accommodation changes and sedimentation. The "relative sea-level curve" is a composite of three "eustatic" sea-level curves (although this could include other, non-eustatic mechanisms, as discussed in Chap. 10), integrated with a smooth subsidence curve. The coloured areas of the composite curve indicate intervals of time when accommodation is being generated, and parasequences are deposited. Examples of parasequences are indicated by the *arrows* (Van Wagoner et al., 1990). AAPG © 1990. Reprinted by permission of the AAPG whose permission is required for further use

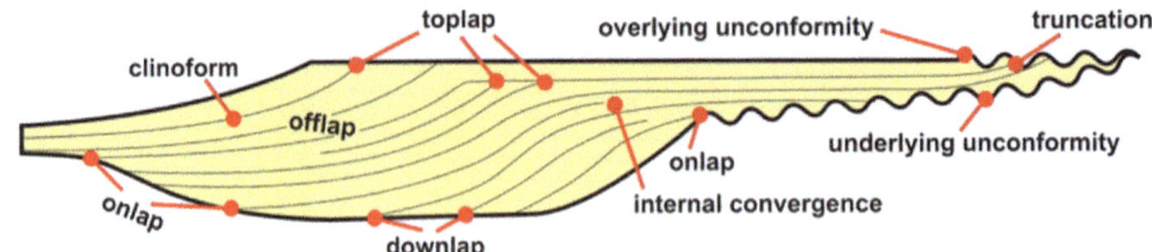



**Fig. 2.4** Sequence architecture, showing common characteristics of "seismic reflection terminations" (redrawn from Vail et al., 1977)

Such considerations begin to be important as the resolution of the geological timescale improves at a global scale."

Various architectural, or geometric, characteristics record the lateral shift in depositional environments in response to sea-level change and subsidence (Figs. 2.4 and 2.5). *Onlap* typically takes place at the base of the succession, recording the beginning of a cycle of sedimentation. *Offlap* develops when the rate of sedimentation exceeds the rate of accommodation generation. An offlap architecture may predominate in settings of high sediment supply. *Toplap* represents the abrupt pinch-out of offlapping units at the shelf-slope break. This develops when there is a major difference in accommodation generation between the shelf and slope, for example when wave, tide, or storm processes inhibit or prevent accumulation on the shelf. Sediment transported across the shelf is eventually delivered to the slope, a process termed *sediment bypass*. Toplap my represent abrupt thinning rather than truncation, with a thick slope unit passing laterally into a condensed section on the shelf. Discrimination between truncation and condensation may then depend on seismic resolution. *Downlap surfaces* may develop as a result of progradation across a basin floor, and they also develop during a transition from onlap to offlap. They typically develop above flooding surfaces, as basin-margin depositional systems begin to prograde seaward following the time of maximum flooding. The

dipping, prograding units are called *clinoforms* (after Rich, 1951), and they lap out downward onto the downlap surface as lateral progradation takes place. The word *lapout* is used as a general term for all these types of stratigraphic termination.

The broad internal characteristics of stratigraphic units may be determined from their *seismic facies*, defined to mean an areally restricted group of seismic reflections whose appearance and characteristics are distinguishable from those of adjacent groups (Sangree and Widmier, 1977). Various attributes may be used to define facies: reflection configuration, continuity, amplitude and frequence spectra, internal velocity, internal geometrical relations, and external three-dimensional form.

Figure 2.6 illustrates the main styles of seismic-facies reflection patterns (Mitchum et al., 1977a). Most of these are best seen in sections parallel to depositional dip. Parallel or subparallel reflections indicate uniform rates of deposition; divergent reflections result from differential subsidence rates, such as in a half-graben or across a shelf-margin hinge zone. Clinoform reflections comprise an important class of seismic-facies patterns. They are particularly common on continental margins, where they commonly represent prograded deltaic or continental-slope outgrowth. Variations in clinoform architecture reflect different combinations of depositional energy, subsidence rates, sediment supply, water depth and sea-level

**Fig. 2.5** A later diagram of sequence architecture (from Vail, 1987), which incorporates the concept of initial onlap followed by progradation across a downlap surface. AAPG © 1987. Reprinted by permission of the AAPG whose permission is required for further use

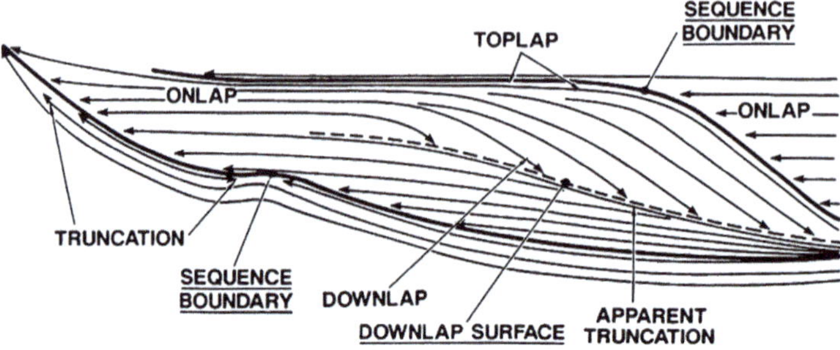

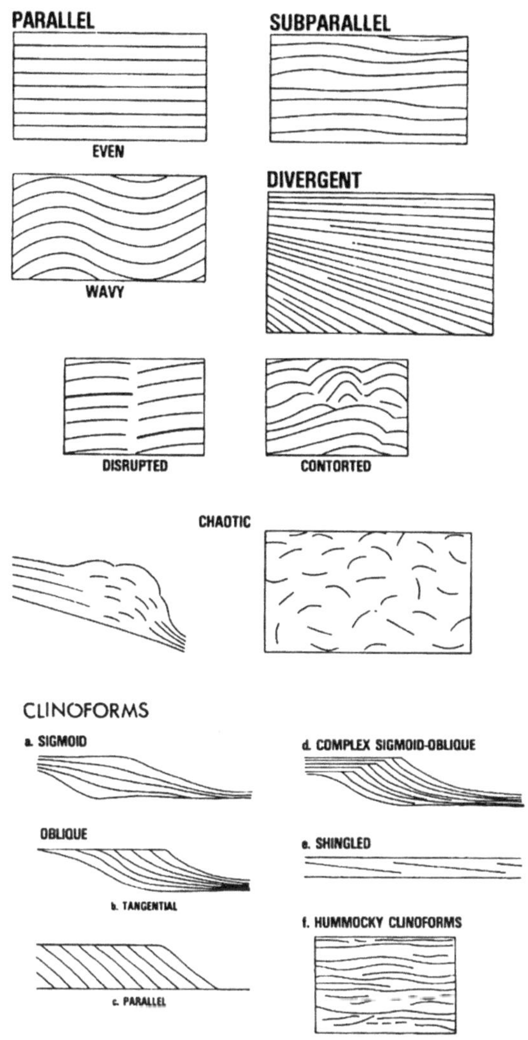



**Fig. 2.6** Typical seismic reflection patterns, illustrating the concept of seismic facies (Mitchum et al., 1977a). AAPG © 1987. Reprinted by permission of the AAPG whose permission is required for further use

position. *Sigmoid clinoforms* tend to have low depositional dips, typically less than 1°, whereas *oblique clinoforms* may show depositional dips up to 10°. *Parallel-oblique clinoform* patterns show no topsets. This usually implies shallow water depths with wave or current scour and sediment bypass to deeper water, perhaps down a submarine canyon that may be revealed on an adjacent seismic cross section. Many seismic sequences show complex offlapping stratigraphy, of which the complex sigmoid-oblique clinoform pattern in Fig. 2.6 is a simple example. This diagram illustrates periods of sea-level still-stand, with the

development of truncated topsets (*toplap*) alternating with periods of sea-level rise (or more rapid basin subsidence), which allowed the lip of the prograding sequence to build upward as well as outward. Mitchum et al. (1977a) described the *hummocky clinoform* pattern as consisting of "irregular discontinuous subparallel reflection segments forming a practically random hummocky pattern marked by nonsystematic reflection terminations and splits. Relief on the hummocks is low, approaching the limits of seismic resolution. The reflection pattern is generally interpreted as strata forming small, interfingering clinoform lobes building into shallow water," such as the upbuilding or offlapping lobes of a delta undergoing distributary switching. Submarine fans may show the same hummocky reflections. *Shingled clinoform* patterns typically reflect offlapping sediment bodies on a continental shelf.

Chaotic reflections may reflect slumped or contorted sediment masses or those with abundant channels or cut-and-fill structures. Disrupted reflections are usually caused by faults. Lenticular patterns are likely to be most common in sections oriented perpendicular to depositional dip. They represent the depositional lobes of deltas or submarine fans.

A *marine flooding surface* is a surface that separates older from younger strata, across which there is evidence of an abrupt increase in water depth. These surfaces are typically prominent and readily recognizable and mappable in the stratigraphic record. Each of the heavy, arrowed lines within the lower, retrogradational part of the sequence shown in Fig. 2.5 are marine flooding surfaces, as are the heavy lines in Fig. 2.7b. The *maximum flooding surface* records the maximum extent of marine drowning, and separates transgressive units below from regressive units above (the dashed line extending obliquely across the centre of the cross-section in Fig. 2.5 is a maximum flooding surface). It commonly is a surface of considerable regional stratigraphic prominence and significance. It may be marked by a widespread shale, or by a condensed section, indicating slow sedimentation at a time of sediment starvation on the continental shelf, and may correspond to a downlap surface, as noted above. The prominence of this surface led Galloway (1989a) to propose that sequences be defined by the maximum flooding surface rather than the subaerial erosion surface. We discuss this, and other alternative concepts, in Sect. 2.4.

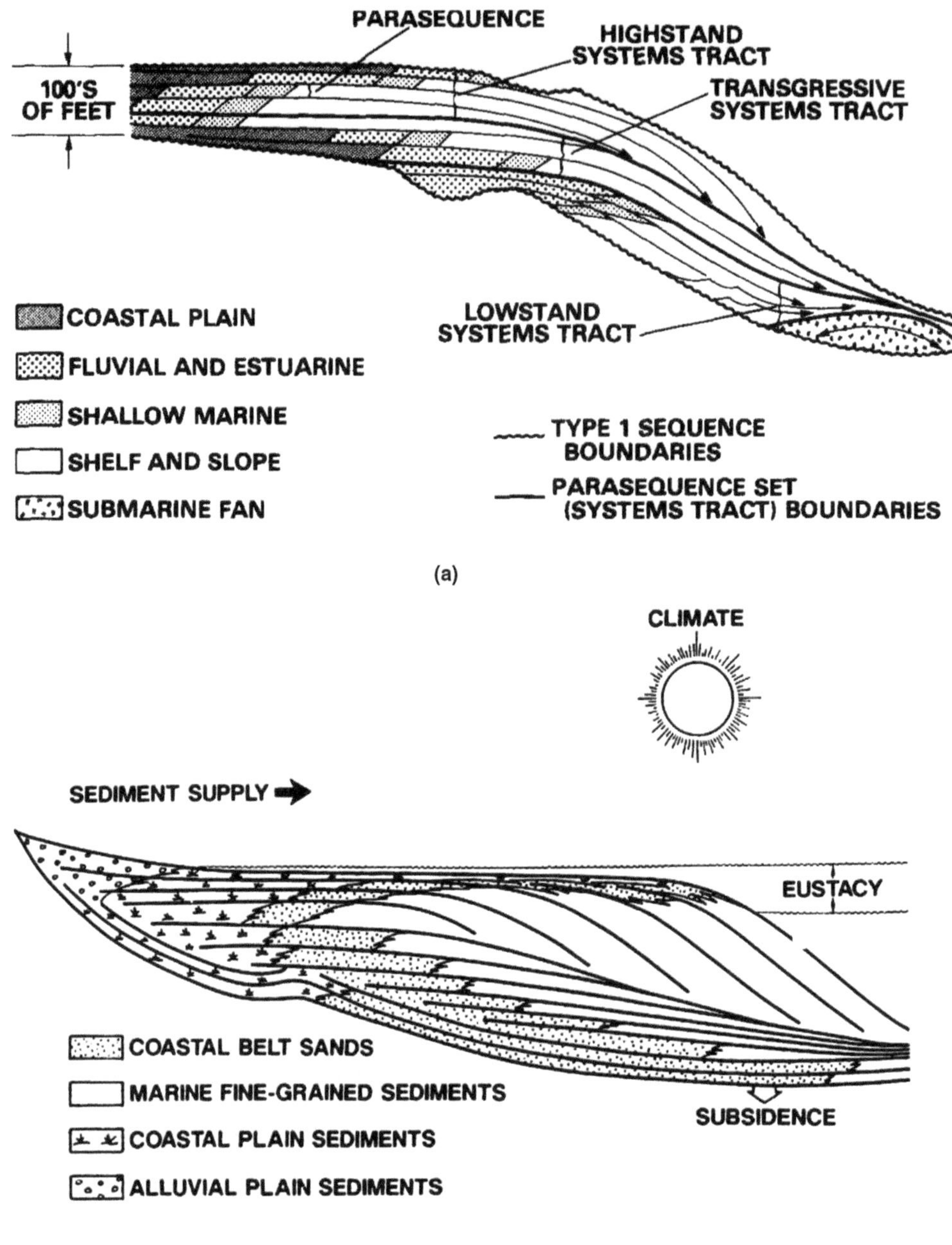



**Fig. 2.7** Diagram of sequences, sequence sets, and composite sequences. (**a**) Parasequences are the shoaling-upward successions, bounded by flooding marine surfaces (the *heavy lines*). (**b**) Sequences are composed of parasequences, which stack into lowstand, transgressive, and highstand sequence sets to form composite sequences (Mitchum and Van Wagoner, 1991)

Sequences may consist of stacked facies successions, each of which shows a gradual upward change in facies character, indicating a progressive shift in local depositional environments. The small packages of strata contained between the heavy lines in Fig. 2.7a are examples of these component packages of strata.

Van Wagoner et al. (1987) erected the term *parasequence* to encompass "a relatively conformable succession of genetically related beds or bedsets bounded by marine flooding surfaces and their correlative surfaces ... Parasequences are progradational and therefore the beds within parasequences shoal upward."

As Walker (1992) pointed out, "parasequences and facies successions … are essentially the same thing, except that the concept of facies succession is broader." However, other types of facies succession occur within sequences (e.g., channel-fill fining-upward successions), and the term parasequence is therefore unnecessarily restrictive. Many such successions are generated by autogenic processes, such as delta-lobe switching, and channel migration, that have nothing to do with sequence controls, and to include them in a term that has the word "sequence" within it may be misleading. Walker (1992) recommended that the term parasequence not be used. Catuneanu (2006, pp. 243–245) pointed out numerous problems with the concept of the parasequence, including the imprecise meaning of the term "flooding surface" (which it is now recognized, may have several different meanings) and the potential confusion with surfaces generated by autogenic processes. He recommended using the term only in the context of progradational units in coastal settings. I suggest that the term be abandoned altogether. We return to this point at the end of the chapter.

### 2.2.3 Depositional Systems and Systems Tracts

The concept of the *depositional-system* and basin-analysis methods based on it were developed largely in the Gulf Coast region as a means of analyzing and interpreting the immense thicknesses of Mesozoic-Cenozoic sediment there that are so rich in oil and gas. A depositional system is defined in the Schlumberger online Oilfied Glossary as

> The three-dimensional array of sediments or lithofacies that fills a basin. Depositional systems vary according to the types of sediments available for deposition as well as the depositional processes and environments in which they are deposited. The dominant depositional systems are alluvial, fluvial, deltaic, marine, lacustrine and eolian systems.

The principles of depositional-systems analysis have never been formally stated, but have been widely used, particularly by geologists of the Bureau of Economic Geology at the University of Texas (notably W. L. Fisher, L. F. Brown Jr., J. H. McGowen, W. E. Galloway and D. E. Frazier). Useful papers on the topic are those by Fisher and McGowen (1967) and Brown and Fisher (1977). Textbook discussions are given by Miall (1999, Chap. 6) and Walker (1992). The concept of depositional episode was developed by Frazier (1974) to explain the construction of Mississippi delta by progradation of successive delta lobes (Fig. 1.14).

Posamentier et al. (1988, p. 110) defined a *depositional system* as "a three-dimensional assemblage of lithofacies, genetically linked by active (modern) or inferred (ancient) processes and environments." A *systems tract* is defined as

> A linkage of contemporaneous depositional systems … Each is defined objectively by stratal geometries at bounding surfaces, position within the sequence, and internal parasequence stacking patterns. Each is interpreted to be associated with a specific segment of the eustatic curve (i.e., eustatic lowstand-lowstand wedge; eustatic rise-transgressive; rapid eustatic fall-lowstand fan, and so on), although not defined on the basis of this association (Posamentier et al., 1988).

Elsewhere, Van Wagoner et al. (1987) stated that "when referring to systems tracts, the terms lowstand and highstand are not meant to imply a unique period of time or position on a cycle of eustatic or relative change of sea level. The actual time of initiation of a systems tract is interpreted to be a function of the interaction between eustasy, sediment supply, and tectonics." There is clearly an inherent, or built-in contradiction here, that results from the use in a descriptive sense of terminology that has a genetic connotation (e.g., transgressive systems tract implies transgression). We return to this problem below.

Systems tracts are named with reference to their assumed position within the sea-level cycle, and these names incorporate ideas about the expected response of a basin to the changing balance between the major sedimentary controls (accommodation and sediment supply) during a base-level cycle. There are four standard systems tracts. These are the *highstand, falling-stage, lowstand,* and *transgressive systems tracts.* Each is illustrated here by a block diagram model with summary remarks outlining the major sedimentary controls and depositional patterns prevailing at that stage of sequence development (Fig. 2.8). Other terms have been used by different workers, but these four systems tracts and their bounding surfaces provide a useful, easy-to-understand model from which to build interpretive concepts.

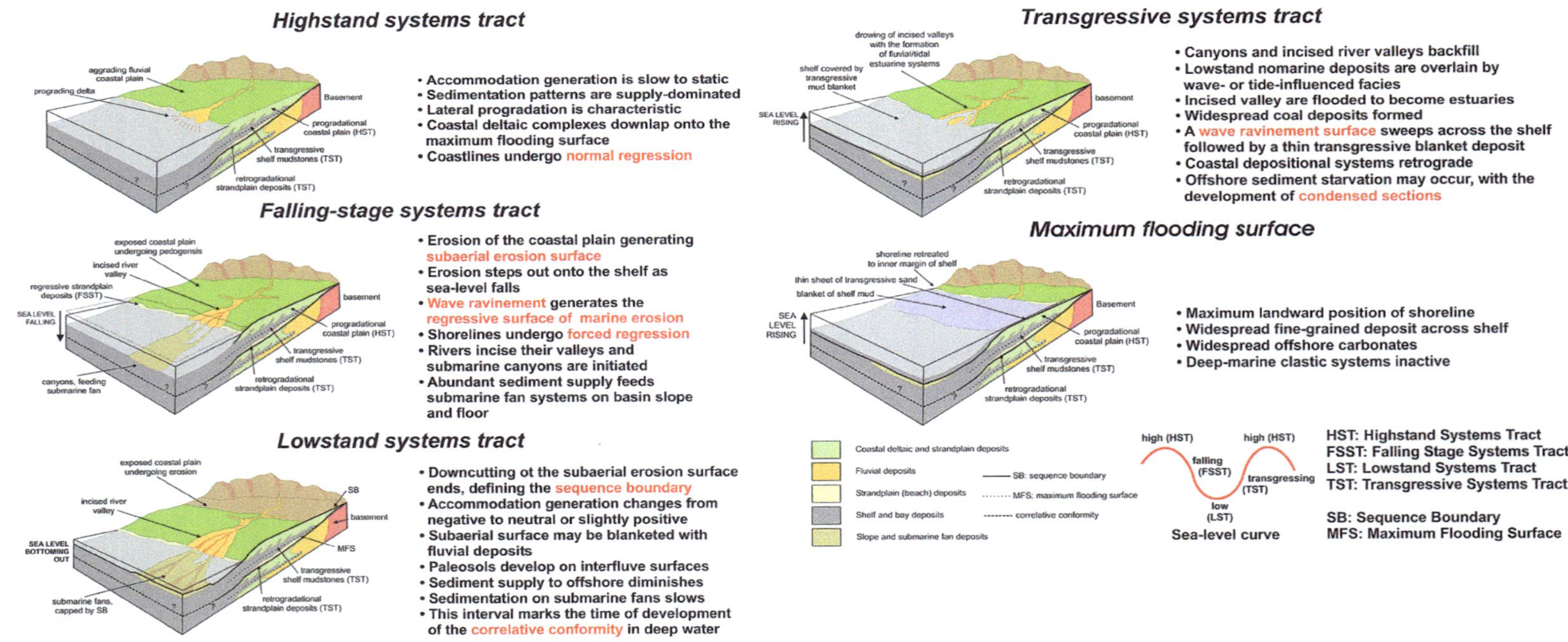



**Fig. 2.8**  Relationship between seismic architecture and depositional systems, highlighting the major depositional controls. In this example, each of the *heavy curved lines* represents a minor flooding surface, and the packages of strata between them have been defined as parasequences (Vail, 1987)

The linking of systems tracts to stages of the sea-level cycle is unfortunate, because it has now been repeatedly demonstrated that the geometric and behavioural features that supposedly characterize each systems tract and its link to the sea-level cycle are not necessarily diagnostic of sea level, but may reflect combinations of several factors. For example, on the east coast of South Island, New Zealand, different stretches of coast are simultaneously undergoing coastal progradation, reflecting a large sediment supply, and coastal retreat and transgression, because of locally high wave energy (Leckie, 1994). Andros Island, in the Bahamas, currently exhibits three different systems-tract conditions. Lowstand conditions characterize the eastern (windward) margin of the island, facing the deep-water channel, the Tongue of the Ocean. Transgressive conditions occur along the northwest margin, where tidal flats are undergoing erosion, whereas on the more sheltered leeward margin, along the southwest edge of the Andros coastline, highstand conditions are suggested by tidal-flat progradation (Schlager, 2005, Fig. 7.6).

## 2.3  Sequence Models in Clastic and Carbonate Settings

In this section, a brief overview of sequence models is provided for the main areas and styles of deposition, marine and nonmarine clastics, and carbonates. A much more complete treatment of this topic forms the core of the book by Catuneanu (2006). In addition, many useful review articles and books provide additional insights into particular areas or themes (e.g., Emery and Myers, 1996; Schlager, 1992, 1993, 2005; Posamentier and Allen, 1999; Hunt and Gawthorpe, 2000; Yoshida et al., 2007).

### 2.3.1  Marine Clastic Depositional Systems and Systems Tracts

The *highstand systems tract (HST)* corresponds to a period when little new accommodation is being added to the depositional environment. As shown in the relative sea-level curve at the lower right corner of Fig. 2.8, base-level rise is in the process of slowing down as it reaches its highest point, immediately prior to commencement of a slow fall. If the sediment supply remains more or less constant, then Rd>Rs at this point (the upper of the three conditions shown in Fig. 1.13). The most characteristic feature of this systems tract is the lateral progradation of coastal sedimentary environments. Major coastal barrier-lagoon and deltaic complexes are the result. *Normal regression* is the term used to describe the seaward advance of the coastline as a result of the progressive addition of sediment to the front of the beach or the delta systems, developing a broad *topset* environment (the *undathem* of Rich, 1951; see Fig. 1.10). This is in contrast to the condition of forced regression, which is described below.

Where the terrigenous sediment supply is high, delta systems may largely dominate the resulting sedimentary succession, as shown in the accompanying example of the Dunvegan delta, Alberta (Fig. 2.9). The allomember boundaries in this diagram indicate times of relative low sea level, followed by flooding. Each allomember boundary is overlain by a mudstone representing the maximum flooding surface, over which delta complexes prograded. Sedimentary environments characteristically include coastal mangrove swamps, and may include significant peat swamps, the sites of future coal development. The numbered subdivisions of each allomember indicate individual deltaic shingles. Subsurface mapping may indicate that shingles of this type shifted laterally as a result of delta switching, in a manner similar to the Mississippi delta and the Yellow River delta. This points to potential confusion in terminology, because upward-shoaling successions, such as those illustrated in Fig. 2.9 correspond to the Van Wagoner et al. (1987) definition of *parasequence*. We return to this point at the end of the chapter (Sect. 2.4).

The thickness of highstand shelf deposits depends on the accommodation generated by marine transgression across the shelf, typically a few tens of metres, up to a maximum of about 200 m. Where the shelf is narrow or the sediment supply is large, deltas may prograde to the shelf-slope break, at which point deltaic sedimentation may extend down slope into the deep basin (Porebski and Steel, 2003). High-amplitude clinoforms may result, including significant volumes of sediment-gravity-flow deposits.

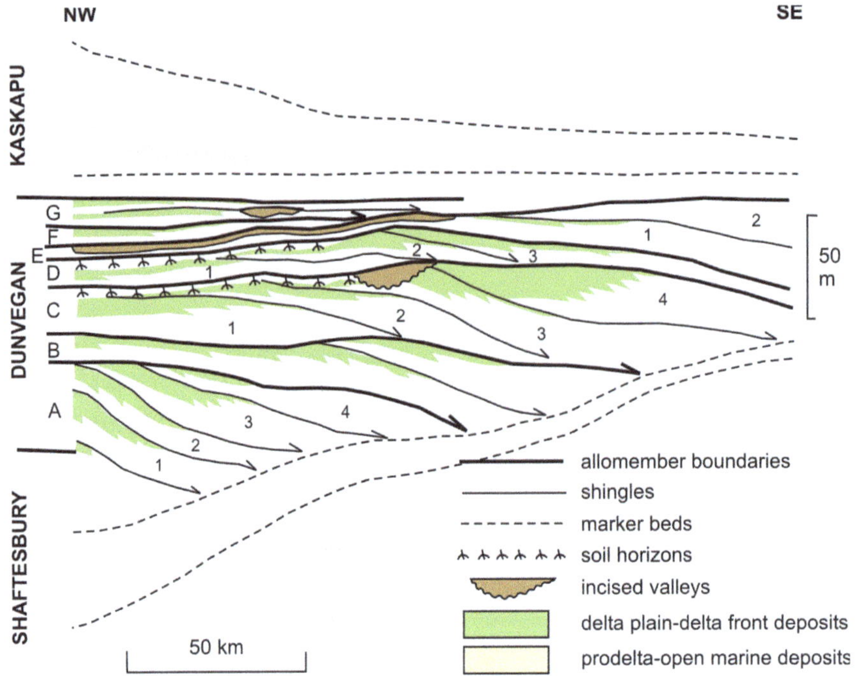



**Fig. 2.9** The Dunvegan alloformation of northwest Alberta is dominated by highstand deposits, reflecting its origins as a deltaic complex. Each shingle represents an individual delta lobe, terminated by a flooding surface (indicated by a downlapping *half-arrow*). Adapted from Bhattacharya (1991)

*The falling-stage systems tract (FSST).* A fall in base level from the highstand position exposes the coastal plain and then the continental shelf, to subaerial erosion. River mouths retreat seaward, and under most conditions, river valleys incise themselves as they continually grade downward to progressively lower sea levels. Deeply incised paleovalleys may result. In the rock record, many of these show evidence of multiple erosional events (Korus et al., 2008), indicating repeated responses to autogenic threshold triggers (Schumm, 1993) or perhaps to minor cycles of base-level change. Significant volumes of sediment are eroded from the coastal plain and the shelf, and are fed through the coastal fluvial systems and onto the shelf. Eventually, these large sediment volumes may be tipped directly over the edge of the shelf onto the continental slope, triggering submarine landslides, debris flows, and turbidity currents. These have a powerful erosive effect, and may initiate development of submarine erosional valleys at the mouths of the major rivers or offshore from major delta distributaries. Many submarine canyons are initiated by this process, and remain as major routes for sediment dispersal through successive cycles of base-level change. The FSST is typically the major period of growth of submarine fans.

The falling base level causes basinward retreat of the shoreline, a process termed *forced regression* by Plint (1991) (Fig. 2.10). The occurrence of forced regression, as distinct from normal regression, may be detected by careful mapping of coastal shoreline sandstone complexes. Fall of sea level causes water depths over the shelf to decrease, increasing the erosive power of waves and tides. This typically leads to the development of a surface called the *regressive surface of marine erosion (RSME)*, which truncates shelf and distal coastal (e.g., deltaic) deposits that had been formed during the preceding highstand phase (Fig. 2.10b). The first such surface to form, at the commencement of a phase of sea-level fall, is termed the *basal surface of forced regression.* Some specialists used this surface as the basis for sequence definition, as discussed in Sect. 2.4. Given an adequate sediment supply, especially if there are pauses during the fall of sea level (Fig. 2.10c), shoreface sand accumulates above the RSME, forming what have come to be informally termed *sharp-based sandstone bodies* (Plint, 1988). These are internally identical to other coastal, regressive sandstone bodies, except that they rest on an erosion surface instead of grading up from the fine-grained shelf sediments, as in the initial coastal

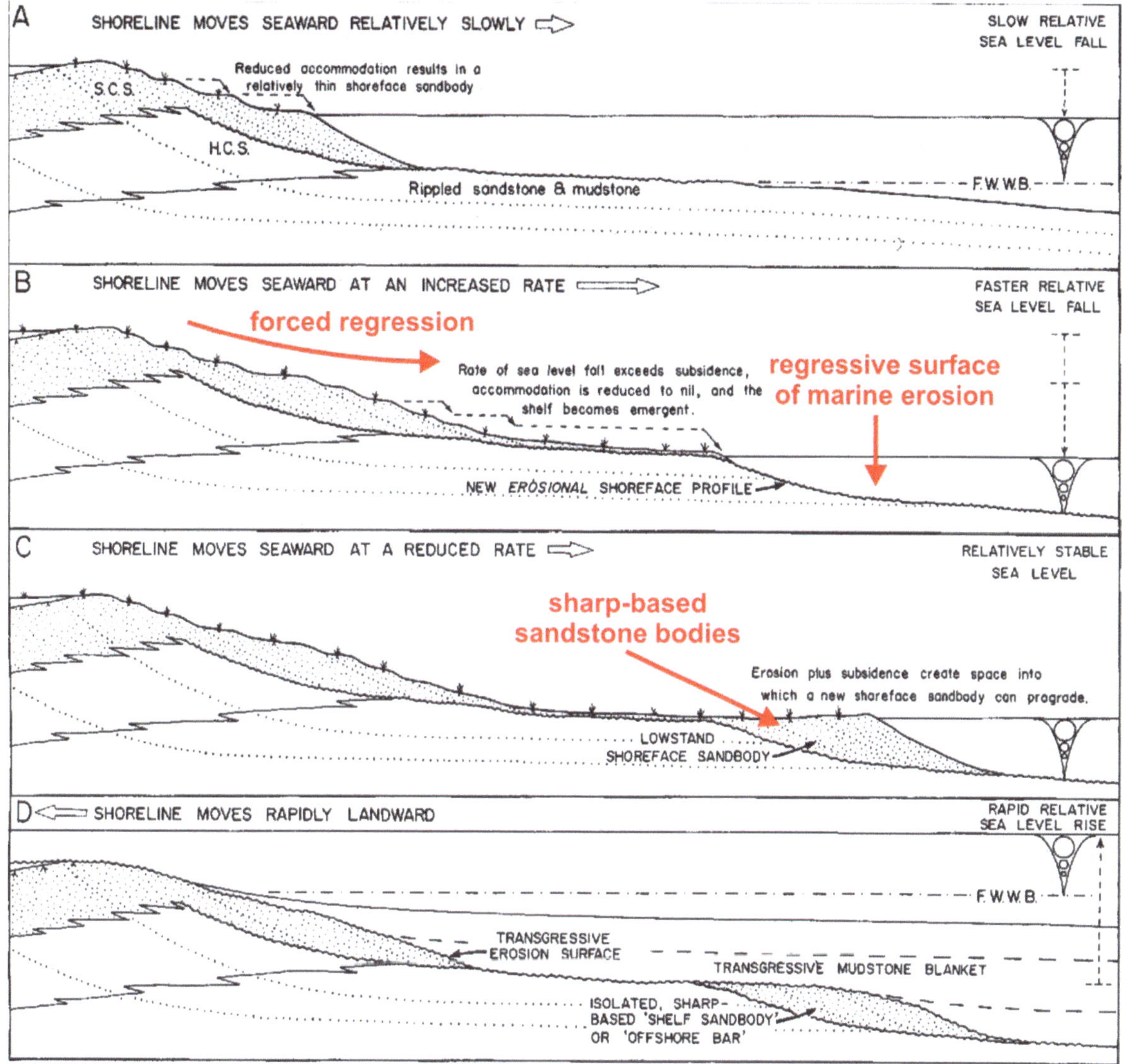



**Fig. 2.10** The process of forced regression, and the development of the regressive surface of marine erosion and "sharp-based sand bodies". Original diagram from Plint (1988)

sands shown in Fig. 2.10a (which are the product of normal regression). Repeated pulses of sea-level fall punctuated by stillstand may develop several offlapping surfaces of marine erosion. Shelf-margin deltas may form where the mouths of major river systems regress to the shelf-slope break during forced regression (Porebski and Steel, 2003).

As noted above, the falling-stage is typically the interval during the sea-level cycle when the sediment supply to the continental shelf and slope is at its greatest. Most sediment accumulation on submarine fans occurs during this and the next phase, the lowstand (discussed below). Most of the early sequence

models (e.g., Posamentier et al., 1988; Posamentier and Vail, 1988) showed submarine fans resting on a basal sequence boundary, but this configuration now seems unlikely. On the continental shelf and coastal plain, the sequence boundary is an erosion surface representing the lowest point to which erosion cuts during the falling stage of the base-level cycle. As sea-level fall slows to its lowest point, sediment delivery from the newly exposed coastal plain and shelf will gradually diminish. Sedimentation on submarine fans will correspondingly slow down, and the deposits may show a gradual upward decrease in average grain size. Sedimentation there may virtually cease once the next phase of sea-level rise commences, and the rivers feeding sediment to the slope become flooded (transgressive systems tract). The sequence boundary, therefore, is likely to be contemporaneous with the middle to upper part of the submarine-fan succession, possibly with the top of it. However, there is unlikely to be an actual mappable break in sedimentation at this level, and it may be difficult to impossible to locate the position in the section corresponding to the turn-around from falling to rising sea level. This horizon is, therefore, what Vail et al. (1977) called a *correlative conformity*, although his original application of the term was to the fine-grained sediments formed in deep water beyond the submarine-fan wedge, out in the deep basin where it was assumed sedimentation would be continuous throughout a sea-level cycle.

The *sequence boundary (SB)* marks the lowest point reached by erosion during the falling stage of the sea-level cycle. On land this is represented by a subaerial erosion surface, which may extend far onto the continental shelf, depending on how far sea level falls. The sequence boundary cuts into the deposits of the highstand systems tract and is overlain by the deposits of the lowstand or transgressive systems tracts. It is therefore typically a surface where a marked facies change takes place, usually from a relatively lower-energy deposit below to a high-energy deposit above. Mapping of such a surface in outcrop or in the subsurface, using well logs, is facilitated by this facies change, except where the boundary juxtaposes fluvial on fluvial facies. In such cases, distinguishing the sequence boundary from other large-scale channel scours may be a difficult undertaking.

The *lowstand systems tract (LST):* This systems tract represents the interval of time when sea-level has bottomed out, and depositional trends undergo a shift from seaward-directed (e.g., progradational) to landward-directed (e.g., retrogradational). Within most depositional systems there is little that may be confidently assigned to the lowstand systems tract. The initial basal fill of incised river valleys, and some of the fill of submarine canyons are deposited during this phase. Volumetrically they are usually of minor importance, but they may be of a coarser grain size than succeeding transgressive deposits. In parts of the incised valley of the Mississippi River, for example (the valley formed during Pleistocene glacioeustatic sea-level lowstands), the basal fill formed during the initial post-glacial transgression is a coarse braided

stream deposit, in contrast to the sandy meandering river deposits that form the bulk of the Mississippi river sediments. The episode of active submarine-fan sedimentation on the continental slope and deep basin may persist through the lowstand phase.

There may be a phase of normal regressive sedimentation at the lowstand coastline. On coastal plains, the lowstand is a time of stillstand, when little erosion or sedimentation takes place. Between the major rivers, on the interfluve uplands, this may therefore be a place where long-established plant growth and soil development takes place. Peat is unlikely to accumulate because of the lack of accommodation, but soils, corresponding in time to the sequence boundary, may be extensive, and the resulting paleosols may therefore be employed for mapping purposes (e.g., McCarthy et al., 1999; Plint et al., 2001).

*Transgressive systems tract (TST):* A rise in base level is typically accompanied by flooding of incised valleys and transgression across the continental shelf (Fig. 2.8). Base-level rise exceeds sediment supply, leading to retrogradation of depositional systems (Rd<Rs in Fig. 1.13c), except that at the mouths of the largest rivers sediment supply may be sufficiently large that deltas may continue to aggrade or prograde.

Flooded river valleys are estuaries; they typically provide ample accommodation for sedimentary accumulation. In estuarine successions, the upward transition from lowstand to transgressive systems tract in estuaries and other coastal river systems is commonly marked by the development of wave- or tide-influenced fluvial facies, such as tidal sand bars containing sigmoidal crossbedding or flaser bedding. The sedimentology of this environment has received much attention (Fig. 2.11), because of the potential for the development of stratigraphic sandstone traps, in the form of valley-fill ribbon sands. Studies of ancient paleovalley fills have shown that many are complex, indicating repeated cycles of base level change and/or autogenic changes in sediment dispersal (Korus et al., 2008).

On the continental shelf the most distinctive feature of most transgressive systems tracts is the development of a widespread *transgressive surface (TS)*, a flooding surface covered with an equally widespread marine mudstone. A transgressive conglomeratic or sandy lag may blanket the flooding surface. Offshore, rapid transgression may cut the deep-water environment off from its sediment source, leading to slow

**Fig. 2.11** Depositional model for estuaries (Reinson, 1992; Dalrymple et al., 1994)

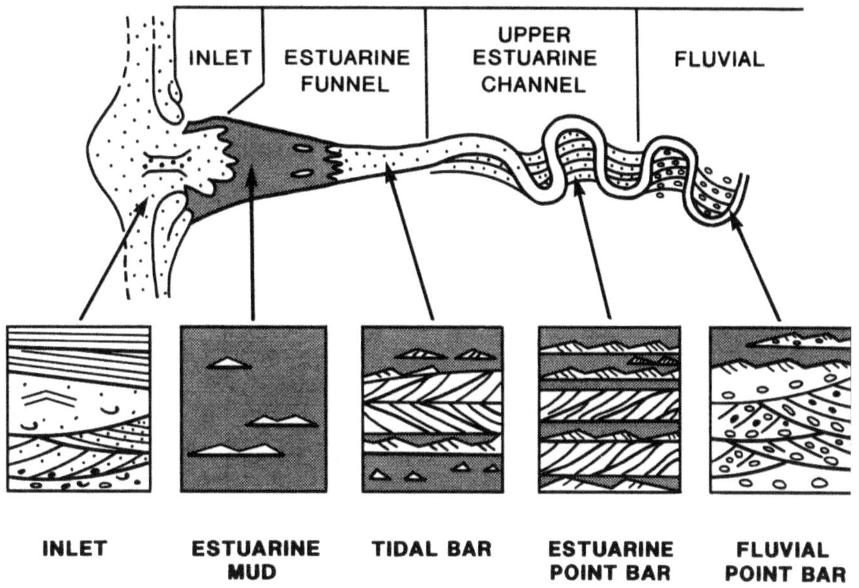



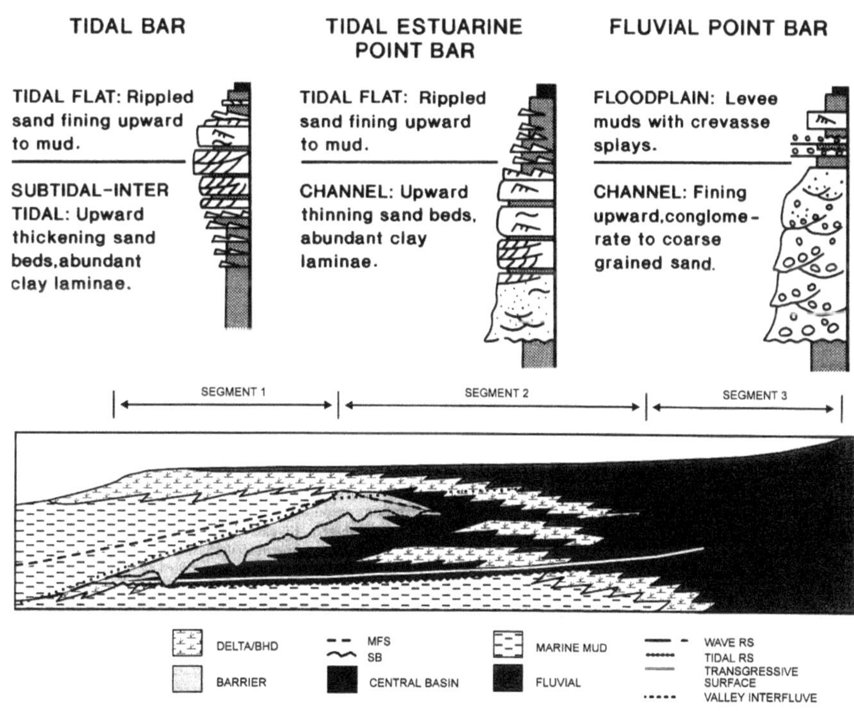



sedimentation, and the formation of a *condensed section*. This is commonly a distinctive facies, consisting of concentrated shell or fish fragments, amalgamated biozones, and a "hot" (high gamma-ray) response on well-logs, reflecting a concentration of radioactive clays (Loutit et al., 1988). Significant volumes of clastic sediment deposited on the shelf may be reworked during transgression. Posamentier (2002) documented numerous complexes of shelf sand ridges constituting parts of shelf transgressive systems tracts that were formed by vigorous wave and tide action. Offshore, limestones may be deposited, such as the

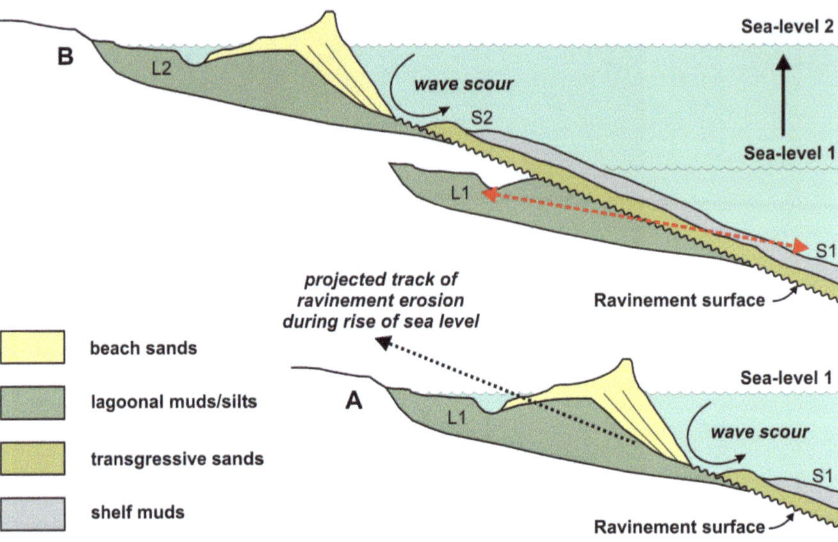



**Fig. 2.12** The process of transgressive ravinement. The *dashed red line* indicates the time correlation between shelf sediments deposited on the ravinement surface and contemporaneous lagoon deposits that are truncated by ravinement erosion as sea-level rises. Adapted from Nummedal and Swift (1987)

several Jurassic and Cretaceous limestones and chalks in the Western Interior Seaway (Greenhorn Limestone, Austin Chalk).

In the nearshore setting, wave erosion during transgression is usually the cause of ravinement, with the development of a diachronous *ravinement surface* (Fig. 2.12). The juxtaposition of marine shelf sediments, above, over coastal shoreline or lagoonal sediments below, creates a prominent surface which should not be confused with a sequence boundary. A ravinement surface marks an upward deepening, the opposite of the facies relationships at most sequence boundaries. In some cases, ravinement erosion may cut down through lowstand deposits and into the underlying highstand systems tract, and in such cases the ravinement surface becomes the sequence boundary (Nummedal and Swift, 1987).

Peat may be deposited on the coastal plain and in deltaic settings at any time during a cycle of base-level change. However, the thickest and most widespread coals are now known to be those formed from peat accumulated during transgression, because of the accommodation provided by rising base level, during a time when clastic influx into the coastal plain is "held back" by the landward-advancing shoreline (Bohacs and Suter, 1997; see Fig. 2.13).

The *maximum flooding surface (MFS)* marks the end of the phase during which the difference between the rate of sea-level rise and the rate of sediment supply is at its greatest (Fig. 2.8). Sea-level rise continues beyond this point, but as the rate of rise slows, sediment input begins to re-establish progradation at the shoreline, and this defines the transition into the highstand systems tract. The offshore shale formed around the time of the MFS is an excellent mapping marker, because of its widespread nature and distinctive facies. In areas distant from the shoreline, where clastic sediment supply is at a minimum, the MFS is commonly marked by calcareous shale, marl or limestone. In some studies, sequence mapping is accomplished using this surface in preference to the sequence boundary, because of its more predictable facies and its consistent horizontality.

The preceding paragraphs constitute a set of useful generalizations. However, there are many exceptions and special cases. For example, consider the ultimate fate of the clastic sediment flux on continental margins during cycles of sea level change. In the traditional model (Posamentier et al., 1988), on which this section is largely based, coastal plain complexes, including deltas, typically accumulate during highstand phases, following a period of coastal plain transgression and flooding, and basin slope and plain deposits, including submarine fans, accumulate during the sea-level falling stage and lowstand. However, these generalizations do not necessarily apply to all continental margins. As Carvajal and Steel (2006, p. 665) pointed out,

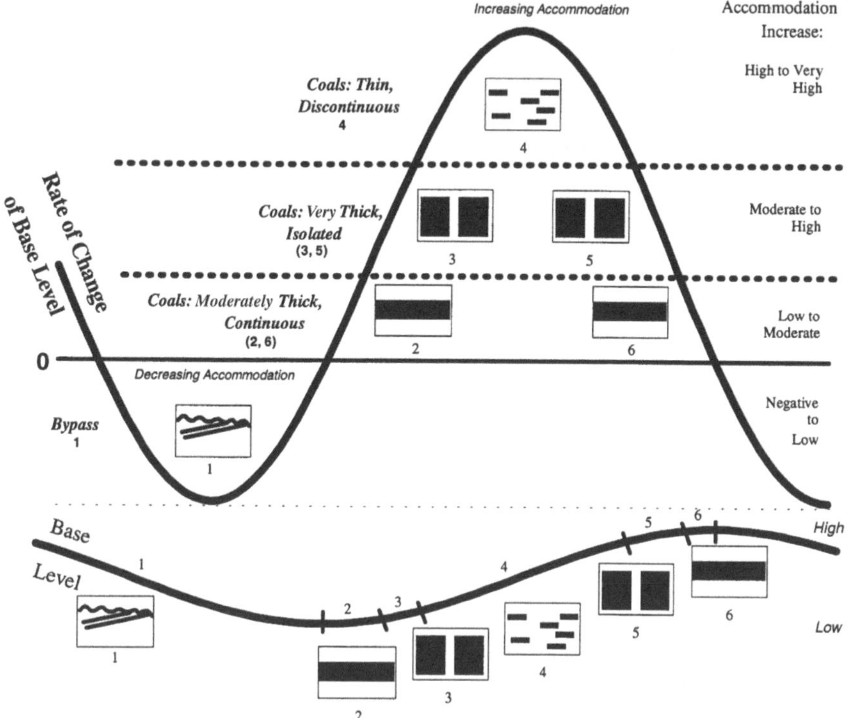



**Fig. 2.13** The dependency of the lateral extent and thickness of coal seams on the rate of change of base level (Bohacs and Suter, 1997). AAPG © 1997. Reprinted by permission of the AAPG whose permission is required for further use

This model has been challenged using examples from narrow shelf settings (e.g., fans in the California Borderland, Gulf of Corinth, and Mediterranean Sea; see Piper and Normark, 2001; Ito and Masuda, 1988) or extremely high supply systems (e.g., Bengal Fan; Weber et al., 1997). In these cases slope canyons extending to almost the shoreline may receive sand from littoral drift or shelf currents during rising sea level. In addition, deltas may easily cross narrow shelves and provide sand for deep-water deposits under normal supply conditions during relative sea-level highstand. It has also been postulated that in moderately wide (tens of kilometers) to wide shelf (hundreds of kilometers) settings, significant volumes of sand can be bypassed to deep-water areas at highstand through shelf-edge deltas (Burgess and Hovius, 1998; Porebski and Steel, 2006). Nonetheless, documenting such delivery either in the modern or ancient has been difficult (except for suggestions from studies at the third order time scale, e.g., McMillen and Winn, 1991), biasing researchers to interpret ancient deep-water deposits preferentially following the lowstand model. Thus, focus on this lowstand model has tended to cause us to overlook (1) the dominant role that sediment supply may play in deep-water sediment delivery, and (2) how such supply-dominated shelf margins can generate deep-water fans even during periods of rising relative sea level.

Covault et al. (2007) similarly noted the development of submarine fans on the California borderland at times of sea-level highstand. The connection of canyon and fan dispersal systems to the littoral sediment supply is the key control on the timing of deposition in this setting.

In addition to the physiographic variations noted here, which complicate the relationship between the base level cycle and systems-tract architecture and development, it is quite possible for episodic changes in systems-tract development at continental margins to have nothing to do with sea-level change at all (Part III of this book). To cite two examples, in the case of the modern Amazon fan, the marked facies variations mapped by the ODP bear no relation to Neogene sea-level changes, but reflect autogenic avulsion processes on the upper fan (Christie-Blick et al., 2007). White and Lovell (1997) demonstrated that in the North Sea basin, peaks in submarine fan sedimentation occurred at times of regional uplift of the crust underlying the British Isles, as a consequence of episodes of magma underplating, resulting in increased sediment delivery to the marine realm (Sect. 10.2.2; Fig. 10.12).

Note, in closing, the caveats at the end of Sect. 2.2.3 regarding the possible confusion between the terminology of systems tracts (highstand, falling stage, etc.) and the actual state of the sea-level cycle which they represent.

### 2.3.2 Nonmarine Depositional Systems

The early sequence model of Posamentier et al. (1988) and Posamentier and Vail (1988) emphasized the accumulation of fluvial deposits during the late highstand phase of the sea-level cycle, based on the graphical models of Jervey (1988). The model suggested that the longitudinal profile of rivers that are graded to sea level would shift seaward during a fall in base level, and that this would generate accommodation for the accumulation of nonmarine sediments. This idea was examined critically by Miall (1991a), who described scenarios where this may and may not occur. The response of fluvial systems to changes in base level was examined from a geomorphic perspective in greater detail by Wescott (1993) and Schumm (1993), and the sequence stratigraphy of nonmarine deposits was critically reviewed in an important paper by Shanley and McCabe (1994). An extensive discussion of the sequence stratigraphy of fluvial deposits was given by Miall (1996, Sect. 11.2.2 and Chap. 13), and only a brief, updated summary of this material is given here.

Shanley and McCabe (1994) discussed the relative importance of downstream base-level controls versus upstream tectonic controls in the development of fluvial architecture. In general, the importance of base-level change diminishes upstream. In large rivers, such as the Mississippi, the evidence from the Quaternary record indicates that sea-level changes affect aggradation and degradation as far upstream as the region of Natchez, Mississippi, about 220 km upstream from the present mouth. Farther upstream than this, source-area effects, including changes in discharge and sediment supply, resulting from tectonism and climate change, are much more important. In the Colorado River of Texas base-level influence extends about 90 km upstream, beyond which the river has been affected primarily by the climate changes of Neogene glaciations. Blum (1994, p. 275), based on his detailed work on the Gulf Coast rivers, stated "At some point upstream rivers become completely independent of higher order relative changes in base level, and are responding to a tectonically controlled long-term average base level of erosion." The response of river systems to climate change is complex. As summarized by Miall (1996, Sect. 12.13.2), cycles of aggradation and degradation in inland areas may be driven by changes in discharge and sediment load, which are in part climate dependent. These cycles may be completely out of phase with those driven primarily by base-level change.

The elements of a generalized sequence model for coastal fluvial deposits are shown in Fig. 2.14. The sequence boundary is commonly an incised valley eroded during the falling stage of the base-level cycle. This valley is filled with fluvial or estuarine deposits during the lowstand to transgressive part of the cycle, with the thickness and facies composition of these beds determined by the balance between the rates of subsidence, base-level change and sediment supply. Away from the incised valley, on interfluve areas, the sequence boundary may be marked by well-developed paleosol horizons (McCarthy et al., 1999; McCarthy, 2002). It is a matter of debate whether the fluvial fill of an incised valley should be assigned to the

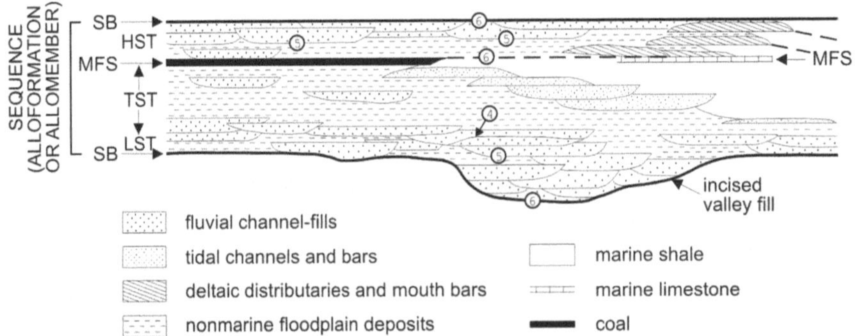



**Fig. 2.14** Sequence model for coastal and shallow-marine deposits. Changes in fluvial facies are based on the nonmarine sequence models of Wright and Marriott (1993), Shanley and McCabe (1994), and Gibling and Bird (1994). Standard systems-tract abbreviations (MFS, etc.; explained in text) and bounding surface rankings (numerals in *circles*) are shown (see Miall, 1988, 1996)

lowstand or the transgressive systems tract. The shape of the sea-level curve and the timing of these deposits relative to this curve are usually not knowable, and so this is a somewhat hypothetical argument.

Transgression is commonly indicated by the appearance of abundant tidal influence in the fluvial succession. Sigmoidal crossbedding, tidal bedding (wavy, flaser and lenticular bedding), oyster beds and brackish to marine trace fossils are all typical indicators of tidal-marine environments. The transition from fluvial to tidal is typically diachronous, and the filling of the incised valley changes from aggradational to retrogradational. Inland from tidal influence, the change from the lowstand to the transgressive phase may be marked by a change in fluvial style or by the development of coal beds (Fig. 2.13). Coal commonly occurs during an initial increase in accommodation, before this is balanced by an increase in clastic supply. Within the valleys of major rivers, the increase in accommodation can result in more loose stacking of channel sand bodies and greater preservation of overbank fines. Changes in fluvial style are also common, with braided rivers typifying lowstand systems and anastomosed or meandering rivers common during times of high rate of generation of accommodation, as during the transgressive phase of the base-level cycle.

A highstand systems tract develops when base-level rise slows, and the rate of generation of accommodation space decreases to a minimum. There are two possible depositional scenarios for this phase of sequence development. Retrogradation of the river systems during transgression will have led to reduced slopes, and a low-energy landscape undergoing slow accumulation of floodplain deposits, limited channel aggradation, and closely-spaced, well-developed soil profiles (Shanley and McCabe, 1994). Given no change in source-area conditions, however, the sediment supply into the basin will continue, and vigorous channel systems will eventually be re-established. Under these conditions, channel bodies will form that show reduced vertical separation relative to the TST, leading to lateral amalgamation of sandstone units and high net-to-gross sandstone ratios (Wright and Marriott, 1993; Olsen et al., 1995; Yoshida et al., 1996). Basinward progradation of coastal depositional systems leads to downlap of deltaic and barrier-strandplain deposits onto the maximum flooding surface. A good nonmarine example of this was described by Ray (1982), who mapped

the progradation of an alluvial plain and deltaic system into lake deposits.

It seems likely that the HST will be poorly represented in most nonmarine basins, because the highstand is usually followed by the next cycle of falling base level, which may result in the removal by subaerial erosion of much or all of the just-formed HST deposits. A minor increase in the sand-shale ratio immediately below the sequence boundary may be the only indication of the highstand phase, as in the Castlegate Sandstone of Utah (Olsen et al., 1995; Yoshida et al., 1996).

Care must be taken to evaluate all the evidence in interpreting such data as net-to-gross sandstone ratios. Changes in this parameter may not always be attributable to changes in the rate of generation of accommodation space. Smith (1994) described a case where an increase in the proportion of channel sandstones in a section seems to have been related not to changes in the rate of generation of accommodation space, but to increased sediment runoff resulting from increased rainfall. In the case of sequences driven by orbital forcing mechanisms, where both base-level change and climate change may be involved, unravelling the complexity of causes and effects is likely to be a continuing challenge. In the model of Shanley and McCabe (1994) a greater degree of channel amalgamation is shown in the TST than in the HST, the opposite of that shown in the model of Wright and Marriott (1993). Shanley and McCabe (1994) suggested that where rising base level is the main control on the rate and style of channel stacking, the rate of generation of accommodation is small during transgression in inland areas while the coastline is still distant, and increases only once transgression has brought the coastline farther inland where the effect of base-level rise on the lower reaches of the river produces a more rapid increase in accommodation. In this model the rate of generation of accommodation is greater during the highstand than during transgression, and results in low net-to-gross sandstone ratios. However, this line of reasoning omits the influence of upstream factors, and must therefore not be followed dogmatically. One must also be cautious in using systems-tract terminology derived from marine processes for the labeling of nonmarine events. There may be a considerable lag in the transmission of a transgression upstream to inland positions by the process of slope reduction, aggradation and tidal invasion. The inland reaches of the river will not "know" that a transgression is occurring, and it is questionable, therefore, whether the deposits formed inland during the initial stages of the marine transgression should be included with the TST.

Currie (1997) suggested an alternative terminology for the standard systems tract terms used for marine basins because, obviously, terms that include such words as transgressive, highstand, etc., are inappropriate for basins that are entirely nonmarine. For falling stage and lowstand deposits Currie (1997) suggested the term *degradational systems tract*, for transgressive deposits, *transitional systems tract*, and for highstand deposits, *aggradational systems tract*. These terms provide analogous ideas regarding changes in accommodation and sediment supply and their consequences for depositional style.

Holbrook et al. (2006) introduced the useful concepts of *buttresses* and *buffers* to account for longitudinal changes in fluvial facies and architecture upstream from a coastline. A buttress is some fixed point that constitutes the downstream control on a fluvial graded profile. In marine basins this will be marine base level (sea level). In inland basins it will be the lip or edge of a basin through which the trunk river flows out of the basin. The buffer is the zone of space above and below the current graded profile which represents the range of reactions that the profile may exhibit given changes in upstream controls, such as tectonism or climate change, that govern the discharge and sediment load of the river. For example, tectonic uplift may increase the sediment load, causing the river to aggrade towards its upper buffer limit. A drop in the buttress, for example as a result of a fall in sea level, may result in incision of the river system, but if the continental shelf newly exposed by the fall in sea level has a similar slope to that of the river profile, there may be little change in the fluvial style of the river. In any of these cases, the response of the river system is to erode or aggrade towards a new dynamically maintained equilibrium profile that balances out the water and sediment flux and the rate of change in *accommodation*. The zone between the upper and lower limits is the buffer zone, and represents the available (potential) *preservation space* for the fluvial system.

Blum (1994) demonstrated that nowhere within coastal fluvial systems is there a single erosion surface that can be related to lowstand erosion. Such surfaces are continually modified by channel scour, even during transgression, because episodes of channel incision may reflect climatically controlled times of low sediment load, which are not synchronous with changes in base level. This is particularly evident landward of the limit of base-level influence. Post-glacial terraces within inland river valleys reveal a history of alternating aggradation and channel incision reflecting climate changes, all of which occurred during the last post-glacial rise in sea level. A major episode of valley incision occurred in Texas not during the time of glacioeustatic sea-level lowstand, but at the beginning of the postglacial sea-level rise, which commenced at about 15 ka (Blum, 1994). The implications of this have yet to be resolved for inland basins where aggradation occurs (because of tectonic subsidence), rather than incision and terrace formation. However, it would seem to suggest that no simple relationship between major bounding surfaces and base level change should be expected.

Figure 2.15 shows a model of fluvial processes in relationship to glacially controlled changes in climate and vegetation, based on Dutch work. These studies, and those in Texas, deal with periglacial regions, where climate change was pronounced, but the areas were not directly affected by glaciation. Vandenberghe (1993) and Vandenberghe et al. (1994) demonstrated that a major period of incision occurred during the transition from cold to warm phases because runoff increased while sediment yield remained low. Vegetation was quickly able to stabilize river banks, reducing sediment delivery, while evapotranspiration remained low, so that the runoff was high. Fluvial styles in aggrading valleys tend to change from braided during glacial phases to meandering during interglacials (Vandenberghe et al., 1994). With increasing warmth, and consequently increasing vegetation density, rivers of anastomosed or meandering style tend to develop, the former particularly in coastal areas where the rate of generation of new accommodation space is high during the period of rapidly rising base level (Törnqvist, 1993; Törnqvist et al., 1993). Vandenberghe (1993) also demonstrated that valley incision tends to occur during the transition from warm to cold phases. Reduced evapotranspiration consequent upon the cooling temperatures occurs while the vegetation cover is still substantial. Therefore runoff increases, while sediment yield remains low. With reduction in vegetation cover as the cold phase becomes established, sediment deliveries increase, and fluvial aggradation is reestablished.

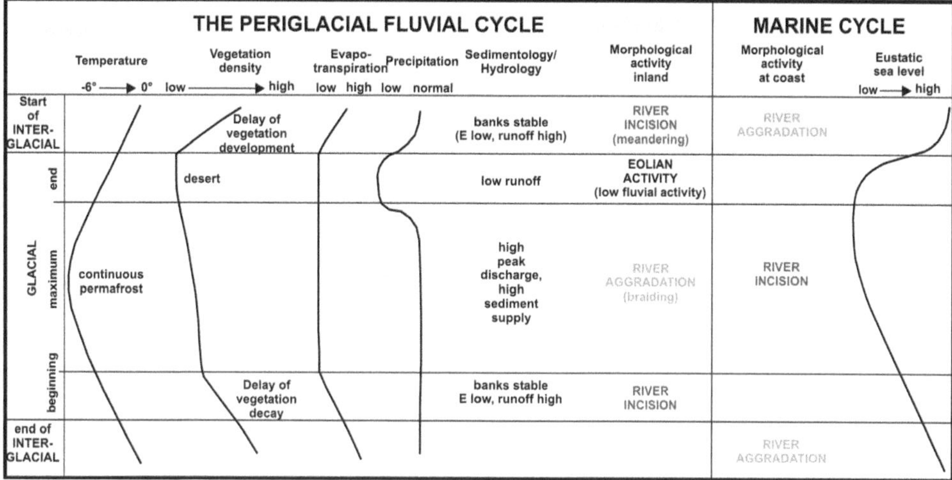



**Fig. 2.15** Relationship among temperature, vegetation density, evapotranspiration, precipitation, and sedimentary processes in river systems during glacial and interglacial phases, and the relationship to the contemporaneous marine cycle. Based on work in the modern Rhine-Meuse system (Vandenberghe, 1993), with the marine cycle added

**Fig. 2.16** Longitudinal NW-SE section, approximately 250 km in length, through the mid-Cretaceous Dakota Group, from Colorado into the northeast corner of New Mexico. The internal architecture consists of a series of unconformity-bounded sandstone sheets that reflect "deposition during repetitive valley-scale cycles of aggradation and incision" (Holbrook et al., 2006, p. 164)

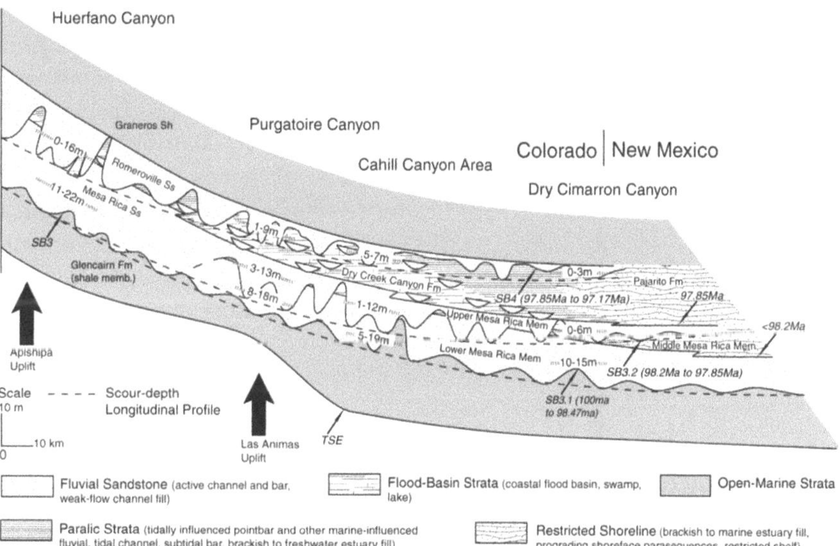



It is apparent that fluvial processes inland and those along the coast may be completely out of phase during the climatic and base-level changes accompanying glacial to interglacial cycles. Within a few tens of kilometres of the sea, valley incision occurs at times of base level lowstand, during cold phases, but the surface may be modified and deepened during the subsequent transgression until it is finally buried. Inland, major erosional bounding surfaces correlate to times of climatic transition, from cold to warm and from warm to cold, that is to say during times of rising and falling sea level, respectively.

The Dakota Group of northeast New Mexico and southeast Colorado provides a good example of an internally architecturally complex fluvial unit generated by a combination of upstream tectonic controls and downstream sea-level cycles (Fig. 2.16: Holbrook et al., 2006). At the coastline, progradation and retrogradation creating three sequences were caused by sea-level cycles on a $10^5$-year time scale. Each of these sequences can be traced updip towards the west, where they are composed of repeated cycles of aggradational valley-fill successions and mutually incised scour surfaces. These cycles reflect autogenic channel shifting within the limited preservation space available under conditions of modest, tectonically-generated accommodation. This space is defined by a lower buffer (in the Holbrook et al., 2006 terminology) set by maximum local channel scour, and an upper buffer set by the ability of the river to aggrade under the prevailing conditions of discharge and sediment load.

### 2.3.3  Carbonate Depositional Systems

Carbonate and clastic depositional systems respond very differently to sea-level change (Sarg, 1988; James and Kendall, 1992; Schlager, 1992, 1993). The differences between carbonates and clastics were not understood at the time the original Exxon sequences models were developed by Vail et al. (1977) and Vail (1987).

Figure 2.17 illustrates a typical carbonate platform, consisting of a wide, carbonate-dominated shelf with a fringing barrier reef, scattered bioherms or patch reefs in the platform interior, and a marginal clastic belt, the width of which depends on the clastic supply delivered to the coast from river mouths, and the strength of the waves and tides redistributing it along the

coast. Figure 2.18 illustrates the differences between the responses of carbonate and clastic systems to sea-level change, which are described briefly here. At times of sea-level lowstand, terrigenous clastics bypass the continental shelf, leading to exposure, erosion, and the development of incised valleys. Submarine canyons are deepened, and sand-rich turbidites systems develop submarine-fan complexes on the slope and the basin floor. Carbonate systems essentially shut down at times of lowstand, because the main "carbonate factory", the continental shelf, is exposed, and commonly undergoes karstification. A narrow shelf-edge belt of reefs or sand shoals may occur, while the deep-water basin is starved of sediment or possibly subjected to hyperconcentration, with the development of evaporite deposits. Evaporites may also develop on the continental shelf during episodes of sea-level fall, when reef barriers serve to block marine circulation over the shelf.

During transgression of a clastic system, incised valleys fill with estuarine deposits and eventually are blanketed with marine shale. There may be a rapid landward translation of facies belts, leaving the continental shelf starved of sediment, so that condensed successions are deposited. By contrast, transgression of a carbonate shelf serves to "turn on" the carbonate factory, with the flooding of the shelf with warm, shallow seas. Thick platform carbonate successions develop, reefs, in particular, being able to grow

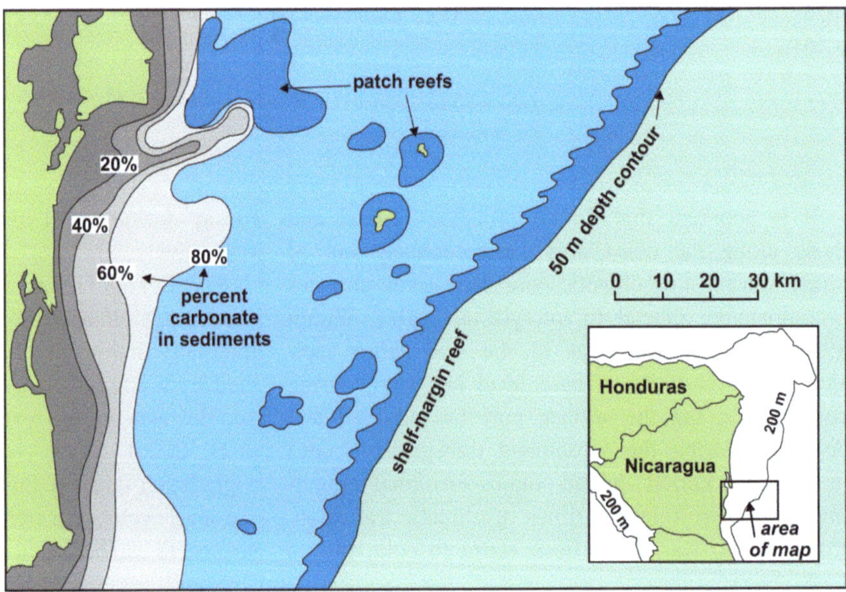


**Fig. 2.17**  A typical carbonate platform. The continental shelf off Nicaragua (Roberts, 1987)

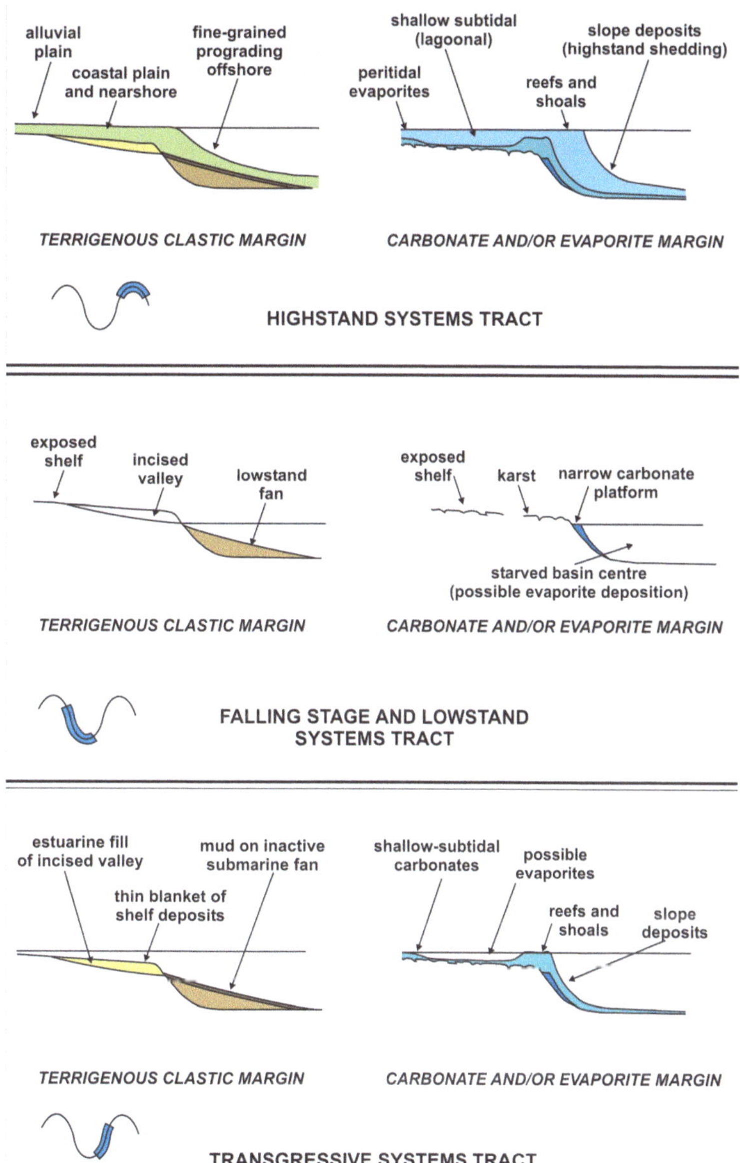

**Fig. 2.18** Differences in the response to base level change between carbonate and clastic systems (adapted from James and Kendall, 1992)



vertically at extremely rapid rates as sea level rises. At times of maximum transgression, the deepest part of the shelf may pass below the photic zone, leading to cessation of carbonate sedimentation and development of a condensed section or hardground. The resulting surface is termed a *drowning unconformity* (Schlager, 1989).

Carbonate and clastic shelves are most alike during times of highstand. The rate of addition of sedimentary accommodation space is low, and lateral progradation is therefore encouraged, with the development

of clinoform slope architectures. Autogenic shoaling-upward cycles are common in both types of environment (e.g., terrigenous deltaic lobes, tidal carbonate cycles). Schlager (1992) stated:

> Prograding [carbonate] margins dominated by offshore sediment transport most closely resemble the classical [siliciclastic] sequence model. They are controlled by loose sediment accumulation and approach the geometry of siliciclastic systems (e.g., leeward margins . . .).

Carbonate platforms generate sediment at the highest rate during highstands of sea level, when platforms are flooded and the carbonate factory is at maximum productivity. The sediment volume commonly exceeds available accommodation, and the excess is delivered to the continental slope, where it may provide the sediment for large-scale progradation by carbonate talus and sediment-gravity flows. This process, termed *highstand shedding* is exemplified by the architecture of the Bahamas Platform (Schlager, 1992). It is the converse of the pattern of siliciclastic sedimentation, within which, as already noted, sediment is fed to the continental slope most rapidly at times of low sea level.

Many ancient shelf deposits are mixed carbonate-clastic successions, containing thin sand banks or deltaic sand sheets interbedded with carbonate platform deposits (Dolan, 1989; Southgate et al., 1993). Galloway and Brown (1973) described an example from the Pennsylvanian of northern central Texas, in which a deltaic system prograded onto a stable carbonate shelf (Fig. 2.19). The term *reciprocal sedimentation* has been used for depositional systems in which carbonates and clastics alternate (Wilson, 1967). In the example given in Fig. 2.19, deltaic distributary channels are incised into the underlying shelf carbonate deposits. Widespread shelf limestones alternate with the clastic sheets and also occur in some interdeltaic embayments. Carbonate banks occur on the outer shelf edge, beyond which the sediments

thicken dramatically into a clinoform slope-clastics system. This association of carbonates and clastics reflects regular changes in sea-level, with the carbonate phase representing high sea-level and the clastic phase low sea-level. The deltaic and shelf-sand sheets and the slope clinoform deposits represent lowstand systems tracts, while the carbonate deposits are highstand deposits. During episodes of high sea-level, clastics are trapped in nearshore deltas, while during low stands much of the detritus bypasses the shelf and is deposited on the slope (arrows in Fig. 2.19).

The reciprocal-sedimentation model should, however, be used with caution. Dunbar et al. (2000) reported on the results of a detailed sampling program across the Great Barrier Reef, from which they concluded that through the glacioeustatic sea-level cycles of the last 300 ka, the maximum rate of siliciclastic sedimentation on the continental slope occurred during transgression, not during the falling stage. They explained this as the result of very low fluvial slopes across the shelf that became exposed during the sea-level falling stage and lowstand, resulting in sediment accumulation and storage on the shelf during this phase of the sea-level cycle. The sediment stored there was mobilized and transported seaward by vigorous marine processes during the subsequent transgressive phases.

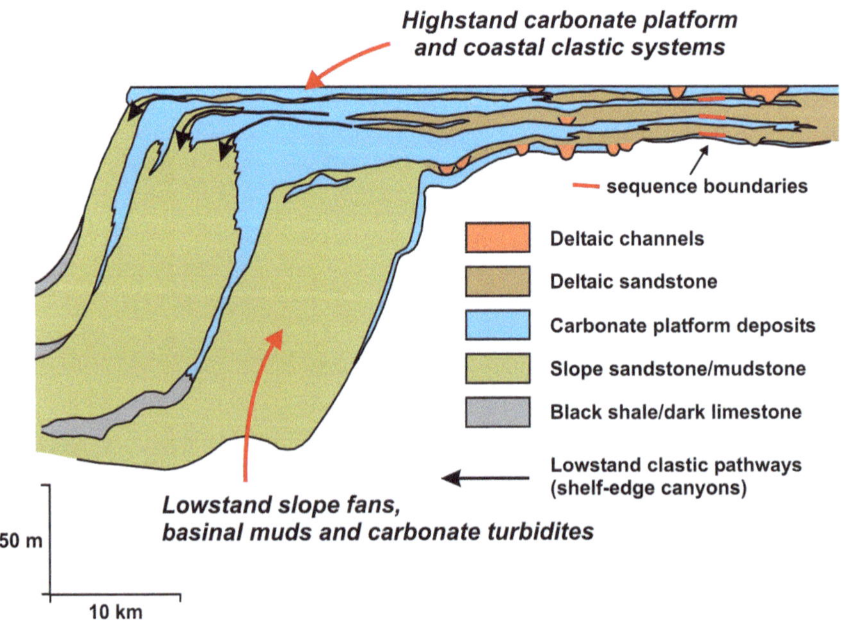


**Fig. 2.19** The stratigraphic architecture of the late Paleozoic continental margin, central Texas, an example of "reciprocal sedimentation" (Galloway and Brown, 1973)

### 2.3.3.1 Breaks in Sedimentation in Carbonate Environments

Submarine erosion, and other processes, can generate breaks in sedimentation without any change in sea level. This is particularly the case in carbonate sediments, which are very sensitive to environmental change, and may develop drowning unconformities (Schlager, 1989, 1992). They may be architecturally similar to lowstand unconformities, and care must be taken to interpret them correctly. They may, in fact, represent an interval of slow sedimentation, with many small hiatuses and interbedded with thin condensed sections, indistinguishable on the seismic record from actual unconformable breaks because of limited seismic resolution. Schlager (1992) stated:

> Drowning requires that the reef or platform be submerged to subphotic depths by a relative rise that exceeds the growth potential of the carbonate system. The race between sea level and platform growth goes over a short distance, the thickness of the photic zone. Holocene systems indicate that their short-term growth potential is an order of magnitude higher than the rates of long-term subsidence or of third-order sea level cycles ... This implies that drowning events must be caused by

unusually rapid pulses of sea level or by environmental change that reduced the growth potential of platforms. With growth reduced, drowning may occur at normal rates of rise.

Schlager (1992) pointed to such environmental changes as the shifts in the El Niño current, which bring about sudden rises in water temperature, beyond the tolerance of many corals. Drowning can also occur when sea-level rise invades flat bank tops, creating shallow lagoons with highly variable temperatures and salinities, plus high suspended-sediment loads due to coastal soil erosion. Oceanic anoxic events, particularly in the Cretaceous, are also known to have caused reef drowning. Schlager (1992) suggested that two Valangian sequence boundaries in the Haq et al. (1987, 1988a) global cycle chart may actually be drowning unconformities that have been misinterpreted as lowstand events. He also noted the erosive effects of submarine currents, and their ability to generate unconformities that may be mapped as sequence boundaries but that have nothing to do with sea-level change. The Cenozoic sequence stratigraphy of the Blake Plateau, off the eastern United States, is

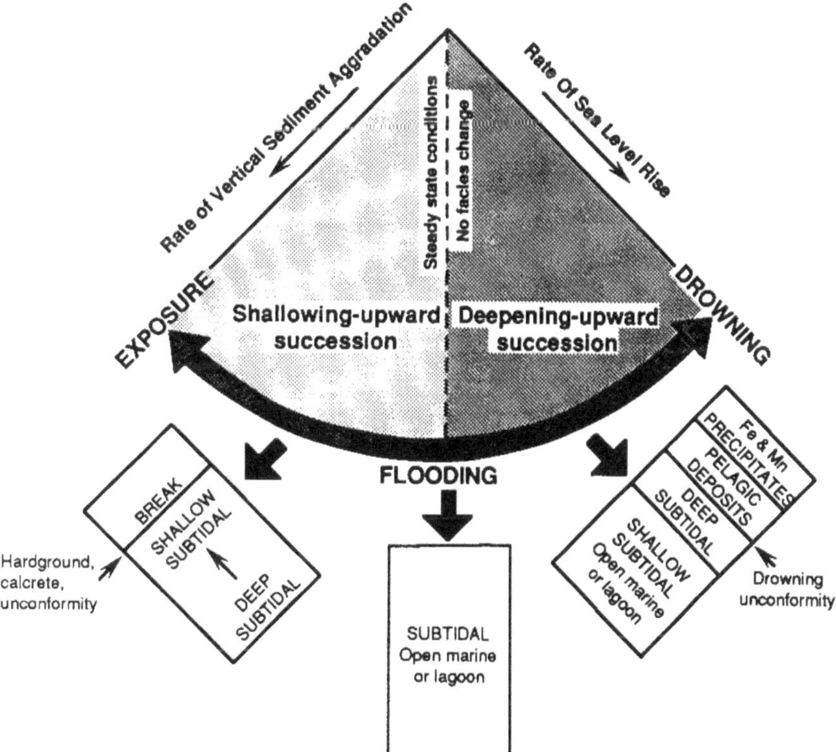



**Fig. 2.20** The types of sequence that develop, depending on variations in the rate of relative sea-level rise and carbonate sediment production (Jones and Desrochers, 1992)

dominated by such breaks in sedimentation that do not correlate with the Exxon global cycle chart, but have been interpreted as the result of erosion by the meandering Gulf Stream.

### 2.3.3.2 Platform Carbonates: Catch-up Versus Keep-up

The style of carbonate sequence-stratigraphy on the platform is mainly a reflection of the balance between sea-level rise and carbonate production, as summarized in a review by Jones and Desrochers (1992). Where sea-level fall exceeds subsidence rates, exposure occurs, and karst surfaces may develop (Figs. 2.20 and 2.21). Rapid trangression, on the other hand, leads to the development of condensed sections, and may shut down the carbonate factory, resulting in a *drowning unconformity*. Nutrient poisoning and choking by siliciclastic detritus may also shut down the carbonate factory at times of high sea level, and this can also lead to the development of drowning unconformities, which may be mistaken for sequence boundaries (e.g., Erlich et al., 1993). James and Bourque (1992) argued that poisoning and choking were the processes most likely to cause a shut down of reef sedimentation, because studies have indicated that under ideal conditions vertical reef growth is capable of keeping pace with the most rapid of sea-level rises.

Figure 2.21 illustrates the main variations in platform architecture that develop in response to changes in the controls noted above. Drowning during a rapid rise in relative sea level is typically followed by backstepping. A slightly less rapid transgression may lead to *catch-up* architecture. Here the sea-floor remains a site of carbonate production, and as sea-level rise slows, sedimentation is able to catch up to the new sea level. Vertical aggradation characterizes the first stage of the catch-up, but lateral progradation may occur late in the cycle, when the rate of generation of new accommodation space decreases. Shoaling-upward sequences are the result.

A balance between sea-level rise and sediment-production rates will lead to *keep-up* successions, in which cyclicity is poorly developed. Eventually, typically at the close of a cycle of sea-level rise, carbonate production may exceed the rate of generation of accommodation space. This can lead to lateral progradation, and highstand shedding of carbonate detritus onto the continental slope.

As with coastal detrital sedimentation, autogenic processes may generate successions that are similar to those that are inferred to have formed in direct response to sea-level change. Pratt et al. (1992) summarized the processes of autogenesis in peritidal environments, where shallowing-upward cycles are characteristic, as a result of short-distance transport and accumulation of carbonate detritus. Lateral progradation and vertical aggradation both may occur. Rapid filling of the available accommodation space may lead

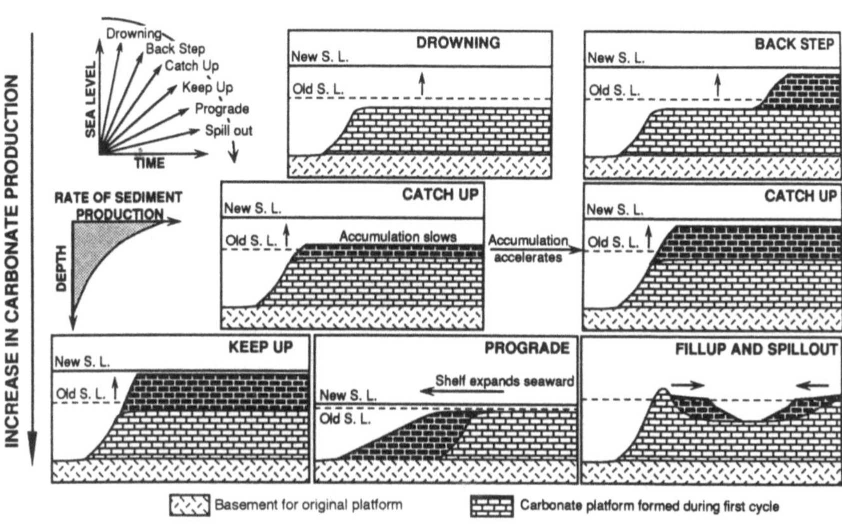


**Fig. 2.21** The styles of carbonate-platform architecture and their dependence on the balance between the rate of sea-level rise and carbonate productivity (Jones and Desrocher, 1992)

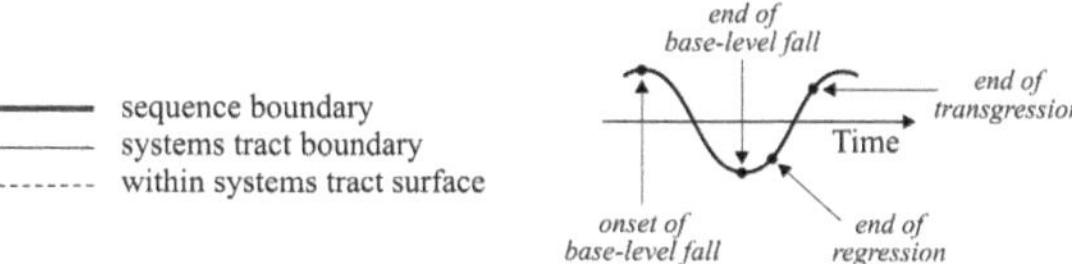

| Events \ Sequence model | Depositional Sequence II | Depositional Sequence III | Depositional Sequence IV | Genetic Sequence | T-R Sequence |
|---|---|---|---|---|---|
|  | HST | early HST | HST | HST | RST |
| *end of transgression* |  |  |  |  |  |
|  | TST | TST | TST | TST | TST |
| *end of regression* |  |  |  |  |  |
|  | late LST (wedge) | LST | LST | late LST (wedge) |  |
| *end of base level fall* |  |  |  |  |  |
|  | early LST (fan) | late HST (fan) | FSST | early LST (fan) | RST |
| *onset of base level fall* |  |  |  |  |  |
|  | HST | early HST (wedge) | HST | HST |  |

sequence boundary
systems tract boundary
within systems tract surface



**Fig. 2.22** Comparison of the five major ways in which sequences have been defined (Catuneanu, 2006, Fig. 1.7). Depositional sequence II refers to the work of Posamentier et al. (1988); Depositional sequence III is that of Van Wagoner et al. (1990); Depositional sequence IV is the definition of Hunt and Tucker (1992); Genetic sequence is the preferred approach of Galloway (1989a) and T-R sequence represents the work of Embry and Johannessen (1992)

to a shut-down of the carbonate factory until relative sea-level rises enough to stimulate its reactivation. The alternative to autogenesis is the invocation of a form of rhythmic tectonic movement to generate the required sea-level change, or Milankovitch mechanisms. As Pratt et al. (1992) noted, the scale and periodicity of tectonism required for metre-scale cycles has not yet been demonstrated.

## 2.4 Sequence Definitions

One of the commonest complaints about sequence stratigraphy is that it is "model-driven." Catuneanu (2006, pp. 6–9) summarized the various approaches that have been taken to defining sequences, and argued the case that the differences between the various models is not important, so long as sequences are described properly with reference to a selected standard model, with correct and appropriate recognition of systems tracts and bounding surfaces. His comparison diagram is reproduced here as Fig. 2.22, and the suite of important surfaces that are used in sequence and systems-tract definition is shown in Fig. 2.23. It should be noted that in each of the sequence definitions shown in Fig. 2.22, a similar set of systems tracts is shown in much the same relationship to each other. Exceptions include the T-R sequence, which makes use of a simplified definition of systems tracts, and such differences as that between the "late highstand" of depositional sequence III and the "falling stage" of depositional sequence IV. The major difference between the sequence models is where different workers have chosen to place the sequence boundary.

In the original Exxon model (Vail et al., 1977) the sequence boundary (commonly abbreviated as *SB* on diagrams) was drawn at the subaerial unconformity surface, following the precedent set by Sloss (1963), an approach which readily permits the sequence framework to be incorporated into an allostratigraphic terminology, at least for coastal deposits, where the subaerial erosion surface is readily mapped. Offshore

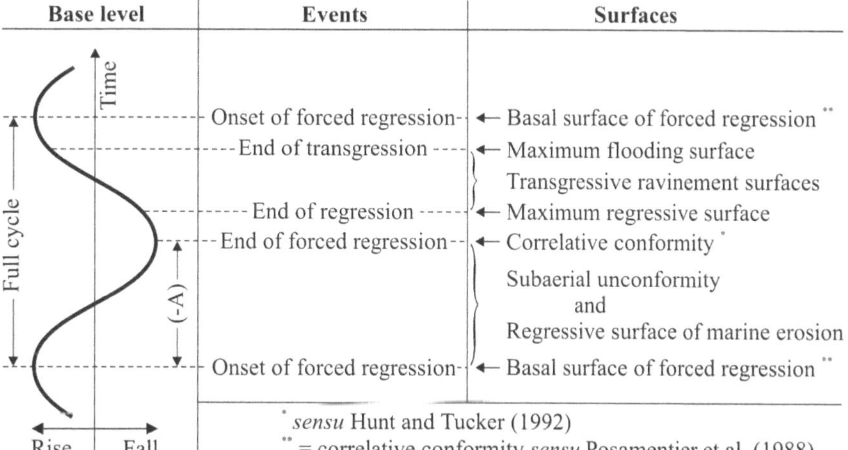



**Fig. 2.23** Stratigraphic surfaces used in the definition of sequences and systems tracts, and their timing, relative to the cycle of base-level change (Catuneanu, 2006, Fig. 4.7)

may be a different story. The first model did not recognize the falling-stage systems tract. The highstand of one sequence was followed directly by the lowstand of the next sequence, with the sequence boundary falling between the two systems tracts. With the recognition of the process of forced regression (Plint, 1988), the forced regressive deposits could be assigned either to a "late highstand" or to an "early lowstand." Based on assumptions about the changing rate of sea level fall, the sequence boundary—the coastal equivalent of the subaerial erosion surface—was placed at the *basal surface of forced regression* (assumed commencement of forced regression). This placement of the sequence boundary is the basis for what Catuneanu (2006) refers to as "depositional sequence II" (Fig. 2.22). The problem with this definition is addressed below. In addition, the early Exxon work defined several different types of sequence-bounding unconformity. Vail and Todd (1981) recognized three types, but later work (e.g., Van Wagoner et al., 1987) simplified this into two, termed type-1 and type-2 unconformities, based on assumptions about the rate of change of sea level and how this was reflected in the sequence architecture. In the rock record, they would be differentiated on the basis of the extent of subaerial erosion and the amount of seaward shift of facies belts. However, Catuneanu (2006, p. 167) pointed out the long-standing

confusions associated with these definitions and recommended that they be abandoned. These types are not discussed further in this book. Schlager (2005, p. 121) recommended the separate recognition of a third type of sequence boundary, one which forms "when sea level rises faster than the system can aggrade, such that a transgressive systems tract directly overlies the preceding highstand tract often with a significant marine hiatus. . . . Marine erosion frequently accentuates this sequence boundary, particularly on drowned carbonate platforms."

Hunt and Tucker (1992) were amongst the first (since Barrell!) to point out that during sea-level fall, subaerial erosion continues until the time of sea-level lowstand, with the continuing transfer of sediment through clastic delivery systems to the shelf, slope and basin, and with continuing downcutting of the *subaerial erosion surface* throughout this phase. The age of the *subaerial erosion surface,* therefore, spans the time up to the end of the phase of sea-level fall, a time substantially later than the time of initiation of forced regression. The use of the *basal surface of forced regression* as a sequence boundary, as in "depositional sequence II" is, therefore, not an ideal surface at which to define the sequence boundary, although, as Catuneanu (pers. com., 2009) reports, it is commonly a prominent surface on seismic-reflection lines. In fact, as Embry (1995) pointed out (see Fig. 2.24), there

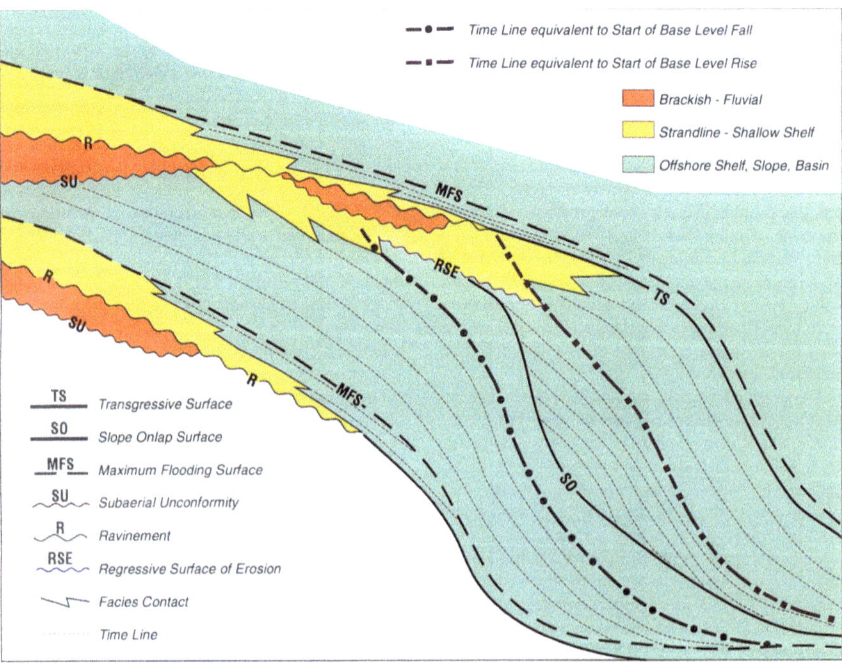


**Fig. 2.24** Schematic cross-section through an ideal continental-margin sequence, showing the relationships between the major surfaces (Embry, 1995, Fig. 1.1)

is no through-going surface associated with forced regression that can be used to extend the subaerial erosion surface offshore for the purpose of defining a sequence boundary. He argued that "from my experience I have found that the most suitable stratigraphic surface for the conformable expression of a sequence boundary is the transgressive surface" (Embry, 1995, p. 4). This meets his criterion—one which all stratigraphers would agree with—that "one of the main purposes of sequence definition [is] a coherent genetic unit without significant internal breaks" (Embry, 1995, p. 2). His preferred definition of sequences, the T-R sequence, places the sequence boundary at the TR surface, at the end of the phase of regression and the time of initial transgression (Figs. 2.22 and 2.24). There is, of course, a delay in time between the end of downcutting of the subaerial erosion surface during the falling stage, and the flooding of the same surface during transgression. The results of the two processes may coincide in the rocks, which is why this surface may provide a good stratigraphic marker; but it is important to remember that the surface is not a time marker, but represents a time gap, with the gap decreasing in duration basinward.

Highlighting the timing of development of the subaerial erosion surface by Hunt and Tucker (1992) also served to highlight the inconsistency of assigning the main succession of submarine fan deposits on the basin floor to the lowstand systems tract, as shown in the Vail et al. (1977) and Posamentier et al. (1988) models, in which these deposits are shown resting on the sequence boundary. Notwithstanding the discussion of the Hunt and Tucker (1992) paper by Kolla et al. (1995), who defended the original Exxon models, this is an inconsistency that required a redefinition of the standard sequence boundary. It is now recognized that in the deep offshore, within submarine-fan deposits formed from sediment delivered to the basin floor during a falling stage, there may be no sharp definition of the end of the falling stage nor of the turn-around and subsequent beginning of the next cycle of sea-level rise and, therefore, no distinct surface at which to draw the sequence boundary. The boundary here is a correlative conformity, and may be very difficult to define in practice.

An alternative sequence model, termed the genetic stratigraphic sequence (Fig. 2.22), was defined by Galloway (1989a), building on the work of Frazier (1974). Although Galloway stressed supposed

philosophical differences between his model and the Exxon model, in practice, the difference between them is simply one of where to define the sequence boundaries. The Exxon model places emphasis on subaerial unconformities, but Galloway (1989a) pointed out that under some circumstances unconformities may be poorly defined or absent and, in any case, are not always easy to recognize and map.

Galloway's (1989a) preference is to draw the sequence boundaries at the maximum flooding surface, which corresponds to the highstand downlap surfaces. He claims that these surfaces are more prominent in the stratigraphic record, and therefore more readily mappable. Galloway's proposal has not met with general acceptance. For example, Walker (1992) disputed one of Galloway's main contentions, that "because shelf deposits are derived from reworked transgressed or contemporary retrogradational deposits, their distribution commonly reflects the paleogeography of the precursor depositional episode." Galloway (1989a) went on to state that "these deposits are best included in and mapped as a facies element of the underlying genetic stratigraphic sequence." However, as Walker (1992) pointed out, most sedimentological parameters, including depth of water, waves, tides, basin geometry, salinity, rates of sediment supply, and grain size, change when an unconformity or a maximum flooding surface is crossed. From the point of view of genetic linkage, therefore, the only sedimentologically related packages lie (1) between a subaerial unconformity and a maximum flooding surface, (2) between a maximum flooding surface and the next younger unconformity, or (3) between a subaerial erosion surface and the overlying unconformity (an incised-valley-fill) (Walker, 1992, p. 11).

However, some workers have found Galloway's use of the maximum flooding surface much more convenient for sequence mapping, for practical reasons. For example, it may yield a prominent gamma-ray spike in wireline logs (Underhill and Partington, 1993a), or it may correspond to widespread and distinctive goniatite bands (Martinsen, 1993), or it may provide a more readily traceable marker, in contrast to the surface at the base of the lowstand systems tract, which may have irregular topography and may be hard to distinguish from other channel-scour surfaces (Gibling and Bird, 1994). In nonmarine sections it may be hard to find the paleosol on interfluves that correlates with the sequence-bounding channel-scour

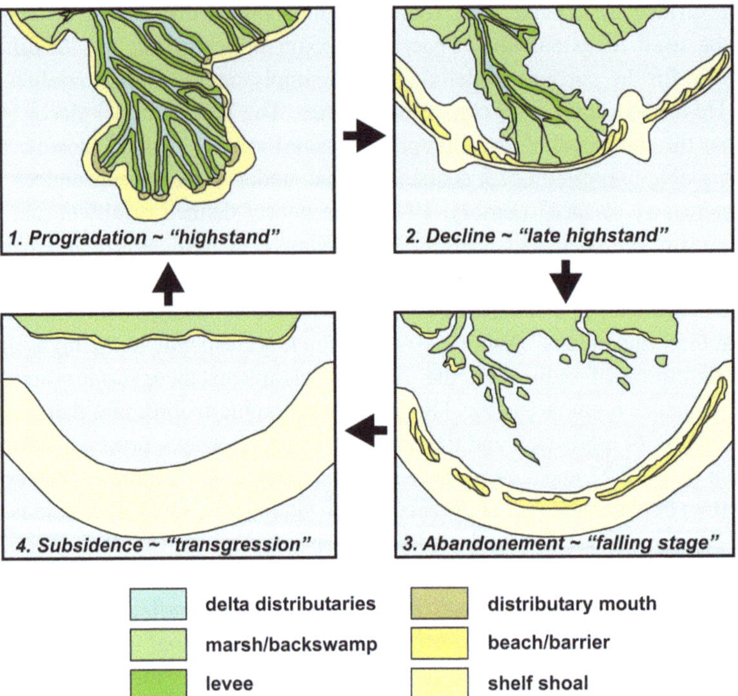



**Fig. 2.25** The development and abandonment of delta lobes in a river dominated, Mississippi-type delta (e.g., see Fig. 7.5), based on detailed analysis of the Mississippi delta system by Boyd and Penland (1988). In *stage 1*, progradation develops an upward-shoaling deltaic succession. Abandonment, followed by subsidence (resulting from compaction) cause the upper layers of the succession to be reworked (*stage 2*), resulting in the development of an extensive barrier island system (*stage 3*). Finally, the deposit undergoes transgression and is covered by marine shale (*stage 4*). Repetition of this succession of events when a new delta lobe progrades back over the older deposit results in shoaling upward successions bounded by transgressive flooding surfaces, that is, parasequences. In this case, however, they are clearly of autogenic origin. Systems-tract designations for each of the four stages are indicated in parenthesis

surface (Martinesen, 1993). In some studies (e.g., Plint et al., 1986; Bhattacharya, 1993) it has been found that ravinement erosion during transgression has removed the transgressive systems tract, so that the marine flooding surface coincides with the sequence boundary.

Finally, a word about the *parasequence*. Proposed originally as part of a hierarchy of terms, the bed, bedset, parasequence, parasequence set, and sequence (Van Wagoner et al., 1990; see Table 4.2), the parasequence has become a source of confusion, as noted above (Sect. 2.2.2). The shingles that compose many deltaic successions (e.g., see Fig. 2.9) are shoaling-upward successions bounded by flooding surfaces, and therefore fit the definition of parasequences. They are commonly the product of a process of autogenic delta switching, as illustrated in Fig. 2.25. This has been demonstrated to be the case in the example of the Dunvegan delta illustrated in Fig. 2.9 (Bhattacharya, 1991). The shingles and their bounding flooding surfaces are therefore local in distribution, and their development has little, if anything, to do with the allogenic mechanisms that generate sequences. However, to apply to these successions a term that contains the word "sequence" in it is inevitably to introduce the implication that they are allogenic in origin and constitute regionally correlatable units. The correct interpretation clearly depends on good mapping to determine the extent and correlatability of each shingle, andit would seem advisable not to use a term in a descriptive sense that carries genetic implications. As noted earlier (Sect. 2.2.2), this author recommends abandoning the term parasequence altogether.

# Chapter 3

# Other Methods for the Stratigraphic Analysis of Cycles of Base-Level Change

## Contents



As discussed in several places in this book, one of the principal problems with the assessment of causality is that there are no absolute reference frames for the calibration of sea-level change. The vertical movements of sea level can only be measured by the stratigraphic record left on Earth's crust, but the crust itself is in constant vertical motion, driven by mantle and lithospheric processes (Burton et al., 1987; Sahagian and Watts, 1991). Much of this book deals with the strategies earth scientists have evolved in their attempts to attack and overcome this problem.

## 3.1 Introduction

A variety of supplementary methods that have evolved for assessing regional and global sea-level changes directly from the stratigraphic record. These include the following:

- Analysing and correlating facies cycles.
- The mapping of changes in stratigraphic volumes or areas covered by successions of a specified age range as an indication of transgression and regression and changing rates of subsidence.
- The use of hypsometric curves to plot broad changes in continental elevation and eustatic sea-level changes.
- The study of paleoshorelines.
- Graphical and numerical methods for the documentation and analysis of the metre-scale cycles that are very common in some parts of the stratigraphic record (e.g., certain lacustrine and shelf-carbonate successions).

## 3.2 Facies Cycles

Vertical changes in lithofacies and biofacies have long been used to reconstruct temporal changes in depositional environments and, with the aid of Walther's Law, to interpret lateral shifts in these environments. Such an approach comprises the basis of the method of facies analysis, as discussed at length in many textbooks (e.g., Miall, 1999, Chap. 4) and review articles (e.g., Wanless, 1991). A single example of lithofacies and biofacies analysis will suffice. Other examples are illustrated elsewhere in this report.

Heckel (1986) presented a basic depositional model for the so-called "Kansas cyclothem", and for the environments of deposition occurring on a gently sloping tropical shelf (Fig. 3.1). Beds representing each of these environments are extraordinarily widespread, indicating shifts of environments of hundreds of kilometres. Black, phosphatic shales with conodonts indicate the deepest marine environments. Skeletal and algal wackestones and grainstones were deposited in shallow marine settings, whereas sandstones and


A.D. Miall, *The Geology of Stratigraphic Sequences,* 2nd ed.,
DOI 10.1007/978-3-642-05027-5_3, © Springer-Verlag Berlin Heidelberg 2010

**Fig. 3.1** (**a**) The typical Kansas cyclothem, showing the interpretation in terms of transgression and regression. (**b**) Model for deposition on a gently sloping tropical shelf, showing the position of rock types that become superimposed with transgression and regression (Heckel, 1986)

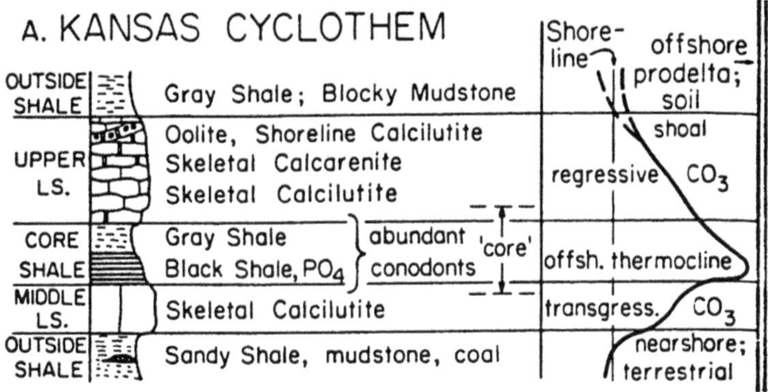


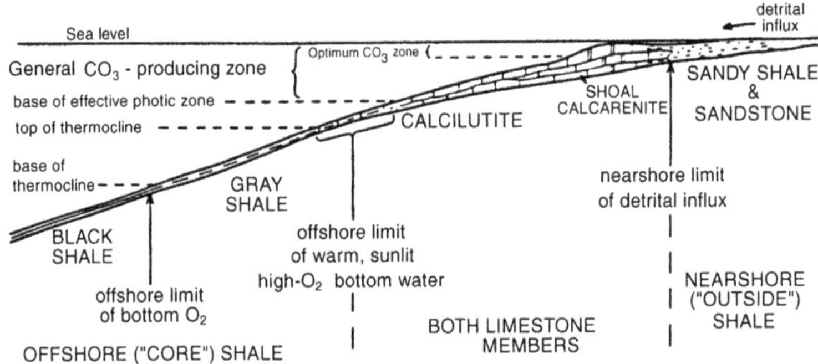


coals indicate nonmarine and marginal-marine settings. The vertical arrangement of these facies indicates transgression and regression. As illustrated in Fig. 11.18, Heckel (1986) and Boardman and Heckel (1989) were able to develop a sea-level curve for part of the Pennsylvanian by correlating these cycles and their contained facies changes across the U. S. mid-continent from Texas to Iowa.

In the foreland basin of the U. S. Western Interior, Weimer (1986) listed the following criteria for recognizing sea-level changes in the stratigraphic record:

1. Progradational shoreline deposits with incised drainage, with overlying marine shale.
2. Valley-fill deposits (of incised drainage system) overlying marine shale:

   A. root zones at or near base of valley-fill sequence.
   B. Paleosol on scour surface.

3. Unconformities within the basin:

   A. Missing faunal zone (except where absent for paleoecological reasons).
   B. Missing facies in a normal regressive succession (e.g., shoreface or delta-front sandstone).
   C. Paleokarst with regolith or paleosoil.
   D. Concentration of one or more of the following on a scour surface: phosphate nodules, glauconite, recrystallized shell debris to form thin lenticular limestone layers.

4. Thin, widespread coal layer overlying marine regressive delta-front sandstone deposits (indicating rising sea level).

A spectrum of short- and longer-term sea-level changes was documented, mainly in Europe and North America, by McKerrow (1979), based on paleobathymetric interpretations of brachiopod communities (Fig. 3.2). In the Ordovician and Silurian, marine shelf faunas consisted mainly of brachiopods, trilobites, corals, stromatoporoids and bryozoans. These comprise what is informally termed the shelly fauna.

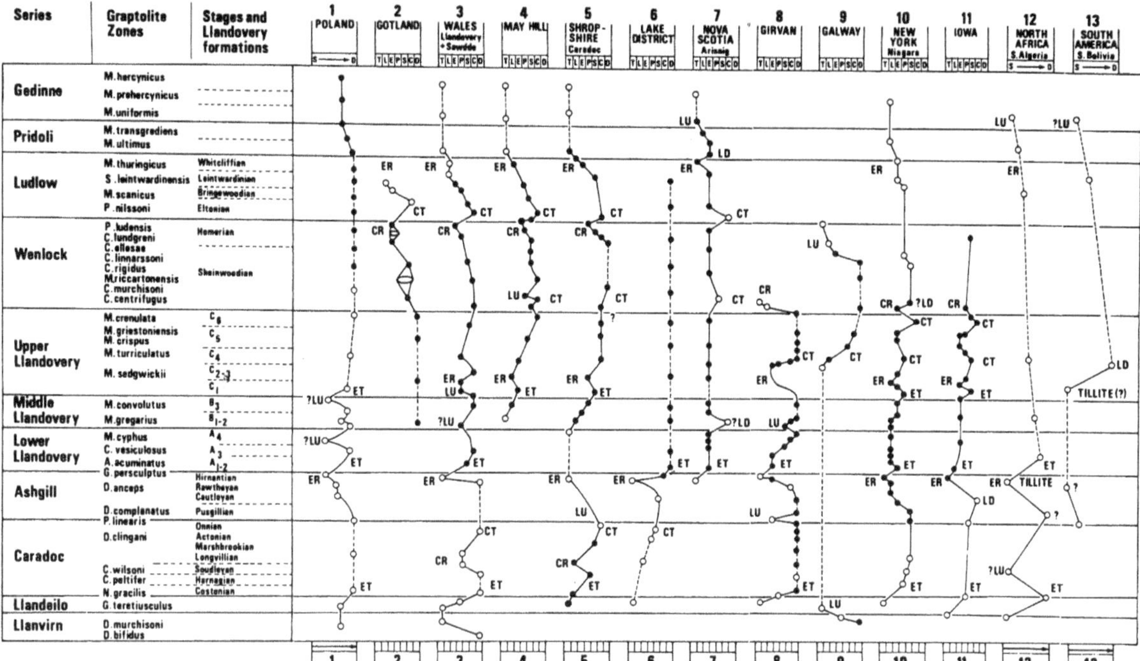

**Fig. 3.2** Depth changes on three continents during the Ordovician and Silurian. *Symbols* at head of column: S->D, shallow->deep; TLEPSCD, transgression, *Lingula, Eocoelia, Pentamerus, Stricklandia, Clorinda,* regression (see text for explanation). Symbols on graphs: E, eustatic; C, continental; L, local; T, transgression; R, regression; U, uplift; D, deepening. *Open circles* indicate uncertainty of depth and/or age. *Dashed lines* indicate terrestrial, unfossiliferous, or very deep environments that do not yield good depth control (McKerrow, 1979)

It contrasts with the graptolitic fauna of deeper-water, continental-slope, and abyssal oceanic environments.

Several authors have attempted to subdivide lower Paleozoic shelf faunas into depth-controlled communities. Ziegler (1965) and Ziegler et al. (1968) recognized five brachiopod-dominated assemblages in the early Silurian, which he named after typical genera. They are *Lingula, Eocoelia, Pentamerus, Stricklandia,* and *Clorinda,* in order of increasing water depth. These communities map out in bands parallel to the shore in shelf sequences in Wales, the Appalachian Basin, New Brunswick, and Iowa (McKerrow, 1979). The communities are not related to distance from shore as the shelf width varies from 5 to more than 100 km, and they are not related to sediment character, as each community occurs in various rock types. However, Cant (1979) has observed that storms can redistribute shallow-water fossil assemblages into deep water, and so some caution must be used in interpreting these data.

Ziegler's faunal differentiation has been established for the Upper Ordovician and the remainder of the

Silurian in a few areas. Sea-level changes over the shelf should be accompanied by lateral shifts in these communities, which should be recognizable in vertical sections through the resulting sediments. McKerrow (1979) used this approach, plus supplementary facies data, to construct depth-change curves for the Middle Ordovician to Early Devonian in 13 locations in Europe, Africa, North America and South America. The results are shown in Fig. 3.2. Some of the depth changes can be correlated between many of the regions examined and may reflect eustatic sea-level changes. Others are more local in scope and were probably caused by regional tectonic events.

McKerrow (1979) distinguished two types of eustatic depth change, slow and fast. Slow changes occurring over a few millions or tens of millions of years include the rise in sea-level during the Llandovery and the fall in the Ludlow and Pridoli. The slow rise and fall during the Silurian took about 40 million years to complete. Fast changes include the rapid rise in latest Llandeilo and earliest Caradoc time, a short-lived fall at the end of the Ashgill, and a rise and fall at the beginning of the late Llandovery. These rapid changes were of 1–2 million years duration.

Examples of sequence analysis of stratigraphic sections based on facies studies are given in Figs. 5.12, 6.1, 7.12, 7.15, and 7.17. Sequence boundaries typically are identified at unconformities, or at transgressive surfaces, where deepening took place.

In the subsurface, it may be necessary to rely on petrophysical logs to define sequences. The various components of the sequences may have distinctive log characteristics ("log motifs"), which aid in sequence definition, especially if the logs can be calibrated against one or more cores through the succession. For example, condensed sections commonly are revealed by gamma-ray spikes. Cant (1992) and Armentrout et al. (1993) described the methods of analysis. Armentrout et al. (1993) used grids of seismic lines and suites of petrophysical logs tied to the seismic cross-sections to correlate Paleogene deposits in the North Sea, and demonstrated that errors in correlation of up to 30 m could occur when markers were traced around correlation loops. In this area wells are up to tens of kilometres apart, and such large errors are not expected in mature areas such as the Alberta Basin or the Gulf Coast.

## 3.3 Areas and Volumes of Stratigraphic Units

In principle, the rising and falling of sea level should be recorded by transgressive and regressive deposits, leaving a record of shifting strandlines and of onlap and offlap. A simple measurement technique for tracking these events is to document the changes in the area of the basin or craton underlain by units of successive age. Sloss (1972) employed this procedure to compare the stratigraphic record of the Western Canada Sedimentary Basin and the Russian Platform (Fig. 3.3). As he noted, "the area covered by a given stratigraphic unit is determined by the maximum area of original deposition less the area of post-depositional erosion of sufficient magnitude to remove the unit" (Sloss, 1972, p. 25). The record of transgression tends to be better preserved than that of regression, because earlier transgressive deposits are covered by later deposits and thereby preserved, whereas during regression there is the tendency for the deposits to be exposed and eroded. Also, "the discovery of an isolated fault block, or diatreme xenolith, or glacial erratic, can shift the purported extent of late-cycle seas by hundreds of kilometres and alter the supposed maximum elevation of

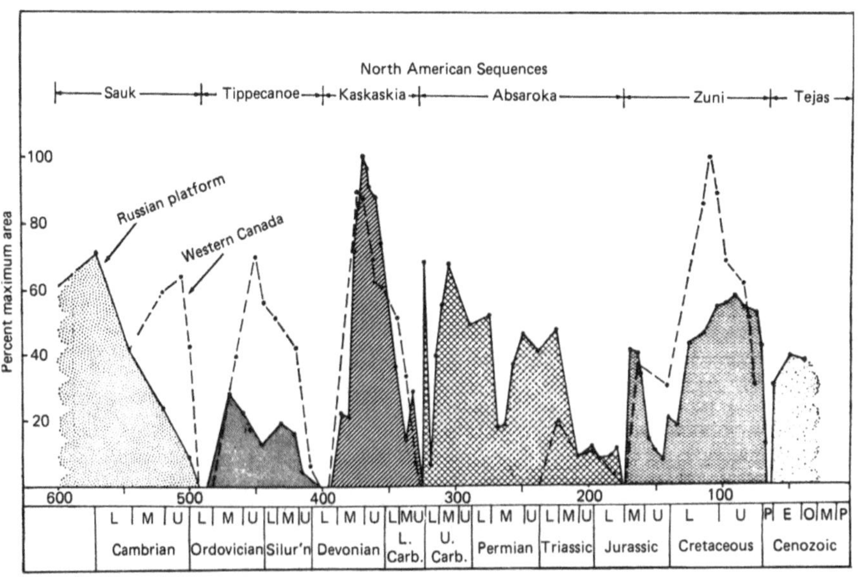


**Fig. 3.3** Comparison of the areal extent of a series of Phanerozoic time slices on the Russian Platform and in the Western Canada Sedimentary Basin. The maps were derived by measuring the area covered by the deposits in a series of standard map areas contained in paleogeographic atlases for the two areas (Sloss, 1972)

sea level by tens to hundreds of metres" (Sloss, 1979, p. 462). Given these constraints, nevertheless, Sloss (1972, 1979) found considerable similarities between the stratigraphic sequences of the Western Canada Sedimentary Basin and the Russian Platform

In order to reduce the error inherent in the analysis of thin feather-edge cratonic remnants, Sloss (1979) turned to intracratonic and pericratonic basins, as representing loci of more continuous subsidence.

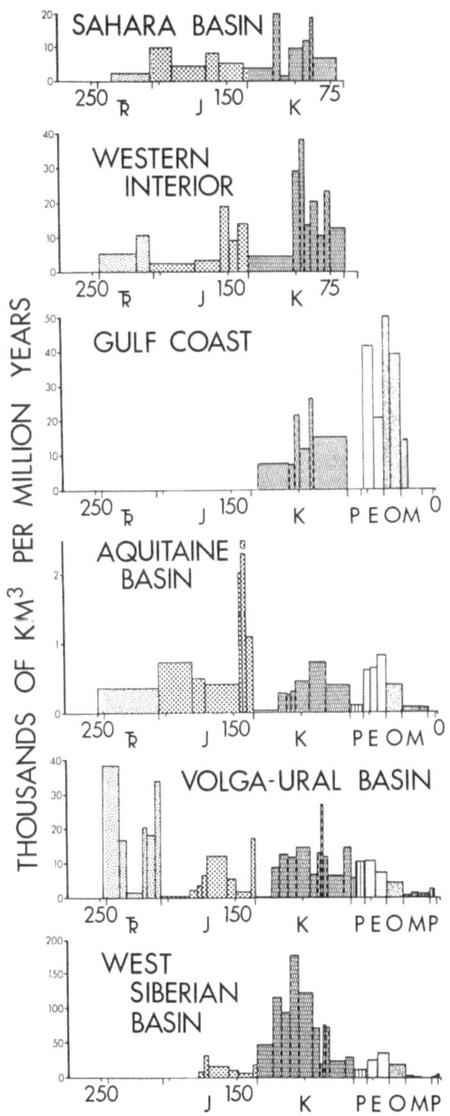

Fig. 3.4 Volume of sediment preserved per unit time in six Mesozoic-Cenozoic basins. TR=Triassic, J=Jurassic, K=Cretaceous, P=Paleocene, E=Eocene, O=Oligocene, M=Miocene (Sloss, 1979)

Figure 3.4 shows the volume of sediment preserved per unit time in six Mesozoic-Cenozoic basins in various tectonic settings. The four basins that yielded data on the Triassic-Early Jurassic period indicated a marked acceleration of subsidence in the Late Triassic or Early Jurassic, and there are also peaks in the mid Jurassic and mid- to Late Cretaceous. These data are combined into a single plot in Fig. 3.5, in which the volume/rate data for each basin have been normalized as a percent of the median rate for each basin and plotted at successive 10 million years increments. The smoothed trend shows a series of cycles about 50 million years long, which are presumably the result of interregional or global processes.

A more recent application of these techniques was presented by Ronov (1994), based on Russian paleogeographic atlases. This work generated a global "first-order" sea-level curve similar to that of Vail et al. (1977), but the method is not suitable for application to more detailed studies because of the coarse scale of stratigraphic resolution in the published atlases.

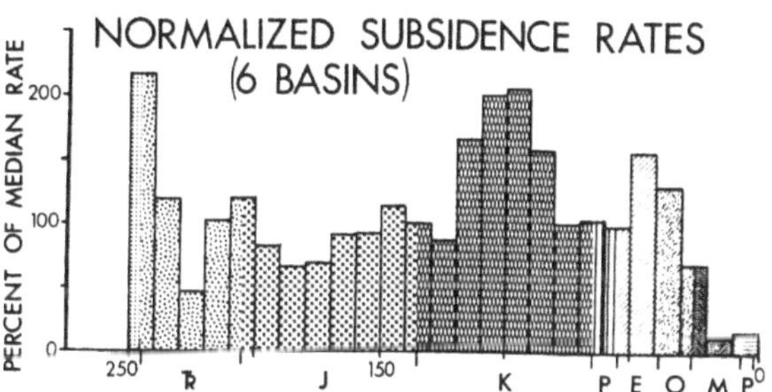

Fig. 3.5 Average Mesozoic-Cenozoic subsidence rates of the six basins shown in Fig. 3.4 (Sloss, 1979)

## 3.4 Hypsometric Curves

A hypsometric (or hypsographic) curve is defined as "a cumulative frequency profile representing the statistical distribution of the absolute or relative areas of the Earth's solid surface (land and sea floor) at various elevations above, or depths below, a given datum, usually sea level" (Bates and Jackson, 1987). Using a planimeter and an equal-area projection, the amount the sea advanced across a continent during any particular time interval can be derived from paleogeographic maps and compared with the curve derived from the

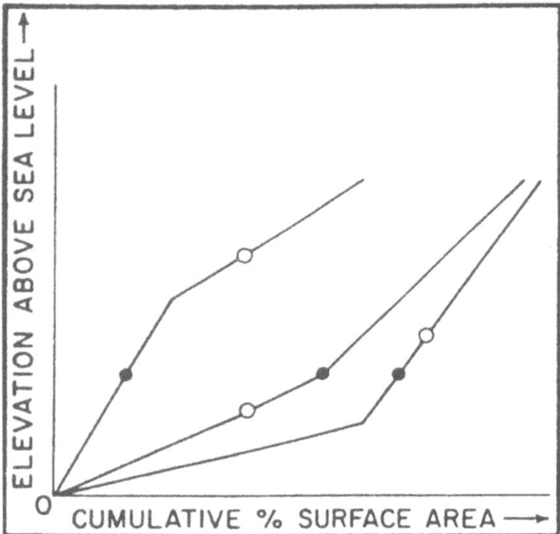

Fig. 3.6 Hypothetical hypsometric curves and various points showing percentages of continental area flooded during a transgression. *Solid points*: these all fall at the same elevation, indicating a transgression due to sea-level rise without subsequent change in continental hyposmetries. *Open circles*: transgression followed by substantial change in continental hypsometries (Bond, 1978)

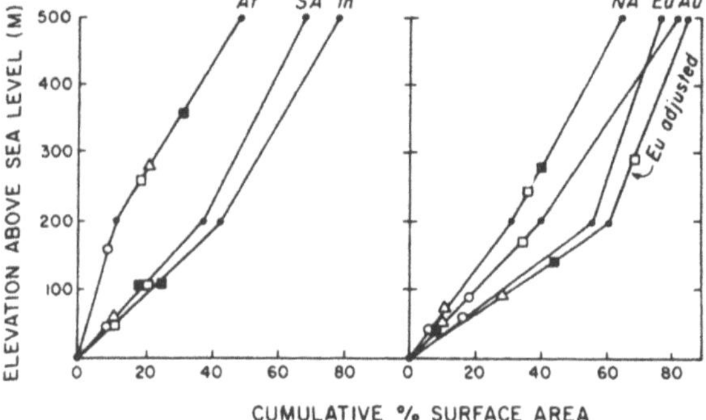

Fig. 3.7 Percentage of flooding plotted on continental hypsometric curves. Af: Africa, Au: Australia, Eu: Europe, In: India, NA: North America, SA: South America. "Europe adjusted" is the curve for Europe with the area south of the Alpine collision zone excluded. Albian: open squares, Late Campanian to early Maastrichtian: *solid squares*, Eocene: *triangles*, Miocene: *open circles* (Bond, 1978)

present continent (Burton et al., 1987). Bond (1976) developed a method of using hypsometric curves for distinguishing between sea-level changes and vertical motions of large continental surfaces. The principal is illustrated in Fig. 3.6. Application of the technique is rendered difficult by the large number of generalizations that must be incorporated. The same difficulties arise as those encountered by Sloss (1972, 1979) in using map distributions, and there is the difficulty in allowing for changes in continental hypsometry over

geological time because of plate-tectonic effects (e.g., crustal stretching and thickening).

Bond (1976) was able to demonstrate differences that developed in continental elevations between the major continental blocks during the mid-Cretaceous to Miocene, and also estimated actual (eustatic) sea-level changes during that time. The results are shown in Figs. 3.7 and 3.8. In Fig. 3.7, the percentages of the continental areas of several continents flooded during several successive time periods are indicated. Assuming no change in continental hypsometry since the time indicated, the rise in sea level required to bring about this percentage of flooding may be read off the graph. These estimates are then replotted in Fig. 3.8a. Clusters of three or four continental points occur at each time period, the close correlation between them suggesting eustatic sea-level

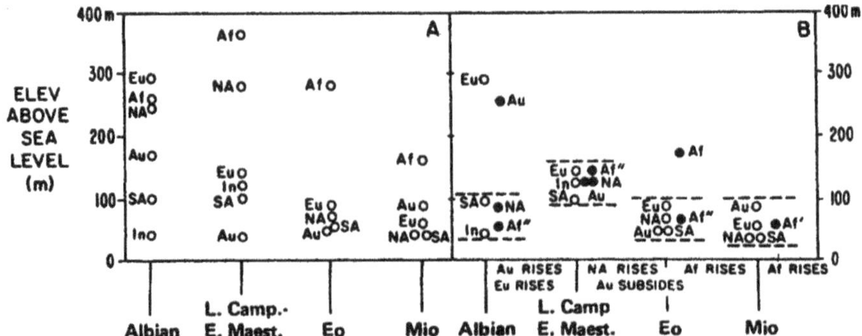

Fig. 3.8 (a) Sea-level elevations read off the curves in Fig. 3.17. (b) After corrections (*black circles*, corrections are explained in the text), clusters are apparent which suggest eustatic sea-level changes (Bond, 1978)

changes. However, Africa appears as an anomaly in the Campanian-Maastrichtian, Eocene and Miocene columns, and there is a scatter of points in the Albian column. Bond (1978) suggested that Africa has been uplifted since the Miocene. A 90-m lowering of the Africa point, to the middle of the Miocene cluster (point Af') does not restore the Africa point to the middle of the Eocene cluster, suggesting an Eocene-Miocene uplift in the order of 210 m (point Af''). Other corrections are also indicated in Fig. 3.8b. Bond (1978) concluded that the final clusters indicate overall generalized changes in global sea level, a rise from the Albian to a maximum of about 150 m above present in the Campanian-Maastrichtian, followed by a gradual fall. It is also clear from these data that the continents have moved vertically independently of each other through Phanerozoic time.

More recent work on hypsometric curves was summarized by Burton et al. (1987). They concluded that although the technique is useful in providing general estimates of long-term changes in sea level there are too many uncertainties to permit its use for determining detailed, short-term changes. A long-term sea-level curve for the Paleozoic was published by Algeo and Seslavisnsky (1995) based on analysis of recent paleogeographic syntheses. They demonstrated significant differences in the flooding histories of the major continents, and used the results to refine and calibrate the long-term trends indicated by the work of Hallam (1984) and Haq et al. (1987, 1988a). Some causes for the variation in continental elevations over time are discussed in Chap. 9.

## 3.5 Backstripping

Backstripping is a technique for performing detailed analysis of subsidence and sedimentation of a basin. The initial purpose of this type of analysis was to reveal tectonic driving mechanisms of basin subsidence. The analysis consists of progressively removing the sedimentary load from a basin, correcting for compaction (plus lithification, if necessary), paleobathymetry, and changes in sea-level, and calculating the depth to basement. The load may be fitted to an Airy-type or flexural subsidence model, depending on the tectonic setting of the basin, and the residual subsidence that is revealed can then be related to thermal behavior and changes

with time in crustal properties. The technique was developed by Sleep (1971) and was first explored in detail by Watts and Ryan (1976). Steckler and Watts (1978) applied the McKenzie (1978) stretching model to offshore stratigraphic data from the continental margin off New York, and Sclater and Christie (1980) applied the techniques to an analysis of the North Sea Basin, in papers that have become standard work on the subject.

A subsidence curve can be predicted from a knowledge of the tectonic setting of a basin. Departures from the curve can then be interpreted in terms of one or more of the "corrections" that are applied during the analysis, in particular, tectonic events and changes in water depth. Water depth in part represents a balance between subsidence and sedimentation rate, but is also affected by sea-level change. If accurate estimates of water depth during sedimentation can be determined, changes in sea level may then be isolated. For deep-water sediments this is difficult because of the imprecision of paleoecological and other methods of estimating this parameter (Dickinson et al., 1987). However, for shallow-water sediments, such as shelf clastics and platform carbonates, depth corrections are small enough that major changes in sea level may become apparent.

The procedure for backstripping a sedimentary basin starts with the division of the stratigraphic column into increments for which the thickness and age range can be accurately determined (Fig. 3.9). These time slices are then added to the basement one by one, calculating the original decompacted thickness and bulk density, and placing its top at a depth below sea-level corresponding to the average depth of water in which the unit was deposited. Figure 3.9 illustrates schematically what happens when a basement surface is loaded with sediment and water. In order to accurately reconstruct subsidence history, these events are deconstructed into discrete steps $t_0$ to $t_4$. Note how at each time step each unit is reduced in thickness as it is loaded by overlying sediments. Also note changes in water depth with time, due to changes in elevation of the sea floor, sea level and sedimentation.

Subsidence history means the changing elevation with time of the basement surface. An accurate reconstruction of the changing position of this surface, therefore, requires us to be able to reconstruct these three factors, (1) compaction, (2) water depth, and (3) sea level.

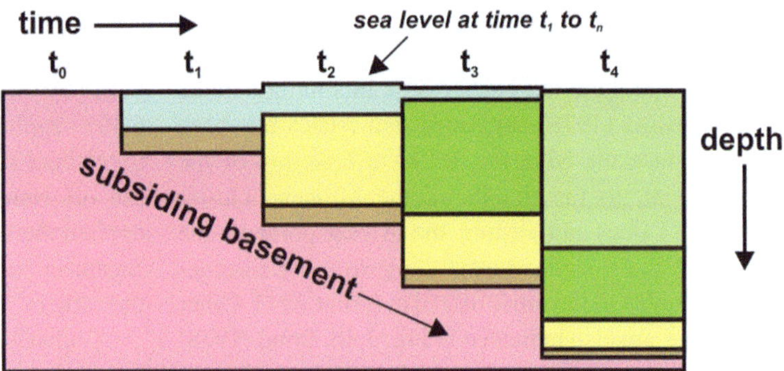



**Fig. 3.9** Time slices in the subsidence history of a basin. Note that water depth, sedimentation rate, and compaction all vary with time. Each increment of sediment is compacted beneath the weight of successive increments, and this effect must be removed for each time slice in the backstripping procedure (from Mayer, 1987)

Compaction may be estimated by measuring changes in porosity with depth. Porosity measurements may be made in the laboratory from core plugs, and they may also be estimated from petrophysical logs. Figure 3.11 illustrates a set of measurements of this parameter that were made in the 1970s, when quantitative studies of subsidence history were first conducted. Note the substantial reduction of porosity with depth, from >50% to about 10% at depths of 4 km. Compaction varies with lithology, being substantial in mud-dominated successions, and considerably less in sandstone and limestone successions, particularly the latter, which may be lithified soon after deposition.

Water depth during sedimentation is difficult to estimate. Benthic-fossil assemblages, trace-fossil assemblages and some sedimentary facies are depth-dependent (Fig. 3.12), but the indications are imprecise.

Sea-level changes are even more difficult to estimate. As discussed in Chap. 1, throughout the history of geological research there has been a search for regularity and cyclicity in geological processes, which has included several attempts to reconstruct sea-level change through the geological past. The most recent and most well known of these was the attempt by Peter Vail of Exxon Corporation to develop a sea-level curve based on analysis of reflection-seismic records. His curve was popular and widely used for more than a decade, but has now been shown to be suspect. This is discussed in Chap. 12.

Given the necessary corrections and adjustments, the isostatic subsidence caused by the weight of the first sedimentary unit can then be calculated, and the depth to the surface on which the sediment was deposited is calculated with only the weight of the water as the basement load (Fig. 3.10). The second unit is then added and adjusted in the same way. The thickness and bulk density of the first unit are adjusted in accordance with the depth of burial beneath the second unit, and so on up the column. An example of the procedure is shown in Fig. 3.13. In this figure, only two layers, a and b, are shown. The first step (column 3) is to remove layer b. The top of layer (a) is then positioned according to the estimate of the water depth at

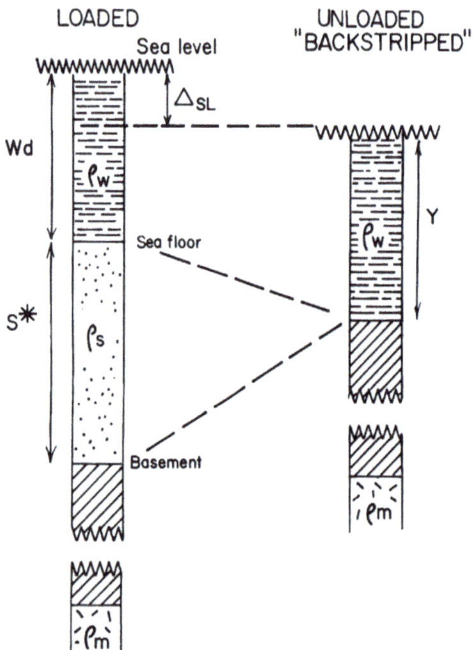



**Fig. 3.10** The basic relationships between a loaded sedimentary section and an unloaded, or backstripped, section. Wd=water depth during deposition of given sedimentary unit, $S^*$=total sediment thickness, $\rho_m$, $\rho_s$, $\rho_w$=density of mantle, sediment and water, $Y$=depth of water with no sediment load, $\Delta_{SL}$=incremental eustatic change in sea-level (from Steckler and Watts, 1978)

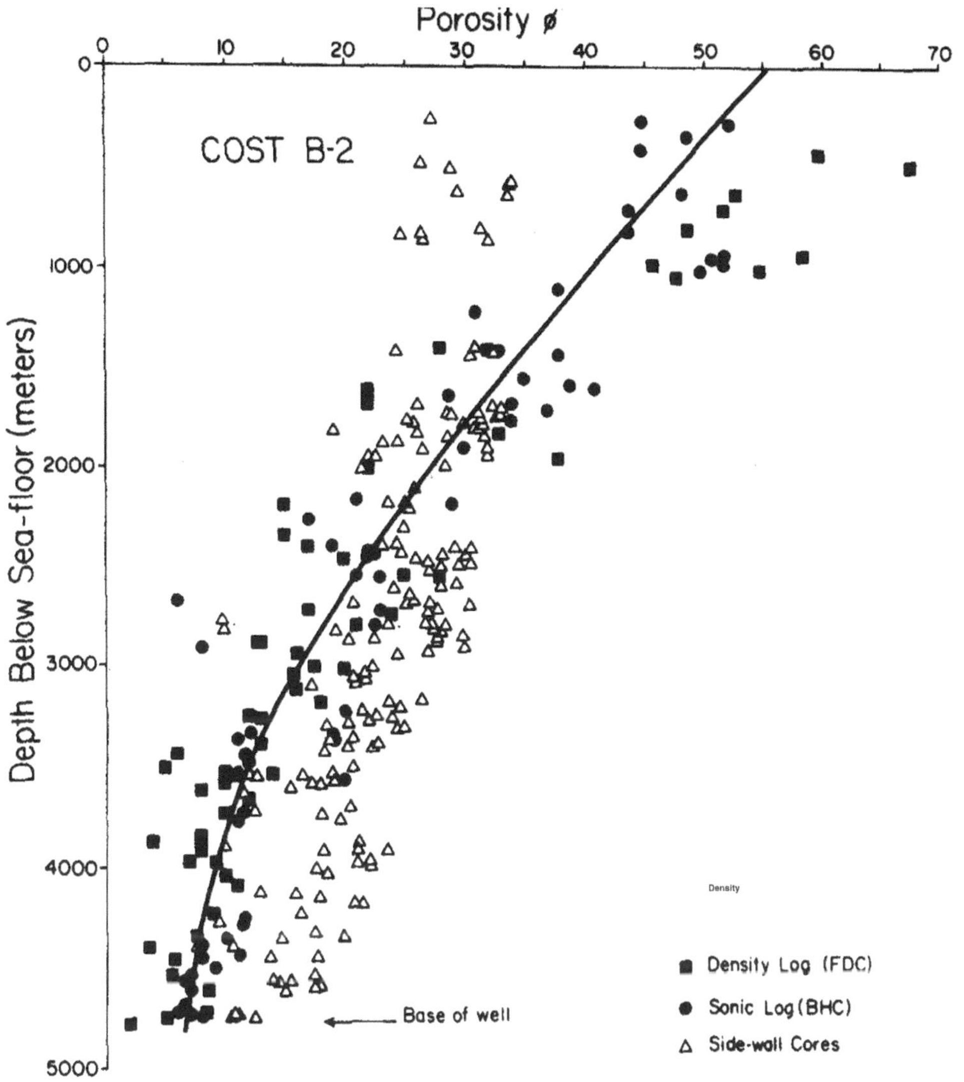

**Fig. 3.11** Variations in porosity with depth in an offshore test well, drilled on the continental shelf off New York (Watts, 1981). AAPG © 1981. Reprinted by permission of the AAPG whose permission is required for further use

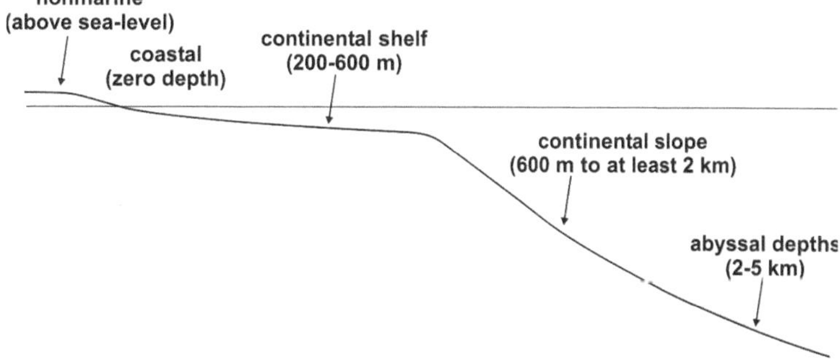


**Fig. 3.12** Depth zones at a continental margin, as indicated by benthic fossil assemblages

**Fig. 3.13** The backstripping procedure, showing the steps required to determine the subsidence due to unit (**a**). This can be broken down into the subsidence due to the sediment load, $S_s$, and the subsidence due to the tectonic driving force, $S_t$ (Célérier, 1988)

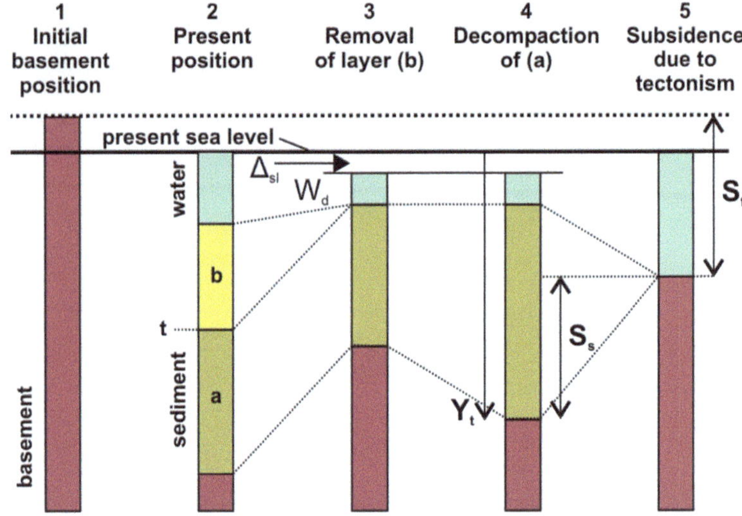


the time of deposition ($W_d$) and also correcting for the estimated difference in sea-level relative to the present day ($\Delta_{sl}$). Unit (a) is then decompacted according to the estimated or measured changes in porosity and the base of the unit moved down to the appropriate level (column 4 in Fig. 3.13). The level of the basal contact then indicates the depth from the original surface which the basement had reached ($Y_t$) at time t, at the end of the deposition of unit (a). The compacted thickness of this layer ($S_s$) corresponds to the sediment load, and the remainder of the subsidence, $S_t$, then corresponds to the subsidence due to the tectonic driving force. Repeating this procedure for each layer in turn corresponds to the incremental unloading from $t_4$ to $t_0$ shown in Fig. 3.9.

An example of the application of this method to the study of sea-level change was given by Bond and Kominz (1984). These authors were concerned with evaluating the tectonic evolution of the early Paleozoic Cordilleran miogeocline of western Canada—the former continental margin. An outcome of their analysis was a model of the geometry of the continental margin based on flexural subsidence, following the methods of Watts (1981) (Fig. 3.14). A restored cross-section of the margin shows that the flexural model does not predict a basin as deep as that indicated by the cross-section. Also, the sediment wedge extend much further onto the craton than can be explained by a flexural-subsidence model (Fig. 3.14). Bond and Kominz (1984) suggested that thermal subsidence may have been underestimated, and could account for the

greater-than-predicted thickness, and that a regional transgression caused by eustatic sea-level rise during the Middle Cambrian to Early Ordovician must have been responsible for depositing the thin sediment blanket that extends for several thousand kilometres onto the craton.

Bond and Kominz (1991a) and Kominz and Bond (1991) subsequently reported a detailed comparison of early-mid Paleozoic subsidence curves for various continental-margin and intracratonic basins in North America. They suggested that sea-level changes could be isolated by studying the subsidence history of Iowa, a central, highly stable area of the craton which is unlikely to have been affected by any tectonism during this period. This analysis identified three episodes of sea-level rise (Fig. 3.15). Their so-called "Iowa baseline curve" was then subtracted from other curves derived from basinal settings. The plots that resulted could still not be fitted to exponential McKenzie-type subsidence curves, as would have been expected from the tectonic setting of the basins. Additional subsidence is indicated, which the authors related to long-wavelength flexural subsidence induced by intraplate stress, a mechanism discussed in Chap. 10. This result is of considerable importance, because intraplate stress has been proposed as a mechanism that may be capable of generating stratigraphic architectures similar to those arising from eustatic sea-level changes over large continental areas (Cloetingh, 1988). This observation adds to the questions about the origins and significance of the Exxon global cycle charts (Chap. 12).

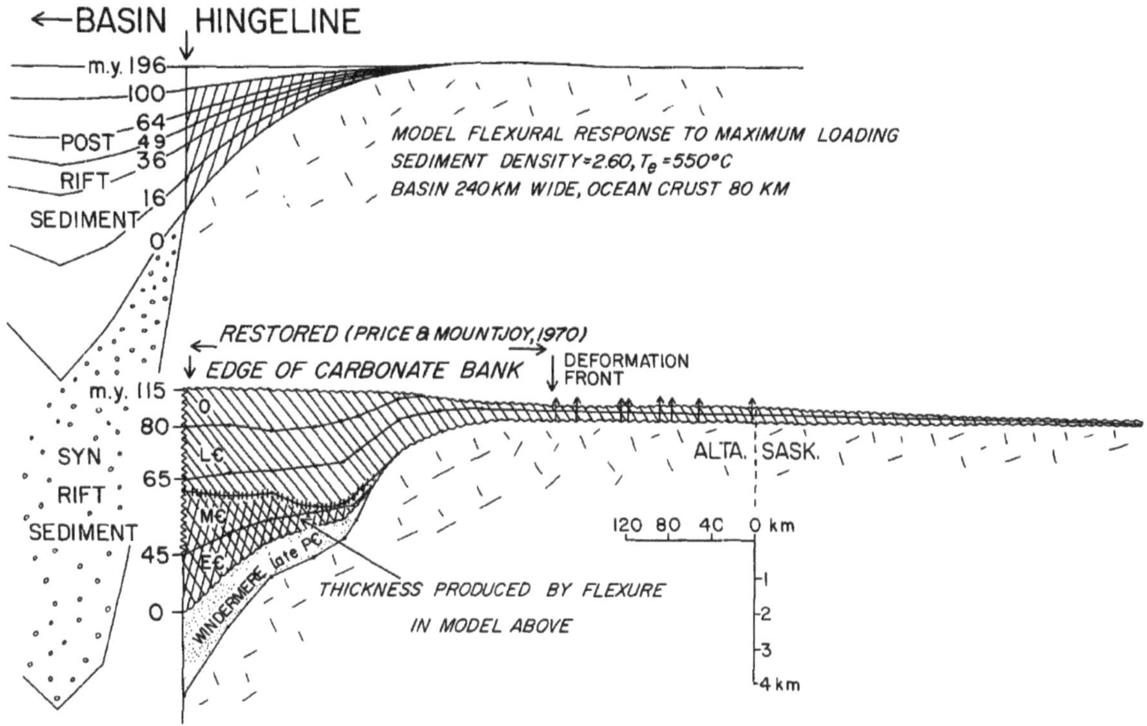


**Fig. 3.14** The Cambrian-Ordovician continental margin of western Canada. Comparison of a restored cross-section (*below*) with a hypothetical model based on flexural loading of stretched crust (*above*). Inverted Vs indicate well locations (Bond and Kominz, 1984)

As discussed below, one of the best places to study sea-level changes is at paleoshorelines, where stratigraphic units thin against stable cratonic areas. This approach guided the choice of Iowa as a "baseline" in the work of Bond and Kominz. Sahagian and Holland (1991) followed a similar approach in their use of stratigraphic data from the Russian platform to develop a eustatic curve by backstripping procedures. By avoiding areas of active subsidence, the need for complex corrections of doubtful accuracy is eliminated. However, even stable continental interiors are subject to vertical motions. The work of Gurnis (1988, 1990, 1992) and Russell and Gurnis (1994) has demonstrated that the entire surface of Earth, including continents and the floor of the oceans, is subject to broad, gentle epeirogenic movements driven by thermal effects of deep-seated mantle processes. Gurnis refers to this characteristic of Earth's surface as *dynamic topography*, a topic discussed at greater length in Chap. 9. The important point to note here is that this work demonstrates that no single location can

be used as a reference location for assessing eustatic sea-level variations.

Attempts have been made to extract additional local and regional detail from one-dimensional vertical profiles using more elaborate backstripping methods. The techniques were described by Bond et al. (1989) and Bond and Kominz (1991b), and were also illustrated by Osleger and Read (1993), and consist of a series of reductions of the primary data. The first step, termed R1 analysis, comprises construction of the decompacted, delithified subsidence curve following procedures summarized above. Next, an exponential subsidence curve is fitted to the R1 curve using least-squares methods. This step is, of course, designed for use where tectonic subsidence is driven by thermal relaxation mechanisms acting over a scale of tens of millions of years, and builds in the assumption that the subsidence follows a simple exponential path. The R2 curve consists of the residuals derived by extracting the exponential curve from the R1 curve. The resulting R2 curve represents the external changes in

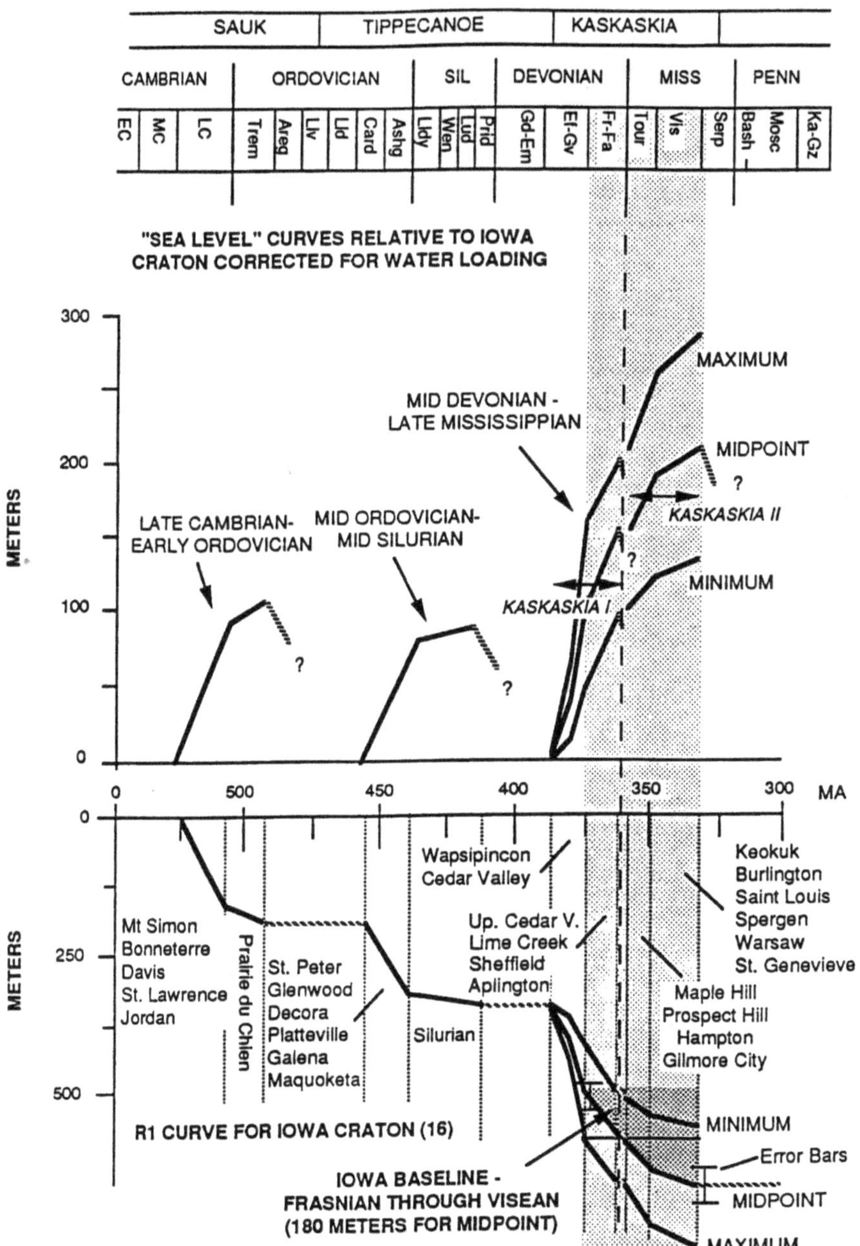


**Fig. 3.15** Subsidence curve for Iowa (*lower curve*), from which a sea-level curve (*upper curve*) has been derived by correcting for sediment and water loading (Bond and Kominz, 1991a)

accommodation superimposed on the thermal subsidence (Fig. 3.16). These changes may be of tectonic and/or eustatic origin, and typically have wavelengths in the million-year range. Comparison of R2 curves from different sections may yield important information on the extent of specific accommodation events. For example, Fig. 3.17 illustrates a set of R2 curves derived for Cambrian sections on the passive western margin of North America. Most of the peaks and troughs can be correlated, suggesting that they are of eustatic origin. A further reduction, termed R3 analysis, fits a polynomial to the R2 curve and plots the residuals. The result is a reflection of the local departures from high-frequency changes in accommodation

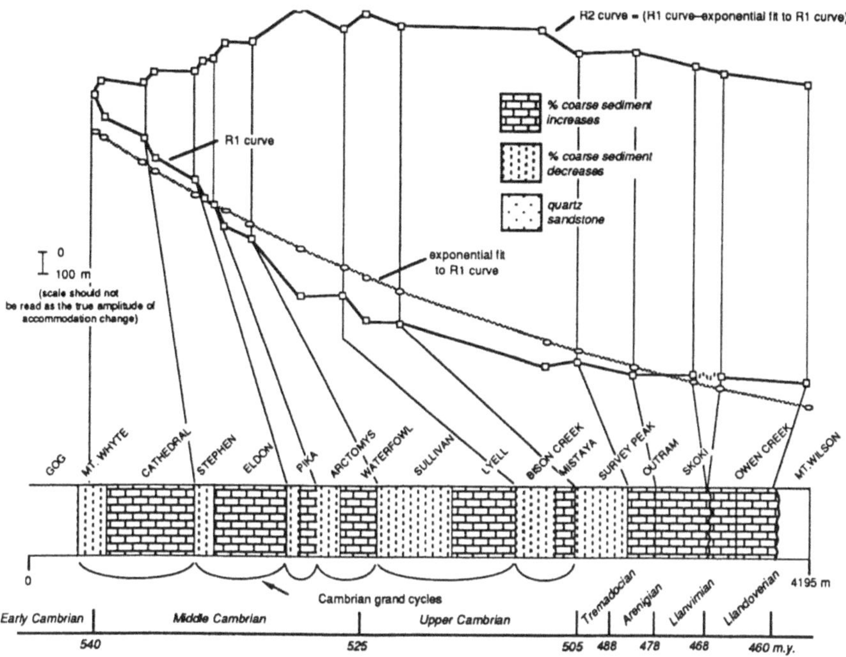



**Fig. 3.16** R1 and R2 curves for Cambrian succession at Mount Wilson, Alberta, which consists of a series of "grand cycles" (Bond and Kominz, 1991b)

space, that may be useful in clarifying differences between curves resulting from changes in facies.

An evaluation of the many sources of error in these methods, including inaccurate subsidence curves and imprecision in correlation of the sections, indicates that the form of the R2 and R3 curves provides a useful general indication of the changes in accommodation with time, but that they cannot yield accurate measurements of the magnitudes of accommodation events (Bond et al., 1989). It seems likely that these methods would only work well for carbonate sediments because, given conditions suitable for the "carbonate factory" to develop, carbonate facies and thicknesses more are sensitive to changes in water depth, whereas in the case of clastic sediments external factors of sediment supply and current dispersion affect resulting thicknesses, and would tend to complicate higher-order data reduction (certainly at the level of R3 analysis).

An alternative approach is to study sediments that are always formed at or near sea level, namely shallow-water carbonates. A special application of backstripping procedures was reported by Lincoln and Schlanger (1991). They documented unconformities and solution surfaces that occur in carbonates comprising the platform on which Enewetak and Bikini Atolls

are built in the South Pacific Ocean. Gradual subsidence of the atolls, driven by the weight of the sediment and the thermal subsidence of the oceanic crust beneath, has preserved a stratigraphic record extending back to the Eocene. Periodic sea-level falls exposed the atoll surfaces, leading to diagenetic changes and the development of karst surfaces. Unless significant erosion takes place during such intervals of exposure, these surfaces are preserved upon subsequent sea-level rise, and provide a record of sea-level history. Figures 3.18 and 3.19 illustrate the principals in the use of atoll stratigraphy to document sea-level change, and Fig. 3.20 illustrates the stratigraphy of the drill holes that were used in their study. Part of the record may be lost to erosion, and reconstruction of the curve depends on the ability to date the sediments accurately and to determine subsidence histories and rates of erosion. Paleodepth corrections must be made for carbonate sediments that are not deposited close to sea level. Lincoln and Schlanger (1991) were able to construct a sea-level curve that tracks the main variations in the Haq et al. (1987, 1988a) global cycle chart reasonably well, but the finer-scale events on the million-year time scale are not detectable (Fig. 3.21).

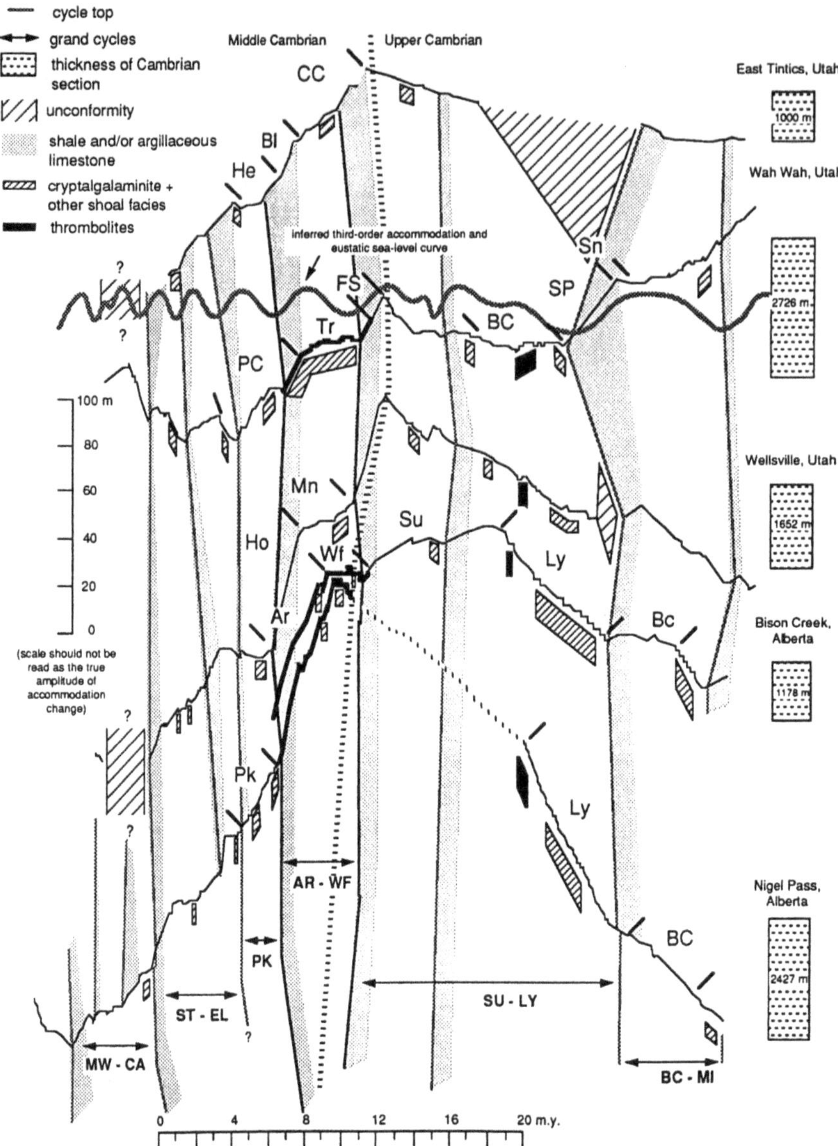


**Fig. 3.17** Correlation of R2 curves for Cambrian sections in Utah and Alberta. The fact that most of the peaks and troughs in these curves can be correlated suggests that the accommodation events are of eustatic origin Two-letter abbreviations refer to stratigraphic names, the details of which are not necessary for the purpose of this book. The relevant stratigraphic names in Alberta are given in Fig. 3.16 (Bond and Kominz, 1991b)

## 3.6 Sea-Level Estimation from Paleoshorelines and Other Fixed Points

A different approach has been to attempt to locate fixed points, such as shorelines, at specific points in time by detailed study of the local geology, choosing a stable area to eliminate tectonic effects as much as possible, and making every possible allowance for other possible complications by carrying out local corrections. The cratonic interior of North America has been chosen as a reference frame for several important studies of this type, because of its location above the the stable Canadian Shield. For example, Sleep (1976) studied a Cretaceous paleoshoreline in Minnesota, and concluded that sea level was approximately 300 m higher than at present during the Late Cretaceous. Kominz

**Fig. 3.18** The origin of solution unconformities within atoll stratigraphies. (**a**) Stage 1, shallow-marine sediments accumulate during a relative rise in sea level; (**b**) Stage 2, during a fall in sea level primary aragonite and high-magnesium calcite dissolve, while low-magnesium calcite is precipitated in the fresh-water zone (Ghyben-Herzberg lens). Karst surfaces may develop. (**c**) Stage 3, during a subsequent rise in sea level the karst surface is preserved as a solution unconformity (Lincoln and Schlanger, 1991)

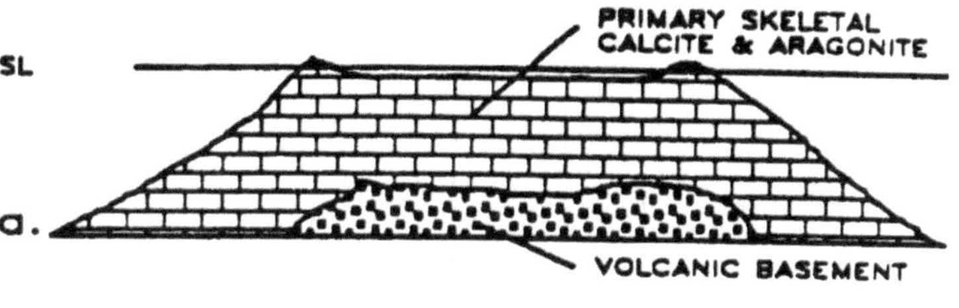


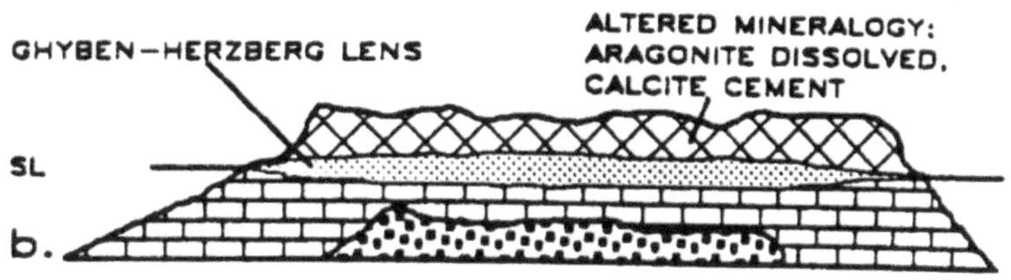

and Bond (1991) chose Iowa as a "baseline" source of stratigraphic data for their calculations of long-term Paleozoic subsidence and sea-level change (Chap. 9). Other research focusing on the study of shorelines on stable cratons has been reported by Sahagian (1987, 1988).

One of the most detailed studies to date of this type is that reported by McDonough and Cross (1991). The starting point of their analysis is explained in the following statement:

> Advantages of this method [the study of paleoshorelines] include its direct measurement of historical sea level elevation at a point in time; its potential for achieving increased temporal resolution; its requirement of fewer assumptions; its capability for testing and falsifying those assumptions, postulates and results; and its potential as the basis for independent evaluation of assumptions used in other approaches. This method makes a single, simplifying assumption: the paleoshoreline has not moved vertically since deposition (or that postdepositional movement can be reconstructed and calculated). Vertical movement of a paleoshoreline may be caused by postdepositional tectonic movement, lithospheric compensation to surface

or other loads, and sediment compaction. This assumption is most likely to be satisfied where strata containing a paleoshoreline were deposited high on the margin of a tectonically stable craton of old, cold lithosphere, and where the stratigraphic section is sufficiently thin that sediment compaction is minimized. These two conditions ... are most likely achieved during former high sea levels when continents were flooded to a maximum extent.

These authors returned to Minnesota to collect stratigraphic data along the thin edge of the Cretaceous sedimentary cover, where a single progradational unit could be traced and dated to within approximately 100 ka. This is important because, as Wise (1974) pointed out, if data from many locations of varying ages are used in paleoshoreline calculations there is a tendency to generalize the results and overestimate the height of eustatic rises of sea level. Earlier work of Sleep (1976) and Sahagian (1987) did not adhere to this guideline, and their results reveal a wide range of estimates.

In Minnesota the single shoreline unit traced by McDonough and Cross (1991) in outcrop and cores

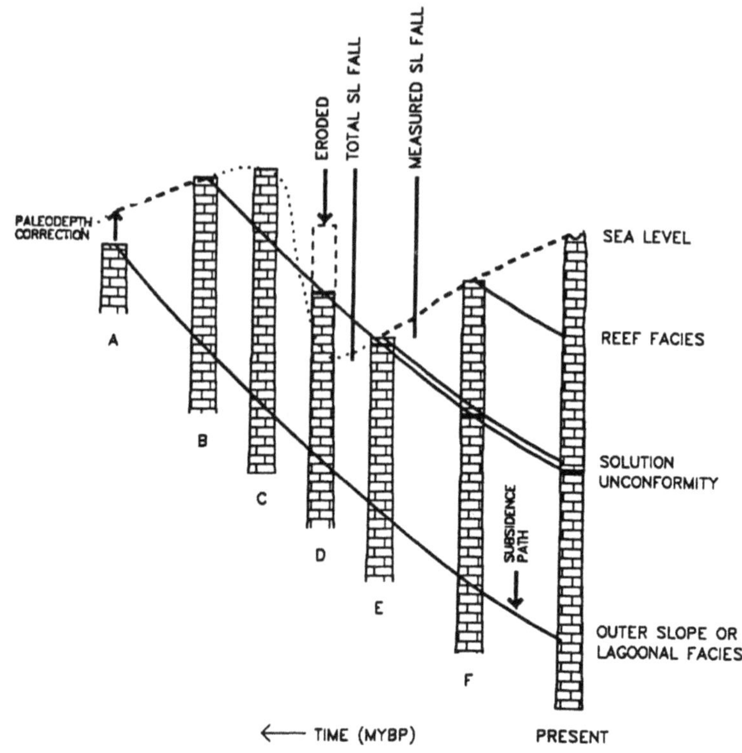



**Fig. 3.19** Model for the development of atoll stratigraphy with sea-level changes. *Sloped lines* display subsidence paths of hypothetical atoll from time A to present. *Dashed-dotted line* is sea-level curve. Dashed portion is that part of the curve which can be reconstructed from the atoll stratigraphy, dotted portion represents the record lost to erosion. (Lincoln and Schlanger, 1991)

varied in present elevation from 266 to 286 m above present sea-level. Minimal corrections for sediment compaction were required because of the location of these beds above very thin older Cretaceous sediments, which, in turn, rest on crystalline basement. Post-Cretaceous movement is thought to have been limited to loading and flexure related to Pleistocene glaciation. Calculations were carried out to eliminate these effects. Possible movements related to dynamic topography effects were not considered. The corrected, average value is 276±24 m above modern sea level. The magnitude of the sea-level rise demonstrated by this work is similar to that calculated by other workers for the Cretaceous (Fig. 3.22), although the wide range of values resulting from this earlier work reflects the broad time range and geographic extent covered by the earlier studies.

A stratigraphic succession may contain many precise records of sea-level elevation, such as shoreline positions. Franseen et al. (1993) and Goldstein and Franseen (1995) showed how these could be used to constrain sea-level curves. The position of sea level at any time may be revealed by stratigraphic tran-

sitions from marine to nonmarine strata, or by evidence of near-sea-level facies such as tidal-flat or beach deposits, reefs, or surfaces of subaerial exposure overlain by marine deposits. The identification of such stratigraphic records provides a series of *pinning points*, the relative elevations of which yield a series of quantitatively fixed points on a sea-level curve. The method is analogous to the use of onlap seismic terminations (Vail et al., 1977), but makes use of a wider variety of outcrop indicators. Corrections must be made for compaction and tectonic tilting, for example by using geopetal evidence to indicate tilts due to differential compaction.

Figure 3.23 illustrates a stratigraphic cross-section in which thirty pinning points have been identified. The sea-level curve constructed from these points is shown in Fig. 3.24. Gaps in this curve indicate time periods for which a reconstruction is not possible because of a loss of the stratigraphic record to erosion. The determination of a few of the individual pinning points is described below to provide some indication of the types of geological reasoning used in the construction of this curve.

**Fig. 3.20** Stratigraphy of carbonate succession beneath Enewetak Atoll, South Pacific Ocean. Note the correlation of solution unconformities, and the alternation of altered and unaltered intervals (Lincoln and Schlanger, 1991)

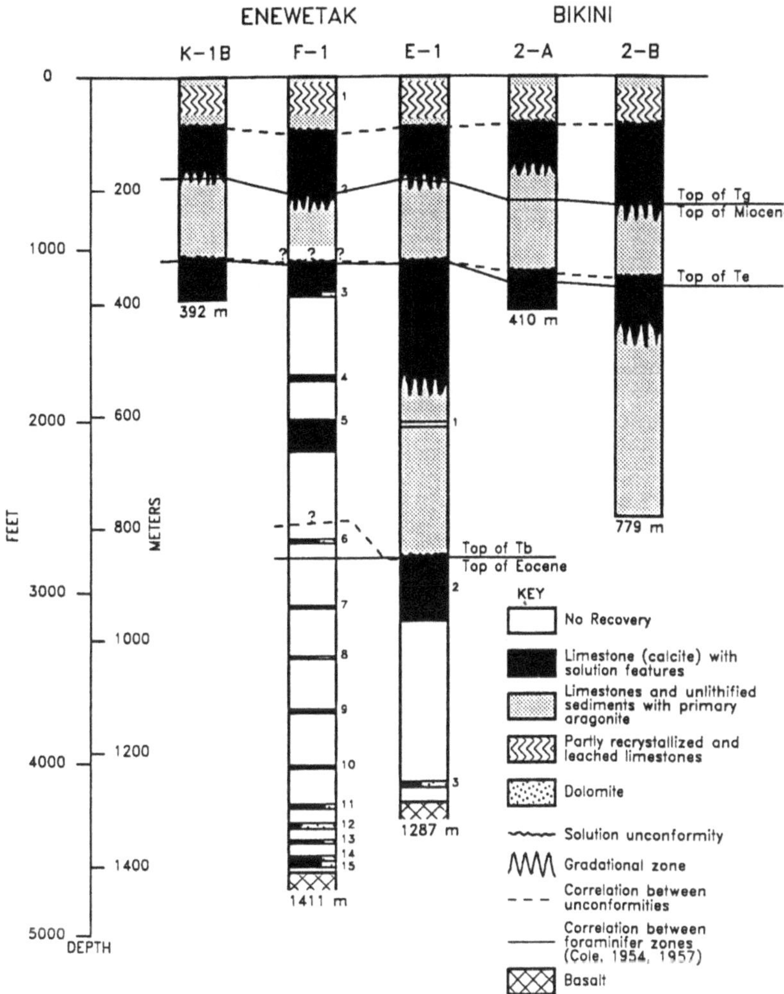


- Pinning point 1: volcanic basement with evidence of subaerial exposure (fissures, spheroidal weathering) overlain by marine deposits.
- Pinning points 7, 8: exposure features on the upper surface of unit DS2 can be traced downslope to their most distal point, interpreted as points on a falling and subsequently rising leg of the sea-level curve.
- Pinning point 10: reef facies encrusted with Porites, indicating a reef-crest environment.

## 3.7  Documentation of Metre-Scale Cycles

Regular successions of metre-scale cycles are common in some successions of shallow-marine carbonate and fine-grained clastic rocks. They may be autogenic or allogenic in origin. Their thicknesses commonly vary in a systematic manner, suggesting the influence of a long-term control on thickness variation. Fischer (1964), in a study of Triassic carbonate cycles in the Calcareous Alps, devised a method of plotting cycle thickness as a means of objectively displaying these thickness variations. This plotting technique has come to be termed the Fischer plot. There has been considerable renewed interest in these cycles and the use of Fischer plots as a means of documenting cyclicity in the Milankovitch band (see Sect. 4.1 for an explanation of this term). Recent studies in which the method has been used include Goldhammer et al. (1987), Read and Goldhammer (1988), Osleger and Read (1991, 1993), and Montañez and Read (1992). Excellent discussions of the method, its advantages and pitfalls, have

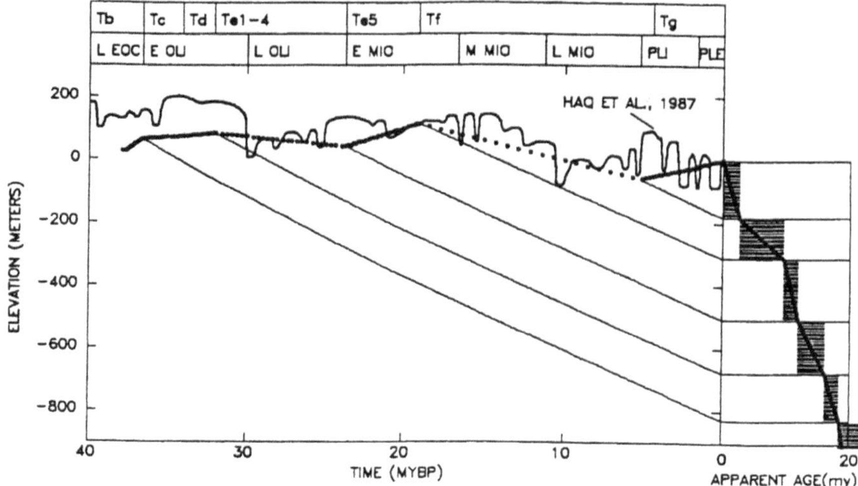


**Fig. 3.21** Depth-age profile for Enewetak, interpolated every 6.6 m from biostratigraphic boundaries, is plotted on the right side of the diagram. The data are backtracked (backstripped) using calculated subsidence and flexure parameters. *Solid circles* indicate the calculated positions of sea-level since the late Eocene (Lincoln and Schlanger, 1991)

been provided by Sadler et al. (1993), Drummond and Wilkinson (1993a), and Boss and Rasmussen (1995).

The basis of the Fischer plot is a zig-zag diagram in which net subsidence is plotted against cycle thickness (Fig. 3.25a). The slope of the subsidence plot is determined from the total thickness of the stratigraphic section divided by the elapsed time for the section. A constant subsidence rate is assumed. The line formed by joining all the cycle tops (heavy line in Fig. 3.25a) rises and falls as an irregular wave train reflecting the changing accommodation space in the depositional environment (Fig. 3.26). A rising slope indicates a succession of thick cycles, suggesting an increase in the rate of generation of accommodation space, such as is brought about by a rise in sea level. A fall in the curve, indicating a succession of thin cycles, suggests a decrease in the rate of generation of accommodation space and a fall in sea level. However, the limitations of the plot need to be borne in mind. A constant rate of subsidence is assumed, and each cycle is assigned the same duration. Neither assumption may be valid. However, considerable information may be

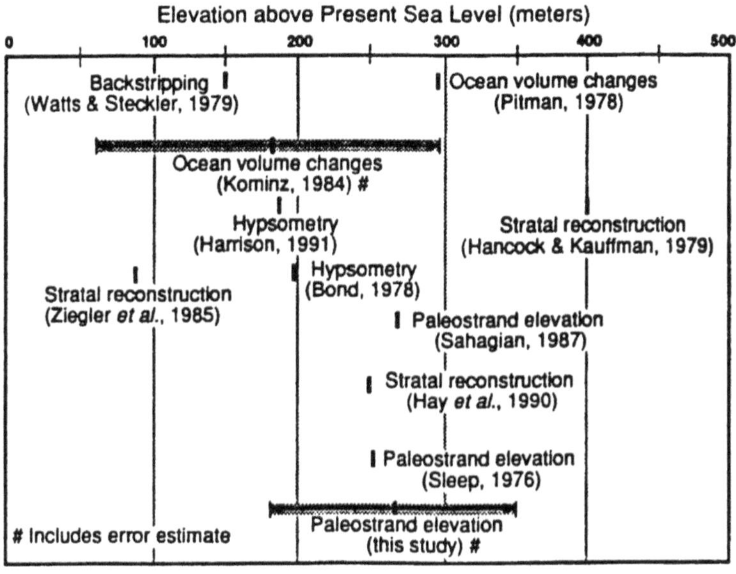


**Fig. 3.22** Ranges in estimates of Cretaceous sea-level resulting from use of different methods of estimation (McDonough and Cross, 1991)

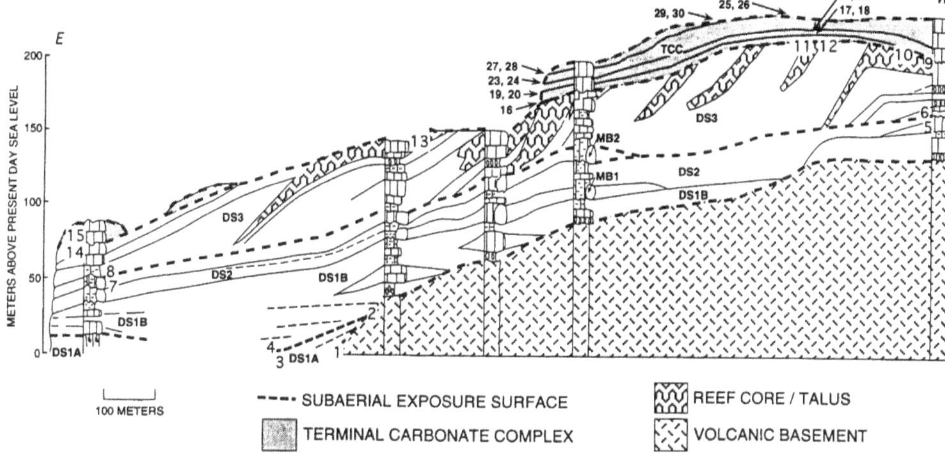


**Fig. 3.23** Stratigraphic cross-section through an Upper Miocene succession in southeast Spain, showing the location of thirty identified pinning points (Goldstein and Franseen, 1995)

derived from the plots if allowance is made for these limitations.

It is important to be clear about what is actually being plotted. As pointed out by Sadler et al. (1993), the vertical axis is in fact a plot of the departure of the individual cycle thickness from mean thickness, and

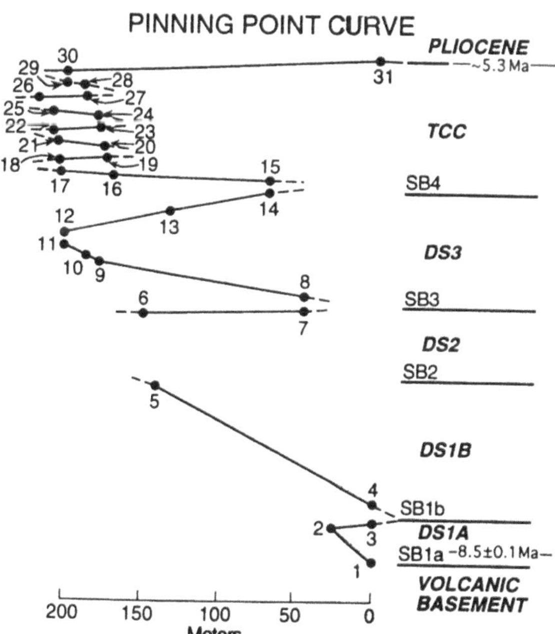


**Fig. 3.24** Sea-level curve constructed using the pinning points shown in Fig. 3.23. *Scale* indicates height of sea-level relative to pinning point #1 (Goldstein and Franseen, 1995)

the horizontal axis should be labelled cycle number, to remove any confusion about its relationship to time (Fig. 3.25c). Because the plots are simply a method of portraying the distribution of total thickness over a given section, the total positive and negative departures from the mean cycle thickness sum to zero, and the graphs always finish at the same vertical position as they start. It is therefore important to base Fischer plots on lengthy sections, or serious distortions may appear, as illustrated in the short segments of the curve replotted in Fig. 3.26b. Sadler et al. (1993) recommended that a plot be based on a minimum of at least fifty cycles, in order to avoid such difficulties, and to overcome possible statistical distortions of a small sample base. Because of all the qualifications on the shape of the plot, Sadler et al. (1993) prefer to describe the form of the curving trace in terms of "waves" rather than "cycles". Drummond and Wilkinson (1993a) demonstrated that "waves" constructed from short runs of cycle thickness could not be statistically distinguished from random noise.

Practical problems in the operational definition of cycles, reflecting the ambiguity that is common in the stratigraphic record, may lead to difficulties in the definition of cycles in practice. Figure 3.26a illustrates this point. A succession has been subdivided using two different approaches to cycle definition, one that generates relatively thicker cycles, and one that generates thinner cycles. The form of the plot is different, although broad trends of rise and fall are similar in the two plots.

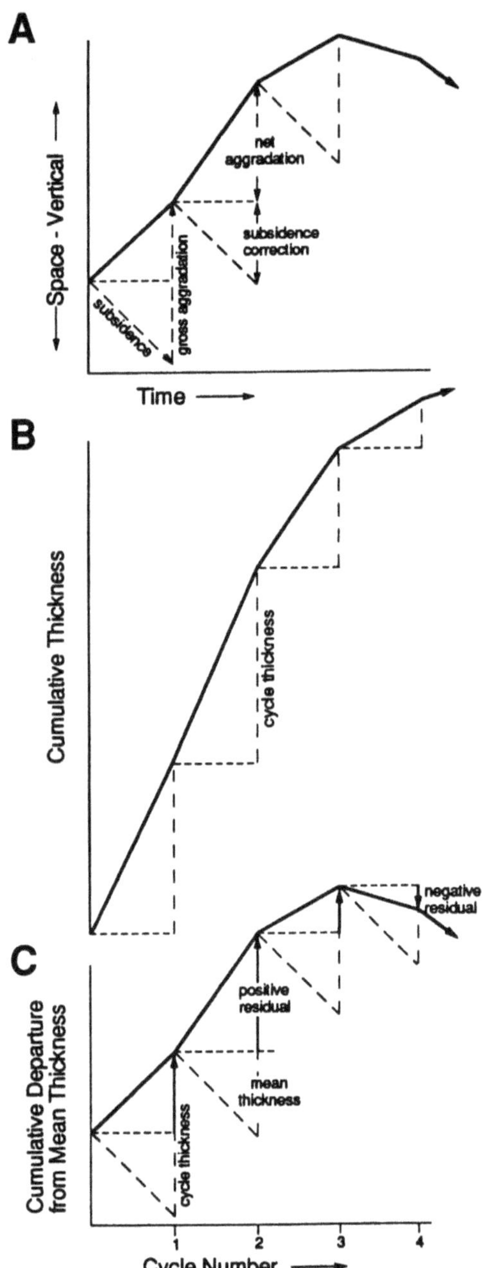



**Fig. 3.25** (**a**) The basis for the Fischer plot; (**b**) replotting the data as a cumulative thickness plot; (**c**) The same information, labelled in a more objective way (Sadler et al., 1993)

What does the Fischer plot reveal about changes in the rate of generation of accommodation space? Although the rise and fall of the "wave" is a reflection of long-term changes, there are a number of important problems and limitations that need to be considered:

1. The plot cannot account for time that is not represented in the section because of a fall in sea level that exposes the depositional environment (what have been termed "missed beats", as explained in Chap. 7). As shown by Drummond and Wilkinson (1993a) where the long-term rate of sea-level fall is greater than the rate of subsidence, long intervals of time (and many cycles) will be missing from the section, resulting in severe distortions of the record.
2. Plots cannot be constructed for intervals of the section that are noncyclic.
3. The plots are one-dimensional, and cannot differentiate between cycles of local extent that may be of autogenic origin, and more regionally extensive cycles of allogenic origin (e.g., Pratt et al., 1992).
4. Distortions may also be introduced by major changes in facies, because sedimentation rates may vary (Pratt et al., 1992).
5. It can be shown that the plots are always asymmetric, because thin cycles are more common than thick cycles, and so falling legs of the wave are longer, and therefore flatter than rising legs (Sadler et al., 1993).
6. As noted by Drummond and Wilkinson (1993a), the plots do not allow for compaction. Differential compaction, reflecting systematic variations in lithology could, theoretically, generate Fischer plots similar to those shown in Fig. 3.26.

Boss and Rasmussen (1995) constructed Fischer plots one cycle in length for the Holocene sedimentary record of the Bahamas carbonate platform, using a seismic transect, and demonstrated that there is no relationship between cycle thickness and accommodation space (depth through the water column and Holocene deposits to the top of the Pleistocene).

Aside from the possibility of differential compaction, regular or episodic changes in accommodation space may be the result of changes in sea level or changes in susbsidence rate, or some combination of both. As discussed in Chap. 10, tectonic mechanisms operate over a wide range of time scales, including the 10- to 100-ka time scale typical of Milankovitch cycles. Therefore, although the Fischer plot provides an objective representation of preserved cycle thicknesses, for all the reasons stated above it offers only a crude representation of changes in accommodation space. Nonetheless, comparisons and correlations between plots for various stratigraphic sections may

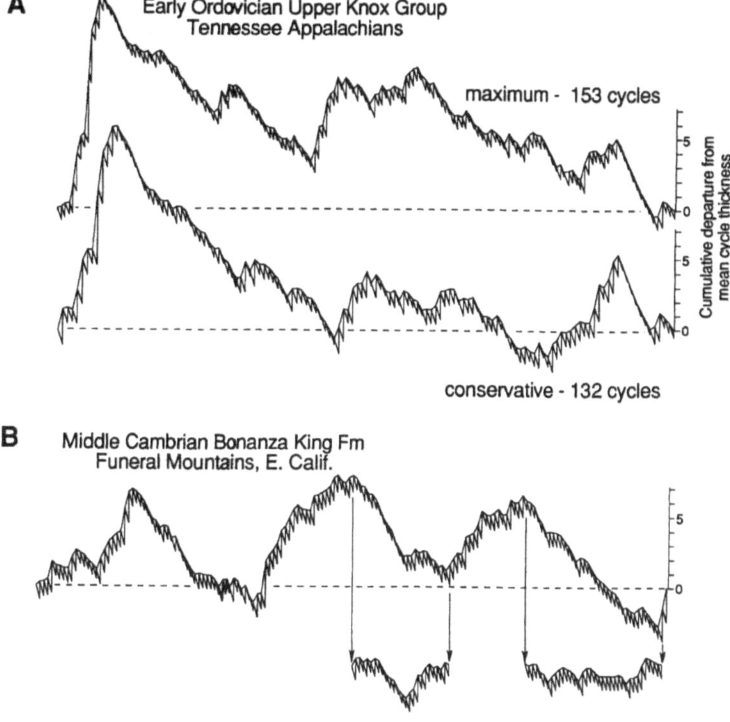



**Fig. 3.26** Examples of Fischer plots for peritidal Paleozoic carbonate successions. (**a**) The same succession measured using two different criteria for cycle definition. (**b**) Another section, in which two legs of the curve have been replotted as if they constituted the entire data succession (Sadler et al., 1993)

provide useful insights. Examples were offered by Osleger and Read (1993).

Figure 3.27 illustrates three methods for determining changes in accommodation space. The "paleobathymetry" curve was constructed using qualitative estimates of water depth based on facies interpretations, of the type discussed in Sect. 3.5. A Fischer plot is shown in the middle, and to the right is the "R3" curve derived from subsidence analysis, using the method described in Sect. 3.5. The three curves are quite similar. Exact correlations cannot be expected, because the stratigraphic section and the paleobathymetry curve are plotted against thickness, whereas the Fischer plot uses cycle number in the vertical axis, and the R3 curve is plotted against decompacted thickness.

Klein and Kupperman (1992) and Klein (1994) proposed a method for evaluating the relative contributions of tectonic subsidence and sea-level change in the generation of late Paleozoic cyclothems. Tectonic subsidence is derived by plotting a backstripping curve and dividing total subsidence (corrected for compaction and loading) by the number of cycles present. Water depth changes are estimated by sedimentological methods and the two results compared.

## 3.8 Integrated Tectonic-Stratigraphic Analysis

A complete basin analysis study incorporates all available sources of information, drawing on seismic, well, or outcrop data, depending on availability. A sequence analysis of the type discussed in Chap. 2 may be combined with the type of analysis summarized in this chapter. To organize the analysis it may be useful to establish a checklist of tasks to be completed, and follow what Catuneanu (2006, pp. 63–71) called the "workflow".

This is the sequence of procedures suggested by Vail et al. (1991):

1. Determine the physical chronostratigraphic framework by interpreting sequences, systems tracts and parasequences and/or simple sequences on outcrops, well logs and seismic data and age date with high resolution biostratigraphy.
2. Construct geohistory, total subsidence and tectonic subsidence curves based on sequence boundary ages.
3. Complete a tectonostratigraphic analysis including:

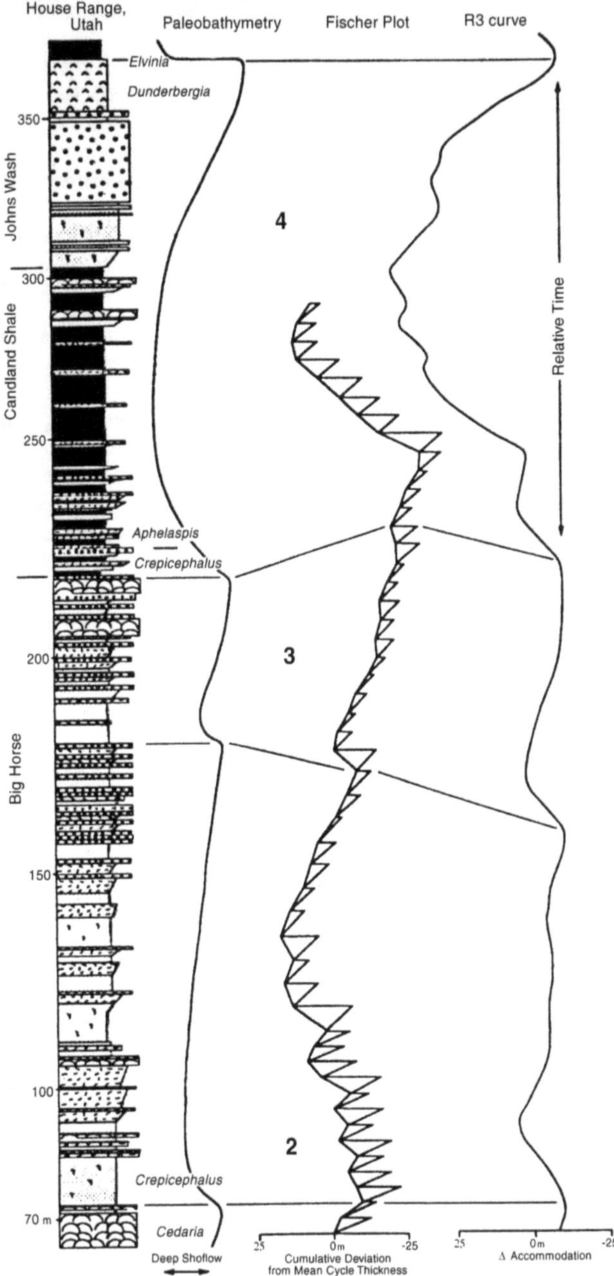



**Fig. 3.27** Comparison of three methods for determining variations in accommodation space, as applied to a Cambrian section in Utah (Osleger and Read, 1993)

A. relate major transgressive-regressive facies cycles to tectonic events.

B. relate changes in rates on tectonic subsidence curves to plate-tectonic events.

C. assign a cause to tectonically enhanced unconformities.

D. relate magmatism to the tectonic subsidence curve.

E. map tectono-stratigraphic units.

F. determine style and orientation of structures within tectono-stratigraphic units.

G. simulate geologic history.

4. Define depositional systems and lithofacies tracts within systems tracts and parasequences or simple sequences.

5. Interpret paleogeography, geologic history, and stratigraphic signatures from resulting cross-sections, maps and chronostratigraphic charts.

6. Locate potential reservoirs and source rocks for possible sites of exploration.

Step one, the "physical chronostratigraphic framework" builds from the original hypothesis of Vail et al. (1977, p. 96) that the global sequence framework constitutes "an instrument of geochronology" (see Chap. 1). As discussed in Part IV, there are strong reasons for abandoning this practice. The establishment of the sequence framework should be a strictly empirical process.

It is suggested that subsidence and maturation analysis (Step 2) be carried out at the conclusion of the detailed analysis, rather than near the beginning, as proposed in the Exxon approach. The reason is that a complete, thorough analysis requires the input of a considerable amount of stratigraphic data. Corrections for changing water depths and for porosity/lithification characteristics, which are an integral part of such analysis, all require a detailed knowledge of the stratigraphic and paleogeographic evolution of the basin.

Catuneanu (2006, p. 63) suggested that the workflow commence with a determination of the type of basin under study. The tectonic setting of the basin determines its subsidence pattern, and a knowledge of this may assist in the prediction of depositional systems and their spatial relationships. Such information may be invaluable in carrying out the preliminary analysis of regional seismic data.

A modified workflow is summarized below. The pattern of work will depend on the nature of the project (regional seismic, exploratory well data, detailed development project, outcrop work) and will involve moving back and forth between several of these steps (particularly between 2, 3, 4 and 5) as the analysis become more refined.

1. Determine tectonic setting of basin, and establish regional style of subsidence and structural deformation.

2. If 2-D regional seismic data are available, analyze and document stratal surfaces and seismic terminations. Incorporate lithostratigraphic well correlations, biostratigraphic and any other information regarding relative or absolute ages.

3. Develop an allostratigraphic framework.

4. Make use of facies data from core and/or outcrop to determine depositional environments and systems. Develop a suite of possible stratigraphic models (e.g., shelf carbonate, slope submarine fan, coastal deltaic, etc.) that conform with available stratigraphic and facies data.

5. Compile provenance and paleocurrent data, by location and by stratigraphic unit.

6. Establish a regional sequence framework with the use of all available chronostratigraphic information.

7. Determine the relationships between sequence boundaries and tectonic events, e.g., by tracing sequence boundaries into areas of structural deformation, and documenting the architecture of onlap/offlap relationships, fault offsets, unconformable discordancies, etc.

8. Establish the relationship between sequence boundaries and regional tectonic history, based on plate-kinematic reconstructions.

9. Explore other possible scenarios for sequence generation, such as climate change, and its effects on sediment supply depositional environments and sea levels.

10. Refine the sequence-stratigraphic model. Subdivide sequences into depositional systems tracts and interpret facies.

11. Construct regional structural, isopach, and facies maps, interpret paleogeographic evolution, and develop plays and prospects based on this analysis.

12. Develop detailed subsidence and thermal-maturation history by backstripping/geohistory analysis.

This book is concerned primarily with the interpretation of sequences. Parts II and III focus on Steps 7−10. Part IV discusses the issue of chronostratigraphy and correlation, and attempts to provide a framework for the use of chronostratigraphic data in the establishment and interpretation of the sequence framework (Steps 2−6).

# Part II
# The Stratigraphic Framework

Abundant stratigraphic data have been amassed since AAPG Memoir 26 was published in 1977, setting off the current high level of interest in sequence stratigraphy. Some of this consists of regional seismic surveys, but a great deal of work has also been carried out with well records and outcrops. The purpose of this section of the book is to present some of the basic stratigraphic data, with a minimum of interpretation, as a basis for discussion of mechanisms and controls that constitutes much of the remainder of the book.

The wealth of data now available has demonstrated that the original concept of a simple rank ordering of sequences, from first- to fifth- (and higher?) order, is too simplistic. Although sequence-generating mechanisms have natural episodicities or periodicities, the time periods (wavelengths) of these processes overlap, and do not support the use of this simple classification (Table 4.1). A subdivision of sequences according to episodicity is used for descriptive purposes in this part of the book; it forms the basis for the subdivision into three chapters; but no genetic interpretations are implied by this subdivision.

Most of the current interest and controversy regarding sequence stratigraphy focuses on sequences of a few-million-years duration, and less. For this reason, considerably more space is devoted to these cycles (Chaps. 6 and 7).

It has been necessary to make a judicious selection of examples from the wealth of data now available in the published record. The reader is urged to turn to the original publications wherever possible for the details of regional setting, stratigraphy, facies, etc.

- Sequential ordering of beds is found in nearly all stratigraphic successions on various scales. It is thought to be the result from the combination of regional to global causal factors and modifying local environmental processes (Einsele and Ricken, 1991, p. 611).
- The Exxon group's distinction of first-, second- and third-order eustatic cycles is useful. The first-order cycles are the two Phanerozoic supercycles of Fischer (1984), and general agreement exists that the changes of sea level are genuinely global. A fair measure of consensus also exists that the second-order cycles (10–18 Myr duration) have a eustatic origin as well, but the third-order cycles (1–10 Myr) are more controversial. Many stratigraphic experts independent of the Exxon group are, however, persuaded of their reality, on the basis of correlation over extensive regions … Besides these larger cycles, some would distinguish smaller, fourth- and fifth-order cycles, usually no more than a few metres thick and signifying durations of tens to hundreds of thousands of years. One currently popular idea is that late Paleozoic and other comparable cyclothems are glacioeustatic phenomena under the ultimate control of an orbital forcing mechanism that affects global climate (Hallam, 1992a, p. 204).
- … the discrimination of stratigraphic hierarchies and their designation as $n$th-order cycles may constitute little more than the arbitrary subdivision of an uninterrupted stratigraphic continuum (Drummond and Wilkinson, 1996, p. 1).
- This system of ordering cycles is widely used but has serious shortcomings. By pretending that there exists a hierarchy of cycles, repetitive patterns, which have nothing in common except their duration, are mixed together. Such a classification is just as meaningless as grouping elephants and fleas into an order based on their size (Schwarzacher, 2000, p. 52, in reference to the Vail et al., 1977, sequence hierarchy)
- Orders of stratigraphic sequences are being used loosely and with widely varying definitions. The orders seem to be subdivisions of convenience rather than an indication of natural structure. It is proposed that, at least at time scales of $10^3$–$10^6$ years, sequences and systems tracts are scale-invariant fractal features in which units bounded by exposure surfaces and units bounded by flooding surfaces are about equally likely (Schlager, 2004, p. 185)

# Chapter 4

# The Major Types of Stratigraphic Cycle

## Contents



## 4.1 Introduction

The duration and episodicity of geological events and stratigraphic cyclicity span at least sixteen orders of magnitude, ranging from the repeat time of burst-sweep cycles in turbulent boundary layers ($10^{-6}$ years), to plate-tectonic cycles involving the formation and breakup of supercontinents ($10^9$ years) (Miall, 1991b; Einsele et al., 1991b; Figs. 4.1 and 4.2). The highest-frequency cyclicity of stratigraphic significance is *Milankovitch-band* cyclicity, over time scales of $10^4$–$10^5$ years. Minor cyclicity may be apparent in the geological record as a result of seasonal changes in weather, fluvial discharge, etc., the so-called *calendar band* of cyclicity (Fischer and Bottjer, 1991). Sunspots and other solar processes generate cyclicity on a $10^1$–$10^2$-year scale of cyclicity, the *solar band* of the time scale. These processes and their products are not discussed in this book.

The highest frequency of cyclicity discussed here overlaps with that of autogenic cyclicity, such as channel migration or delta-lobe switching ($10^3$–$10^4$ years). Deltaic cycles are of the upward shoaling type, bounded by the flooding surfaces formed by lobe abandonment, and therefore fit the definition of *parasequences*. This is one of the difficulties with that term, as touched on later.

There are four basic types of stratigraphic sequence (types A to D in Table 4.1). These four types are introduced briefly in this chapter, with detailed discussion and documentation constituting the remainder of Part II. These cycle types reflect the independent operation of at least four type of process, including regional tectonism and various controls on eustasy. Work is still in progress on the analysis of tectonic mechanisms to determine the importance of other processes comparable in scale, duration and areal effect. Important second-order effects triggered or driven by these primary mechanisms include changes in global climate, magmatism, ocean-water circulation patterns, the carbon and oxygen cycles, and biogenesis, all of which have significant, measurable effects on the sediments. Sorting out the various processes and their effects is currently one of the most vigorous areas of basin analysis research.

## 4.2 Sequence Hierarchy

Depositional units constitute a hierarchy of scales, from the small-scale ripple cross-laminae through the various units constituting depositional systems, to the scale of the entire basin fill. This concept of "hierarchy" is discussed in the context of geological time in Chap. 13, and the ideas are summarized in Table 13.1. Important recent papers that deal with the subject of depositional hierarchy are those of Van Wagoner et al. (1990), Miall (1991b), Mitchum and Van Wagoner (1991), Nio and Yang (1991), Drummond


A.D. Miall, *The Geology of Stratigraphic Sequences,* 2nd ed.,
DOI 10.1007/978-3-642-05027-5_4, © Springer-Verlag Berlin Heidelberg 2010

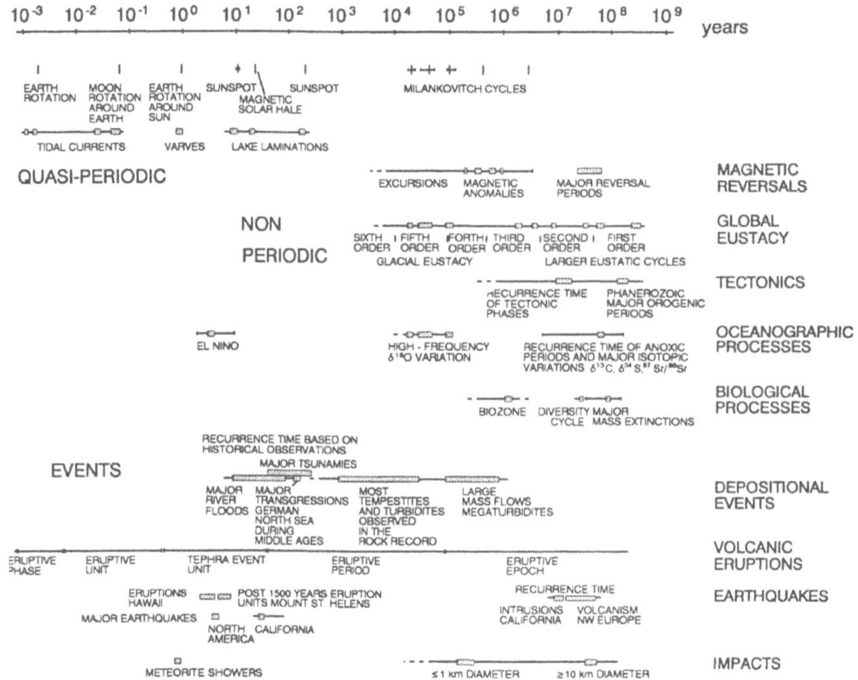



**Fig. 4.1**  Recurrence time of periodic and episodic processes and events in geology (Einsele et al., 1991b)

and Wilkinson (1996) and Schlager (2004). There is a need to relate the hierarchies based primarily on outcrop studies of depositional units and bounding surfaces, to the sequence-stratigraphic system of nomenclature.

An empirical classification of stratigraphic sequences using rank-order designations, based on their duration, was devised by Vail et al. (1977). This classification is widely known, but has become increasingly unsatisfactory as more has been learned about sequences and their generating mechanisms. The terminology is shown in Table 4.1, but is no longer recommended (Schwarzacher, 2000; Schlager, 2004).

Carter et al. (1991) pointed out that although a hierarchy of sequences exists in the stratigraphic record, their distinctiveness in terms of duration or recurrence interval is only approximate, the scales of the sequences (for example, their thicknesses) do not define mutually exclusive ranges, and they do not necessarily nest internally in a logical or ordered pattern. They stated (p. 45):

It is not self-evident that the SSM [sequence-stratigraphic model] at successive orders can be adequately represented as the sum of a set of topologically similar SSMs that comprise parasequences of the next higher order. ... There is also the additional difficulty that in areas of high sediment supply the thickness of sixth or fifth-order sequences may greatly exceed 'typical' third-order sequence thickness ... We conclude that it is unlikely that sequences at all orders correspond to a 'Russian doll' stacked set, whereby the SSM applies at any level and the sequence at that level is viewed as forming from a large number of finer sequences (parasequences) of the next higher order. None the less, some orders of sequence do indeed embrace lower orders, e.g. the major first-order thermo-tectonic cycle that incorporates the complete sedimentary history of the Canterbury Basin ..., which includes examples of second, third, fourth, and probably fifth-order sequences.

The type of internal stacking described in this quote is characteristic of basin fills, as described in many examples in this book (e.g., Fig. 6.1).

Drummond and Wilkinson (1996) carried out a quantitative study of the duration and thickness of stratigraphic sequences and confirmed the opinions of Carter et al. (1991). Their major conclusion was that "discrimination of stratigraphic hierarchies and their designation as *n*th-order cycles may constitute little more than the arbitrary subdivision of an uninterrupted stratigraphic continuum."

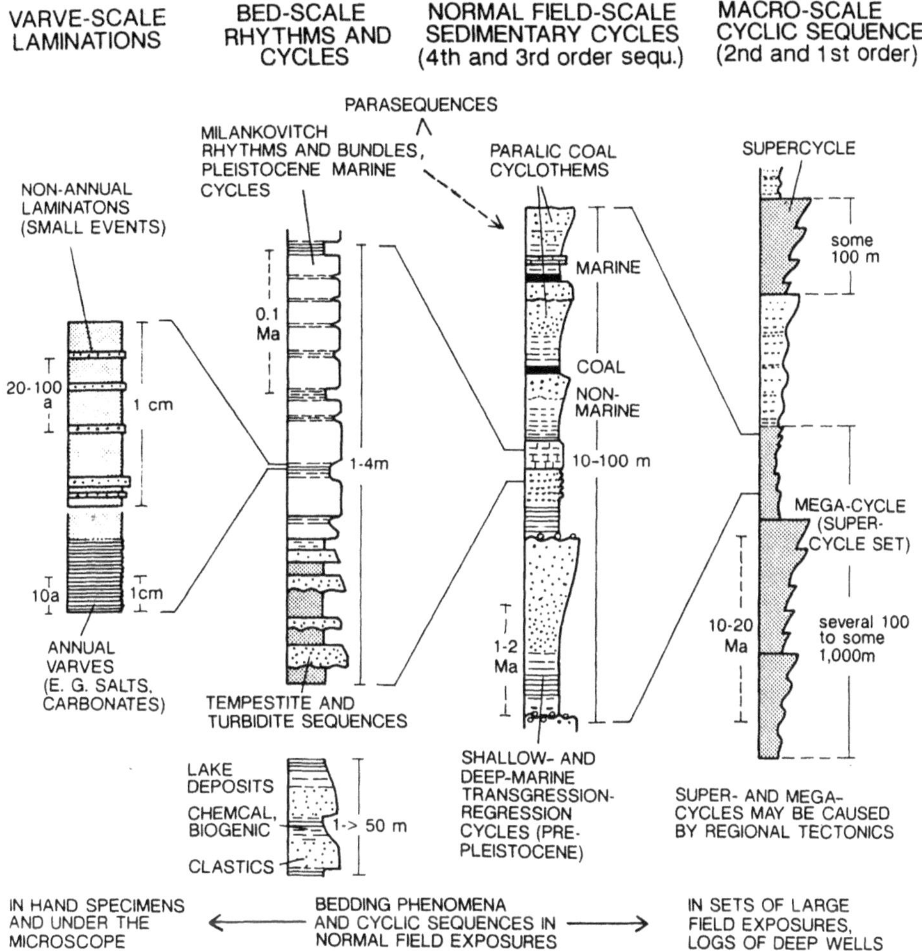


**Fig. 4.2** The scales of cyclic sedimentation in the stratigraphic record. This book is concerned with all but the first of these types (varve-scale laminations) (Einsele et al., 1991b)

Embry (1995) also found the existing hierarchical classification of sequences arbitrary, and he proposed what he claimed is a more objective approach, based on certain key descriptive characteristics of the sequences themselves, such as the amount of deformation and the degree of facies change at the sequence boundary. He objected that a classification based on measured (or assumed) duration is subjective and meaningless. His classification is discussed below.

Schlager (2004, p. 185; see Fig. 4.3 in this book) noted that:

Although the principle of defining orders by duration has been almost universally followed, the actual values used in the definitions scatter widely. [Figure 4.3] shows the definitions used in key publications since the introduction of the concept. In the range of 2nd to 3rd order, the

discrepancies at any one boundary are about one-half order; for shorter categories, they are even larger. Moreover, the values do not seem to converge with time and improving data. [Another figure] plots the durations of sequence cycles of 2nd and 3rd order on the sea-level curve of Haq et al. (1987). The two categories clearly differ in their modes but broadly overlap in range.

Schlager (2004) demonstrated that sequence architecture is fractal in character; that is, sequences are self-similar at a wide range of scales. According to this mathematical model, which is discussed further in Chap. 13, durations of any length may be expected to occur, from thousands of years to hundreds of millions of years. The only reason systematization in sequence classification is possible is because some of these durations correspond to the periodicity of certain natural processes.

**Table 4.1** Stratigraphic cycles and their causes

| Sequence type | Duration (million years) | Other terminology |
| --- | --- | --- |
| A. Global supercontinent cycle | 200–400 | First-order cycle (Vail et al., 1977) |
| B. Cycles generated by continental-scale mantle thermal processes (dynamic topography), and by plate kinematics, including: | 10–100 | Second-order cycle (Vail et al., 1977), supercycle (Vail et al., 1977), sequence (Sloss, 1963) |
|   1. Eustatic cycles induced by volume changes in global mid-oceanic spreading centres | | |
|   2. Regional cycles of basement movement induced by extensional downwarp and crustal loading. | | |
| C. Regional to local cycles of basement movement caused by regional plate kinematics, including changes in intraplate-stress regime | 0.01–10 | 3rd- to 5th order cycles (Vail et al., 1977).<br>3rd-order cycles also termed: megacyclothem (Heckel, 1986), mesothem (Ramsbottom, 1979) |
| D. Global cycles generated by orbital forcing, including glacioeustasy, productivity cycles, etc. | 0.01–2 | 4th- and 5th-order cycles (Vail et al., 1977), Milankovitch cycles, cyclothem (Wanless and Weller, 1932), major and minor cycles (Heckel, 1986) |

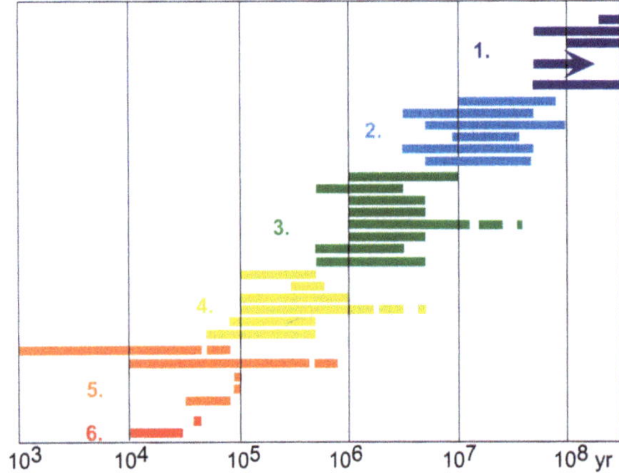



**Fig. 4.3** Duration of orders of stratigraphic sequences as defined by various authors. In each category, oldest publication is on top. Note large differences, particularly in 4th–6th orders (Schlager, 2004)

devised for tidal deposits by Nio and Yang (1990), based on that of Miall. Bounding surface codes for the two systems are given in Table 4.4. However there are substantial disagreements between the two systems in terms of the interpreted time duration of the various units. Sequence nomenclature of the Exxon group of workers has been concerned primarily with "third-order" cycles (group 10 of Table 13.1). These represent depositional units and time spans at least an order of magnitude greater than most of the units with which Miall (1991b) was concerned. However, Van Wagoner et al. (1990) and Mitchum and Van Wagoner (1991) claimed that they can trace widespread high-frequency sequences nested within the third-order cycles through the Gulf of Mexico. Their stratigraphy (and the terminology they have developed to describe it) involves depositional units of the same scale as those classified by Miall (1988, 1991b) and Nio and Yang (1991). Some attempt at a reconciliation of the terminologies is therefore necessary.

Figure 4.5 illustrates a hierarchy of depositional units or lithosomes developed for fluvial deposits (Miall, 1988, 1991b). A very similar field scheme was

Figure 4.6 and Tables 4.2 and 4.3 illustrate the original Exxon nomenclature of the sequence stratigraphers. Sequences, sequence-sets and composite

**Fig. 4.4**  Types of sequence anatomy, reflecting the scaling of sequences to the types of driving processes (Schlager, 2005, Fig. 6.13)

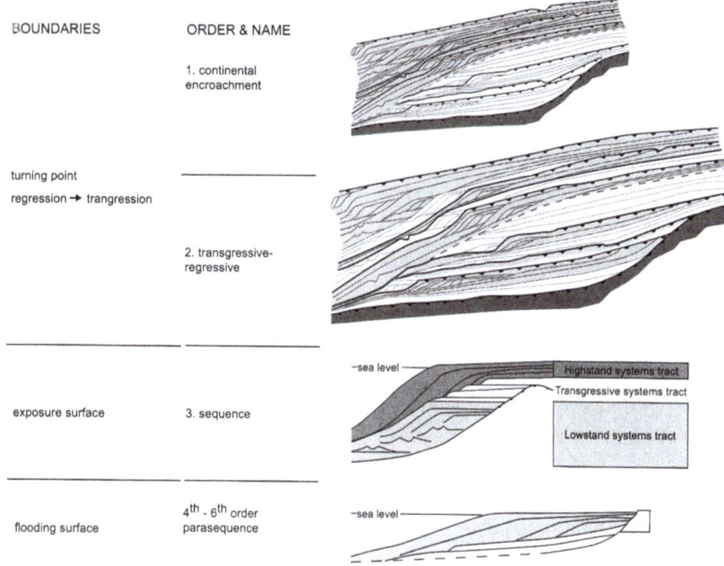


**Fig. 4.5**  The scales of depositional elements in a fluvial system. Circled numbers indicate the ranks of the bounding surfaces. In diagram D the sand flat is shown as being built up by migrating "sand waves". Foreset terminations of these are shown on top of diagram, but the internal crossbedding that results has been omitted for clarity (Miall, 1996)

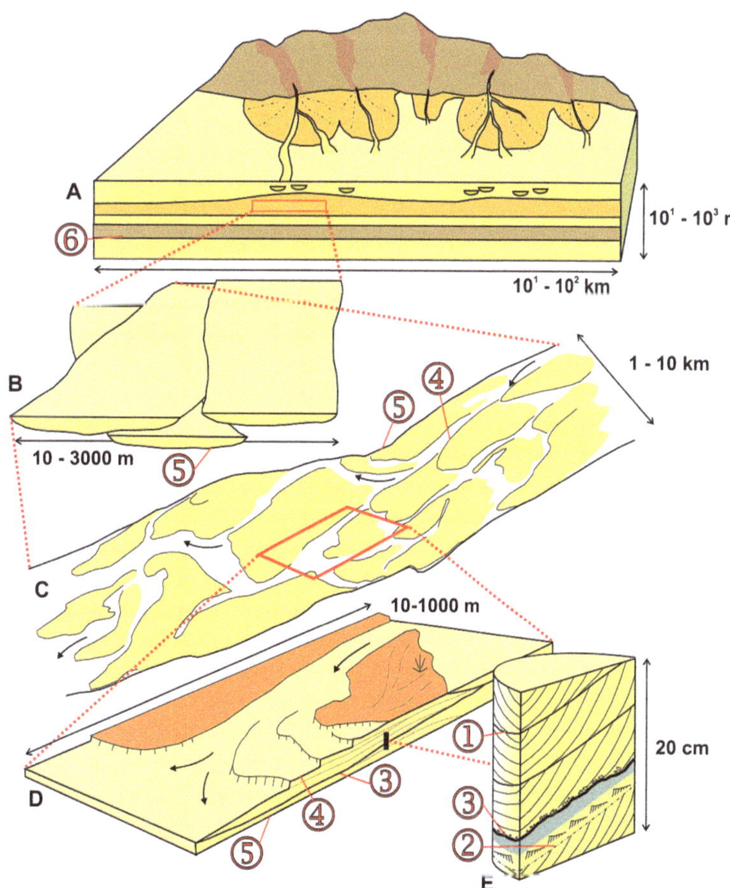

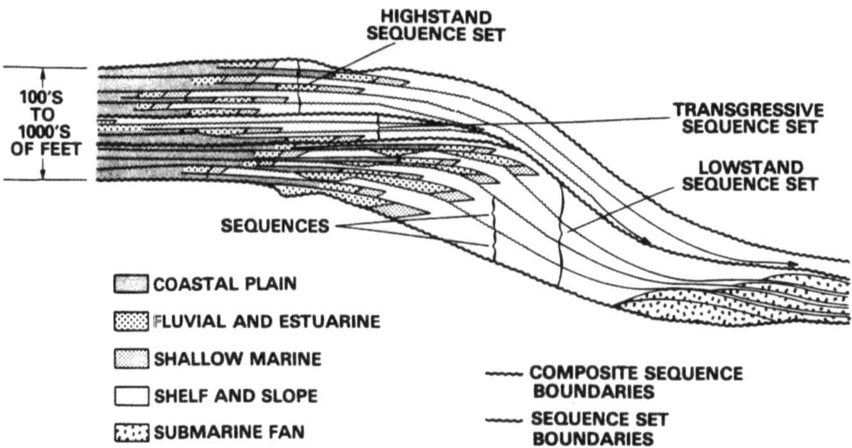



**Fig. 4.6** Diagram of sequences, sequence sets, and composite sequences. Individual sequences, composed of parasequences, stack into lowstand, transgressive, and highstand sequence sets to form composite sequences (Mitchum and Van Wagoner, 1991)

sequences are illustrated in order to demonstrate the multiples scales of relative sea-level change cycle that may combine to generate a stratigraphic architecture. *Parasequences* may be comparable in duration to fifth- or sixth-order sequences. However, the relationship is not necessarily as simple as this. In some studies high-frequency ($10^5$-year) sequences constitute the basic building-blocks of the stratigraphy, and may themselves be subdivided into parasequences. For example, Fig. 2.9 illustrates the $10^5$-year sequences of the Dunvegan Delta, Alberta. Each of the sequences, named informally A–G, has been

**Table 4.2** Definitions and descriptions of terms used in sequence stratigraphy (Van Wagoner et al., 1990)

| Stratal units | Definitions | Range of thicknesses (feet) | Range of lateral extents (sq.miles) | Range of times for formation (years) | Tool Resolution |
|---|---|---|---|---|---|
| Sequence | A relatively conformable succession of genetically related strata bounded by unconformities and their correlative conformities (Mitchum et al., 1977 a,b) | | | | Paleo / Exploration Seismic |
| Para-sequence set | A succession of genetically related parasequences forming a distinctive stacking pattern and commonly bounded by major marine-flooding surfaces and their correlative surfaces | | | | |
| Parase-quence | A relatively conformable succession of genetically related beds or bedsets bounded by marine-flooding surfaces and their correlative surfaces | | | | |
| Bedset | See Table 4.3 | | | | Well log |
| Bed | See Table 4.3 | | | | |
| Lamina set | See Table 4.3 | | | | Core and outcrop |
| Lamina | See Table 4.3 | | | | |

Range of thicknesses (feet): 1000  100  10  1  Inches
Range of lateral extents (sq.miles): 10 000  1000  100  10  1
Range of times for formation (years): $10^6$  $10^5$  $10^4$  $10^3$  $10^2$  10  1

**Table 4.3** Detailed characteristics of lamina, bed, and bedset (Van Wagoner et al., 1990; based on Campbell, 1967)

| Stratal unit | Definition | Characteristics of constituent stratal units | Depositional processes | Characteristics of bounding surfaces |
|---|---|---|---|---|
| Bedset | A relatively comformable succession of genetically related beds bounded by surfaces (called bedset surfaces) of erosion, non-deposition, or their correlative conformities | Beds above below bedset always differ in composition, texture, or sedimentary structure from those composing the bedset | Episodic or periodic. (same as bed below) | (Same as bed below) plus<br>• Bedsets and bedset surfaces form over a longer period of time than beds<br>• Commonly have a greater lateral extent than bedding surfaces |
| Bed | A relatively conformable succession of genetically related laminae or laminasets bounded by surfaces (called bedding surfaces) of erosion non-deposition, or their correlative conformities | Not all beds contain laminaets | Episodic or periodic.<br>Episodic deposition includes deposition from storms, floods, debris flows, turbidity currents.<br>Periodic deposition inclues deposition from seasonal or climatic changes | • Form rapidly, minutes to years<br>• Separate all younger strata from all older strata over the extent of the surfaces<br>• Facies changes are bounded by bedding surfaces<br>• Useful for chronostratigraphy under certain circumstances<br>• Time represented by bedding surfaces probably greater than time represented by bed<br>• Areal extents very widely from square feet to 1000s square miles |
| Laminaset | A relatively conformable succession of genetically related laminae bounded by surfaces (called laminaset surface) of erosion non-deposition or their correlative conformities | Consists of a group or set of conformable laminae that compose distinctive structures in a bed | Episodic, commonly found in wave or current-rippled beds, turbidites, wave-rippled intervals in hummocky bedsets, or cross beds as reverse flow ripples or rippled toes of foresets | • Form rapidly, minutes to days<br>• Smaller areal extent than encompassing bed |
| Lamina | The smallest megascopic layer | Uniform in composition/texture<br>Never internally layered | Episodic | • Forms very rapidly, minutes to hours<br>• Smaller areal extent than encompassing bed |

defined as an allomember, and each may itself be subdivided into suites of offlapping shingles representing $10^4$ years of sedimentation. Each of the shingles is an upward coarsening facies succession, which, therefore, itself fits the definition of a parasequence (Fig. 2.3). Shingle boundaries would be classified as sixth-order bounding surfaces in the Miall (1988) classification, or as "F" surfaces in the Nio and Yang (1991) system. At least three scales of succession therefore need to be accommodated in the nomenclature.

The shingles in the Dunvegan delta have been shown by careful mapping to be localized in distribution and comparable in three dimensional architecture to the lobes of the modern Mississippi delta (Bhattacharya and Walker, 1991; Sect. 2.3.1). They are autogenic in origin. This raises a problem that was not acknowledged by Mitchum and Van Wagoner (1991) and the other Exxon workers: the importance of autogenic processes in the development of facies successions (parasequences). Those successions constituting the "shingles" in Fig. 2.9, and the various lobes of the post-glacial Mississippi Delta (Fig. 7.5) are interpreted to be of autogenic origin. They developed by delta-lobe switching as a result of major channel crevassing and consequent shortening of the transport path to the sea (Bhattacharya and Walker, 1992). The resulting vertical succession is very similar to that of stratigraphic sequences, including the presence of prograding ("constructive") clinoforms similar to highstand systems tracts, and delta-lobe abandonment surfaces followed by "destructive" deposits comparable to the sequence boundaries and transgressive systems tracts of the sequence models

(Fig. 2.25). The successions are comparable in vertical succession and time span to $10^4$-year ("fifth-order") sequences. However, no sea-level change need be invoked to explain delta-lobe switching. Very careful delta mapping is required to distinguish autogenic delta shingles and lobes from "true" high-frequency cycles (e.g., Bhattacharya, 1991; Bhattacharya and Walker, 1991). They key test is whether specific facies successions can or cannot be precisely correlated to those in other, genetically unrelated depositional systems.

Table 4.4 is an attempt to set out the relationships among three types of hierarchical stratigraphic terminology that have been described in the literature. Sequence nomenclature is that of Van Wagoner et al. (1990) and Mitchum and Van Wagoner (1991). These authors defined third-order cycles as those having durations of 1–2 million years, which represents only the shorter end of the time spans traditionally classified as third-order. Fourth- and fifth-order sequences, in their classification, are those with periodicities or episodicities of $10^5$ and $10^4$ years, respectively. The bounding-surface classifications of Miall (1988, 1991b) and Nio and Yang (1991) are based on outcrop architectural descriptions. These classifications were built from the small-scale units upward, by mapping the way in which facies units combine into macroforms (e.g., bars), channels, and larger features. This contrasts with the sequence nomenclature, which represents an attempt to develop an ever more refined subdivision of the larger-scale (regional or global) stratigraphic units. The major overlap and possible source of confusion in the nomenclature systems is at the level of Miall's (1991b) groups 8 and 9

**Table 4.4** Suggested relationships between various hierarchical systems of stratigraphic classification

| Sequence duration (years) | Sequence nomenclature | Bounding surface classifications Miall[a], Nio and Yang[b] | Proposed allostratigraphic terminology |
|---|---|---|---|
| $10^6$ | 3rd-order sequence | 8[a] | Alloformation |
| $10^5$ | 4th-order sequence (or regional parasequence set) | 7[a] | Allomember |
| $10^4$ | 5th-order sequence (or regional parasequnce set) | 6[a], F[b] | Allomember or submember |
| $10^{3-4}$ | Parasequence | 5, 6[a]; E, F[b] | Facies succession |

Bounding surface lettering of Nio and Yang (1991) is inserted here on the basis of a comparison between the architectural features described by these authors and Miall (1988), but does not correspond to their sequence designations. The hierarchical sequence nomenclature shown in column two is based on Vail et al. (1977), but is no longer recommended. Superscripts a and b refer to the nomenclature of Miall and of Nio and Yang, respectively, in column 3.

(Table 13.1). High-frequency ("fourth-" and "fifth-order") sequences and large-scale autogenic facies successions may have similar vertical thicknesses and lithologic profiles, and represent similar time spans, but they are quite different in three-dimensional distribution and have quite different origins. It should be borne in mind that modern work is now suggesting that although natural hierarchies of sequences are common in the rock record, the durations of the constituent sequences are variable, and there is no genetic basis for a formal rank ordering.

An appropriate system of non-genetic stratigraphic nomenclature might be the solution to the confusion. A perfect system does not, at present, exist, but the system of allostratigraphy devised by the North American Commission on Stratigraphic Nomenclature (1983) is a useful first approach. Allostratigraphic nomenclature is discussed in Sect. 1.7, and some of the terms and suggested relationships to sequence and bounding-surface nomenclature are shown in Table 4.4.

Embry (1993, 1995) objected to the rank-order classification of sequences (first- to sixth-order) proposed by Vail et al. (1977). This classification is not based on any descriptive characteristics of the sequences themselves except their duration, which may be poorly known, and has led to classifications of sequences in the global cycle chart that Embry (1993, 1995) regards as arbitrary. Embry's (1993, 1995) proposed solution is to define a five-fold classification of sequences based on six descriptive criteria:

1. The areal extent over which the sequence can be recognized;
2. The areal extent of the unconformable portion of the boundary;
3. The degree of deformation that strata underlying the unconformable portion of the boundary underwent during boundary generation;
4. The magnitude of the deepening of the sea and the flooding of the basin margin as represented

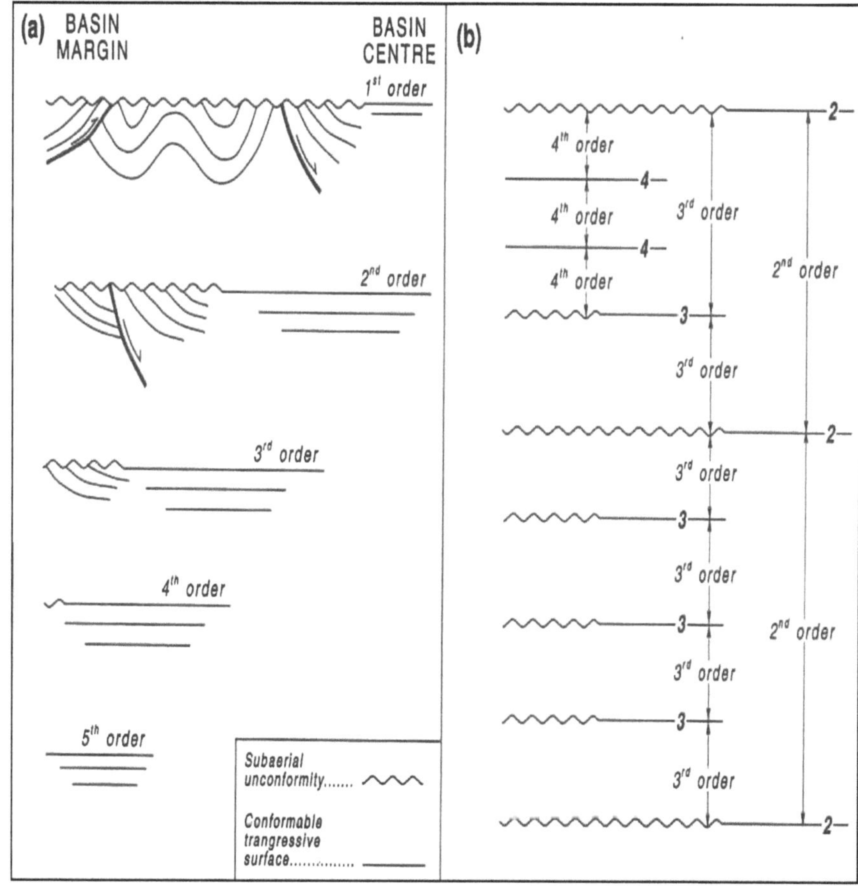


**Fig. 4.7** (a) A Schematic classification of stratigraphic sequences, according to the proposal by Embry (1993, 1995). As shown in the hypothetical sequence diagram (b) Sequences cannot contain a sequence boundary with the same or lower order than its highest order boundary, and the order of a sequence is equal to the order of its highest order boundary (Embry, 1993, 1995)

by the nature and extent of the transgressive strata overlying the boundary;

5. The degree of change of the sedimentary regime across the boundary;

6. The degree of change of the tectonic setting of the basin and surrounding areas across boundary.

The application of these criteria to a sequence classification is illustrated in Fig. 4.7. There are two problems with this classification. One is that it implies tectonic control in sequence generation. Sequences generated by glacioeustasy, such as the upper Paleozoic cyclothems and those of Neogene age on modern continental margins, would be first-order sequences in this classification on the basis of their areal distribution (they are potentially global in extent), but fifth-order on the basis of the nature of their bounding surfaces (because of the absence of tectonism in the generation of the sequence boundary). The second problem is that the classification requires good preservation of the basin margin in order for deformation at the sequence boundary to be properly assessed. This may not always be available, and it is possible that sequences of first to third order in this scheme could be mistakenly assigned to the fourth order if their uplifted and eroded marginal portions are not preserved. However, this classification has certainly clarified relationships in basins such as the Sverdrup Basin of the Canadian Arctic, and there are undoubtedly other areas where this particular type of objective descriptive approach could be useful.

In this book, a formal classification of sequences according to "order" or level of deformation is avoided as far as possible, because such classifications are increasingly being shown to be arbitrary. However, it is useful to refer to sequences in a simple descriptive sense as being of low or high frequency, or to make reference to their periodicity or episodicity in terms of the order of magnitude of the cycle frequency (e.g., $10^7$-year cycles for what were earlier termed "second-order" cycles). The work of Drummond and Wilkinson (1996) and Schlager (2004) would appear to suggest that there is a continuum of sequence types spanning the geological time scale from $10^4$ to $10^7$ year periodicities. Nevertheless, the geological record comprises four basic types of sequence, which are introduced briefly here, below.

## 4.3 The Supercontinent Cycle

A very long-term eustatic cycle is generated as a result of the assembly of supercontinents and their subsequent rifting and dispersal. The complete cycle takes 400–500 million years, and the process has been underway since at least 2 Ga. About four complete cycles may have occurred. The assembly of a supercontinent inhibits radiogenic heat loss from the core and mantle, leading to thermal doming, rifting, and breakup of the continent, so that a period of vigorous seafloor spreading ensues. The process of "subduction-pull" by downgoing cold oceanic plates enhances and may speed supercontinent fragmentation (Heller and Angevine, 1985; Gurnis, 1992; Burgess, 2008). There are significant consequences for global climate and sea-level. These mechanisms are discussed in more detail in Chap. 9.

Major cycles of sea-level rise and fall during the Phanerozoic are illustrated in Fig. 4.8. Vail et al. (1977) referred to these as first-order cycles. They include the two extended periods of maximum marine transgression in the Late Cambrian to Mississipian, and the Cretaceous, and a period of maximum regression in the Pennsylvanian to Jurassic. A glance at the geological map of North America confirms the importance of these broad changes. The Canadian Shield is flanked by Cretaceous rocks resting unconformably on a Cambrian or Ordovician to Devonian sequence over wide areas of the craton, from the Great Lakes region across the Prairies into the Beaufort-Mackenzie region and the Arctic Platform. Rocks of Pennsylvanian to Jurassic age are largely confined to intracratonic basins and the mobile belts flanking the cratonic interior.

Worsley et al. (1984, 1986), Hoffman (1989, 1991), Dalziel (1991), and Rogers (1996) argued that there is fragmentary evidence for several comparable cycles during the Precambrian, back to at least 2 Ga, possibly 3 Ga (earlier work on this subject is summarized by Williams, 1981). They reviewed the evidence for four major global episodes of thermotectonic activity comparable in magnitude and extent to the combined Caledonian-Acadian and Hercynian-Appalachian events of the Devonian to Permian. These orogenic episodes may indicate supercontinent assembly. They were followed by the intrusion of major dyke swarms, which probably formed during the initial rifting of the supercontinent. Evolutionary milestones in

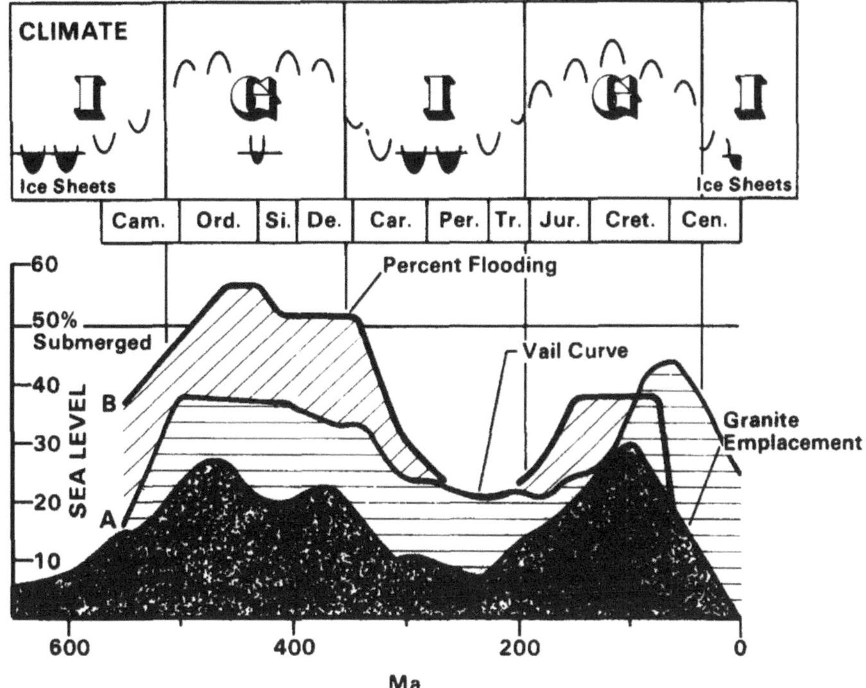


**Fig. 4.8** Supercontinent cycles during the Phanerozoic, including sea-level change (from Vail et al., 1977), percent flooding of the continents (from Fischer, 1981), and volume of granite emplacement (from Engel and Engel, 1964). The generation of the sea-level curve is discussed in Chap. 12 (Vail et al., 1977). This diagram, which was compiled by Worsley et al. (1984), also shows fluctuations in global climate that accompanied these changes. I, icehouse; G, greenhouse climate (discussed in text)

the biologic realm also seem to have postdated the major orogenic episodes, possibly indicating the explosion of biotic diversity in newly flooded shelf seas. Continental flooding is caused by eustatic sea-level rise driven by increased rates of seafloor spreading 50–100 million years after rifting, and by the stretching and thermal subsidence of new continental margins.

Stratigraphic evidence for these Precambrian cycles has yet to be assembled. There are many thick sequences of Precambrian strata around the world, but they have been much disturbed by tectonism, and accurate dating is difficult to achieve. Much work remains to be done in this area.

## 4.4 Cycles with Episodicities of Tens of Millions of Years

It has now been satisfactorily demonstrated that the original six cycles of Sloss (1963) can be recognized and correlated with comparable cratonic cycles in other continents. Sloss (1972, 1979) showed that a similar sequence chronology could be recognized in Europe and Russia. In his 1972 paper, Sloss reported on an analysis of detailed isopach and lithofacies maps of the Western Canada Sedimentary Basin and the Russian Platform. The data source included 29 Canadian maps (McCrossan and Glaister, 1964) and 62 Russian maps (Vinogradov and Nalivkin, 1960; Vinogradov et al., 1961). Each map was divided into a grid with intersection points spaced about 60 km apart, and the thickness and lithofacies were recorded for each point. From these data, the areal extent and volume of each of the mapped units could be calculated and compared. This approach is subject to possibly serious error because of the high probability of intersequence and even intrasequence erosion. The presence of a single isolated outlier or fault block beyond the edge of the main basin can change the interpreted former area of extent of a map unit by hundreds of square kilometres. Nevertheless, the data from the two areas show remarkable similarities (Fig. 4.9), and the detailed statistical documentation confirms that Sloss's six sequences can be recognized in two widely separated continents that

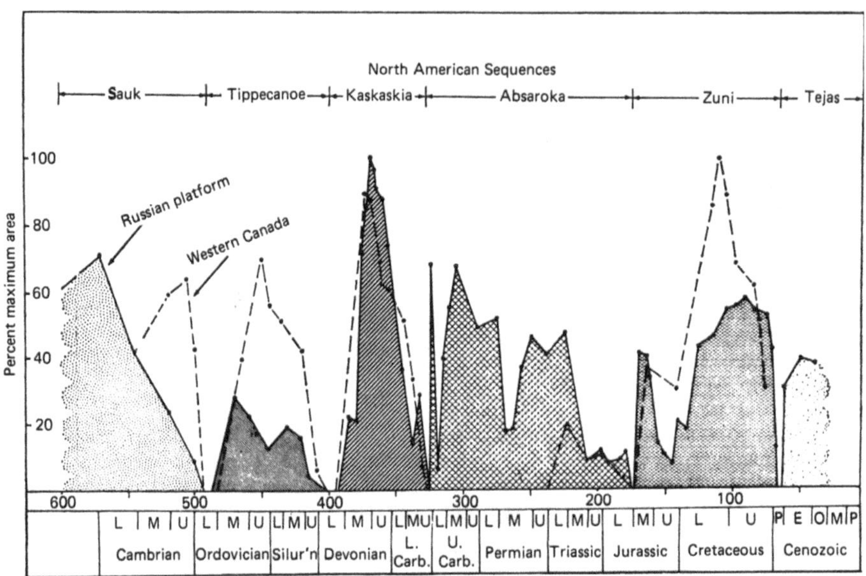



**Fig. 4.9** Areas of preservation of units in western Canada and the Russian platform showing relationship to the six sequences of Sloss (1963). L, M, U = Lower, Middle, Upper; P = Paleocene, E = Eocene, O = Oligocene, M = Miocene, P = Pliocene (Sloss, 1972)

formerly would have been assumed to have undergone a quite different geological history.

These Phanerozoic cycles are clearly global in scope. They have been attributed to eustatic changes in sea level in response to volume changes of oceanic spreading centres, an interpretation first suggested by Hallam (1963), in his summary of Late Cretaceous and Cenozoic events, and subsequently elaborated by Pitman (1978). However, eustatic sea-level changes cannot explain all the features of the major sequences. They commonly are separated by angular unconformities, and in many cases, the sequences contain thick continental deposits. A passive rise in sea level would terminate widespread nonmarine deposition, and, therefore, additional processes must be involved. Geophysical studies have demonstrated that the crust flexes on a regional scale in response to convergent, divergent and transcurrent plate movements, a process termed *intraplate stress* (Sect. 10.4). Detailed back-strippping and other studies, of the type outlined in Chap. 3, show that the crust is constantly undergoing gentle vertical motions and broad continental tilts. This form of *epeirogeny* has recently been termed *dynamic topography* (Sect. 9.3.2). Broad, regional cycles of change in basement elevation occur in response to thermal changes in the crust and mantle, crustal thickening and thinning, and sediment loading.

## 4.5 Cycles with Million-Year Episodicities

Detailed stratigraphic and facies studies of well dated Phanerozoic strata have yielded a wealth of information concerning cyclicity over time scales of 1–10 million years. Vail et al. (1977) termed these third-order cycles. Examples of sequence-stratigraphic data are documented in Chap. 6. The intent is not to provide an exhaustive catalog of either the literature or the results, but to impart an overview of the range of types of stratigraphic cycle that have been described in various tectonic settings, and to provide a flavor of the kinds of research currently being conducted in this area. There is increasing evidence for the occurrence of the same type of cyclicity during the Precambrian (Christie-Blick et al., 1988), but the difficulty of achieving accurate chronostratigraphic correlations in Precambrian sediments has retarded the application of cyclic concepts to Precambrian stratigraphy (Catuneanu and Eriksson, 1999).

Several different techniques have been used in attempts to reconstruct a chronostratigraphic record of these stratigraphic cycles. The techniques of Vail and his co-workers (1977) and Haq et al. (1987, 1988), as applied to analyses of the Mesozoic and Cenozoic sequence, are discussed in Chap. 12. Sloss-type outcrop-area plots have been used by several workers, such as in Hallam's (1975) analysis of the Jurassic record. Detailed stratigraphic reconstructions, emphasizing lithofacies or biofacies data, or both, constitute the third main method of reconstructing sea-level change. Papers in this category are too numerous to list. The recent work of Graciansky et al. (1998), which attempted to synthesize much of these data, is discussed in Chap. 12. A few workers have attempted to reconcile the results of these various techniques in order to come up with average or compromise sea-level curves (e.g., Hallam, 1981, 1984, 1992a). There are no reliable methods of quantifying the averaging procedures, and so the amplitudes of the curves (the amounts of sea-level change) cannot be taken too literally.

A single example of million-year episodicity is illustrated here, that of the Early Cretaceous to modern record of the continental shelf of New Jersey

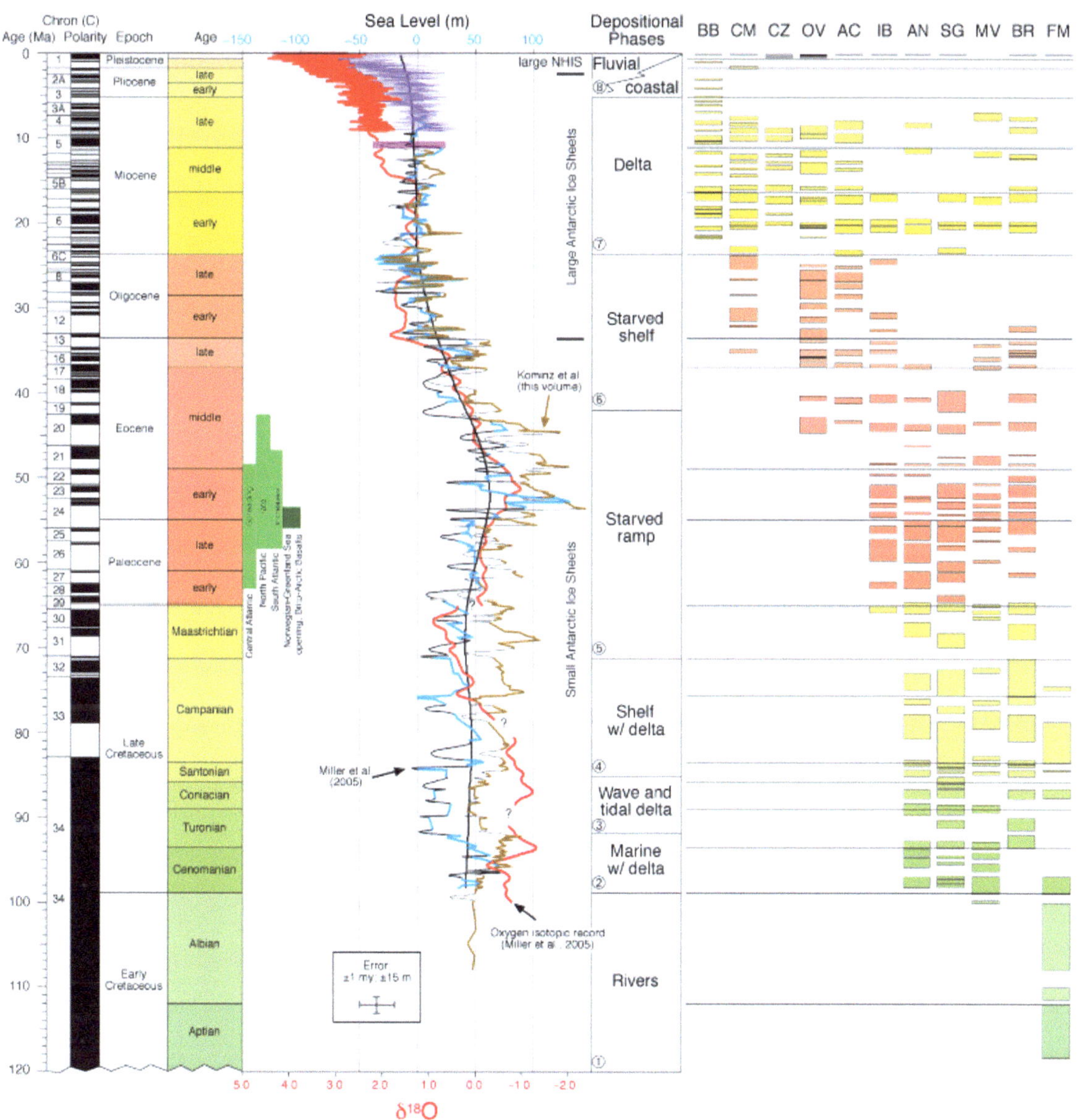



**Fig. 4.10** Sequence stratigraphy of the New Jersey continental margin, from Browning et al. (2008), based on data from eleven core holes (columns at right, identified by two-letter location codes). A composite sea-level curve is shown in brown, and is compared to earlier sea-level and oxygen isotope curves obtained from the same area by Miller et al. (2004, 2005a)

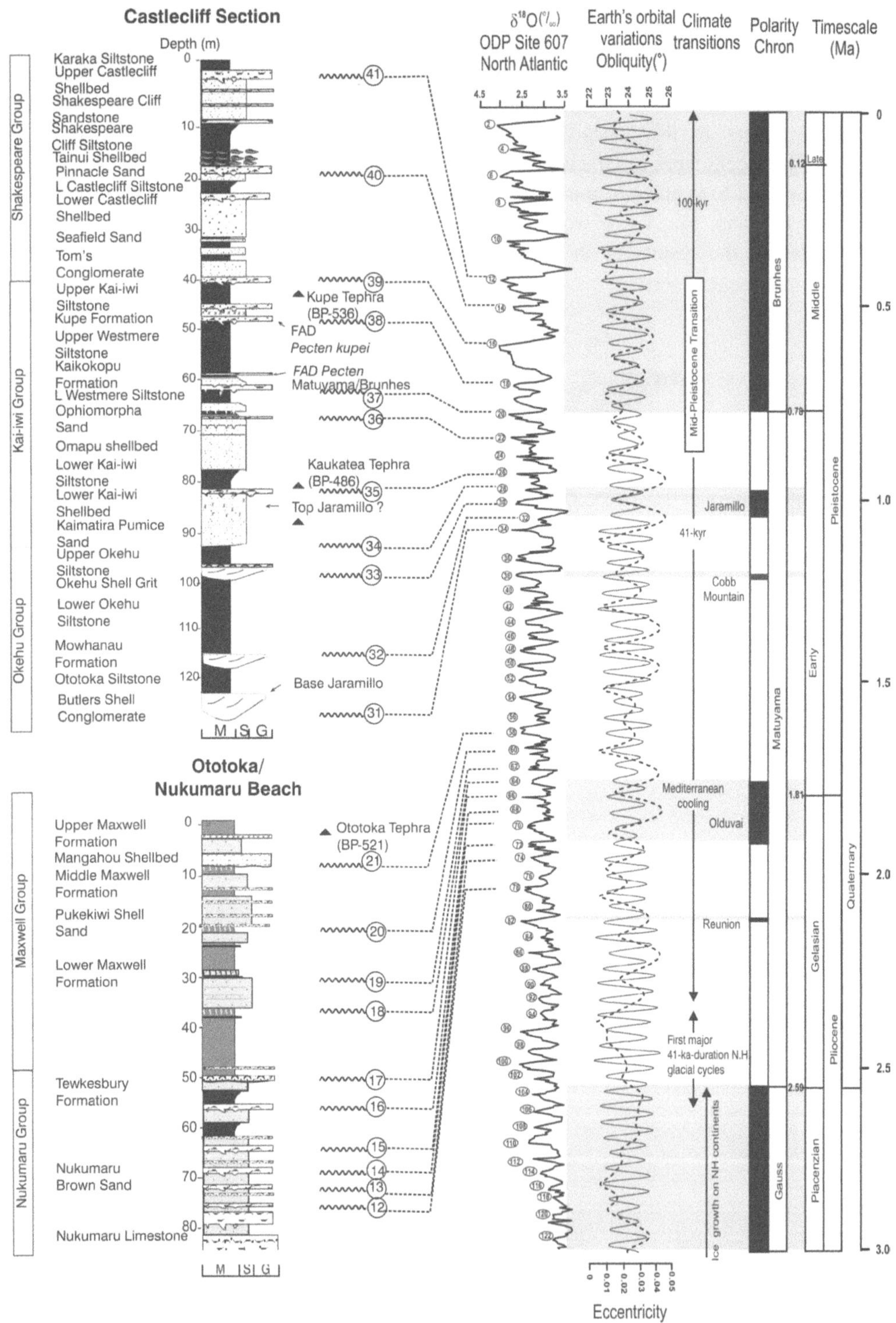



**Fig. 4.11** Composite stratigraphic columns for the Nukumaru and Castlecliff coastal sections, Wanganui Basin, New Zealand, showing lithostratigraphy, sequence stratigraphy, and correlations with the oxygen isotope timescale (Naish et al., 2005). This particularly complete Pliocene to Pleistocene section serves as a regional standard for geological time (Turner et al., 2005)

(Fig. 4.10). This particularly well dated composite section illustrates many of the scientific points that form the thesis of this book, and is, therefore, one to which we return many times in subsequent chapters.

Currently the origins of these cycles, including whether they are regional or global in scope, remains to be conclusively determined. Partial success has been achieved in correlating selected sections around the globe with each other but, as discussed in Sect. 14.6, the evidence for global eustasy is still very limited. Several tectonic mechanisms for the development of regional (continent-wide) cycles have been identified, as discussed in Chap. 10.

## 4.6  Cycles with Episodicities of Less Than One Million Years

There is a complete spectrum of cycles with durations down to those of a few-thousand years duration. The classification of these cycles that is commonly used, into those of fourth-order rank (0.2–0.5 million years duration) and those of fifth-order rank (0.01–0.2 million years) is one of convenience, but is now increasingly seen as arbitrary. As described and illustrated in Chap. 7, there are several distinct types of cycle that can be classified on the basis of their sedimentological character and tectonic and climatic setting, and this range of types suggests that there is more than one, possibly several, interacting, generative mechanism. Deposits formed in areas distant from detrital sediment sources, such as pelagic deposits, carbonate shelves and some lakes, commonly show successions of sequences that may be traced through wholesale facies changes, or between areas having different tectonic histories, indicating that they are not autogenic in origin. The useful, informal descriptive term metre-scale cycle is commonly used to refer to such sequences.

Many of the cycles are of climatic origin, caused by changes in the amount of solar radiation received by the earth and its global distribution that are, in turn, brought about by irregularities in the earth's motions about the sun, a process termed *orbital forcing*. Such processes are also referred to as *Milankovitch processes* (or mechanisms) after the Serbian mathematician who was the first to establish the quantitative theoretical basis for the concept (Chap. 11). Glacioeustasy is the single most important result, but the importance of what is termed here "non-glacial Milankovitch cyclicity" (Sect. 11.2.5) is being increasingly recognized throughout the geological record. The Neogene glaciation generated suites of clastic cycles throughout the world in most tectonic settings. An

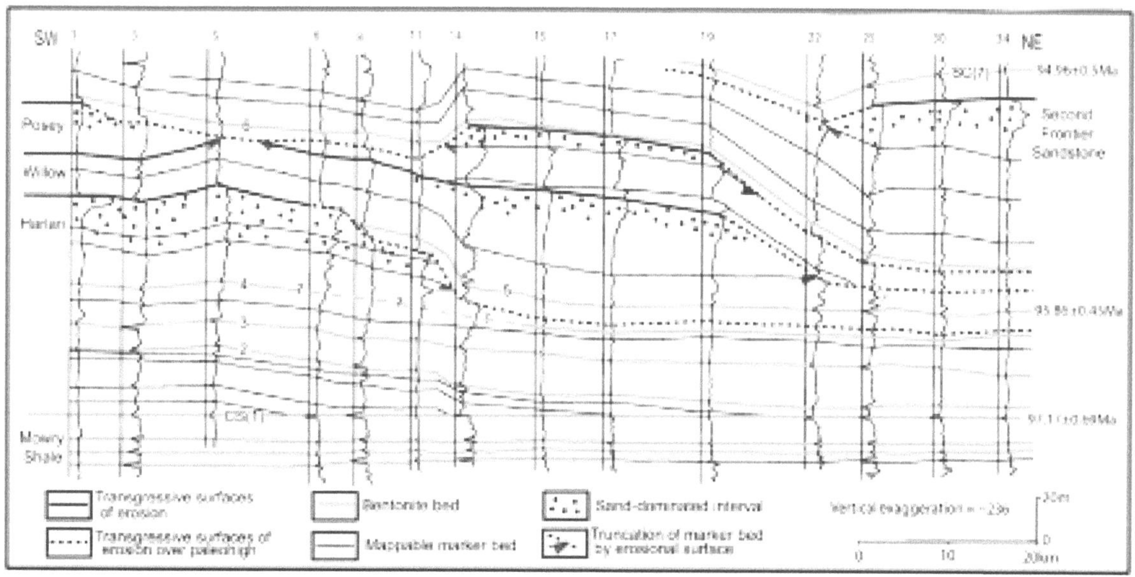

**Fig. 4.12** Resistivity well-log cross-section showing distribution of seven regionally recognized bentonites (*thick gray lines*), sandy portions of allomembers (*dotted pattern*), and tectonically driven erosional surfaces (*dashed lines*) (Vakarelov et al., 2006)

example is given in Fig. 4.11. The well-known Upper Paleozoic cyclothems of the northern hemisphere are also interpreted as glacioeustatic in origin. Other types of climatic cycle are discussed in Chap. 7. There is growing evidence for the occurrence of glacioeustasy during the Cretaceous and early Cenozoic, as discussed in Chaps. 11 and 13.

Detailed mapping of tectonically-active basins is providing abundant evidence of the importance of local to regional tectonism as a generator of high-frequency sequence stratigraphies. There are many such examples from ancient foreland basins. Cycles of loading and unloading due to contractional tectonism, crustal flexure, uplift and erosion, may generate changes in accommodation and resulting cycles of sedimentation and erosion over time scales as short as a few thousand years. An example is illustrated in Fig. 4.12, additional examples are described in Chap. 7, and the mechanisms are discussed in Chap. 11.

In the Western Interior Basin of North America, there is excellent evidence from the Cretaceous stratigraphic record of both climatic and tectonic forcing of accommodation and sedimentation. In different regions, at different times during the Cretaceous, the evidence appears to indicate one or the other as the dominant process, and determining the ultimate allogenic controls is one of the current themes of active research in this basin. Evidence from the New Jersey coast and from some ODP records is providing increasing evidence of glacioeustasy during the Cretaceous and Paleogene (Sects. 11.3.3 and 14.6.4), and there is also evidence from the ancient record of possible "sub-Milankovitch" cyclicity, that is, with cycle frequencies in the $10^3$-year range (Sect. 14.7.2).

In tectonically active basins, distinguishing tectonic mechanisms from the products of orbital forcing offers particular challenges, as discussed in Sect. 11.4.

# Chapter 5

# Cycles with Episodicities of Tens to Hundreds of Millions of Years

## Contents



## 5.1 Climate, Sedimentation and Biogenesis

It is becoming increasingly clear that global climate, oceanic circulation, sedimentation patterns, biotic diversity, and evolutionary trends are all linked to variations in sea level (Fischer, 1984; Worsley and Nance, 1989; Worsley et al., 1984, 1986, 1991; Veevers, 1990; Rogers and Santosh, 2004). These linkages can be traced through the long-term cycle of supercontinent assembly and dispersal (Fig. 5.1), and many of the same changes can also be observed on shorter time scales. For example, certain trends in sedimentation patterns are related to the position and direction of change of sea level relative to continental margins, and there is some evidence (touched on below) for such variations occurring with episodicities of millions to tens of millions of years.

During supercontinent assembly and dispersal, individual continental plates may undergo latitudinal drift, which carries them through various climate belts. This leads to long-term changes in climate and consequent changes in sedimentary styles, of the type described by Cecil (1990) and Perlmutter and Matthews (1990). This is discussed further in Chaps. 8 and 9.

Flooding of the continents during the fragmentation phase of the supercontinent cycle causes widespread dispersion of marine clastic sediments, followed by chemical sedimentation as the supply of clastics from thermally subsiding continental fragments decreases. Evaporites may be abundant in the newly formed rifted basins and nascent oceans. Clastic sedimentation on the continents decreases to a minimum during the maximum-dispersal phase. At times of continental assembly, large volumes of terrigenous detritus are produced by erosion of newly emergent orogens, and are deposited as clastic wedges in inland basins (e.g., in foredeeps) and along continental margins. During the stasis phase, the presence of large, mountain-rimmed continental-interior basins may lead to climatic extremes in the continental interiors, and continental sedimentation may be dominated by eolian or glacial facies.

The depositional systems tracts described in Chap. 2 are dependent on sea level over the long and short term. Erosional and depositional events in submarine canyons and fans are markedly affected by sea level. During high stands of the sea, continental detritus tends to be trapped along the shorelines in deltas and coastal plain complexes (Shanmugam and Moiola, 1982). G. deV. Klein (oral communication, 1982) suggested that the major tidalite sequences of the Phanerozoic correspond to periods of high sea level, when shelf widths and tidal effects were at a maximum.

Shelf progradation rates may be high at times of high sea level; Carbonate platforms may undergo what Schlager (1991) called "highstand shedding"; but the deep ocean tends to be starved. Clinoform


A.D. Miall, *The Geology of Stratigraphic Sequences,* 2nd ed.,
DOI 10.1007/978-3-642-05027-5_5, © Springer-Verlag Berlin Heidelberg 2010

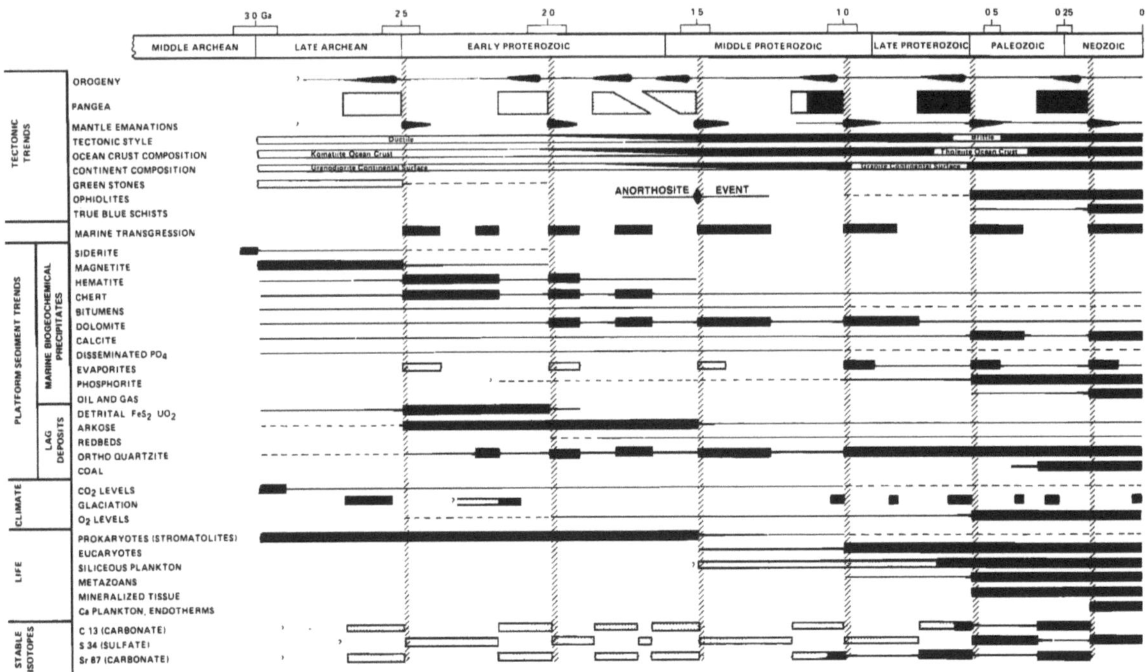

**Fig. 5.1** Summary of late Archean to Recent trends in tectonism, platform sedimentation, climate, life, and marine platform stable isotopes. Trends are shown as follows: abundant, intense or heavy: *solid bars*; common or moderate: dashed or solid lines (*dotted* where speculative). Proposed supercontinent fragmentation events are indicated by *vertical hatched bars* (Worsley et al., 1986)

shelf-slope architecture and offlapping sequences result. Conversely, during periods of low sea level, canyon erosion is active and much detritus is fed directly to the upper reaches of submarine canyons. The result is deep-marine onlap and/or basinward progradation of thick fan sequences. Because of the increased sediment supply to the deep oceans and increased thermohaline circulation (see below) during periods of low sea level, contourites are also likely to be more common at these times. Shanmugam and Moiola (1982) attempted to document the major occurrences of thick submarine fan and contourite deposits in the stratigraphic record, and they claimed that most correspond to periods of minimum sea level on the Vail et al. (1977) curves. Given the widespread controversy regarding these curves (Part IV) correlation with the curve is no longer regarded as significant, but the general relationships between depositional systems and sea-level change remain useful.

Several studies have suggested that many physical, chemical, and biological events in the continents and oceans are correlated with each other and that they change cyclicly over periods of $10^6$ to $10^7$ a. Funnell (1981) termed this autocorrelation. The subject was explored in depth by Fischer and Arthur (1977), who studied the Mesozoic and Cenozoic record, and later by Leggett et al. (1981), who carried out a similar analysis for the early Paleozoic. Berner et al. (1983) are credited with the elucidation of a long-term cycle of atmospheric change related to changing rates of sea-floor spreading; a concept now termed the *BLAG hypothesis,* based on the initial letters of the authors (Ruddiman and Prell, 1997, p. 9; Ruddiman, 2008, pp. 71–72). Worsley et al. (1986, 1991) and Worsley and Nance (1989) provided a compilation of current data and ideas.

The main control on autocorrelation appears to be long-term sea-level change, which is partly controlled by rates of sea-floor spreading (Pitman, 1978). During periods of high sea level, the world's oceans tend to be warm, with much-reduced latitudinal and vertical temperature gradients. Elevated $CO_2$ levels are attributed to high rates of volcanic outgassing. Although the area of exposed continent is relatively low, the rate of release of calcium and phosphorus by weathering remains high because of elevated surface temperatures, so that the rate of burial of $CO_2$ as carbonate remains constant, and the $CO_2$ content of the atmosphere rises, increasing the climatic greenhouse effect. Oceanic circulation is relatively sluggish, leading to increased stratification and severe oxygen depletion at depth. Deposition of organic-rich sediments in the deep ocean becomes widespread, the carbonate compensation depth rises and faunal diversity increases. These are the times of global anoxic events when widespread black shales are deposited around the world. Increased nutrient supply caused by high rates of mid-ocean volcanism encourages planktonic growth, leading to high general rates of biotic activity, and this may also, in part, account for the development of an unusual number of prolific oil source beds during the Cretaceous (Larson, 1991).

During periods of low sea level, global climates are more variable, and the oceans are, in general, cooler and better oxygenated because of better circulation. The $CO_2$ content of the atmosphere is reduced by greater combination with calcium and phosphorus, and burial as carbonates and phosphates, reflecting increased supply of these elements from continental weathering. The climatic greenhouse effect is therefore reduced. Many faunal niches in shelf regions are destroyed by subaerial exposure. Submarine erosion may be intensified. Initiation of these episodes of lower sea level may be a cause of biotic crises, in which faunal diversity is sharply reduced by major extinctions (Newell, 1967).

Fischer and Arthur (1977) encapsulated these faunal variations by applying the terms *polytaxic* and *oligotaxic* to periods of, respectively, high and low faunal diversity. The corresponding climatic milieus are referred to as *greenhouse* type when sea levels are high and climates are globally uniform and relatively warm, and *icehouse* type when variable climates accompany times of low sea level. Fischer and Arthur (1977) and Fischer (1981) demonstrated that since the Triassic the oceans have fluctuated between these two modes during an approximately 32-million years-long cycle. These cycles correlate reasonably well with the Sloss-type cycles discussed in Sect. 3.3, although there does not seem to be much support for the idea of a precise 32-million years regularity to the cycles. The climate changes are crudely cumulative, and can be correlated in general terms with the supercontinent cycles of sea-level change, as shown in Fig. 4.7.

Major glaciations occurred during the periods of icehouse climate in the late Precambrian, Late Devonian-Permian, and mid- to late Cenozoic. The causes of these long-term regional glaciations are complex (Eyles, 1993, 2008; Eyles and Januszczak, 2004), but widespread uplift as a result of continental collision and/or rifting seem to be major causes. A shorter glacial episode that occurred in the Late Ordovician to Early Silurian does not fit this pattern. Eyles (2008) suggested that regional uplift, a dynamic topography effect resulting from mantle heating, may have helped to trigger this glacial episode.

Carbonate sedimentation trends reflect the long-term oscillations from icehouse to greenhouse climates. Carbonate production is increased during times of high global sea level and greenhouse climate because of the greater extent of shallow shelf seas. A compilation by James (1983) showed that reefs tend to be more abundant at times of high global sea level (the long-term highs identified in Fig. 4.7). Lumsden (1985) found that the distribution of dolomite in deep-sea sediments follows the same trend; again, because of the greater production of carbonate sediment when areas of shelf seas are high. Sandberg (1983) suggested that low-Mg calcite is the dominant calcium carbonate mineral precipitated during times of high sea level and greenhouse climates, such as during the Ordovician-Devonian and Jurassic-Cretaceous, whereas high-Mg calcite and aragonite are dominant at times of low sea level and icehouse climates. Mackenzie and Pigott (1981) and Worsley et al. (1986) documented parallel changes in carbon and oxygen distribution and isotopic composition (Fig. 5.2).

## 5.2 The Supercontinent Cycle

### 5.2.1 The Tectonic-Stratigraphic Model

Evidence for two supercontinent cycles in the Phanerozoic, and several more in the Precambrian was assembled by Worsley et al. (1984, 1986), who synthesized a vast and diverse data base. Rogers and Santosh (2004) have updated and extended these ideas. The major elements of the Worsley model are illustrated in Fig. 5.2, and are discussed below.

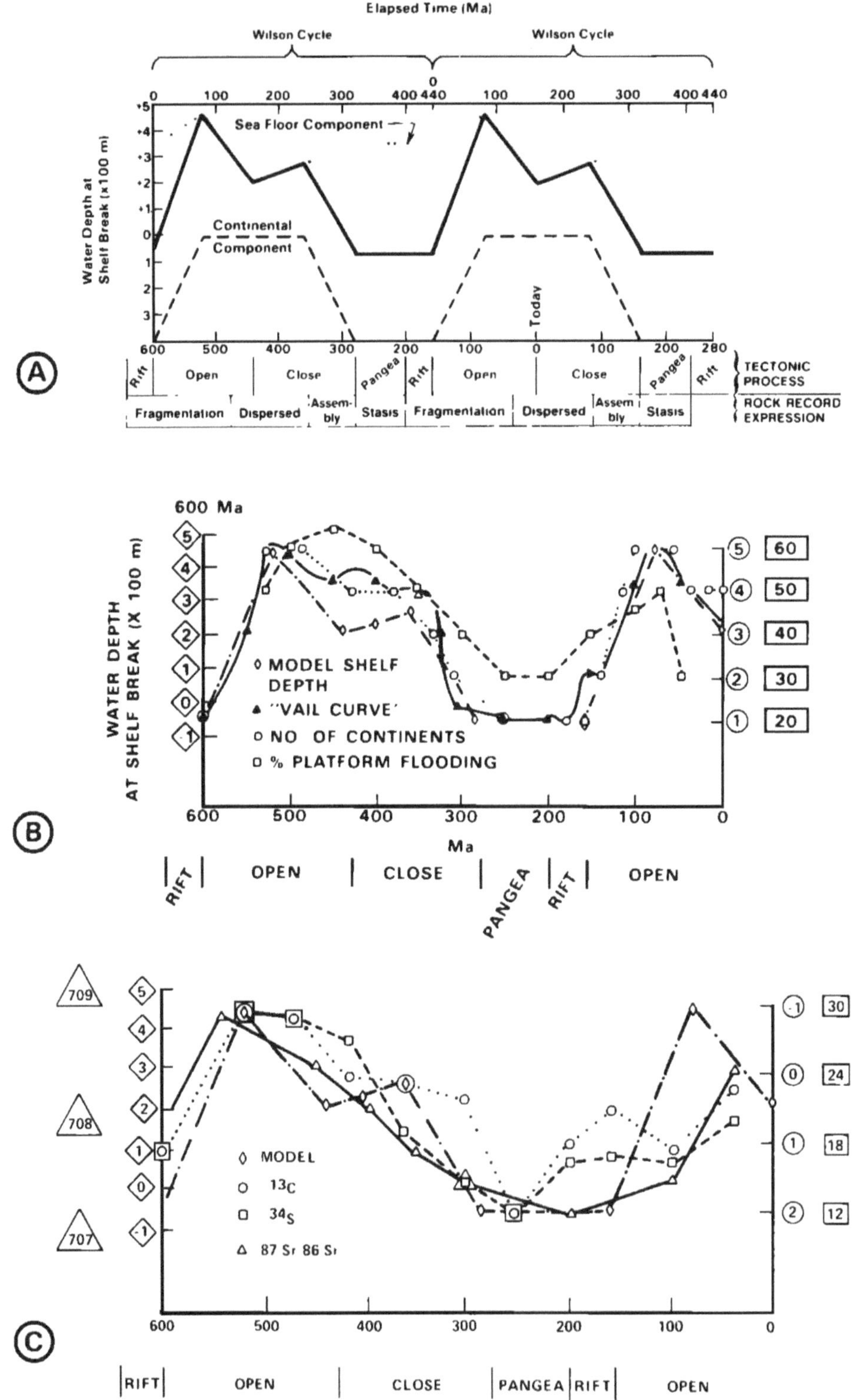



**Fig. 5.2** Summary of the supercontinent assembly-fragmentation cycle, and its effects on the Phanerozoic record. (**a**) Tectonic components of the model, showing two complete assembly-fragmentation cycles; (**b**) Comparison of the model with the long-term eustatic curve of Vail et al. (1977), platform flooding, and number of continents; (**c**) Trends in stable isotopes (Worsley et al., 1986)

The cyclic model has four tectonic phases: the *fragmentation phase* is the interval during which the supercontinent undergoes rifting and dispersion, over a period of approximately 160 million years. This is a period of active generation of oceanic crust. The average age of the crust therefore decreases. As demonstrated in Chap. 9, this is a cause of rising global sea level, because of the buoyancy of young oceanic crust and the presence of long, active, thermally-expanded, sea-floor spreading centres. Collisional orogeny is at a minimum during this period, whereas arc magmatism is active, with the emplacement of felsic magma, and continental mafic magmatism is at a peak, with the emplacement of mafic dyke swarms and mafic volcanism accompanying the formation of rift systems.

During the *maximum dispersal phase* mature, stretched continental margins face wide, aging oceans, as in the case of the Atlantic Ocean and its bordering continents at present. The world oceanic crust reaches its maximum average age about 200 million years after initial rifting. Atlantic-type oceans may begin to subduct, at which time the presence of old, cold oceanic crust entering the subduction zone leads to subduction-hinge roll-back, and the development of backarc spreading. Continental magmatism (as distinct from arc magmatism) is at a minimum during this period, and global heat flow is also at a minimum. Sea levels fall during this phase.

The *assembly phase* closes old Atlantic-type oceans, so that the average age of the oceanic crust increases and global heat flow increases to intermediate values. Convergent tectonism and its accompanying felsic magmatism increase globally, resulting in increased continental relief, decreased continental area, and a corresponding increase in ocean-basin volume, with resulting low global sea levels. Terrane-collision events may be abundant within the subducting oceans, while backarc spreading may occur along the trailing edges of the converging continents.

The *supercontinent stasis phase* is characterized by epeirogenic uplift of the new supercontinent, as heat builds up beneath it. Oceanic crust is at an intermediate age, and the combination of elevated continental crust and intermediate oceanic depths leads to the phase of maximum eustatic sea-level fall. Collisional orogeny is at a minimum, whereas subduction continues around the margins of the supercontinent.

## 5.2.2 The Phanerozoic Record

Many events in the Phanerozoic have long been suspected to be of global importance. Some of the more important of these are summarized here (Figs. 5.1 and 5.2). Worsley et al. (1984, 1986) postulated two supercontinent cycles during the Phanerozoic. Veevers (1990) divided the most recent of these into five stages, as noted below and in Fig. 5.3, column VIII.

Late Proterozoic glaciation, the evidence for which is widespread in Greenland and Scandinavia, may be related to the formation of regional ice caps on rift margins elevated by thermal doming prior to continental separation and the breakup of a late Precambrian supercontinent (Eyles, 1993; Eyles and Januszczak, 2004). I find these explanations more convincing for Late Proterozoic glaciation (see also Allen and Etienne, 2008), than the Snowball Earth concept of Hoffman and Schrag (2000). The Cambrian transgression may reflect the subsequent increase in the rate of sea-floor spreading as continental dispersal accelerated (Matthews and Cowie, 1979; Donovan and Jones, 1979). It has been suggested that the late Precambrian Sparagmite sequence of Norway was formed in rifts representing the incipient Iapetus (proto-Atlantic) Ocean (Bjørlykke et al., 1976). The exceptionally high sea level stand during the Ordovician might then relate to the rapid widening of the Iapetus Ocean (and probably other world oceans).

Sea-level lowering occurred during the Caledonian-Acadian and Hercynian-Appalachian suturing of Pangea between the Devonian and Permian (Schopf, 1974). The presence of this supercontinent over the south pole is thought to have been the cause of increasingly continental-type climates, leading to the Late Devonian-Permian glaciations of Gondwana (Crowell, 1978; Caputo and Crowell, 1985; Eyles, 1993; Fig. 11.17) and the widespread Pennsylvanian to Jurassic eolian facies of the United States and Europe (Kocurek, 1988a).

Kominz and Bond (1991) and Bond and Kominz (1991a) documented anomalously large subsidence events in North American cratonic basins and on the Cordilleran and Appalachian continental margins during the Late Devonian and Early Mississippian. They suggested that basin deepening and arch uplift may have been caused by intraplate stresses associated with plate convergence toward a region of mantle

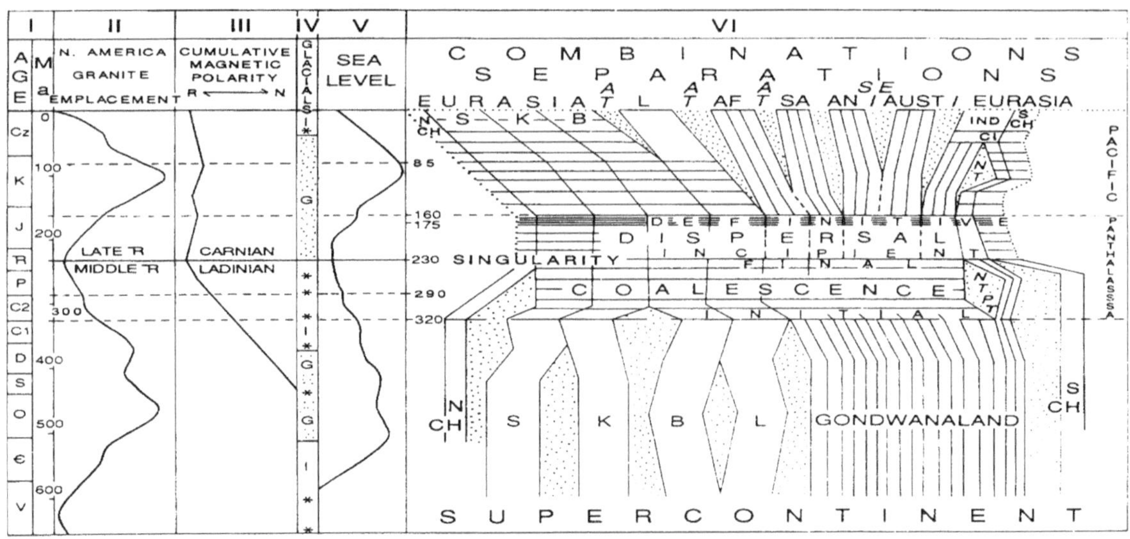

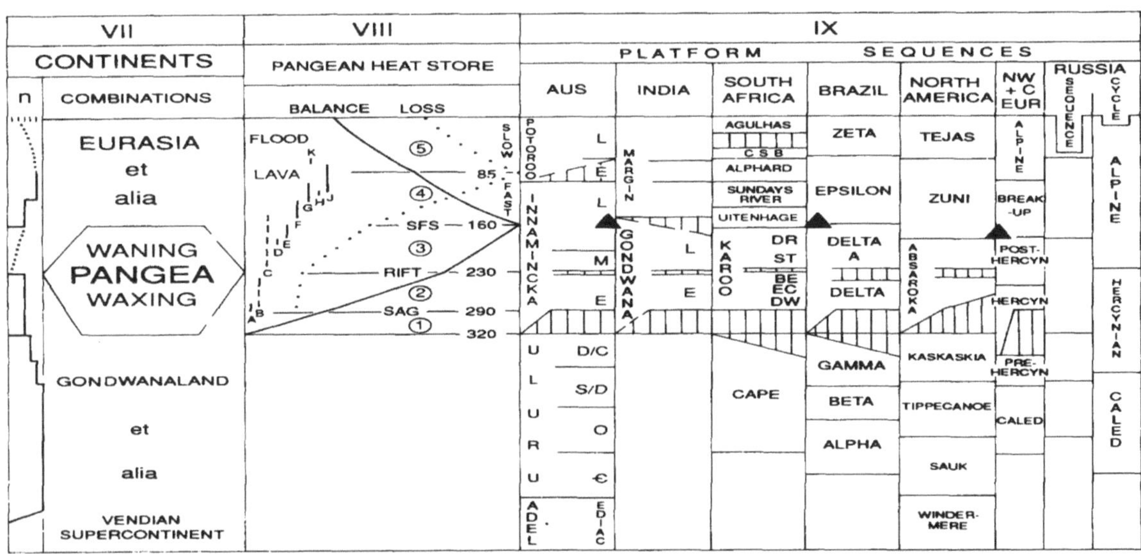

**Fig. 5.3** Phanerozoic tectonism, schematic continental assembly and rifting events, and summary of long-term stratigraphic cyclicity in the major continental areas. Column VI shows, schematically, the assembly and rifting of Pangea. Abbreviations of continental fragments are, from left to right: N CH=north China, S=Siberia, K=Kazakhstania, B=Baltica, L=Laurentia, Af=Africa, SA=South America, AN=Antarctica, Aust=Australia, IND=India, S CH=south China, CI=Cimmeria. Abbreviations of oceans are: AT=Atlantic (north, central, south), SEI=southeast Indian, NT=Neotethys, PT=Paleotethys. Numbered stages in column VIII are discussed in text (Veevers, 1990)

downwelling during Pangea assembly. This mechanism is discussed in greater detail in Chap. 9.

Worldwide transgressions during the Jurassic and Cretaceous probably reflect the progressive splitting of Pangea (Figs. 9.5 and Fig. 9.6). North America and Africa rifted apart in the mid-Jurassic, South America and Africa in the Early Cretaceous; North America and Britain split in the mid-Cretaceous, as did Africa and Antarctica; India and Madagascar rifted apart in the Late Cretaceous, and the North Atlantic split extended northward between Greenland and Scandinavia in the early Paleocene (summary and data sources in Bally and Snelson, 1980; Uchupi and Emery, 1991). Larson (1991) stated that, "during the mid-Cretaceous, starting in earliest Aptian time (124 Ma), there were eruptions from an extraordinary upwelling of heat and deep-mantle material in the form of one or several very large plumes." Mantle processes are discussed in Chap. 9.

Veevers (1990) subdivided the Pennsylvanian to Cenozoic history of Pangea into five stages (Fig. 5.3). The assembly of Pangea was completed by Permian time, including the Ouachita orogeny, which sutured Gondwana to Laurasia. Stage 1 (Late Carboniferous, 320–290 Ma, the platform stage) corresponds to a widespread stratigraphic gap on the continents, caused by the thermal uplift accompanying continental assembly. Stage 2 (Permian-mid Triassic, 290–230 Ma, the sag stage) is represented by the early Gondwana stratigraphic sequences, and marks the local thinning of the Pangea lithosphere during its initial stretching, with the formation of broad basins or sags. Stage 3 (Late Triassic-Late Jurassic, 230–160 Ma, the rifting stage), is represented by ocean-margin rift successions marking the initial break-up of Pangea. Stage 4 (Late Jurassic-Late Cretaceous, the spreading stage), marks the time of maximum rate of continental dispersal, with high global sea levels and corresponding widespread shelf sedimentation, and the development of thick continental-margin successions on the borders of Atlantic-type oceans. Stage 5 (Late Cretaceous-present, 85-0 Ma), a phase during which the rate of sea-floor spreading and subduction slowed, corresponds to the end of the dispersion phase of Worsley et al. (1986), and the beginning of the continental assembly phase, with major collisions occurring along the Alpine-Himalayan belt.

Applying this set of stages to the Phanerozoic Earth, we can demonstrate two main periods of high sea level during the ocean-opening stage, the mid-Paleozoic, from the Ordovician to the Devonian (the dispersal of *Rodinia*), and again during the Cretaceous (the dispersal of *Pangea*) (Fig. 4.8). During both these periods, much of the cratonic interior of Europe, Asia and other major continents was covered by vast shelf seas, whereas during the intervening *Pangea* phase (late Paleozoic and Triassic) sea levels were at an all-time low. This is reflected in the broad sequence stratigraphy of the continental interiors, which, in North America, is characterized by onlapping, mainly marine shelf deposits of Ordovician to Devonian age and of Cretaceous age, separated by a major regional unconformity representing the Permian to mid-Jurassic interval. Only at the margins of the craton are deposits of this age range present (Fig. 5.4).

There is currently considerable discussion regarding the nature of possible earlier (Precambrian) supercontinent cycles. This is beyond the scope of the present book (see Rogers and Santosh, 2004). A speculative reconstruction of Rodinia was offered by Hoffman (1991), who suggested that at around 700 Ma Laurentia was situated at the centre of a supercontinent that subsequently dispersed and "turned inside-out" to form Gondwana around the end of the Precambrian, and subsequently Pangea in the late-Paleozoic. Other work on this problem is referenced in Sect. 4.2 and Chap. 9.

## 5.3 Cycles with Episodicities of Tens of Millions of years

### 5.3.1 Regional to Intercontinental Correlations

Sloss (1963, 1972) established the six Indian-name sequences in North America (Fig. 1.12; Fig. 5.5), and demonstrated their correlation with similar sequences on the Russian platform (Fig. 4.9). Vail et al. (1977) referred to these as *second-order cycles*, or *supercycles*. They are now commonly called "Sloss sequences." This section briefly reviews other work on this type of stratigraphic sequence.

Even within North America, the recognition of the Sloss sequence record is commonly not at all clear. Figure 5.6 provides a compilation table of Paleozoic stratigraphy of the ancient western continental margin of north America, and Fig. 5.7 is a stratigraphic cross-section through the margin of the Franklin Basin in Arctic Canada. In both cases, the chronostratigraphic positions of Sloss sequence boundaries are indicated. The Sauk I, II and III boundaries and the Kaskaskia I–II boundaries appear to correspond to stratigraphic contacts in Nevada, Alberta and British Columbia, but the other Paleozoic sequence boundaries do not appear to correlate to stratigraphic contacts in these areas, and in other parts of western North America there is little correspondence to the Paleozoic Sloss sequence record. In Arctic Canada the record is even more clearly one of control by local tectonism (Fig. 5.7). Syntheses of Arctic tectonism and sedimentation (e.g., Trettin, 1991) indicate the extent and importance of local tectonic episodes, particularly

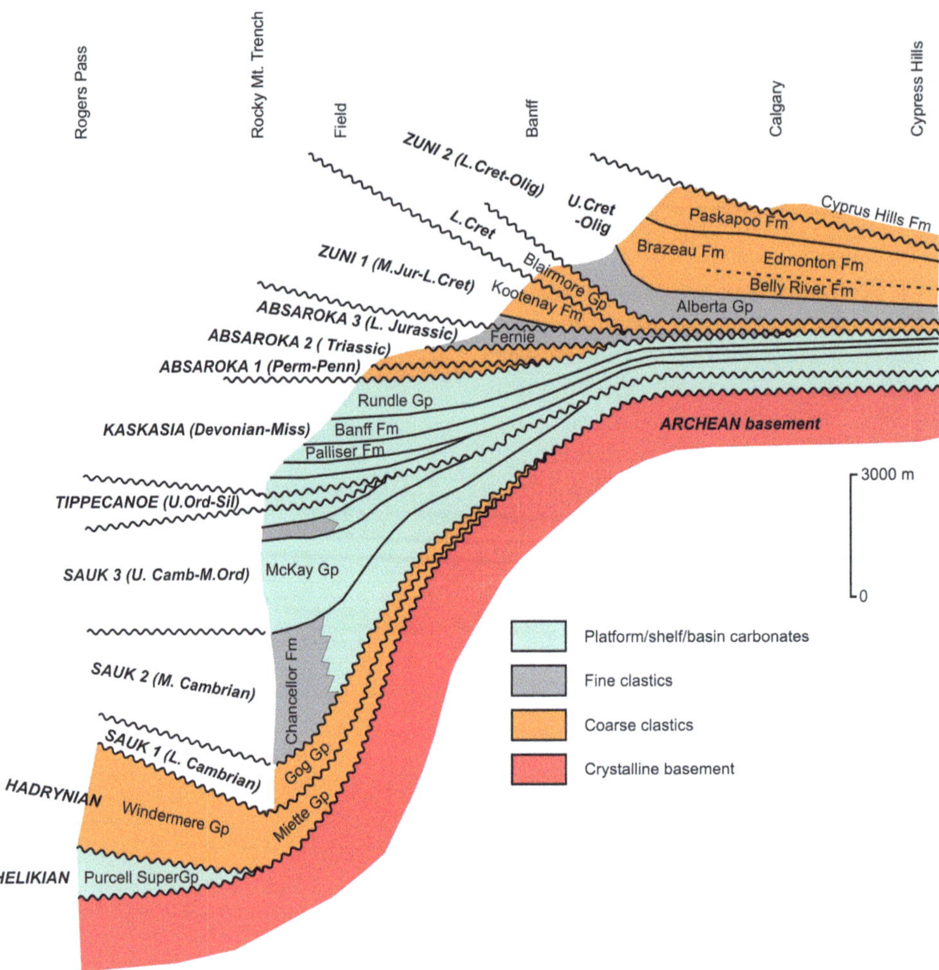


**Fig. 5.4** Generalized stratigraphic cross-section through the Western Canada Sedimentary Basin, showing the lithostratigraphy and Sloss-sequences. Adapted from Price et al. (1972)

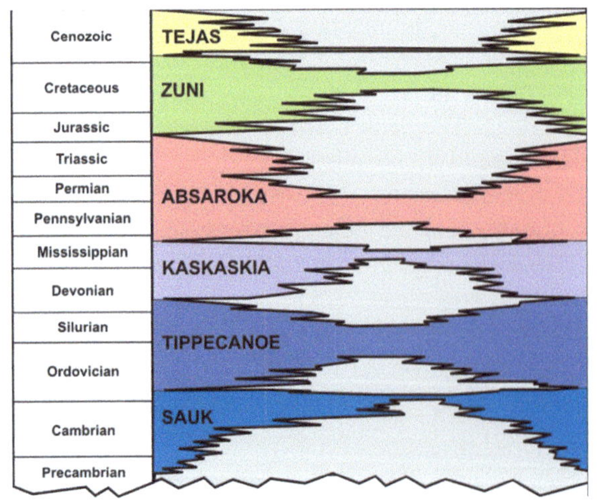


**Fig. 5.5** The "Sloss sequences". Redrawn from Sloss (1963)

in the Cornwallis-Prince of Wales Islands area, which was affected by a crustal ramp dislocation in the Late Silurian-Early Devonian (a far-field effect of the Caledonian collision of Greenland and Scandinavia, see Miall and Blakey, 2008).

Stratigraphic relationships and isopach patterns of the Sloss sequences indicate the influence of epeirogenic uplift, subsidence, tilting and warping of the craton, throughout the Phanerozoic. In the original definition of his six sequences, Sloss (1963) provided several cross-sections illustrating the angular unconformity between successive sequences. One of these has been redrawn and is reproduced here as Fig. 5.8. This clear evidence of tectonism contained within the original definition of the sequences is important, because it was largely forgotten and set aside in the enthusiasm for the model of global eustasy during the 1980s and 1990s (Sloss, 1991).

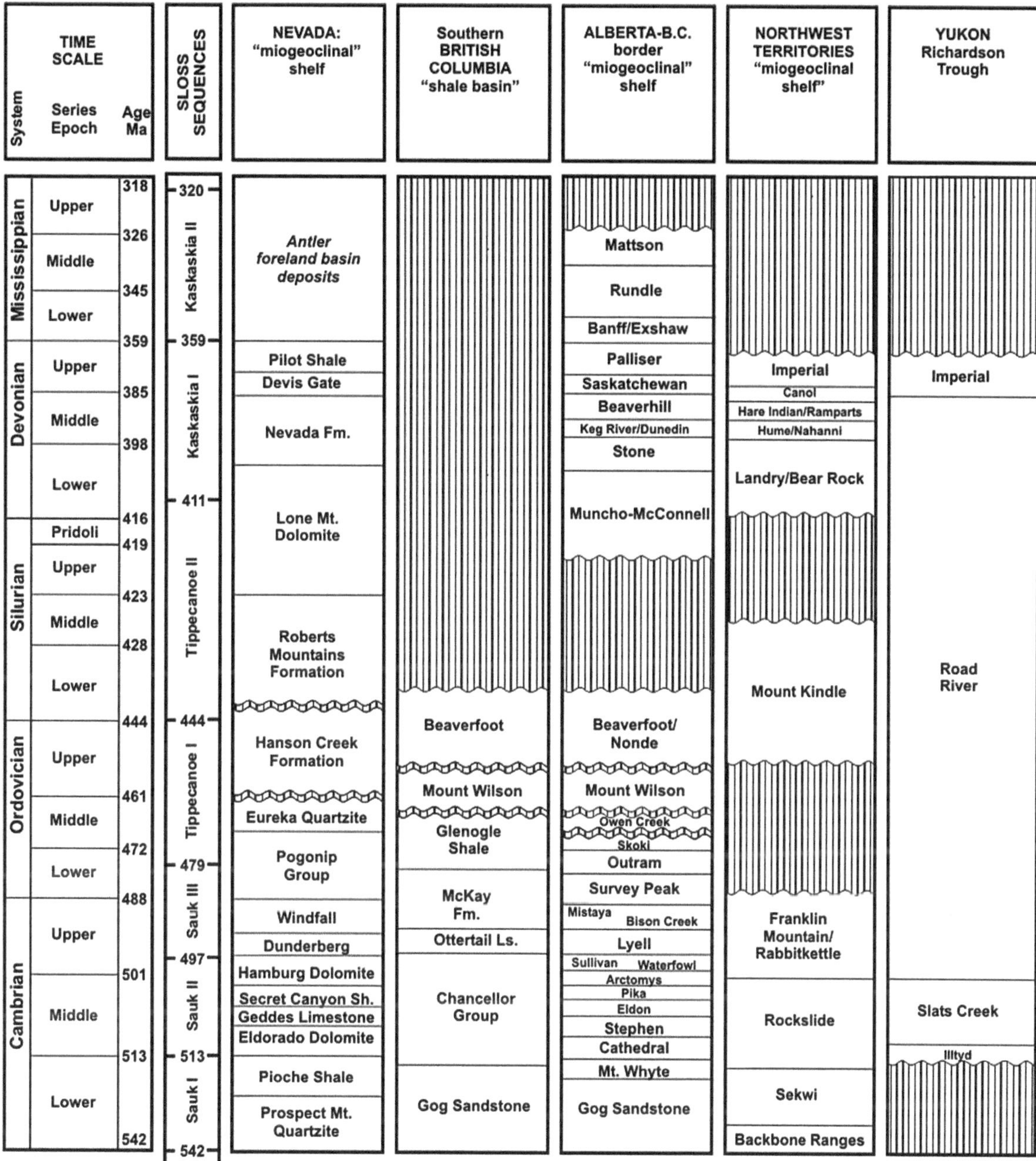


**Fig. 5.6** Correlation table of the stratigraphy of the western Paleozoic continental margin of North America (Laurentia), showing selected stratigraphic names. Sources: Time scale from www.stratigraphy.org; Sloss Sequences from Sloss (1988b). Regional columns complied by Miall (2008)

In a major updating of the sequence concept as applied to North American geology, prepared for the decade of North American Geology Project of the Geological Society of America, Sloss (1988b) generated a series of maps showing variations in subsidence rates represented by each of the sequences across the continental United States. Two of his maps are reproduced here. Figure 5.9 shows the subsidence pattern for the Sauk Sequence (Middle Cambrian-Lower Ordovician). Figure 5.10 illustrates the subsidence pattern for the Absaroka-I sequence (Pennsylvanian-Lower Permian). The Sauk map indicates significant

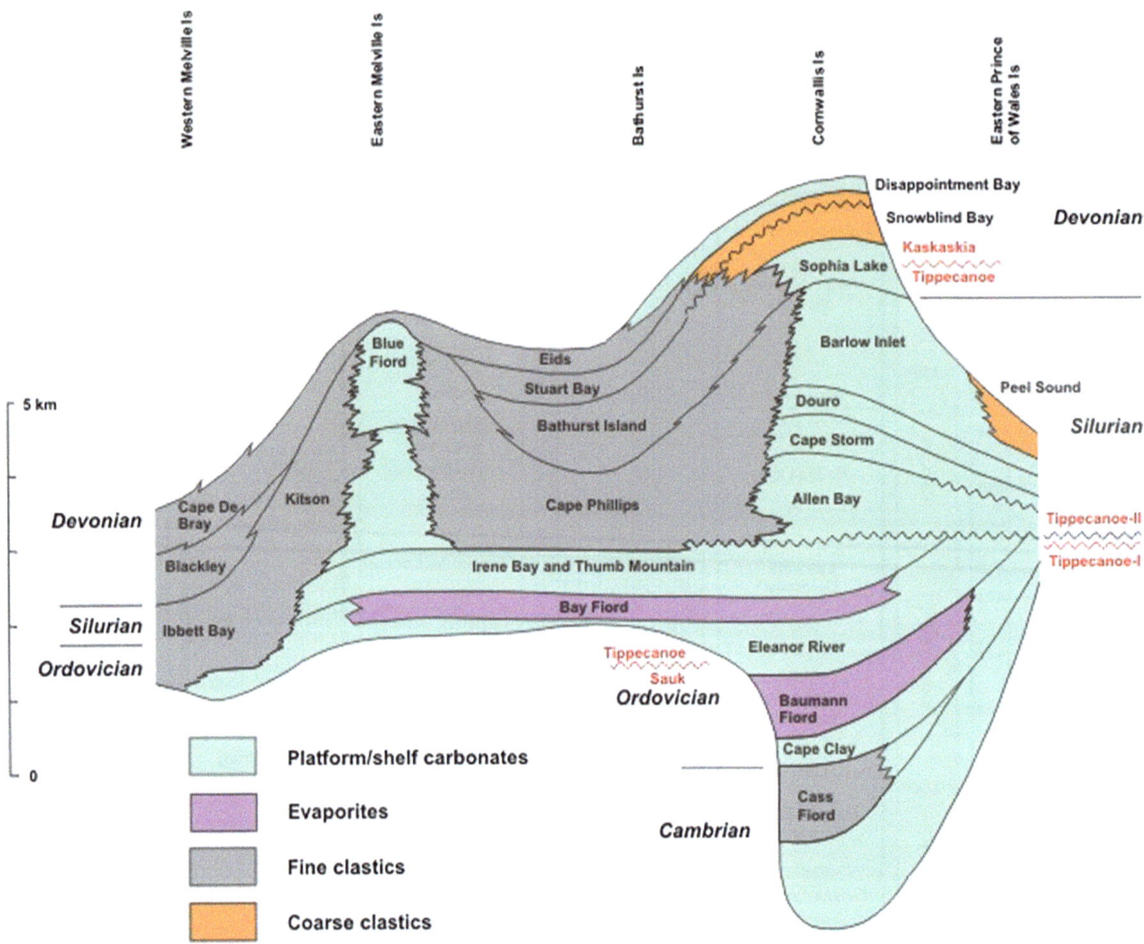



**Fig. 5.7** Stratigraphic cross-section across the southern margin of the Franklin Basin, Arctic Canada. Adapted from Trettin (1991)

rates of subsidence on the eastern and western continental margins, with a central area (labeled with the number 9 in Fig. 5.9) indicated as an area of no net subsidence. The lack of Sauk deposits across this central part of the continent could be attributed to erosion, except that many of the maps of Lower Paleozoic stratigraphy show a similar pattern. This belt, extending northeast-southwest across the cratonic interior, has been called the Transcontinental Arch, a term that appears to have originated with Eardley (1951). It may never have functioned as a single, coherent, transcontinental structure, but isopach patterns confirm that it defines an area that tended to be structurally positive throughout the early Paleozoic.

During the mid- to late Paleozoic, regional tectonic trends changed, and the Absaroka-I map shown in Fig. 5.10 reveals a quite different pattern of regional epeirogeny, with rapid subsidence occurring along the southern margin of the continent, together with the clear influence of differential uplift and subsidence associated with the development of the Ancestral Rocky Mountains (Blakey, 2008). The reasons for these differences are discussed in Sect. 9.3.2.

Turning briefly to the Mesozoic-Cenozoic, Fig. 5.11 provides a table of regional correlations of the Jurassic to Paleogene stratigraphy of the Western Interior Seaway of North America, upon which the Sloss sequence boundaries have been superimposed. It is important to establish the broad patterns of distribution of these sequences, as a basis for the discussion later in this book (Chap. 9.) concerning our modern understanding of the processes that generated the Sloss sequences. The Zuni-I–II sequence boundary corresponds, in some areas, to a very significant

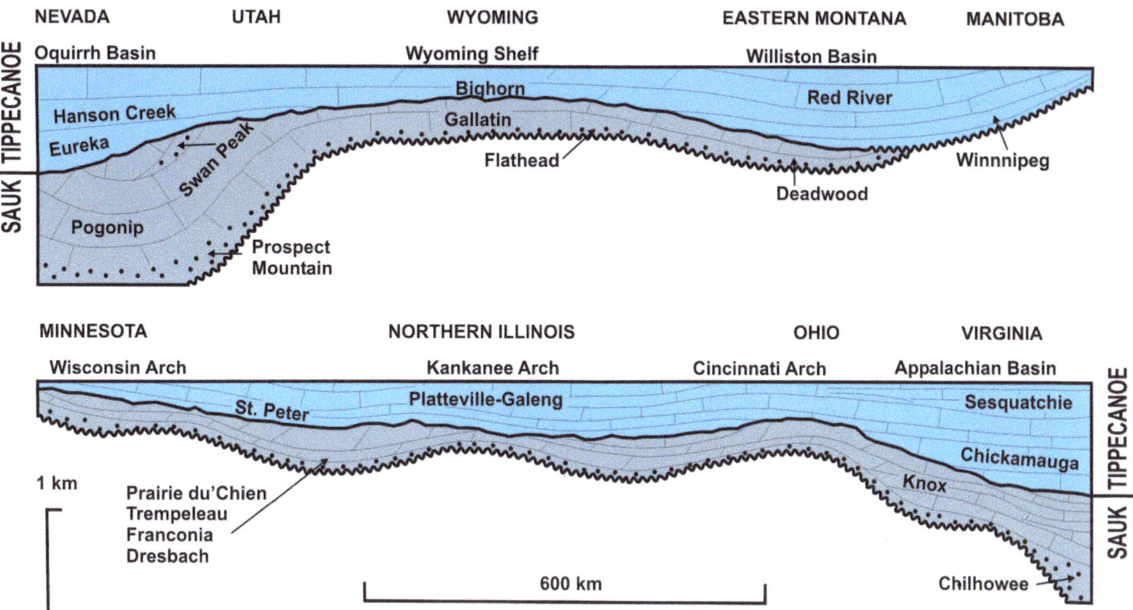


**Fig. 5.8** Stratigraphic reconstructions of the Sauk and Tippecanoe sequences through the cratonic interior of North America. Redrawn from Sloss (1963)

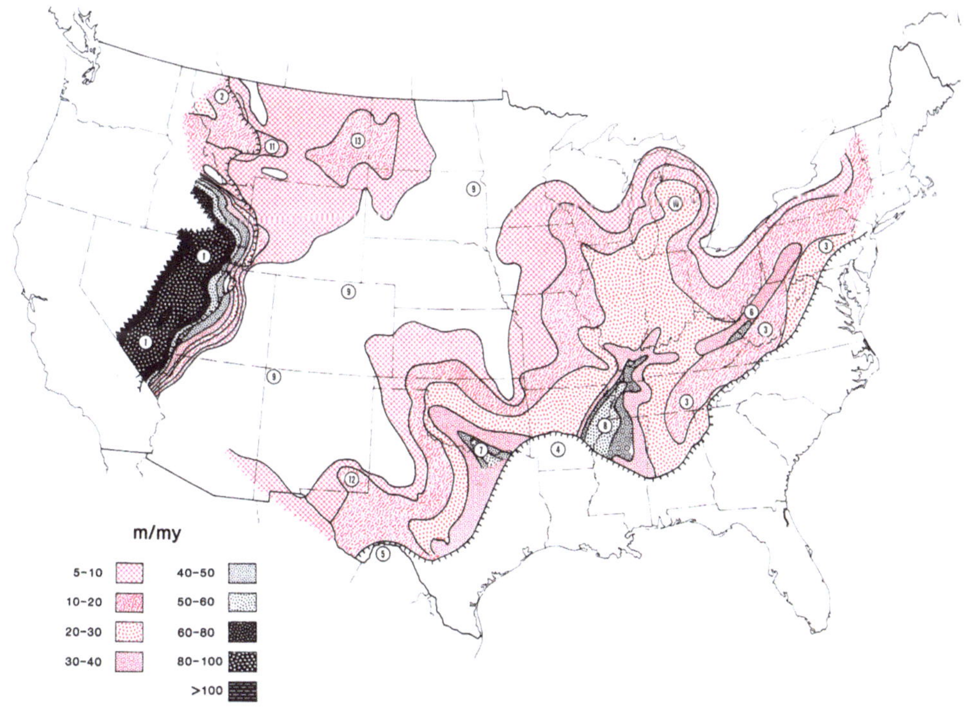


**Fig. 5.9** Subsidence pattern for the Sauk Sequence (Middle Cambrian-Lower Ordovician) (Sloss, 1988b). the Transcontinental Arch, indicated by the numeral "9", was a persistent feature of cratonic architecture through much of the early Paleozoic

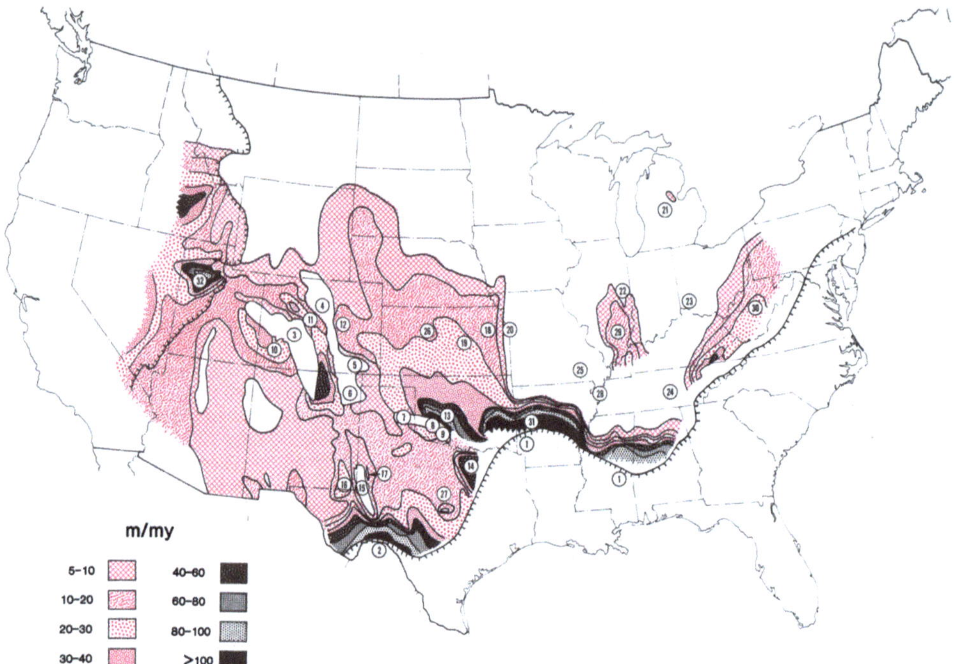



**Fig. 5.10** Subsidence pattern for the Absaroka-I sequence (Pennsylvanian-Lower Permian) (Sloss, 1988b). Note the absence of the Transcontinental Arch, and the appearance, instead, of a complex pattern of basins and highs in the south-western part of the United States

regional unconformity. However, it has been argued that this unconformity is tectonic in origin, related to patterns of orogenic activity along the Cordilleran Mountains. Beaumont et al. (1993) attributed the development of this important regional unconformity to "post-orogenic exhumation", following an episode of terrane collision along the active western continental margin (see also Sect. 5.3.2, below). The Zuni II–III sequence boundary is not clearly recognizable from this regional stratigraphic synthesis. The end of the Zuni sequence appears to have occurred during a period of active clastic-wedge progradation from the Rocky Mountains—the North Horn, Paskapoo and Ravenscrag formations are all nonmarine deposits derived from uplift and erosion of the Rocky Mountains (Miall et al. 2008).

A continuing problem of sequence interpretation is to disentangle the local from the regional. Fig. 5.12 illustrates a stratigraphic synthesis of the so-called "Grand Cycles" of the southern Canadian Rocky Mountains (see Chap. 6. for further discussion of these), within which two Sloss sequence boundaries occur. The lowermost of these, the Sauk I–II boundary,

clearly coincides with a major regional stratigraphic gap corresponding to at least one major biozone missing in the Middle Cambrian. The Sauk II–III boundary, while it appears to correlate with one of the Grand Cycle boundaries, does not appear to constitute a major sedimentary break in this area. As discussed in Sect. 14.3, it is commonly very difficult to unravel the chronostratigraphic significance of unconformities. The major break at the Sauk I–II unconformity may indicate that two separate cycles of sea-level change, of different frequency, were in phase during the Middle Cambrian, whereas the minor Sauk II–III boundary suggests that no such process was operating at that time, during the early Late Cambrian.

Soares et al. (1978) reported a stratigraphic analysis of the three major intracratonic basins in Brazil, namely, the Amazon, Parnaiba, and Parana basins, all of which contain successions spanning most of the Phanerozoic. Their interpretation of the geomorphic behavior of these basins is given in Fig. 5.13. They recognized seven sequences, which correlate reasonably closely with those of Sloss, as shown in Fig. 5.14.

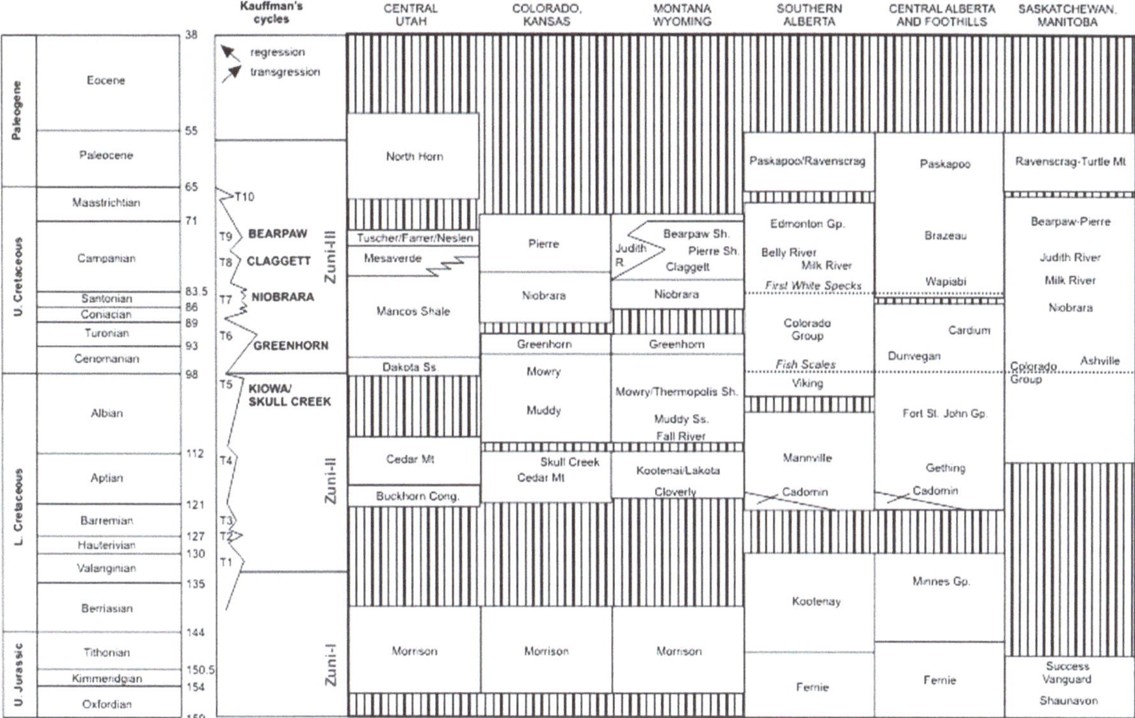

Fig. 5.11 Generalized stratigraphic table for the Western Interior Basin, showing only the most well-known of the numerous lithostratigraphic names and the most widespread of the regional unconformities. Derived from numerous sources. Kauffman's cycle nomenclature is from Kauffman (1984), Kauffman and Caldwell (1993). Sloss sequence boundaries added from Sloss (1988b)

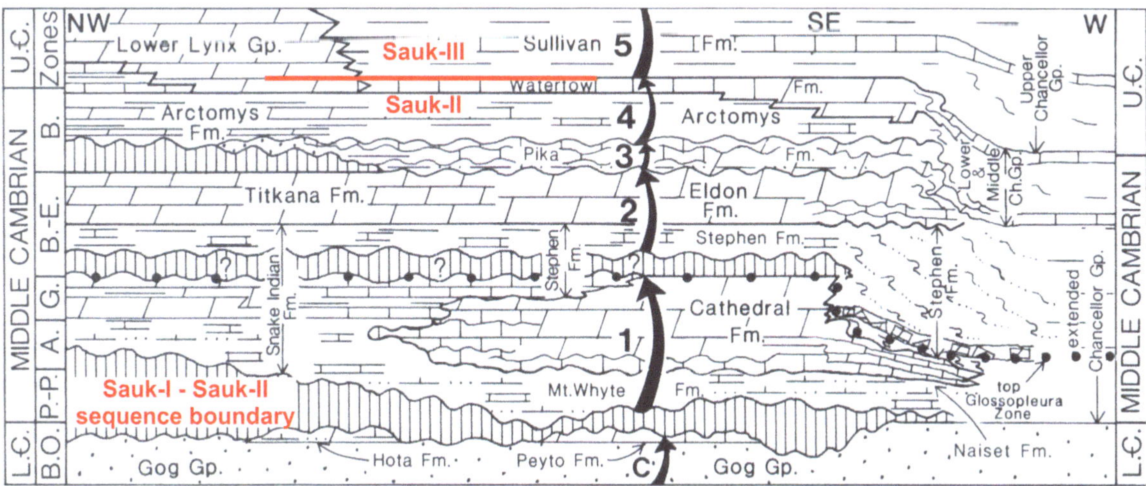


Fig. 5.12 Schematic stratigraphic cross-section of the Cambrian rocks of the western continental margin of Canada, extending from Jasper National Park southeastward to Banff National Park, and then westward across the Kicking Horse Rim. Six of the Grand Cycles of Aitken (1966, 1978) are highlighted, and the Sloss sequence boundaries have been added. Adapted from Fritz et al. (1992)

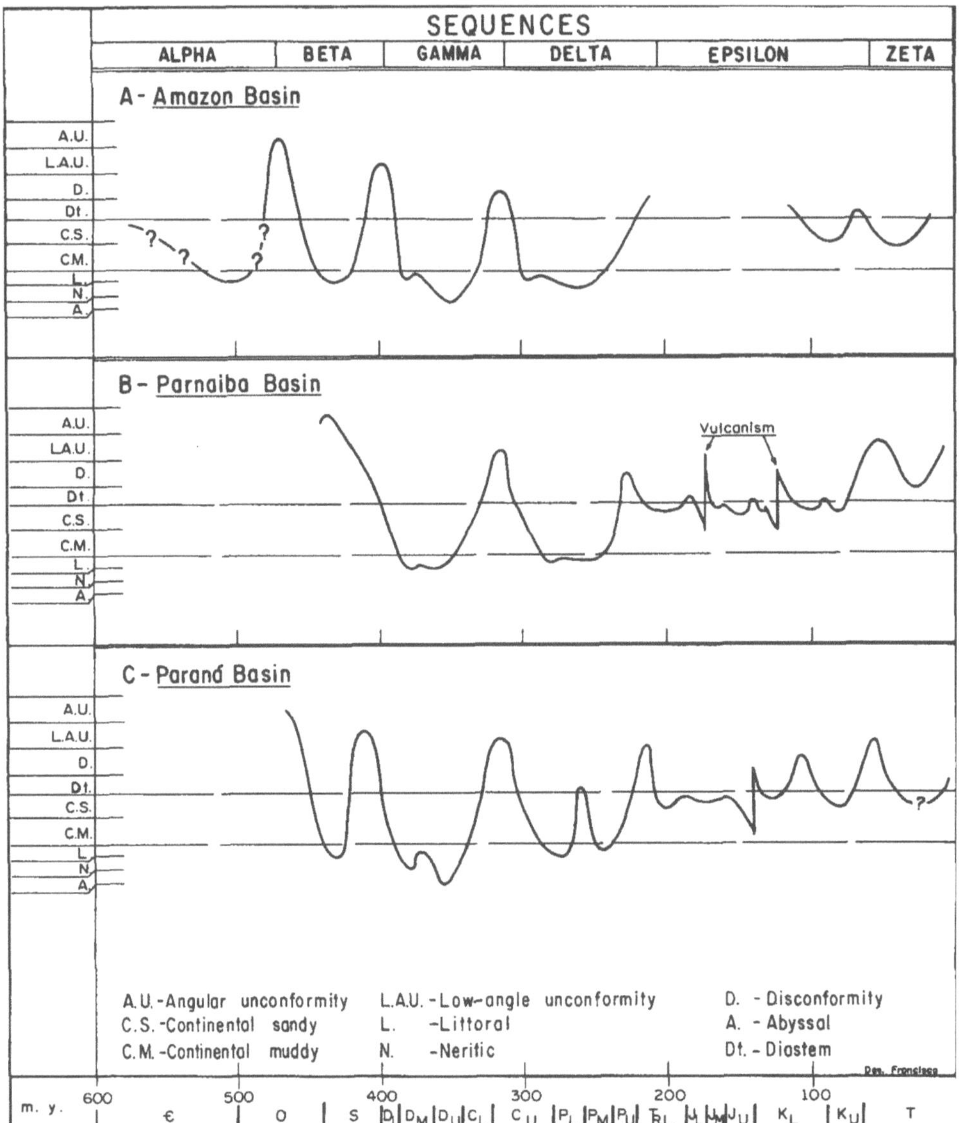


**Fig. 5.13** Geomorphic expression of oscillatory movements in three Brazilian basins (Soares et al., 1978)

Soares et al. (1978) described an epeirogenic cycle consisting of five phases, which explained stratigraphic events in each of the Brazilian sequences. Note that a tectonic control is indicated for this cycle.

1. Initial rapid basin subsidence with development of nonmarine facies and numerous local unconformities.
2. Basin subsidence slower, with deepening basin centres, marine transgression and differentiation of central marine and marginal nonmarine facies belts.
3. Development of intrabasin uplifts and local downwarps, much local facies variability.
4. Renewed basinwide subsidence, time of maximum transgression, generally fine-grained deposits.
5. Broad cratonic uplift, return to nonmarine sedimentation.

In summary, the detailed stratigraphic relationships revealed within North America, and the nature of the regional correlations of the Sloss sequences

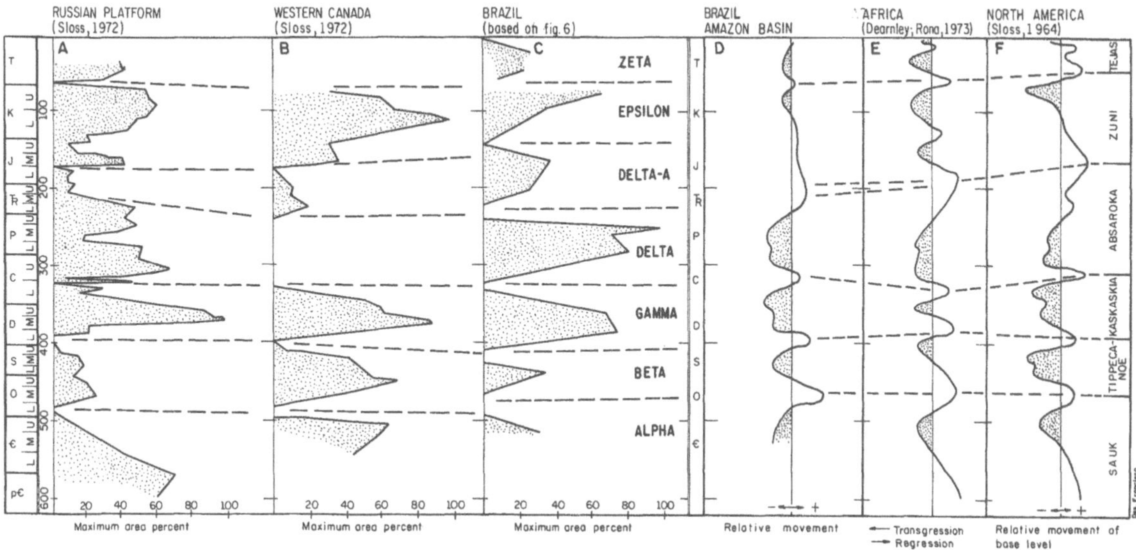


**Fig. 5.14**  Correlation of sequences of oscillatory movement in North American, Brazilian, European, and African cratons. *A*, *B*, and *C* are based on preserved sediments, and *D*, *E*, and *F*, are based on relative base-level movements (Soares et al., 1978)

to South America and the Russian Platform, suggest that while there appears to be an overriding process, probably eustatic sea-level change, that has caused similar patterns of sequences to develop in at least three continents, the details of the sequences show that there are significant local variations in preservation and in the ages of the sequence boundaries. Epeirogenic deformation was also clearly a factor in the generation of these sequences, as is discussed in Chap. 9. Plate-margin effects related to orogenic episodes are discussed in the next section.

Recent stratigraphic work has provided important new insights into the anatomy of cratonic sequences. Runkel et al. (2007, 2008) examined the Sauk Sequence in Wisconsin. The sequence is a little over 100 m thick in this area, and developed on a ramp dipping gently off an intracratonic upwarp. Very detailed biostratigraphic work using trilobites, brachiopods and conodonts permits a subdivision of the succession into some 31 zones, averaging about 0.6 million years in duration (using stage-boundary ages posted at www.stratigraphy.org), and the stratigraphic architecture this has revealed has yielded some fascinating insights. The sequence consists of four subsequences (Fig. 5.15). Each one developed by clinoform progradation of coastal and nearshore deposits, which extended the coastline as much as 200 km into

the slightly deeper basinal area. Water depths are estimated to have been less than 100 m in the basin, and depositional slopes 0.1 m/km, or less. Subsidence rates were <0.01 m/ka. The clinoform configuration helps to explain how water depths were adequate to maintain the fully marine, open circulation conditions that are indicated by facies studies of these and other Paleozoic cratonic units.

## 5.3.2 *Tectonostratigraphic Sequences*

In many basins, tectonic influences are indicated by the presence of angular unconformities, faults that terminate at sequence boundaries, and changes in isopach patterns and in detrital petrology, indicating changes in regional sediment transport directions. Such lines of observational evidence have long constituted a set of criteria by which stratigraphic successions have been subdivided. For example, many of Levorsen's (1943) "layers of geology" consist of packages of strata (what we now call sequences) separated by angular unconformities (e.g., Fig. 1.9). These criteria were usefully summarized by Embry (1997) as a guide to distinguishing sequence boundaries generated by tectonism from those caused by eustatic sea-level changes, and may also be employed to distinguish tectonic influence

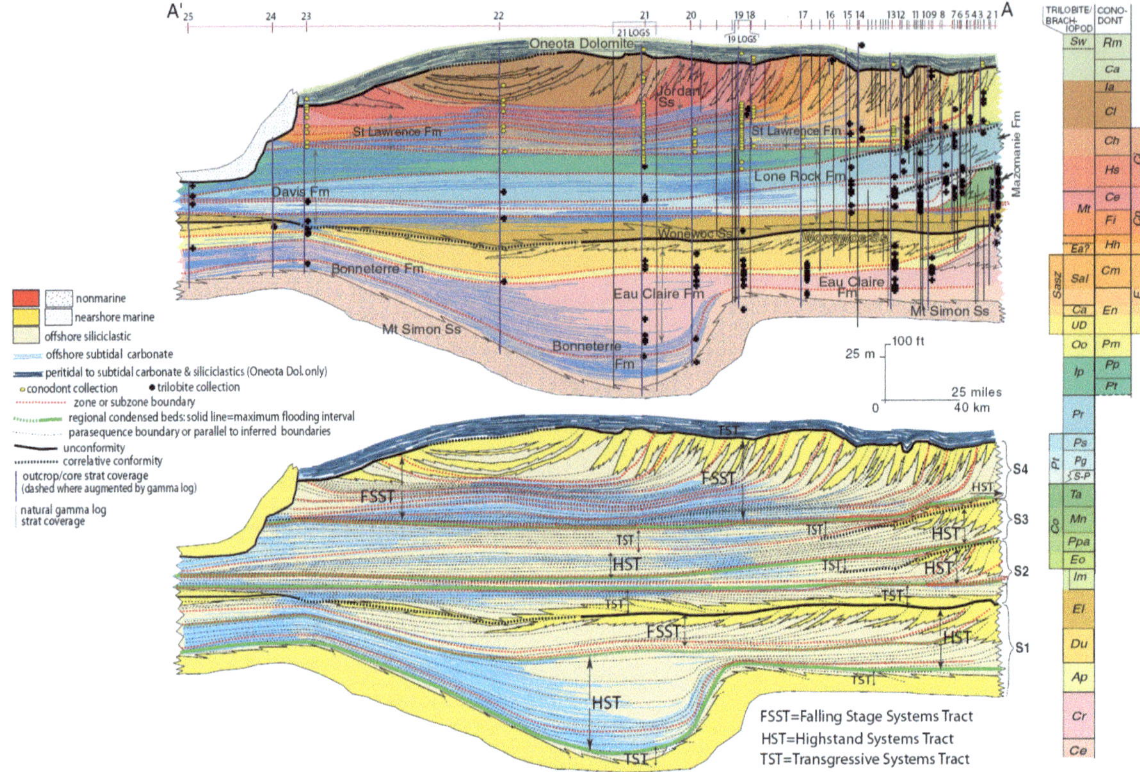

**Fig. 5.15** Transect from intracratonic uplift (*right*) to basin (*left*), Wisconsin, showing the detailed stratigraphy of the Sauk Sequence. *Upper* diagram shows the biostratigraphy according to the key at *right*; *lower panel* shows the same transect ornamented according to sedimentary environments (legend at *left*) (Runkel et al., 2007, Fig. 6)

from orbital forcing as a primary driving mechanism in high-frequency sequence generation. We discuss this further in Chap. 10.

Major structural features, such as unconformities, and other patterns of internal structural deformation, are commonly used as a means to subdivide regional stratigraphies into tectonic sequences, sometimes called *tectonostratigraphic sequences*, that commonly span millions to tens of millions of years. They typically reflect major steps in the plate-tectonic evolution of the area, such as the transition from rifting to thermal subsidence on extensional continental margins—the so-called *breakup unconformity*, and the migrating regional erosion surface associated with forebulge erosion in front of a developing fold-thrust belt. Regional reflection-seismic data are particularly useful in establishing such broad internal subdivisions.

The margins of the North Atlantic Ocean provide a good example of the development of tectonos-

tratigraphic sequences. On the Grand Banks, east of Newfoundland (Figs. 5.16 and 5.17), there were five broad phases of basin subsidence (Tankard and Welsink, 1987; Welsink et al., 1989): (1) The first phase, of Late Triassic-early Jurassic age, lasted some 20–30 million years, and resulted in the deposition of a redbed complex capped by evaporites. (2) Movement on major northwest-southeast faults, including the Newfoundland fracture zone and the parallel transfer faults on the Grand Banks initiated the fragmentation of the shelf platform. The Murre Fault, and others of this suite, which were initiated at this time, came to define the major structure of the Jeanne'd'Arc basin (Fig. 5.16). (3) A phase of slow thermal subsidence during the Early and Middle Jurassic led to the deposition of a monotonous mudstone-carbonate succession (epeiric basin phase in Fig. 5.17). (4) Africa began to separate from Nova Scotia in the Middle Jurassic, at about 175 Ma, and this led to about 40 km of continent-continent displacement along the

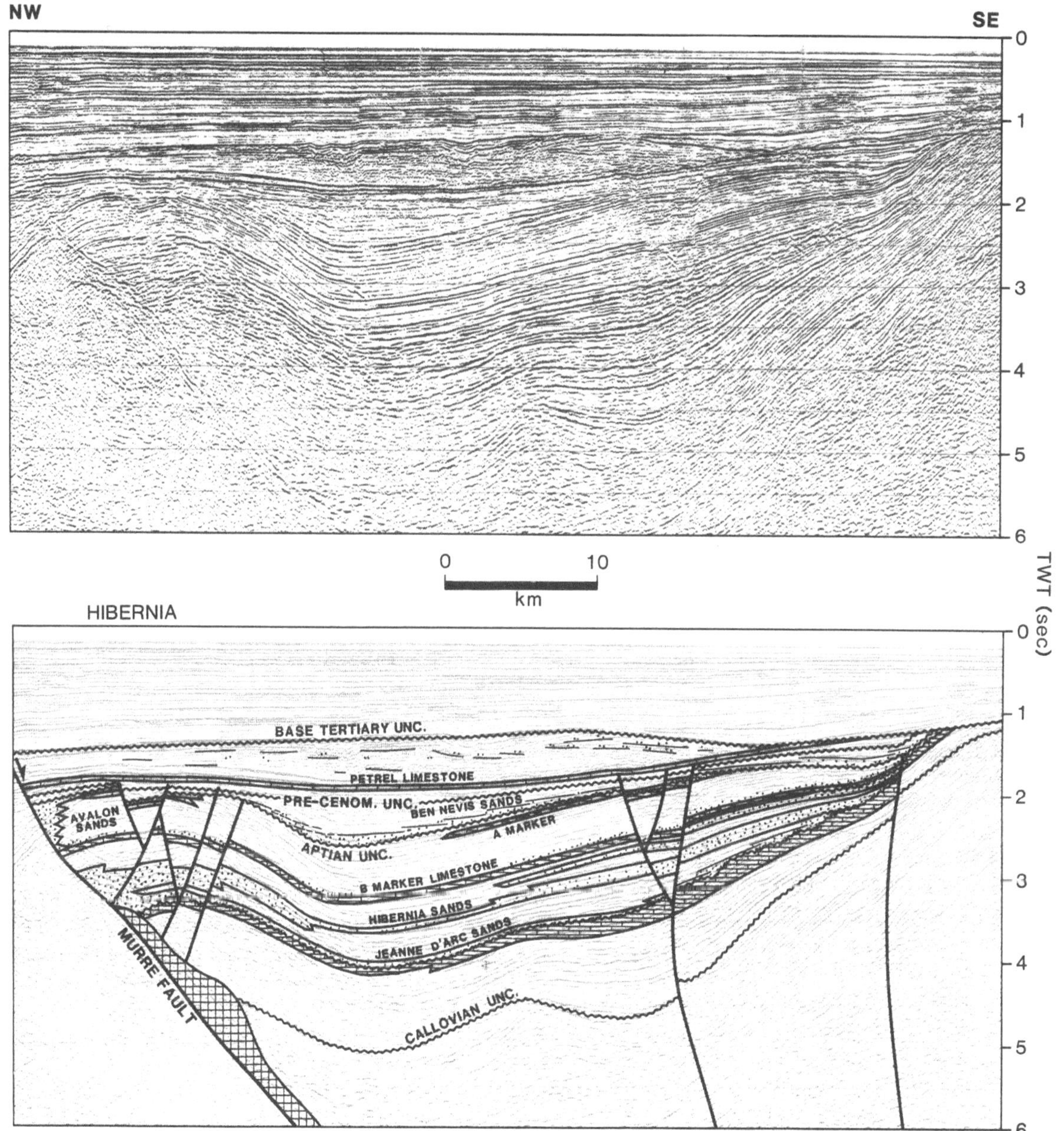


**Fig. 5.16** Seismic line oriented northwest-southeast through the Hibernia field, Jeanne d'Arc Basin, off Newfoundland. The field is a roll-over trap adjacent to the Murre Fault, at *left*. Salt that has risen along the Murre fault is shown by the pattern of squares (Tankard et al., 1989). AAPG © 1989. Reprinted by permission of the AAPG whose permission is required for further use

Newfoundland fracture zone and the transfer faults until Valanginian time. Synrift deposits developed in the Jeanne d'Arc and other basins. This phase of movement came to end with the separation of Iberia at about 115 Ma. (5) During the post-rift phase, which followed, oceanic crust began to develop off the northeast margin of the Grand Banks, and crustal extension led to the development of a suite of crossing faults oriented northeast-southwest. This led into the post-rift thermal subsidence phase of Fig. 5.17.

The Grand Banks is but one of several large areas of marginal continental crust that underwent several

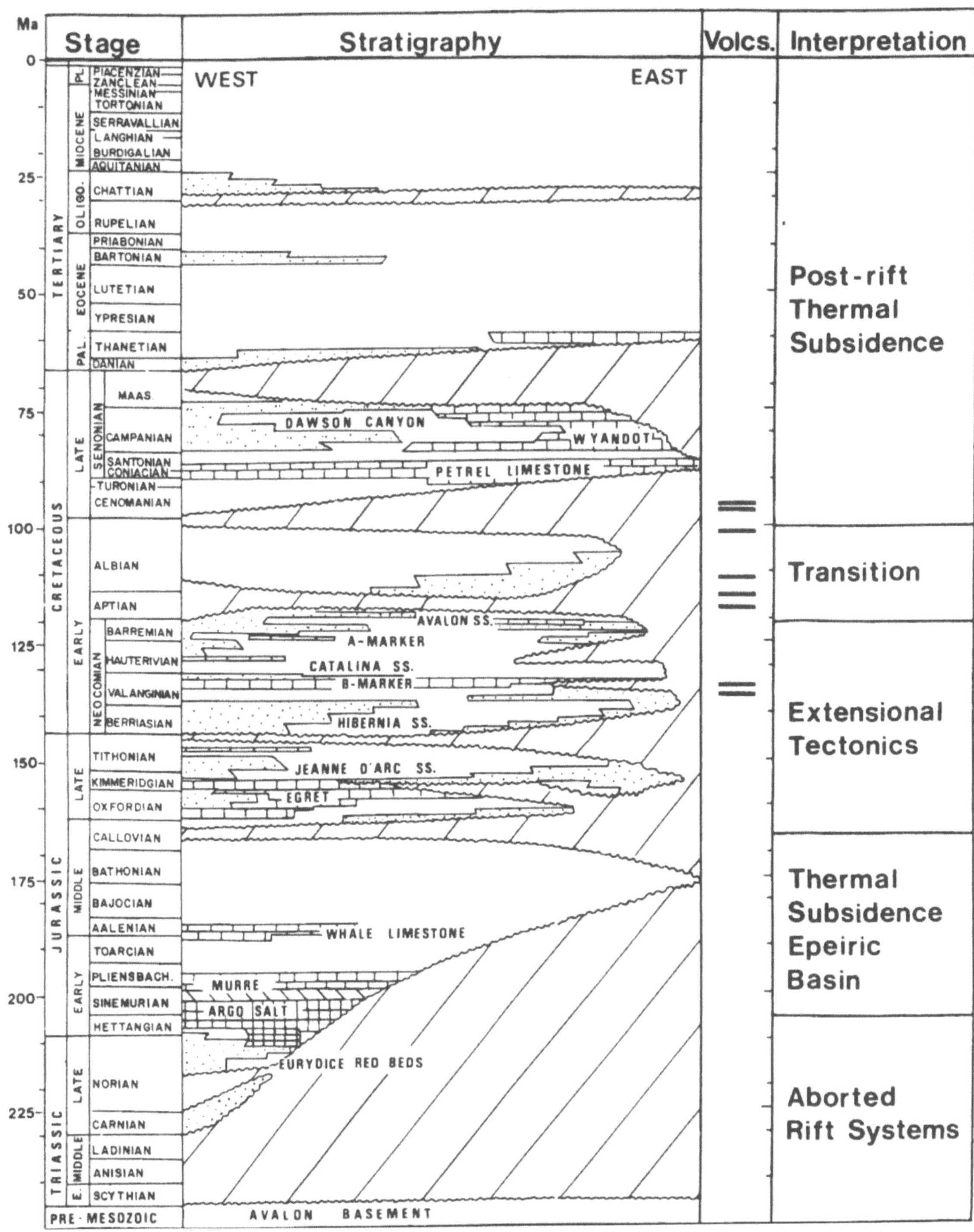


**Fig. 5.17** Tectonostratigraphic column for the Jeanne d'Arc Basin, Grand Banks, Newfoundland, showing major episodes in basin evolution, and prominent unconformities (Welsink and Tankard, 1988). AAPG © 1988. Reprinted by permission of the AAPG whose permission is required for further use

phases of rifting, followed by thermal subsidence, as the Atlantic Ocean formed and grew in size. Many comparative studies of this complex series of events have been carried out. For example, Sinclair et al. (1994) compared the structural and stratigraphic histories of the Jeanne d'Arc basin, the Porcupine Basin, southwest of Ireland, and the Moray Firth Basin, off northeast Scotland (Fig. 5.18). Differences

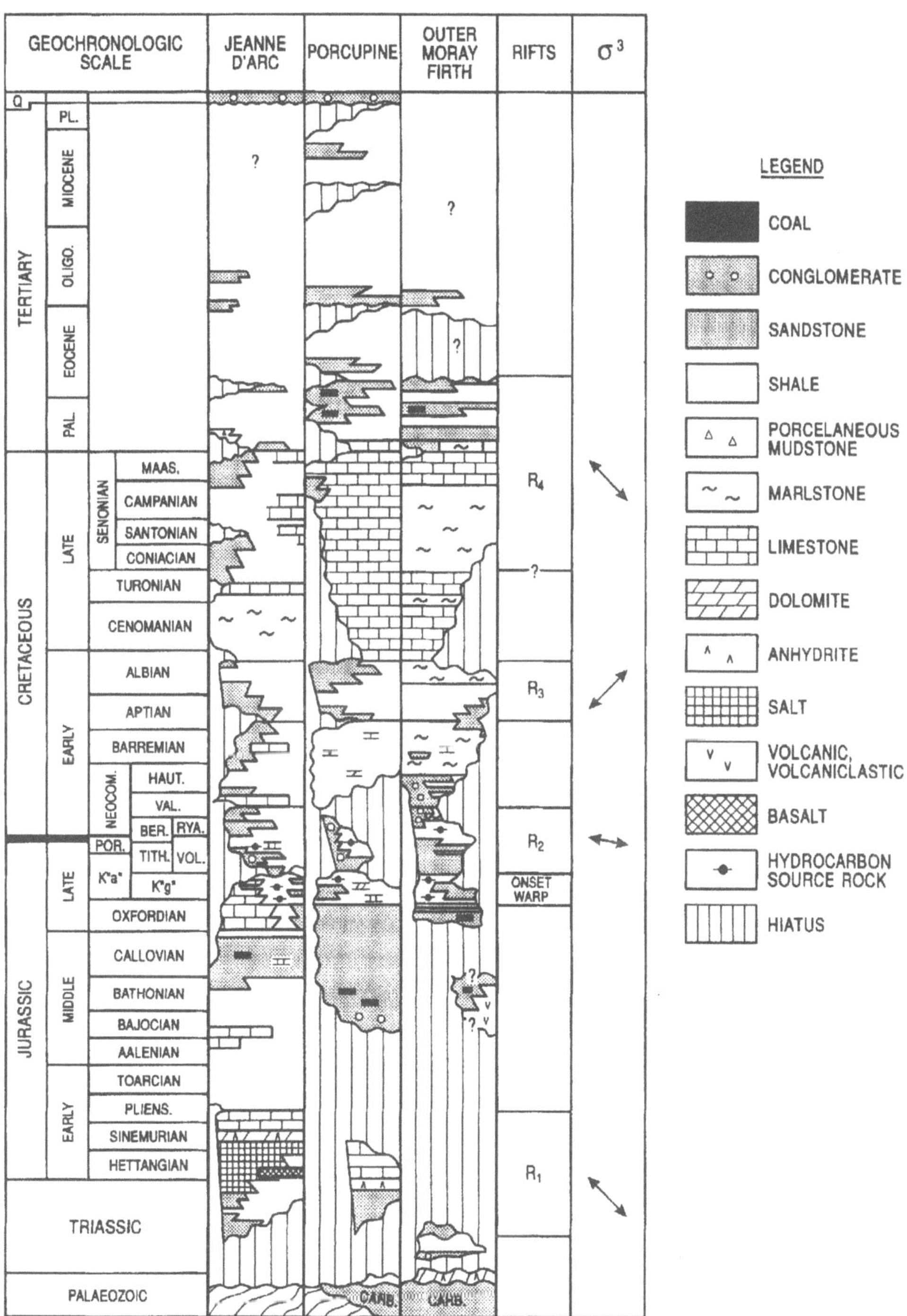


**Fig. 5.18**  Comparative stratigraphic columns for three Atlantic-margin sedimentary basins (Sinclair et al., 1994)

in the sequence histories between these areas relate to local modest differences in the structural response to tectonism, and variations in the local sediment supply. As discussed in Sect. 10.4, regional changes in intraplate stress patterns appear to have occurred simultaneously through the North Atlantic area as a result of adjustments in the spreading patterns (shifts in the poles of small-circle rotation of the adjoining plates), as evidenced by what appear to be contemporaneous deformational episodes of different style in different locations.

Two other examples of regional seismic-stratigraphic analyses of extensional continental margins are illustrated in Figs. 5.19 and 5.20. Examples of unconformable sequence boundaries that are clearly related to regional tectonism are particularly clearly seen in the Brazilian example (Fig. 5.20). In this continental-margin section off northern Brazil, Petrobras (1988) identified three sequences. The first, of Aptian age (Sequence I), comprises sediment deposited during the rift phase of Atlantic opening. Component seismic unit 1 consists of fluvial and lacustrine sediments, and unit 2 consists of sediment deposited during the first marine transgression. Sequence II was deposited during the beginning of the thermal subsidence phase, with the unconformity at the sequence I/II boundary corresponding to the breakup unconformity. Unit 3, of Albian age, consists of onlapping restricted and shallow-marine

sediments. Unit 4, of Turonian-Santonian age, is a thick slope clastic succession showing deep-marine onlap. Sequence III was also deposited during the thermal subsidence phase. It is subdivided into six units spanning the Campanian to present. Units 5–7 contain prograding platform-edge carbonate deposits, and the remaining units are platform carbonates.

A chart comparing the stratigraphies on the same three extensional continental margins, the Beaufort Sea, Grand Banks, and Brazil (Santos Basin) is shown in Fig. 5.21. Based on the kind of structural architectures shown in Figs. 5.16, 5.17, 5.18, 5.19, 5.20 and 5.21, the sequence boundaries have been interpreted as the product of extensional faulting, salt movement, and other mechanisms. The "orogenic wedges" shown in the Beaufort Sea column refer to contractional tectonism that occurred along the southwest margin of the sea, in Yukon and northeast Alaska, whereas the sequences shown in Fig. 5.19 were generated by deltaic and continental-margin progradation across the south- to southeastern margin of the sea, which was characterized by extensional tectonism.

It may be noted that the sequence boundaries are almost all of different ages in the three basins. At the time this chart was published by Hubbard (1988), the global eustasy model was widely accepted as the basis for interpreting and correlating sequence stratigraphies worldwide, and this work, based on regional seismic analysis, was therefore quite controversial.

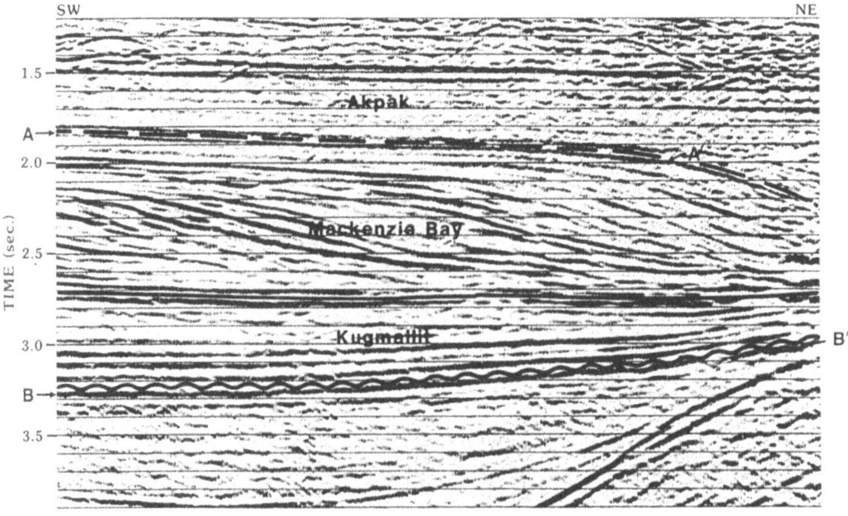



**Fig. 5.19** Seismic expression of sequence boundaries in Beaufort-Mackenzie Basin, with the names of the major sequences. From top to bottom these boundaries have been dated at 11, 25, and 36 Ma (Dixon and Dietrich, 1988, 1990)

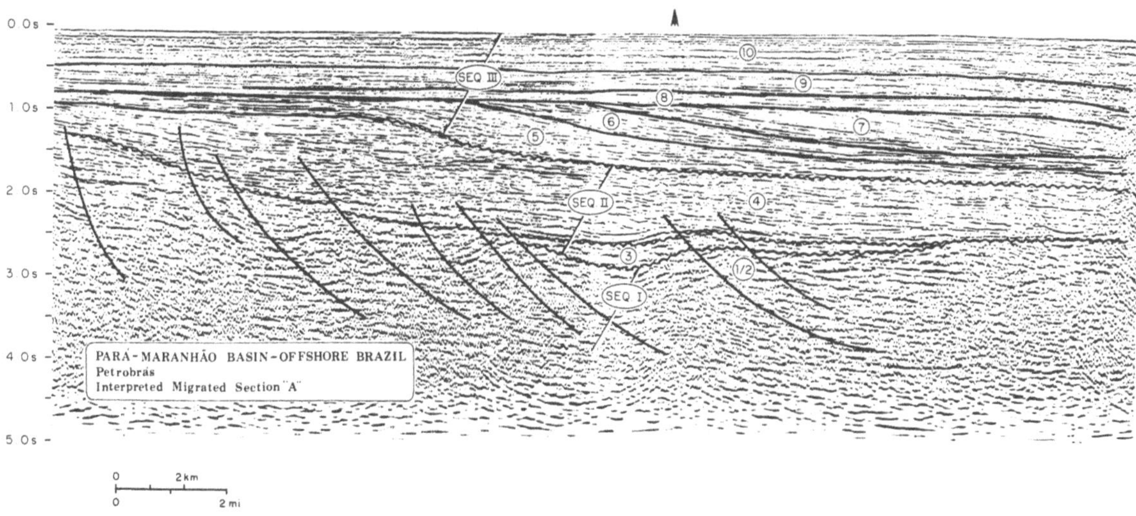


**Fig. 5.20** Interpreted seismic section from offshore northern Brazil, showing three sequences and their component seismic-facies units 1–10 (Petrobras Exploration Department, 1988). AAPG © 1988. Reprinted by permission of the AAPG whose permission is required for further use

Van Wagoner et al. (1990, p. 50), while accepting that the tectonic episodes identified in this analysis were important, nevertheless claimed that these were merely "tectonically enhanced" sequence boundaries that were fundamentally of eustatic origin. This subject is discussed at length in Chap. 12.

On convergent plate margins it is to be expected that tectonism will be the dominant control on the long-term development of basin architecture. For example, Seyfried et al. (1991) mapped a series of regional unconformities in Cretaceous-Cenozoic sections of Costa Rica that are clearly related to arc tectonism and volcanism (Fig. 5.22). The spacing of these unconformities indicates the occurrence of tectonic episodes tens of millions of years apart. Seyfried et al. (1991) stated that a cycle of compression, uplift, erosion, unconformity, subsidence (tilting) and basin-filling has occurred in Costa Rica and Nicaragua three times since the late Eocene, and suggested a control by long-term intraplate stress. Similarly, in the Andean backarc basin of Argentina, changes in tectonic regime over intervals of 15–65 million years since the Triassic are considered to have been responsible for the main variations in stratigraphic architecture there (Hallam, 1991; Legarreta and Uliana, 1991).

In the Western Canada Sedimentary Basin, five clastic sequences ("clastic wedges") span the mid-Jurassic to mid-Cenozoic (Fig. 5.23). Crustal loading occurred as a series of pulses, as successive terranes arrived at and were obducted onto the western continental margin (see summary in Miall et al., 2008). These events are represented in the basin as a series of clastic wedges. Westerly sediment sources associated with contractional tectonism appeared for the first time in the Late Jurassic, including the Mesocordilleran Geanticline of Nevada. The first major clastic wedge, constituting the Morrison and Kootenay formations, continued into the Berriasian, but much of the Berriasian to Barremian (Neocomian) interval is represented by a regional unconformity throughout the Western Interior Basin. This period corresponds to a "magmatic lull" in the Cordillera. The base of the Cretaceous section, of late Berriasian or Aptian age, typically consists of a sheet of coarse, fluvial gravels, throughout much of the Western Interior Basin. Provenance studies of the foreland-basin strata indicated that following the regional mid-Cretaceous episode of tectonic quiescence, erosion tapped into oceanic-arc and related rocks, and syndepositional continental magmatic rocks of Quesnellia, far to the west of the orogenic front. Examination of the ages of basal Mannville conglomerates deposited at this time, and reconstruction of the subsidence histories, suggest that a new phase of flexural loading and subsidence commenced shortly after deposition, initiating a new "constructive" phase of development of the Cordilleran orogen.

At least two Cretaceous cycles of transgression occurred in northern Canada, but marine waters did

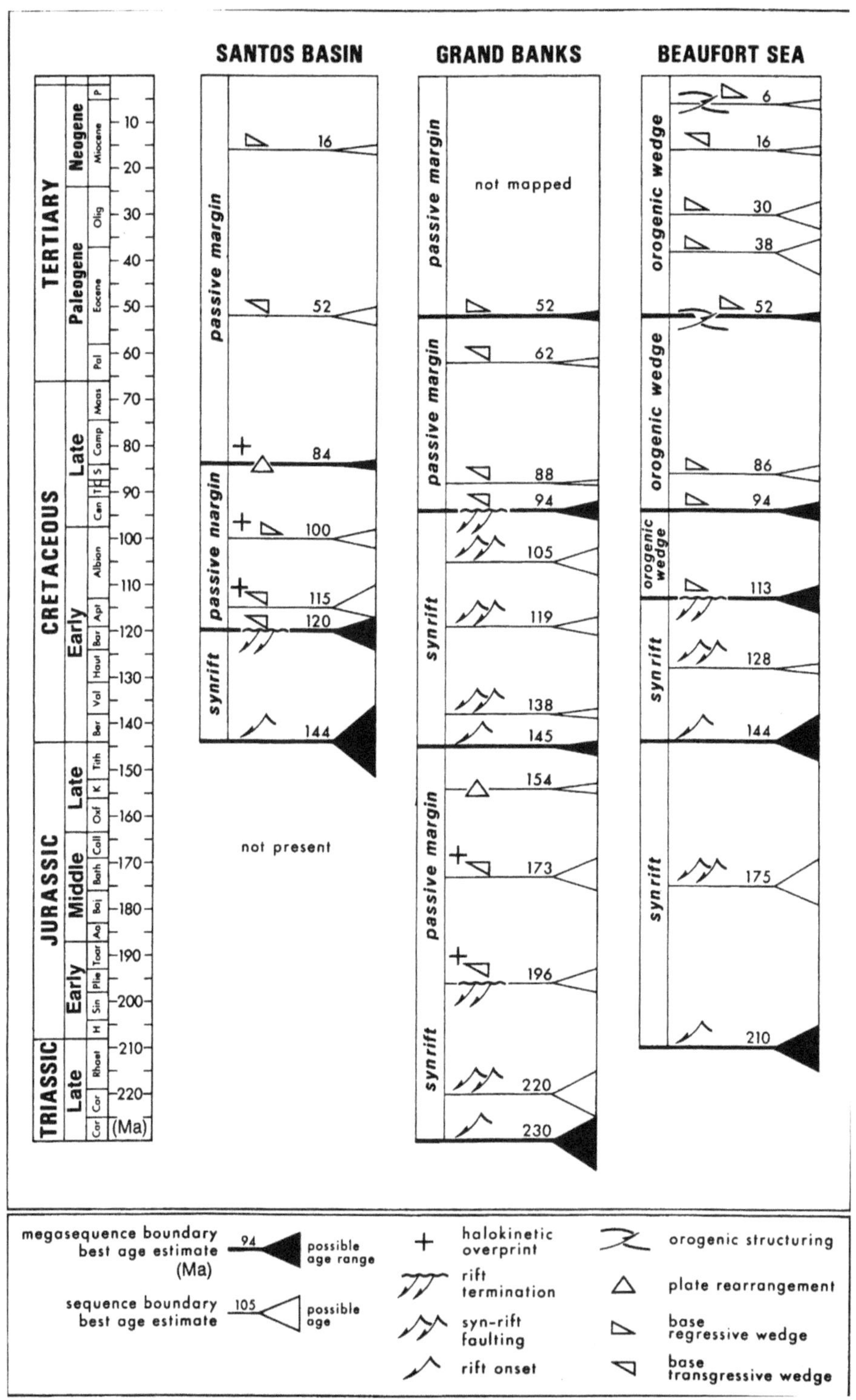


**Fig. 5.21** Estimated ages of sequence boundaries in three extensional-margin basins. The Santos basin is in South America, and is not discussed in this book (Hubbard, 1988)

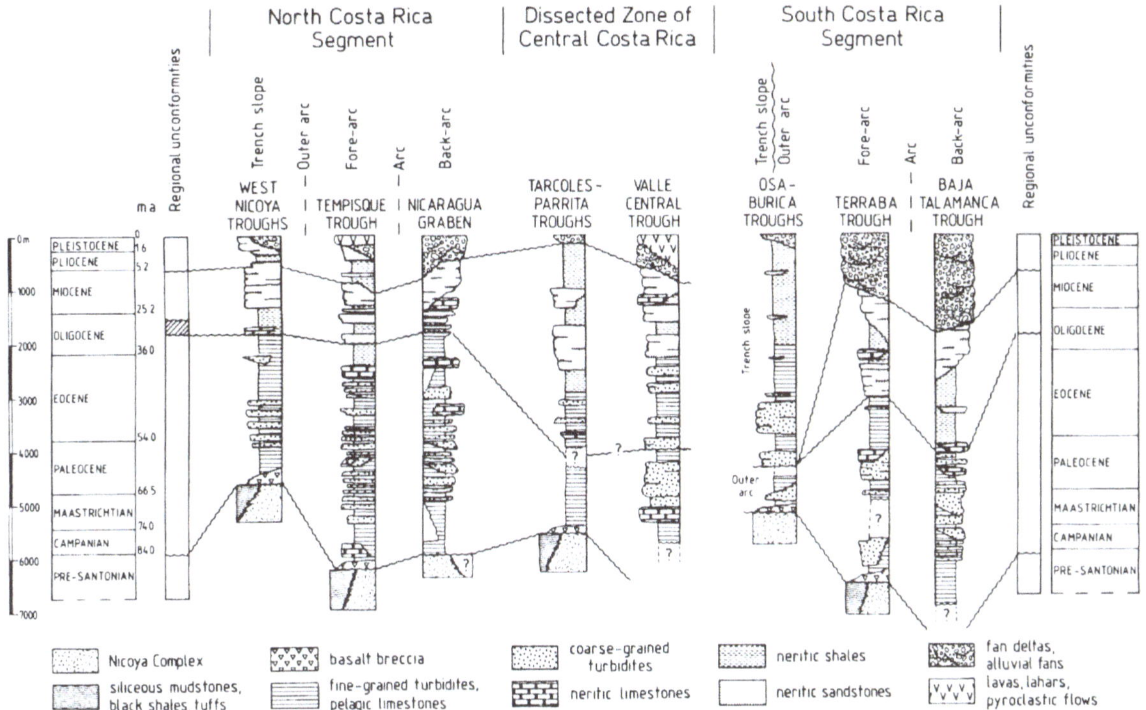



**Fig. 5.22** Stratigraphic correlation of Cretaceous-Cenozoic sections in Cost Rica. Note the subdivision of the section into "sequences" by the presence of regional unconformities. However, these are angular unconformities developed in response to regional convergent tectonism (Seyfried et al., 1991)

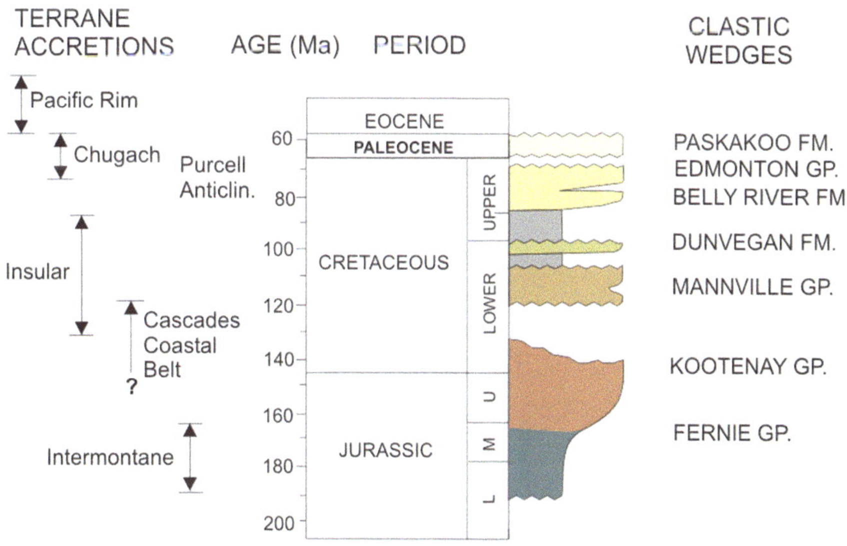



**Fig. 5.23** The six clastic wedges of the Western Canada foreland basin, shown as a function of time, and the times of accretion of allochthonous terranes. Also shown is the period of development of the Purcell anticlinorium (PA). Adapted from Stockmal et al. (1992)

not extend southward into the Western Interior Basin until the Aptian. During the Aptian-earliest Albian interval, most of the Western Interior Basin was occupied by fluvial and estuarine systems assigned to such units as the Mannville Group in Alberta-Saskatchewan, and the Kootenai Formation of Montana. The Upper Cretaceous stratigraphy of the Western Interior Basin is characterized by the deposits of several major marine transgressions. Gaps in the stratigraphic record are numerous; some represent millions of years, although most are less than one million years in duration. Eustatic sea-level changes were probably partly responsible for this stratigraphic architecture, but regional and local tectonic processes were also important (see summary by Miall et al., 2008).

## 5.4 Main Conclusions

1. Many of the broad characteristics of the global stratigraphic record can be related to the changes in continental scale, crustal thickness, climate, latitudinal position and eustatic sea level that result from the assembly and dispersal of supercontinents over a 200–400 million years time period.

2. Cratonic sequences of about 10–100 million years duration can be traced and correlated among several of the earth's major continental interiors, including the interior of the United States and Canada, Russia and Brazil. These are now commonly called *Sloss sequences*, after L. L. Sloss, who first identified and named the North American sequences and devoted much of his life's work to their analysis and interpretation.

3. Sequences tens of millions of years in duration may be the result of regional tectonism, including the effects of long-term changes in plate-kinematic patterns. These have been termed *tectonostratigraphic sequences*. They are the stratigraphic response to plate rifting, fragmentation and thermal subsidence, and to episodic contractional tectonism. They may be correlatable with each other within areas as large as two or three adjacent tectonic plates, the tectonics of which were dominated by the same major events, such as a large-scale plate rifting or collision event.

4. On the $10^{7-8}$-year time scale, regional stratigraphy is typically a product of both global processes (eustatic sea-level change, long-term climate change) and regional tectonism related to plate kinematics.

# Chapter 6

# Cycles with Million-Year Episodicities

## Contents



A wealth of stratigraphic data has accumulated for cycles having durations and episodicities of a few million years. They have been recorded in a wide variety of Phanerozoic basins in many different tectonic settings. They constitute the main basis of the Exxon global cycle charts, where they are termed "third-order cycles" (Haq et al., 1987, 1988a). A small selection of these is described in this chapter in order to illustrate stratigraphic patterns and their reflection of tectonic setting. Sequence concepts have also been applied to the study of the Precambrian record (e.g., Christie-Blick et al., 1988), but this work is not discussed here because at this time the record is fragmentary and regional correlations are very limited.

## 6.1 Continental Margins

It is now routine practice to subdivide stratigraphic successions into sequences, on various scales, based on the recognition and correlation of key surfaces and facies packages. Figures 6.1 and 6.2 consist of a set of diagrams constructed to illustrate the Upper Cretaceous stratigraphy of Oman. Three scales of "accommodation cycles" have been recognized in this data set. The thickest of the cycles spans much of the Cenomanian stage, and corresponds to a $10^6$-year sequence. It can be divided into two higher-order sets of sequences, of the type discussed in Chap. 7.

### 6.1.1 Clastic Platforms and Margins

The Atlantic and Gulf Coast continental margins of the United States are classic examples of extensional continental margins (Fig. 6.3). Their stratigraphy and structure were influential in the development of plate-tectonic basin models for continental margins, and much research has been carried out there to investigate the tectonic history and petroleum potential of the major basins. Techniques for backstripping and for modeling of the geophysical controls of flexural and thermal subsidence were first developed using Atlantic-margin data (see Miall, 1999, Chap. 7 for summary). In more recent years, a major research project has been undertaken to thoroughly analyze the sequence stratigraphy of the Atlantic margin off the coast of New Jersey. This project, led by K. G. Miller of Rutgers University, is discussed below in greater detail, and in Sect. 14.6.1.

One of the suggestions made by the Vail-Haq-Hardenbol/Exxon school of seismic stratigraphy was that "Neogene stratigraphic successions from a number of continents are characterized by very similar stratal geometries" (Bartek et al., 1991, p. 6753;


A.D. Miall, *The Geology of Stratigraphic Sequences*, 2nd ed.,
DOI 10.1007/978-3-642-05027-5_6, © Springer-Verlag Berlin Heidelberg 2010

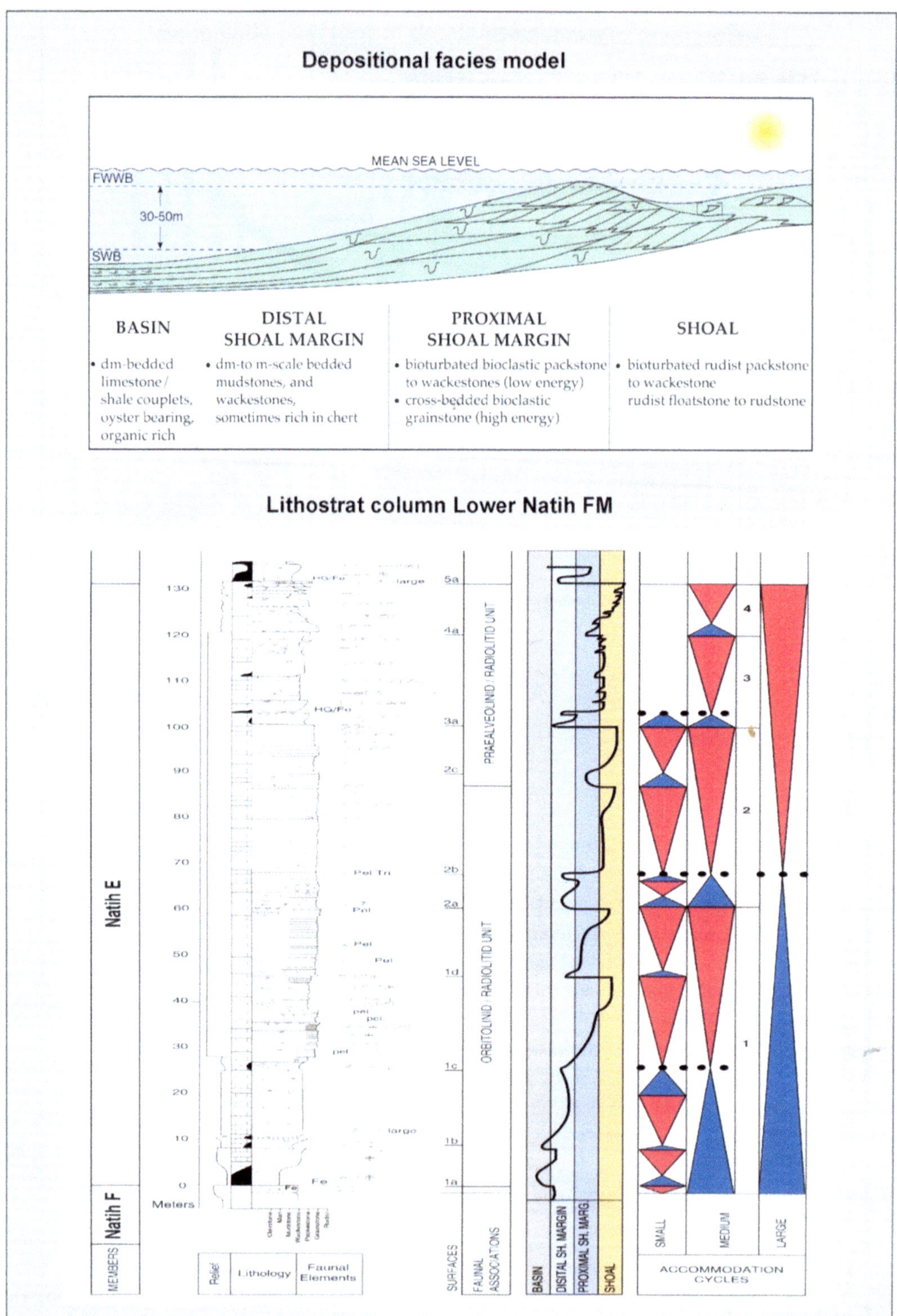


**Fig. 6.1** The subdivision of a stratigraphic section into "accommodation cycles" at several different scales, showing a relative sea-level curve (at centre of *lower diagram*) and its relationship to coastal depositional systems (*top diagram*). From a study of Upper Cretaceous deposits in Oman (Veeken, 2007, Fig. 4.34)

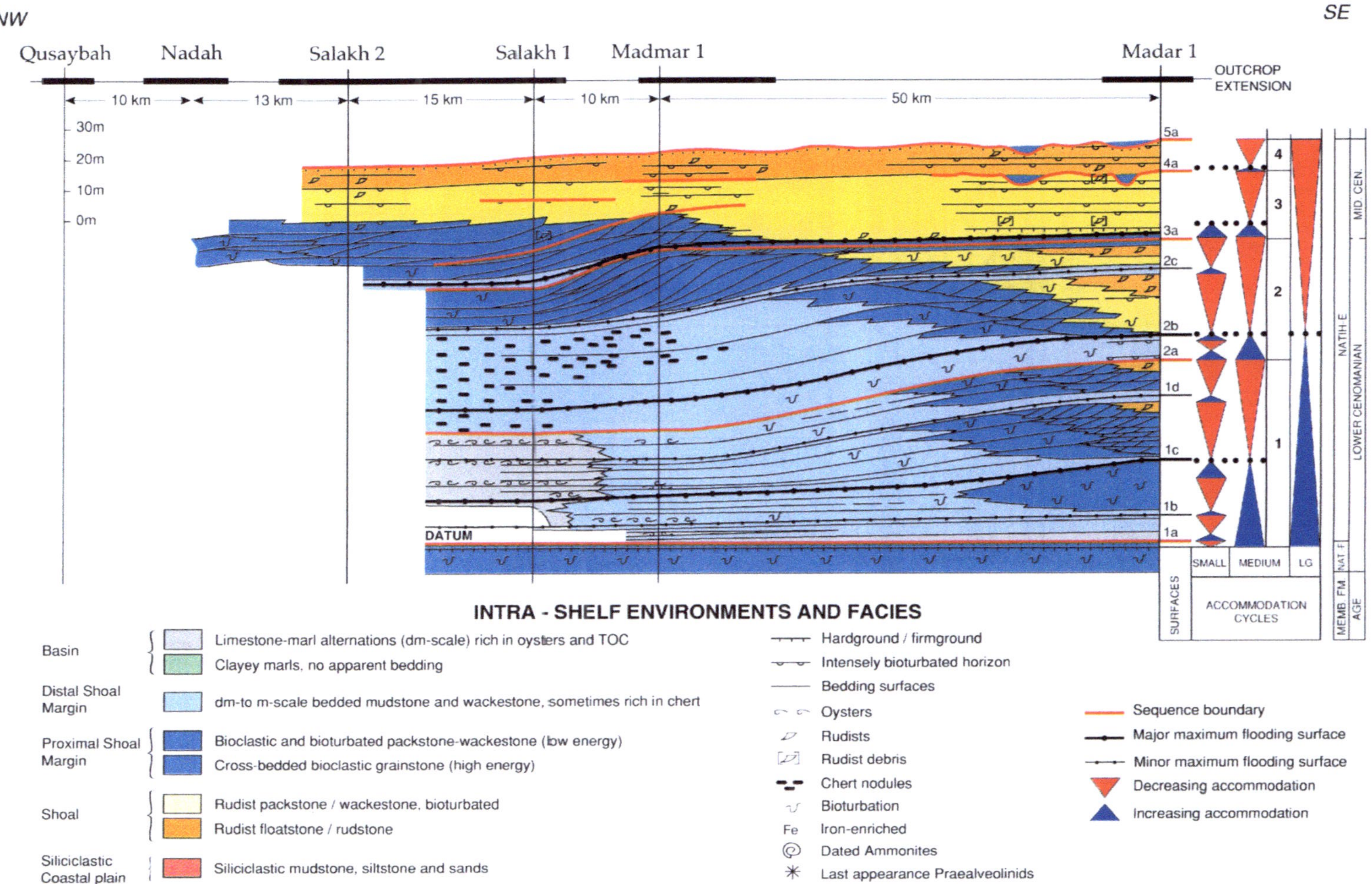



**Fig. 6.2** High-resolution sequence stratigraphy of the same section in Fig. 6.1, showing the lateral and vertical relationships of the various facies assemblages. From a study of Upper Cretaceous deposits in Oman (Veeken, 2007, Fig. 4.35)

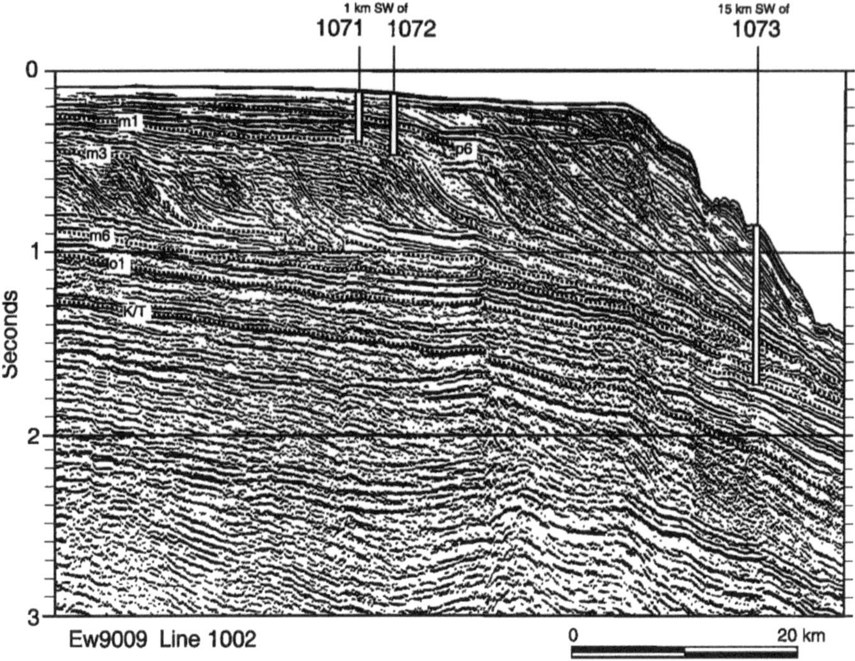



**Fig. 6.3** Reflection-seismic line across the outer New Jersey continental shelf and slope, showing the Cretaceous-Tertiary boundary (K/T) and some of the Cenozoic sequence boundaries that can be differentiated along this section. These are part of a data set that defines a complete succession of $10^6$-year sequences along this part of the North American margin (see Fig. 4.10) (Miller et al., 1998, Fig. 3)

**Fig. 6.4** Schematic cross-section of a generalized, hypothetical continental margin, based on assessment of a global record of seismic-reflection data by Bartek et al. (1991). This section purports to illustrate what the authors term "the global stratigraphic signature of the Upper Paleogene and Neogene. This version of the illustration is from McGowran (2005, Fig. 5.21)

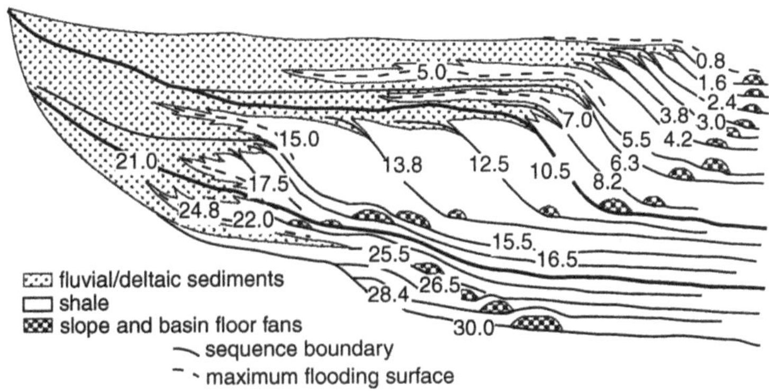



see Fig. 6.4). It was thought that this indicated a global control of continental-margin stratigraphy by eustatic sea-level change. The signature included 12 recognizable "third-order" elements, beginning with these three: "(1) Lower Oligocene landward thickening, (2) upper Oligocene wedge, which laps out at or near the shelf margin and thickens basinward, (3) basal lower Miocene flooding," and so on (Bartek et al., 1991, p. 6753). As discussed in Chap. 14, subsequent detailed chronostratigraphic correlation of some individual continental margin sections does not lend support to this characterization of Neogene stratigraphy. While there is good agreement in sequence boundary ages between some individual successions, paradoxically, the Vail Neogene "signature" of Fig. 6.4 does not correlate with these events (Fig. 14.34). Nevertheless, cross sections such as Fig. 6.4, as constructed for individual basin-fill stratigraphies, are an invaluable guide to the long-range development of the basin fill. For example, this diagram suggests a long-term transgression from 30 to 21 Ma, during which the shelf margin underwent significant backstepping, and a long-term period of high sediment supply from 15 Ma to the present day, during which time the continental margin underwent both substantial progradation and aggradation.

The lack of "fit" of the sequence boundary ages of Fig. 6.4 with other, independently dated continental margin sections (Fig. 14.34) is probably a consequence of the method by which the ages of the events and the procedure by which correlation was carried out. As discussed at length elsewhere in this book (Sect. 1.2, Chap. 12), the Vail-Haq-Hardenbol/Exxon "School" of sequence stratigraphy adopted a "deductive" approach to the issue of sequence classification and correlation, and this has led to problems that limit the usefulness of the work.

Some of the details of the Atlantic-margin project of K. G. Miller are discussed here. This project has focused on the New Jersey margin, but also includes a detailed analysis of an outcrop section in Alabama, which provides some interesting comparisons. The St. Stephens Quarry in Alabama was one of the sites used in the early Exxon work as a reference section for Paleogene stratigraphy, in particular, for documentation of the Eocene-Oligocene boundary (Baum and Vail, 1988). As Miller et al. (2008a) noted, this is potentially an important section, because the Eocene-Oligocene boundary has long been thought to coincide with a major drop in sea level following a build-up in Antarctic ice volumes, and there has existed a need to document this transition in detail. Modern oxygen-isotope data spanning the Cenozoic (Miller et al., 2005a, b) do, in fact, show such a drop (Fig. 11.16). Given controversies surrounding the sequence stratigraphy and correlation of this section, Miller et al. (2008a) undertook a thorough re-evaluation of the succession, based on a continuous core obtained at the site.

Figure 6.5 illustrates the sequence stratigraphy and correlations of the Alabama section. The precise positioning of the sequence boundaries was accomplished using calcareous nannofossils and planktonic foraminifera, tied into the time scale of Berggren et al. (1995). The method of dating is described in Sect. 14.6.1. As discussed there, potential errors of $\pm 0.5$ million years may be expected with this type

of data set at this level in the geological column. Given that, a reasonably close comparison is apparent between the Alabama and New Jersey sections, in Fig. 6.5. A virtually continuous section in Alabama from 31.7 to 33.9 Ma contains two breaks that correlate with longer hiatuses in the New Jersey section, and the other unconformities in the Alabama section partially overlap those in New Jersey. The two areas are approximately 1,500 km apart and located in different, but tectonically related structural provinces. Does this indicate that the correlations pass the test for eustasy? It can be observed, here, that two sequence boundaries shown in Fig. 6.5, at 33.5 and 33.9 Ma, appear to coincide with sudden increases in $\delta^{18}O$, which would seem to confirm a relationship to ice build-up. We return to this question in Chap. 14.

The complete section on the New Jersey continental margin (coastal plain, shelf and slope) spans the Lower Cretaceous to present, and exhibits a set of $10^6$-year sequences (Fig. 4.10). A portion of the section is shown in Fig. 6.6. The synthesis was constructed from seismic and borehole data, calibrated using a series of drill cores subjected to lithofacies and biofacies analysis, biostratigraphy and other methods of dating, including strontium and oxygen isotope data and magnetostratigraphy (see Sect. 14.6.1). As discussed above (see also discussion of Fig. 6.6 by Browning et al., 2008, p. 241), the Eocene-Oligocene interval illustrated in Fig. 6.6 spans a transition into cool, icehouse climates, with a substantial buildup of ice postulated to have occurred on Antarctica (see Chap. 11). Facies analysis of drill cores through this interval document a transition from a carbonate ramp accumulating carbonate-rich clays to a prograding siliciclastic margin, commencing in the middle Eocene. In Fig. 6.6 it can be seen that carbonate sediments occur only in the lowermost sequence (E8). The total time span represented by the preserved sequences illustrated in Fig. 6.6 represents only 32% of the total elapsed time encompassed by this diagram (middle Eocene to late Oligocene), as measured at the $10^6$-year scale.

Figure 6.6 includes two versions of a calculated sea-level curve constructed using the backstripping methods summarized in Sect. 3.5. As noted by Kominz et al. (2008, p. 211), such curves can, initially, only be assumed to relate to local sea-level changes. Careful construction of local curves must precede any attempt to develop global curves, a topic discussed at

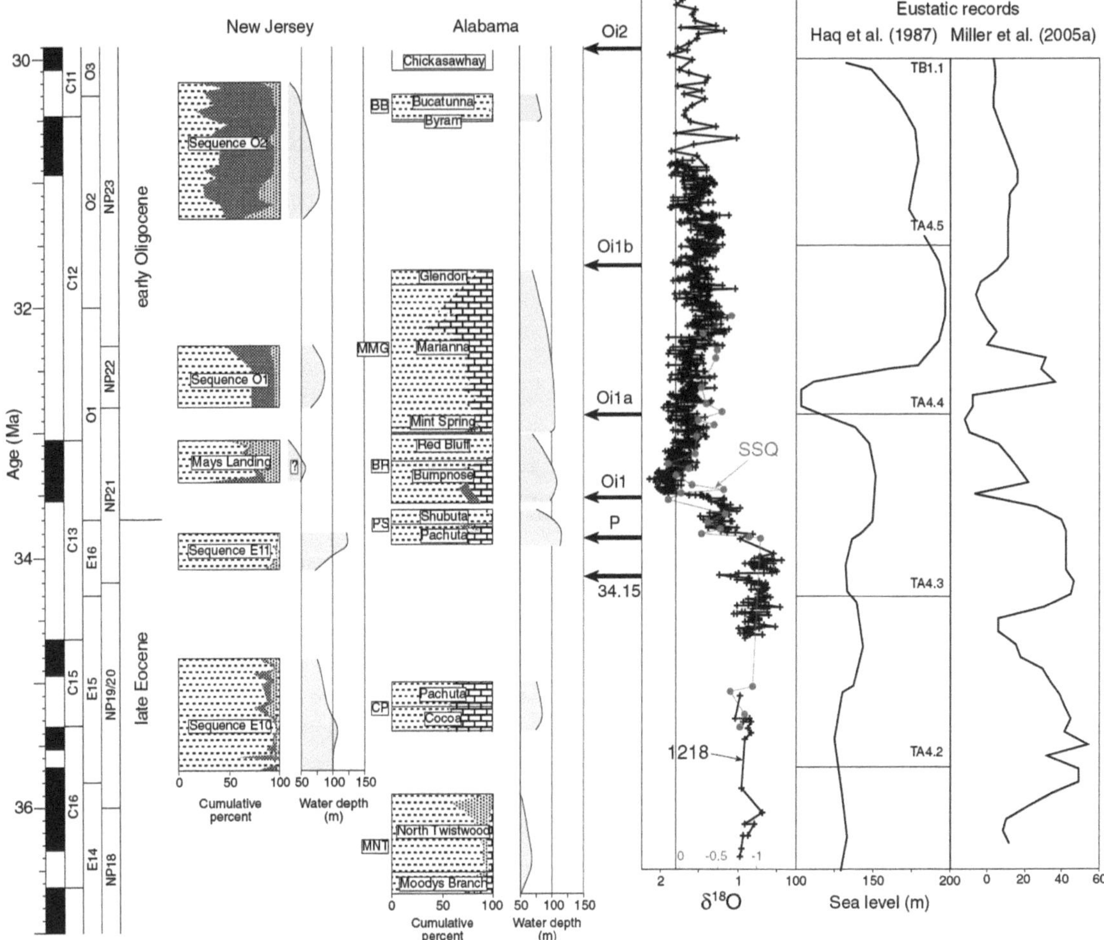



**Fig. 6.5** The sequence record of an outcrop section in Alabama, compared with contemporaneous strata in New Jersey, and correlated against a standard oxygen isotope curve and published sea-level curves. The *rectangles* for each sequence indicate the assigned ages of both tops and bottoms. The internal ornamentation indicates lithofacies composition. Water depths, determined from biofacies data, are shown in the small graphs to the *right* of each *rectangle*. *Arrows* labeled Oi2, Oi1b, etc., refer to sequence boundaries in the New Jersey sections. This diagram is from the Alabama study (Miller et al., 2008a, Fig. 8); the New Jersey data are from Miller et al. (1998). The methods of chronostratigraphic correlation used in the construction of this diagram are discussed with reference to the New Jersey work of K. G. Miller and his colleagues in Chap. 14

length in Chap. 14. This is because basement elevation is controlled by local to regional tectonic effects (thermal and flexural subsidence, dynamic loading, etc.; see Chap. 10) and by sediment loading, as well as by the absolute elevation of the sea. Location-specific sea-level curves must also take into account changing water depths during sedimentation and post-depositional sediment compaction effects (Miall, 1999, Chap. 7; Allen and Allen, 2005). This is why detailed sedimentological studies (for this case study, reported by Browning et al., 2008) are essential.

### 6.1.2 Carbonate Cycles of Platforms and Craton Margins

Carbonate sediments are sensitive indicators of changing sea level for various reasons, including the following three: Firstly, the "carbonate factory" that develops within warm, shallow continental shelves that are free of clastic detritus can produce carbonate sediment at a rate that normally is rapid enough to keep up with the most rapid of sea-level changes. Secondly, within such shallow-water carbonates, depth changes are indicated

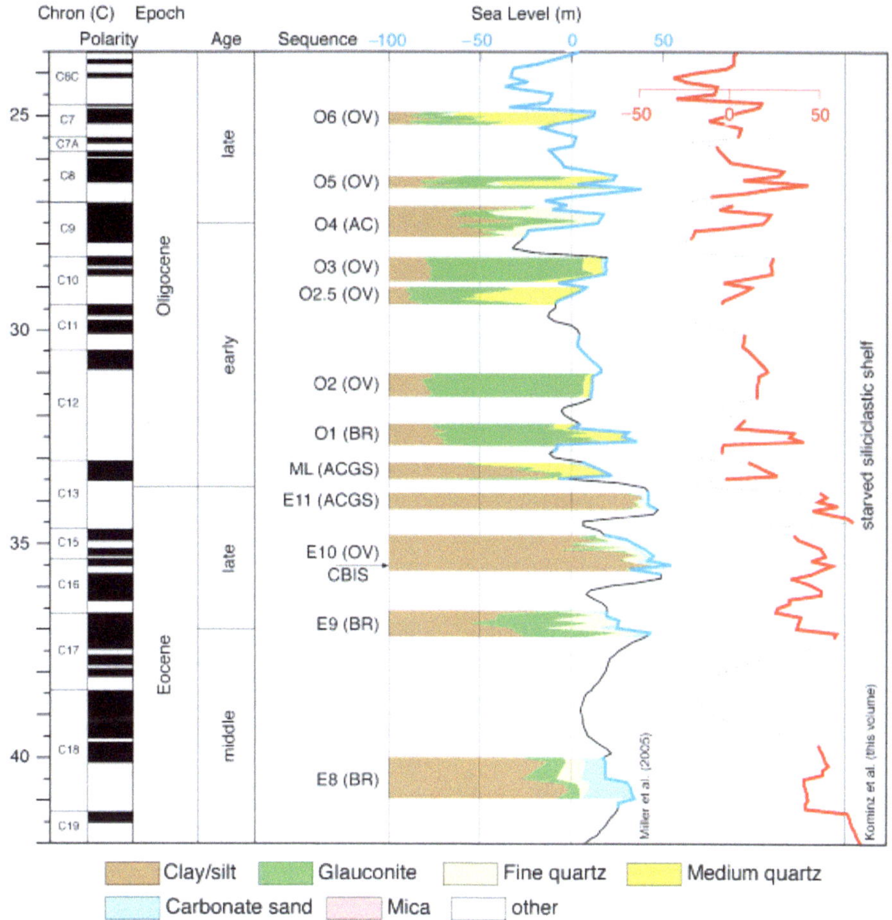



**Fig. 6.6** Eocene-Oligocene sequence stratigraphy of New Jersey. The sequence terminology (E8, O1, etc.) is based on detailed studies of drill cores from the New Jersey coastal plain, the names of which are indicated by abbreviations (BR, OV, etc.). The sea level *curves* represent different version of back-stripping reconstructions of the New Jersey data by Miller et al. (2005a) and Kominz et al. (2008). From Browning et al. (2008, Fig. 11)

by a variety of depth-sensitive facies characteristics. Thirdly, within carbonate platforms most carbonate sediments are deposited where they are produced. The complications of sediment redistribution by autogenic processes that characterize clastic sediments (e.g., the development of delta lobes) therefore are less extreme. For these reasons stratigraphic sequences are commonly well developed in carbonate rocks. Hierarchies of sequences may be present, reflecting the integration of several different generative mechanisms (Sarg, 1988; Goldhammer et al., 1990; Schlager, 1992a). It is important to note, however, that sequence boundaries can develop as a result of submarine erosion and environmental change, as well as in response to sea-level change (Sect. 2.3.3).

Prograding continental margins provide many examples of $10^6$-year carbonate sequence stratigraphy, and these are well displayed in regional reflection-seismic records. Eberli and Ginsburg (1988, 1989) and Eberli et al. (1997) described an excellent example of this, revealing the dramatic lateral growth by progradation of the Bahama Platform. Other examples were given by Sarg (1988), Mitchum and Uliana (1988), Epting (1989) and Tcherepanov et al. (2008). Bosellini (1984) and Sarg (1988) provided outcrop examples. Many detailed case studies are contained in the book edited by Loucks and Sarg (1993).

Figure 6.7 illustrates the architecture and sequence stratigraphy of Miocene to Recent carbonate buildups in offshore Sarawak. The buildups are gradually

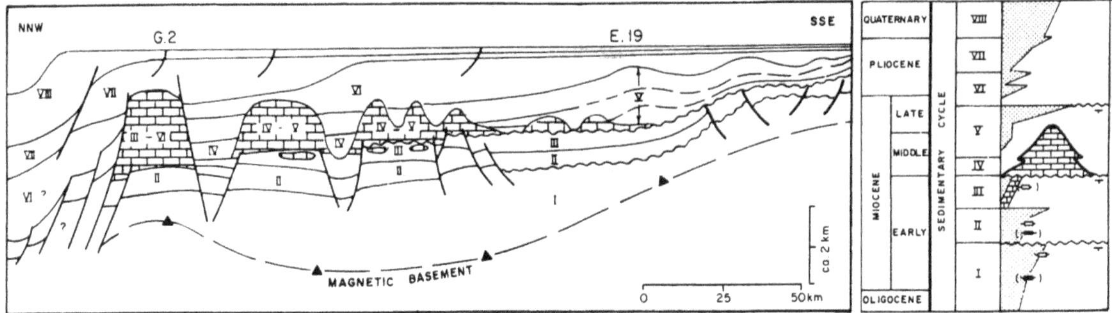

**Fig. 6.7** Simplified stratigraphic cross-section showing relationship between carbonate buildups and depositional cycles, offshore Sarawak. The carbonate buildups of cycles III–VI are buried by deltaic clastics of cycles V–VIII. Seaward stepping of the carbonate buildups and deltaic progradation are continuing to the present day (Epting, 1989). AAPG © 1989. Reprinted by permission of the AAPG whose permission is required for further use

extending seaward, and are being buried by progradation of deltaic clastics. This is a very common pattern in continental-margin carbonate shelves (e.g., Devonian of the Alberta Basin, modern Gulf of Papua, as discussed and illustrated below). In the Sarawak and Gulf of Papua examples, the most recent reef buildups are not yet drowned by deltaic progradation and are still actively developing.

The Bahamas Bank is one of the more thoroughly studied carbonate margins. Some of the research carried out there by R. N. Ginsburg and G. Eberli is summarized in Figs. 6.8, 6.9, and 6.10. The Bahamas platform was built on fault blocks that formed initially during the mid-Jurassic, although the present topography of the bank, consisting of flat-topped carbonate platforms cut by deep oceanic channels, is thought to have originated in the Late Cretaceous. The platform has expanded laterally toward the Florida margin by spectacular lateral clinoform progradation (Figs. 6.8 and 6.9). Windward margins are steep and show chaotic seismic facies. Leeward margins evolved from steep, fault-bounded margins into bypass margins and into low-angle accretionary slopes during the Miocene (Fig. 6.9). The present configuration of the continental slopes around the platform and inter-island channels is partly the result of carbonate progradation and partly the result of increased submarine erosion, probably reflecting the gradual acceleration of thermohaline oceanic circulation as the climate became cooler with increasing latitudinal temperature gradients, during the Cenozoic.

Eberli and Ginsburg (1989) recognized multiple sequence boundaries in the Cretaceous-Cenozoic carbonate cover (Figs. 6.8 and 6.9), reflecting the interplay between vertical motions of the platform and eustatic sea-level changes. Architectural details of the sequences clearly record relative sea-level changes (Fig. 6.10). Two cored holes subsequently located

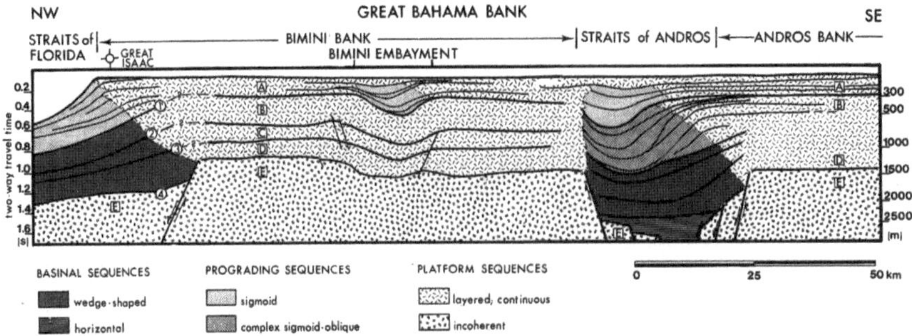

**Fig. 6.8** Schematic cross-section across the northwest margin of Great Bahama Bank and margin of Andros Bank, based on seismic-reflection data. Patterns indicate variations in seismic facies. The margin facing the Straits of Florida and that on the SE side of the Straits of Andros are typical leeward platform margins. That on the NW side of the Straits of Andros is a windward margin. The evolution of the leeward margins is shown in greater detail in Fig. 6.7 (Eberli and Ginsburg, 1989)

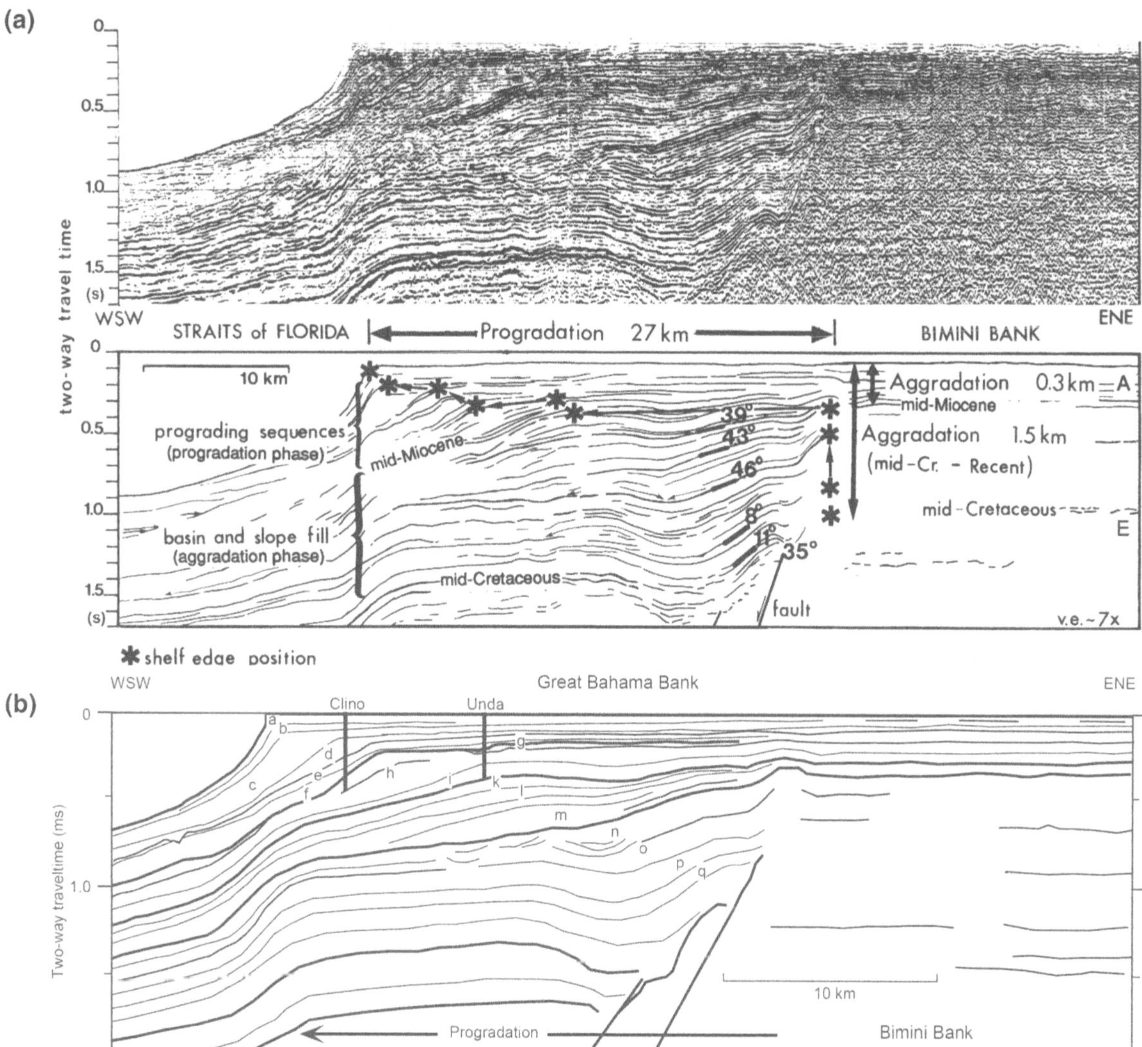



**Fig. 6.9** West margin of Bimini Bank, showing details of margin evolution. (**a**) Seismic section and angles documenting the decrease in continental slope through time. *Asterisks* denote position of shelf edge (Eberli and Ginsburg, 1989). (**b**) Interpretation of the section based on the analysis of two core holes drilled along the line of section (Eberli et al., 1997). The ages of the sequences a–q are provided in Fig. 13.34

on and near the centre of the seismic line shown in Fig. 6.9b (within the topsets and foresets of the platform) drilled down to a Middle Miocene level at a maximum depth of 678 m, and revealed much about the stratigraphy, sedimentology, fluids and other features of the platform rocks (Eberli et al., 1997). Analyses indicate the importance of "highstand shedding" as the primary control on the depositional evolution of the Neogene western Great Bahama Bank:

> The principal source of sediment to the slope is the extensive offbank transport of suspended, fine-grained banktop 'background' sediment during periods of sea-level highstands when the entire platform was submerged providing the bulk, more than 80%, of the slope sediment. During sea-level falls, the supply of fine-grained sediment to the slope environment is reduced or completely stopped. These deposits of reworked margin-derived material form thin intercalations in the 'background' sediment. Factors controlling the thickness, composition, and diagenesis of the deposits, and the formation of discontinuity surfaces are (1) the morphology of the platform, hardgrounds may develop at the base (ramp morphology) or at the top of the lowstand deposits (flat-topped platform), (2) the frequency and amplitude of sea-level changes, and (3) the water depth and distance to the margin (Eberli et al., 1997, p. 35).

**Fig. 6.10** Architectural details of carbonate sequence margins reveal changes in accommodation. Abundant sediment generation during a highstand in relative sea-level results in a trajectory of the facies belts sloping obliquely upward (diagrams **a**, **c**). An intervening episode of low relative sea level causes a downward shift in facies belts and reduction in sediment supply from a narrow coastal strip, resulting in a topset termination to the reef facies body that records the lowstand event (Eberli and. Ginsburg, 1989, Fig. 8)

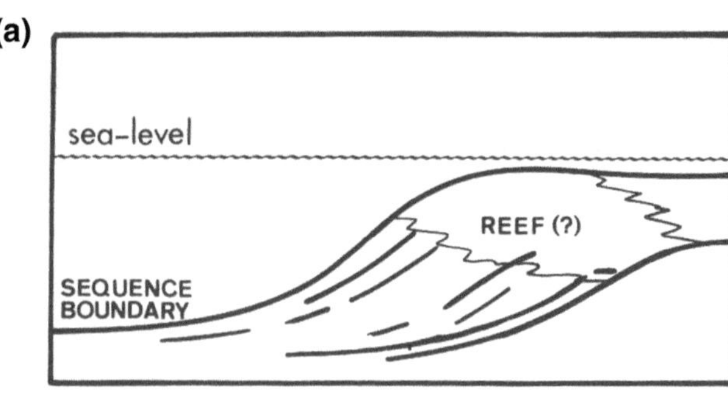


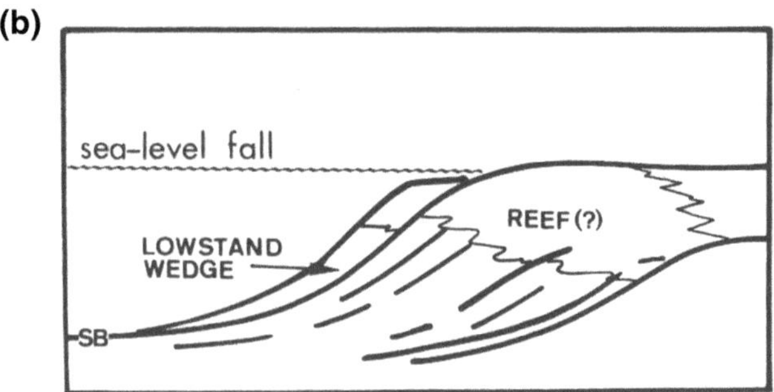


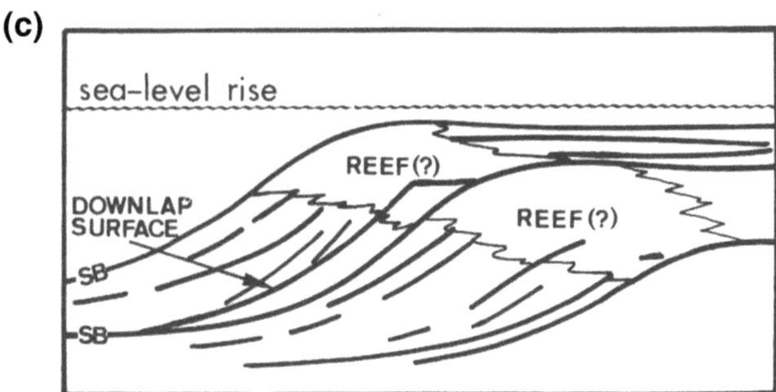


A suite of ODP cores drilled mainly on the lower foresets and bottomsets along the seismic line of Fig. 6.8 provided additional details. They showed that much of the lower foreset deposits consists of carbonate turbidites, while the bottomset deposits include contourites (Betzler et al., 1999). Depositional slopes on the clinoforms reach a maximum of 47°, but this is exceptional. Most slopes are less than 4° (Betzler et al., 1999), a point that is obscured by the large vertical exaggerations characteristic of seismic displays.

These cores reveal three scales of sequence cyclicity, large scale cycles 60–170 m thick, a medium scale of cyclicity tens of metres thick, and a small-scale cyclicity on a metre scale (Eberli et al., 1997). Major sigmoid reflections within the progradational deposits define boundaries between individual sequences, formed at sea-level lowstands. They are estimated to represent cycles in the order of 1–2 million years duration (sequences a–q in Fig. 6.9b). Larger scale "megasequences" spanning 10–20 million years are identified by major bounding surfaces (letters in boxes in Fig. 6.8).The highest-frequency sequences represent sea-level cyclicity on a 20-ka time scale, according to Betzler et al. (1999). Three major progradational episodes, of late Miocene, late early Pliocene and latest Pliocene age, are considered to indicate sea-level lowstands (Eberli et al., 1997). The major lowstand unconformity dated as Late Pliocene-basal Pleistocene has been interpreted as correlating to the global lowstand that records the build-up of continental ice cover in the northern hemisphere.

## 6.1.3 Mixed Carbonate-Clastic Successions

A recent seismic survey of the Gulf of Papua provides some high-resolution detail of $10^6$-year sequence episodicity in that basin (Tcherepanov et al., 2008). This basin was initiated by break-up of Gondwana (middle Triassic-middle Jurassic) and modified by spreading of the Coral Sea (Late Cretaceous-Paleocene). It became part of a peripheral foreland (proforeland) basin following the collision of Australia with the Banda arc in the Oligocene. A dominantly carbonate succession accumulated there from Eocene time until near the end of the Miocene, at which time it was uplifted and exposed. A subsequent rise in relative sea level resulted in flooding, backstepping and drowning of the carbonate platform, after which a major phase of clastic progradation commenced, sourced from the rising Papuan fold belt to the north. The relatively thin Eocene-Late Miocene carbonate succession and the overlying prograding clastic wedge are well displayed in reflection seismic data (Figs. 6.11 and 6.12). The extent and architecture of the

carbonate succession do not appear to relate to the foreland basin setting, whereas the subsequent prograding clastic wedge is clearly a product of tectonic uplift of a clastic source to the north. The gradual burial of the reefs is well seen in Fig. 6.12, showing a very similar succession of events to that which occurred in offshore Sarawak (Fig. 6.7).

Stratigraphic data, including well-log correlation, strontium-isotope stratigraphy and biostratigraphy have been used to divide the carbonate succession into ten sequences, averaging about 2.7 million years in duration (Fig. 6.11). Well cuttings and cores indicate a range of sedimentary environments from shelf to upper bathyal. Isochron maps of these sequences (Fig. 6.13) permit a reconstruction of the evolving architecture of the carbonate-dominated succession, which, in turn, throws light on the relationship between the development of accommodation and the sediment supply. These trends are summarized in Fig. 6.14, as follows: 1. Late Oligocene-early Miocene aggradation, backstepping and partial drowning; 2. late early Miocene-early middle Miocene aggradation; 3. middle Miocene downward shift in facies belts; 4. late middle Miocene lateral progradation; 5. late Miocene-early Pliocene flooding and aggradation. There is some correspondence of these events with the "Global stratigraphic signature" of the Upper Paleogene and

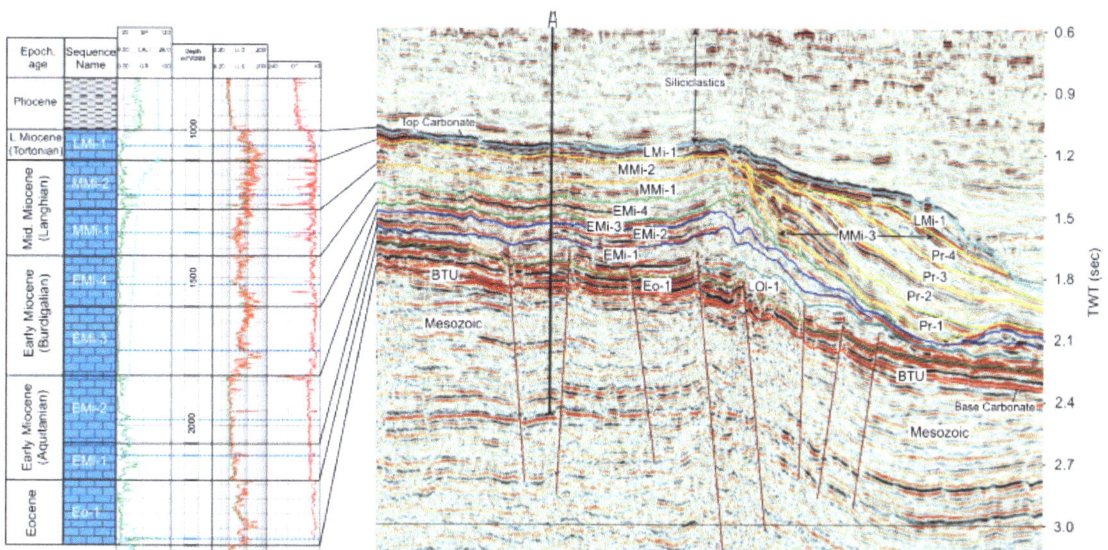


**Fig. 6.11** Seismic profile across the continental shelf, Gulf of Papua, showing stratigraphic ties to an exploration well. Eight sequences of Eocene to late Miocene age have been defined based on this type of data (Tcherepanov et al., 2008, Fig. 4a)

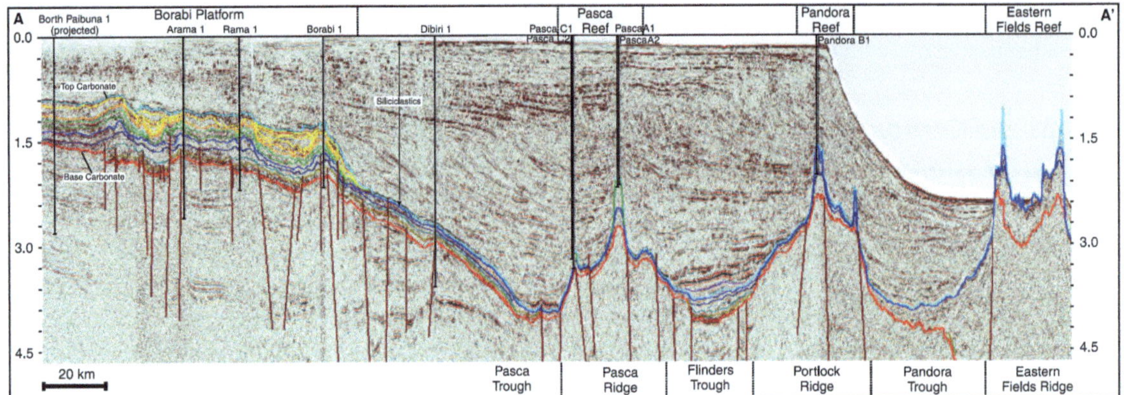


**Fig. 6.12** An Eocene-middle Miocene carbonate succession (defined by "top carbonate" and "base carbonate" underlying the Gulf of Papua is overlain by clastics of Late Miocene to modern, in age. These have prograded from the continental margin, leaving only the Eastern Fields and a few other small reef bodies exposed. The reefs have been largely drowned by rising sea level, and only the tip of the Eastern Fields reef is still active (Tcherepanov et al., 2008, Fig. 5A)

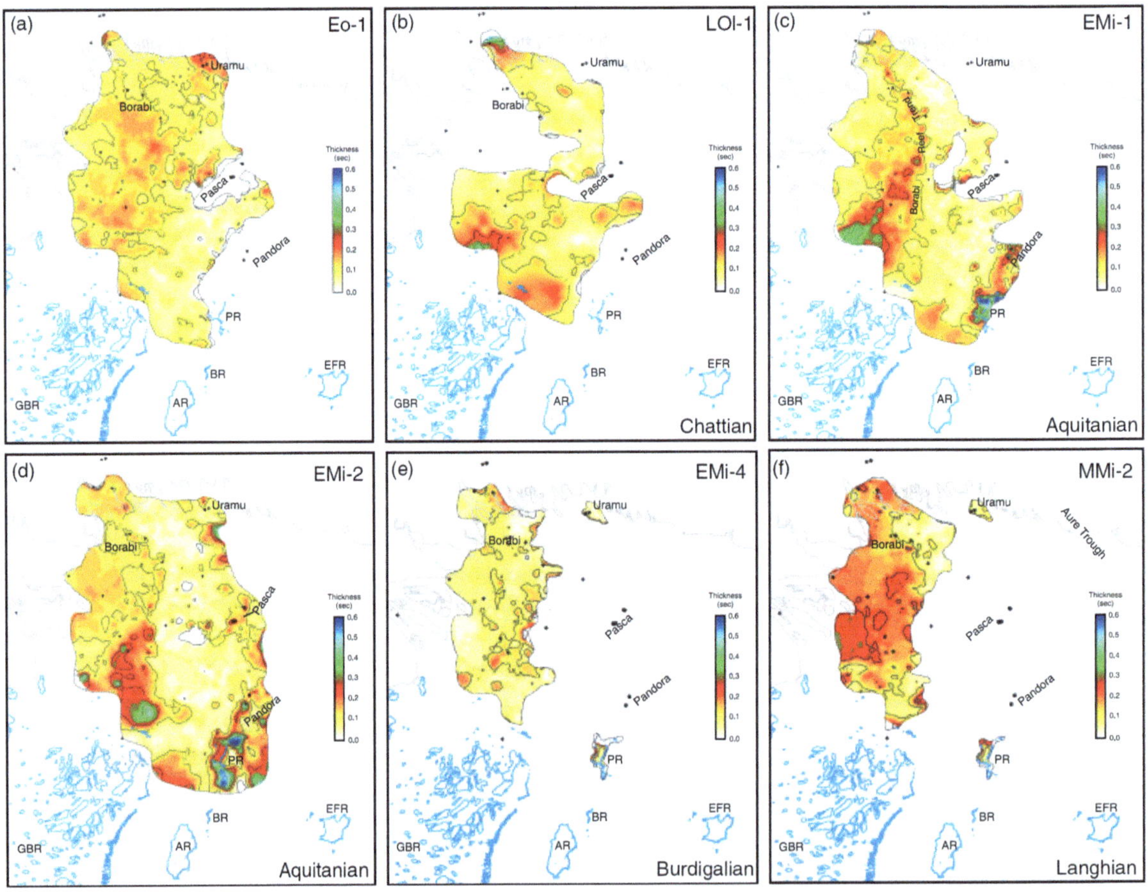

**Fig. 6.13** Isochron maps of selected carbonate sequences, Gulf of Papua, showing their extent and distribution. Sequence nomenclature (*top right* corner of each map) is as shown in Fig. 6.11 (Tcherepanov et al., 2008, Fig. 7)

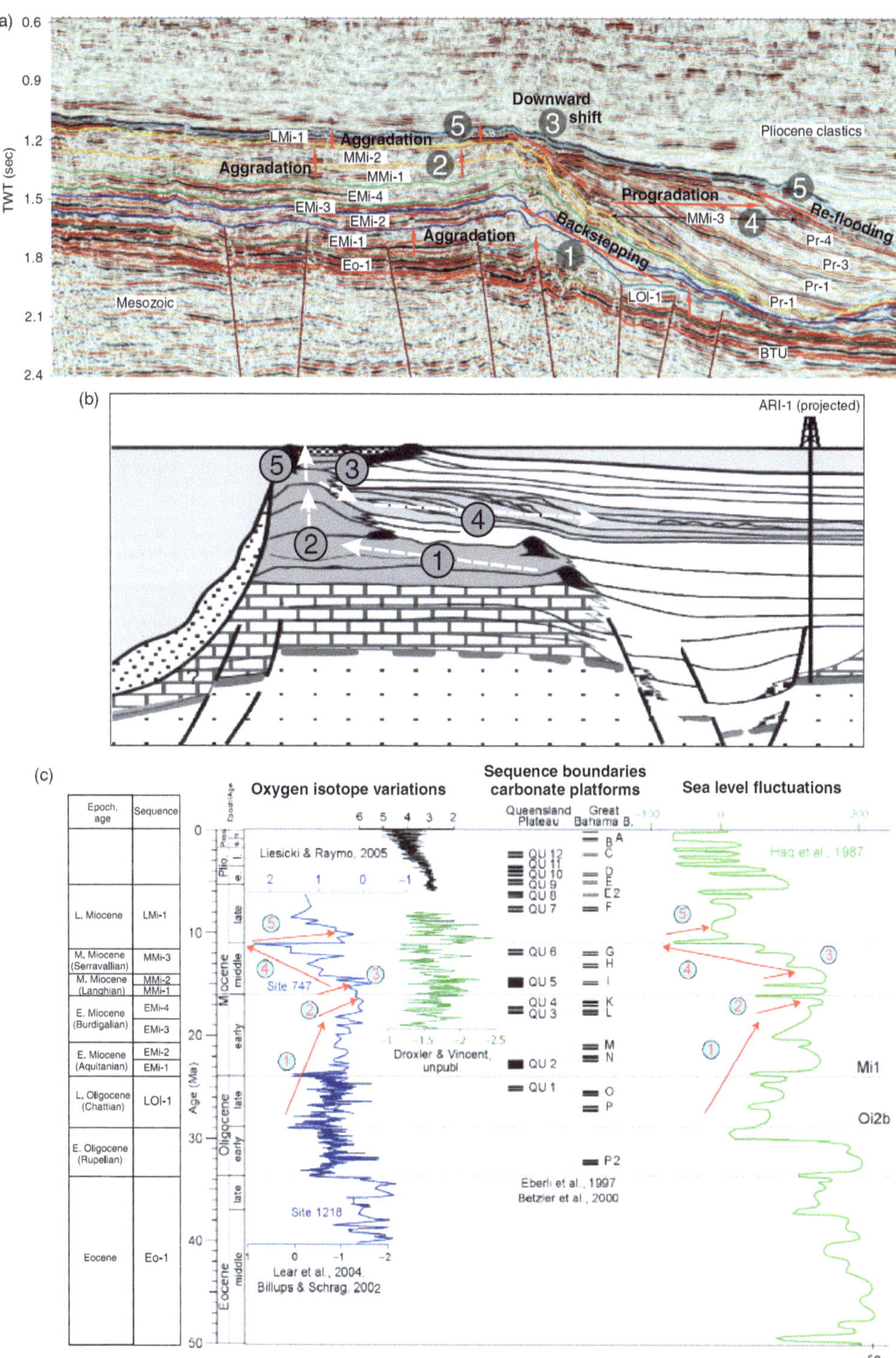



**Fig. 6.14** (**a**) Typical seismic section, Gulf of Papua, showing the major architectural features of the stratigraphy. (**b**) As seismic synthesis diagram for the West Maldives (Beloposky and Droxller, 2004). Aggradation, progradation, backstepping, and other architectural features of the stratigraphy are clearly definable from the seismic data in these two locations, but they appear to define a series of changes in accommodation with longer-term episodicities than the events in Queensland and Great Bahama (from Betzler et al., 2000), with which these events are compared in diagram (**c**) This figure is from Tcherepanov et al. (2008, Fig. 10). The Queensland-Bahamas correlations are shown in Fig. 13.34 and are discussed in Sect. 13.6.3

Neogene, described by Bartek et al. (1991; illustrated here as Fig. 6.4), but as noted in Sect. 14.6.3, global correlation of this "signature" with other areas remains poor.

A well known and much studied example of reefal carbonates is the Upper Devonian succession of Alberta. These rocks span the late Givetian to Fammenian, a period spanning more than 25 million years. The Givetian to late Frasnian part of the succession discussed here has been described in detail in the Atlas of the Western Canada Sedimentary Basin (Oldale and Munday, 1994; Switzer et al., 1994). A more recent sequence-stratigraphic reinterpretation and synthesis has been published by Potma et al. (2001).

A schematic summary of the succession is shown in Fig. 6.15. The geology shown in this diagram provides an illustration of the problem with the traditional "order" sequence terminology. Oldale and Munday (1994, p. 155), in referring to the Beaverhill Lake Group, stated:

> Two second-order depositional phases are recognized within the strata … a transgressive 'reefal' phase (a

term introduced by Stoakes, 1988) and a regressive 'basin-fill' phase. Each phase exhibits a distinctive style of deposition and consists of genetically related depositional cycles (parasequences). The phases are bounded by an unconformity or a surface of nondeposition (disconformity) and can be equated to a depositional 'sequence' utilizing the sequence stratigraphic concept. Each cycle reflects a third-order depositional sequence.

They identified three reef cycles within the transgressive phase and "numerous" shale and argillaceous carbonate cycles comprising the regressive phase. However, Potma et al. (2001) referred to the entire succession from the base of the Gillwood Sand to the base of the Gramina Siltstone Formation as a "second-order sequence", and they subdivided the Beaverhill Lake Group into just three "third-order" sequences (Fig. 6.15). The average duration of the nine sequences identified by Potma et al. (2001) is about 3 million years, and so their characterization is consistent with Vail et al.'s (1977) original definition of third-order sequences, but as discussed in Sect. 4.2, the "order" classification has ceased to have any useful meaning.

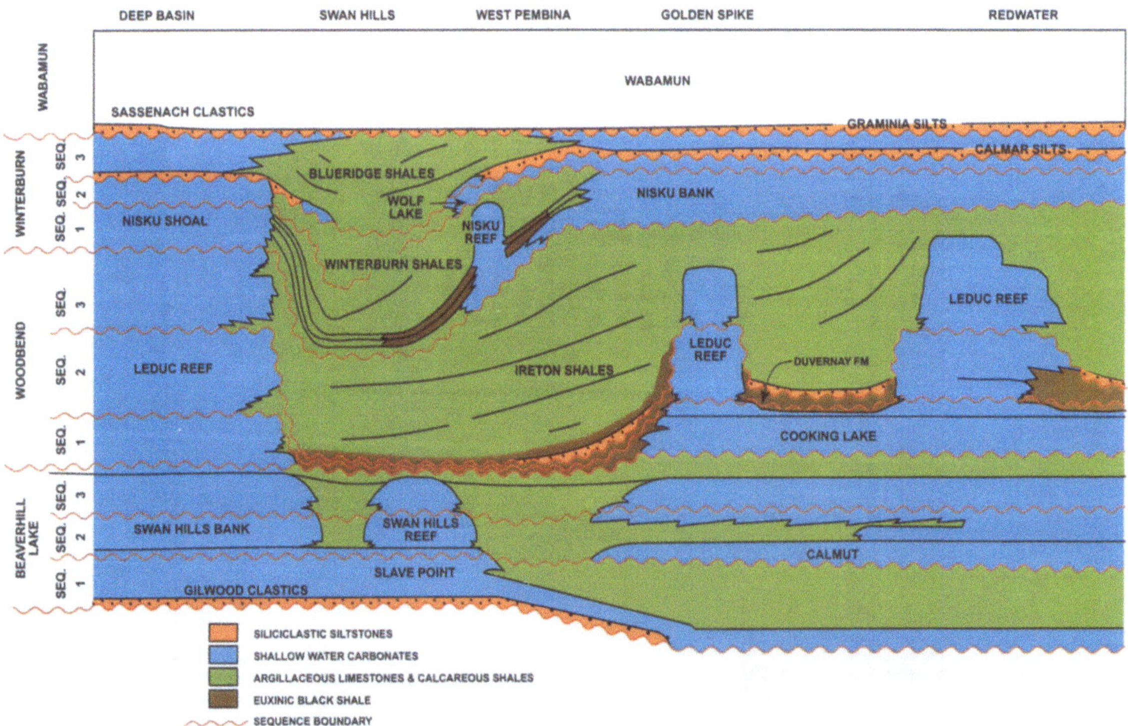


**Fig. 6.15** Schematic sequence stratigraphic cross-section of the Late Givetian to Frasnian succession of central Alberta. Line of section is oriented southeast to northwest Potma et al., 2001, Fig. 2)

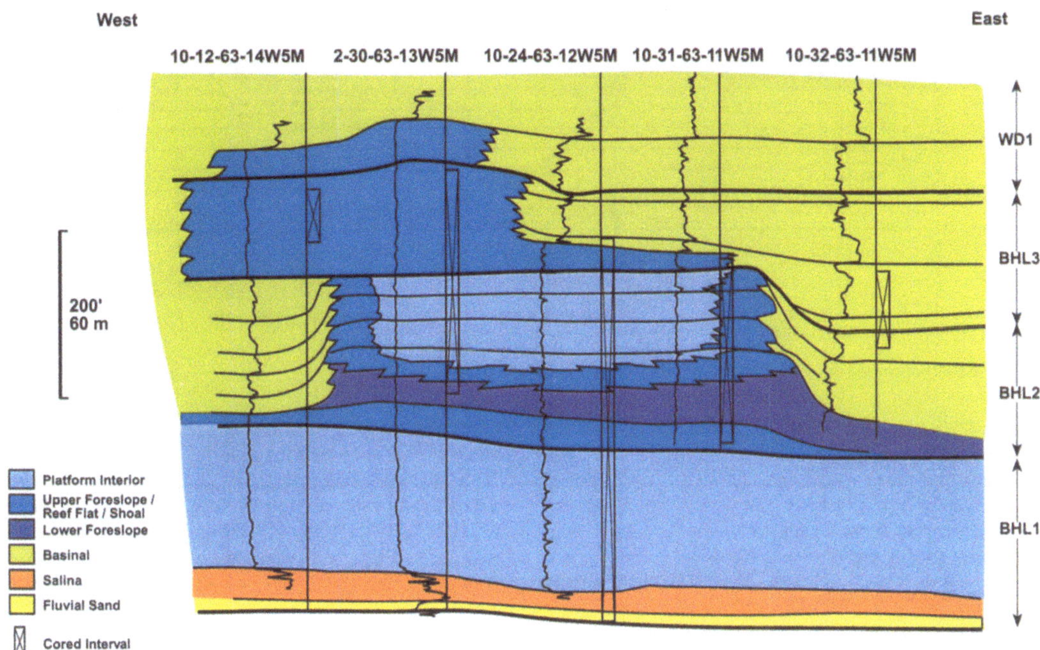


**Fig. 6.16**  Stratigraphy of the Judy Creek West Pool Alberta. Sequences (BHL1, etc.) are as shown in Fig. 6.15 (Potma et al., 2001, Fig. 8)

Sequence BLH1 commences with fluvial sandstone of the Gillwood Formation, interfingering with platform carbonate and evaporites to the east, and resting on a major regional unconformity (Figs. 6.15 and 6.16). The carbonates consist of *Amphipora* grainstones and boundstones. The BLH1-BLH2 boundary is interpreted as a regional unconformity, and is followed by the reef buildups of the Swan Hills Formation. These consist of atolls dominated by the stromatoporoid *Amphipora*, with well-defined foreslopes that can be mapped based on detailed wireline log correlation. Each of the reefs consist of stacked small-scale shallowing-upward cycles capped by a marine flooding surfaces. Potma et al. (2001, p. 47) termed these "parasequence sets" and indicated that they are "extensively correlatable." Each of the parasequence sets may be locally subdivided into metre-scale parasequences, particularly within the lagoonal facies. Sequence BLH3 indicates a westward backstepping of the reef facies into shallower water environments, closer to a regional upwarp, the West Alberta Ridge. This suggests a gradual, long-term deepening during deposition of the Beaverhill Lake Group.

The base of Woodbend sequence 1 is locally marked by the presence of charophyte oogonia, the reproductive part of fresh or brackish-water plants. This is interpreted to represent a fall of sea level of about 5 m. The main part of sequence WD1 consists of the widespread Cooking Lake Platform and the basal Leduc reef, between which the fine-grained clastics of the Ireton Formation prograded from clastic sources to the northeast. The platform has a near-vertical western margin, with a relief of about 100 m. The Leduc "pinnacle reefs" that subsequently developed at the edge of the platform built the platform margin relief to a maximum of more than 300 m, and wireline log correlations suggest that much of that relief was present during deposition, as a major topographic contrast between reef and basin.

Additional detail of WD2 is shown in Fig. 6.17. Outcrop data from the Rocky Mountains show that the sequence consists of seven shoaling-upward parasequence sets representing peritidal cycles.

Figure 6.18 is a model that has been developed to explain the development of the reef and related systems tracts in the Upper Devonian of Alberta. Throughgoing karst surfaces—difficult to pinpoint except where core is available and the geologist knows what to look for—indicate sea-level lowstands. These are commonly present within the reef deposits, and

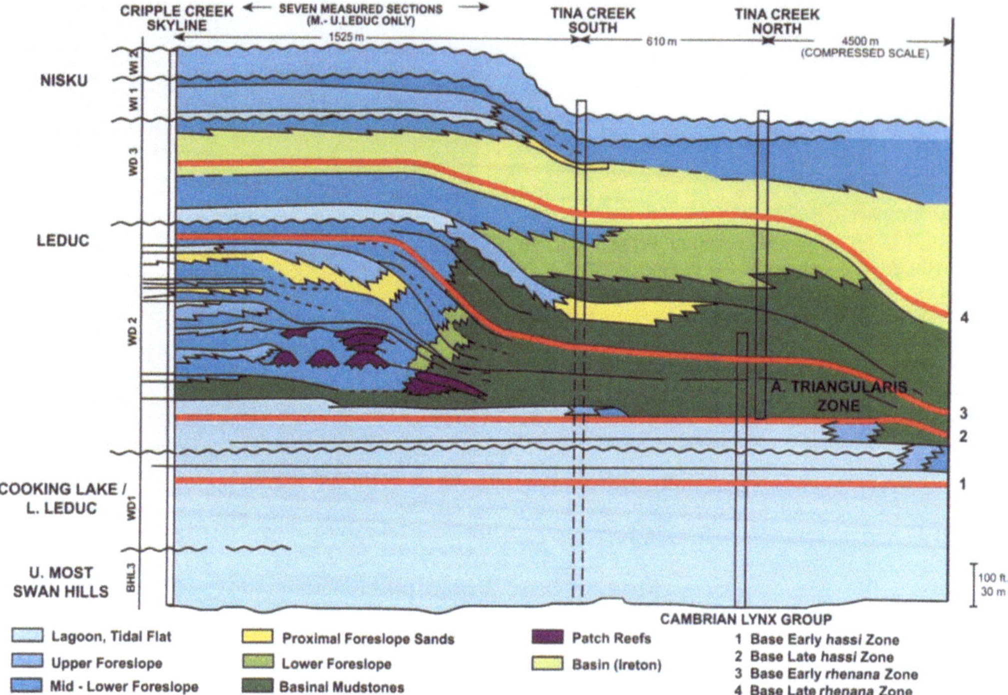


**Fig. 6.17** Sequence stratigraphy of the Cripple Creek reef margin, Alberta, constructed from outcrop and subsurface data (Potma et al., 2001, Fig. 13)

are useful as indicators of sequence boundaries. Potma et al. (2001, pp. 81–82) explain reef development as follows:

> The pattern of sedimentation in the reef complexes is similar for each sequence. A drop in sea level initiates the base of the sequence. The reef complex is exposed and, due to the steep sides of the complex, there are few suitable sites for the reestablishment and growth of carbonate producing organisms downslope. Carbonate production on the atoll effectively ceases. Extensive tidal flats, some dissolution and occasional green shale beds, generally mark the surface. We interpret that the shales were sourced from the Canadian Shield to the northeast, and deposited by (fluvial) by-pass channels, through the exposed carbonate platform margins, into the basin. They form concentrations on the reef platforms due to the lack of dilution by carbonate production during depositional hiati. These shales, and cemented micritic tidal flats, typically form impediments to vertical fluid flow in the reefs. Minor dolomitization of the exposed reef complexes can also occur at this time.
>
> Coarse siliciclastic deposition into the basin is also greater during sea level lowstand, due to fluvial by-pass of the basin-fringing carbonate banks. The major sequence boundary at the base of the third Woodbend sequence has lowstand siliciclastics associated with it …. We correlate these quartz sands to time-equivalent exposure surfaces in the Redwater, Leduc and Golden Spike reef complexes.
>
> Above the sequence boundary, as relative sea level begins to rise in the transgressive portion of the sequence, the complex is again flooded, and carbonate production is re-established. Generally, this occurs somewhat landward of the pre-existing edge of the underlying reef complex. Initially, the sediments commonly consist of tidal flat deposits, but display more open marine character as relative sea level rise accelerates. The transgressive systems tract consists of a series of backstepping, well-circulated lagoon to shoal cycles. These units generally have good primary reservoir quality. The retrogradation creates an accretionary high, forming the locus of progradation during the early stages of the highstand systems tract.
>
> During highstand of sea level, aggradation, progradation or retrogradation of the reef can occur. The first cycle of the highstand is commonly characterized by strong progradation that begins at topographic highs on the underlying transgressive systems tract and then expands seaward toward the platform margin of the pre-existing sequence. The lower foreslope lithofacies at its base are thicker, and these downlap onto the maximum flooding surface of the transgressive systems tract. This part of the section may form a significant vertical permeability barrier at the edges of large reef complexes or over the entire base of base of smaller reefs …, and adds to the tight nature of the rocks adjacent to sequence boundaries.

**Fig. 6.18**  Schematic evolution of reef systems tracts, Devonian, Alberta (Potma et al., 2001, Fig. 30)

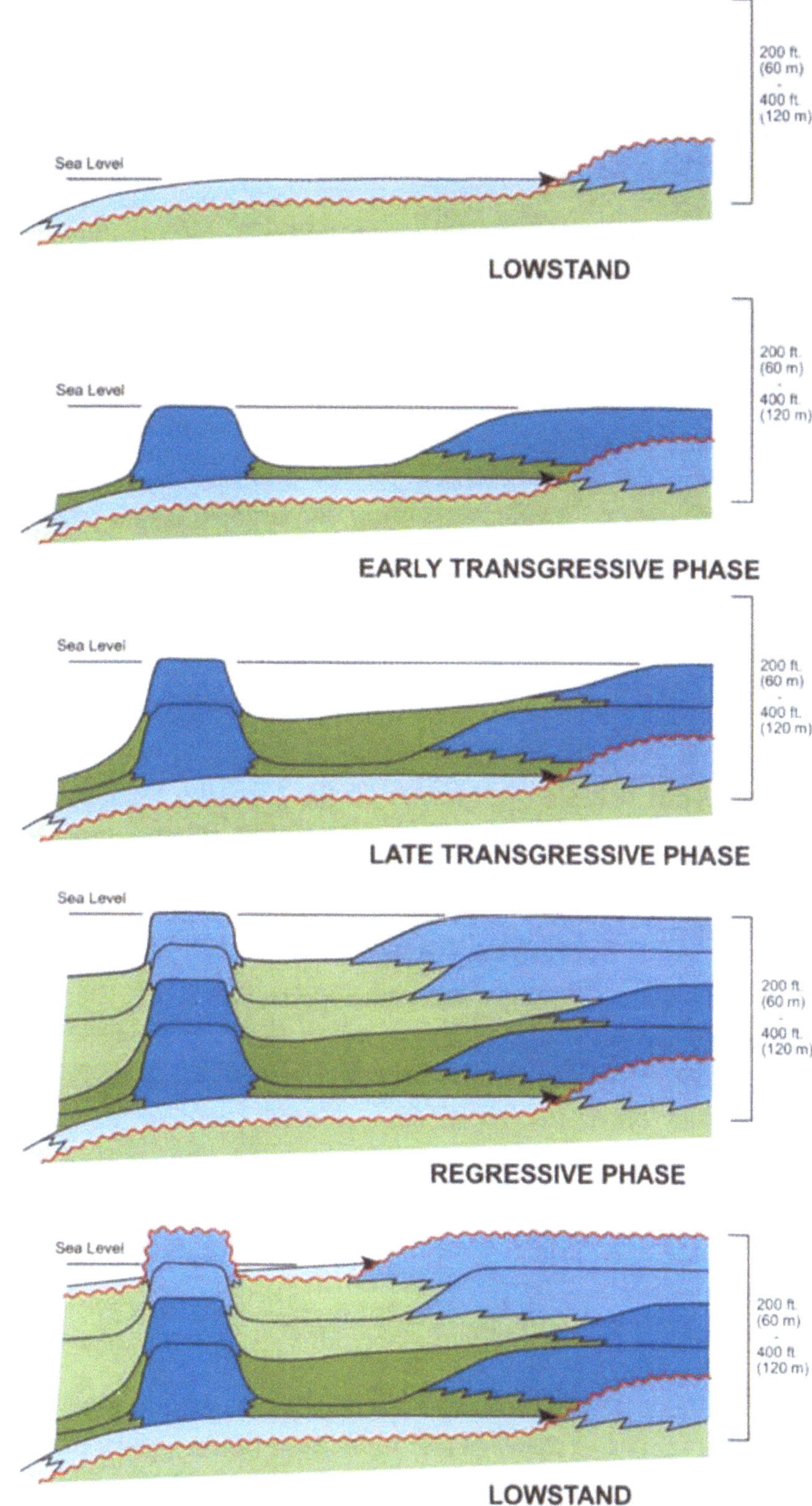

## 6.2 Foreland Basins

### 6.2.1 Foreland Basin of the North American Western Interior

During the Cretaceous, the Western Interior of the United States and Canada formed a vast epicontinental seaway along a foreland basin extending from the Arctic Ocean to the Gulf Coast. The basin was asymmetric, with more rapid subsidence and sedimentation occurring along the western flank of the basin, adjacent to the fold-thrust belt of the Sevier orogen (Fig. 6.19; DeCelles, 2004; Miall et al., 2008). Up to 5 km of sediments accumulated during the Cretaceous. They constitute a classic "clastic wedge" (Fig. 6.20), as this term was defined by Sloss (1962). Weimer (1960) was the first to recognize that the Upper Cretaceous section constitutes a succession of large-scale transgressive-regressive cycles with $10^6$-year episodicities. Figure 6.21 illustrates Weimer's (1986) most recent synthesis of the stratigraphy and age of Cretaceous cycles in the Western Interior Seaway, including some of the key stratigraphic names from the Rocky Mountain and other basins. Major interregional unconformities and their ages in Ma are indicated on this diagram. They do not correlate in any particularly obvious way with the sequence boundaries in

the Exxon charts, unless allowance is made for errors of up to 1 or 2 million years, in which case they all correlate. The relationship between transgressive-regressive cycles and tectonism in the foreland basin was discussed by Fouch et al. (1983) and Kauffman (1984), and is considered at some length in Sect. 10.3.3.1.

Major regressive sandstone wedges within the succession include the Ferron, Emery, Blackhawk, Castlegate and Price River sandstones (Fig. 6.20). Details of the lowermost of these wedges are illustrated in Figs. 6.20 and 6.22, based on the work of Ryer (1984) and Shanley and McCabe (1991). In the latter, the sequences constitute alluvial-coastal plain facies successions ranging from 16 to 180 m in thickness. As shown in Fig. 6.23, some of these sandstone wedges are capable of even further subdivision. These high-order cycles are discussed in Chap. 7. Suffice it to note here that the foreland basin clastic wedge illustrated at the increasing scales of Figs. 6.20, 6.22 and 6.23 displays three scales of sequence cyclicity.

Another well studied cycle in the foreland-basin clastic wedge is that of the Gallup Sandstone and associated beds in San Juan Basin, New Mexico (Fig. 6.24; Molenaar, 1983). The Gallup Sandstone is of Coniacian-Turonian age, and represents approximately 1 million years of sedimentation. The sequence stratigraphy of these rocks has been described by

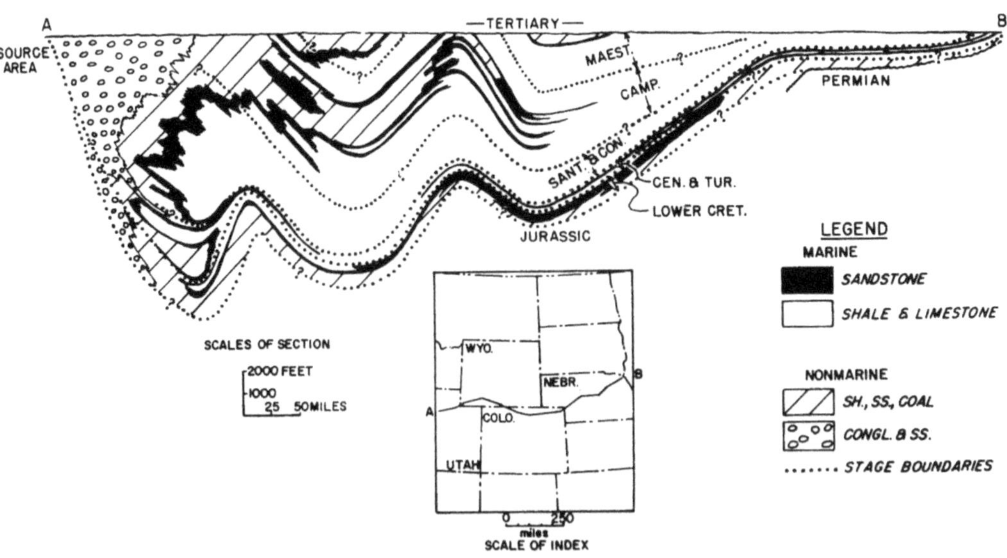


**Fig. 6.19** Diagrammatic restored cross-section through the Upper Cretaceous rocks of the Western Interior Seaway, flattened on a datum at the base of the Tertiary (Weimer, 1970)

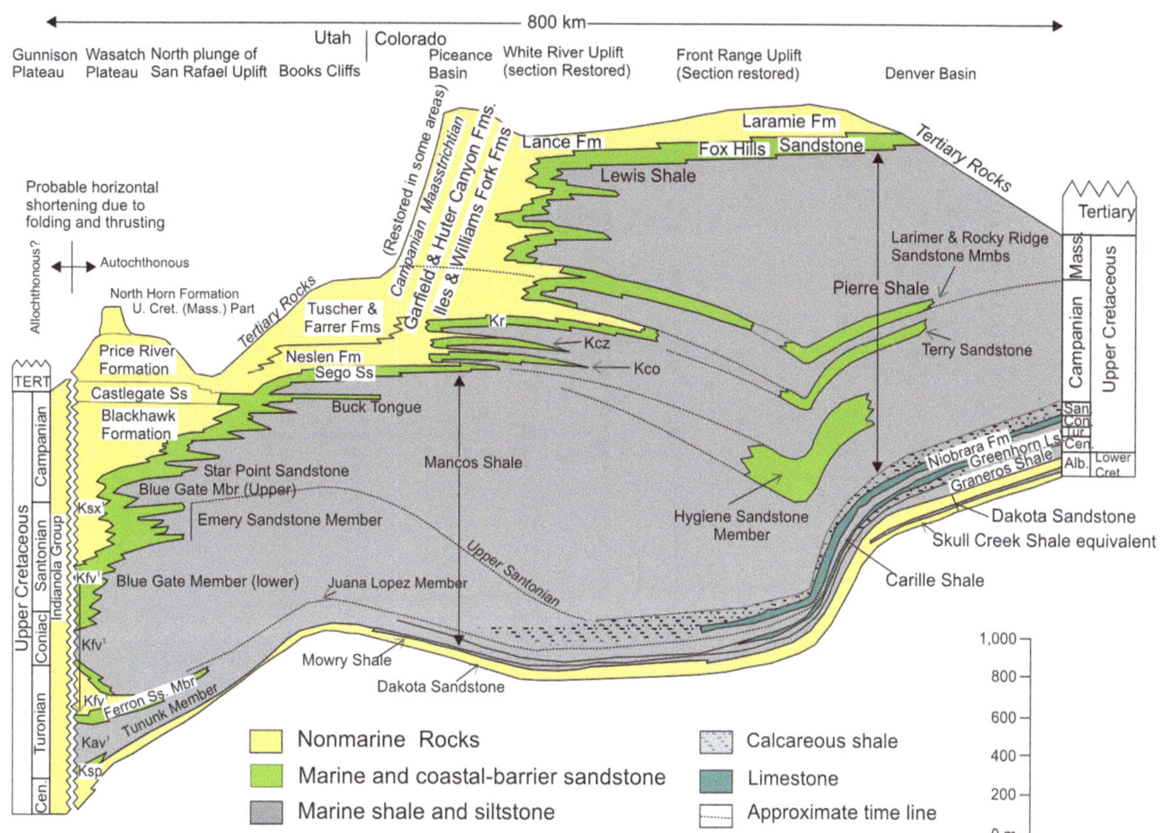


**Fig. 6.20** Enlarged portion of Fig. 6.19, showing details of the stratigraphy of the Upper Cretaceous clastic wedge of Utah and Colorado (Molenaar and Rice, 1988)

Nummedal and Swift (1987), Nummedal et al. (1989), Nummedal (1990), and Nummedal and Molenaar (1995). The Gallup Sandstone can be subdivided into component higher-frequency cycles. It includes component members formed in a variety of shelf, coastal, and nonmarine environments.

A more recent study led by D. Nummedal focused on the identification and mapping of "megasequences" in Wyoming that are clearly of tectonic origin (Liu et al., 2005). These sequences are discussed briefly, here and tectonic mechanisms are considered in depth in Sect. 10.3.3.1. Mapping by R. L. Armstrong, P. G. DeCelles, and many others since the 1960s has unraveled a close relationship between episodes of thrust faulting and uplift along the Sevier fold-thrust belt and the development of coarse, source-proximal conglomerates derived directly from those uplifts. Eastward, the conglomerates pass into nonmarine and shallow-marine coastal plain deposits. Commonly the conglomerates are themselves cut and displaced by the

faults, which is further evidence of the genetic relationship between tectonism and sedimentation. Figure 6.25 summarizes this relationship, as it has been documented in northeastern Utah and Southern Wyoming, and Fig. 6.26 is a synthesis of the stratigraphy of these sequences. Figure 6.27 is a chronostratigraphic chart which synthesizes the stratigraphy, facies and timing of these sequences in relationship to regional and global events.

The definition of the sequences shown in Figs. 6.26 and 6.27 did not automatically emerge from a synthesis of the regional stratigraphy, but required a search for and a recognition of the key indicators of changing regional accommodation that are now know to characterize sequence generation. This is a good illustration of the "genetic" nature of sequence stratigraphy—application of sequence concepts can greatly facilitate synthesis and interpretation, but only if there are existing concepts and models that are appropriate for the field case under study.

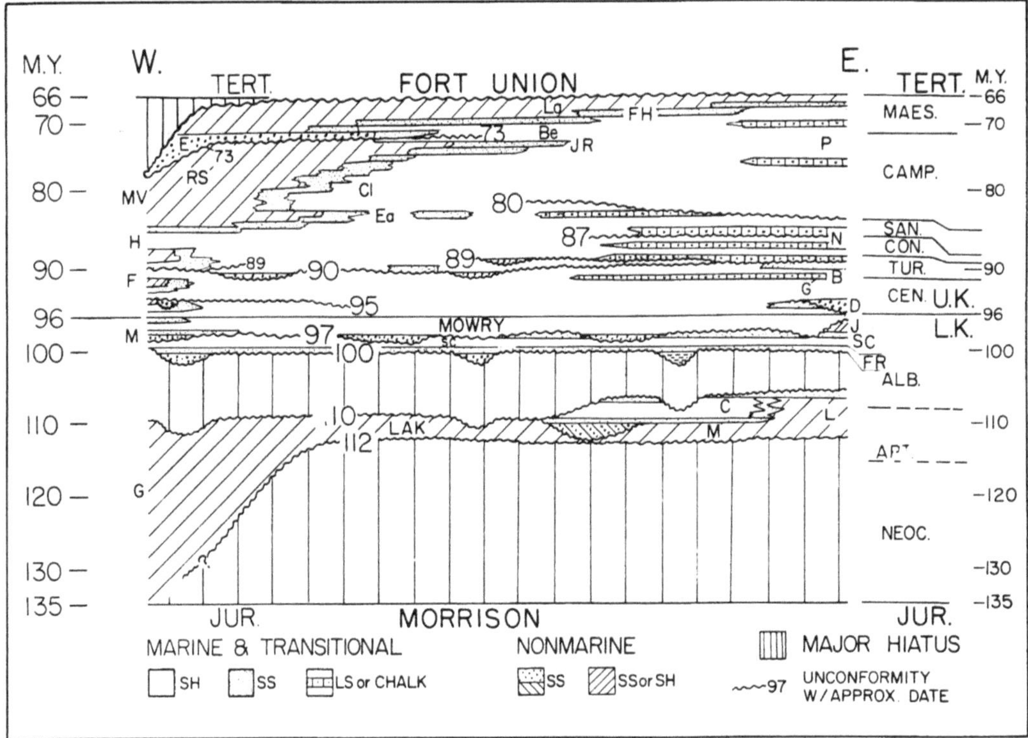


**Fig. 6.21** Diagrammatic west–east cross-section through the Western Interior Seaway of the Rocky Mountains, from Wyoming-Montana in the west to eastern Colorado-Black Hills-eastern Alberta in the east, showing stratigraphic positions and approximate dates of major transgressive units and interregional unconformities. Formations or groups to the west are: G, Gannett; SC, Skull Creek; M, Mowry; F, Frontier, H, Hilliard, MV, Mesaverde; RS, Rock Springs; E, Ericson; Ea, Eagle; Cl, Claggett; JR, Judith River; Be, Bearpaw; FH, Fox Hills; La, Lance. To the east formations are L, Lytle; LAK, Lakota; FR, Fall River; SC, Skull Creek; J and D sands of Denver basin; G, Greenhorn; B, Benton; N, Niobrara; P, Pierre; M and C, McMurray and Clearwater of Canada (Weimer, 1986)

This illustrates both the strengths and the pitfalls of the method. In the case under study here, sequence mapping was facilitated by the appropriate choice of datum for the construction of the stratigraphic synthesis (Fig. 6.26). This is not a mundane methodological issue, but may become a key to the elucidation of stratigraphic relationships. The datum, in this case, was placed at the base of the Canyon Creek Member of the Ericson Formation, which clarifies the progradational nature of the units above, and introduces as little distortion as possible to the complex tectono-stratigraphic relationships of the units below. The five sequences into which the succession has been divided were recognized in the basis of "an iterative process, in which regional unconformities and/or surfaces across which there was a demonstrable rapid change in subsidence regime were the chosen boundaries" (Liu et al., 2005, p. 493). Correlation of the units westward, through the facies change into the syntectonic conglomerates, was also an important criterion. Using accommodation cycles (such as those summarized in Chap. 2 of this book) as a model for interpretation, lithostratigraphic units that have long been known in this area could be assigned their appropriate position in the succession of systems tracts. Specific marine shales could then be identified as transgressive deposits or representing maximum flooding intervals, upward-coarsening transitions from coastal-plain sandstone to coarse conglomerate could be assigned to highstand systems tracts, and so on.

A sequence subdivision of the mid-Jurassic to Paleocene fill of the Alberta foreland basin is shown in Fig. 6.28. The thick, coarse, predominantly nonmarine clastic wedges, including the Kootenay,

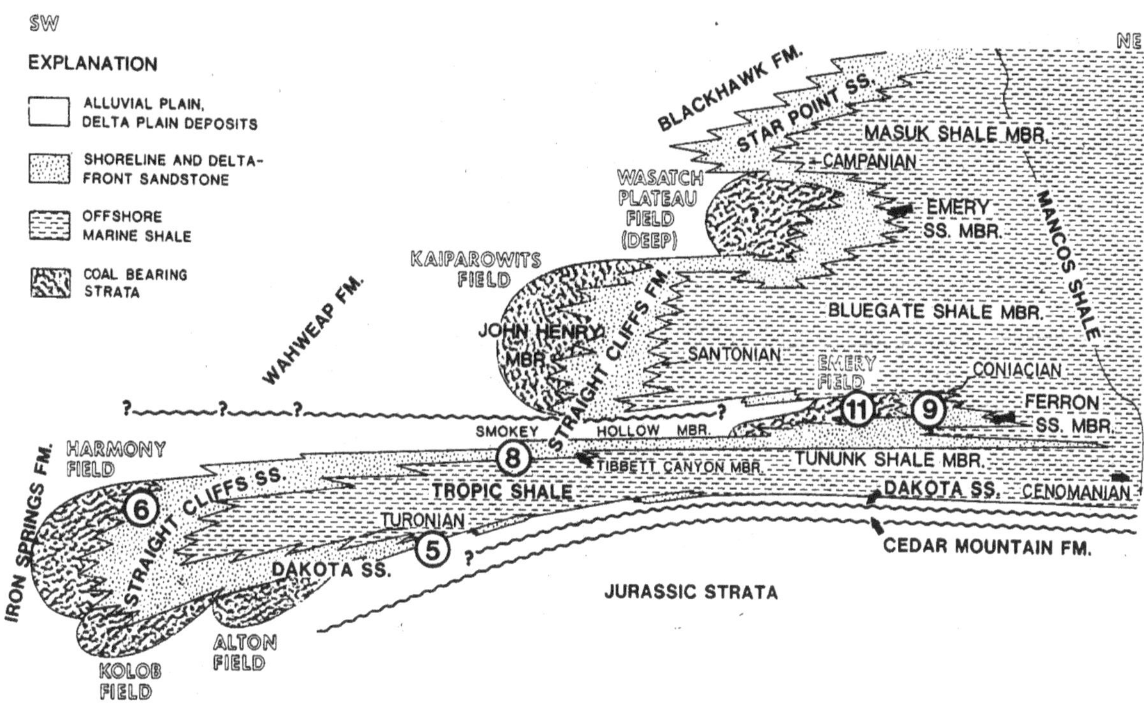



**Fig. 6.22** Diagrammatic cross-section through the Ferron Sandstone and equivalent beds, southwestern Utah, showing the major (third-order) clastic cycles (Ryer, 1984)

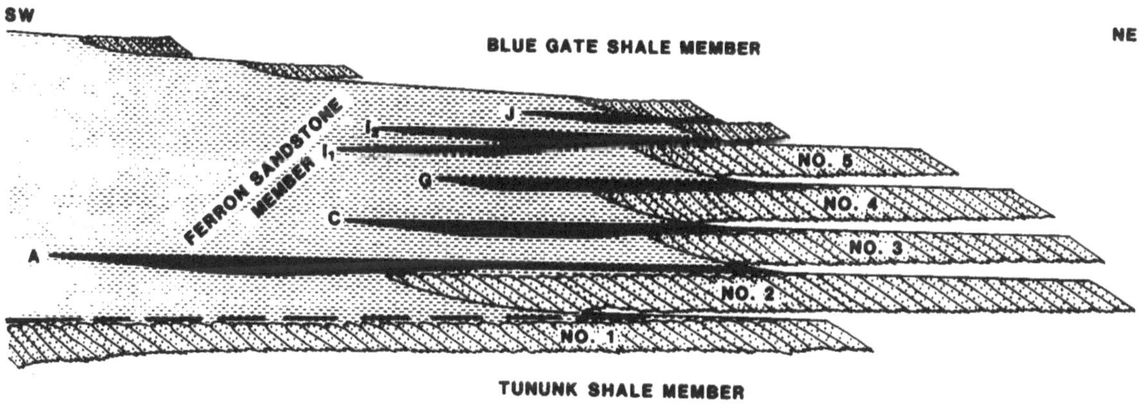



**Fig. 6.23** Schematic cross-section of the Ferron Sandstone, central Utah. The stratigraphic position of these beds within the foreland-basin clastic wedge is shown in Fig. 6.22 (Ryer, 1984)

Mannville, Belly River-Edmonton and Paskapoo, each span several million years and would be classified as second- or third-order sequences using the Vail et al. (1977) classification. The relationship of these sequences to the orogenic development of the Cordillera is discussed in Chap. 10. The Mannville represents much of the Aptian and Albian stages, totaling 12–14 million years, which places it within the second-order classification of Vail et al. (1977). Note, however, that in Fig. 6.28 the Miannville is classified as one of several third-order sequences, and can be subdivided into a suite of constituent fourth-order sequences, additional details of which are shown in Fig. 6.29. Once again the "order" classification does not seem to provide much in the way of useful insights into the origins of these sequences, especially given

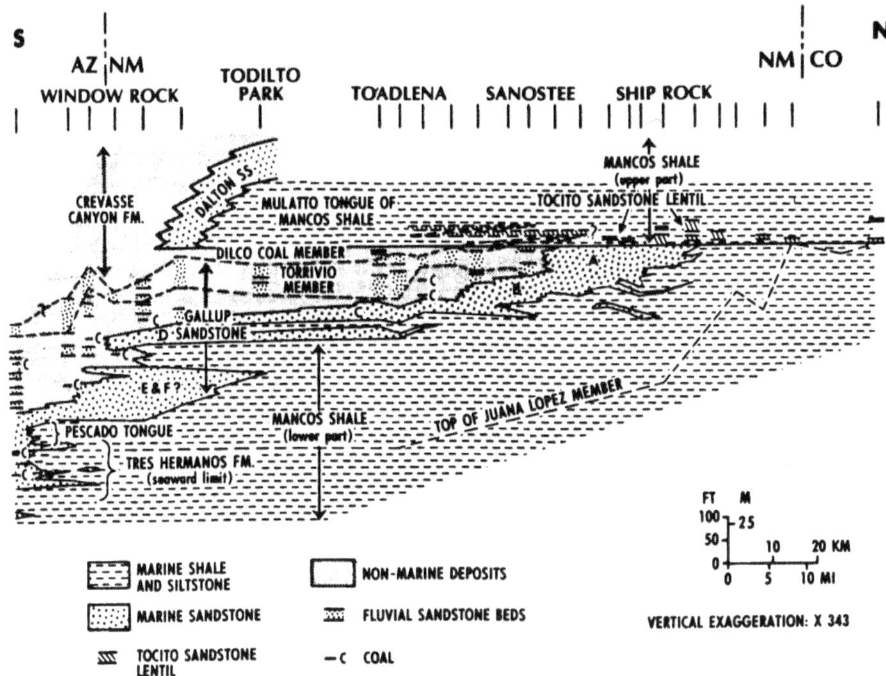


**Fig. 6.24** Summary of the sequence stratigraphy of the Gallup Sandstone and associated strata, San Juan Basin, New Mexico (Nummedal, 1990)

that, as shown in Fig. 6.28, some of the unconformities between the sequences may represent longer time spans than the sequences, and therefore fall into a higher order in the classification!

The Mannville can be subdivided into a lower suite of nonmarine to estuarine units, commencing with the Cadomin Conglomerate (Fig. 6.29). This is a thin but very widespread unit, with equivalents such as the Burro Canyon, Cloverly and Lakota in the United States, indicating aggradation of a lowstand deposit following a long and widespread regional unconformity (Heller and Paola, 1989) This lower suite of units constitutes the lowstand to transgressive systems tracts of the overall Mannville sequence. The Falher and Notikewin formations constitute the highstand systems tract of the Mannville sequence. Nonmarine coastal-plain deposits of the Upper Mannville pass north-westward, in northwestern Alberta and northeastern British Columbia, into a coastal zone where marine-to-nonmarine facies transitions permit a subdivision into a suite of high-frequency sequences constituting the Falher Formation. Prograding shoreface sandstones and conglomerates are capped by flooding surfaces in a landward (SE) direction, and downlap northwestward onto the maximum flooding surface of the Mannville

sequence. We discuss these high-frequency sequences in Chap. 7.

### 6.2.2 Other Foreland Basins

The Alpine foreland basins of western Europe have been intensively studied in recent years, aided by reflection-seismic data, magnetostratigraphic and refined biostratigraphic correlation, structural mapping and modern sedimentological methods (Mascle et al., 1998). A great deal has been learned from this work about the relationship between sedimentation and tectonics. We touch here briefly on one of these recent studies, that of the early-middle Eocene (Ypresian-Lutetian) succession in the Tremp-Ager sub-basin in the southern Pyrenees (Nijman, 1998). This is classified as a piggyback basin, but it correspond to the deepest and widest part of the Ebro foreland basin on the southern flank of the Pyrenean fold-thrust belt. The basin fill rests on, is cut by, and is overlain by thrust faults.

A paleogeographic block diagram of the basin is shown in Fig. 6.30. It shows that the basin filled longitudinally, from east to west, with sediment

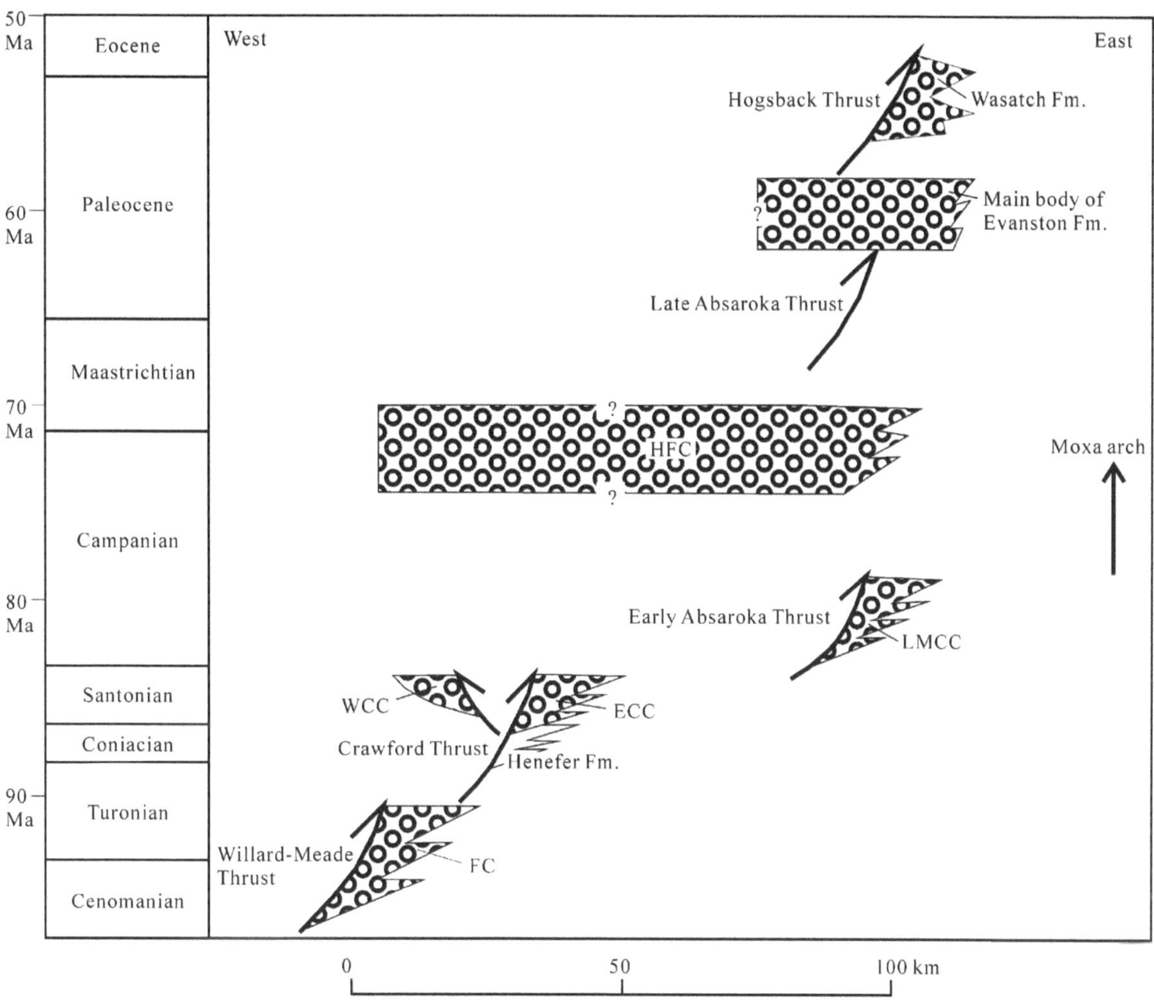


**Fig. 6.25** The time-space relationship between thrust fault episodes and synorogenic conglomerate deposition in northeastern Utah and southern Wyoming. HFC, Hams Fork Conglomerate; LMCC, LittleMuddy Creek Conglomerate; WCC, Weber Canyon Conglomerate; ECC, Echo Canyon Conglomerate; FC, Frontier Conglomerate (Liu et al., 2005, Fig. 3)

derived in part from the northeast, and in part from the southeast. Deposition was in part contemporaneous with and in part modulated by tectonism, the nature of which is discussed further in Sect. 10.3.3.3. Environments varied down dip from alluvial fan to alluvial plain, to deltaic, to submarine fan in the west.

Cyclicity in the basin fill is present at three nested scales, two of which are shown in Fig. 6.31. Three "megasequences," UM-C, UM-D and UM-E, represent major cycles in the Upper Montanya Group and range up to 200 m in thickness. These may be divided into "sequences." Megasequence UM-D is divisible into six such sequences (Fig. 6.31). These range from 20 to 40 m in thickness and are compared by Nijman (1998, p. 141) to parasequences. He noted that the fluvial Castisent Sandstone, a lateral equivalent of the deltaic Middle Montanya (Fig. 6.32), contains a high-order cyclicity. The various subenvironments exhibited by these deposits are indicated by the colours in Fig. 6.31. This is also summarized in Fig. 6.32, which provides a chronostratigraphic classification of the sequences, based on marine micropaleontology. This shows that during the 11.5 million-year time span of the Ypresian-Lutetian, nine megasequences developed, indicating that they average 1.3 million years in duration. Nijman (1998) noted that

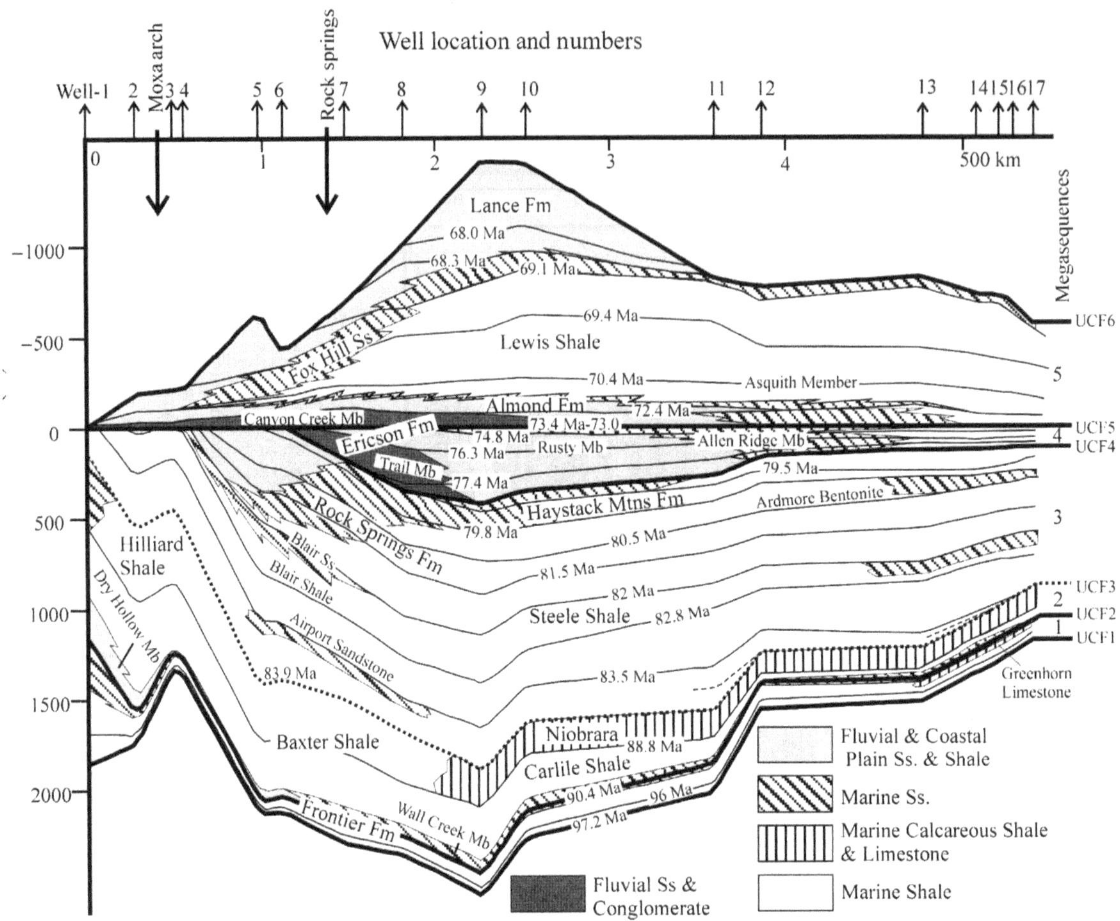


**Fig. 6.26** The stratigraphy of Upper Cretaceous (mid-Cenomanian-Maastrichtian) "megasequences" in southern Wyoming. UCF=unconformity (Liu et al., 2005, Fig. 4)

there is no correspondence between thicknesses, age ranges and rates of sedimentation calculated for each of the megasequences, which suggests a non-regular generating mechanism. He also noted a general correspondence of some of the ages of the megasequence boundaries with sea-level events in the Haq et al. (1987) sea level curve (Fig. 6.32). It is a curious feature of some research literature, to which we return in Sect. 10.3.3.3, that basins which are clearly filled under the influence of active syndepositional tectonism, are nonetheless simultaneously interpreted with reference to the global cycle chart, as though sequence boundaries can be generated simultaneously by global sea-level change and regional tectonism.

Analysis of the Appalachian foreland basin (Ettensohn, 1994, 2008) has demonstrated the existence of stratigraphic packages very similar to those of the Alberta foreland basin and the Pyrenean basin. Figure 6.33 illustrates the "tectophases" of Devonian and early Mississippian age. These four phases span about 66 million years and therefore average 16.5 million years in duration. They contain within them constituent cycles that exhibit similar facies successions. Each cycle commences with a carbonate or calcareous sandstone unit resting on an unconformity. This passes up into a black shale which, in most cases, onlaps the carbonate to rest on a major pre-Devonian unconformity. The shale then passes up into a thick clastic succession of sandstones and shales that together constitute the well-known Catskill "delta." The five cycles that constitute the third tectophase in Fig. 6.33 span the mid-Givetian to late Famennian, and therefore average about 5.5 million years in duration.

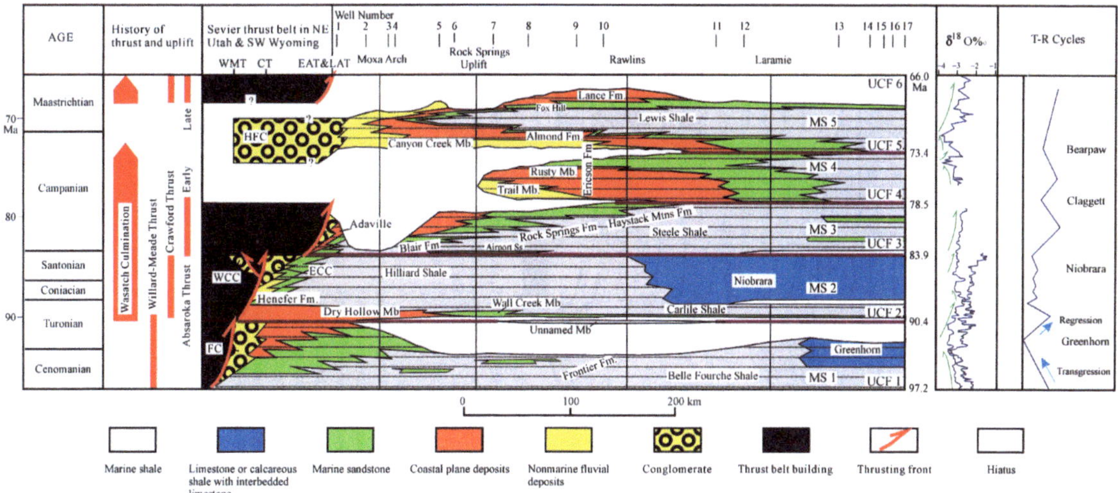

**Fig. 6.27** Correlation chart for the Upper Cretaceous "megasequences" of southern Wyoming, showing facies variations, relationship to thrusting episodes and, at *right*, the oxygen isotope curve from Abreu et al. (1998) and the T-R cycles of Kauffman (1984). WMT, Willard-Meade thrust; CT, Crawford thrust; EAT, Early Absaroka thrust; LAT, Late Absaroka thrust. MS = megasequence, UCF = unconformity (Liu et al., 2005, Fig. 5)

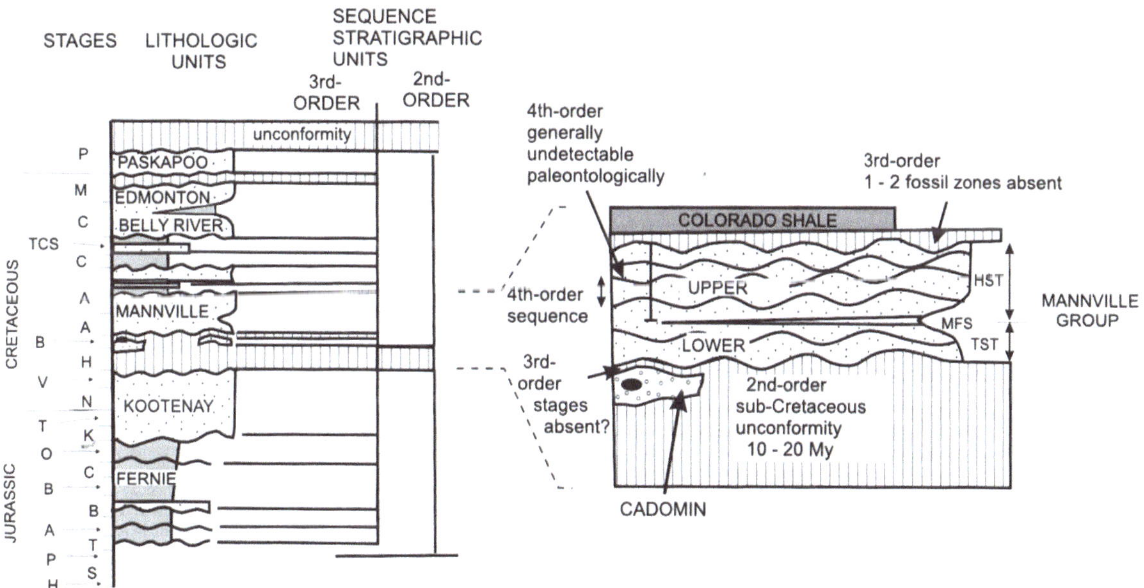

**Fig. 6.28** A sequence subdivision and classification of the clastic wedges of the Alberta Basin (Cant, 1998, Fig. 1)

## 6.3 Arc-Related Basins

### 6.3.1 Forearc Basins

Forearc and backarc basins occur within convergent continental margins, so-called "active margins", a term which emphasizes the importance of tectonism in controlling stratigraphic architectures. Several recent studies of arc-related basins have examined the basin fills from the perspective of sequence stratigraphy, and in many cases major differences with the stratigraphic styles of extensional and rifted margins, and even with foreland basins, have become apparent. There is little

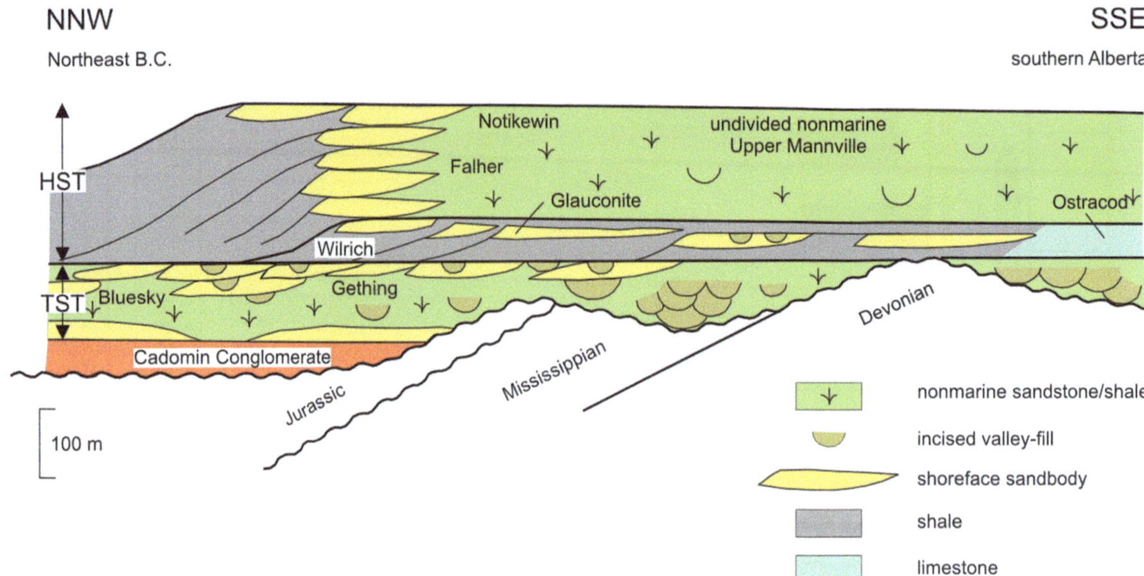



**Fig. 6.29** Composite, simplified cross-section of the Mannville Group of Alberta, oriented along the axis of the basin, The group represents a single large-scale sequence, spanning 12–14 million years, which can be subdivided into a suite of higher-order sequences, each of which is marked by a regressive shoreline sandbody and an onlap surface (adapted from Cant, 1996)

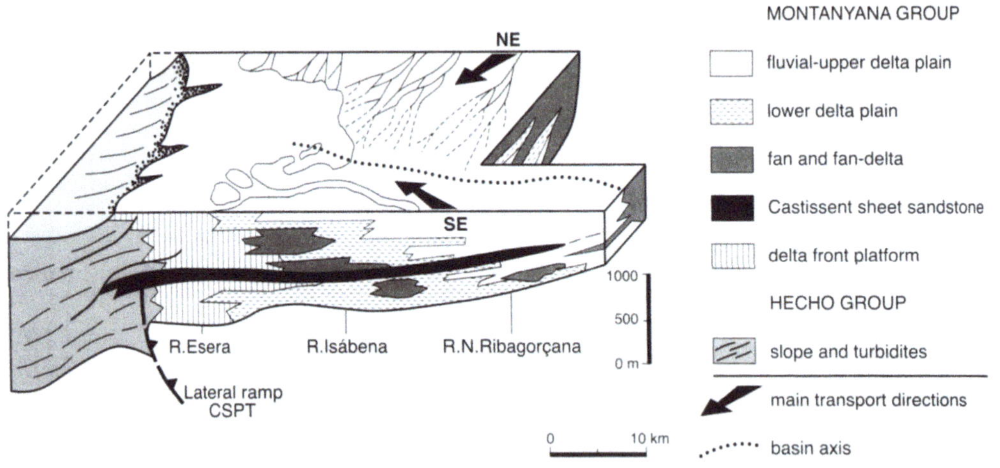



**Fig. 6.30** Block diagram of the Tremp-Ager basin, northern Spain, during the Eocene. The front panel of the diagram is oriented west–east, showing that the axis of the basin runs approximately ESE-WNW, and is filled, in part, by progradation in that direction. The source of alluvial detritus from the northeast is from the Pyrenean uplift, that from the southeast, from the Iberian hinterland (Nijman, 1998, Fig. 5)

clear evidence for the existence of cycles caused by $10^6$-year eustatic sea-level cycles. This contrasts with the record of $10^{4-5}$-year cyclicity, including that of glacioeustatic origin which is locally prominent in arc-related basins and has been mapped and documented in detail, for example, within the Japanese islands (Figs. 7.21 and 7.22), and North Island New Zealand (Fig. 4.11).

Several studies of arc-related basins have been carried out in Nicaragua and Costa Rica (Seyfried et al., 1991; Schmidt and Seyfried, 1991; Kolb and Schmidt, 1991; Winsemann and Seyfried, 1991). In general,

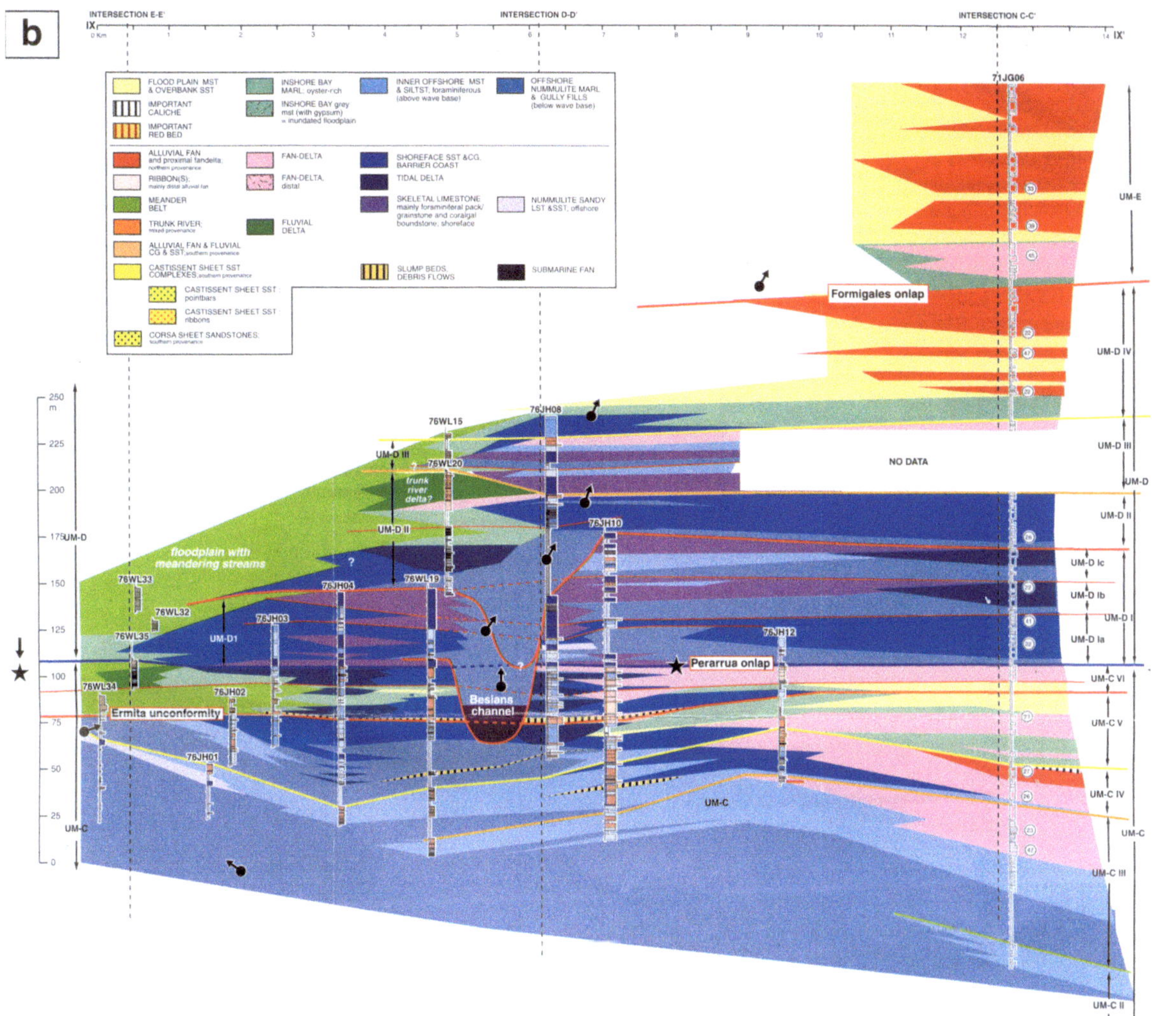



**Fig. 6.31** South–north stratigraphic reconstruction across the Tremp-Ager basin, northern Spain, showing the various depositional environments, contact relationships and, at *right*, the subdivision of the succession into sequences and megasequences. The *arrows* attached to *small black circles* shown the gradual shift in the position of the basin axis, as a result of the changing balance in sediment supply from the north and the south (Nijman, 1998, Fig. 4b)

they demonstrate the importance of convergent tectonism as a dominant control in the development of stratigraphic architecture. An example of a sequence-stratigraphic framework of Miocene age in a fore-arc basin in Nicaragua was provided by Kolb and Schmidt (1991; Fig. 6.34). Fan deltas developed during episodes of sea-level fall, as pyroclastic and alluvial deposits spread across an exposed shelf. Coastlines retrogressed far inland during periods of sea-level rise. Correlations of these and other sections in Central America with the Exxon global cycle chart seem forced and unconvincing. Biostratigraphic evidence for the correlations is extremely limited. In most cases, it is evident that folding, faulting, tilting and tectonic uplift and subsidence are the major sedimentary controls (Seyfried et al., 1991; Schmidt and Seyfried, 1991). Seyfried et al. (1991) documented the presence of regional angular unconformities that can be mapped for distances of 900 km, as discussed in Sect. 5.3.2 (Fig. 5.22).

**Fig. 6.32** Composite chronostratigraphic cross-section through the Montanyana Group in the Tremp-Ager basin, showing the biostratigraphic zonation and interpreted ages of the sequences. The development of this basin, including the shift in the basin axis (shown here) is discussed in Sect. 10.3.3.3. (Nijman, 1998, Fig. 13)

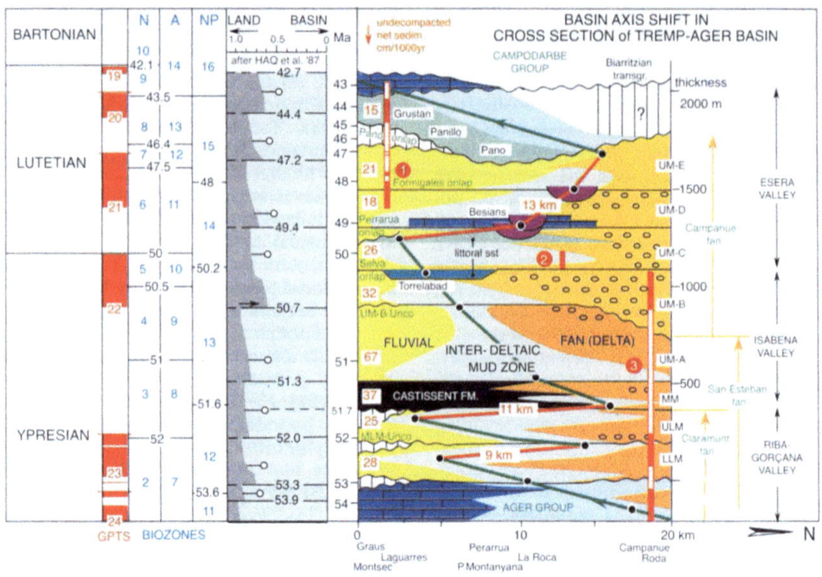

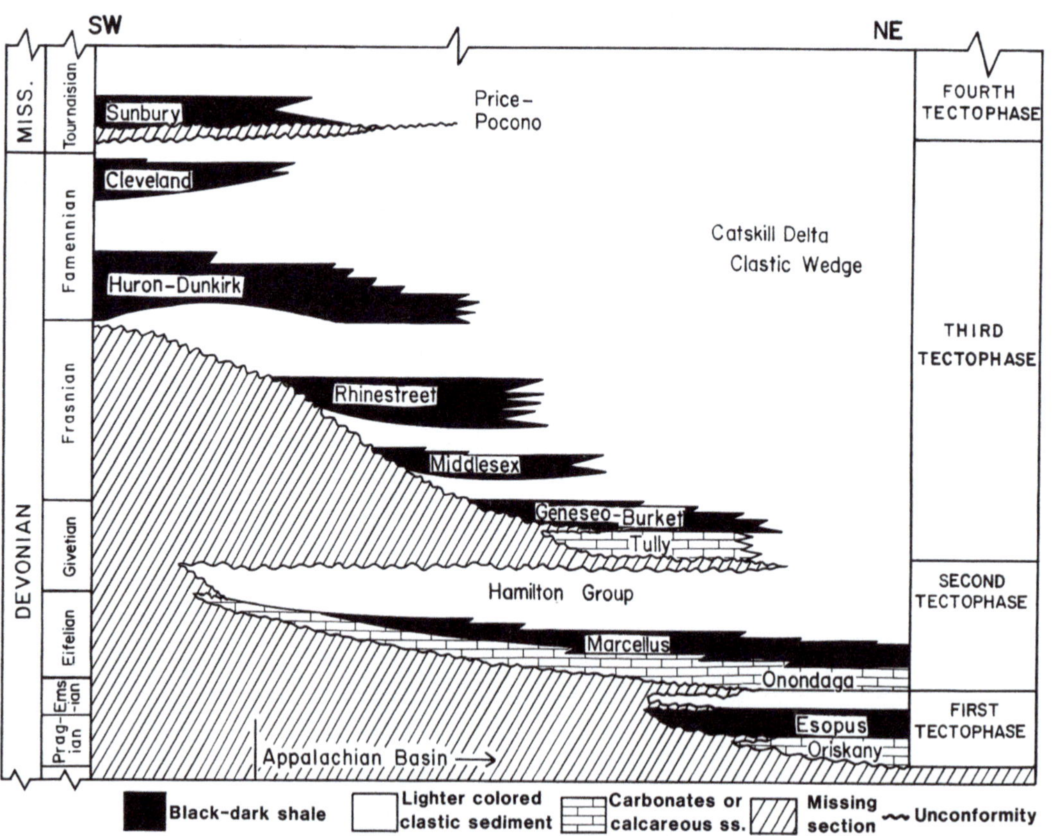



**Fig. 6.33** Composite stratigraphic section from east-central New York to north-central Ohio, across part of the Appalachian foreland basin, showing how the Devonian stratigraphic succession may be subdivided into "tectophases" and cycles (Ettensohn, 1994, Fig. 13)

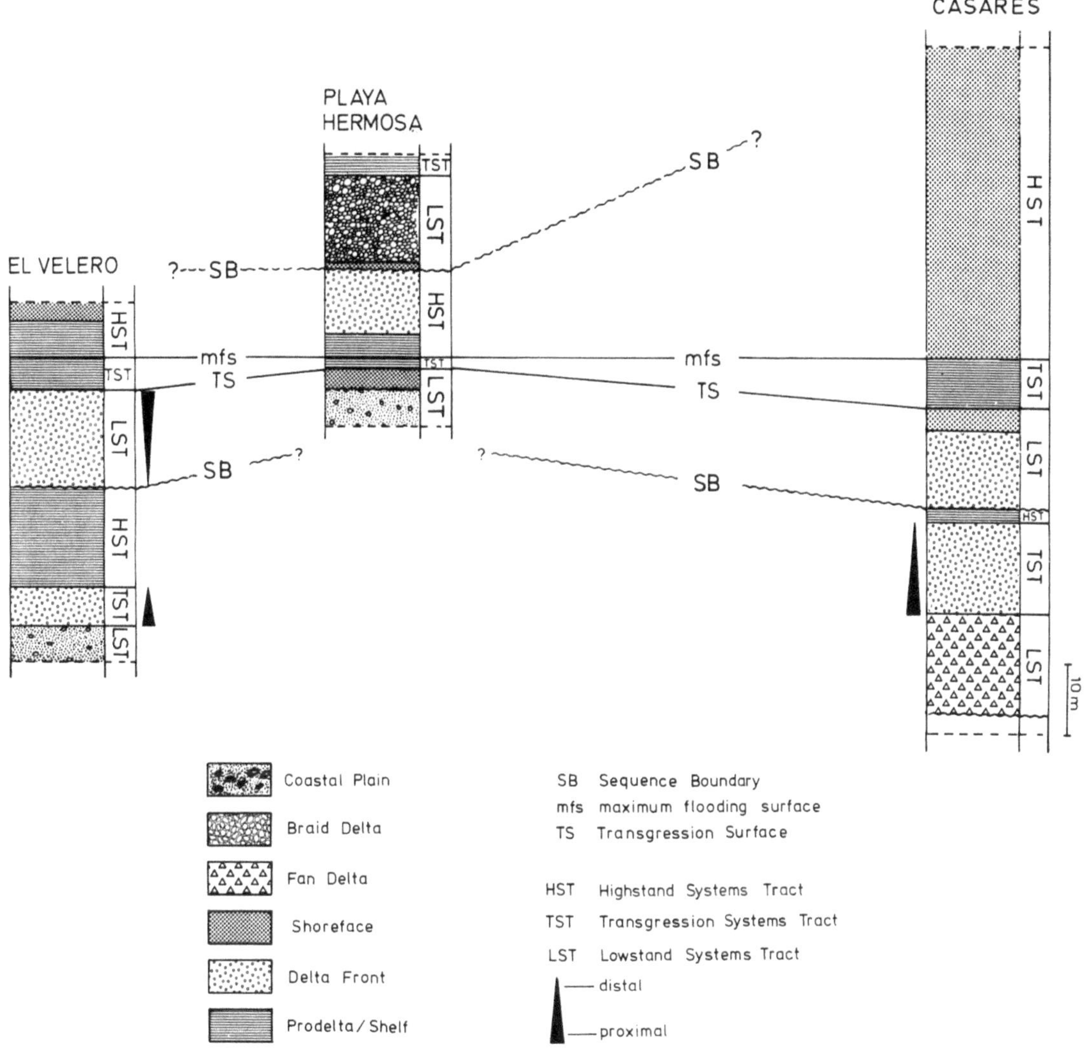



**Fig. 6.34**   Sequence framework, Miocene beds of southwestern Nicaragua (Kolb and Schmidt, 1991)

In some cases, volcanic control of the sediment supply is the critical factor. Thus Winsemann and Seyfreid (1991, pp. 286–287) stated

The formation of depositional sequences in the deepwater sediments of southern Central America is strongly related to the morphotectonic evolution of the island-arc system. Each depositional sequence reflects the complex interaction between global sea-level fluctuations, sediment supply, and tectonic activity. Sediment supply and tectonic activity overprinted the eustatic effects and enhanced or lessened them. If large supplies of clastics or uplift overcame the eustatic effects, deep marine sands were also deposited during highstand of sea level, whereas under conditions of low sediment input, thin-bedded turbidites were deposited even during lowstands of sea level.

These tectonic and sediment-supply considerations are of paramount local importance, as discussed in Chap. 10.

A Jurassic-Cretaceous forearc basin succession in the Antarctic Peninsula is dominated by major facies changes at intervals of several millions of years, suggesting a control by "third-order" tectonism (Butterworth, 1991). One major, basin-wide stratigraphic event, an abrupt shallowing, followed by a transgression, gave rise to a unit named the Jupiter Glacier Member (JGM in Fig. 6.35), consisting of deepening-upward shelf deposits. Butterworth (1991) suggested correlation with a major eustatic low near the Berriasian-Valanginian boundary indicated on the

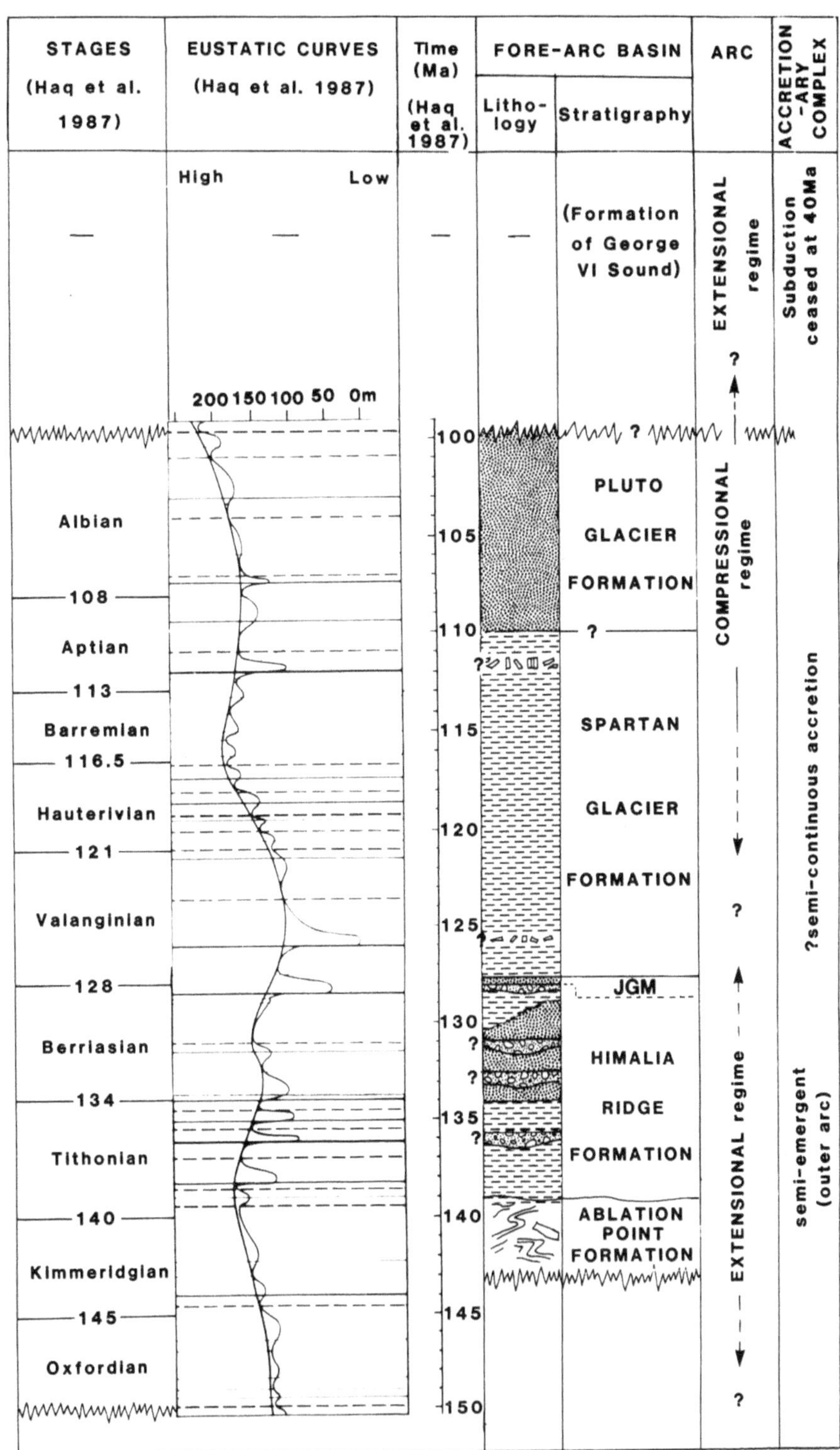


**Fig. 6.35** Stratigraphy of the southern Antarctic Peninsula, compared to the Exxon global cycle chart (Haq et al., 1987, 1988a). One stratigraphic event, a shallowing that gave rise to the Jupiter Glacier Member (JGM), may correlate with the Berriasian-Valanginian eustatic low on the cycle chart (Butterworth, 1991)

Exxon chart (Fig. 6.35). However, this seems somewhat fortuitous. It is the only such correlation that can be proposed for this basin. Butterworth (1991) discussed the possibility of tectonic control of this particular event and suggested (p. 327): "It is unlikely that many eustatically-controlled sea-level fluctuations will be recognized from forearc basins, unless it is possible to demonstrate that there were prolonged periods of tectonic quiescence."

A stratigraphic synthesis of the Cenozoic deposits of Cyprus by Robertson et al. (1991) also resulted in tentative (and rather tenuous) correlations of some stratigraphic events with the Exxon global cycle chart (Fig. 6.36). The evidence indicates that subduction and strike-slip deformation had a dominant effect on sedimentation patterns in this area. During the late Miocene (Messinian) plate-tectonic events led to the isolation and desiccation of the entire Mediterranean basin, an event that is clearly recorded in Cyprus (Fig. 6.36), at a time when global sea levels underwent several fluctuations, according to Haq et al. (1987, 1988a).

### 6.3.2 Backarc Basins

The tectonic evolution of backarc basins may be similar to that of extensional margins, especially along the interior, cratonic flanks of the basin. Legarreta and Uliana (1991) found that the Neuquén Basin, a backarc basin flanking the Andes in Argentina, underwent an "exponential thermo-mechanical subsidence" pattern, following an early Mesozoic thermal event. Sediment-supply conditions along the cratonic flank of a backarc basin are also likely to be comparable to those of extensional margins. For this reason these basins may show stratigraphic patterns comparable to those on Atlantic-type margins, including the presence of major carbonate suites (Sect. 6.1.2), relatively mature clastics, and a sequence architecture containing evidence of cyclicity with $10^6$–$10^7$-year episodicities. Two studies of Andean basins confirm that this is the case. Legarreta and Uliana (1991) described a sequence stratigraphy that they correlated directly with the Exxon chart, while Hallam (1991), summarizing his own work plus that of various South American geologists, developed his own regional sea-level curve that contains transgressive-regressive cycles with $10^6$–$10^7$-year frequencies.

## 6.4 Cyclothems and Mesothems

A unique type of high-frequency cyclicity characterizes the Carboniferous and Lower Permian strata of much of the American midcontinent, northwest Europe and the Russian platform (Ross and Ross, 1988; Heckel, 1990, 1994). As discussed in Chap. 11 there is general agreement that these cycles are glacioeustatic in origin. The Carboniferous-Early Permian corresponds to the time when major continental ice caps were forming and retreating throughout the great southern Gondwana supercontinent (Caputo and Crowell, 1985; Veevers and Powell, 1987; Eyles, 1993, 2008), while the areas where cyclothems and mesothems occur lay close to the late Paleozoic paleoequator. The term *cyclothem* was proposed by Wanless and Weller (1932) following their study of these cycles in the Upper Paleozoic rocks of the American midcontinent. They exhibit a periodicity in the $10^4$–$10^5$-year time band. The term cyclothem, and the cycles to which it applies, are discussed in Chap. 11.

Ramsbottom (1979) noted that unusually extensive transgressive and regressive beds forming the sequence boundaries between some of the cyclothems enable groups of about four or five of them to be combined into larger cycles showing a $10^6$-year periodicity, and he proposed the term *mesothem* for these. Moore (1936) and Wagner (1964) termed groups of cyclothems *megacyclothems*, but Heckel (1986) showed that these are higher-order cycles than the mesothems discussed here. Holdsworth and Collinson (1988) used the term *major cycle* for Ramsbottom's mesothems. They compare in duration to the "third-order" cycles of Haq et al. (1987, 1988a). Ramsbottom (1979) identified nearly forty such cycles in the Carboniferous of northwest Europe, and showed that their average duration ranged from 1.1 million years in the Namurian to 3.6 million years in the Dinantian. A chronostratigraphic chart of these cycles as they occur in Britain is shown in Fig. 6.37, and a more detailed Wheeler diagram, showing the main lithostratigraphic components of the Namurian mesothems of part of northern England, is given in Fig. 6.38.

In the Namurian, each mesothem consists of a muddy sequence at the base containing several cyclothems, followed by one or more sandy cyclothems. The lower, muddy parts of the mesothems

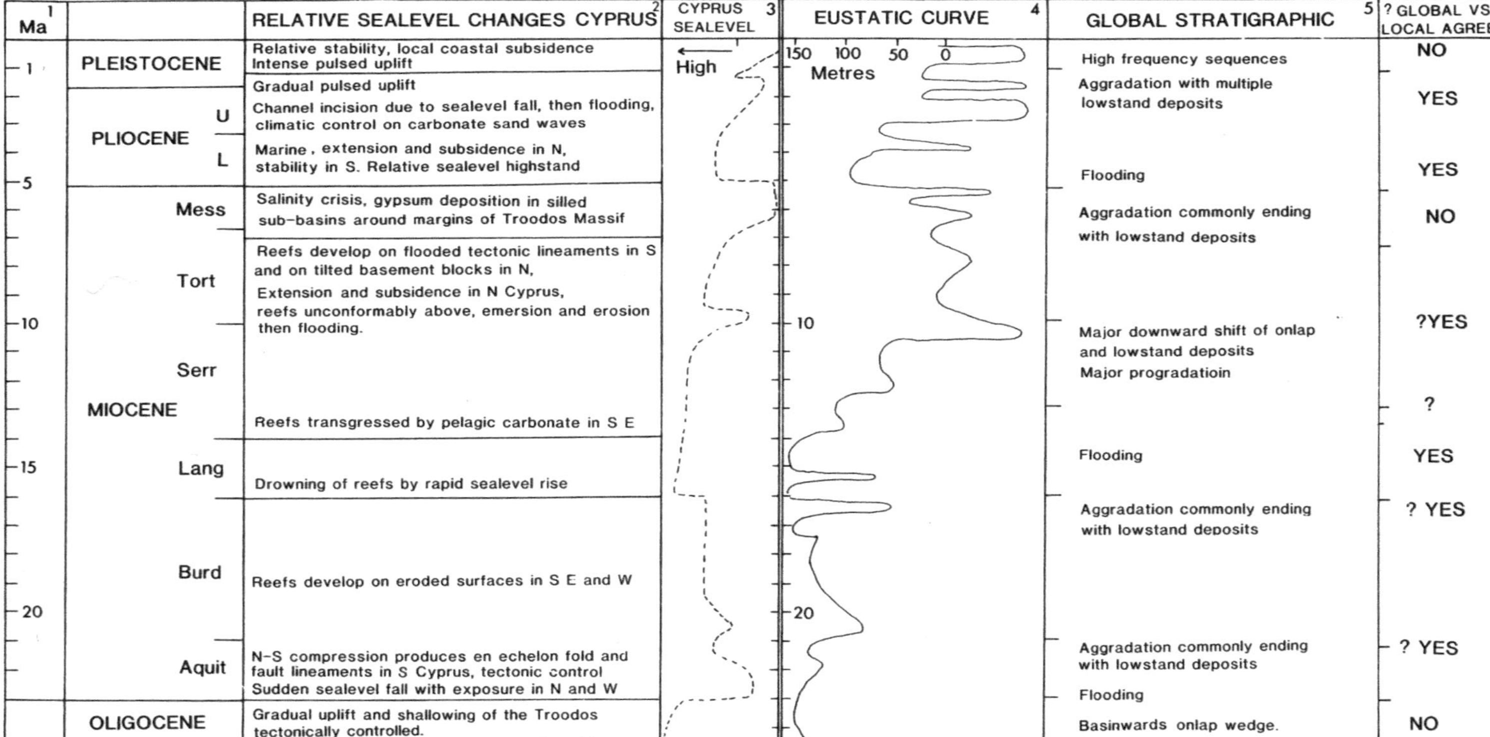


**Fig. 6.36** Summary of stratigraphic events in Cyprus, with an interpreted sea-level curve. This is compared to the Exxon global cycle chart (Haq et al., 1987, 1988a). From Robertson et al. (1991)

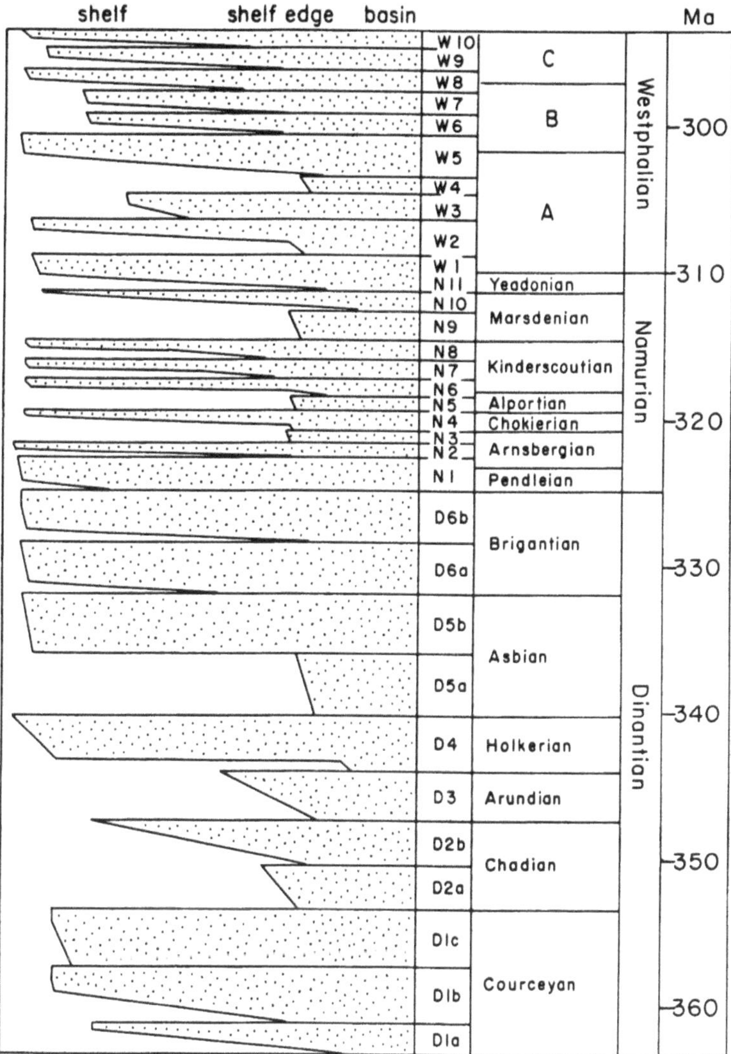

**Fig. 6.37** Mesothems in the Carboniferous succession of northern England (Ramsbottom, 1979)

are broadly transgressive, although the transgressions were slow and pulsed. The original concept of Ramsbottom (1979) was that each cyclothemic transgression reaching further than its predecessor out from the basin on to the shelf., although there has been considerable discussion about the validity of this concept, based on detailed stratigraphic documentation (e.g., Holdsworth and Collinson, 1988). Basal beds in each cycle may contain evidence of high salinities, reflecting the isolation of individual small basins at times of low sea level. Marine beds containing distinctive goniatite faunas are supposedly more extensive at the transgressive base of mesothems than in the component cyclothems, indicating more pronounced (higher amplitude) eustatic rises corresponding to the

mesothemic cyclicity. The regression at the end of each mesothem appears to have occurred rapidly. The sandy phase commonly commenced with turbidites and is followed by thick deltaic sandstones (commonly called Grits in the British Namurian; see Fig. 6.38). The cycles may be capped by coal. On the shelf each mesothem is bounded by a disconformity (Fig. 6.38), but sedimentation probably was continuous in the basins. Deltaic progradation was rapid, approaching the growth rate of the modern Mississippi delta.

Ramsbottom (1979) noted that the Namurian mesothems were of the shortest duration, and many do not extend up onto the shelf. This stage spans the Mississippian-Pennsylvanian boundary, which is designated as the boundary between the Kaskaskia and

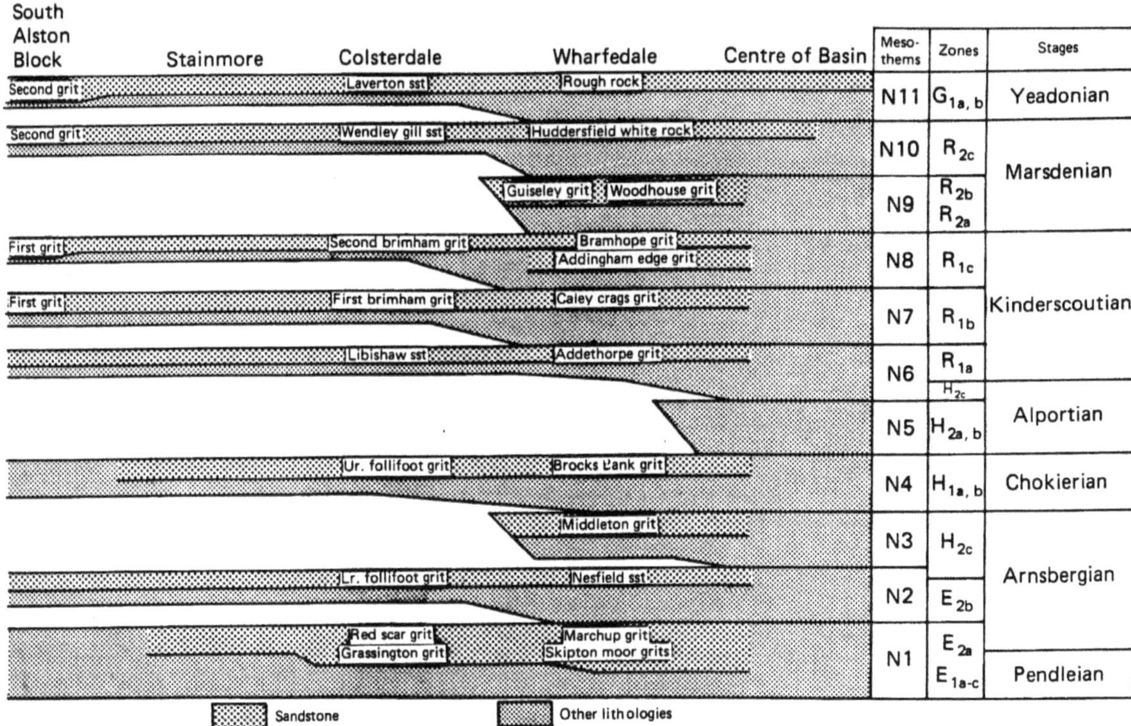


**Fig. 6.38** Chronostratigraphic section of the Namurian rocks of part of northern England, showing the main lithostratigraphic units (Ramsbottom, 1979)

Absaroka sequences in North America, a time at which sea level was at a long-term eustatic lowstand (Sloss, 1963; Figs. 4.8, 5.5).

A subsurface analysis of the Middle Pennsylvanian cyclothem record in Kansas reveals an unmistakeable "mesothem" pattern, which Youle et al. (1994) attributed to $10^6$-year eustatic cycles. They noted that higher-frequency cyclothems form a transgressive set in which deep-water deposits become more important upwards, and extend successively further onto the craton. The first few cycles overstep each other to onlap Mississippian basement. The highstand sequence set shows a distinctly progradational pattern, recording the long-term gradual drop in average sea level.

The Carboniferous succession in northern England, the type area for the mesothem model, contains evidence for considerable local tectonism, and for lithostratigraphic complexity resulting from local autogenic controls, such as delta-lobe switching. As a result, mesothemic cyclicity is not everywhere apparent, and Holdsworth and Collinson (1988) provided a critique of the mesothem model based on detailed local studies. In places the evidence does not support the simple mesothem model of Ramsbottom (1979) (see also Leeder, 1988). However, identification of a comparable form of cyclicity in the Late Paleozoic cyclothem succession of Kansas, as well as in other parts of Europe, as noted below, indicates that the mesothem concept should not be discarded. Groups of cycles are more likely to be mappable where tectonic influences are minor, as in Kansas.

Hampson et al. (1999) discussed the cyclothem model as applied to the Upper Carboniferous, coal-bearing succession of the Ruhr Basin, Germany. They refer to groupings of cyclothems that have been termed mesothems, but do not make it clear whether the mesothem concept can be applied to their rocks. By contrast, Cózar et al. (2006) made explicit reference to the mesothem concept, in their study of Lower Carboniferous (Visean) deposits in southwestern Spain. They used foraminiferal biostratigraphy to date sections through a mixed carbonate-siliciclastic succession, and suggested a

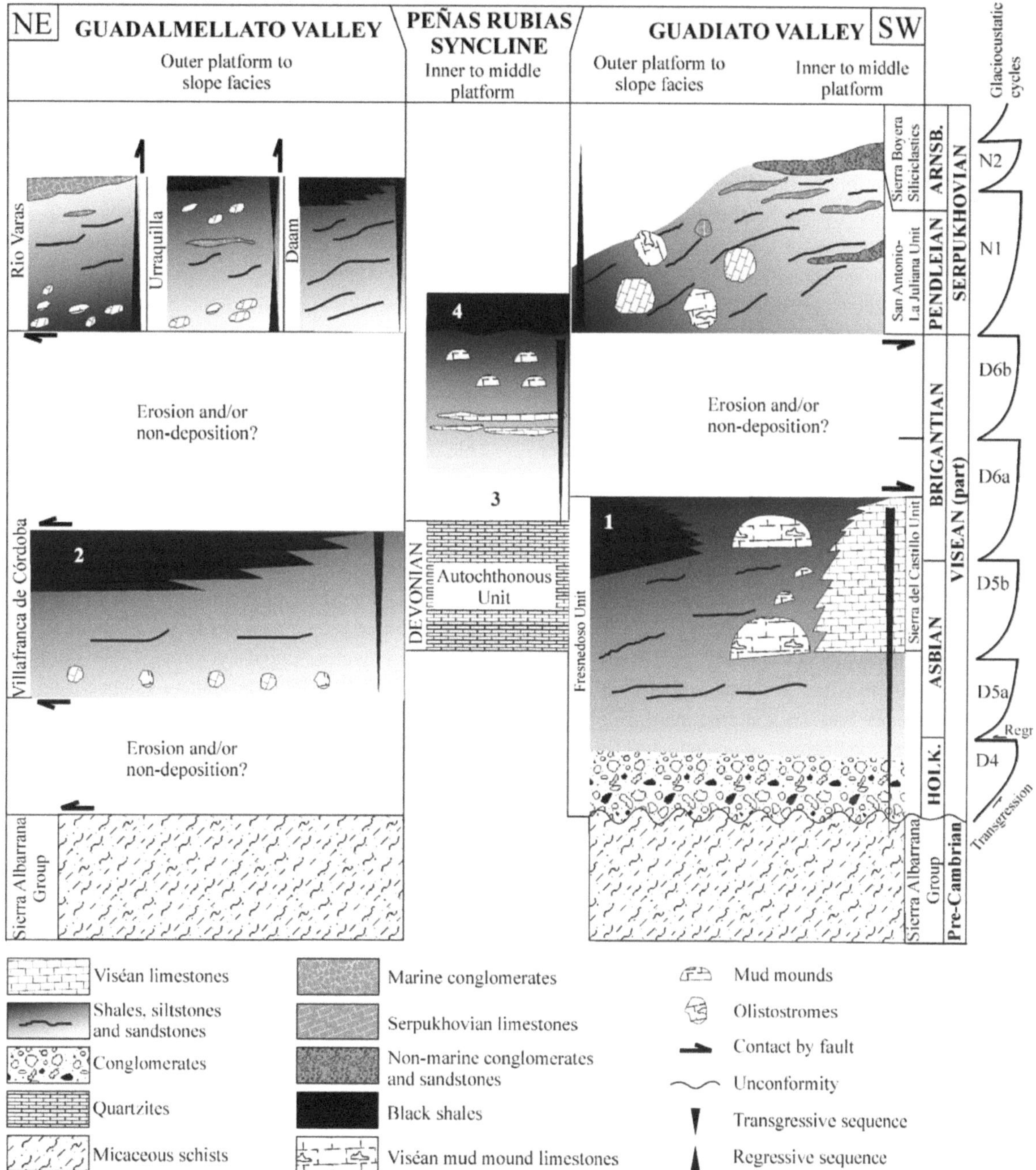


**Fig. 6.39** Lithostratigraphic correlation of the upper Viséan to Serpukhovian rocks in the Guadalmellato and Guadiato valleys, Spain. 1–4: Brigantian age of these black shales. Glacioeustatic cycles (mesothems) defined by Ramsbottom (1979) for the Holkerian to Serpukhovian are shown in the *right hand column*. (Cózar et al., 2006, Fig. 4)

correlation with Ramsbottom's (1979) mesothem scheme (Fig. 6.39).

Ross and Ross (1988) carried out a worldwide comparison of these upper Paleozoic cyclic deposits, and proposed detailed correlations between cratonic sections in the United States, northwest Europe and the Russian platform. They recognized about 60 cycles in the Lower Carboniferous to Lower Permian stratigraphic record. They suggested that similar cyclic successions may occur in contemporaneous continental-margin sections, but that because most of these have been deformed by post-Paleozoic plate-tectonic events the record from these areas is less well known. In view of the scepticism surrounding the mesothem concept, and the difficulties in correlation based on limited chronostratigraphic data, such inter-continental correlation may be regarded as premature.

## 6.5 Conclusions

1. Stratigraphic sequences with $10^6$-year episodicities are common in rifted and extensional continental-margin basins and on cratonic platforms. Their architecture comprises repeated transgressive-regressive packages of siliciclastic or carbonate deposits, of tabular form on continental shelves, or comprising prograding clinoform slope wedges. Mixed carbonate-siliciclastic sequences formed by reciprocal sedimentation are common. These sequences can readily be interpreted using standard systems-tract models.

2. Sequences of comparable duration are also common in foreland basins and can also be interpreted using the systems-tract approach. Siliciclastic deposits are dominant, particularly thick alluvial deposits in proximal settings. Transgressive carbonate successions may occur on the distal ramp and forebulge of foreland basins. Evidence for tectonic control is common.

3. In forearc basins, stratigraphic evolution is dominated by tectonic subsidence, faulting, and volcanism.

4. Where backarc basins border a continental margin, the tectonic and stratigraphic history are comparable to those of extensional continental margins, and commonly contain well-developed records of $10^6$-year carbonate or clastic stratigraphic sequences.

5. A distinctive type of cyclicity occurs in upper Paleozoic rocks of the northern hemisphere. They are of glacioeustatic origin, resulting from the great glaciation of the Gondwana supercontinent. High-frequency cycles, called cyclothems, are the most prominent type of stratigraphic sequence (and are discussed in Chap. 7), but groupings of these into $10^6$-year mesothems or major cycles can be recognized. Some detailed stratigraphic studies have thrown doubt on the mesothem concept as applied to the Upper Paleozoic record of northwest Europe, whereas some evidence for the concept has been obtained from the US midcontinent.

# Chapter 7

# Cycles with Episodicities of Less than One Million Years

## Contents



## 7.1 Introduction

Cycles of $10^4$–$10^5$ years duration can be grouped into five main types (Sects. 7.2, 7.3, 7.4, 7.5, and 7.6), reflecting their stratigraphic composition, tectonic setting, and age. Many of these cycles are thought to have been generated by global climate changes driven by orbital forcing, the so-called *Milankovitch mechanisms*, as described in Chap. 11. The term derives from the name of the Yugoslavian mathematician who was the first to provide the mathematical basis for the theory of astronomical forcing (Milankovitch, 1930, 1941). However, there is increasing evidence of the tectonic origin of some of the clastic cycles in foreland basins (Sect. 7.6), as discussed in Chap. 10.

It has been common practice to subdivide high-frequency sequences into those of fourth order, with durations in the $10^5$-year range, and those of fifth-order, with episodicities of $10^4$ years. This twofold subdivision has been based largely on interpretations that invoke low- and high- frequency Milankovitch mechanisms, respectively. However, this classification should now be abandoned. There is increasing evidence for other sequence-generating mechanisms that do not readily fall into simple temporal classifications, and it has now been demonstrated that sequence thicknesses and durations have log-normal distributions that lack significant modes (Drummond and Wilkinson, 1996). Schlager (2004) argued that sequence durations have a fractal distribution. This topic is discussed further in Sect. 4.2

As demonstrated by Schwarzacher (1993), orbital forcing includes a periodicity of 2.035 million years (related to orbital eccentricity), but few examples of this have been described in the literature. A model for Cretaceous cyclicity that includes this periodicity is discussed in Chap. 11. Most examples of orbital periodicities demonstrated from the rock record fall between the 413 ka eccentricity period and the 19 ka precession period. As the examples in this chapter demonstrate, a range of episodicities has been calculated for cycles in the $10^5$-year time band in the geological record, from a few hundred-thousand years to over 1 million years. No preferred time ranges or clustering of these periods around particular frequencies has emerged from the analysis, which could indicate one of two possibilities: (1) the driving forces are not characterized by temporal regularity, or (2) preferred frequencies do in fact exist (such as harmonics of the orbital-forcing process) but are obscured by inaccuracies in the chronostratigraphic control.

This book does not deal with cycles in the so-called *solar band* or *calendar band* of geological time. The solar band refers to cyclicity in the $10^1$–$10^2$ year range, including the sun-spot cycle and its possible geological effects (e.g., El Niño current changes). The calendar


A.D. Miall, *The Geology of Stratigraphic Sequences*, 2nd ed.,
DOI 10.1007/978-3-642-05027-5_7, © Springer-Verlag Berlin Heidelberg 2010

band refers to cyclicity relating to earth's seasonal rhythms (freeze-thaw, spring run-off, varves, etc.) and the tidal and other effects driven by the moon (Fischer and Bottjer, 1991).

The discussion in this chapter is provided in order to illustrate some well-documented examples of high-frequency sequence stratigraphy, and to demonstrate, briefly, how such stratigraphic syntheses throw light on depositional, tectonic processes and other allogenic processes, and are an aid to petroleum development. These examples, amongst others, are used as a basis for a discussion of driving mechanisms in Chaps. 10 and 11.

## 7.2 Neogene Clastic Cycles of Continental Margins

Marine clastic cycles of Neogene age are widespread, occurring particularly in continental shelf and slope settings. They have been identified by detailed outcrop studies, and by analysis of high-quality seismic-reflection data from offshore regions. They have also been documented in deep-sea sediments, using DSDP data. Modern multidisciplinary studies of the stratigraphic record have been able to show correlations of the cycles with the oxygen-isotope record, indicating a direct correlation between sea level and ocean temperatures. These sequences are clearly of glacioeustatic origin.

### 7.2.1 The Gulf Coast Basin of the United States

The Gulf of Mexico is one of the most intensively studied regions of the world, and the Neogene sequence stratigraphy of this area is now well known. Alluvial, deltaic and shallow-marine deposits of the Mississippi valley and delta area were first the subject of detailed study by Fisk (1939, 1944), who recognized the relationships among river terraces, erosional valleys and depositional surfaces and attributed them to glacial changes in sea level. Early attempts at sequence-stratigraphic modelling were carried out by Frazier (1974; see Fig. 1.14). Fisher and McGowen (1967) documented Cenozoic depositional systems and demonstrated that the major sediment-delivery pathways that supply the Gulf Coast Basin have been

in place since the early Cenozoic. Suter et al. (1987) and Boyd et al. (1989) interpreted the stratigraphy of the Lousiana shelf, including the lobes of the Mississippi delta, in terms of seven Quaternary stratigraphic sequences. Laterally equivalent deposits on the Texas Gulf Coast were described by Morton and Price (1987). The downdip equivalents, comprising the deposits of the Mississippi fan, have been studied by Feeley et al. (1990) and Weimer (1990), who recognized at least 13 sequences representing the last 3.5 million years of geological time. Several studies by W. E. Galloway have provided syntheses of Gulf Coastal Plain stratigraphy (Galloway, 1989b, 2008).

Figure 7.1 illustrates the main tectonic elements of the Texas coast, and the distribution of the major late Pleistocene and Holocene depositional systems. Similar depositional systems have been in place since early in the Cenozoic, subject to repeated cycles of sea-level change, which have caused transgressions and regressions shifting shorelines by locally more than 150 km. The total thickness of the Jurassic to Recent sediment pile reaches a maximum of more than 20 km beneath the modern continental shelf (Fig. 7.2). Near the Texas-Louisiana border, Pleistocene deposits alone exceed 3.5 km in thickness, and have extended the continental shelf seaward by nearly 50 km as a result of aggradation and progradation (Fig. 7.2). Fluvial sediment supply from the continental interior was vast, and supplied individual deltaic systems up to 70 km across and 180 m thick. Tectonic influences on sediment supply can be recognized in the variations in thickness of individual deltaic complexes along strike around the Gulf Coast (Galloway, 1989b, 2008). On a local scale, the effects of autogenic delta-lobe switching, and local tectonic disturbances caused by growth faulting and by salt diapirism render regional sequence recognition and correlation difficult. As demonstrated below, modern seismic methods, including horizontal seismic sections and seismic geomorphology, permit highly refined local reconstructions of depositional architecture and the nature of depositional controls and cyclic processes. Presented here are just a few examples of high-frequency sequence cyclicity spanning the Mesozoic to Plio-Pleistocene.

Shallow, high-resolution seismic studies, together with core data, have provided a chronostratigraphic framework for the mid-Pleistocene to Holocene stratigraphy of the continental shelf, which ties the sequence record to glacioeustatic sea-level changes

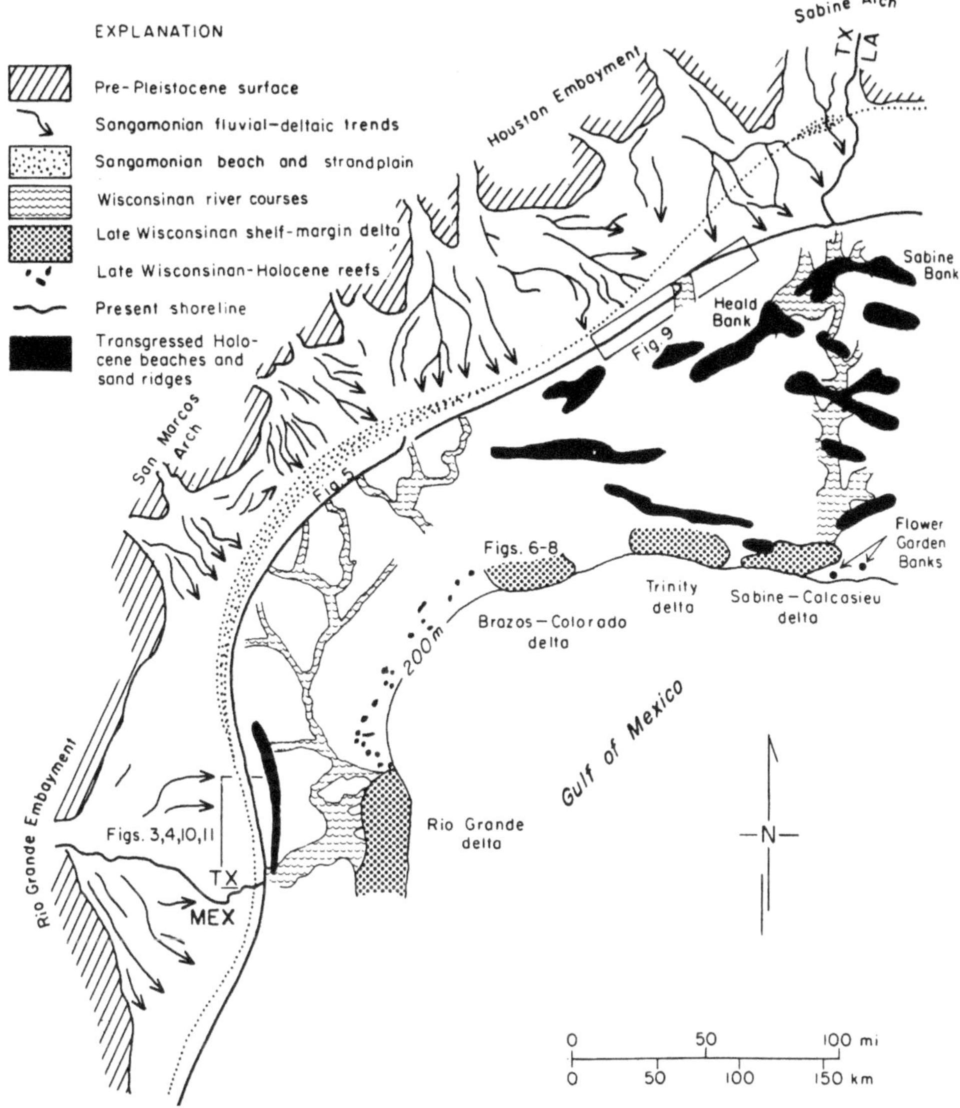



**Fig. 7.1** Tectonic elements and regional distribution of late Pleistocene and Holocene depositional systems, Texas coast (Morton and Price, 1987)

reflected by the oxygen-isotope curve (Suter et al., 1987; Thomas and Anderson, 1994; Fig. 7.3). Example of high-resolution seismic-reflection cross-sections through these deposits are shown in Fig. 7.4. Fluvial systems formed networks of incised valleys crossing the continental shelf during sea-level lowstands (Fig. 7.4). Near the shelf margin, these incised valleys have erosional relief of nearly 60 m. and are locally more than 10 km wide. At the same time, shelf-margin deltas formed. The position of some these is shown in Fig. 7.1. The incised valleys were subsequently filled

by episodes of aggradation during the transgressive to highstand phases (Fig. 7.4). The scale and rate of sea-level cycles suggests a classification of the sequences into fourth- and fifth-orders (Fig. 7.3).

Nine sequences have developed on the Louisiana shelf during the last 300 ka (Fig. 7.3). Sequence 1 is interpreted to represent the lowstand associated with the Illinoian glacial stage. Sequences 5 and 6 correspond to the post-Wisconsinan sea-level rise since 18 ka. The Mississippi delta complex developed during this final stage of sea-level rise (Fig. 7.5). The

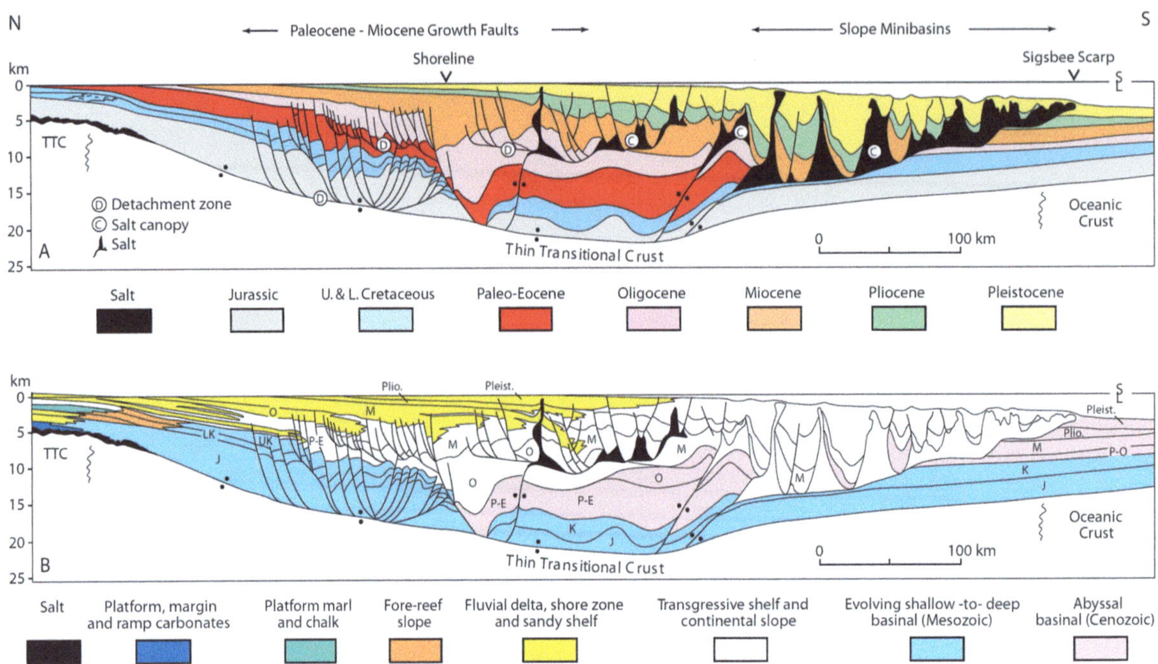



**Fig. 7.2** North–south cross-section through the Gulf of Mexico continental margin from central Louisiana to the base of the continental slope, showing the chronostratigraphy (*top*) and depositional facies (*below*) (Galloway, 2008, Fig. 6)

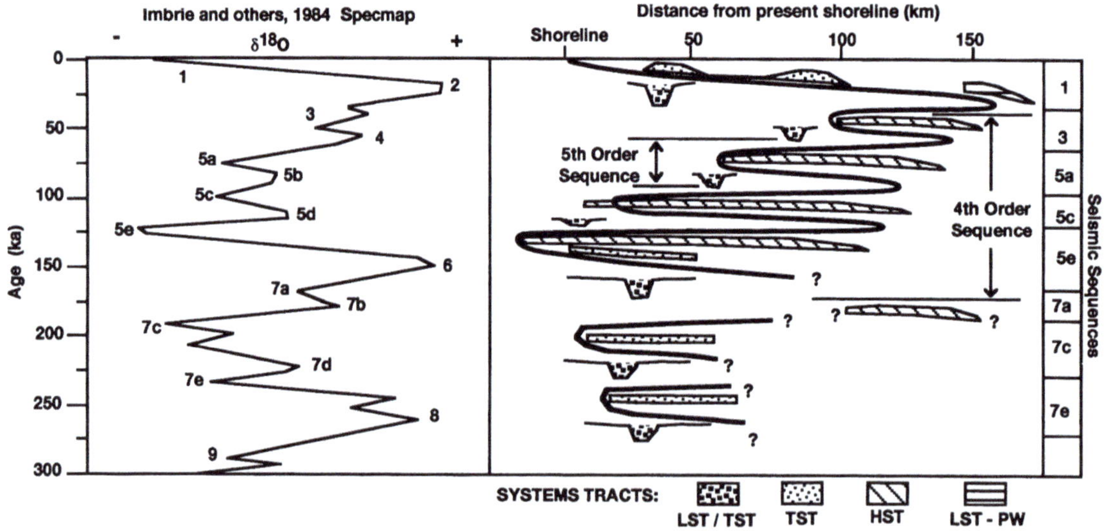



**Fig. 7.3** Correlation of the sequence stratigraphy of the Gulf-Coast continental margin with the oxygen isotope curve of Imbrie et al. (1984) (Thomas and Anderson, 1994, Fig. 8)

first three lobes (numbered 1–3 in Fig. 7.5) are interpreted as transgressive systems tracts. They backstep onto the shelf because sediment supply to the delta was unable to keep pace with rising sea level. The culmination of the sea-level rise occurred at about 3–4 ka, and resulted in the retreat of the coastline to the mouth of the Mississippi alluvial valley. The St. Bernard lobe represents the first of four highstand systems tracts that have formed since this sea-level highstand. Delta-lobe switching between the various positions has taken

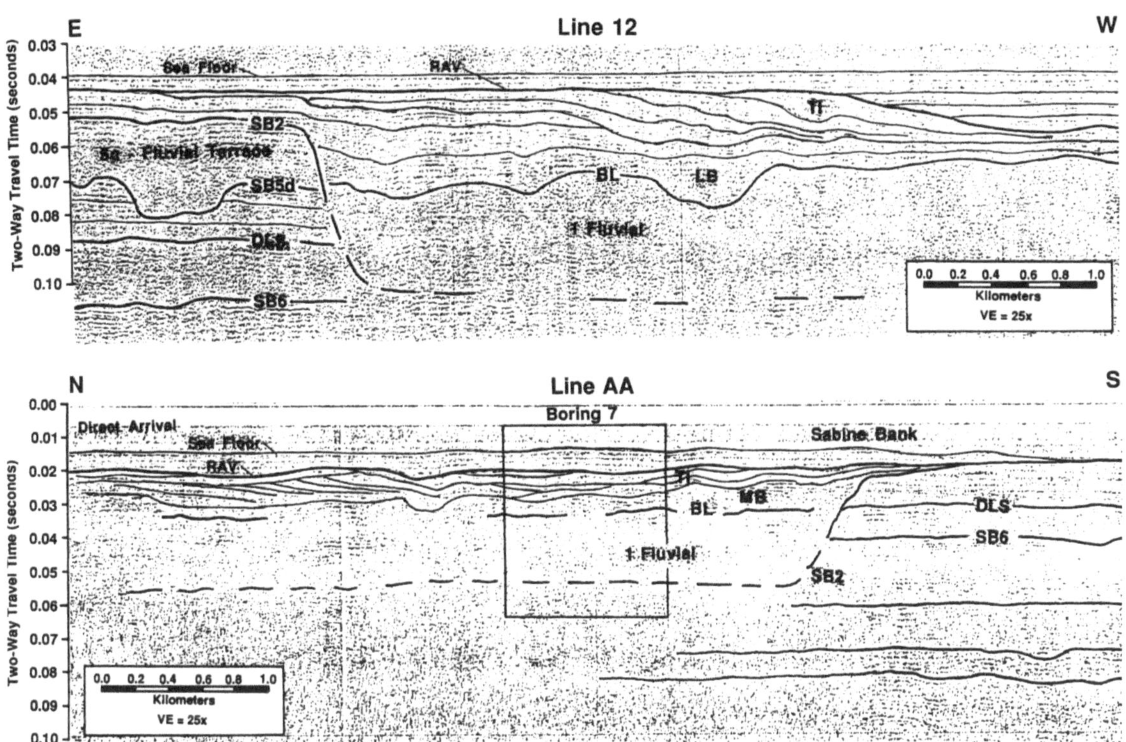


**Fig. 7.4** High-resolution seismic profiles through incised valley systems on the continental shelf near Houston. BL: bay line, MB: middle bay, TI: tidal inlet, AV: ravinement surface, DLS: downlap surface, SB: sequence boundary, with numbers corresponding to those shown in Fig. 7.3 (Thomas and Anderson, 1994, Figs. 16 and 17)

place because of autogenic processes relating to the flattening of the river slope as the lobes prograded seaward (see Fig. 2.25).

The Mississippi fan (Fig. 7.6) is the most well studied fan deposit in the world. It is also one of the largest, consisting of more than 3.5 km of sediment deposited during the last 6.5 million years. More than 2 km, constituting about 70% of the fan volume, has been deposited since 800 ka. Seismic and sonar surveys and the results of DSDP leg 96 have enabled a very detailed sequence analysis to be compiled (Feeley et al., 1990; Weimer, 1990). Weimer (1990) documented 17 seismic sequences within the fan succession, consisting mostly of lowstand channel-levee depositional systems (Figs. 7.7, 7.8, 7.9). Feeley et al. (1990) correlated the sequence stratigraphy with the oxygen-isotope curve (Fig. 7.10). However, the actual chronostratigraphic evidence for the correlations derived from the fan sediments themselves appears to be limited, consisting of a few biostratigraphic picks in several key wells. The correlations shown in Fig. 7.10 are therefore based on the standard Exxon method of

pattern matching, constrained by correlation of a few proven fixed points.

An interesting study by Roof et al. (1991) demonstrated that not all Gulf sediments are controlled by processes related to orbital forcing. An ODP core obtained from the west Florida carbonate ramp slope spanning the last 5.4 Ma of sedimentation along the Gulf margins was analyzed for cyclic frequencies, with quite mixed results. Although some intervals of the core could clearly be related to orbital control of such parameters as sediment supply (from the distant Mississippi delta) and changes in biogenic productivity, related to changing water temperatures, other factors intervened, such as shifts in surface and bottom currents in response to slow changes in oceanic circulation following closure of the Isthmus of Panama.

In the remainder of this section we examine briefly two case studies of high-frequency sequence stratigraphy in the Gulf Coast basin that demonstrate the value of detailed log and core analysis and, where possible, a close correlation between high-resolution

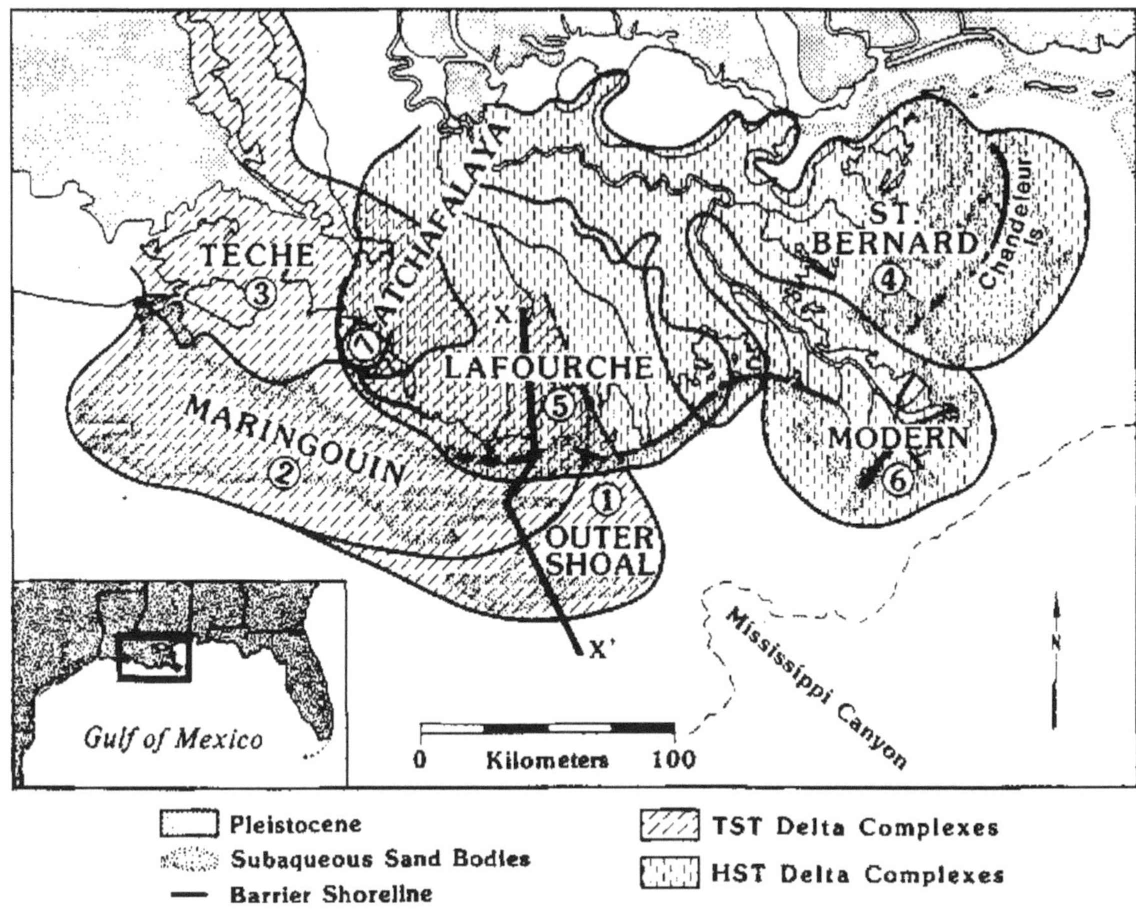



**Fig. 7.5** The delta lobes of the Holocene Mississipi River. Lobe 1 is dated at about 9 ka, and represents the first of three trangressive systems tracts formed following the end of the Wisconsinan glacial stage. Lobes 5–7 are highstand systems tracts (Boyd et al., 1989)

reflection-seismic data (including the use of horizontal sections extracted from three-dimensional seismic data), well-log and core data.

Zeng and Hentz (2004) employed seismic-geomorphic methods to interpret horizontal-seismic sections through Miocene depositional systems in offshore Louisiana. In such horizontal sections, and in conjunction with well data, reflection amplitudes may be interpreted in terms of depositional environments and sedimentary facies, and then this information may be used to improve interpretations of conventional 2-D seismic sections tied to well logs. In earlier work, the upper lower through upper Miocene succession in Starfak and Tiger Shoal fields had been subdivided into 10 third-order sequences and 59 fourth-order sequences. Many of these are shown in Fig. 7.11, and Fig. 7.12 is an example of well-log correlations through the area. "Limited well calibration in the two-field area shows an excellent tie between seismic amplitude and log lithology that can be readily applied (extrapolated) to areas in the seismic volume where there are no well data. This extrapolation enables identification of the horizontal distribution of lithofacies, with (1) strongly negative amplitude (red) indicating thick sandstones, (2) weakly negative amplitude denoting thin or shaly sandstones, and (3) positive amplitude (black) corresponding to shale beds. On all three stratal slices, incised-valley fills are defined by strongly negative (red) amplitudes that indicate well-developed sandstones (30–130 ft [10–40 m])" (Zeng and Hentz, 2004, p. 166). Shale/mudstone-filled portions of incised-valley fills are commonly not easily distinguished from laterally adjacent (but older) shaly highstand systems tract strata. Most incised-valley fills

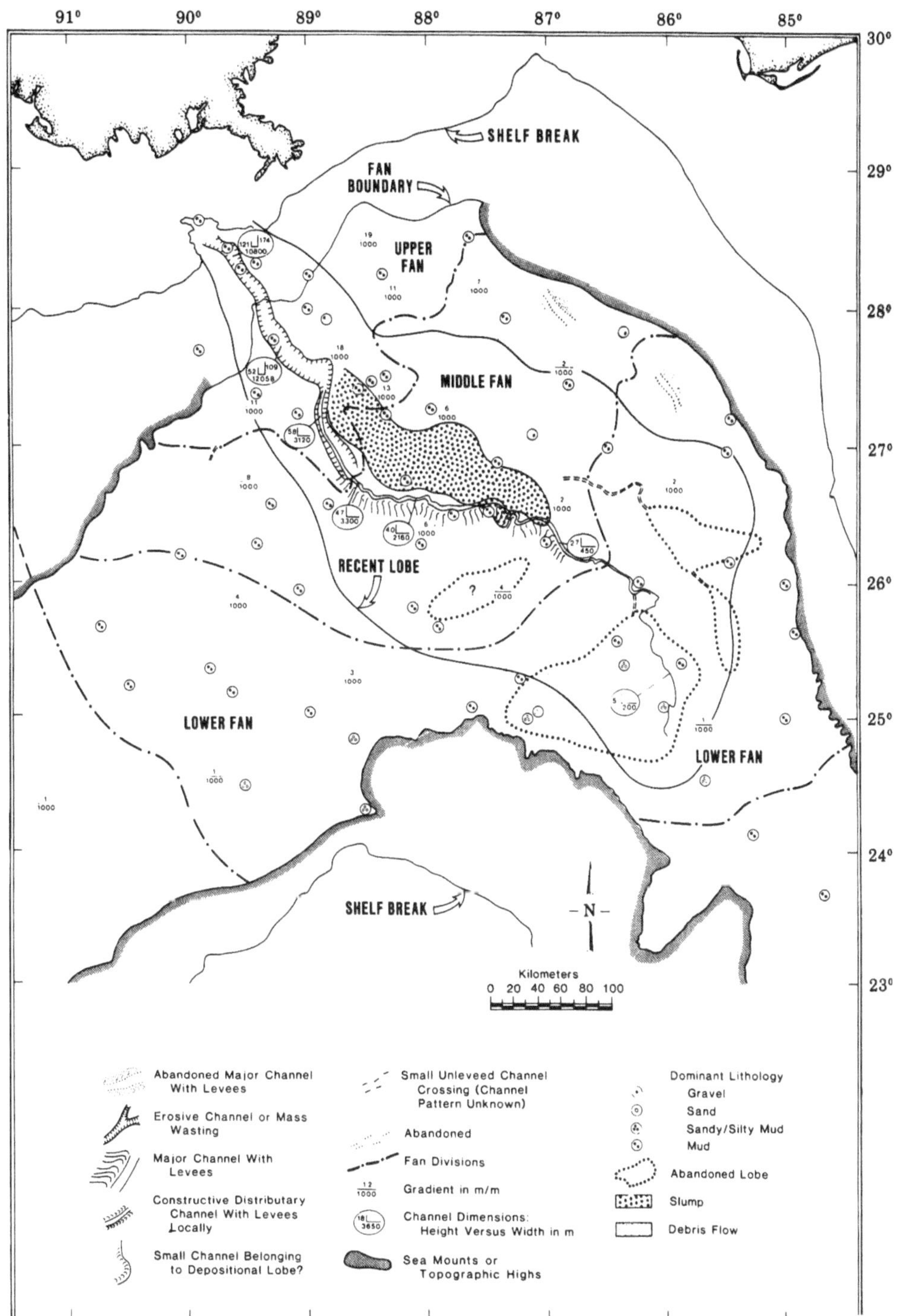



**Fig. 7.6** Map of the recent Mississippi fan lobe, showing the position of the central channel. Note that the feeder canyon is not currently opposite the active delta, which accounts for the fact that the fan is not actively accumulating sediment at the present time (Bouma et al., 1985)

in the area are the largest curvilinear, dip-oriented, high-amplitude features in map view (1–50 km wide), exhibiting low to moderate sinuosity (Fig. 7.13). The thickness and width of these incised valleys are comparable to those observed in nearby Quaternary incised valley systems.

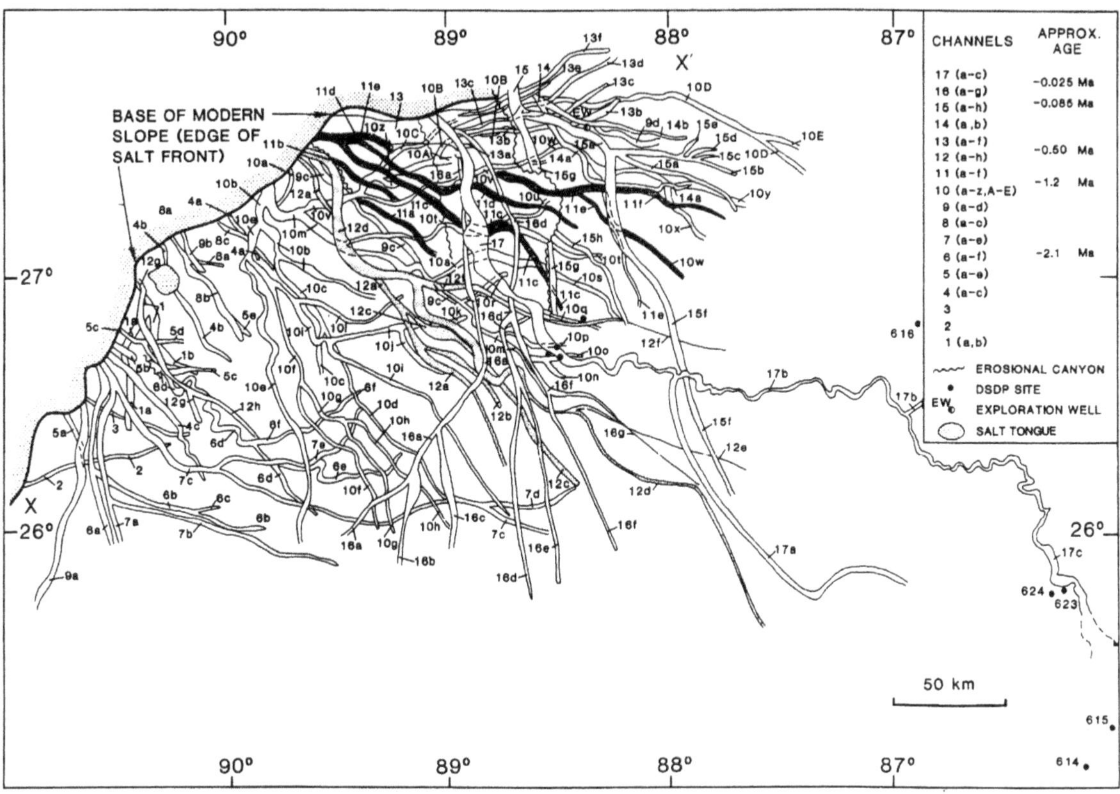



**Fig. 7.7** Locations of the channel valleys of the 17 channel-levee systems comprising the Mississippi fan. They are numbered in order from oldest to youngest (Weimer, 1990)

The mature, super-giant East Texas field has been the most productive oil field in the US lower 48 states and the second largest in the country (Ambrose et al., 2009). Declining production led to a re-appraisal of the stratigraphy and sedimentology of the reservoir units, and a few of the results are reported here. Primary hydrocarbon accumulation in the field is in west-dipping Woodbine sandstones that are truncated on the west flank of the Sabine uplift by a subregional unconformity below the Austin Chalk that formed during the Late Cretaceous after uplift and erosion. Impermeable calcareous siltstones of the Austin form the seal throughout the field. A shale unit at the base of the Eagle Ford Group is correlated to a widespread anoxic event that has been dated as 91.5 Ma, and this provided the main basis for tying the section into the Haq et al. (1987) global cycle chart (Fig. 7.14). The dubious value of this particular practice is discussed in Chap. 12.

A sequence-stratigraphic analysis was carried out on some 225 well logs distributed across the central and south-central parts of the East Texas Basin. A maximum of 14 "fourth-order" sequences was identified within the Woodbine succession in the central part of the basin. Sequence boundaries, transgressive surfaces and maximum flooding within the sequences were interpreted primarily from gamma-ray signatures (Fig. 7.14), supplemented by core data. The Woodbine succession is dominated by lowstand incised-valley fills, which occur as low-gamma-ray zones as much as 45 m thick, forming blocky log responses with abrupt horizontal basal erosion surfaces. Sequence boundaries at the base of the incised-valley fills were correlated with those at the tops of adjacent upward coarsening cycles, which represent older highstand systems tracts into which the valley fills were cut.

The new, detailed sequence-stratigraphic analysis throws new light on the evolution of the basin and

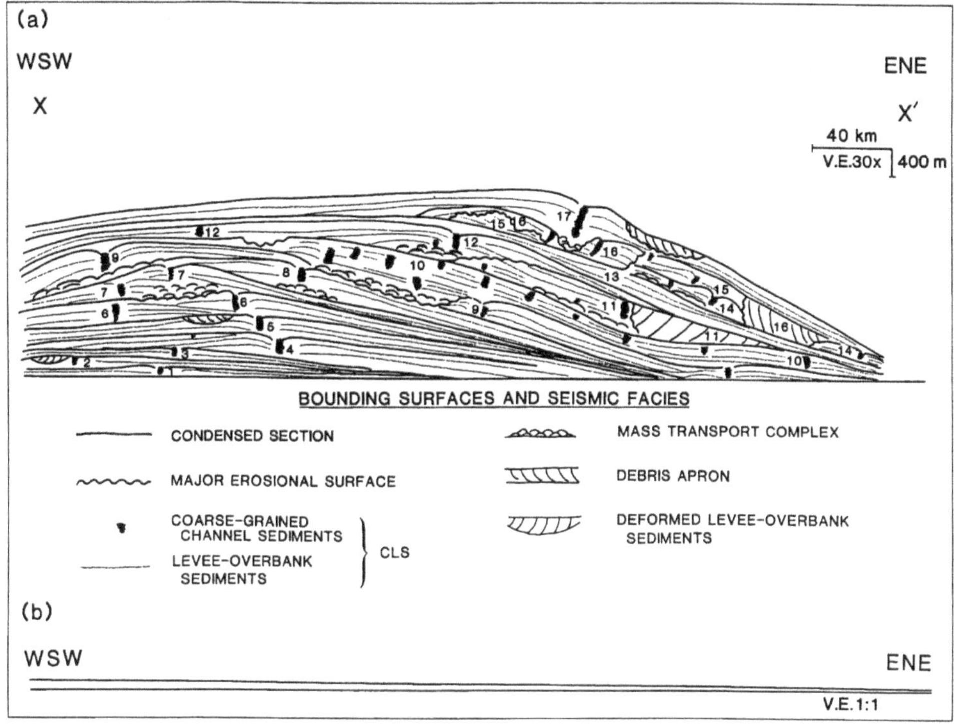


**Fig. 7.8** Schematic strike-oriented cross-section across the Mississippi fan, showing the mounded pattern of the 17 channel-levee systems. Note the true vertical scale in the diagram at *bottom* (Weimer, 1990)

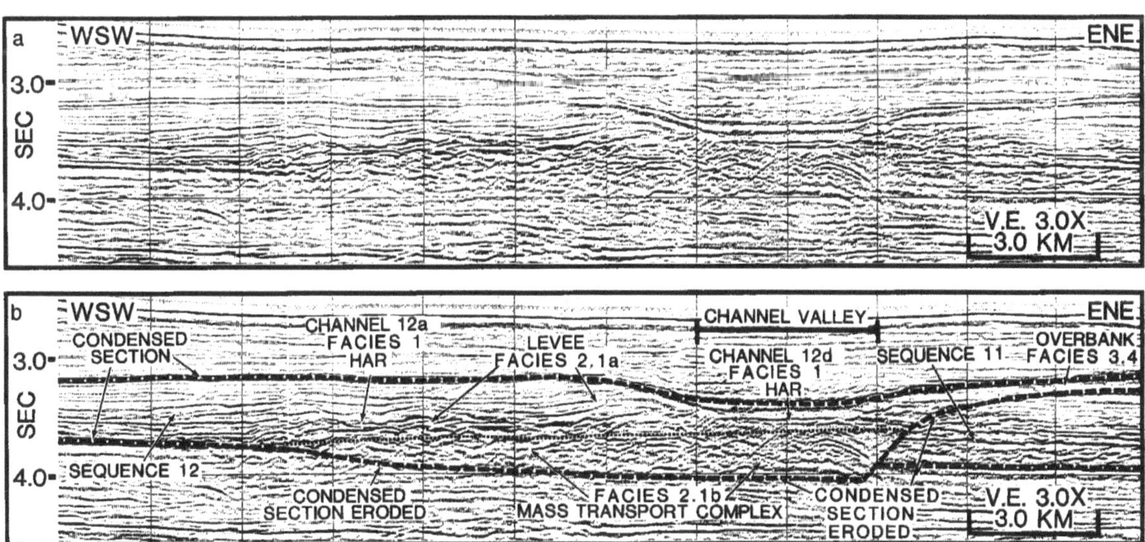


**Fig. 7.9** Example of a strike-oriented seismic profile across the Mississippi fan, showing the seismic facies character of some of the channel-levee systems. *Upper section* is uninterpreted, *lower section* shows position of two sequence boundaries, defined by condensed sections (*dashed lines*) (Weimer, 1990)

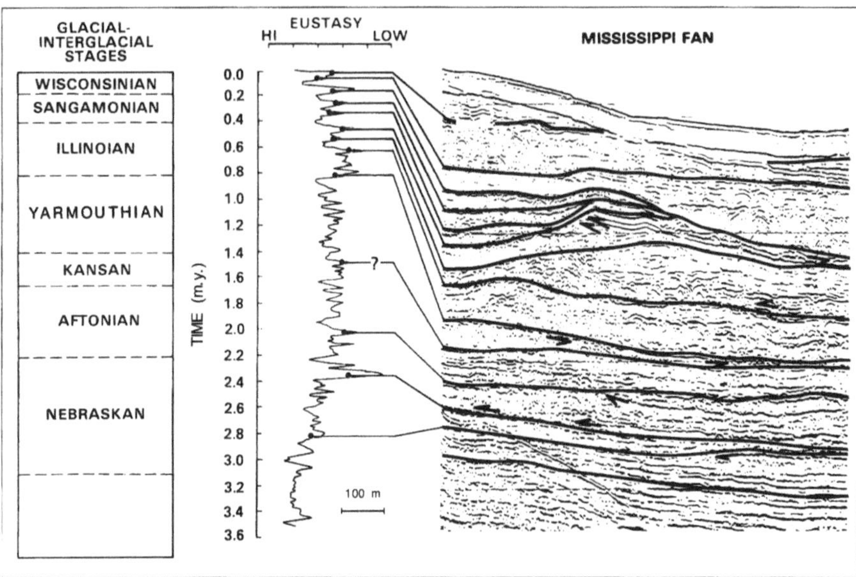


**Fig. 7.10** Correlation of the Mississippi fan sequences (*right*) with the oxygen-isotope curve for the late Pliocene-Pleistocene (Feeley et al., 1990)

the flanking Sabine Uplift. It can clearly be demonstrated (see Fig. 7.15) that the first five fourth-order sequences, bounded by sequence boundaries 10–50, are truncated by the Austin Chalk. The remaining sequences (6–14, bounded at the top by SB 150) show depositional thinning and pinch-out on the updip flank of the Sabine Uplift. They are overlain conformably by the widespread Eagle Ford marine shale, and the entire succession was then uplifted, tilted further westward, and eroded prior to deposition of the Austin Chalk. This analysis clarifies the tectonic evolution of the Sabine Uplift, showing that it was not an active uplift during the first phase of Woodbine deposition (sequences 1–5).

The new stratigraphic analysis presented by Ambrose et al. (2009) also provides the basis for the development of enhanced recovery projects, which rely on a detailed knowledge of the reservoir architecture. High-resolution sequence correlations permit detailed subdivision of the Woodbine succession for the purpose of developing detailed environmental interpretations of the reservoir units, including incised-valley fills, fluvial and deltaic channels and bars, from which reliable net-sandstone isopachs can be constructed.

### 7.2.2 Wanganui Basin, North Island, New Zealand

Several studies of the continental shelf and slope of New Zealand have demonstrated a high-frequency sequence stratigraphy in the deposits, and a close correlation of the Neogene succession with the oxygen-isotope record (Fulthorpe and Carter, 1989; Kamp and Turner, 1990; Carter et al., 1991; Fulthorpe, 1991). These studies also illustrate several points regarding lower-order cycles and the global cycle chart, as discussed by Carter et al. (1991). During the Neogene, New Zealand underwent active uplift along the transpressive Alpine Fault, and abundant sediment was shed into basins situated in various tectonic settings on the continental shelves surrounding the North and South islands. The Canterbury Basin, on South Island, is located in an extensional, passive-margin setting, while the Wanganui basin is situated in a backarc setting, behind the arc of the North Island Axial Ranges (Fig. 7.16). In the Canterbury Basin the beds above an initial syn-rift fill constitute the transgressive systems-tract of a major long-term thermo-tectonic cycle. Transgression reached a max-

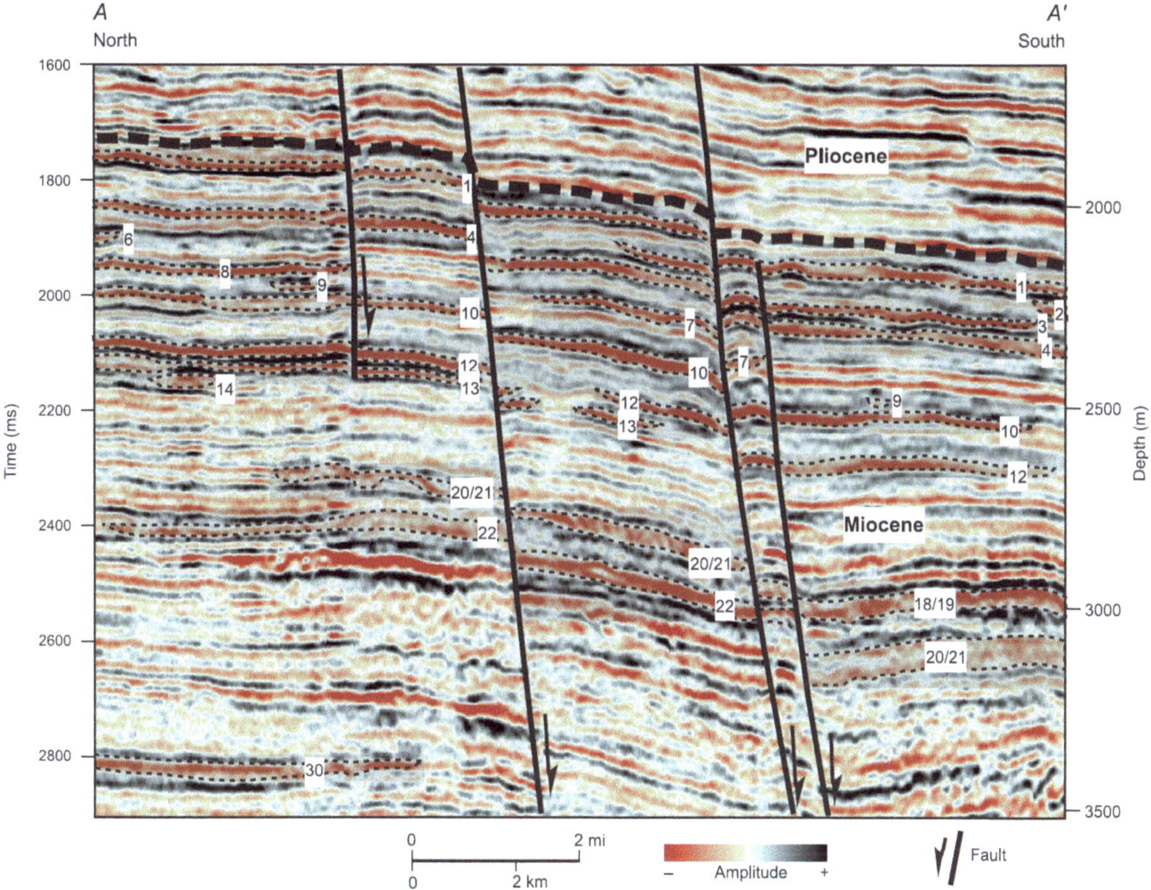


**Fig. 7.11** North–south dip seismic section across the continental shelf off central Louisiana, showing representative Miocene seismic reflection configurations in the Vermilion Block 50, Tiger Shoal area. Parallel to divergent facies are the dominant seismic facies in the formation. Interpreted lowstand incised-valley deposits (labeled 1–30) in this study are seismically thin, commonly terminating in continuous seismic events, and are not distinctively different from the surrounding sediments in seismic characteristics. Interpretation is not repeatable without geomorphologic information from the horizontal dimension (Zeng and Hentz, 2004, Fig. 1). AAPG © 2004. Reprinted by permission of the AAPG whose permission is required for further use

imum during the Oligocene, with the deposition of a condensed succession of pelagic to hemipelagic limestone. Transpression along the Alpine fault then led to increased sediment supply and the development of a highstand progradational phase in the Miocene.

According to Carter et al. (1991) the mid-Oligocene sea-level high indicated by the Canterbury Basin condensed section is a "spectacular mismatch with the predicted 29 Ma lowstand" of the global cycle chart of Haq et al. (1987, 1988a). However, Loutit et al. (1988, p. 192), discussing the same basin, did not see it this way. They stated that "the New Zealand Oligocene provides a spectacular example of the effects of subsidence and sea-level movements on stratal geometry." While this is true, it does not explain how they reconcile the relative sea-level changes revealed by the Canterbury Basin stratigraphy with the global cycle chart. We do not discuss this controversy further, but refer the reader to Chap. 12, where a general discussion of the Exxon global cycle chart is presented.

The Wanganui Basin of North Island, New Zealand (location shown in Fig. 7.16), contains a particularly complete Pliocene-Pleistocene sedimentary record. The stratigraphy of the Wanganui Basin has received focused attention by a team of New Zealand geologists, led by T. R. Naish, because of its importance in the

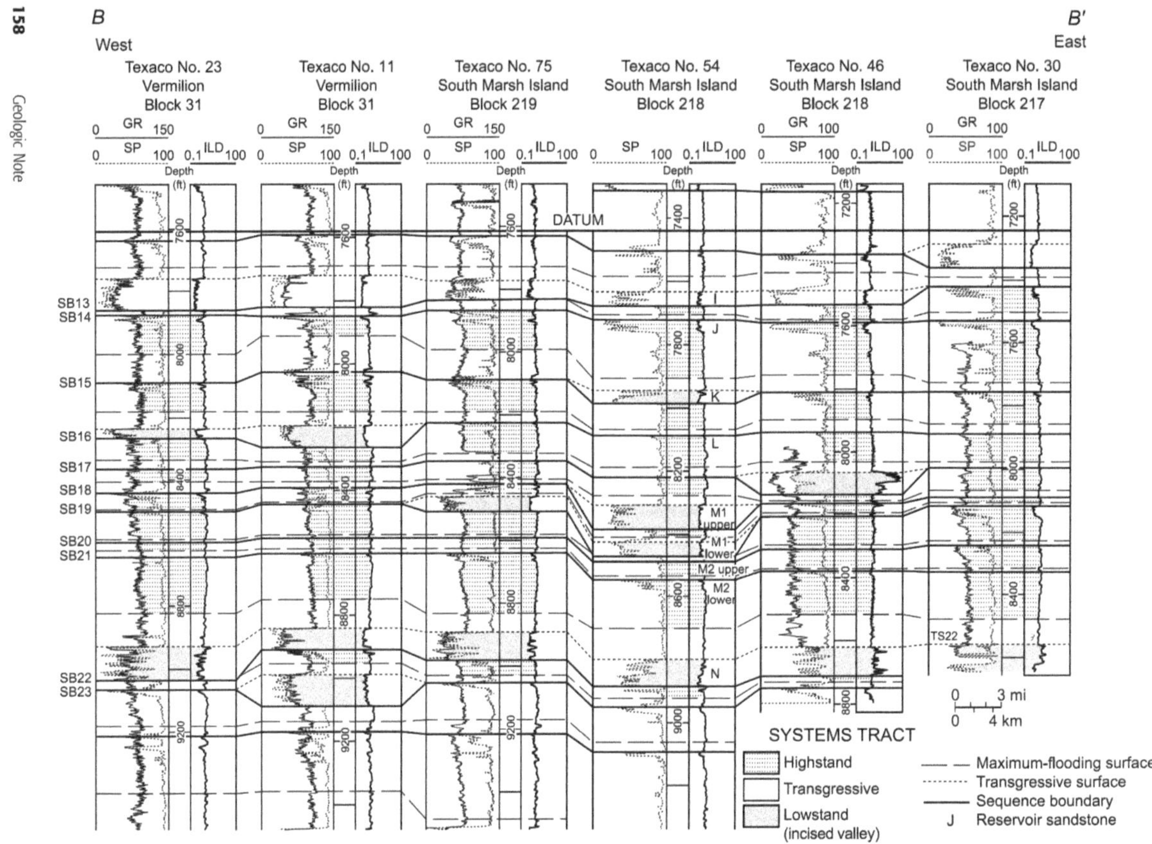



**Fig. 7.12** West–east well cross section of fourth-order sequences 23 through 13 (third-order sequences 4 and 5) with datum at the fourth-order maximum-flooding surface above SB12. Fourth-order sequence-stratigraphic surfaces were interpreted primarily from wire-line-log patterns and faunal data (Zeng and Hentz, 2004, Fig. 3). AAPG © 2004. Reprinted by permission of the AAPG whose permission is required for further use

development of the global time scale for the last 3 million years of earth history (the succession has become one of the best documented sections through the Pliocene to Recent sedimentary record), and because of the light a detailed sedimentological analysis can thrown on the relationships between sea level, sediment supply and systems-tract development (Naish and Kamp, 1995, 1997). A special issue of the Journal of the Royal Society of New Zealand (Naish et al., 2005) contains a suite of paper discussing the Canterbury and Wanganui basin stratigraphies and other papers. The reader is referred, in particular, to Naish et al. (2005).

Coastal cliff sections near the town of Wanganui have been interpreted in terms of $10^5$-year cyclicity by Kamp and Turner (1990) and by Carter et al. (1991), while correlative cycles also are present in offshore sediments, as revealed by seismic surveys (Carter et al., 1991). The succession is characterized by stacked cycles showing 41 and 100 ka periodicity. The exposed sediments represent transgressive and highstand deposits, and can be subdivided into sequences 5–22 m thick. Figure 4.11 summarizes the onshore stratigraphic section and its correlation with the oxygen isotope record. The sequence boundaries represent significant hiatuses. In fact, as revealed by the correlations with the oxygen-isotope record (Fig. 7.17), it is of interest to note that in sum the hiatuses represent about as much time as that represented by sedimentation. Even-numbered oxygen-isotope stages correspond to times of sea-level lowstand (glacial stages). Regionally, the succession shows a broad facies change from clastic-rich (siltstone and sandstone

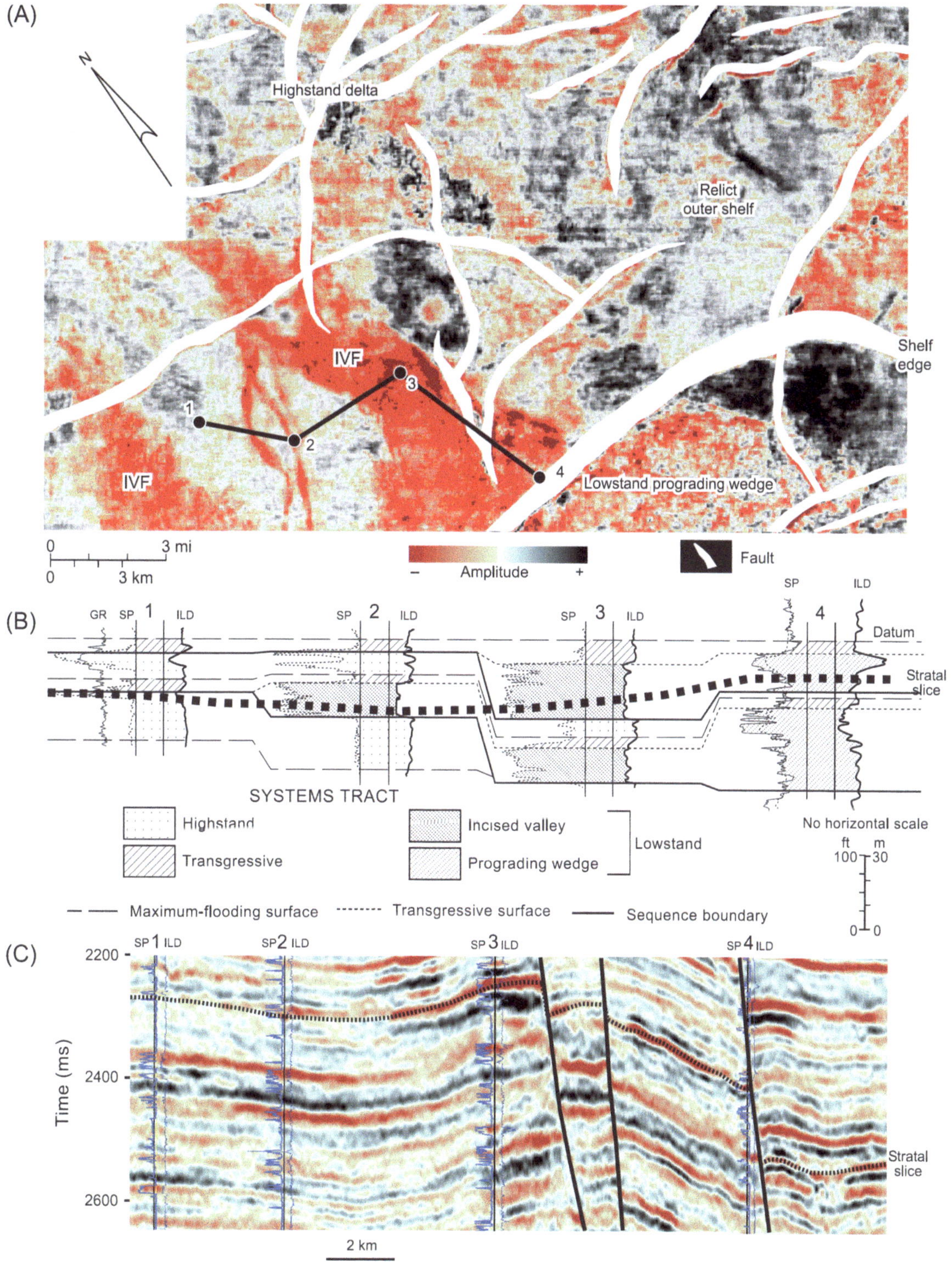



**Fig. 7.13** An example of vertical and horizontal seismic sections integrated with well-log data, showing incised valley systems and related strata. Systems tracts on stratal slice and wire-line-log facies patterns in fourth-order sequence set 18/19 Louisiana outer shelf/upper slope. (**a**) Amplitude stratal slice. (**b**) Well cross section with datum at maximum-flooding surface without horizontal scale. (**c**) Cross-well seismic section with the stratal slice labeled. IVF = incised valley fill (Zeng and Hentz, 2004, Fig. 14). AAPG © 2004. Reprinted by permission of the AAPG whose permission is required for further use

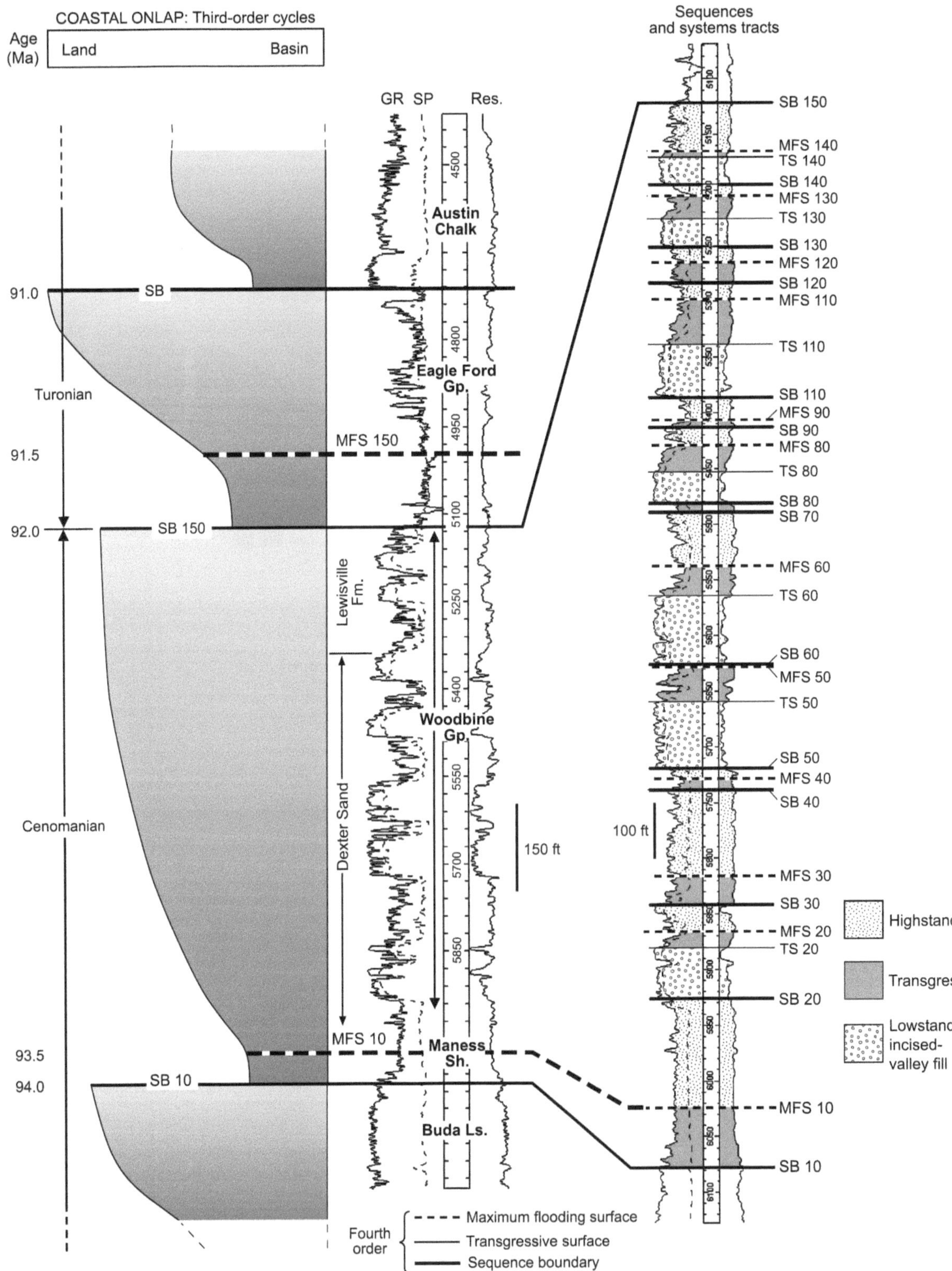


**Fig. 7.14** Sequence stratigraphy of the Upper Cretaceous Woodbine Group in the East Texas Field. The log at the *right* is an expanded version of the log through the Woodbine Group. Correlations with the Haq et al. (1987, 1988a) "global cycle chart" are shown at *left*, a practice that is discussed in Chap. 12 (Ambrose et al., 2009, Fig. 4). AAPG © 2009. Reprinted by permission of the AAPG whose permission is required for further use

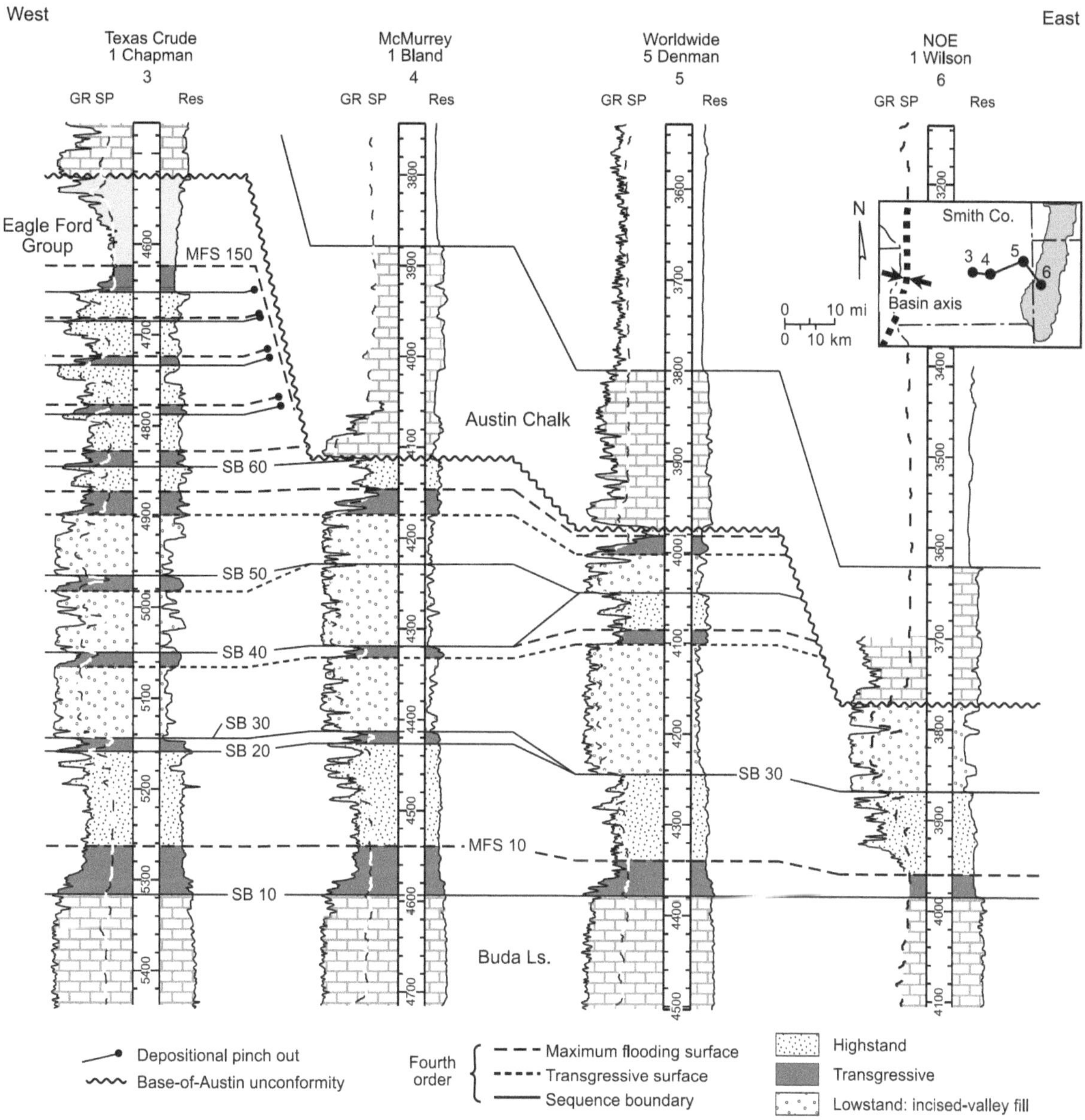



**Fig. 7.15** Details of sequence-stratigraphic and systems-tract framework of the Woodbine Group in the eastern East Texas Basin–western Sabine uplift area. The beds exhibit a regional dip the west, and thin eastward up the flank of the Sabine Uplift. This cross-section is drawn to emphasize stratigraphic relationship, using as a datum sequence boundary SB-10. Two particularly salient features are illustrated: (1) only the oldest five fourth-order sequences S1–S5 (SB 10–60) are truncated by the base-of-Austin unconformity, and (2) all younger Woodbine sequences depositionally pinch out below the shaly third-order transgressive systems tract (capped by maximum flooding surface MFS-150) of the lowermost Eagle Ford Group. No horizontal scale. GR = gamma ray; SP = spontaneous potential; Res = resistivity. NOE = T.C. Noe Oil Account (Ambrose et al., 2009, Fig. 6). AAPG © 2009. Reprinted by permission of the AAPG whose permission is required for further use

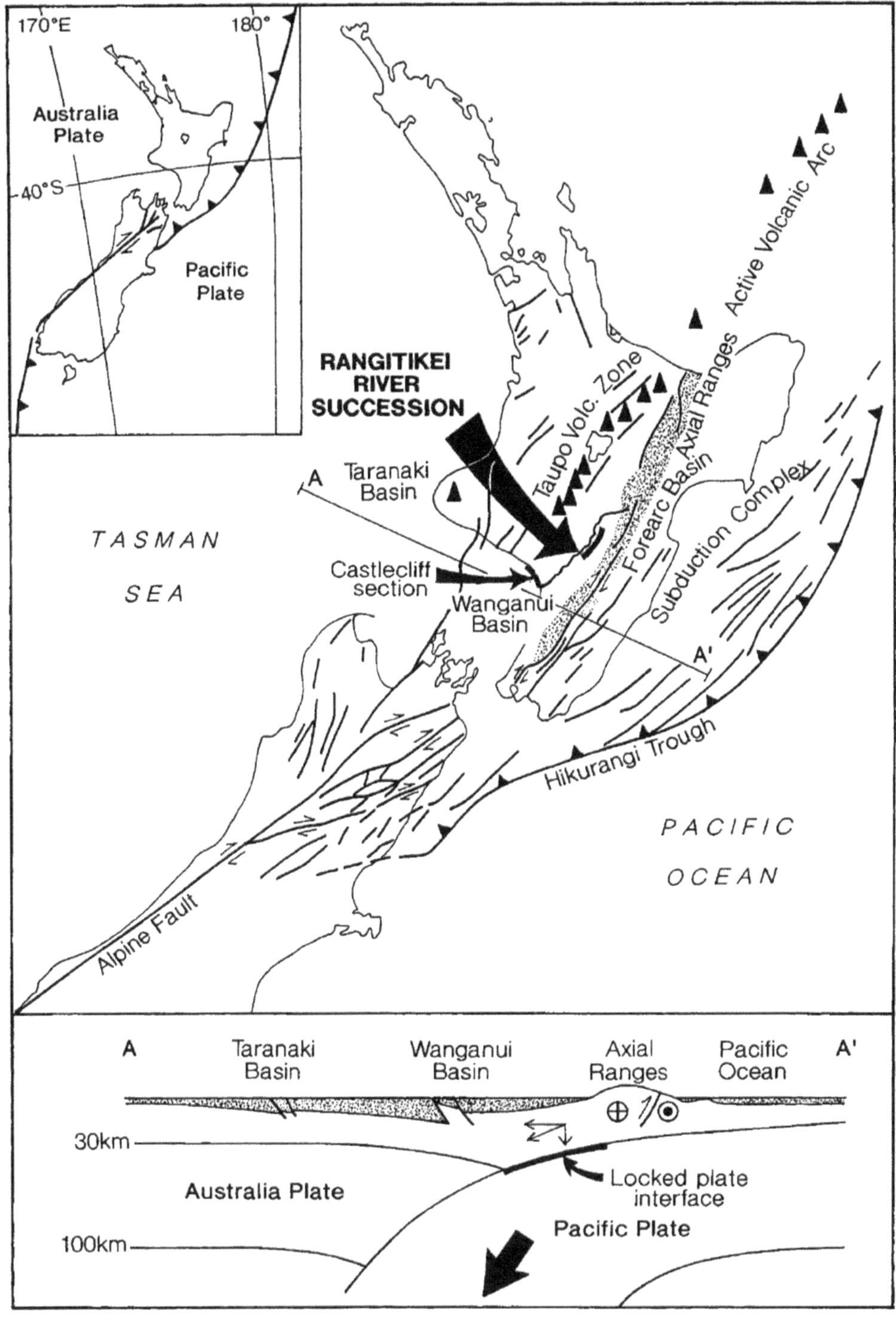


**Fig. 7.16** Plate-tectonic setting of the Wanganui Basin, North Island, New Zealand (Naish and Kamp, 1997, Fig. 1)

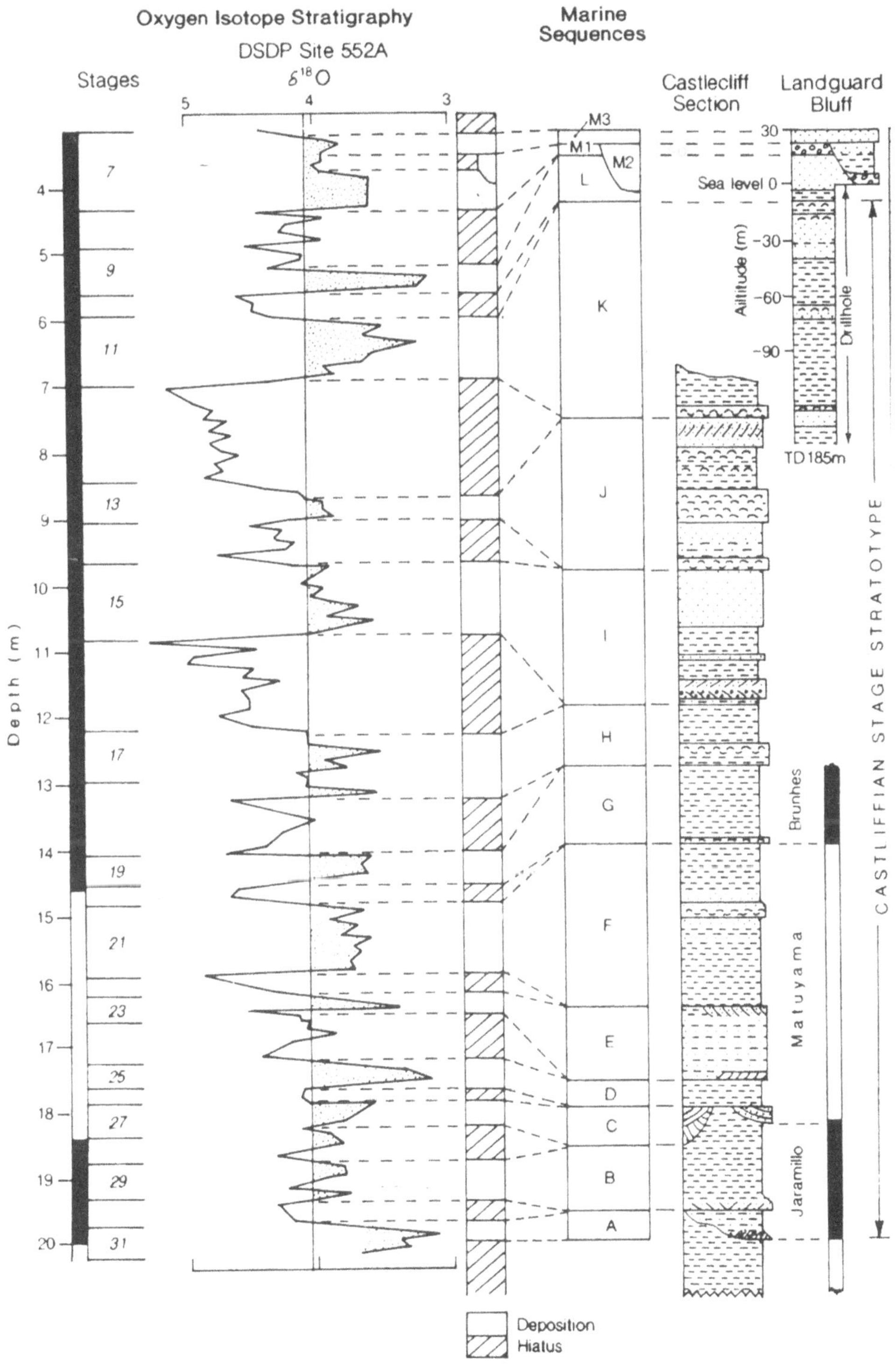


**Fig. 7.17**  Correlation of the Castlecliff and Landguard Bluff sections, North Island, New Zealand, with oxygen-isotope stratigraphy derived from DSDP data, and with nearby coastal terraces (Kamp and Turner, 1990)

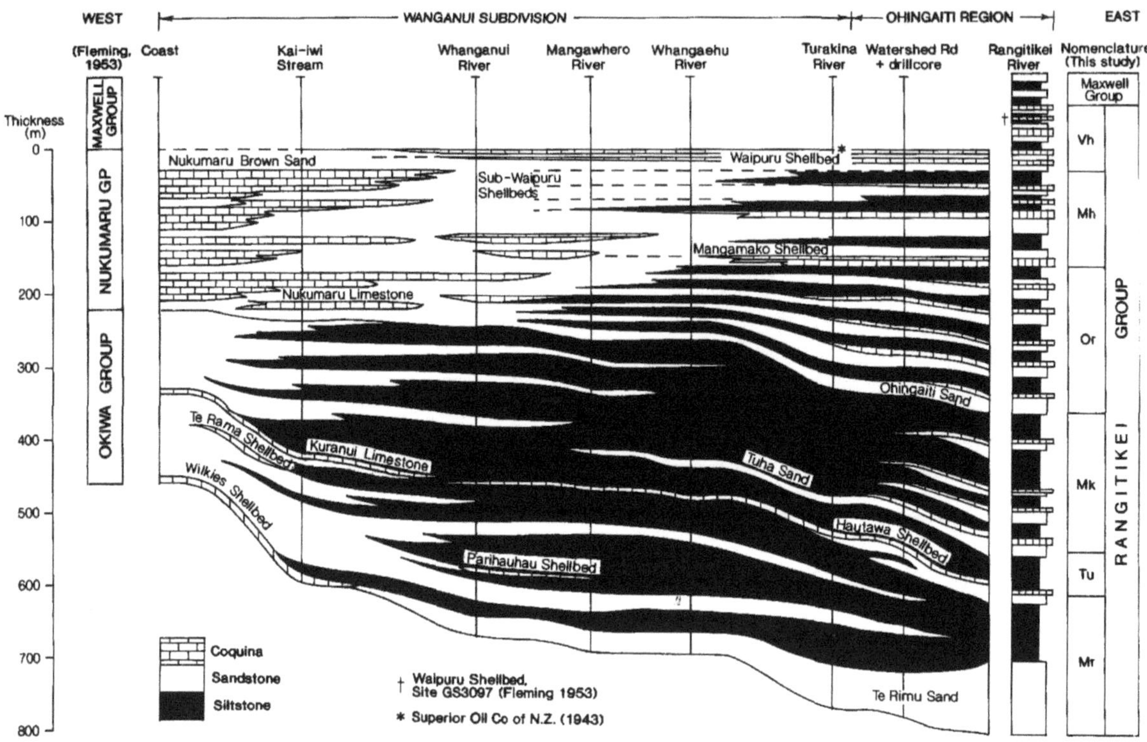

**Fig. 7.18** Schematic cross-section through the Wanganui Basin, showing east–west facies change from coarse-grained, siliciclastic sands and bioclastic lenticular limestones in the west to finer grained sandy siltstones and siltstones in the east (Naish and Kamp, 1995, Fig. 10)

dominated) with minor shell-bed lenses in the east, to thicker and more continuous coquina limestone units in the west (Fig. 7.18). The succession can be followed into prograding clinoform cycles at the shelf margin (Carter et al., 1991).

The conceptual sequence stratigraphic model is as follows (abbreviated slightly from Naish et al., 2005, pp. 96–97):

1. A basal sequence boundary (SB) consisting of an unconformity, which is coincident with the transgressive surface of erosion (TSE).
2. A transgressive systems tract (TST), corresponding mainly to a condensed fossiliferous deepening upwards interval comprising shell-beds. Commonly an onlap shell-bed containing shallow water infaunal molluscs is overlain by a backlap shell-bed comprising significantly more *in-situ* epifaunal components (molluscs, barnacles, brachiopods, and bryozoans) within a siltstone matrix of offshore shelf affinity. These superposed, or vertically gradational, shell-bed units are termed "compound shell-beds". Onlap shell-beds may be overlain by sand rich shoreline facies or heterolithic laminated and bedded intertidal facies.
3. A downlap surface (DLS).
4. A highstand systems tract (HST) comprising an aggradational interval of bioturbated shelf siltstone, or sandier shoreline inner shelf facies, depending on position on the palaeoshelf.
5. A regressive systems tract (RST) that marks a transition from fine sandy siltstone or fine sandstone facies of inner shelf affinity in the upper HST to strongly progradational, "forced" regressive shoreface-foreshore-intertidal facies assemblage. In some sequences, marginal marine swamp/lacustrine and coastal plain facies are preserved in the top of RSTs below the sequence boundary.

The facies composition of the sequences varies from coastal plain to outer shelf. This has been recognized by the definition of some 11 sequence "motifs", named after characteristic outcrops locations (Fig. 7.19). Detailed sedimentology and numerical modeling of

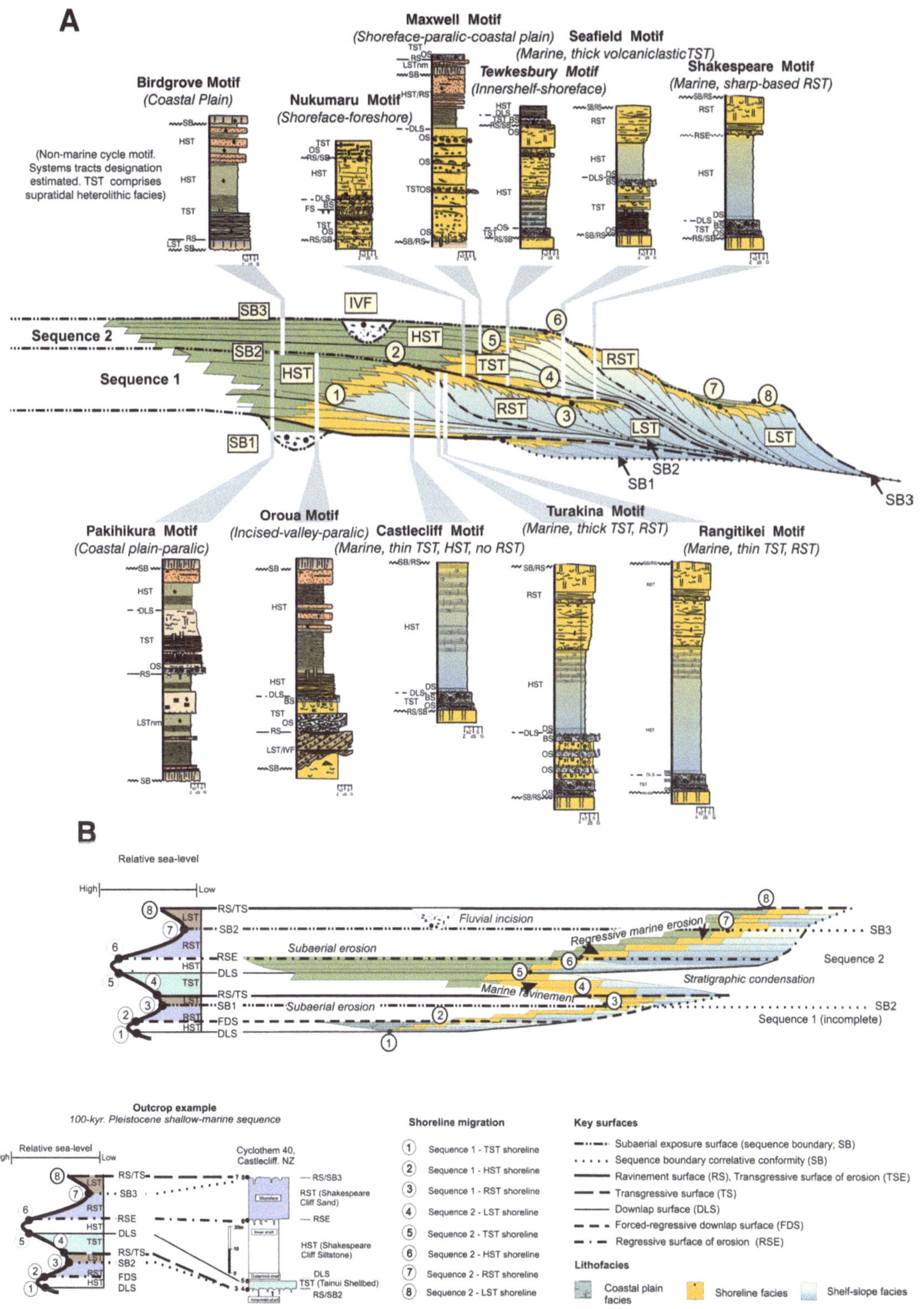


**Fig. 7.19** Conceptual (**a**) sequence stratigraphic and (**b**) chronostratigraphic models for Milankovitch-scale Wanganui Basin depositional sequences. The stratigraphic architecture and timing of development of systems tracts and key surfaces are shown with respect to a relative cycle of sea level. Outcrop motifs of Wanganui sequences are illustrated in the context of a 2D coastal plain to outer shelf progradational setting (Naish et al., 2005, Fig. 6)

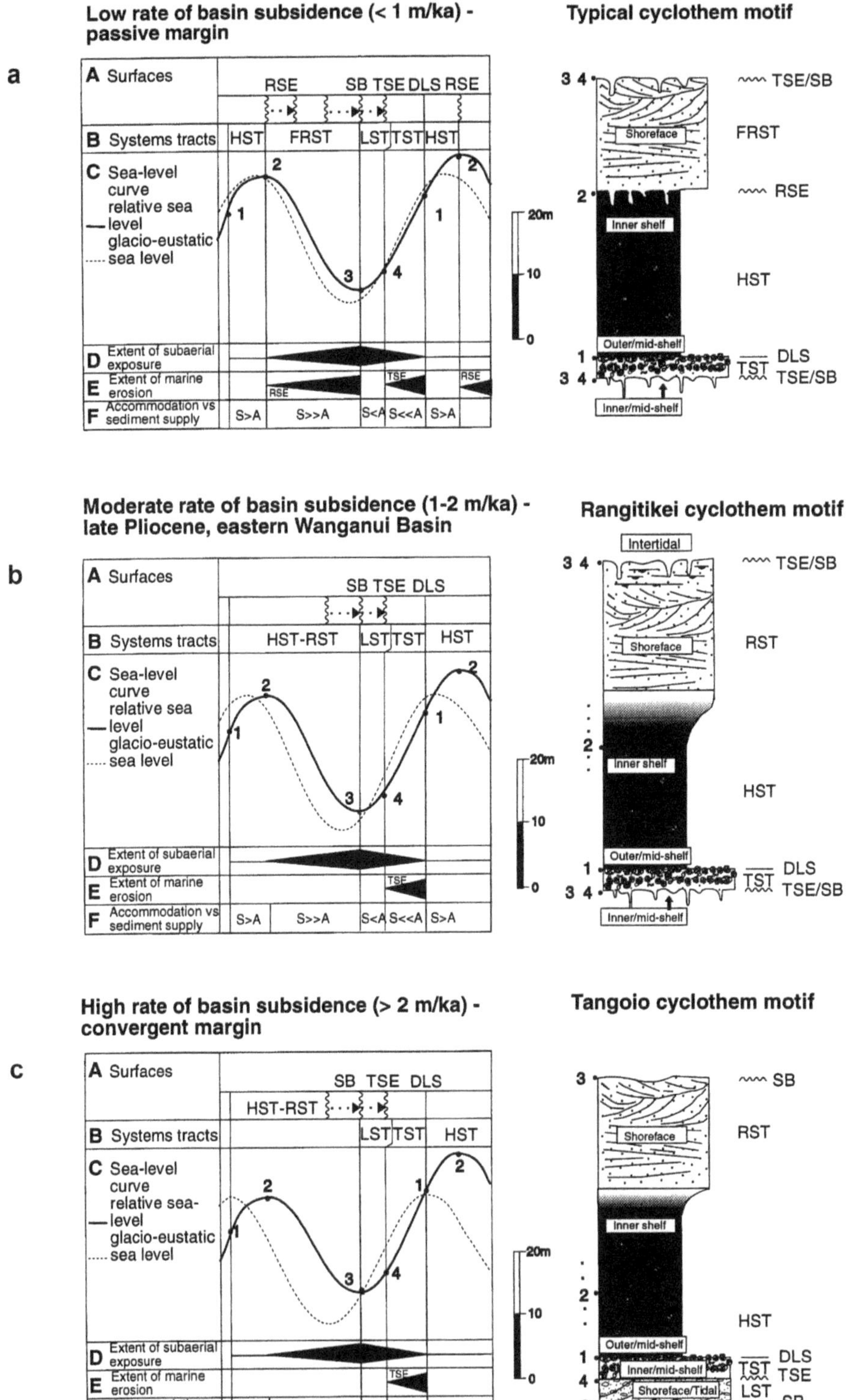


**Fig. 7.20** The timing of development of systems tracts and key surfaces in Wanganui cyclothems, showing variations in response to changes in subsidence rate. HST = highstand systems tract, RST = regressive systems tract, FRST = forced regressive systems tract, LST = lowstand systems tract, TST = transgressive systems tract, RSE = regressive surface of erosion, TSE = transgressive surface of erosion, DLS = downlap surface, SB = sequence boundary (Naish and Kamp, 1997, Fig. 10)

the sequences enabled Naish and Kamp (1997) to present a discussion of the relationships between systems tracts and key surfaces within sequences, and their timing relative to the sea-level curve, a relationship that has formed the basis for heated discussions about sequence classification (a discussion taken up at length by Catuneanu, 2006; Catuneanu et al., 2009). Three of the sequences motifs are shown against curves of relative and actual (glacioeustatic) sea-level change in Fig. 7.20, illustrating how the development of systems tracts and key surfaces correlate to specific points on the curve of changing accommodation in response to variations in the rate of subsidence.

### 7.2.3 Other Examples of Neogene High-Frequency Cycles

Neogene high-frequency cycles have been well documented in forearc and backarc settings on the continental margins of Japan (Ito, 1992, 1995; Ito and O'Hara, 1994; Masuda, 1994). Examples of $10^4$-year cyclicity are illustrated in Fig. 7.21. This succession consists of six sequences spanning the 0.4–0.8 Ma period,

and estimated to have durations ranging from 45,000 to 50,000 years. Each sequence can be subdivided into systems tracts, comprising mainly progradational slope and shelf deposits. The refined stratigraphic correlations indicated in this outcrop study were made possible by the presence of numerous volcanic ash beds, which can be correlated on the basis of their geochemical signatures and fission-track ages. Ito (1992) showed how seawater temperatures fluctuated during sedimentation, and correlated these data with the oxygen-isotope record (Fig. 7.22). He interpreted the cycles to be of glacioeustatic origin (Chap. 10).

As discussed in Chap. 2, sediment delivery to the continental slope and basin plain takes place predominantly during the falling stage of the base-level cycle, and this accounts for the major stages of growth of submarine fans on the sea floor. In some cases, sequences are dominated by the falling-stage systems tract, including significant submarine fan deposits, as in the examples illustrated here in Figs. 7.23, 7.24, and 7.25 from the continental margin of Borneo, Indonesia.

Reflection-seismic data shows that on the eastern continental margin of Borneo, Indonesia, the

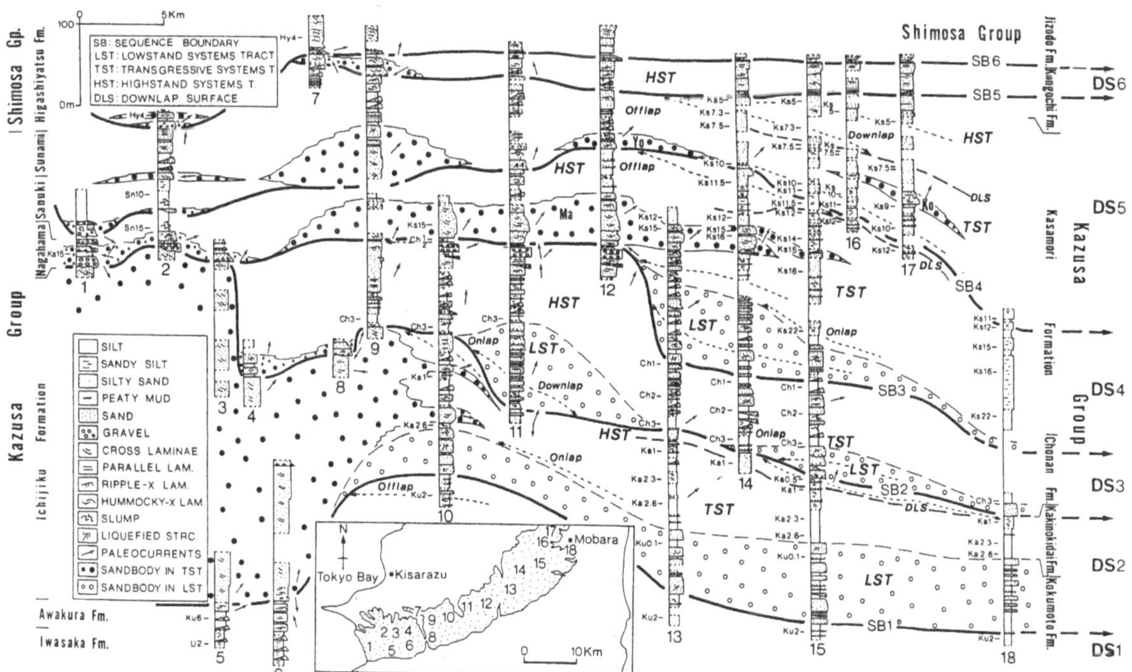

**Fig. 7.21** Stratigraphic cross-section of the upper part of the Kazusa Group in Boso Peninsula, Japan, illustrating glacioeustatic cycles. Codes to *left* of each column indicate names of volcanic ash layers used for chronostratigraphic correlation. Six stratigraphic sequences are recognized, spanning the period 0.4–0.8 Ma (Ito, 1992)

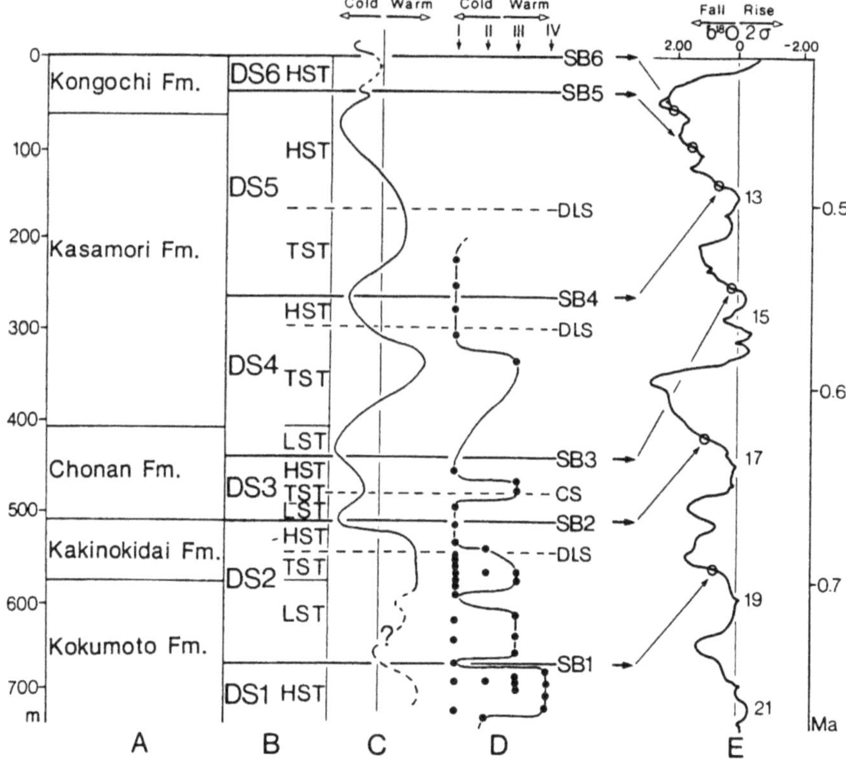



**Fig. 7.22** Correlation of the Kazusa Group, Boso Peninsula, Japan (columns A, B), with the oxygen isotope record (column E). Also shown are temperature records derived from benthic molluscs (column C) and planktonic molluscs (column D). Compiled by Ito (1992)

last four cycles to be deposited by glacioeustatic sea-level change are dominated by progradational shelf-margin, slope and basin-floor deposits. The ages of the sequences are shown against the oxygen-isotope curve at the lower right of Fig. 7.23. This curve shows the asymmetric nature of the sea-level cycle, consisting of a slow falling stage (gradual cooling and ice formation; increasing $\delta O^{18}$ values) and a rapid rising stage. Note the carbonate buildups formed during sea-level rise. These are drowned by rapidly rising sea level, or by deltaic clastic influxes during the subsequent fall. The geometry of the sequences—gently-dipping shelf reflectors, a distinct shelf margin, and prograding clinoforms beyond the shelf edge—is generated by the adjustment of shelf and slope depositional processes to the slowly falling sea level. The flooding surfaces (numbered P1 to P3) dip gently seaward because they are developed over the deposits formed during the preceding slow fall.

Figures 7.24 and 7.25 provides examples of the power of modern, three-dimensional reflection-seismic techniques to reveal the detailed internal architecture of slope and fan deposits. The three cross-sections also provide excellent illustrations of the standard three-fold subdivision of submarine fans into upper, middle and lower. The inner fan is characterized by a single large channel with well-developed levees; the mid fan by a rougher topography, corresponding to the pattern of small distributary channels, and the outer fan by a smooth, convex-up stratigraphy, consistent with an unchannelized surface crossed by sheet-like turbidites.

What is becoming a classic field study of high-frequency carbonate sequence stratigraphy is the Upper Miocene platform carbonate succession of Mallorca (Pomar, 1993; Pomar and Ward, 1999; Schlager, 2005; Fig. 7.26). The complete section represents about 2 million years, and can be subdivided into four sequences and numerous subsequences or parasequences. This is a good example of what Schlager (2005, p. 137) terms a "supply-dominated" succession. The platform prograded about 20 km during its brief life. The internal architecture of the platform reveals numerous "sigmoids" resulting from rapid progradation, with minor aggradation. Backstepping, and any other evidence of deepening-upward have not

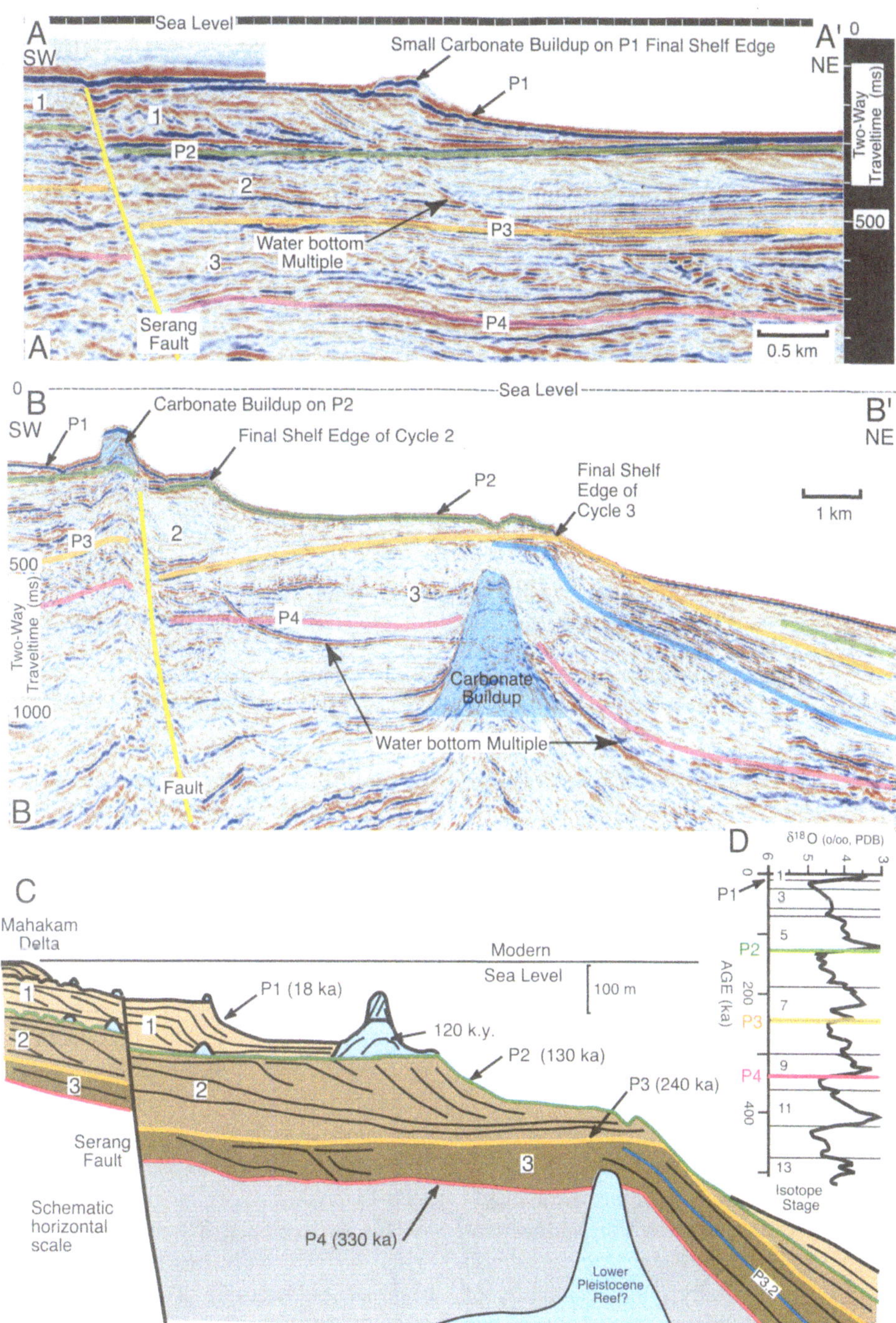


**Fig. 7.23** Two seismic cross-sections and a reconstructed cross-section through sequences formed by glacioeustatic sea-level changes on the east coast of Borneo, Indonesia. Timing is indicated by the $\delta O^{18}$ curve at *right*. Diagrams in Figs. 7.24 and 7.25 show slope-to-basin deposits at the right-hand end of these sections (Saller et al., 2004). AAPG © 2004. Reprinted by permission of the AAPG whose permission is required for further use

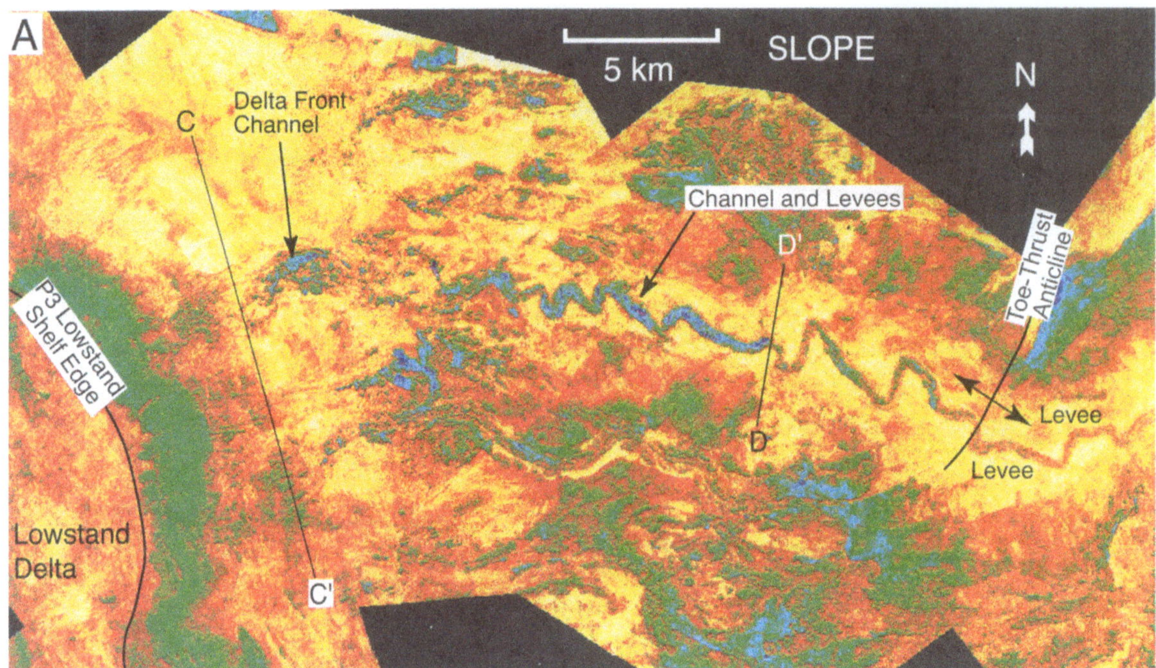


**Fig. 7.24** Horizontal seismic section through channel and levee deposits on the continental slope, Borneo. Cycle 3 of Fig. 7.23 (Saller et al., 2004). AAPG © 2004. Reprinted by permission of the AAPG whose permission is required for further use

been observed. Sequence bounding surfaces are erosional, with abundant evidence of terrestrial exposure. Flooding events consist of beds of open-shelf limestone onlapping tongues of reef debris in the clinoforms. These rarely reach up to the crest of the marginal reef.

### 7.2.4 The Deep-Marine Record

In their introduction to a volume entitled "Cyclostratigraphy: approaches and case studies", Fischer et al. (2004, p. 10) stated "The pelagic and hemipelagic facies of bathyal depths are attractive for various reasons. Depth and distance from the geological turmoil associated with lands generally imply continuity of accumulation, and limited fluctuations in facies and in accumulation rates. Pelagic faunas also offer optimal biostratigraphic control."

As detailed in Sect. 14.7, the Neogene pelagic sedimentary record is now being used to construct a high-resolution time scale by correlating measured and numbered beds in continuous pelagic successions to

the specifics of the orbital control on depositional conditions, working back, cycle by cycle from the present day. This is a type of stratigraphy in which orbital control is expressed by changes in chemical and biogenic sedimentary processes, rather than by glacioeustasy. The Pliocene Trubi Marls of Sicily are the classic field location (Fischer et al., 2009). There, the cyclicity is expressed in the form of metre-scale quadruplets of grey marl-limestone to beige marl-limestone. At intervals of about 20 m the quadruplets become highly calcareous and may lose marl members (Fig. 7.27). Each of these scales of cyclicity corresponds to a type of orbital frequency, as discussed in Chap. 11 and Sect. 14.7.

This is, in principal, only a qualitative extension of the body of work that has accumulated since the landmark publications of Emiliani (1955) and Hays et al. (1976) on the use of oxygen-isotope stratigraphy to document the climatic cyclicity of Neogene glacial to interglacial fluctuations in ice cores and deep-sea drill cores. As a result of convergent tectonics in the Mediterranean region, the Neogene deep-marine record of the paleo-Mediterranean (or late Tethys, if you prefer) has been uplifted and is superbly exposed

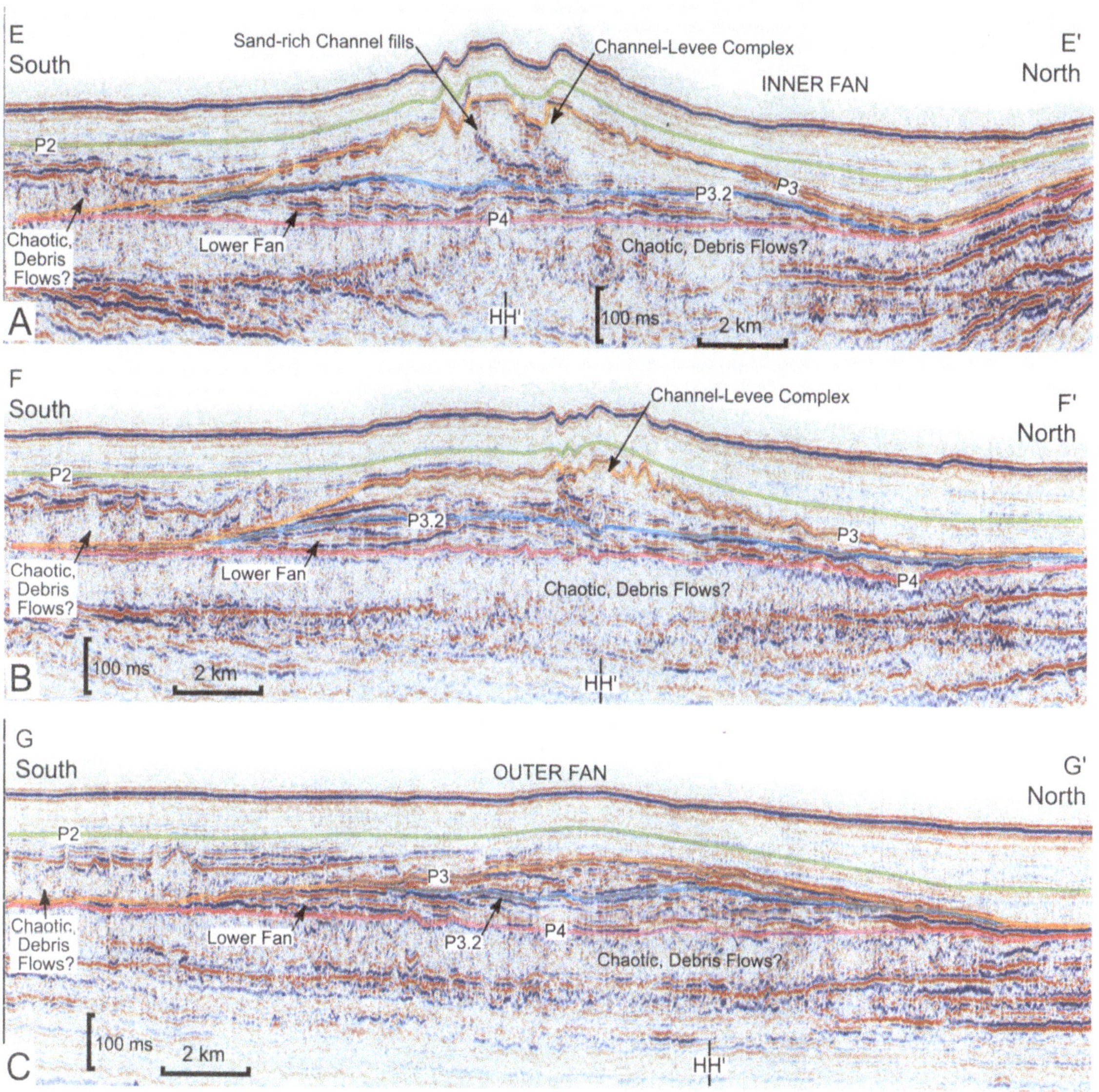

**Fig. 7.25**  Three strike-sections across the submarine fan formed on the basin floor, downstream from the deposits illustrated in Fig. 7.24 (Saller et al., 2004). AAPG © 2004. Reprinted by permission of the AAPG whose permission is required for further use

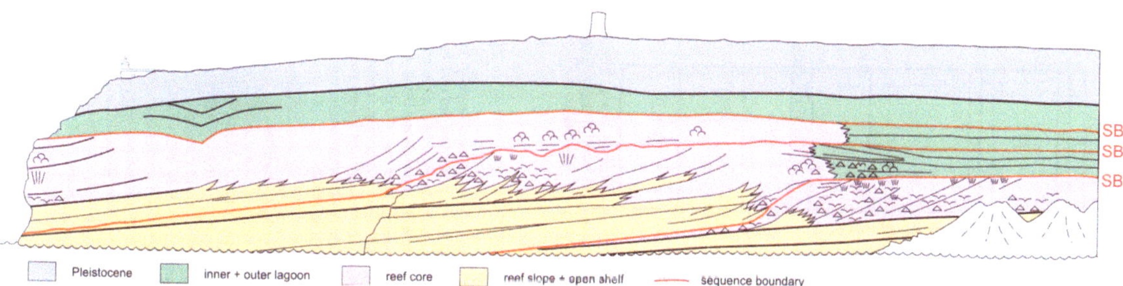


**Fig. 7.26**  The platform carbonate succession at Cap Blanc, Mallorca. The portion of the cliff shown here is about 480 m long, and 90 m high. The complete cliff section extends for over 2 km (Schlager, 2005, Fig. 7.42, after Pomar, 1993)

**Fig. 7.27** The classic Trubi Marl succession (Pliocene) of Capo Rosello, Sicily. The cycles are numbered in the *lower* photograph. g = grey marl layers, b = beige marl layers. Photographs by F. Hilgen (in Fischer et al., 2009, Fig. 12)

**Fig. 7.28** Correlation of three partial composite sections through upper Miocene cyclic successions in southern Italy and Malta. Cycles are numbered from 1 to 160 from *top to bottom* (numbers to *right* of column). Letters and numbers in *circles* to *left* of columns are bioevents. A 3-m scale is shown at *bottom left*. Correlations are shown to the orbital frequencies at *right*. These are discussed in Chap. 14 (Jaccarino et al., 2004, Fig. 5)

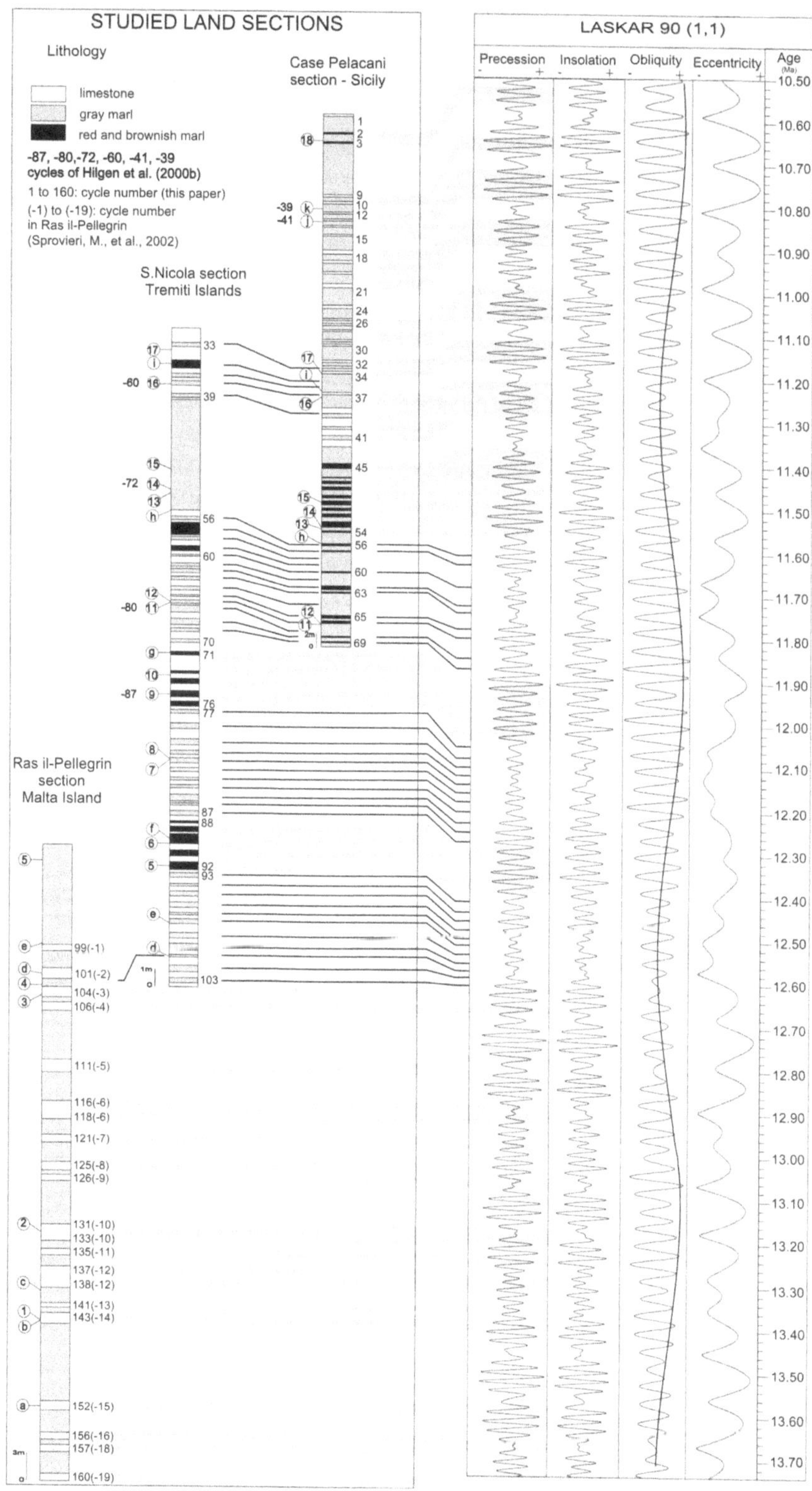

in several coastal regions around this sea, especially in southern mainland Italy and Sicily. The stratigraphy has been documented by meticulous cycle counts, together with rigorous and systematic analysis of cyclic parameters, such as colour, chemical composition, or petrophysical response. Cyclicity has been demonstrated by advanced statistical analyses, and this has helped to build a convincing case that these warm-water, deep-marine successions hold a virtually complete record of Neogene climatic history, unblemished by diagenesis or tectonism.

As a second example we illustrate here the work to correlate Middle Miocene sections in Italy and Malta, reported by Jaccarino et al. (2004). The cycles consist of marls with rhythmic alternations of calcareous and/or organic content. Correlations of partial sections were accomplished using a series of planktonic foraminiferal bioevents and abundance curves. As shown in Fig. 7.28, a composite section has been constructed from three main locations consisting of 160 cycles, averaging about 1 m in thickness, and spanning the 10.50–13.71 Ma time interval. Average cycle duration is therefore in the range of 20,000 years but, as discussed in Chap. 11, the orbital driving forces that generate this type of section result in complex cyclic patterns of three or four or more interacting lengths and durations.

## 7.3 Pre-neogene Marine Carbonate and Clastic Cycles

A wide variety of $10^4$–$10^5$-year carbonate and clastic cycles has been documented in the pre-Neogene sedimentary record in various tectonic settings (Fischer, 1986; Fischer and Bottjer, 1991; Dennison and Ettensohn, 1994; de Boer and Smith, 1994a; D'Argenio et al., 2004a). These are best preserved on carbonate shelves and in deep marine and lacustrine environments, away from clastic influxes, because the autogenic redistribution of detrital sediment tends to obscure regional cyclic patterns. However, high-frequency cyclicity has now also been suggested for many other types of stratigraphic succession, even including some nonmarine strata, as discussed in this section.

Amongst the most well studied cyclic successions in North America is the Permian Capitan reef complex of the Guadalupe Mountains of southern New Mexico

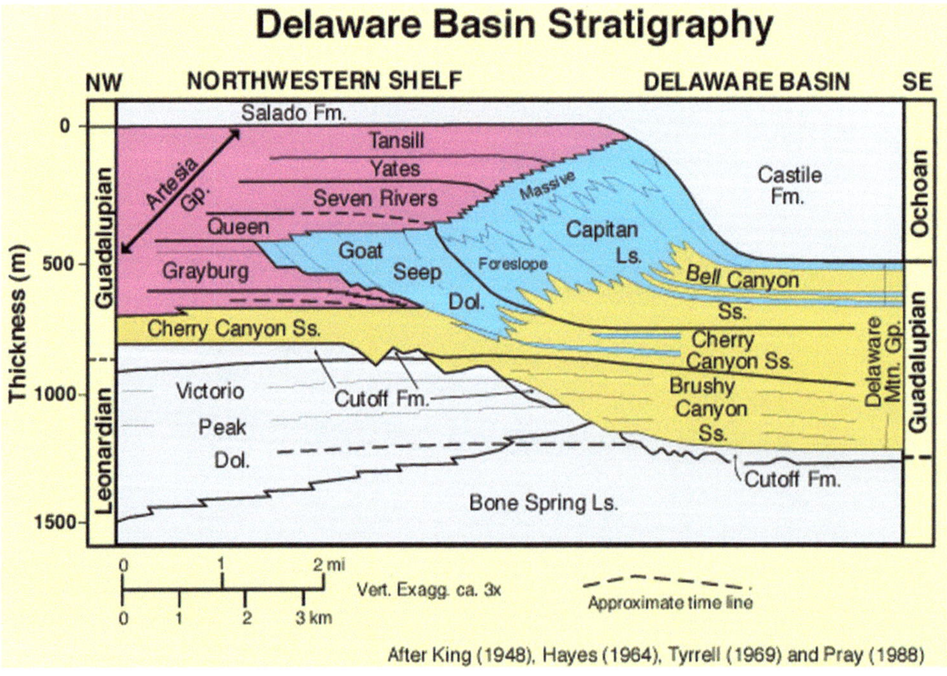


**Fig. 7.29** The stratigraphic relationships between shelf, slope and basin, Capitan reef area, Guadalupe Mountains, Southern New Mexico and West Texas (Scholle, 2006)

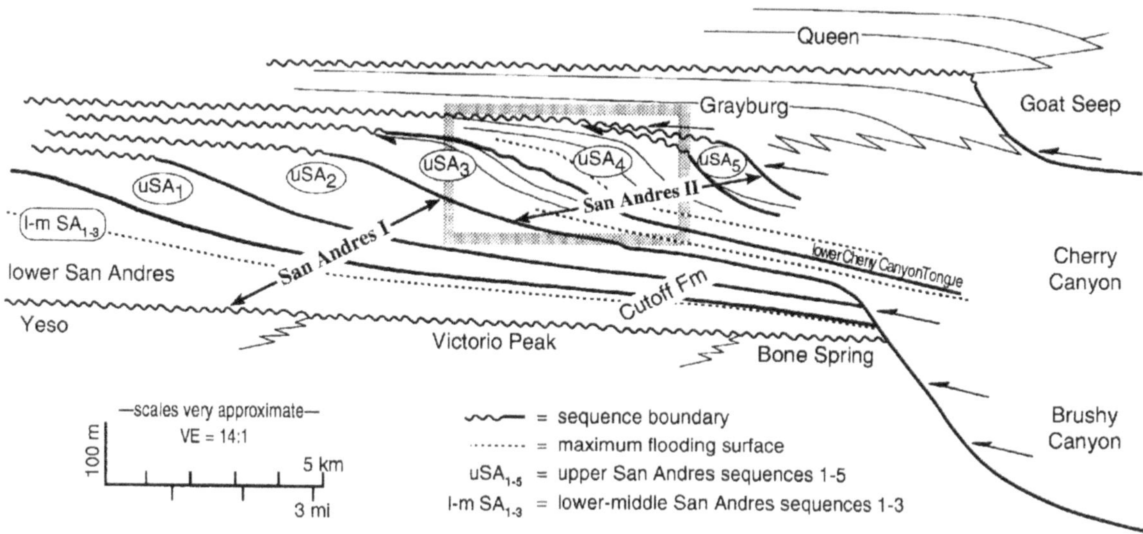



**Fig. 7.30** Stratigraphy of the shelf-slope transition of the Sand Andres and related units, Capitan reef complex, Guadalupe Mountains, New Mexico. San Andres I and II have been interpreted to represent "third-order" sequences (Sonnenfeld and Cross, 1993, Fig. 4). The area shown in the *box* is enlarged in Fig. 7.31. AAPG © 1993. Reprinted by permission of the AAPG whose permission is required for further use

and adjacent areas of Texas. Numerous workers have described the clinoform stratigraphy of this reef-to-basin transition, since King (1942, 1948) made these rocks famous. We provide here a single example of the subdivision of the succession into nested cycles (Fig. 7.29). The Capitan reef prograded basinward, developing a spectacular clinoform stratigraphy. Clastic sediments accumulated in submarine fan and other deep marine environments at the foot of the slope. During deposition of the Delaware Mountain Group the relief difference between shelf and basin was 200–500 m. We illustrate here a slightly earlier stage in margin development, the cyclic sediments of the San Andres Formation, during which the clinoforms were 40–100 m high (Sonnenfeld and Cross, 1993). The San Andres Formation constitutes

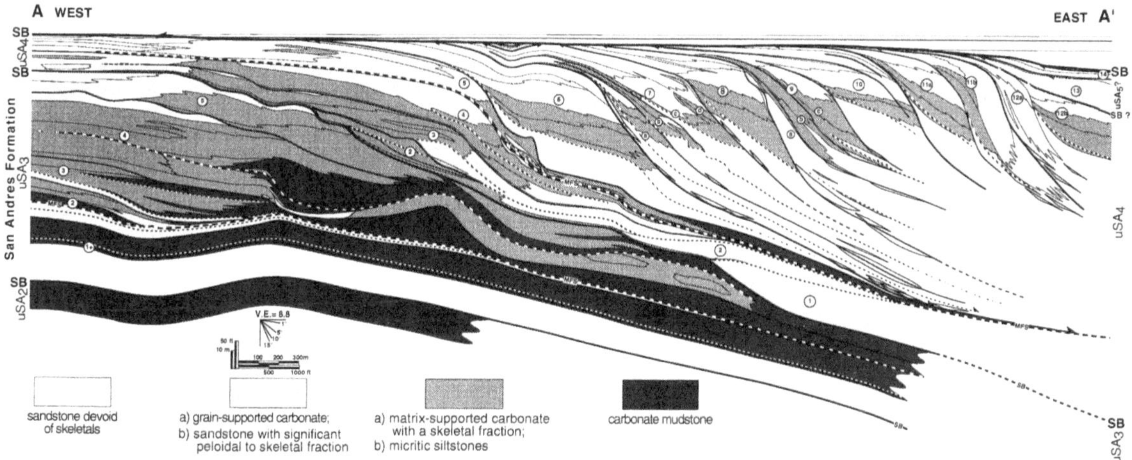



**Fig. 7.31** Enlarged cross-section through the San Andres Formation, showing the details of sequences uSA₂, uSA₃ and uSA₄. Each of these three sequences can be subdivided into high-frequency cycles, as indicated by the numbers n *circles* (Sonnenfeld and Cross, 1993, Fig. 26). AAPG © 1993. Reprinted by permission of the AAPG whose permission is required for further use

at least ten sequences, each estimated to represent 210–400 ka of sedimentation. Six of these are shown in Fig. 7.30, and the details of two of these are enlarged in Fig. 7.31. Sequence uSA$_3$ rests on a tongue of Brushy Canyon Sandstone, which here consists of hummocky cross-stratified sandstone deposited on a ramp. Five high-frequency cycles may be recognized within uSA$_3$, exhibiting a landward-stepping (retrogradational) to aggradational stacking pattern. Facies dislocations across cycle boundaries are interpreted to indicate that they formed by high-frequency changes in sea level. There is then a substantial downward shift of the overlying deposits of cycle 1 of overlying sequence uSA$_4$, which consists of carbonate turbidites, and is confined to the base of the slope. The 13 subsequence cycles account for several kilometres of basinward progradation of the platform margin. Lowstand to transgressive deposits of each cycle consist of fine-grained silty sandstones. These pass upward into prograding highstand packages of bioclastic packstones and wackestones capped by swaly-bedded sandstones. Sequence uSA$_4$ is capped by a karst surfaces, indicating widespread exposure.

A mixed carbonate-clastic suite of Milankovitch cycles in the Permian Yates Formation of the Delaware Basin, west Texas, was described by Borer and Harris (1991). The cycles are prominent in a mid-shelf association, seaward of the inner-shelf evaporite-clastic belt and landward of the shelf-margin Capitan reef belt. The cycles comprise carbonate-clastic couplets. Carbonates consist mainly of algal and peloidal dolomudstones, which yield low gamma-ray readings. Clastics consist of arkosic sandstones and argillaceous siltstones that have high gamma-ray reponses. The alternation between the two main lithofacies types is readily apparent on gamma-ray logs, which reveal the presence of two scales of cyclicity (Fig. 7.32). Clastic intervals were deposited by marine, and possibly eolian processes, during low stands of sea level, and carbonates were formed during transgressions.

The Triassic record of the Dolomites in northern Italy contains a spectacular record of cyclic sedimentation (Goldhammer et al., 1987, 1990; Goldhammer and Harris, 1989; Hinnov and Goldhammer, 1991; Goldhammer et al., 1993), including "nested" cycles of 10$^4$-, 10$^5$- and 10$^6$-year episodicity. These cycles are interpreted to be primarily of eustatic origin, with basin subsidence and autogenic marine processes acting to modify the overall succession (Goldhammer et al.,

1987, 1990; Jones and Desrochers, 1992). The sediments developed on an isolated carbonate platform, named the Latemar Massif. They consist of cyclic, platform deposits passing laterally into progradingd slope deposits. The interpreted 10$^4$-year cycles record abrupt shallowing and exposure, with the development of subtidal platform deposits that underwent exposure and vadose diagenesis. The cycles in the Lower and Upper Cycle Facies average 0.7 m in thickness. In the Lower Platform Facies and the Tepee Facies cycle composition and thickness are different, and Goldhammer and Harris (1989) and Goldhammer et al. (1987, 1990) interpreted this as a result of the overprint of slow (10$^6$-year) eustasy on the higher frequency cycles of sea-level change. During deposition of the Lower Platform Facies the area was undergoing a prolonged long-term rise in sea level. The result was the maintenance of subtidal conditions over the entire platform (~0.5 million years):

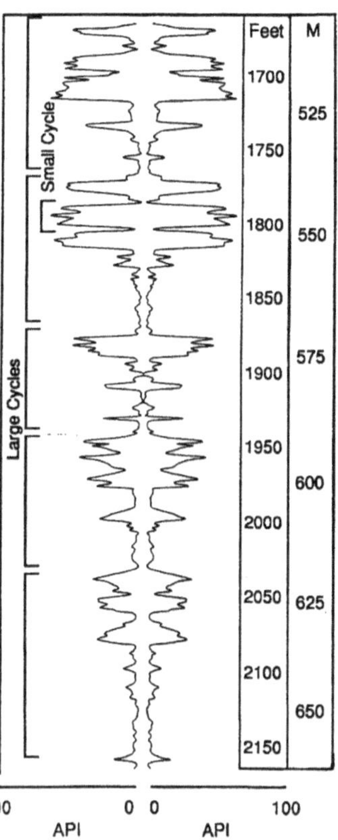


**Fig. 7.32** Composite gamma-ray log through the Yates Formation. Log is repeated in mirror-image to emphasize cyclicity (Borer and Harris, 1991). AAPG © 1991. Reprinted by permission of the AAPG whose permission is required for further use

Subtidal sedimentation is unable to keep up, and thus not every beat of fifth-order high-frequency sea-level can "touch down" on the platform top and subaerially expose the top of the sediment column. These "missed beats" of subaerial exposure result in the formation of thick (average approximately 10 m), fourth-order amalgamated megacycles. During this phase of subtidal carbonate aggradation, subaerial exposure occurs infrequently, and lengthy periods of marine submergence promote abundant syndepositional marine diagenesis (Goldhammer et al., 1990, p. 549).

As discussed in Sect. 14.7.2, there is now an ongoing controversy regarding the age range represented by these limestones, which has important implications for the duration of the cycles. They may not all be "orbitally-forced" in the sense described in Chap. 11.

Other carbonate suites comparable to the Alpine Triassic (in episodicity but not necessarily in facies) include a Mississippian example in Wyoming and Montana (Elrick and Read, 1991) and various Cambrian examples around the margins of the North American continent (Osleger and Read, 1991). Weedon (1986) described centimetre-decimetre-scale limestone-shale couplets In the Lower Jurassic Lias of southern Britain, and discussed their origin in terms of Milankovitch mechanisms. In the Cretaceous Niobrara Formation of Colorado, limestone-shale couplets occur in bundles of 1–12 couplets, each bundle ranging from about 1 to 5 m in thickness. Laferriere et al. (1987) analyzed these deposits in terms of Milankovitch rhythms.

A spectacular clinoform succession is that of the Neocomian in the West Siberia Basin, Russia, which is one of the largest sedimentary basins in the world. This is a broad "sag" basin, comparable in origin to the North Sea Basin. Resting on a Paleozoic basement, basin formation commenced with rifting in the Triassic, followed by slow subsidence through the Jurassic to Cenozoic (Pinous et al., 2001). Seismic-reflection data demonstrated the existence of a widespread clinoform succession that prograded into the basin from the east and west sides during the Early Cretaceous (Fig. 7.33). Clinoform topsets are coastal-plain and shelf deposits, predominantly sandstones. These pass down into slope muds and siltstones, and then into deep-water turbidites, deposited in a series of offlapping submarine fans. A typical well profile is shown in Fig. 7.34. Sixteen sequences (labeled K-1 to K-16, from oldest to youngest) have been identified within the prograding, west-dipping clinoforms on the

east side of the basin (Fig. 7.35). The total thickness of marine sequences ranges from 350 to 400 m in the east, to 700 m in the west. These sequences prograded about 500 km in 9 million years. Each, therefore, represents about 560,000 years. Some may be traced along strike for 500 km, which is regarded as strong evidence for eustatic control of sequence generation (Pinous et al., 2001, p. 1727).

Pelagic successions containing a clear signature of cyclostratigraphic deposition have been recorded in several pre-Neogene successions. As noted by Grippo et al. (2004, p. 57), the Umbria-Marche area of the Apennines, in Italy, contain "a continuous history of pelagic-hemipelagic sedimentation extending from Early Jurassic times into the Miocene." An example is illustrated in Fig. 7.36. This is from a 50–60-m section of Albian coccolith-globigerinid ooze and marl, now compacted into alternating beds of marlstone and limestone. Dark grey to black bands are coloured by carbonaceous organic matter and iron sulphide and are interpreted as the deposits of anoxic intervals. Limestone beds are rich in skeletal remains, and indicate periods of high organic productivity. Some of the limestones are red in colour, as a result of the oxidation of all iron to the ferric state. The causes of these environmental variations are discussed in Chap. 11.

Figure 7.36 illustrates part of a drill core. The grey-scale plot was derived by scanning of photographs of the core with a densitomcter. The individual cycles are interpreted as precessional in origin. They are grouped into "bundles" that reflect a 95-ka eccentricity periodicity, and these, in turn, into 406-ka eccentricity harmonic "superbundles."

## 7.4 Late Paleozoic Cyclothems

The first cycles to be described in the geologic literature were the Mississipian Yoredale cycles of the English Pennines area (Wilson, 1975; Holdsworth and Collinson, 1988). They represent only part of a lengthy and widespread coal-bearing cyclic succession that spans much of the Carboniferous and Permian and extends throughout NW Europe (e.g., Ramsbottom, 1979). The classic Pennsylvanian cyclothems of the American Midcontinent were amongst the first cycles to be described in North America. Wanless and Weller (1932) coined the term cyclothem for them.

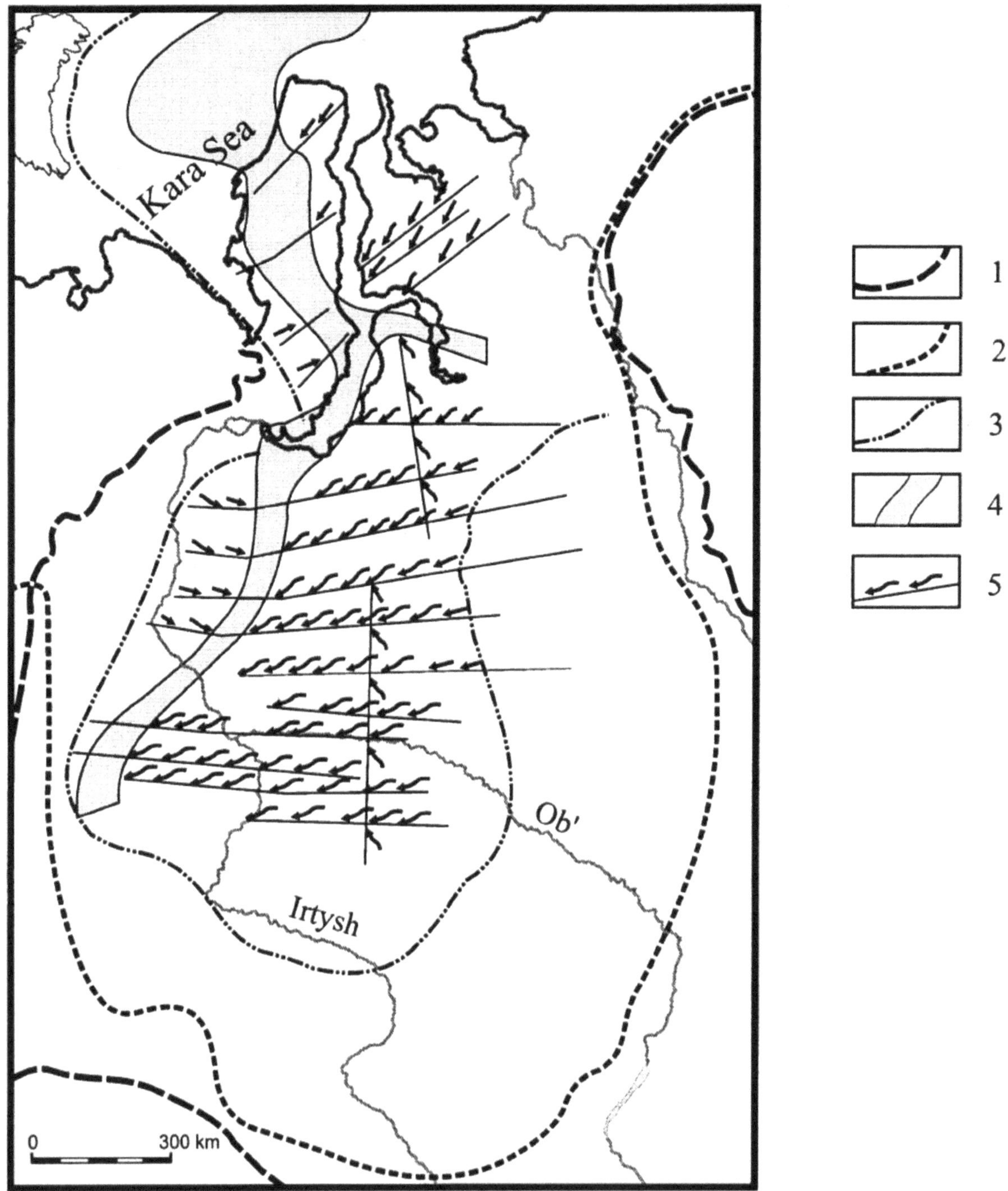



**Fig. 7.33** Generalized map of Neocomian sedimentation of the West Siberian basin. The *arrows* delineate dip directions of the clinoforms. Note the asymmetrical structure of the basin fill. 1 = boundary of the basin; 2 = maximum transgression (Volgian); 3 = approximate shoreline in the beginning of the Valanginian, 4 = terminal channel zone; 5 = seismic lines, dip directions of the clinoforms (Pinous et al., 2001, Fig. 2). AAPG © 2001. Reprinted by permission of the AAPG whose permission is required for further use

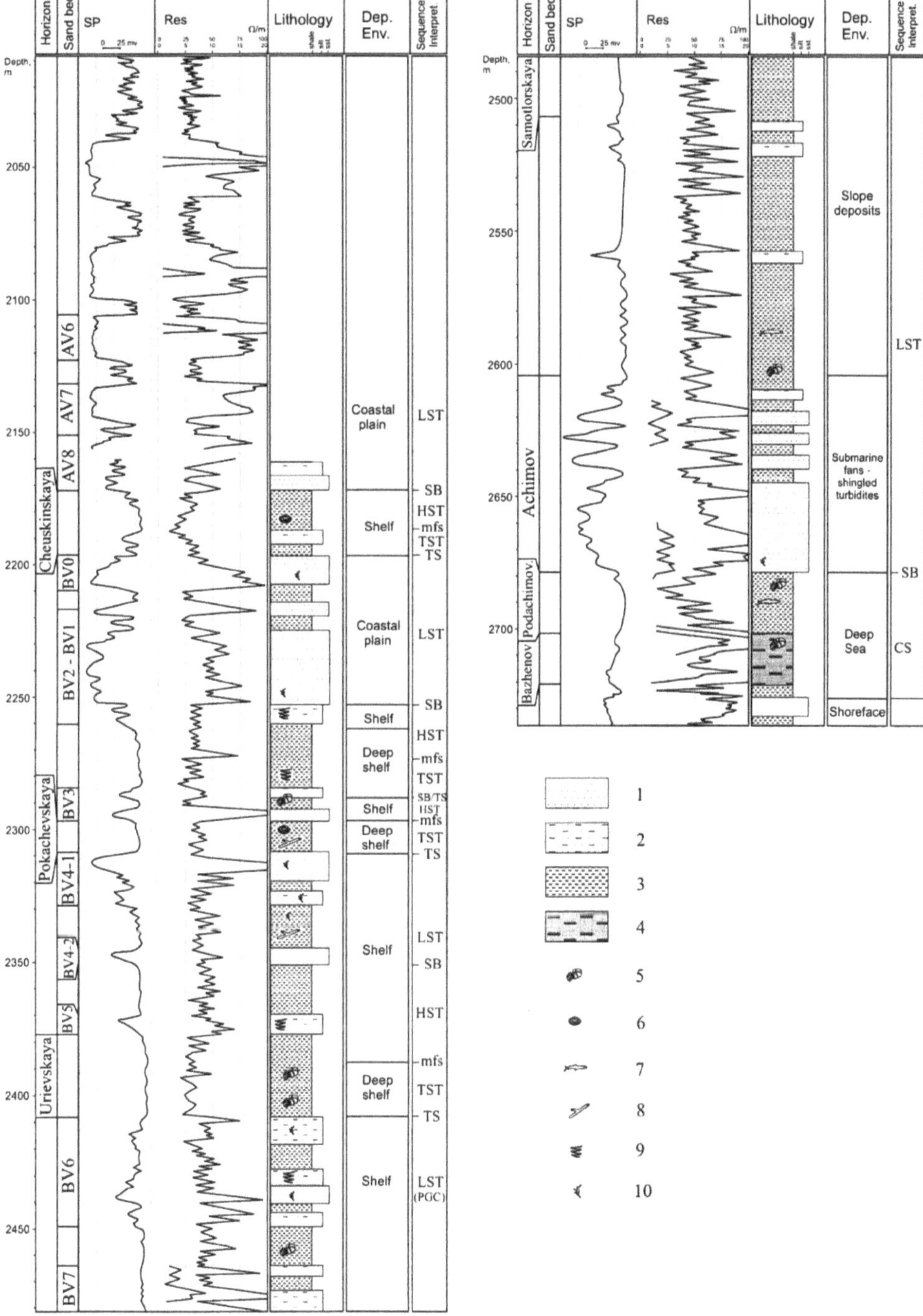







**Fig. 7.34** Stratigraphic subdivision of the Neocomian section of the Pokachevskaya 41 well (western margin of the Nizhnevartovsk arch). LST = lowstand systems tract; TST = transgressive systems tract; HST = highstand systems tract; SB = sequence boundary; TS = transgressive surface; mfs = maximum flooding surface; CS = condensed section; PGC = prograding complex; K-7 = sequence indexes; 1 = sandstone; 2 = siltstone; 3 = shale; 4 = bituminous shale; 5 = ammonites; 6 = bivalves; 7 = fish bones; 8 = chondrites; 9 = *Teichichnus*; 10 = plant detritus (Pinous et al., 2001, Fig. 3). AAPG © 2001. Reprinted by permission of the AAPG whose permission is required for further use

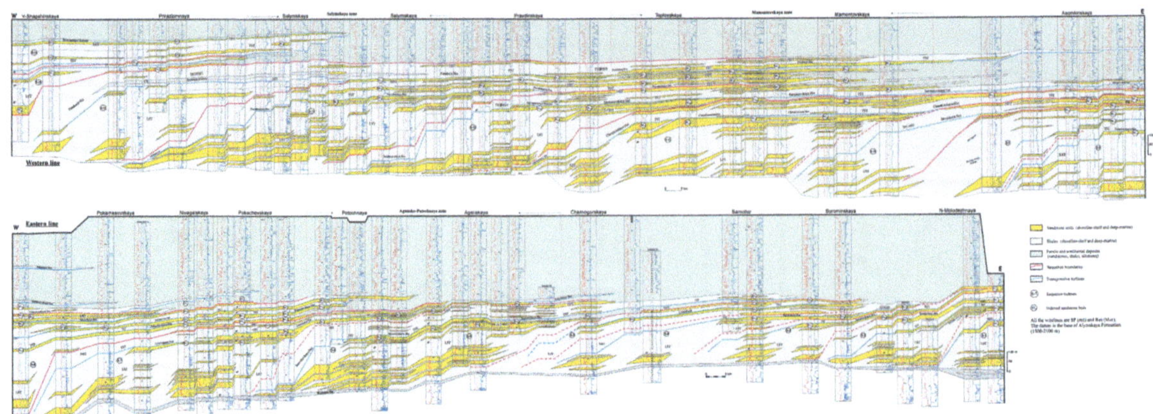

**Fig. 7.35** Regional wire-line correlation line across the Nizhnevartovsk and Surgut arches. The western and eastern lines are joined at the Asomkinskaya 22 well and together represent a continuous line from N-Molodezhnaya to V-Shapshinskaya. All the wire lines are spontaneous potential (*red*) and resistivity (*blue*). The datum is the base of Alymskaya Formation (1,500–2,100 m). 1 = sandstone units (shoreline-shelf and deep-marine); 2 = shales (shoreline-shelf and deep-marine); 3 = paralic and continental deposits (sandstones, shales, siltstones); 4 = sequence boundaries; 5 = transgressive surfaces; 6 = sequence indexes; 7 = indexed sandstone beds (Pinous et al., 2001, Fig. 4). AAPG © 2001. Reprinted by permission of the AAPG whose permission is required for further use

They have also been called Klüpfel cycles, after Klüpfel (1917). These cycles in the northern hemisphere are now interpreted primarily as the product of glacioeustasy (Crowell, 1978); they developed in response to the major Carboniferous-Permian glaciation of the Gondwana supercontinent. Klein and Willard (1989), Klein and Kupperman (1992), and Klein (1994) have also pointed out the importance of regional tectonism in the development of cyclothems, as discussed below. Groups of cyclothems have been termed mesothems (Ramsbottom, 1979; Busch and Rollins, 1984; Sect. 6.4).

Klein and Willard (1989) summarized stratigraphic data indicating that in the North American Interior there are essentially three types of cyclothem, an Appalachian type, an Illinois type, and a Kansas type (Fig. 7.37). The types differ mainly in the carbonate:clastic ratio within each cyclothem. Appalachian-type cyclothems were deposited in a foreland basin close to a major clastic source (the Alleghenian orogen) and are clastic dominated. Illinois-type cyclothems were deposited in an intracratonic basin that was partially "yoked" to the Appalachian foreland basin by the flexural effects of foreland-basin thrust loading. This basin was more distant from the sediment supply, and consequently contains a thinner, finer-grained clastic component. Kansas-type cyclothems were deposited within a cratonic area

a long distance inboard from clastic sources, and are carbonate-dominated. More than 100 such repetitions have been mapped in Kansas (Moore, 1964; Heckel, 1986, 1990, 1994). Some individual beds within the cyclothems can be traced for more than 300 km. Crowell (1978) reviewed ideas first proposed in the 1930s interpreting the cyclothems as the product of transgressions and regressions driven by glacioeustasy. He illustrated a typical Illinois-type cyclothem, and suggested the general range of depositional environments that led to the development of the succession during repeated transgressions and regressions (Fig. 7.38). Wilson (1975) developed generalized facies models for the Midcontinent cyclothems and those in Texas, suggesting how the various types might be related by lateral facies changes (Fig. 7.39).

Figure 7.40 illustrates how a typical cyclothem in Illinois varies as individual components thicken and thin. Thick sandstones and shales represent deltaic units deposited in regions of local subsidence. Nonmarine sandstones may rest on an erosional, channelized unconformity at the top of the underlying marine units. Coals, underclays, black shales, and marine limestones typically retain considerable uniformity over hundreds of kilometers (Wanless, 1964). In cratonic areas away from detrital sources, the nonmarine clastic units may thin to zero, so that the

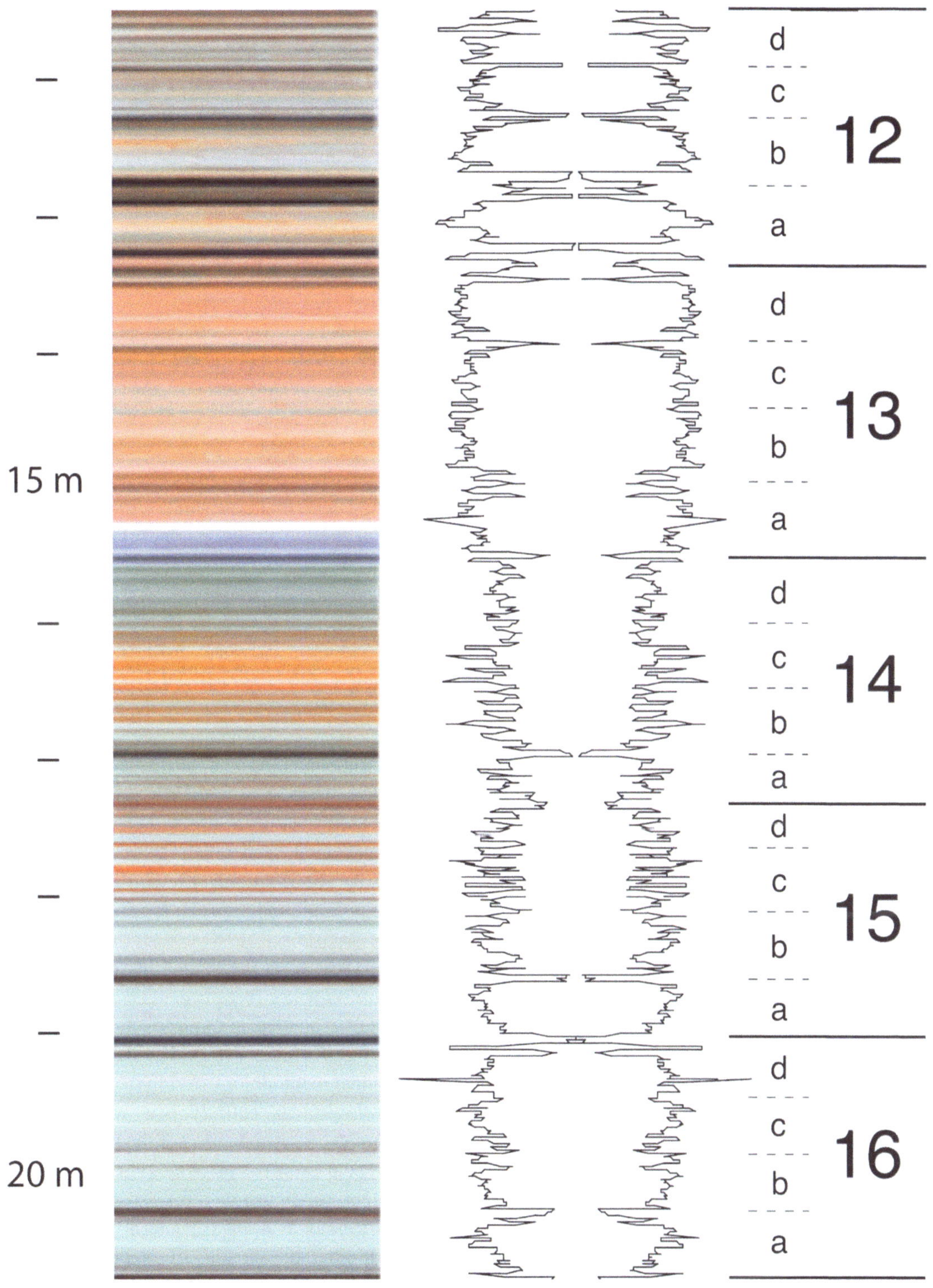


**Fig. 7.36** Photograph of part of a core spanning about 2 million years of section, showing the complex pattern of cyclic bundling revealed b colour differences. At *centre* is a plot of a grey-scale scan (shown twice, in mirror image) showing variation from black mudstone at centre to white limestone at extremes. Numbers refer to "superbundles", and letters refer to "bundles", as discussed in Chap. 11 (Grippo et al., 2004, Fig. 4)

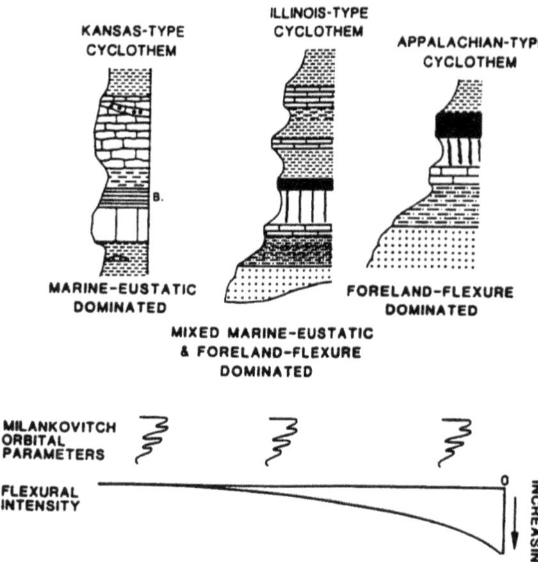


**Fig. 7.37** The three major types of North American cyclothem. *Dashes* = gray shale, *dots* = sandstone, *bricks* = limestone, *black* = coal, *dot-dash pattern* = siltstone, B = black shale (Klein and Willard, 1989)

marine limestones rest on each other, as in the Kansas cyclothems.

The cyclothems formed the basis for an early depositional model (van Siclen, 1958), where they extend southward across the craton margin in central Texas. As noted in Sect. 1.4, this anticipated the development of sequence stratigraphy by some 20 years (Fig. 1.11). Figure 7.41 is an example of the Upper Pennsylvanian stratigraphy of this area. Like many ancient shelf deposits, this area was characterized by a mixed carbonate-clastic assemblage, containing thin sand banks or deltaic sand sheets and carbonate banks (Galloway and Brown, 1973) Deltaic systems prograded onto a stable carbonate shelf (Fig. 7.41). Deltaic distributary channels are incised into the underlying shelf carbonate deposits. Widespread shelf limestones alternate with the clastic sheets and also occur in some interdeltaic embayments. Carbonate banks occur on the outer shelf edge, beyond which the sediments thicken dramatically into a clinoform slope clastic system. This association of carbonates and clastics reflects regular changes in sea level, with the carbonate phase representing high sea level and the clastic phase low sea level. The deltaic and shelf-sand sheets and the slope clinoform deposits represent lowstand systems tracts, whereas the carbonate

deposits are highstand deposits. During episodes of high sea level, clastics were trapped in nearshore deltas, while during lowstands, much of the detritus bypassed the shelf and was deposited on the slope (arrows in Fig. 7.41). Other examples of this style of sedimentation are given by Cant (1992), who noted the use of the term *reciprocal sedimentation* for depositional systems in which carbonates and clastics alternate. Note that the term "reciprocal stratigraphy" has also been used for the pattern of alternating subsidence and uplift observed in foreland basins (Sect. 10.3.3.1).

Heckel (1986) recognized a spectrum of cycle types in the cyclothems of the North American Midcontinent. Major cycles record inundations far onto the craton, which deposited widespread conodont-bearing shales. Minor cycles lack conodont-rich shales and represent minor transgressions onto the craton margin in the southern part of his project area (Kansas and Oklahoma). Heckel (1986) estimated that the major cycles spanned 235–400 ka, and the minor cycles had durations of 40–120 ka. Detailed study of outcrops and cores enabled Heckel (1986) to erect a curve showing regional sea-level changes, as discussed in Chap. 11.

There is by no means universal agreement that cyclothems (and the larger-scale groupings of cyclothems, termed mesothems: see Sect. 6.4) may be correlated over wide areas. For example, George (1978) provided a detailed critique of the work reported by Ramsbottom (1979), based in part on a meticulous examination of the biostratigraphic record. Some geologists reject the concept of the cyclothem entirely. They note the numerous variations in the cyclothem sequence (e.g., Fig. 7.39) and maintain that the concept of an ideal or model cyclothem is a dubious one (e.g., Duff et al., 1967). The development of the facies-model methodology during the 1960s and 1970s gave geologists an entirely new way of analyzing cyclic sequences. Application of the process approach, using vertical profiles and Walther's law, showed that many cyclic relationships are the result of such autogenic processes as point-bar lateral accretion, deltaic progradation, or the shoaling of tidal flats. Ferm (1975) showed that most variations in the cyclothem sequence could be explained by the progradation and abandonment of deltas. He interpreted each cyclothem as a complex clastic wedge, which he termed the Allegheny duck model, because of the fancied resemblance of

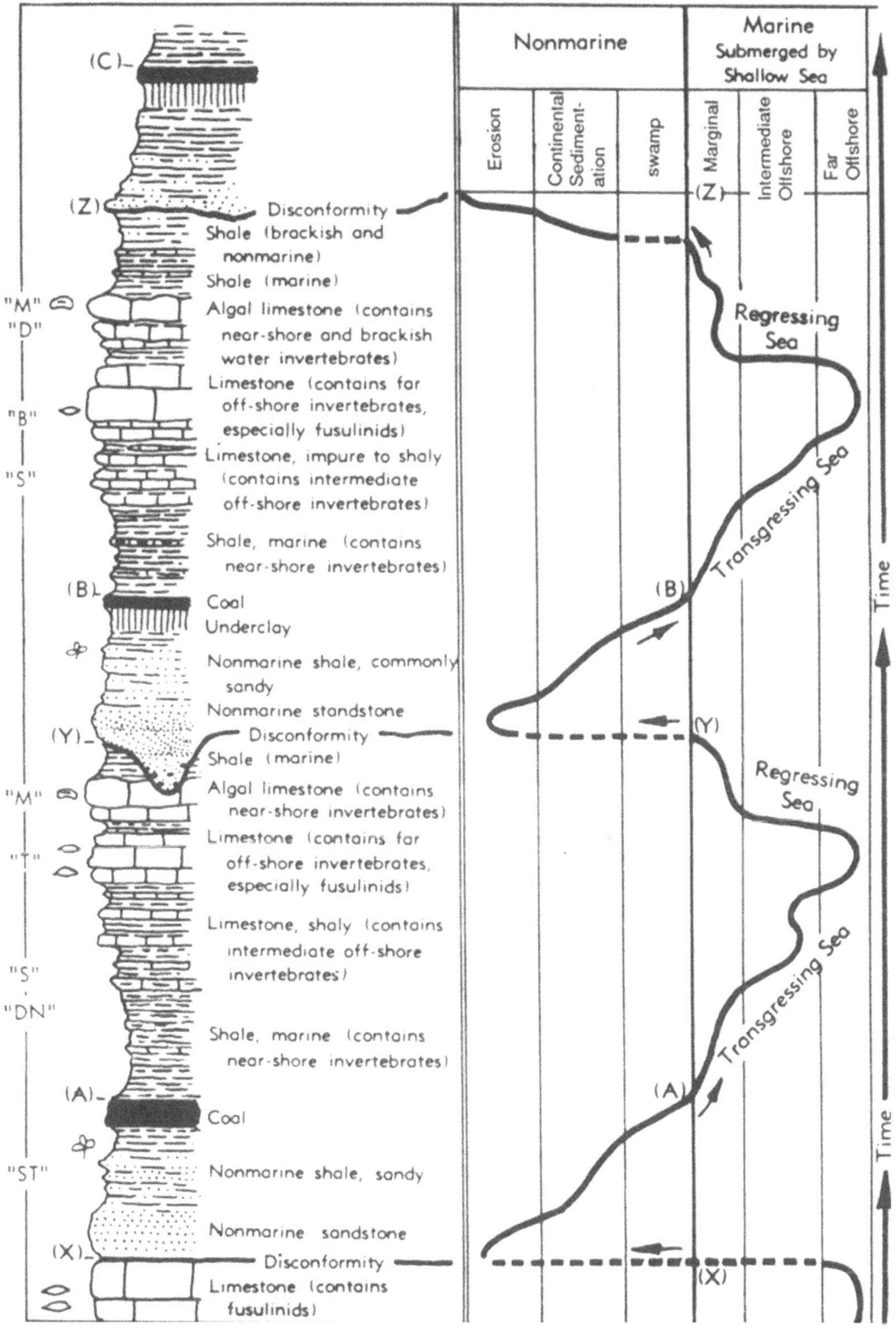


**Fig. 7.38** Two typical Illinois-type Carboniferous cyclothems, showing interpretation in terms of transgression and regression (Crowell, 1978; based on Moore, 1964)

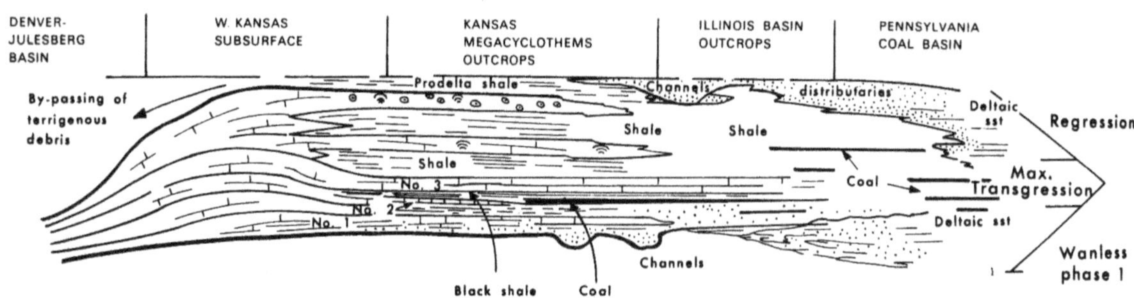


**Fig. 7.39** Stratigraphic model of a single Midcontinent cyclothem from the Appalachians to Kansas, showing the relationships of the three main types of cyclothem illustrated in Fig. 7.37. Each cycle is a few tens of metres thick, and may extend laterally for distances of more than 1,500 km (Wilson, 1975)

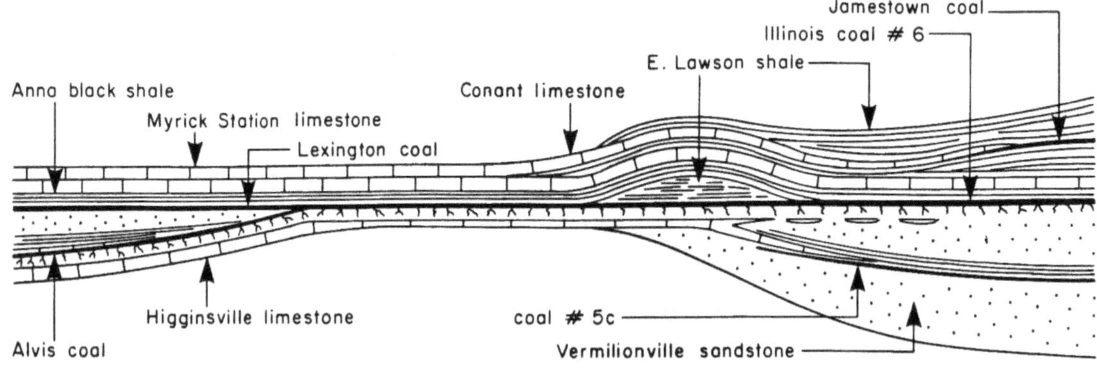


**Fig. 7.40** Generalized cross-section through a cyclothem, in Illinois, illustrating lateral variations (after Wanless, 1964)

the model cross-section to a flying duck. (Ferm also expressed doubts about his own model in the 1975 paper, but it seems a good one to the writer.) That cyclothems can now be explained in modern facies-model terms does not mean that the concept and term cyclothem are no longer valid. The widespread nature of these distinctive sequences, their restriction to rocks of Pennsylvanian and Permian age, and in spite of the doubts of George and others, the fact that individual cyclothems and mesothems can be correlated for considerable distances call for a special interpretation. As Busch and Rollins (1984) noted, and as illustrated in Fig. 7.40, autogenic controls are certainly important on a local scale. In places extra channel-fill or other units may be present, or minor cycles are cut out by channel erosion.

The renewed interest in sequence stratigraphy and eustasy since the late 1970s has revived the credibility of the glacioeustatic model of Wanless and Shepard (1936), and this interpretation is now widely accepted (Crowell, 1978; see Chap. 11). However, considerable debate continues regarding the relative importance of tectonism and climate-change in the generation of these cycles (Dennison and Ettensohn, 1994). For example, Klein and Willard (1989) and Klein and Kupperman (1992) have argued that tectonic influences were important in the development of the clastic-rich Appalachian-type cyclothems and, to a lesser extent, the Illinois-type cyclothems. They suggested that episodic flexural subsidence driven by thrust-sheet loading within the Appalachian foreland basin was a primary control in generating the clastic detritus and in controlling the transgressive-regressive cyclicity of the depositional environment. In Sect. 7.6, other classes of foreland-basin cycles are described which, it has been suggested, may also be tectonically driven, and so this is not a unique idea. The tectonic mechanisms are discussed in Chap. 10. Some

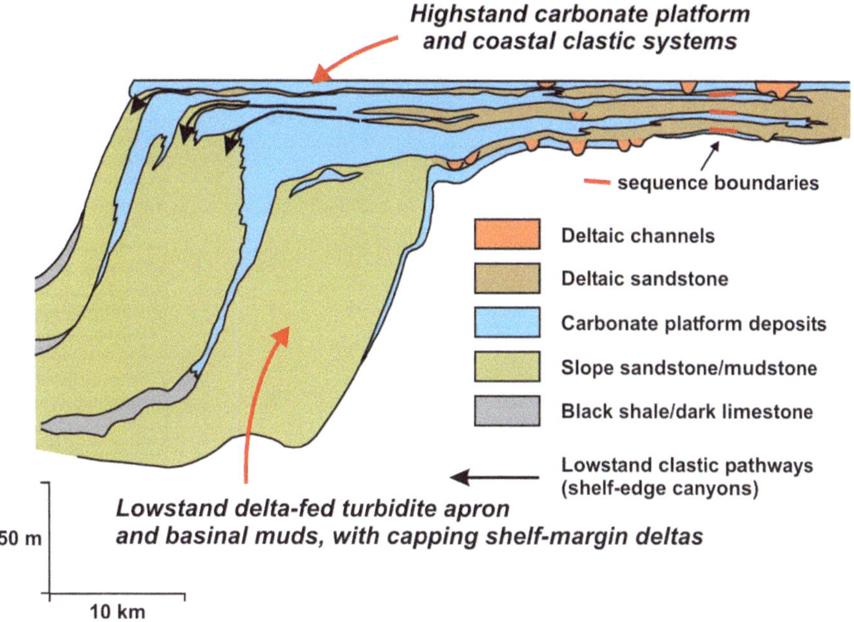


**Fig. 7.41** A mixed carbonate-clastic shelf-slope depositional systems tract, formed as a result of alternating periods of high and low sea level (Galloway and Brown, 1973)

models, such as that of Wilson (1975; see Fig. 7.39), indicate that the three cyclothem types are variants within single mappable stratigraphic entities. But this raises an interesting question: how can carbonate-dominated, eustatically-driven cyclothems within the Kansas craton be correlative with clastic-dominated, tectonically-driven cyclothems of the Appalachian foreland basin, unless eustasy and tectonism were exactly in phase on a $10^5$-year time scale? Klein and Willard (1989) did not examine this question. According to G. deV. Klein (Personal communication, 1995) it was not the intention of their work to suggest a synchroneity between tectonism and climate change. The existing correlation framework for cyclothems in the Appalachian foreland basin and the Midcontinent provides only a partial test of such sychroneity (Heckel, 1994), and debate continues on this point (Sect. 11.3.4).

Recent work in the Western Interior foreland basin has indicated a correlation between some Cretaceous basin-margin clastic sequences and basin-centre chemical sequences (Elder et al., 1994), which raises important questions regarding the nature of the controlling processes. We return to this paper in Sect. 7.6, and the processes are discussed in Chap. 11. The relationship between tectonism and climate change in the Appalachian foreland basin is addressed in more detail in Sect. 10.3.3.2.

## 7.5 Lacustrine Clastic and Chemical Rhythms

Lake sediments are extremely sensitive to changes in temperature, water depth and sediment supply, and are commonly rhythmic, indicating a cyclicity in the external tectonic and climatic controls. The term *metre-scale cycle* is commonly used for this type of sequence. In many cases, Milankovitch mechanisms ($10^4$–$10^5$-year periodicities) are indicated. Amongst the best-studied examples are the Triassic-Jurassic deposits of the Newark Supergroup in eastern North America (Van Houten, 1964; Olsen, 1984, 1986, 1990), and the Eocene Green River Formation of Wyoming (Bradley, 1929; Eugster and Hardie, 1975; Fischer and Roberts, 1991). The Devonian Orcadie Basin of Scotland also contains a cyclic lacustrine succession (Donovan, 1975, 1978, 1980).

The Newark basins are a series of extensional basins developed during the initial, rifting phase of the breakup of Pangea and the opening of the modern Atlantic Ocean. P. E. Olsen (1990) recognized three

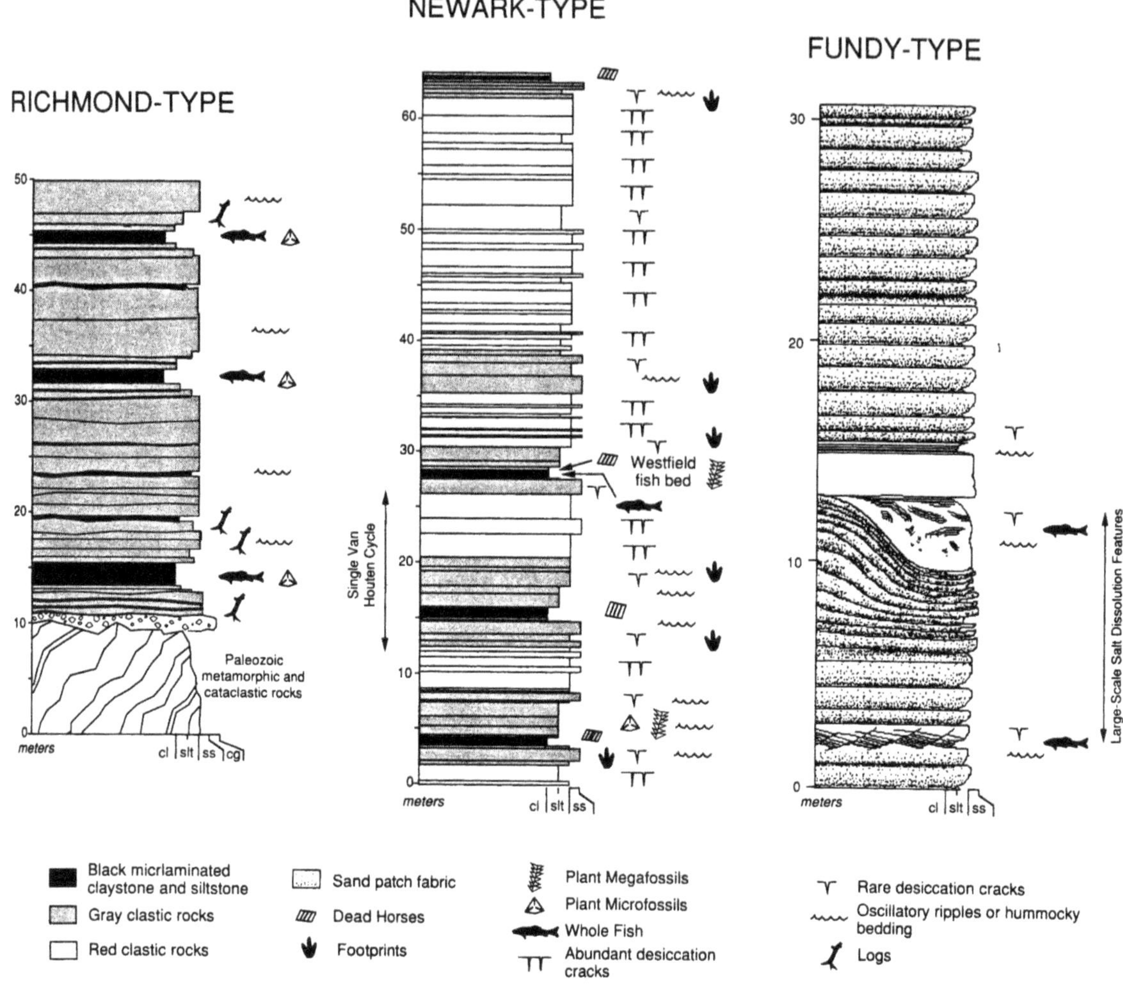



**Fig. 7.42** Three types of cyclic lacustrine succession in Triassic Newark-type basins of eastern North America (Olsen, 1990). AAPG © 1990. Reprinted by permission of the AAPG whose permission is required for further use

types of cyclic succession, which he named after the basins where they are well-known (Fig. 7.42). Sedimentological studies of the Newark-type successions indicate variations in depositional environment from deep water to complete exposure, suggesting cyclic variations in lake level by as much as 200 m, as a result of variations in precipitation. The cycles vary in thickness (and therefore in duration) by at least three orders of magnitude, ranging from 1.5 to 35 m in thickness, and from 20 to 400 ka in estimated duration (Fig. 7.43). P. E. Olsen (1990) called these Van Houten cycles, after the pioneering work on the deposits by Van Houten (1964). P. E. Olsen (1990, p. 213) stated that

Drastic changes in lake level apparently inhibited the buildup of high-relief sedimentary features (other than alluvial fans) within the basin both by wave action during transgression and regression and by the brief time the water was deep. Consequently, lacustrine strata are characterized by extreme lateral continuity and by a tendency for coarse-grained sediment to be absent from deeper water facies and restricted to basin margins ... Large-scale sequence boundaries and large deltas apparently are absent from the main basin fill.

Van Houten (1964) distinguished two main types of cycle (Fig. 7.44), chemical cycles, averaging 3 m in thickness, formed during periods of closed drainage and relatively high evaporation rates, and detrital cycles, averaging 5 m in thickness, that formed during

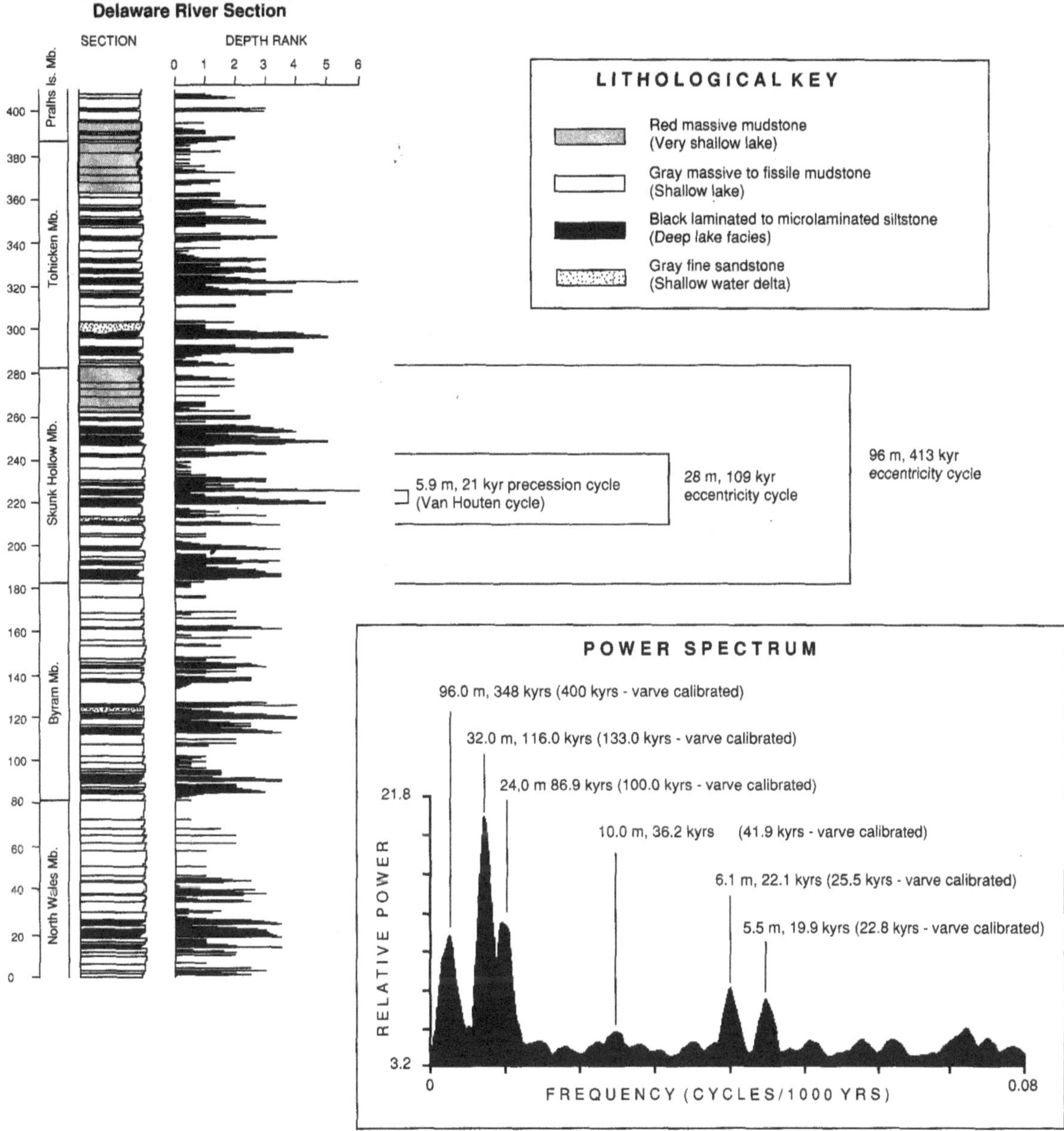



**Fig. 7.43** Section of Newark-type lacustrine facies complex in the middle Lockatong Formation, Newark Basin. The power spectrum was derived by Fourier time-series analysis. Depth ranking indicates a ranking of increasing inferred water depth based on sedimentological analysis. Values in kyrs are cycle periods in thousands of years (Olsen, 1990). AAPG © 1990. Reprinted by permission of the AAPG whose permission is required for further use

more humid periods, when there was probably a run-off flow-through that maintained low concentrations of dissolved sediment. The detrital cycles are coarsening-upward in type, and are interpreted as regressive sequences formed by basin filling following a rapid rise in lake level.

Richmond-type successions (Fig. 7.42) are characterized by significant coals, and by bioturbated and microlaminated siltstones and sandstones containing no evidence of exposure. The successions are thought to indicate relatively humid environments, in which complete basin desiccation was rare. During

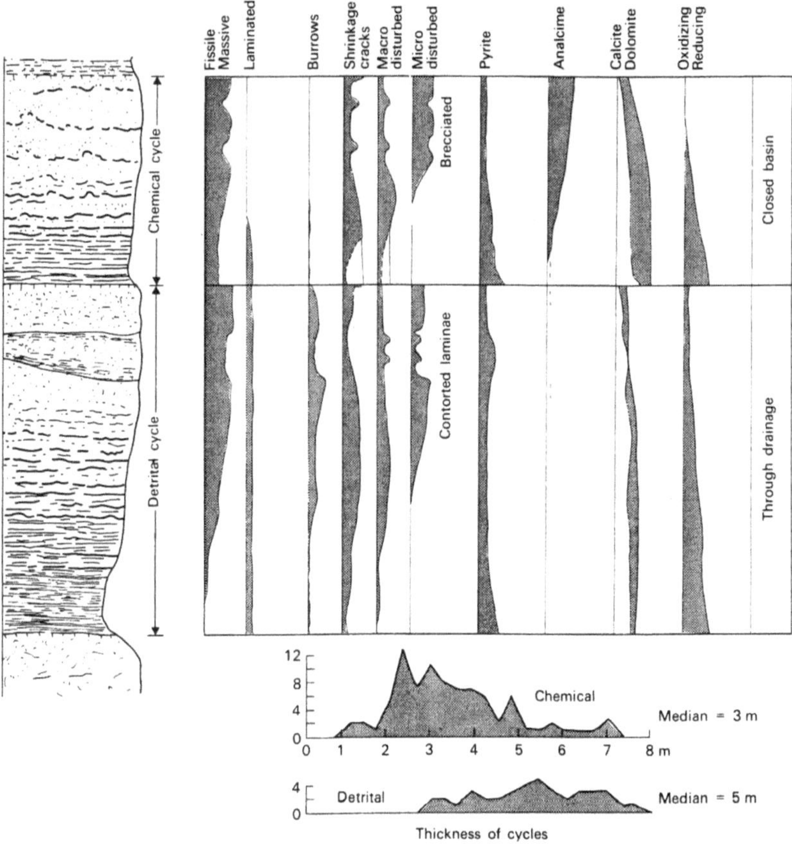


**Fig. 7.44** Model of detrital and chemical cycles in the Lockatong Formation (Triassic), Newark Basin, New Jersey (Van Houten, 1964)

some cycles the lake may have remained deep long enough that high-relief sedimentary features were built, including large, prograding deltas, fan deltas, submarine channels and fans. Large-scale erosional unconformities with deep channeling developed during lake-level lowstands.

Fundy-type successions are "characterized by a cyclicity consisting mostly of what are termed sand-patch cycles … These represent alternations between shallow perennial lakes and playas with well-developed efflorescent salt crusts" (Olsen, 1990, p. 218). Gypsum nodules, salt-collapse structures and eolian dunes are also locally present. Water flow rarely exceeded evaporation in these basins, consequently deep-lake conditions were rarely established, depositional relief remained low, and cycles show considerable lateral continuity. A schematic synthesis of basin stratigraphy and cycle types is shown in Fig. 7.45.

The Green River Formation was deposited in several lacustrine basins. Figure 7.46 illustrates the generalized stratigraphy, showing marginal alluvial complexes draining into an arid basin centre (Fig. 7.47). Annual varves and solar-band cycles are prominent in these deposits (Fischer and Roberts, 1991) but are not discussed here. Milankovitch cycles are well developed (Fig. 7.48), with some cycles traceable for more than 20 km without significant thickness changes. Figure 7.48a shows the basic cycle, consisting of three parts. The oil-shale facies consists of organic-rich dolomitic laminites and breccias, formed by the preservation of algal and fungal matter during relatively wet phases. During drier phases a hypersaline lake developed, forming the trona facies. Halite is also present in some beds of this facies. The marlstone facies consists of thin-bedded dolomitic mudstones with silt-mud laminae, showing evidence of frequent desiccation in a playa environment. The silt-mud laminae represent occasional sheet floods. The Tipton Member is thought to represent a slightly wetter period than the Wilkins Peak Member. The cyclicity is not visible in core, but is picked up by

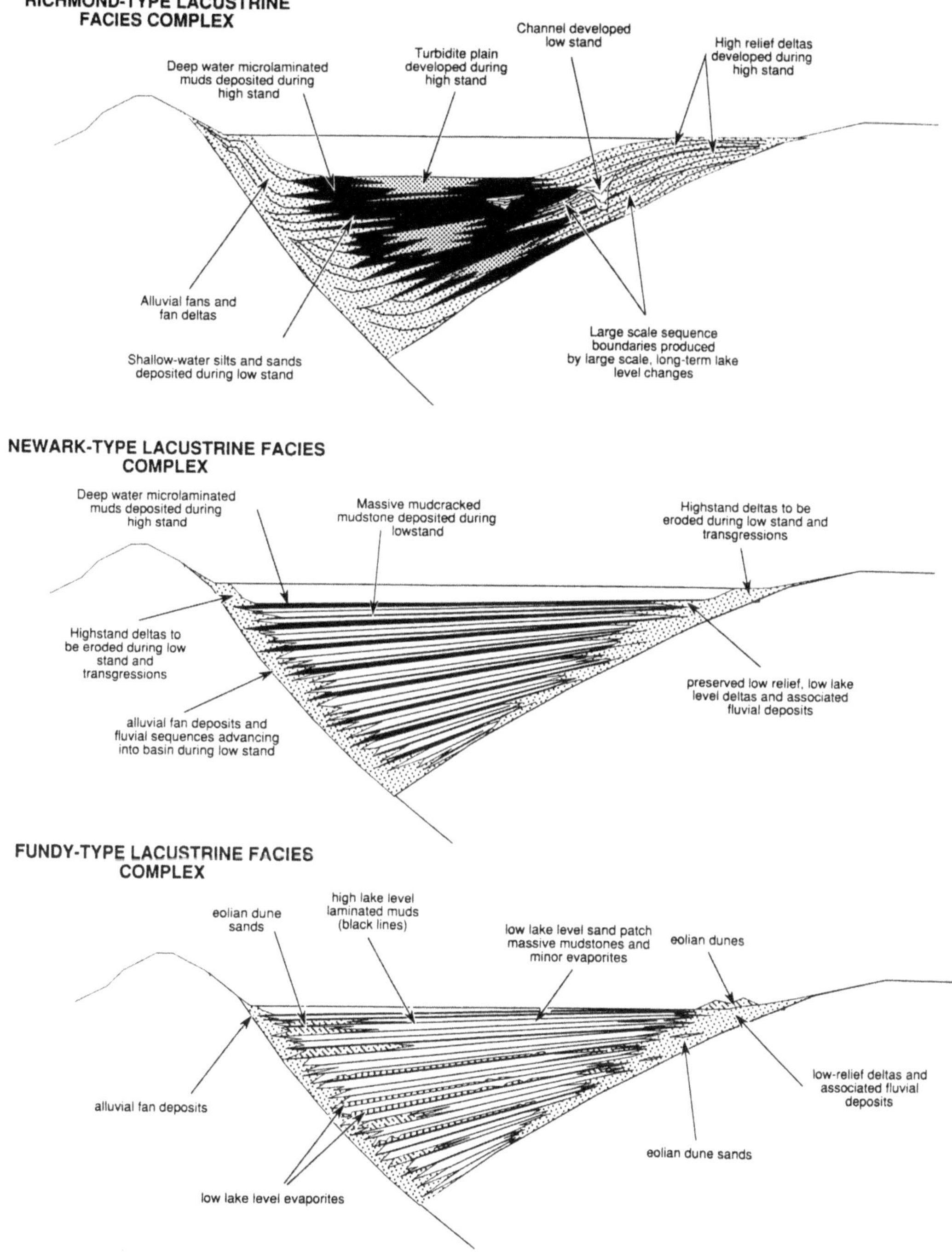


**Fig. 7.45**  Idealized lacustrine facies complexes, Newark basins of eastern North America (Olsen, 1990). AAPG © 1990. Reprinted by permission of the AAPG whose permission is required for further use

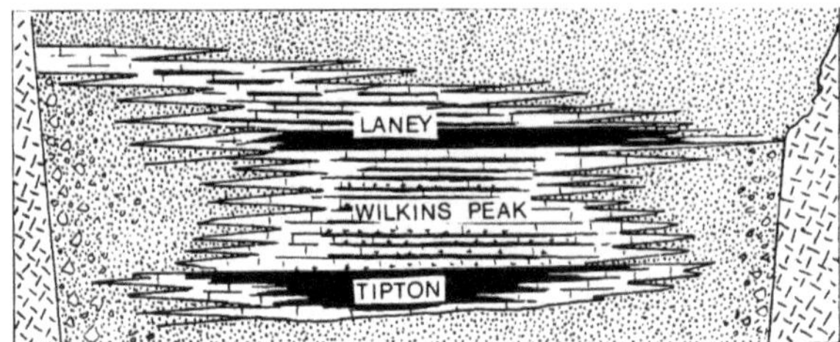


**Fig. 7.46** Generalized stratigraphy of the Green River Formation (Eocene), Wyoming. Marginal mountains shed coarse alluvium into an intermontane basin. Black units are varved, lacustrine oil shales. Vertical scale about 1 km, horizontal scale 200–300 km (Fischer and Roberts, 1991)

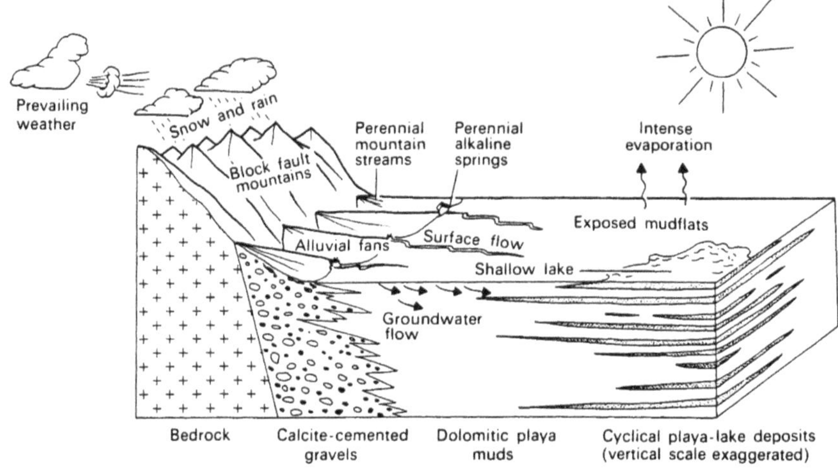


**Fig. 7.47** Schematic block diagram showing depositional framework of the Wilkins Peak Member, Green River Formation (Allen and Collinson, 1986, after Eugster and Hardie, 1975)

gamma-ray and sonic-velocity logs, which record subtle variations in uranium and thorium linked to organic matter, potassium in clay, and carbonate content (Fig. 7.48b). Mean cycle thickness is 2.35 m. In the Wilkins Peak Member cycles average 3.7 m in thickness. Bundles of five cycles indicate a longer-term cyclicity (Fig. 7.48c), of a type discussed in Sect. 11.2.6.

Eolian environments are also sensitive to climate change, and several ancient eolian units display a particular form of climatic cyclicity (Kocurek and Havholm, 1993). One of the most detailed studies to date of this type of high-frequency sequence stratigraphy is that of the Rotliegend Sandstone (Permian) of the North Sea Basin, one of the major gas-bearing reservoir units of that petroleum province. The studies by Yang and Nio (1993) and Yang and Baumfalk (1994) are based entirely on core analysis. Figure 7.49 shows one of their log correlation diagrams. A statistical methodology was used to calibrate gamma ray logs against facies identified in core. Note the lateral transition between a succession dominated by eolian and arid-fluvial facies on the left side of this diagram, and lacustrine and evaporite units on the right. A facies model for the transition between arid and humid environments is shown in Fig. 7.50. This model was used to calibrate and correlate the sequence record in sections such as those shown in Fig. 7.49. Evaporites dominate the lowstand systems tract in basin centre environments, at the same time as erg deposits (dry dunes) extend throughout the basin margins. With increasing humidity lacustrine muds appear, and these may be used as maximum flooding surfaces for the purpose of correlation.

Twelve sequences were identified in the Upper Rotliegend. Two of these, RO2.1, RO2.2, are shown

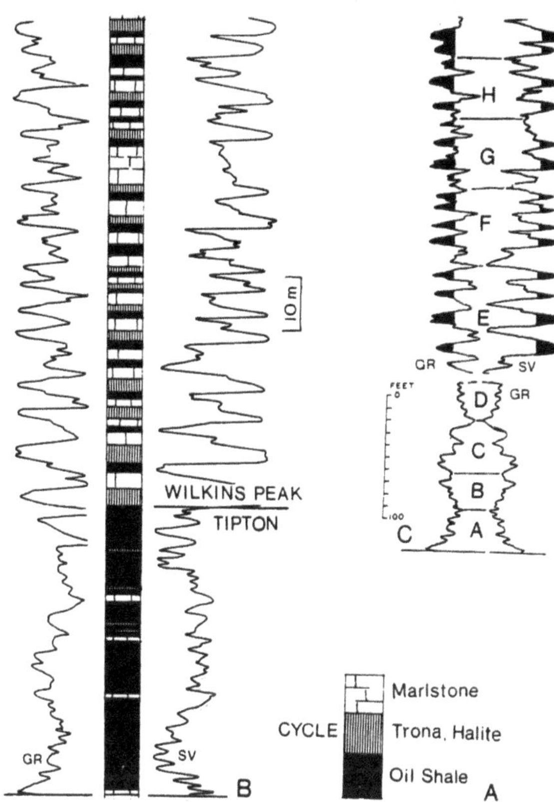



**Fig. 7.48** Cyclicity in the Tipton and Wilkins Peak Members of the Green River Formation, Green River Basin, Wyoming. *A*. Basic desiccation cycle. *B*. Basal 129 m of Green River Formation, showing gamma-ray log (GR) and sonic velocity log (SV). *C*. Log plot illustrating the grouping of cycles into thicker bundles in part of the Tipton Member. *A–D* is a mirror-image GR plot, *E–H* is a GR-SV plot (Fischer and Roberts, 1991)

in detail in Fig. 7.49, and the complete succession is shown in Fig. 7.51. The sequences are typically about 50 m thick. Chronostratigraphic information indicates that these sequences span 10.7 million years, indicating that they have average durations of 0.9 million years. Yang and Nio (1993) performed detailed time-series analysis that suggest a range in durations of 0.7–1.4 million years. Yang and Nio (1993) refer to these as "third-order" sequences, and suggest a clustering into higher-order "supersequences", as shown in Fig. 7.51. We comment elsewhere (Sects. 4.2 and 13.1) on the appropriateness of the "order" designation of sequences.

As a result of their time series analysis Yang and Nio (1993) were also able to report that that there are "Milankovitch"-type frequencies embedded

in the gamma-ray data. These results are shown in Fig. 7.52, and were confirmed by studies of core, which shows the fluctuation from arid to damp environments over a vertical interval of about 8 m. The various Milankovitch frequencies identified in the core are listed in the bottom right of Fig. 7.52, and appear similar to the modern frequencies. However, as discussed in Chap. 11 and Sect. 14.7, the assignment of precise frequencies to the record of the distant geological past is not necessarily a very convincing exercise.

## 7.6 High-Frequency Cycles in Foreland Basins

Shallow-marine to nonmarine clastic sequences with $10^4$–$10^6$-year episodicities are common in the Cretaceous strata of the Western Interior of North America (sequences with $10^6$-year episodicity are discussed in Sect. 6.2). They constitute a distinctive type of cycle, in part because they occur within a foreland basin, with its characteristic pattern of subsidence and sediment supply. Foreland basins are, of course, tectonically highly active, and tectonic influences on sequence development seem to be indicated, as discussed at some length in Sect. 10.3.3. Climatic control has also been suggested for some sequences, including glacioeustatic causes (e.g., Plint, 1991; Elder et al., 1994). Climatic forcing of depositional processes, in the absence of sea-level changes, has also been suggested (Elder et al., 1994; Sageman et al., 1997). The debate about causation has become more complex since evidence has accumulated that there may have been small, temporary ice caps on Antarctica at several times during the Cretaceous (Sect. 14.6.4).

The Western Interior foreland basin is one of the best areas in the world to study the relationships between sedimentation and tectonics, because of the enormous amount of data that has been assembled for this region (Mossop and Shetsen, 1994; Miall et al., 2008, and reference sources therein). Cyclicity on the $10^6$-year time scale is discussed in Sect. 6.2.1. Formal studies of the sequence stratigraphy commenced with the landmark paper by Plint et al. (1986), dealing with a single unit of Upper Cretaceous age in Alberta (discussed below). Since that time, the research group led by A. G. Plint has carried out an extensive research

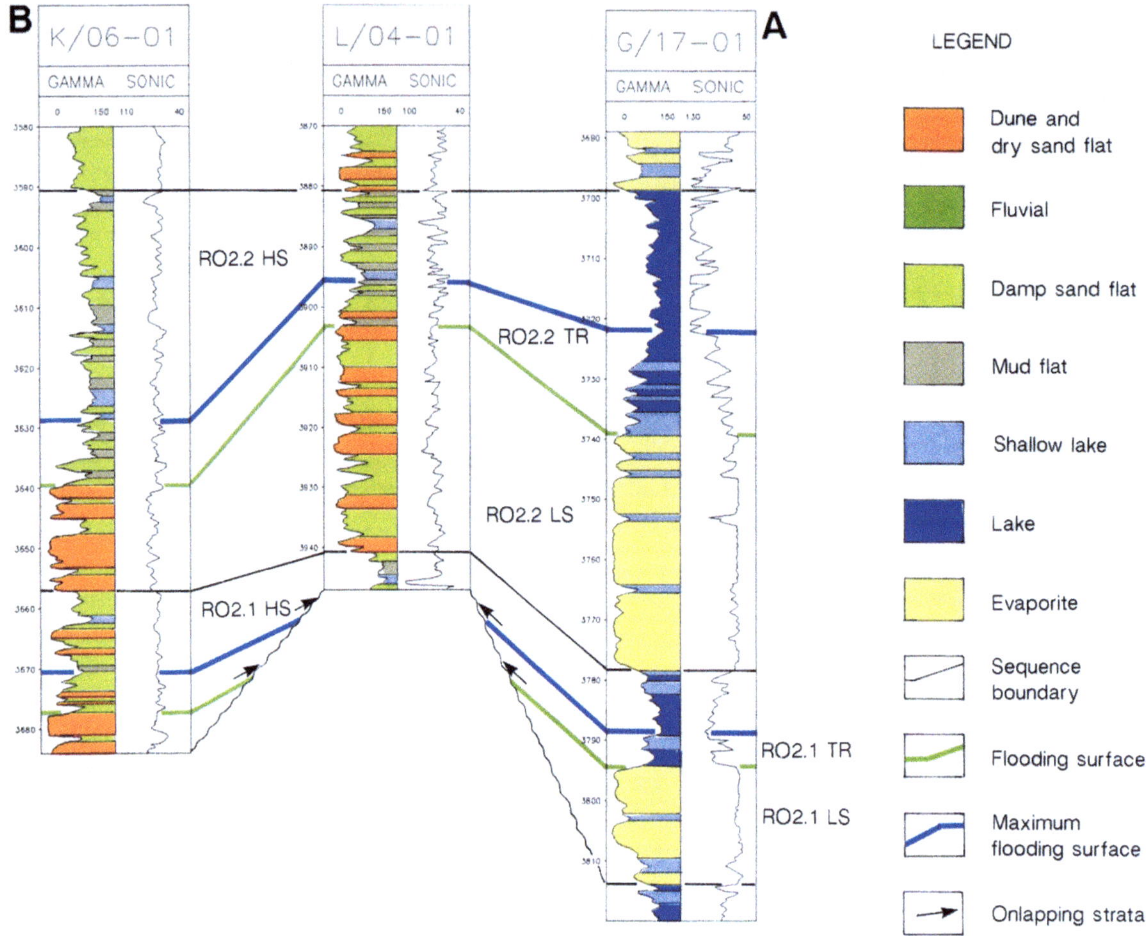



**Fig. 7.49** Stratigraphy of the Rotliegend Sandstone (Permian) of the North Sea Basin, showing log correlation between three wells spaced over a distance of about 120 km off the Dutch coast. Facies analysis is based on core calibrations of gamma-ray logs. Lateral correlations are based on a facies model of climatic cyclicity, shown in Fig. 7.50, and as discussed in the text (Yang and Nio, 1993, Fig. 14). AAPG © 1993. Reprinted by permission of the AAPG whose permission is required for further use

program on the Cretaceous sequence stratigraphy of the northern Alberta subsurface and the outcrop belt of the foothills, extending through the western part of the province. Units displaying well-developed internal high-frequency cyclicity include the Moosebar, Gates/Falher (part of the Mannville Group), Viking, Dunvegan, Kaskapau, Cardium, Muskiki and Bad Heart Formations. As shown by Plint et al. (1992) these units, plus several other clastic tongues in this basin, may be compared to the "third-order" cycles of the Exxon global cycle chart, although it is no longer widely accepted that this episodicity was driven by eustatic sea-level changes on that time scale.

The Upper Cenomanian to Lower Coniacian stratigraphy of the Alberta Basin, in northern Alberta, is shown in Figs. 7.53 and 7.54. The cross-section in Fig. 7.53 shows the dip extent of shoreface sandstones and marine conglomerates, which have the stratigraphic appearance of typical tectonically-derived clastic wedges.

The strongly wedge-shaped units of the Kaskapau indicate rapid flexural subsidence which appears to have limited the progradation of shoreface sandstone from the west. In contrast, the Cardium has much more tabular allomembers which suggest that the rate of flexural subsidence was much lower; this permitted shoreface sandstone and conglomerate to prograde much further into the basin. The thin Erin Lodge, Howard Creek and Josephine Creek sandstones are interpreted as having prograded off an emergent forebulge that was elevated during rapid flexural subsidence during

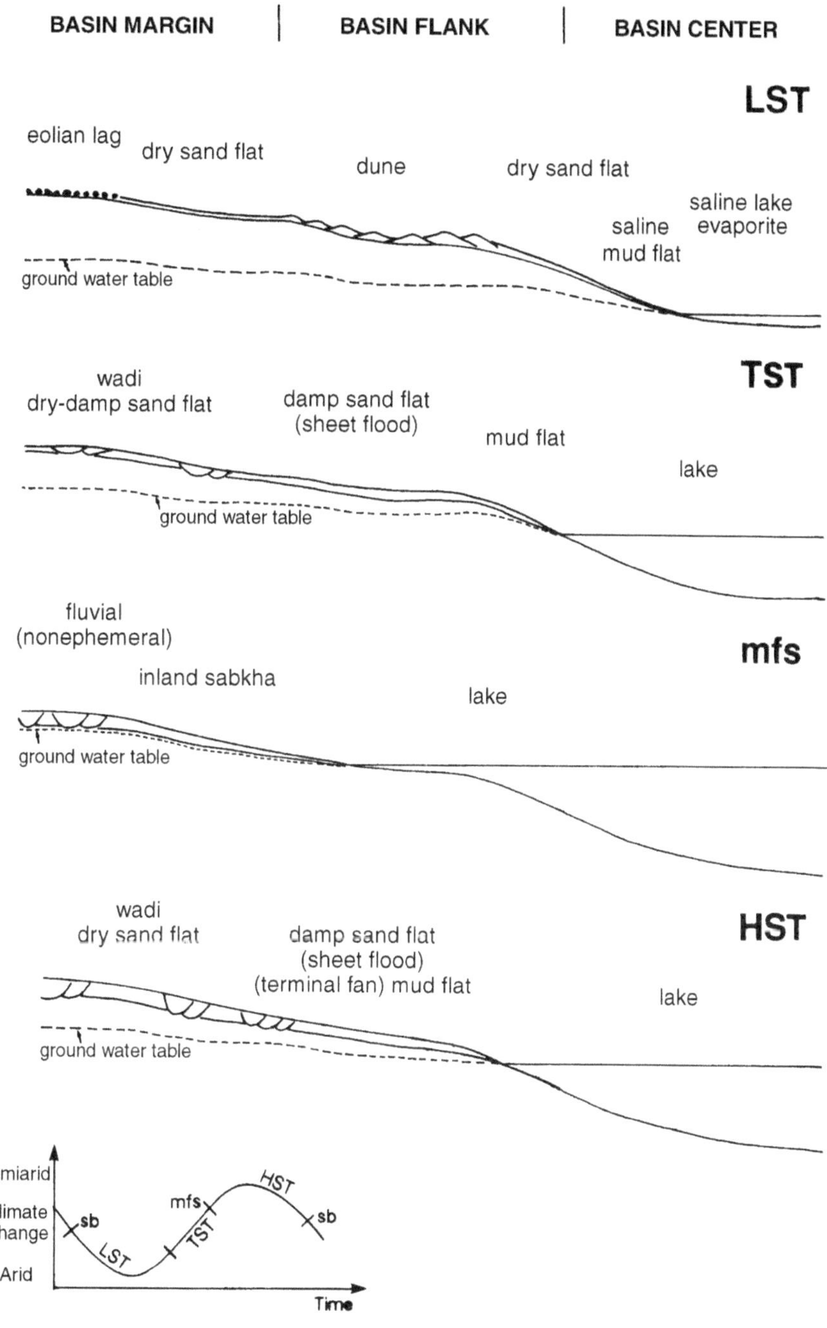



**Fig. 7.50** A facies model for the climatic cyclicity of the Rotliegend Sandstone of the North Sea Basin (Yang and Nio, 1993, Fig. 19). AAPG © 1993. Reprinted by permission of the AAPG whose permission is required for further use

Kaskapau Unit I time. Relative sea-level rise, possibly linked to the Greenhorn transgression, and sediment loading, submerged the forebulge above Unit I, halting erosion. The vertical distribution of conglomerate throughout the entire succession is shown on a logarithmic scale by the bar chart on the left.

Kaskapau Units I and II have the most prominent wedge shape which might indicate the most rapid flexure rate; this may have suppressed gravel delivery to the shoreline. In contrast, Units III, IV and V become gradually less wedge-shaped, possibly indicating a decreased rate of flexure; this might have permitted

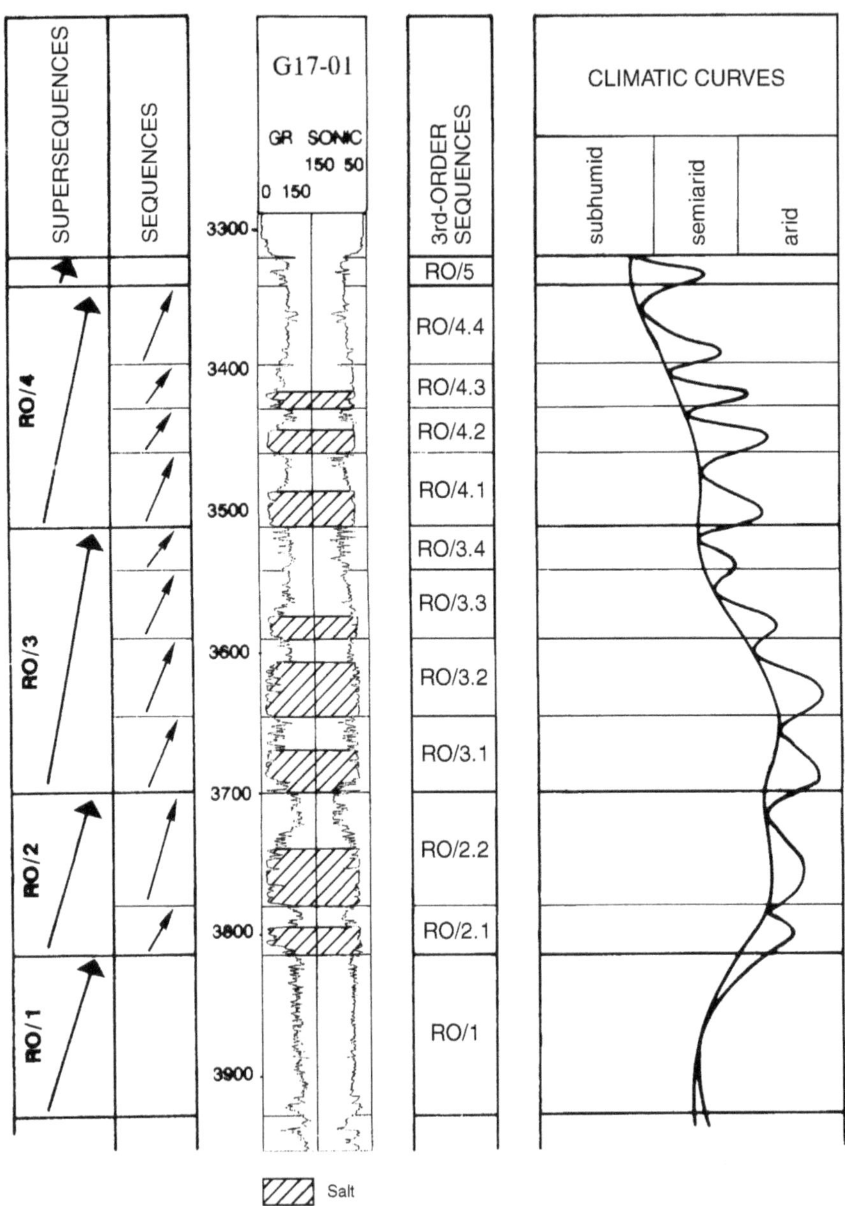


**Fig. 7.51** The sequence and supersequences of the Rotliegend Sandstone of the North Sea Basin (Yang and Nio, 1993). AAPG © 1993. Reprinted by permission of the AAPG whose permission is required for further use

more effective gravel transport to the shoreline, indicated by the upward increase in conglomerate abundance. This trend is dramatically amplified in the overlying Cardium Formation (Varban and Plint, 2008, p. 399).

The Upper Cenomanian to Lower Coniacian stratigraphy may be subdivided into 37 allomembers, spanning 5.5 million years, yielding an average duration of 149 ka for each cycle, although biostratigraphic evidence suggests that some members, notably Cardium allomember C6, lasted longer than others (Fig. 7.54). Varban and Plint (2008, p. 410) estimated the duration of the Kaskapau allomembers as averaging 125 ka. Their distribution suggests a gradual eastward onlap of the forebulge of the foreland basin, a distance of more than 300 km.

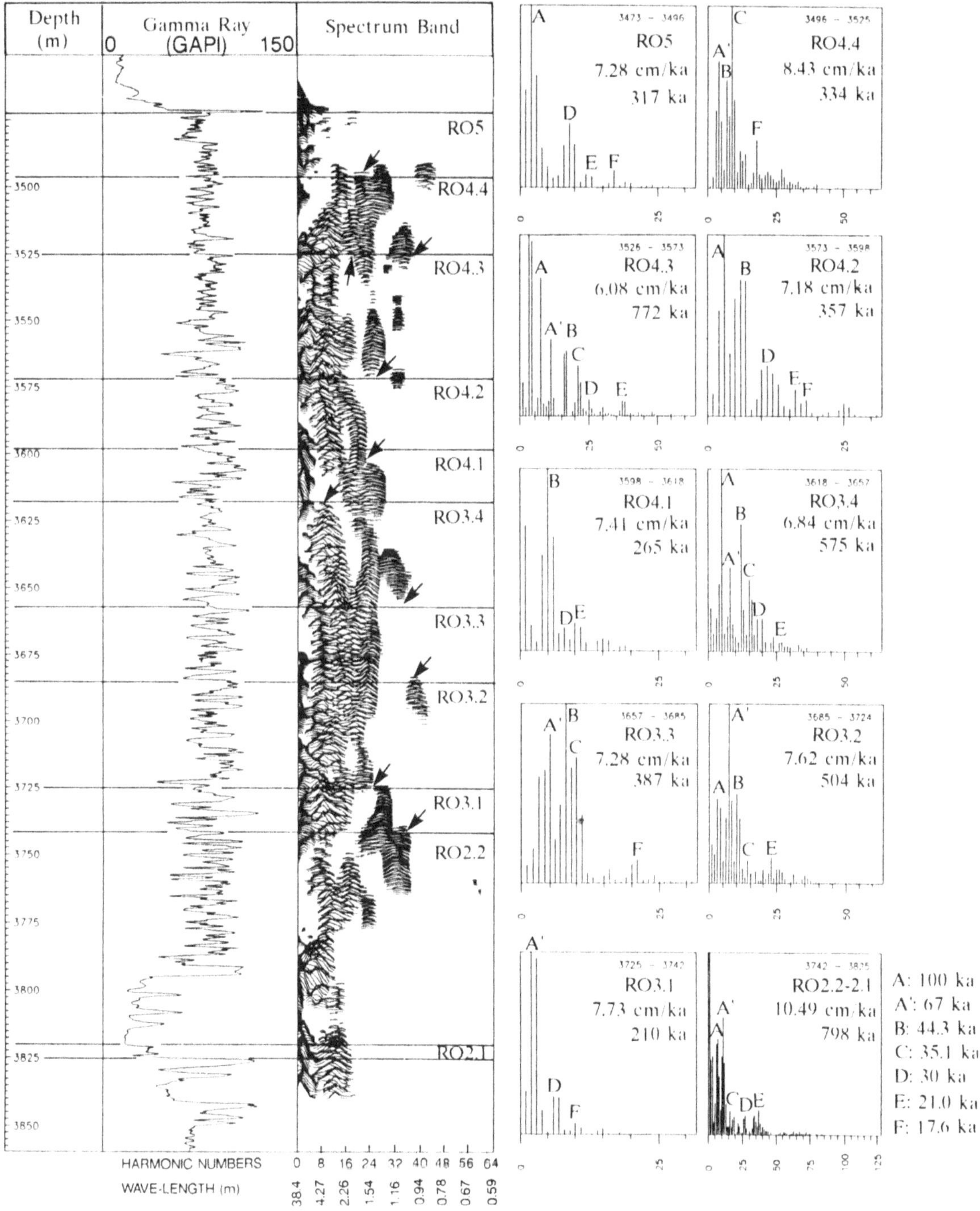


**Fig. 7.52** Spectral analysis of the gammy-ray records through the Rotliegend Sandstone of the North Sea Basin (Yang and Nio, 1993). AAPG © 1993. Reprinted by permission of the AAPG whose permission is required for further use

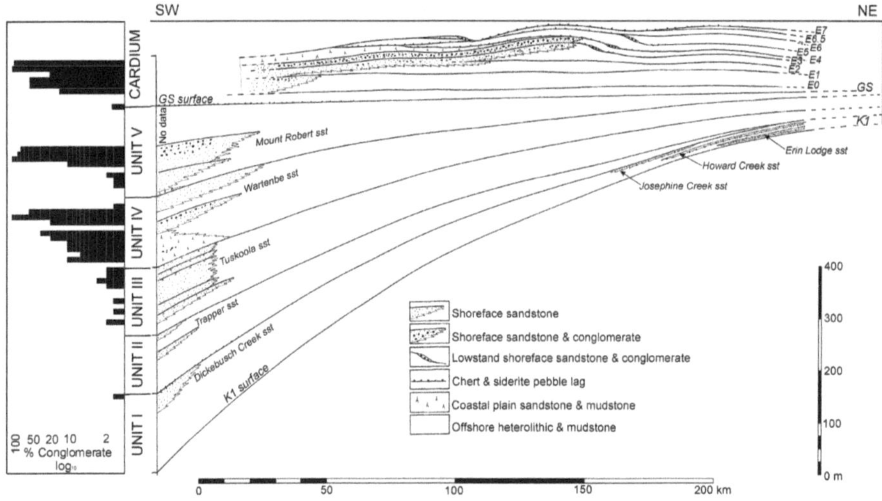


**Fig. 7.53** Dip cross-section, drawn to scale, showing the shape of Kaskapau Formation Units I–V and also the nine allomembers of the overlying Cardium alloformation, defined by the erosion surfaces GS to E7 (Varban and Plint, 2008, Fig. 2)

Although the overall geometry of the Kaskapau and Cardium formations suggests a "clastic-wedge" interpretation, as noted above, there are features of the sedimentology and architecture of some of the component units that suggest eustatic control. Evidence of rapid changes in accommodation that extend without significant modification from foredeep to forebulge is cited by and Plint and Kreitner (2007) and by Varban and Plint (2008). We discuss the interplay of tectonic and eustatic processes in foreland basins in Chap. 10.

The Cardium and Viking formations of Alberta each consist primarily of shelf deposits, and both may be subdivided using allostratigraphic methods based on the recognition and mapping of major bounding erosion surfaces. The Cardium Formation was the first unit in the Alberta basin to be subdivided in this way (Plint et al., 1986), and recognition of the architectural style of the formation constituted a major breakthrough in foreland-basin geology when the 1986 paper was published. His original reconstructed cross-section is shown here as Fig. 7.55. Indeed, the stratigraphic concepts were considered controversial, and some diverging opinions were published (Rine et al., 1987). Subsequent detailed papers on the Cardium Formation include those by Bergman and Walker (1987) and Walker and Eyles (1988). The allostratigraphy of the Viking Formation has been described by Boreen and Walker (1991).

The allostratigraphy of the Cardium Formation, an important hydrocarbon-producing unit in the Alberta Basin, was developed by Plint et al. (1986) as part of a detailed regional surface-subsurface study of the many producing fields in the area. Numerous local studies had, over the years, led to a confusing welter of local informal terminologies for sandstone horizons and marker units within the Cardium Formation and much controversy regarding the depositional environments of the various facies. Routine but meticulous lithostratigraphic correlation of subsurface records led Plint to the recognition that this unit, which is only about 100 m thick, contains at least seven basin-wide erosion surfaces, indicating the occurrence of this many events of erosion and transgression (Fig. 7.55). Some, at least, of the erosion surfaces can be traced for more than 500 km. Not only does this new framework provide a rational basis for basin-wide correlation, but it also throws a wholly new light on the depositional history of the formation. The unit consists largely of a series of basin-wide coarsening-upward cycles capped by sandstone units containing hummocky cross-stratification, overlain in some areas by lenticular conglomerate beds. The cycles, up to the level of the sandstones, are readily interpreted as the product of a shoaling shelf environment, with the sediment surface gradually building up to storm wave base. The problem had always been to fit the conglomerate into

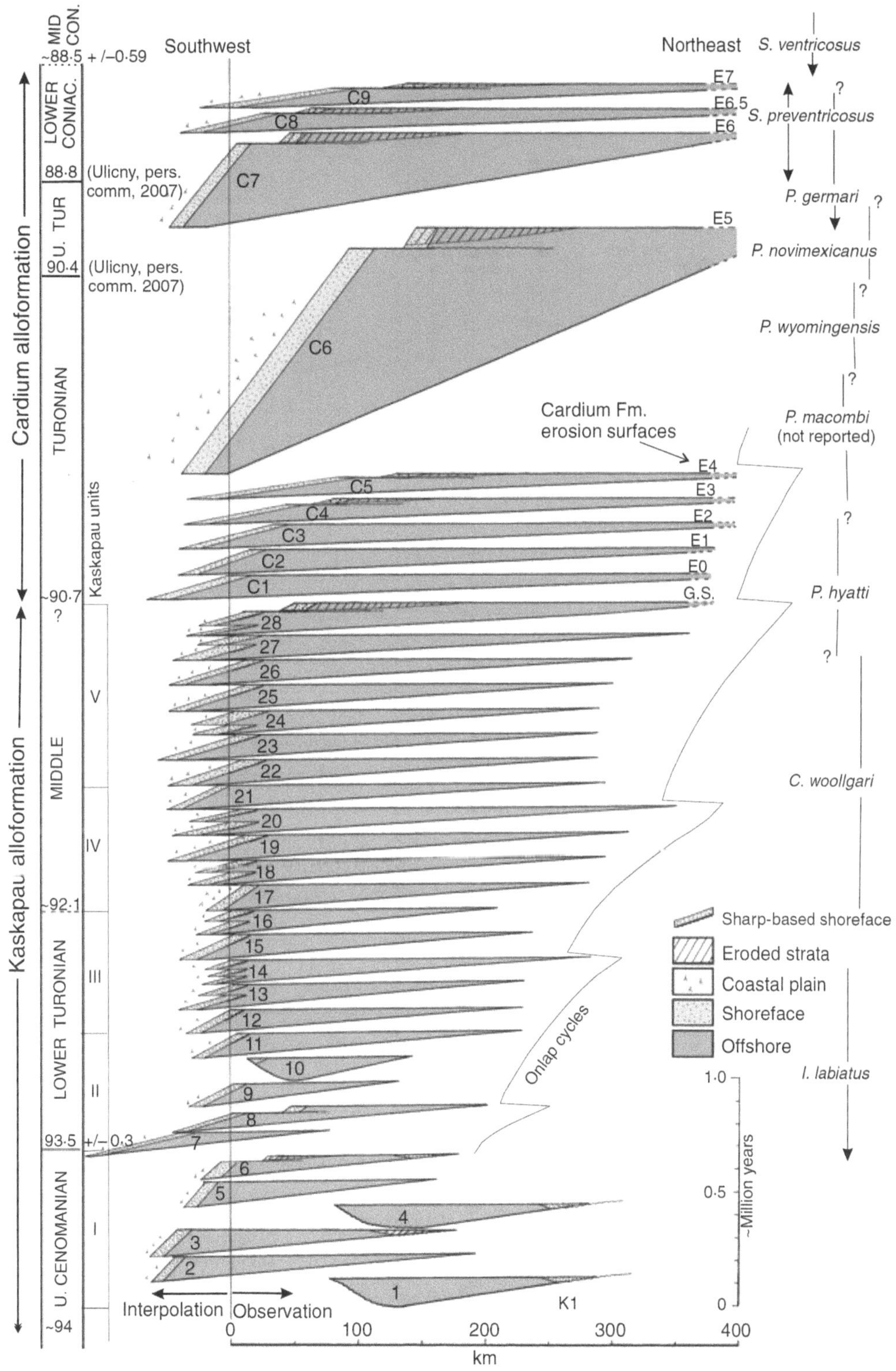


**Fig. 7.54** Chronostratigraphy of the Upper Cretaceous allomembers of northwestern Alberta. "Interpolation" refers to assumed proximal portions of each allomember, now uplifted and removed by erosion (Varban and Plint, 2008, Fig. 5)

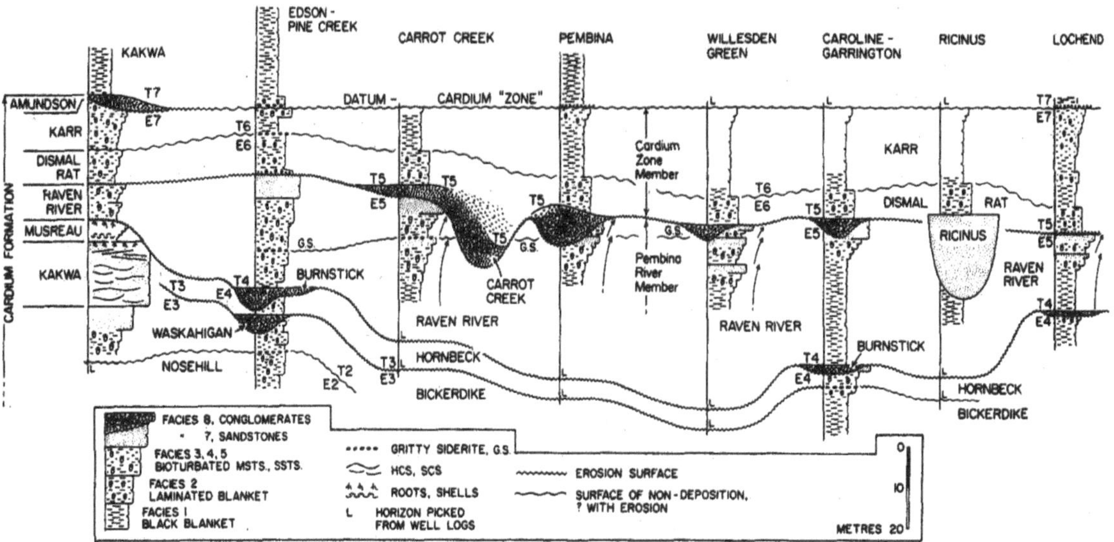

**Fig. 7.55** One of the first example of high-frequency cycles to be identified in a foreland-basin succession: the Cardium stratigraphy of the Alberta Basin. Surfaces of erosion and transgression are numbered E2/T2, etc. HCS, SCS = hummocky and swaly cross-stratification (Plint et al., 1986)

this interpretation. These coarse deposits are tens to several hundreds of kilometers from the basin margin, and the problem of transporting coarse detritus out this far in a shelf setting had been discussed in the Cardium literature for many years. The new interpretation by Plint provided a simple resolution to the problem. The conglomerates are not a conformable cap to the coarsening-upward cycles, as had always been thought. Each conglomerate lens rests on one of the erosion surfaces. They are therefore, in all probability, beach deposits, which originated as fluvial detritus transported basinward during regression of the shoreline, and were concentrated along temporary shorelines during the initial transgression at the beginning of a new cycle of relative sea-level rise. The seven Cardium sequences of Fig. 7.55 have been expanded to nine by later work (Varban and Plint, 2008). They span about 2 million years (Fig. 7.54), therefore the average duration of each sea-level cycle is about 220 ka.

The Dunvegan Formation represents a major delta complex up to 300 m thick that prograded into the Alberta Basin from the northwest over a period of about 1.5 million years (Bhattacharya, 1988, 1991; Bhattacharya and Walker, 1992). It can be subdivided into seven allomembers, which are separated from each other by widespread flooding surfaces (Fig. 7.56). The allomembers have each been mapped over an area in the order of 300,000 $km^2$. Each allomember ranges up to 80 m in thickness, and represents a depositional episode about 200 ka in duration. The bounding marine-flooding surfaces are attributed to "allocyclically controlled relative rises in sea level, probably caused by a rate increase in tectonically induced basin subsidence" (Bhattacharya, 1991). The allomembers consist of shingled clinoform deposits up to 30 m thick, each shingle comprising a heterolithic deltaic complex representing $10^4$ years of sedimentation. The deltas are of river-dominated type and compare in scale and composition (except for being somewhat more sandy) with the delta lobes of the modern Mississippi delta. Offsetting of the shingles within each allomember probably results from autogenic distributary switching, as in the modern Mississippi (see Fig. 2.25).

The early Cretaceous Mannville Group is characterized by nested sequence cyclicity at several time scales. $10^6$-year cyclicity is discussed in Sect. 6.2.1 (see Figs. 6.28 and 6.29). The upper Mannville also displays a pronounced high-frequency cyclicity. Cyclicity in the Gates and Moosebar Formations, and their subsurface equivalent, the Falher Formation,

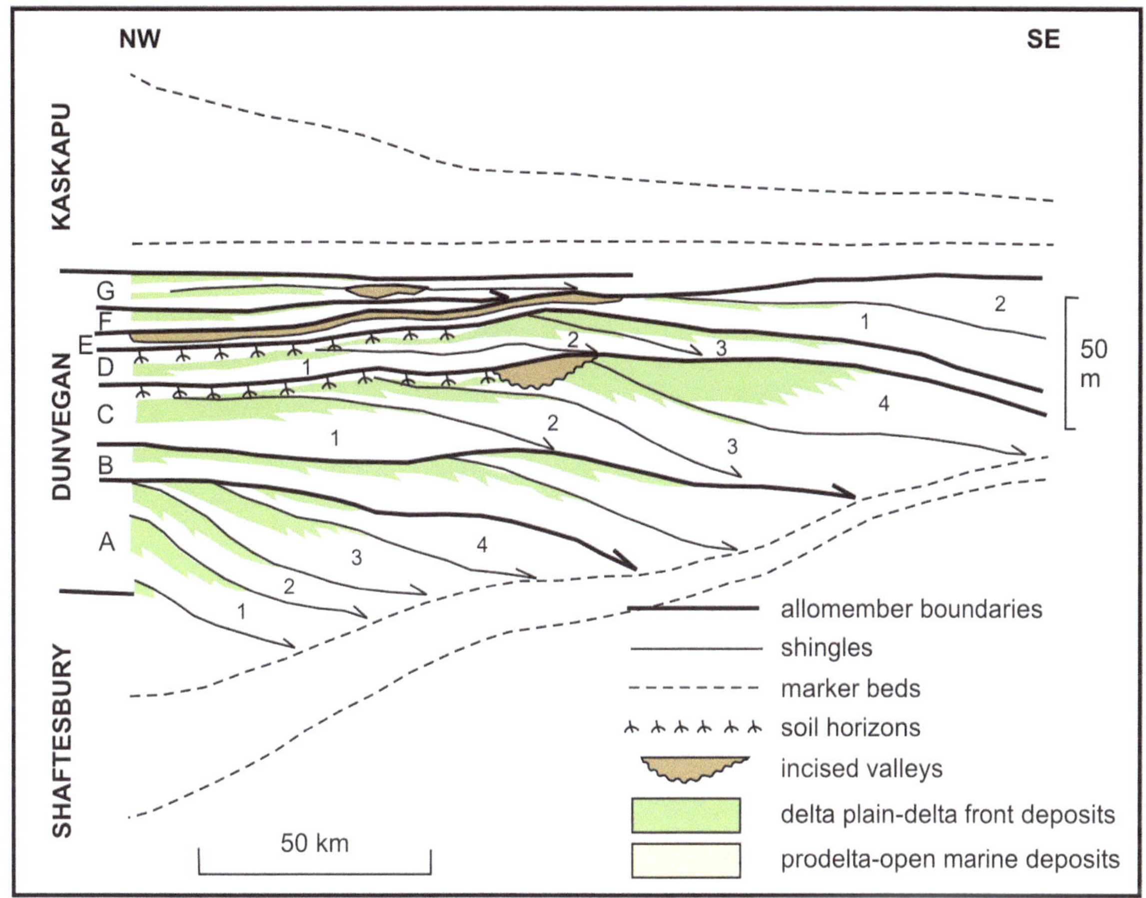



**Fig. 7.56** Schematic regional dip-oriented cross-section of the Dunvegan Formation, west-central Alberta. *Heavy lines* are regional flooding surfaces that subdivide the formation into seven allostratigraphic units. Within each unit separate offlapping shingles can be mapped, as shown by the numbers. Blank units are marine shale. *Arrows* indicate downlap stratigraphic terminations (Bhattacharya, 1991)

of Albian age, was described by Cant (1984, 1995), Leckie (1986), and Carmichael (1988). Seven transgressive-regressive cycles have been mapped within a stratigraphic interval that is interpreted to have lasted for between 3 and 6 million years (Fig. 7.57). Leckie (1986) estimated the cycles to have had durations of between 103,000 and 275,000 years. The cycles consist of thin transgressive deposits, including estuarine sandstones and lag gravels, followed by regressive shoreline and deltaic deposits formed during relative sea-level highstands. Evidence of wave and tide activity is present in these deposits, which form coarsening-upward cycles.

Ryer (1977, 1983, 1984) and Cross (1988) examined coal seams contained within high-frequency cycles in the Rocky Mountain foreland basin. Ryer's work dealt with the coals of the Ferron Sandstone in southern Utah. His generalized cross-section showing the stratigraphic position of the Ferron Sandstone, as a long-term ($10^6$-year) regressive tongue, is given in Fig. 6.22. The detailed lithostratigraphy of this sandstone unit is shown in Fig. 6.23. Ryer (1984) showed that the thickest coal developments occur at times of transgressive maxima and minima on the long-term sea-level cycle, when facies tend to stack vertically. The Ferron examples exemplify a regressive maximum, when $10^5$-year deltaic sandstone cycles

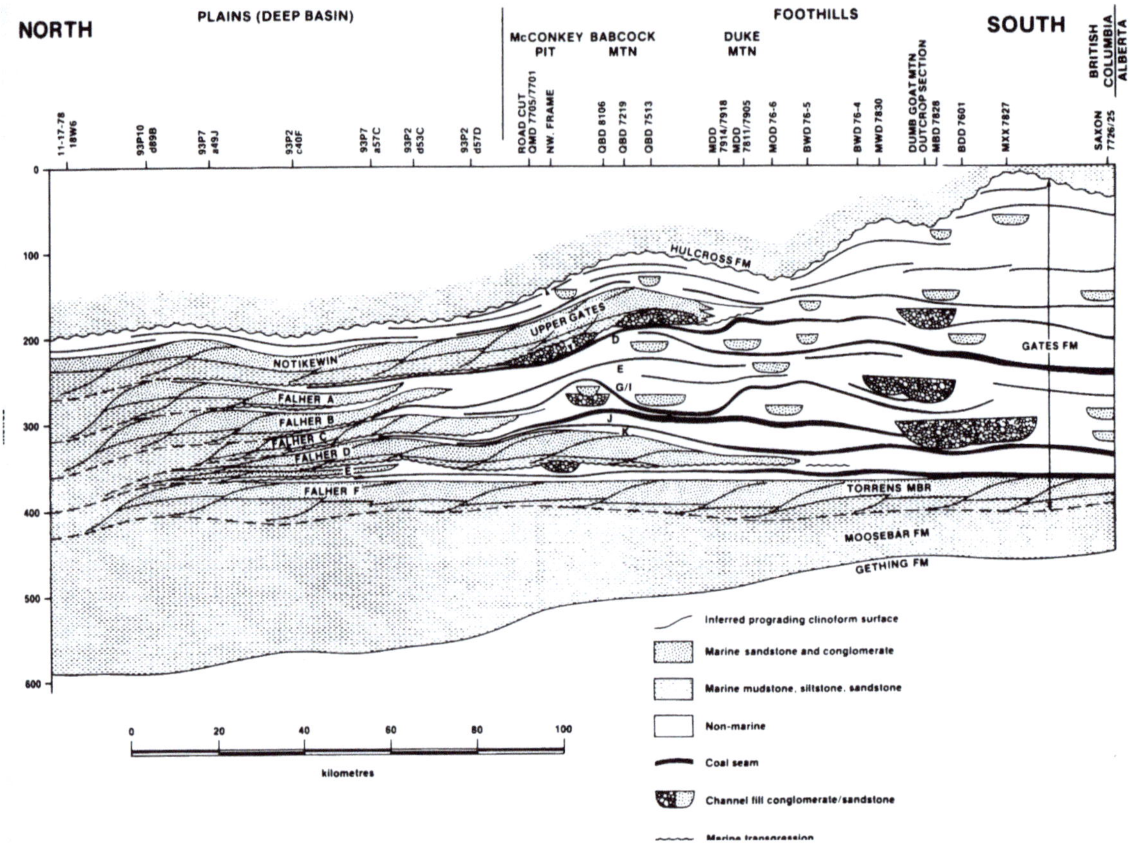

**Fig. 7.57** Regional cross-section through parts of the Gates and Moosebar Formations (Upper Mannville Group, see Fig. 6.29), Northeastern British Columbia (Carmichael, 1988)

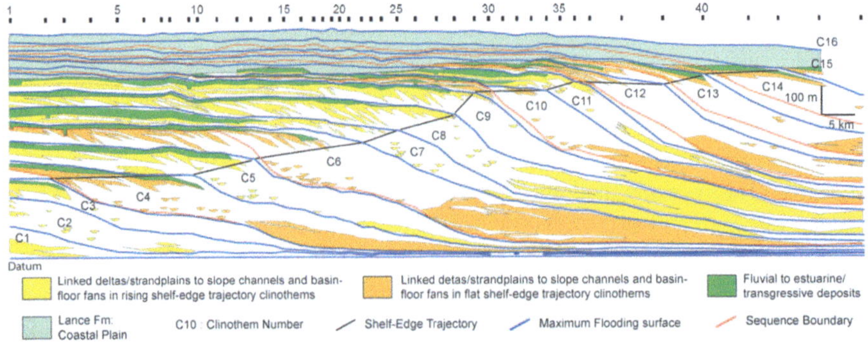



**Fig. 7.58** Clinoforms in the Lance–Fox Hills–Lewis depositional system in southern Wyoming (Carvajal and Steel, 2006, Fig. 2)

prograded far into the basin. A possible comparison with Alleghenian cyclothems may be made (see below).

A spectacular example of clinoform stratigraphy is exhibited in Fig. 7.58. This Maastrichtian succession is described as "the final third-order shoreline regression of the Cretaceous Western Interior Seaway" (Carvajal and Steel, 2006, p. 666). By this time, Laramide fragmentation of the foreland basin had just begun, affecting sediment basinal slopes. Progradation of the clinforms was southward, as a result, contrasting with the eastward sediment transport and progradation that was characteristic of the main phase of basin fill from mid-Jurassic to late Cretaceous time (Miall et al., 2008). Clinoform sets, with amplitudes of up to 430 m "consist of (1) a regressive lower component produced by deltas and/or strandplains crossing the shelf, and in our data set always reaching the shelf edge; (2) a more steeply dipping basinward component created by sediment gravity flows on a long slope below the shelf edge, reflecting an increment of shelf-margin growth; and (3) a transgressive upper component produced by landward-migrating coastal plain, estuary, and barrier lagoon systems." (Carvajal and Steel, 2006, p. 665). The 16 clinoform intervals spanned 1.8 million years, according to the biostratigraphic correlations noted by

Carvajal and Steel (2006), Each sequence therefore represents an average of 113 ka.

A particularly interesting study of $10^5$-year sequences in the Western Interior basin was reported by Elder et al. (1994). They demonstrated that a succession of five Upper Cretaceous clastic strandline parasequences in southern Utah could be correlated with basin-centre limestone-marl sequences in Kansas, some 1,500 km to the east (Fig. 7.59). Clastic sequences, as discussed in this section, are commonly attributed to tectonic mechanisms or to eustatic sea-level changes, whereas chemical cycles are usually explained in terms of orbital-forcing mechanisms, in which sea-level change is not necessarily a requirement. This correlation therefore raises interesting problems of interpretation, that are discussed further in Chaps. 10 and 11. In subsequent studies, Sageman et al. (1997, 1998) explored the nature of the cyclicity by a careful analysis of a drill core through the Bridge Creek Limestone (the limestone-marl cycles in the basin centre). They measured variations in organic carbon and calcium carbonate content through the core and applied time-series analysis to quantitatively investigate the cyclicity embedded in the succession. Several "Milankovitch" frequencies emerged from this analysis.

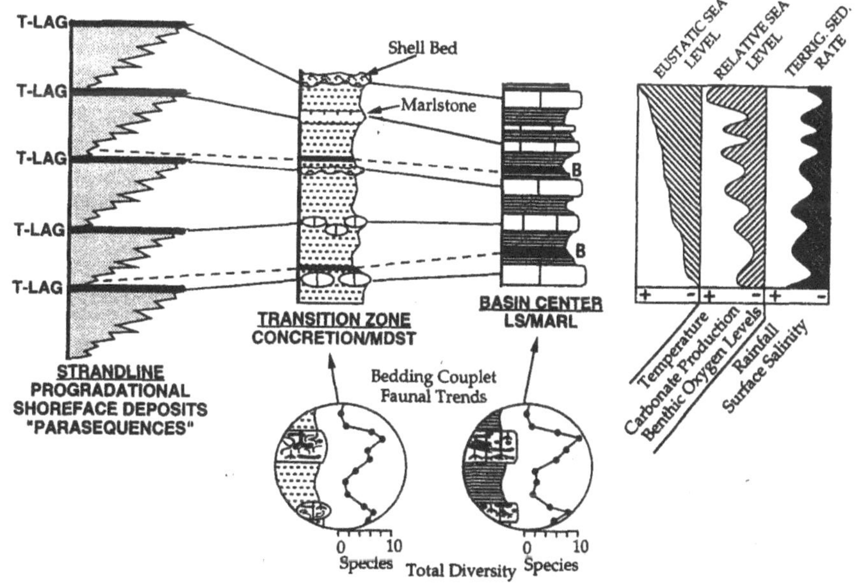



**Fig. 7.59** Correlation of clastic sequence in southern Utah (*left*) with limestone-marl sequences in Kansas. T-LAG = transgressive lag; *dashed lines* with "B" are bentonites (Elder et al., 1994)

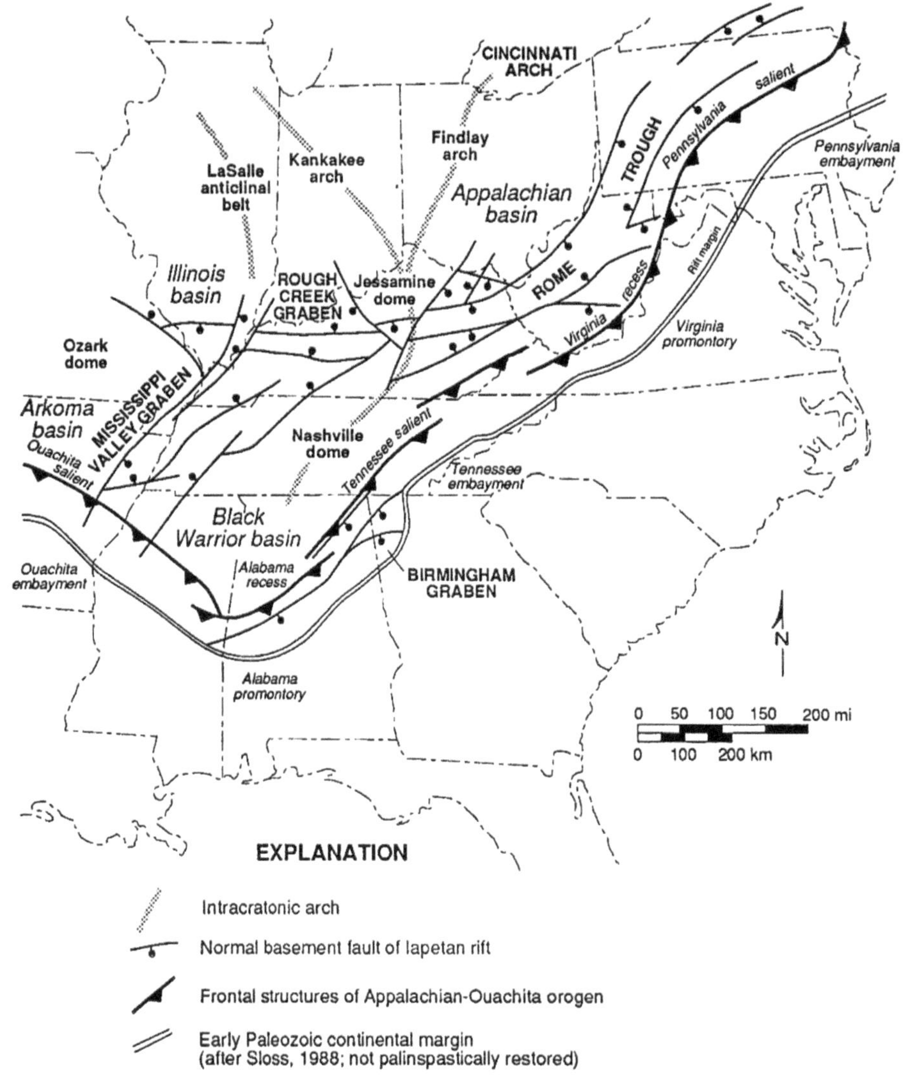



**Fig. 7.60**  Location of the Appalachian foreland basin

High-frequency cyclicity is well-developed in other foreland basins. A single example is presented here. The Appalachian and Black Warrior basins are genetically linked foreland basins developed adjacent to the Appalachian orogen (Fig. 7.60). During the Pennsylvanian, successions of coal-bearing cycles were developed in these basins. These are the same age as the classic cyclothems, discussed in Sect. 7.4, and there has been ongoing debate regarding the relative importance of tectonism versus eustasy as a control on accommodation and sediment supply in the generation of these cyclic successions (e.g., Dennison and Ettensohn, 1994).

Figure 7.61 illustrates a stratigraphic cross-section constructed through the Pennsylvanian cycles of northern Alabama. Correlation of these cycles is based on the recognition of stratigraphic bundles called "coal groups" (Pashin, 1994, p. 90). The cycles range in thickness from 11 m on the northern basin margin to more than 200 m in the deep basin centre. Each contains a basal marine mudstone, and typically coarsens upward into a sandstone. Coal-groups cap each cycle, and consist of successions of mudstone, sandstone, underclay and coal, commonly more than one seam. Pashin (1994) estimated that the cycles each represent approximately 200–500 ka. Given that these deposits

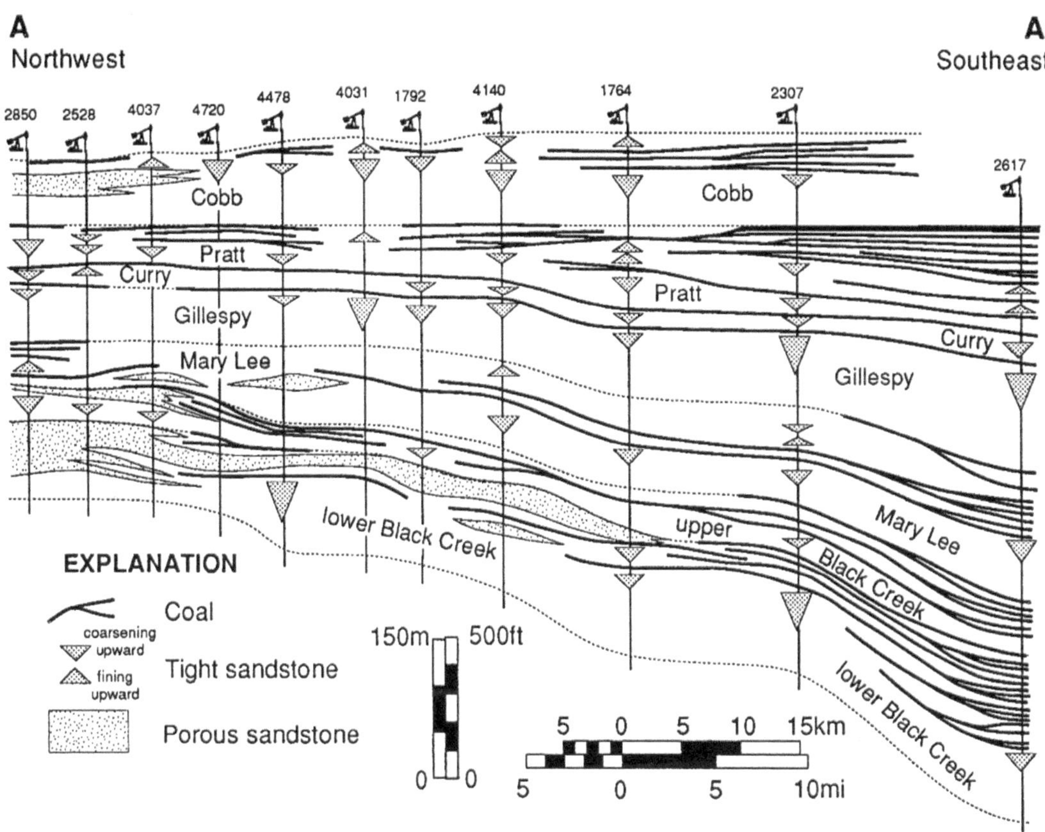



**Fig. 7.61** Stratigraphic cross-section through the Pennsylvanian cycles of northern Alabama. Correlation of these cycles is based on the recognition of stratigraphic bundles called "coal groups" (Pashin, 1994)

are of the same age as the classic cyclothems, an obvious hypothesis is that glacioeustasy is a driving mechanism, with transgression and regression responsible for the development of the stratigraphic succession of each cycle (c.f., Fig. 7.38). But tectonic control is also suggested by the presence of these deposits in an active foreland basin. Isopach maps of the cycles (Fig. 7.62) indicate a shift in the depocentre during the deposition of these cycles from a location in the southeast part of the basin to an area of broad subsidence along the southern margin of the basin. Facies and sand isolith patterns (not shown) suggest that the same broad area of uplift—the orogen to the east and southeast, served as the major sediment source throughout the accumulation of these deposits. Was high-frequency tectonism also involved in the development of these cycles? We return to this question in Chap. 10.

## 7.7 Main Conclusions

1. The Neogene stratigraphic records of continental margins, including the continental shelf, slope, and deep basin, have been intensively studied by reflection-seismic surveying and offshore drilling, including DSDP surveys. Most stratigraphic sections in both carbonate- and clastic-dominated successions are characterized by cycles with $10^4$–$10^5$-year episodicities ("fourth-" and "fifth-order" cycles in the earlier terminology). These can be correlated with the ocean-temperature cycles defined by the oxygen-isotope chronostratigraphic record, and are interpreted to be of glacioeustatic origin.

2. Many stratigraphic sections in the pre-Neogene record also contain prominent Milankovitch cycles. They are particularly prominent, and have been

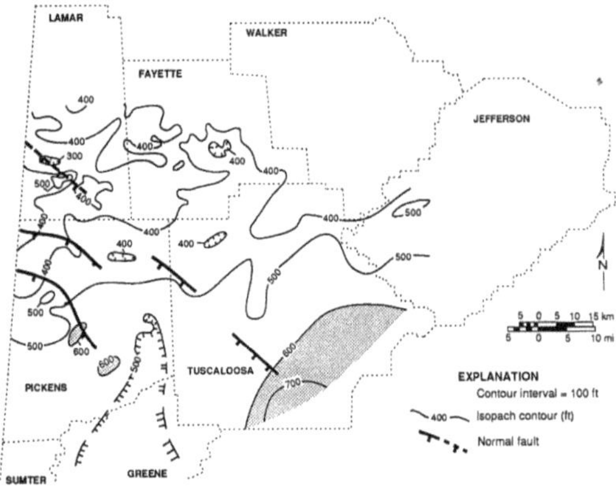

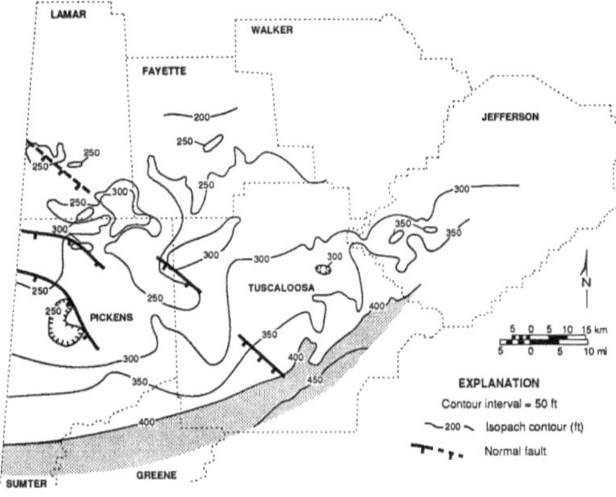



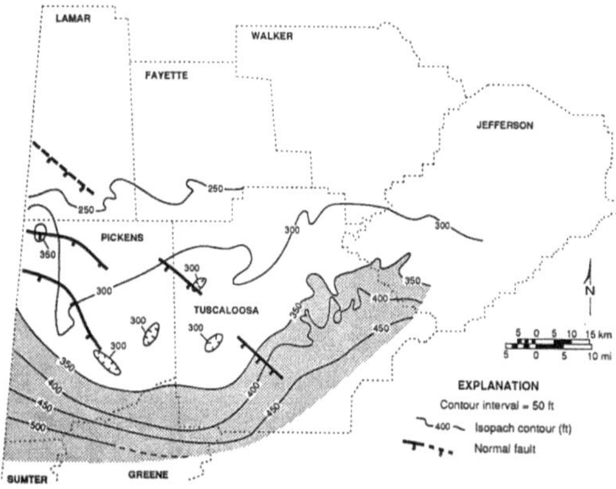

**Fig. 7.62** Isopach maps of selected cycles shown in Fig. 7.61 (Pashin, 1994)

well-described, in various Mesozoic-Cenozoic deep-marine sections around the Mediterranean basin. However, comparable cycles are also present

in many other types of succession, including lacustrine, marginal-marine-evaporite, pelagic-marine and nonmarine suites. There is controversy regarding the origins of many of these cycles, as there is doubt that glacioeustasy can be appealed to during much of Phanerozoic time, when little or no evidence exists for the presence of large continental ice caps anywhere on earth. However, other types of orbital forcing may be responsible (Chap. 11).

3. A particularly well-known type of presumed Milankovitch cycle is the upper Paleozoic record of cyclothems in the midcontinent region of North America and northwest Europe. These cycles contain much of the economic coal deposits of the northern hemisphere and have been widely studied since the 1930s. It has long been generally agreed that these cycles were driven by the late Paleozoic glaciation of the Gondwana super-continent.

4. Many examples of clastic high-frequency cycles have been described from the Cretaceous sedimentary record of the Western Interior of North America, notably in Alberta, Canada, and the Colorado Plateau area of the United States. These cycles were all deposited within a tectonically active foreland-basin setting. Tectonism is known to have been a significant control in the development of the large-scale architecture of the basin fill, and it now seems likely that tectonism also controlled high-frequency cyclicity. Comparable cycles are present in other foreland basins.

5. There is evidence for cyclicity controlled by climate change, and also some circumstantial evidence for modest, high-frequency sea-level changes in some of the Cretaceous successions of the Western Interior Seaway, Some chemical cycles of the deep basin may be correlated with clastic cycles of the basin margin, and Milankovitch mechanisms have been suggested as a contributing or controlling factor.

6. There are similarities between the clastic cycles of the Western Interior foreland basin and the Upper Paleozoic cyclothems of the Alleghenian foreland basin of the eastern United States, suggesting comparable generating causes.

# Part III
# Mechanisms

Although eustasy as the dominant mechanism controlling sequence architecture was an integral part of the Vail/Exxon approach to sequence stratigraphy, L. L. Sloss, on whose work it was originally based, was always convinced that tectonism was an important component of sedimentary controls. With the critiques of the global eustasy model that began to appear in the 1980s and 1990s, research broadened to explore other mechanisms that affect the generation and removal of accommodation, and the sediment supply from which sequences are built. It has been demonstrated that there are several global tectonic mechanisms and astronomical effects that generate eustatic sea-level changes on several different time scales and of varying amplitude. These are discussed in this part of the book. The processes are summarized in Table 8.1 and Fig. 8.1.

Research has, in addition, demonstrated that there is a range of tectonic processes that generate relative sea-level changes on a local to regional scale. The rates and magnitudes of change are variable, and much research remains to be conducted to quantify these effects. However, it is now clear that many forcing functions of different frequencies and amplitudes are likely to be active at the same time, and that they may not be mutually independent; for example, tectonic uplift affects accommodation and, at the same time, may affect climate, and both affect sediment supply. Such processes, acting independently or together, may produce effects that are very similar to those of eustasy, but on a regional to continental scale (Table 8.2, Figs. 8.1, 8.2). This is of critical importance, because it means that sequence stratigraphies recorded in any given basin, however well documented, cannot be assumed to be global, and therefore representative of a global framework of eustatic cycles. In fact, it calls into question the very concept of the global cycle chart based on the tying together of key sections from supposedly representative areas.

The focus of the Exxon models was on eustatic sea-level change as the primary control of sequence architecture. If eustasy is the primary mechanism, then it justifies the use of sequence stratigraphies for the purpose of constructing a global cycle chart, the assumption being that a sequence record anywhere will reflect the same global eustatic controls. Tectonism is treated in the Exxon models as a slow background effect that does not substantially modify the stratigraphic response to eustatic sea-level change. However, sequences of tectonic origin may be expected to vary in age from region to region, because they are generated by plate-tectonic processes and mantle effects that are progressive and diachronous. The only reliable test of a eustatic control is precise correlation. If sequences in widely spaced basins of different tectonic setting are of the same age, they may be assumed to be of eustatic origin. This emphasizes the need for very precise chronostratigraphic control, a subject discussed in Part IV.

Mechanisms, the range of processes that alter sea level and sediment supply, are the subject of this part of the book.

- Nature vibrates with rhythms, climatic and diastrophic, those finding expression ranging in period from the rapid oscillation of surface waters, recorded in ripple mark, to those long-defended stirrings of the deep titans which have divided earth history into periods and eras (Barrell, 1917, p. 745).
- Allocyclic sequences ... are mainly caused by variations external to the considered sedimentary system (e.g., the basin), such as climatic changes, tectonic movements in the source area, global sea level variations, etc. ... Such processes often tend to generate cyclic phenomena of a larger lateral continuity and time period than auto-cyclic processes. The most characteristic effect of some of the allocyclic processes is that they operate simultaneously in different basins in a similar way. Thus it should be possible to correlate part of the allocyclic sequences over long distances and perhaps even from one basin to another (Einsele et al., 1991b, p. 7).
- Sea level is an ill-defined concept that incorporates the effects of isostasy, eustasy, ice volume, passive margin subsidence, lithospheric deformation and flexure, ero-sion and consequent sediment loading, and changes in rate of plate generation and consumption that all serve to alter the elevation of the world seafloor with respect to the elevation of the continental surface. As such, quantitatively defining a uni-versally acceptable definition of sea level to the satisfaction of all has so far proven an intractable problem (Dockal and Worsley, 1991, p. 6805).
- It is unfortunate that a change in water depth is the only parameter that we can infer from facies analysis, because in order to isolate the effects of one of the independent variables (eustasy, tectonics, sedimentation), we are forced to make assumptions about the magnitude and rate of the other two variables (Plint et al., 1992, p. 16).
- In parallel to the eustasy-driven sequence stratigraphy, which held by far the largest share of the market, other researchers went to the opposite end of the spec-trum by suggesting a methodology that favored tectonism as the main driver of stratigraphic cyclicity. This version of sequence stratigraphy was introduced as "tectonostratigraphy" ... The major weakness of both schools of thought is that a priori interpretation of the main allogenic control on accommodation was auto-matically attached to any sequence delineation, which gave the impression that sequence stratigraphy is more of an interpretation artifact than an empirical, data-based method. This a priori interpretation facet of sequence stratigraphy attracted considerable criticism and placed an unwanted shade on a method that otherwise represents a truly important advance in the science of sedimentary geology. Fixing the damaged image of sequence stratigraphy only requires the basic understanding that base-level changes can be controlled by any combination of eustatic and tec-tonic forces, and that the dominance of any of these allogenic mechanisms should be assessed on a case by case basis. It became clear that sequence stratigraphy needed to be dissociated from the global-eustasy model, and that a more objective analysis should be based on empirical evidence that can actually be observed in outcrop or the subsurface (Catuneanu, 2006, p. 5).

# Chapter 8

# Summary of Sequence-Generating Mechanisms

Barrell (1917) was one of the first earth scientists to fully comprehend the full range of dynamic processes affecting the Earth's crust (see quote from his paper at the beginning of this part of the book). These included both orogeny and epeirogeny. *Orogeny* is a term that came into general use in the mid-nineteenth century to encompass the processes of vertical and lateral motion and deformation of the earth's crust involved in the generation of deformed belts and mountain ranges (Bates and Jackson, 1987). *Epeirogeny* is a term coined by Gilbert (1890) to encompass the development of the larger features of continental interiors, including plateaus and basins, predominantly by processes of vertical motion. *Eustasy* was recognized by Suess (1885–1909) and formed the theoretical basis for his studies of global stratigraphy and orogeny (Fig. 1.7). By the 1930s, significance of *glacioeustasy* for understanding parts of the ancient stratigraphic record was recognized by those working on the US Mid-continent cyclothems (Shepard and Wanless, 1935). A more complete discussion of the early search for cyclic or rhythmic mechanisms is presented in Sect. 1.4.

During a period in the 1980s, following the initial success of the new seismic stratigraphy (Vail et al., 1977) there was an over-emphasis on the importance of eustatic sea-level change as the dominant control on stratigraphic architecture but, as summarized in Sect. 1.5, Subsequent research, much of it directly stimulated by questions that Vail's work raised, has uncovered diverse processes that govern the generation and filling of sedimentary accommodation. Dewey and Pitman (1998, p. 14) stated:

> It is our contention that local or regional tectonic control of sea level dominates eustatic effects, except for short lived glacio-eustatic periods and, therefore, that most sequences and third-order cycles are controlled tectonically on a local and regional scale rather than being global-eustatically controlled. We suggest that eustatic sea-level changes are nowhere near as large as have been claimed [by Haq et al., 1987] and have been falsely correlated and constricted into a global eustatic time scale whereby a circular reasoning forces correlation and then uses that correlation as a standard into which all other sequences are squeezed. …. We wish only to caution against the belief that third order sequences can be used as a global eustatic time scale. However, we do believe that Stille's (1924) concepts of global orogenic and eustatic sea-level episodicity need a fresh evaluation in the framework of the assembly and disruption of supercontinents. This evaluation will only come when really fine-scale methods of physical/chemical stratigraphic correlation become widely available for Phanerozoic strata.

Figure 8.1 and Table 8.1 provide a summary of the rates and durations of sedimentary processes. These vary in duration over fourteen orders of magnitude of time, from the short-term processes that develop ripples and laminae under running water (seconds to minutes) to the long-term geological processes that characterize plate-tectonic changes to Earth's crust (tens to hundreds of millions of years). The depositional products of these processes may be grouped conveniently into "groups" that each encompass processes operating within one order of magnitude of time. Sedimentological studies are facilitated by the fact that in the rock record, each of these groups can be defined by their architecture and by the nature of the bounding surfaces that enclose and define them. Figure 4.5 illustrates this point with reference to fluvial deposits, where the numbers correspond to the numbered bounding surfaces that can be defined at each scale at which a fluvial deposit is examined, from a small piece of drill core up to the scale of an entire basin. In Table 8.1, references are provided to work

A.D. Miall, *The Geology of Stratigraphic Sequences*, 2nd ed.,
DOI 10.1007/978-3-642-05027-5_8, © Springer-Verlag Berlin Heidelberg 2010

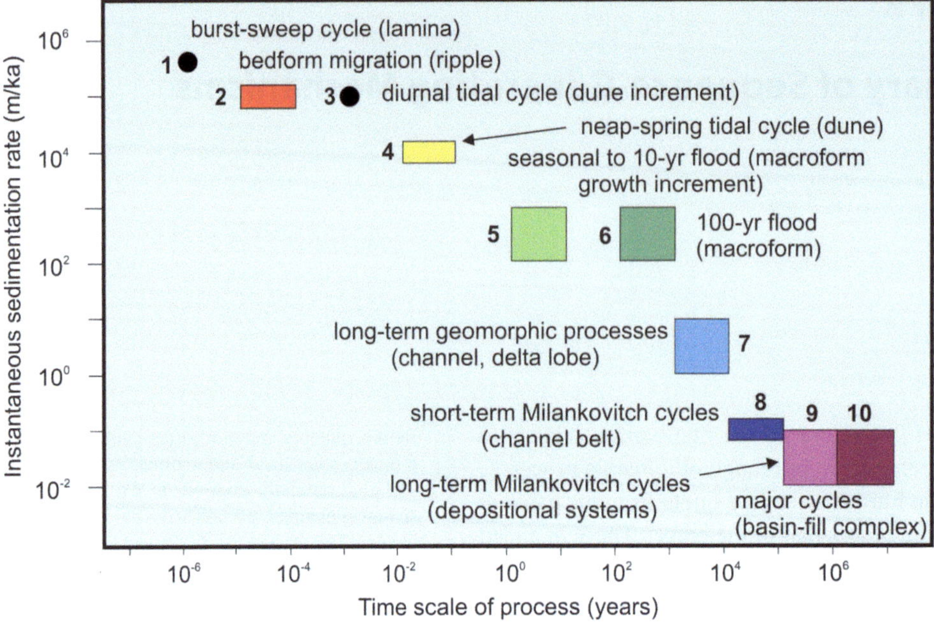


**Fig. 8.1** Rates and durations of sedimentary processes (see also Table 8.1)

that recognized patterns of hierarchical subdivision in other environments.

Sequences represent the product of the longer-term processes shown in Fig. 8.1 and Table 8.1, specifically, groups 8, 9 and 10. This breakdown refers only to rates and durations, and does not help to clarify the nature of the processes or driving mechanisms that form sequences. As discussed in Sect. 4.2, sequences constitute a hierarchy, but not a hierarchy that can be defined with neat mathematical precision. This is because of the multiplicity of processes which, we now recognize, is responsible for generating sequences, some global, some regional to continental in scale. A further layer of complexity stems from the fact that more than one process may be operating at any given time. This was recognized by Barrell (1917), whose chart of changing base level shows oscillations with three different frequencies occurring simultaneously (Fig. 1.3). Dewey and Pitman (1998) provided a brief review of the Pangea supercontinent cycle (Sect. 9.2), within which and upon which a range of processes were superimposed, generating local and regional sea-level changes (Table 8.2).

A summary of our current understanding of sequence-generating mechanisms is provided in Table 4.1.

A graphical summary of the most important geological processes is shown in Fig. 8.2. Here, the time scale is expressed as *recurrence interval*, the meaning of which should not be taken to always indicate rhythmic periodicity, but rather the time scale of episodicity, over which these processes tend to act.

In this diagram, the rectangle labeled "tectonic" refers to the time scale and physical scale of major, long-term tectonic processes, as opposed to the shorter time frame of individual tectonic events—the box labeled "structural". Plate tectonics involves not only changes in the vertical elevation of plate margins, but also changes in ocean-basin configurations and depths. By changing the volumes of the ocean basins, plate tectonics is therefore one of the drivers of eustatic sea-level change (Chap. 9). Because this eustatic process and continental tectonism are outcomes of the same driving force, they are typically contemporaneous, and the resulting changes in accommodation in the affected basins are commonly complex. This is discussed further in Chaps. 9 and 10.

*Intraplate stress* (Fig. 8.2) is now recognized as a significant control on sequence architecture (Sect. 10.4). It is also referred to as *in-plane stress*, and its effects, which may be transmitted for significant

**Table 8.1** Hierarchies of architectural units in clastic deposits

| Group | Time scale of process (years) | Examples of processes | Instantaneous sed. rate (m/ka) | Fluvial, deltaic *Miall* | Eolian *Brookfield, Kocurek* | Coastal, estuarine *Allen[a] Nio and Yang[b]* | Shelf *Dott and Bourgeois,[c] Shurr[d]* | Submarine fan *Mutti & Normark* |
|---|---|---|---|---|---|---|---|---|
| 1. | $10^{-6}$ | Burst-sweep cycle | | Lamina | Grainflow grainfall | Lamina | Lamina | |
| 2. | $10^{-5}$–$10^{-4}$ | Bedform migration | $10^5$ | Ripple (microform) [1st-order surface] | Ripple | Ripple [E3 surface[a]] [A[b]] | [3-surface in HCS[c]] | |
| 3. | $10^{-3}$ | Diurnal tidal cycle | $10^5$ | Diurnal dune incr., react. surf. [1st-order surface] | Daily cycle [3rd-order surface] | Tidal bundle [E2 surface[a]][A[b]] | [2-surface in HCS[c]] | |
| 4. | $10^{-2}$–$10^{-1}$ | Neap-spring tidal cycle | $10^4$ | Dune (mesoform) [2nd-order surface] | Dune [3rd-order surface] | Neap-spring bundle, storm, layer [E2[a], B[b]] | HCS sequence [1-surface[c]] | |
| 5. | $10^0$–$10^1$ | Seasonal to 10 year flood | $10^{2-3}$ | Macroform growth increment [3rd-order surface] | Reactivation [2nd, 3rd-order] surfaces, annual cycle | Sand wave, major storm layer [E1[a], C[b]] | HCS sequence [1-surface[c]] | |
| 6. | $10^2$–$10^3$ | 100 year flood | $10^{2-3}$ | Macroform, e.g. point bar levee, splay [4th-order surface] | Dune, draa [1st-, 2nd-order surfaces] | Sand wave field, washover fan [D[b]] | [facies pack-age (V)[d]] | Macroform [5] |
| 7. | $10^3$–$10^4$ | Long term geomorphic processes | $10^0$–$10^1$ | Channel, delta lobe [5th-order surface] | Draa, erg [1st-order, super surface] | Sand-ridge, barrier island, tidal channel [E[b]] | [elongate lens (IV)[d]] | Minor lobe, channel-levee [4] |
| 8. | $10^4$–$10^5$ | 5th-order (Milankovitch) cycles | $10^{-1}$ | Channel belt [6th-order surface] | Erg [super surface] | Sand-ridge field, c-u cycle [F[b]] | [regional lentil (III)[d]] | Major lobe [turb. stage: 3] |
| 9. | $10^5$–$10^6$ | 4th-order (Milankovitch) cycles | $10^{-1}$–$10^{-2}$ | Depo. system, alluvial fan, major delta | Erg [super surface] | c-u cycle [G[b]] | [ss sheet (II)[d]] | Depo. system [2] |
| 10. | $10^6$–$10^7$ | 3rd-order cycles | $10^{-1}$–$10^{-2}$ | Basin-fill complex | Basin-fill complex | Coastal-plain complex [H[b]] | [lithosome (I)[d]] | Fan complex [1] |

Hierarchical subdivisions of other authors are given in square brackets. Names of authors are at head of each column. [a,b,c,d]: terminology of authors indicated at head of columns. HCS: hummocky cross-stratification. c-u: coarsening-upward. Adapted from Miall (1991b).

**Table 8.2** Rates and magnitudes of processes affecting sea-level*

| Process | Region affected | Type of result | Rate m/ka | Duration million years | Tota possible change (m) |
|---|---|---|---|---|---|
| *Eustatic process* | | | | | |
| Age distribution of earth's oceanic crust | Global | Eustasy | 0.001 | 100+ | 100 |
| Sea-floor ridge volume changes | Global | Eustasy | 0.002–0.01 | 50–100 | 300 |
| Continental shortening, stretching | Global | Eustasy | 0.0006–0.0008 | 45–80 | 38–50 |
| Density changes associated with intraplate stress | Global | Eustasy | 1 | 0.05 | 50 |
| Flooding of inland areas below sea-level | Global | Eustasy | Virtually inst. | Virtually inst. | Up to 10 |
| Continental ice formation | Global | Eustatic fall | 1.5 | 0.1 | 150 |
| Continental ice melting | Global | Eustatic rise | 4–10 | 0.02–0.04 | 80–400 |
| Marine ice-sheet decoupling | Global | Eustatic rise | 30–50 | 0.002 | 6–10 |
| *Processes leading to uplift of continental crust* | | | | | |
| Heating beneath supercontinent | Hemisphere | Uplift of crust | 0.005–0.01 | 100 | 500–1000 |
| Thermal doming accompanying rifting | Rift flanks bulge | Thermal | 0.012 | 16 | 250 |
| Convergent tectonism | Arc-trench systems, collision zones | Uplift of fault blocks, nappes | 0.5[a] 10[b] | 2 0.2 | 1000 2000 |
| Intraplate stress | Entire plates | Modification of flexural deflections | 0.01–0.1 | 1–10 | 100 |
| Unsteadiness in mantle convection | Areas of $10^4$–$10^6$ km$^2$ | Regional warping | 1–10 | 0.01–0.1 | 100 |
| *Processes leading to subsidence of the continental crust* | | | | | |
| Post-rift thermal subsidence of cont. margin | Continental margin | Hinged subsidence | 0.03–0.07[c] 0.005–0.03[d] | 20 200 | 600–1400 2000–4000 |
| Flexural loading | Foreland basin | Basin subsidence | 0.08–1.0 | 2–15 | 1000–4000 |
| Intraplate stress | Entire plates | Modification of flexural deflections | 0.01–0.1 | 10 | 100 |
| Unsteadiness in mantle convection | Areas of $10^4$–$10^6$ km$^2$ | Regional warping | 1–10 | 0.01–0.1 | 100 |

*This table was compiled from numerous sources, as explained in the relevant sections of Chaps. 9, 10 and 11.
[a]Long-term rate.
[b]Short-term rate.
[c]Initial thermal subsidence.
[d]Decay of thermal anomaly.

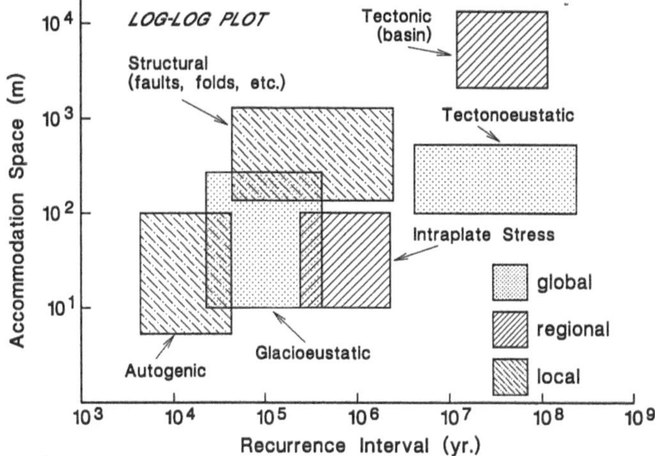


**Fig. 8.2** Estimated ranges in accommodation space and recurrence interval attributable to alternate controls of stratigraphic cyclicity, where accommodation space (per cyclic event) sets limits on thicknesses of individual cycles and recurrence interval (between initiation of successive cycles) governs characteristic frequency of observed cyclicity. The structural field for folds and faults is delimited to embrace the full growth of regional structures. All fields shown are tentative and subject to revision or reinterpretation (tectonic box pertains to full basin evolution). From Dickinson et al. (1994, Fig. 4)

distances within and across tectonic plates, are commonly referred to as *far-field* effects. Another significant process, acting over longer time scales, but not shown in Fig. 8.2, is *dynamic topography* (Sect. 9.3.2).

In Fig. 8.2 "glacioeustatic" refers to the high-frequency recurrence interval of eustatic sea-level changes that result from climatic fluctuations between glacial and interglacial episodes. These are, of course, well known from the Neogene, and have now been extensively documented from the more ancient geological record (Chap. 7). The sedimentary products correspond to groups 8 and 9 of the stratigraphic hierarchy (Fig. 8.1, Table 8.1). Orbital forcing is the driving mechanism. As discussed in Sects. 7.3, 7.4, 7.5 and 7.6, there is extensive geological evidence for non-glacial orbital forcing, that is, orbitally-forced climate change that did not involve glacial freeze-thaw, but did lead to extensive climatic changes that are reflected in the strength and significance of some key sedimentological processes, such as organic productivity and the redox conditions of ocean waters. This is examined at length in Chap. 11.

A further discussion of the hierarchies of geologic time and its implications for the analysis of the ways in which time is stored in the geological record is presented in Chap. 13.

# Chapter 9

# Long-Term Eustasy and Epeirogeny

## Contents



## 9.1 Mantle Processes and Dynamic Topography

The major cause of change in the Earth's crust is the radiogenic heat engine, which drives mantle convection and generates the geomagnetic field. Convection distributes heat and drives plate tectonics. Oceanic and continental crust are subject to heating close to sea-floor spreading centres where new, hot, oceanic crust is generated (causing "ridge push"), and to cooling overlying subduction zones, where cold oceanic crust descends to the mantle (generating "subduction pull"). Widespread cooling and subsidence occur over the downwelling zones where continental fragments converge. The same cooling and subsidence occur above old, cold, downgoing oceanic slabs in subduction zones, and partial melting of the water-saturated rocks as they descend in the subduction zone is what generates arc magmatism. The differential heat distribution and consequent heating and cooling of different parts of the overlying Earth's surface results in broad regional uplifts, downwarps and tilts, because of the effects on crustal densities. These processes maintain what is called *dynamic topography* (Fig. 9.1).

The effects of crustal heating are to cause uplift. This occurs along the flanks of new continental rift systems (e.g., parts of East Africa) and above mantle plumes. Thermal doming beneath supercontinents may elevate the crust by as much as 1 km over periods of 100 million years. Subsidence takes place over cooling areas of the Earth's crust, such as areas of aging oceanic and continental crust distant from spreading centres, and over regions of mantle downwelling.

The concepts of dynamic topography are now being explored with the techniques of a new field, called computational geodynamics, "in which computer models of mantle convection are used in the interpretation of contemporaneous geophysical observations like seismic tomography and the geoid as well as of time-integrated observations from isotope geochemistry" (Gurnis, 1992). Such models are capable of integrating large volumes of detailed stratigraphic data using the backstripping procedures outlined in Sect. 3.5. These developments are of great significance, and their implications for sequence stratigraphy have yet to be fully realized.

Vertical movement of the crust causes relative changes in sea level on a regional or continental scale. This is true *epeirogeny*. However, thermal changes also result in changes in the volume of the ocean basins, which lead, in turn, to eustatic changes in sea level. The same broad crust-mantle processes generate both regional and eustatic effects that may or may not be in phase. The result is a highly complex sequence of sea-level changes, and clear global eustatic signals may not always be present in the stratigraphic record.

Continental-scale vertical movements caused by the process of epeirogeny have been amply demonstrated by stratigraphers (e.g., Sloss, 1963; Bond, 1976, 1978; Burgess, 2008; see Sect. 3.4), but were only explained


A.D. Miall, *The Geology of Stratigraphic Sequences*, 2nd ed.,
DOI 10.1007/978-3-642-05027-5_9, © Springer-Verlag Berlin Heidelberg 2010

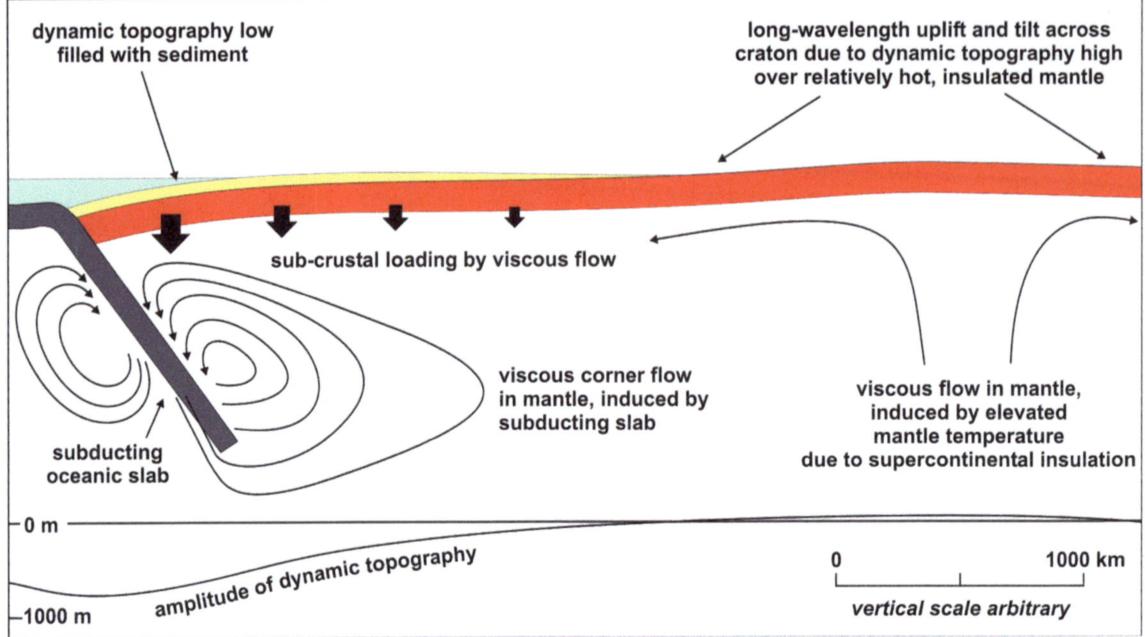


**Fig. 9.1** The generation of dynamic topography and controls on sea-level change and regional vertical crustal motion by thermal effects related to shallow and deep mantle-convection. Widespread heating beneath supercontinents (at *right*) generates continental uplift and an elevated geoid. More localized depression of the geoid is caused by the subduction of cold slabs of oceanic crust (at *left*) (Burgess, 2008, Fig. 6)

recently. At a time when the global eustasy model remained popular, Sloss (1991) maintained that vertical and tilting movements driven by epeirogeny were indicated by the large-scale architecture of the $10^7$-year sequences (the "Sloss sequences"). His cross-sections (e.g., Fig. 5.8) and his maps of subsidence patters of the craton beneath the United States (e.g., Figs. 5.9 and 5.10) make this point, and it is one reinforced by the discussion of the Phanerozoic evolution of the craton by Burgess (2008), who stated (p. 37):

> ... there is a very basic observation that demonstrates that eustasy was not the only contributor to the relative sea-level changes recorded by cratonic sequences. Long-term eustatic oscillations certainly must have contributed to development of the transgressive and regressive sequence elements, but long-wavelength post-depositional tilting of the cratonic strata and angular sequence-bounding unconformities, both ubiquitous features of North American cratonic strata ..., obviously require a tectonic mechanism, and cannot be explained by eustatic change alone. Applying such simple reasoning to North American cratonic stratal patterns often provides a reasonable indication of the degree of tectonic and eustatic influence on relative sea-level.

Only in the last two decades has this research finally provided a theoretical basis for the process of epeirogeny, a process that has commonly been invoked to explain the broad vertical movements of the crust indicated by stratigraphic studies, but which has lacked a basis in the modern theories of plate tectonics.

The processes referred to here are relatively long-term in their effect, and are capable of explaining much of the cyclicity that has been recorded on time scales of tens to hundreds of millions of years. They are the subject of this chapter.

## 9.2 Supercontinent Cycles

Early work by Wilson, Sutton, Bott, Condie, Windley, and others suggested that the Phanerozoic history of the earth (the broad, long-term stratigraphic patterns outlined in Sect. 5.2) can be related to the assembly and breakup of the Pangea supercontinent. Most recent workers have adopted this long-term plate-tectonic cycle as the basis for hypotheses of the earth's dynamics (Anderson, 1982, 1984, 1994; Worsley et al.,

1984, 1986; Gurnis, 1988; Veevers, 1990; Dewey and Pitman, 1998).

The formation of a supercontinent creates a thermal blanket that inhibits convective release of radiogenic heat from the mantle (Fig. 9.1). Changes in the rotation of the Earth's core and in the convective patterns in the mantle may be either the cause or the consequence of these surface events, which also appear to be linked to changes in the earth's magnetic field (Anderson, 1984; Maxwell, 1984). Development of the thermal blanket may be the cause of the eventual breakup of the supercontinent, following establishment of a new pattern of mantle convection. The formation of the thermal blanket beneath a supercontinent leads to heating and regional epeirogenic uplift on a continental scale. Gurnis (1988, 1990, 1992) demonstrated that the uplift rate would be 5–10 m/million years, and could persist for 100 million years, resulting in an uplift of 500 m to 1 km. (This is the first of many processes of sea-level change discussed in this book for which quantitative estimates of rate, duration and magnitude are available. The processes are listed in Table 8.2, which is referred to throughout the remainder of this book). A summary of the process of supercontinent assembly and fragmentation is provided in Sect. 5.2 (see also Fig. 9.2).

It has been argued that dynamic mantle uplift is extremely long-lived. It generates a positive geoid anomaly that survives for $10^8$ years. Crough and Jurdy (1980) demonstrated the existence of a large positive anomaly beneath Africa. Veevers (1990) showed the position of this anomaly beneath a reconstruction of the Pangea supercontinent. The correspondence is remarkably close, and confirms that Africa was at the centre of Pangea. As noted in Sect. 3.4, Bond (1976, 1978) has demonstrated that Africa has undergone anomalous uplift since the early Tertiary. This is too late to have been caused directly by the heating effect, which would have taken place in the late Paleozoic or early Mesozoic following continental assembly. However, it is possible that the uplift relates to intraplate compressive stress generated by the opening of oceans virtually all around the continent. Dispersing continental fragments tend to migrate toward geoid lows, where mantle temperatures are lower, and where relative sea-levels will rise, leading to extensive platform flooding (Gurnis, 1988, 1990, 1992).

The total length of rifting continental margins and of seafloor spreading centres increases during the breakup of a supercontinent and may be accompanied by increased rates of oceanic crust generation, and active subduction, plutonism, and arc volcanism on the outer, convergent plate margins of the dispersing fragments, as suggested by the westward drift of the Americas since the Triassic, and the consequent subduction of tens of thousands of kilometres of paleo-Pacific (Panthalassa) oceanic crust (Engebretson et al., 1985). Spreading rates are episodic, reflecting the structure and behaviour of the mantle convection cells that drive them (Gurnis, 1988). Major eustatic transgressions occur because of the displacement of ocean waters by thermally elevated young oceanic crust and active spreading centers in the new Atlantic-type oceans. These factors were considered by Pitman (1978), Kominz (1984) and Harrison (1990) in the development of a model for cycles of eustatic sea-level changes over time periods of tens to hundreds of millions of years (see next section).

Heller and Angevine (1985) argued that during the first 50–100 million years after initiation of breakup of a supercontinent, the global average age of oceanic crust decreases because of the active development of Atlantic-type oceans. This will lead to a rise in sea-level, without any change in global average spreading rate. Dockal and Worsley (1991) developed a simple isostatic model to quantify the effects of changing age of the earth's oceanic crust, simplifying the earth to a two-ocean system, with an opening Atlantic-type ocean replacing a Pacific-type ocean undergoing consumption. They demonstrated that this effect alone can

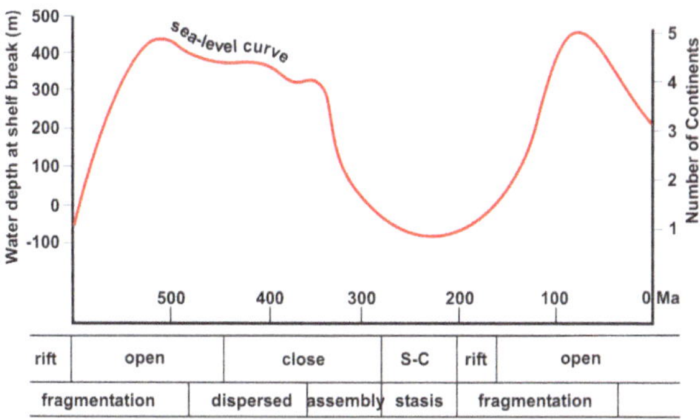


**Fig. 9.2** The supercontinent cycle, showing its relationship to long-term changes in sea level. The "stasis" phase corresponds to the main period of supercontinent assembly (S-C), which was Pangea in the diagram above (simplified from Fig. 5.2)

account for a change in sea level of $\sim$100 m over about 120 million years (Table 8.2).

It has been suggested that the average rate of spreading slows at times of continental assembly, at a time of ridge reordering following major continental collision and suturing events. Collision results in crustal shortening and thickening, which has the effect of increasing the ocean-basin volume. Therefore, at the end of a supercontinent assembly cycle, large areas of old, and therefore cool, and subsided oceanic crust will underlie the world's oceans (Worsley et al., 1984, 1986). All these effects lead to enlargement of the world's ocean basins. Times of low sea-level might therefore be expected to correlate with, or follow, major suturing episodes (Valentine and Moores, 1970, 1972; Larson and Pitman, 1972; Vail et al., 1977; Schwan, 1980; Heller and Angevine, 1985).

The effects of sea-level change on climates, sedimentation and biogenesis are briefly described in Sect. 5.1.

Although eustatic sea level is predicted to fall during continental assembly, Kominz and Bond (1991) documented a synchronous rise in relative sea-level in intracratonic basins and continental margins throughout North America during the middle Paleozoic (Late Devonian-mid Mississippian) (Fig. 9.3), at which time it is postulated that the late Proterozoic supercontinent was dispersing, and Pangea was assembling (Worsley et al., 1984, 1986). Kominz and Bond (1991) attributed the regional rise in sea level to synchronous enhanced subsidence, and argued that this could not have been caused directly by plate-margin processes. Many of the basins are beyond the flexural reach of the continental-margin orogenies that were underway at the time, and some of the data were derived from areas undergoing continental extension where no thermal event has been documented that could explain the timing or rate of subsidence. The synchronous nature of the subsidence (Fig. 9.3) calls for a continental-scale process, and Kominz and Bond (1991) suggested that the cause was intraplate compressive stress resulting from the movement of the North American plate over a region of mantle downwelling during supercontinent dispersion. Kominz and Bond (1991, p. 59) stated:

> ... the converging limbs of the convection system would increase the in-plane compressive stresses at the base of the lithosphere. For a critical level of stress, all preexisting positive and negative lithospheric deflections (i.e., arches and basins) will be enhanced; arches will tend to

move upward and basins will tend to subside. The convection modeling predicts that the maximum compressive stress in a downwelling region is 60 to 70 MPa (Gurnis, 1988), a range that probably is sufficient to reactivate preexisting deflections, assuming a viscoelastic lithospheric rheology.

Subsidence over downwelling mantle is an expression of "dynamic topography", a process described in Sect. 9.3.2. The hypothesis of intraplate stress has been developed by Cloetingh (1986, 1988). It may be responsible for regional changes in relative sea level over time scales of tens of thousands to tens of millions of years, and is discussed at greater length in Sect. 10.4.

In the case of the North American sea-level rise in the Middle Paleozoic, subsidence would have been enhanced by the increase in crustal density overlying a cool downwelling current (the dynamic topography effect). It is not yet clear how much this effect contributed to the overall relative sea-level rise.

As pointed out by Gurnis (1992), there is an ambiguity in attributing causes to long-term sea-level changes. Mantle convection leads to generation of dynamic topographies, which are reflected in the stratigraphic record by their continental-scale effects on relative sea levels. However, the same processes lead to changes in the global average rate of sea-floor spreading, which affect the volume of the ocean basins, and thereby generate eustatic sea-level changes. During times of supercontinent fragmentation, in areas of mantle downwelling, these two processes will be in phase, and therefore additive, which makes it difficult to separate and quantify their effects.

## 9.3 Cycles with Episodicities of Tens of Millions of Years

### 9.3.1 Eustasy

Hallam (1963) suggested that eustatic sea-level oscillations could be caused by variations in oceanic ridge volumes. Later workers (e.g. Russell, 1968; Valentine and Moores, 1970, 1972; Rona, 1973; Hays and Pitman, 1973; Pitman, 1978) applied the increasing knowledge of plate-tectonic processes to suggest that variations in seafloor spreading rates, variations in total ridge length, or both are the cause of the volume

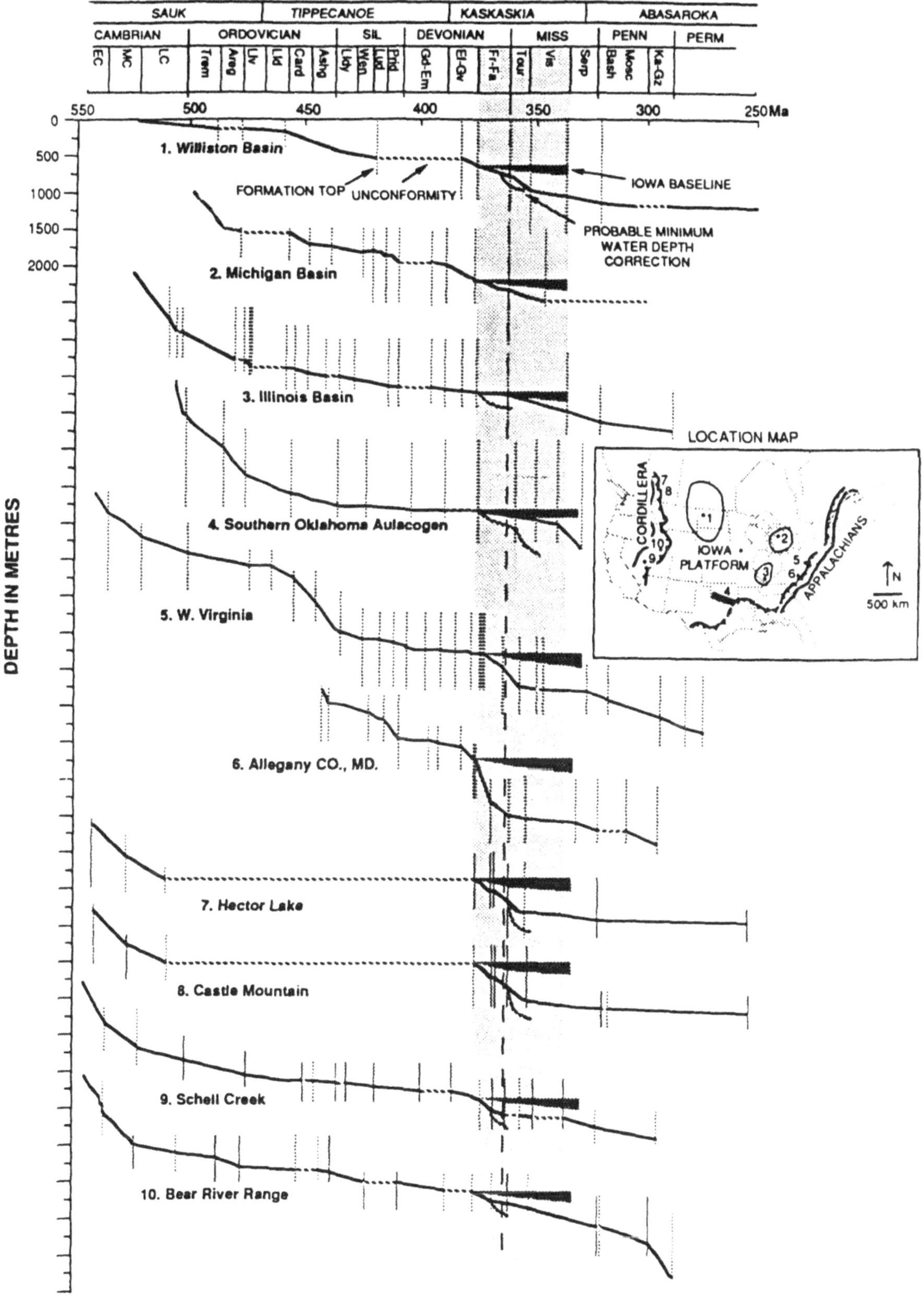


**Fig. 9.3**   Subsidence curves from various locations in North America, showing period of accelerated subsidence in the Late Devonian to middle Mississippian (Kominz and Bond, 1991)

changes. The average age of the oceanic crust also changes, especially during the assembly and dispersal of supercontinents, as noted in the previous section, and this also affects the volume of the ocean basins.

The oceanic lithosphere formed at a spreading center is initially hot, and cools as it moves away from the axis. Cooling is accompanied by thermal contraction and subsidence (Sclater et al., 1971). The age-versus-depth relationship is constant, regardless of spreading history and follows a time-dependent exponential cooling curve (McKenzie and Sclater, 1971), as does the overlying continental crust.

Lowstands of sea level would occur during episodes of slow spreading, during which relatively small volumes of hot oceanic lithosphere are being generated. Conversely, episodes of fast spreading would raise sea levels by increasing the ridge volumes. Using the data of Sclater et al. (1971), Pitman (1978) modeled volume changes in a hypothetical ridge, as shown in Fig. 9.4.

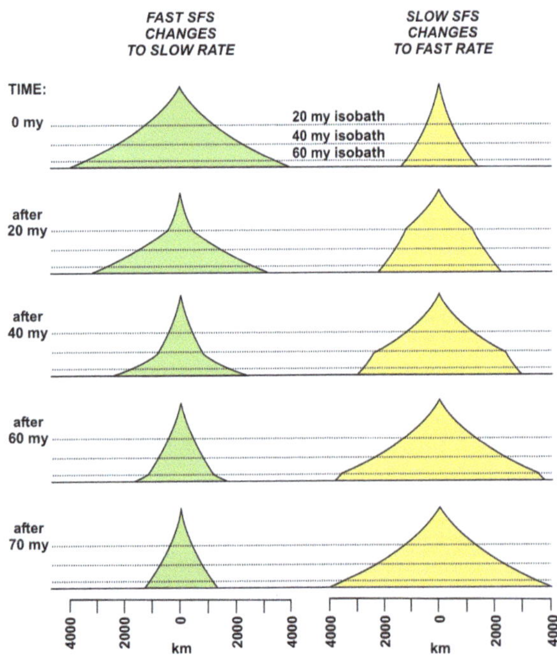


**Fig. 9.4** Profiles of spreading ridges showing the effect of different spreading rates on volume at 20, 40, 60, and 70 million years after initial condition. Right: Profile of a ridge that has been spreading at 20 m/ka for 70 million years and changes to 60 m/ka at time zero. After 70 million years, the ridge has three times its starting volume. Left: Ridge that has been spreading at 60 m/ka for 70 million years and changes to 20 m/ka at time zero. After 70 million years its volume has been reduced to one third (Pitman, 1978, 1979)

The elevation of any part of a ridge can be calculated by converting age to depth, using an appropriate spreading rate. Pitman (1978) showed that a ridge spreading at 60 m/ka will have three times the volume of one spreading at 20 m/ka, provided these rates last for 70 million years (Fig. 9.4). This is the time taken for the oldest (outermost) part of the ridge to subside to average oceanic abyssal depths of 5.5 km, by which time the ridge has achieved an equilibrium profile. The total length of the world midoceanic ridge system is about 45,000 km (Hays and Pitman, 1973; Pitman, 1978), and Pitman (1978) argued that, allowing for the shape of the continental margins, measured spreading rates can account for eustatic sea-level changes up to a maximum rate of 0.01 m/ka. Later compilations (e.g., Pitman and Golovchenko, 1991; Dewey and Pitman, 1998) demonstrated average rates of 0.002–0.003 m/ka for the Jurassic to mid-Tertiary. This is fast enough to generate Sloss-type cycles, those with episodicities of tens of millions of years.

A rise or fall in sea-level is not necessarily the same thing as a transgression or regression. As the rifted margins of a continent move away from a spreading center, they subside as a result of thermal contraction, crustal attenuation, and possibly, phase changes (Sleep, 1971; Watts and Ryan, 1976). Sediments deposited on the subsiding margin cause further isostatic subsidence. The stretched and loaded margin rotates downward around a hinge located at or near the inboard limit of stretched crust. Immediately after rifting and the appearance of oceanic crust, the margins may subside at rates in the order of 0.2 m/ka, decreasing after a few million years to about 0.03–0.07 m/ka and to less than 0.03 m/ka after about 20 million years (Watts and Ryan, 1976; McKenzie, 1978; Pitman, 1978). Average rates of tectonic subsidence on these trailing margins calculated by backstripping procedures are little more than 0.01 m/ka (Pitman and Golovchenko, 1983). Shelf-edge subsidence appears to be always slightly faster than the rate of long-term eustatic sea-level change caused by changes in ridge volumes, and therefore transgressions can actually occur during periods of falling eustatic sea level, if the rate of lowering is sufficiently slow. Conversely, regressions can occur locally during periods of rising sea level if there is an adequate sediment supply. Growth of large deltas (such as that of the Mississippi) following the Holocene postglacial sea-level rise is adequate demonstration of this. However, such local regressions are not relevant to the consideration of global stratigraphic cycles. Except on continental margins underlain by unusually old, rigid crust, which subside slowly, it is unlikely that sea-level changes caused by ridge-volume changes could lower sea-level to beyond the shelf-break (Pitman and Golovchenko, 1983). These tectonic processes that control relative sea level on a regional scale are discussed further in Chap. 10.

The spreading histories of the world oceans are now reasonably well understood (including the recent synthesis of the history of the Arctic regions by Lawver et al., 2002), based on deep-sea-drilling and magnetic-reversal data. Knowledge of worldwide spreading rates enabled Hays and Pitman (1973) to calculate ridge-volume changes and a sea-level curve for the last 110 Ma. A revised version of this curve was calculated for the period 85–15 Ma by Pitman (1978), based on refinements in oceanic data. It showed a gradual drop in sea level of 350 m at an average rate of 0.005 m/ka. Sea levels rose to a maximum during a period of fast spreading between 110 and 85 Ma (Larson and Pitman, 1972), and the subsequent drop reflects slower spreading rates. By calculating the relationship between spreading rates, subsidence rates, and falling sea-level, Pitman (1978) was able to model a major global transgression during the Eocene, as actually documented from stratigraphic evidence by Hallam (1963), and a second transgression during the Miocene. An early Oligocene regression is probably related to the onset of large-scale Antarctic glaciation (Fig. 4.10). Vail et al. (1977) used Pitman's curve to calibrate their chart of relative changes of sea level, in the belief that in this way they were adjusting the curve to show true eustatic sea-level change. They suggested that positive departures from Pitman's curve (where Vail's curve shows a higher sea level than Pitman) are the result of temporary increases in subsidence rates because of sediment loading. Negative departures were attributed to rapid sea-level falls driven by glacioeustasy, a process not factored into Pitman's curve (Vail et al., 1977, p. 92). It is of interest that Vail et al. (1977) mentioned tectonic subsidence as a factor in generating onlap and relative rises in sea level, because the general, indeed overriding importance of this process (discussed in Chap. 11) was been almost completely ignored in subsequent work by the Exxon group, until very recently.

Kominz (1984) reexamined the data on which Pitman's (1978, 1979) curve was based, carrying out her own calculations and incorporating new data, and showed that, because of an incomplete data base, a considerable error must still be allowed for in the development and use of a long-term sea-level curve. She used arbitrary, estimated spreading rates for the Tethyan Ocean because, of course, this ocean has now been completely subducted, whereas during the Mesozoic and early Tertiary it was one of the world's major oceans and its sea-floor spreading history would have had a considerable effect on the eustatic curve. Other errors include inaccurate dating of the sea floor, and missing data; for example the spreading history of the Arctic Ocean was unknown at the time of her synthesis (and is still incompletely understood). Kominz (1984) concluded that the range of error is $\sim$120 m at 80 Ma, decreasing to $\sim$10 m at present. This has important implications for the accuracy of backstripping calculations used to reconstruct basin subsidence histories. More recent work by Kominz and her colleagues on the sea-level record is discussed below and in Sect. 14.6.

A compilation of spreading centres associated with the breakup of Pangea is shown in Figs. 9.5, and 9.6 shows the age of initiation of rifting along each segment of these breakup fractures. The initiation of each ridge would have had an effect on the global average spreading rates, and on the age distribution of the oceanic crust, with consequences for eustatic fluctuations over time periods of tens of millions of years. However, as discussed in Chap. 10, the same events are now also thought to have had a profound effect on regional intraplate-stress regimes, with consequences for regional warping and tilting, and the generation of relative sea-level changes over large continental areas. Rifting and continental separation is also one of the major processes involved in the development of *tectonostratigraphic sequences* (Sect. 10.2).

Estimates of the rate and magnitude of eustatic sea-level changes that can be attributed to volume changes in sea-floor spreading centres were made by Pitman and Golovchenko (1991), based on their earlier work and that of Kominz (1984), and are given in Table 8.2. Dockal and Worsley (1991) examined the effects on the age distribution of the Earth's oceanic crust during the formation and breakup of supercontinents. As noted in the previous section, a two-ocean model, in which a Pacific-type ocean (Panthalassa) is replaced by an Atlantic-type ocean, can account for tens of metres of sea-level change over a time scale of hundreds of

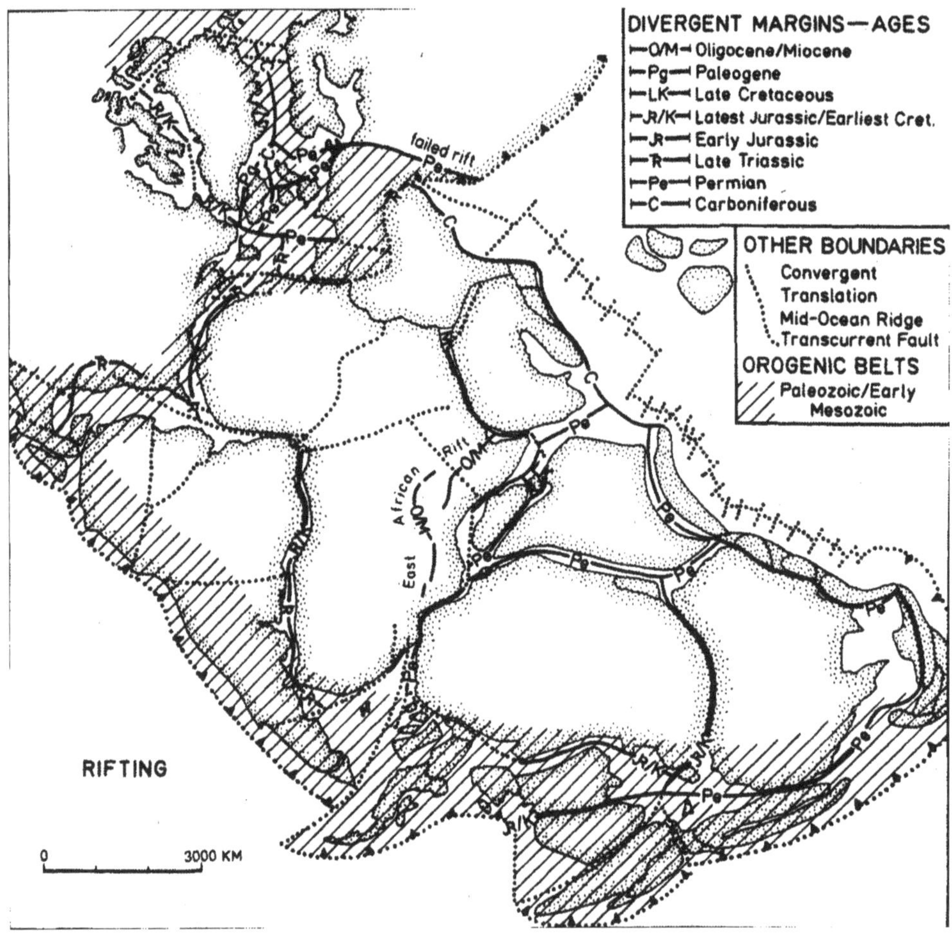


**Fig. 9.5** Paleogeography of Pangea during the earliest Triassic, showing the distribution and ages of initiation of rifting on what became divergent continental margins during the Mesozoic and Cenozoic (Uchupi and Emery, 1991)

millions of years. Preliminary modeling of more complex oceanic systems, such as a two-phase opening of the Atlantic Ocean, indicated that second-order effects would also occur with an amplitude up to about 10 m. The opening of other, smaller oceans (e.g., Labrador Sea, Red Sea) would have had similar, if smaller effects. Dewey and Pitman (1998) made other calculation, including the effects on the ocean-basin volumes of the large-scale continental compression and contraction that occurs during collision and suturing, such as that of India against Asia, and the effects of the filling of large-scale depressions in the Earth's crust, such as that represented by the Mediterranean basin, which desiccated and dried out completely during the Miocene. Their estimates of the effects on eustatic sea level are included in Table 8.2.

Other processes that could possibly affect eustatic sea levels on a time scale of tens to hundreds of millions of years were reviewed by Harrison (1990), who calculated the effects on sea level by relating them to changes in the total volume of the world ridge system. Continental collision increases the thickness of the crust and decreases its area, thus increasing the area and volume of the ocean basins. Major collision and shortening events, such as that between India and Asia, therefore result in a lowering of sea level. The generation and subsequent cooling and subsidence of large oceanic volcanic extrusive masses can be shown to have a modest effect on sea level. Sediment deposited in oceans has an isostatic loading effect which amplifies the subsidence due to crustal aging, but also displaces water. Changes in global average ocean

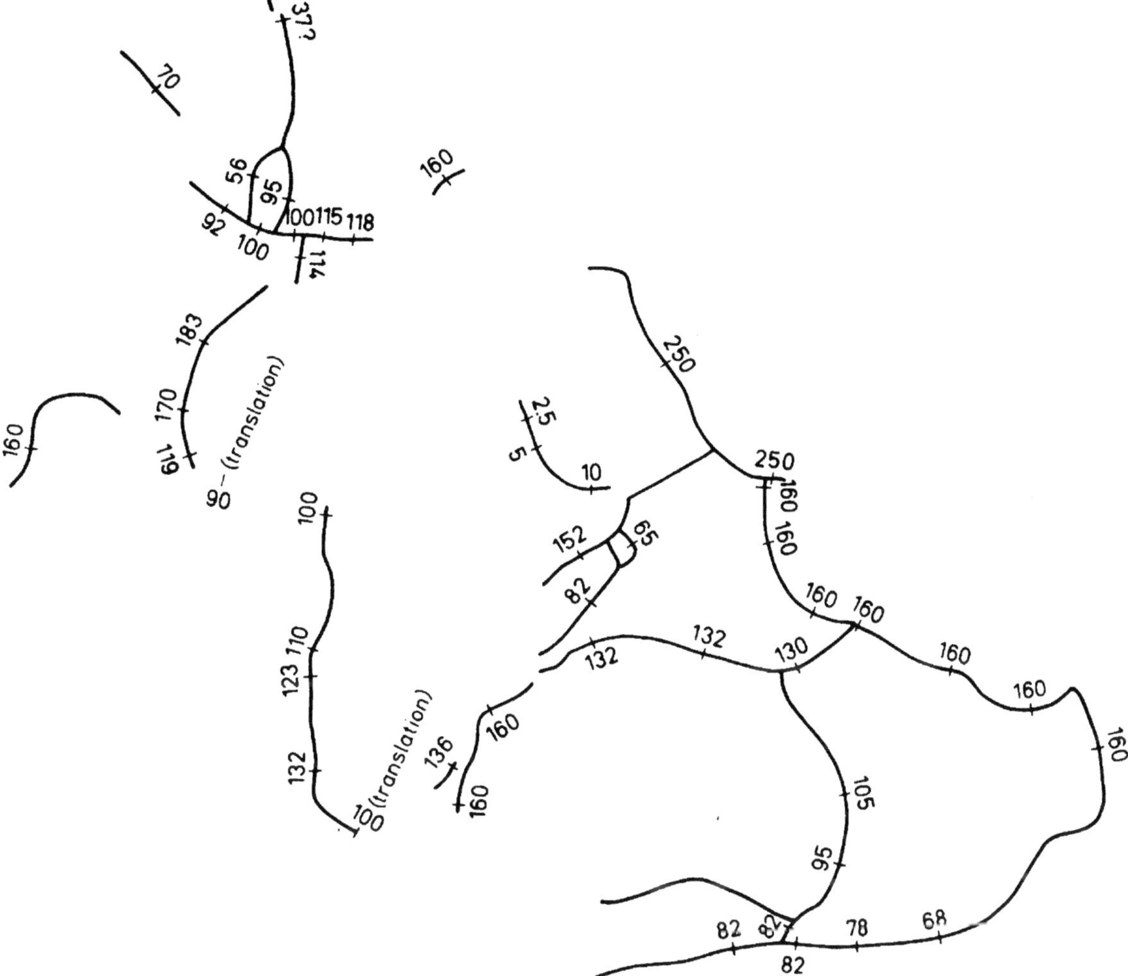


**Fig. 9.6** Divergent continental margins that developed within and around Pangea, showing the age of initiation of seafloor spreading (Uchupi and Emery, 1991)

temperature change the volume of the water through thermal expansion and contraction, without changing its load. Significant temperature changes have occurred as the Earth has cycled between icehouse and greenhouse states. Fairbridge (1961) estimated that a 1° rise in ocean temperatures would lead to a 2 m rise in sea level. Harrison (1990) and Dewey and Pitman (1998) calculated this effect at 10 m.

Modern work on long-term sea-level changes has been discussed by Dewey and Pitman (1998), Müller et al. (2008) and Kominz et al. (2008). As Dewey and Pitman (1998, p. 10) noted:

> ... the rise and fall of sea level can only be viewed rationally in the context of major tectonic episodicity involving the assembly, dismemberment and dispersal of the larger continental masses in "Pangea"-type supercontinent megacycles, where tectonics and sea level are closely interrelated with much buffering, feedback loops and enhancement. This further involves a close relationship between continental distribution, plate kinematics and state of stress, climate, geomorphology, sedimentary facies and basin stratigraphy.

Müller et al. (2008) presented a comprehensive reconstruction of the global age-area and depth-area distribution of ocean floor, including remnants of subducted crust, since the Early Cretaceous (140 Ma), to compute the effects of changes in crustal production, sediment thickness, and ocean-basin depth and area on sea-level fluctuations through time. They made use of recent data regarding the age distribution of oceanic crust, the possible effect of the eruptions associated with large igneous provinces (the effect of which they estimate could raise sea level by as much as 5 m), and the effects of the subduction of oceanic crust. The latter, by cooling the overlying mantle, could cause widespread subsidence of the overlying crust. This is the process known as dynamic topography, and is discussed in the next section.

A range of modern data sets, including those of Müller et al. (2008), was compiled by Kominz et al. (2008), the results of which are shown in Figs. 9.7 and 9.8. The curves shown in Fig. 9.7 were compiled using various methods. Kominz (1984) analyzed the effect of sea-floor spreading on ridge volumes and, thus, eustasy, and indicated the large range of potential error associated with the estimates, based on the knowledge available at that time concerning ocean-floor history,. Two of her plots are provided, the range of sea level, assuming minimum spreading rates (Bio) for the Cretaceous quiet period (green band), and the mean, best estimate result (b.e.) curve. Müller et al. (2008) updated these curves by analyzing ocean volume change including the effect of ridge volume, sediment volume, large igneous province emplacement, and the changing area of the oceans. Ice is not included in their results, which are plotted assuming an ice free world (54 m above today's sea level). Also included in these figures are some older estimates. Bond (1979) studied global continental flooding history and estimated a range of possible sea level for broad periods of time. Sahagian et al. (1996) carried out a detailed analysis of Cretaceous and older sea level on the stable Russian Platform. Watts and Thorne (1984) combined backstripping and forward modelling of the Atlantic coastal stratigraphy to derive their sea level estimates. The Haq et al. (1987) long-term and short-term curves are included, but lie somewhat to the right of the other curves because their estimates for sea-level fluctuations were substantially larger that those of most other workers, despite being calibrated against the best-estimate values of Kominz (1984)— note the coincidence of these two curves at 230 m at

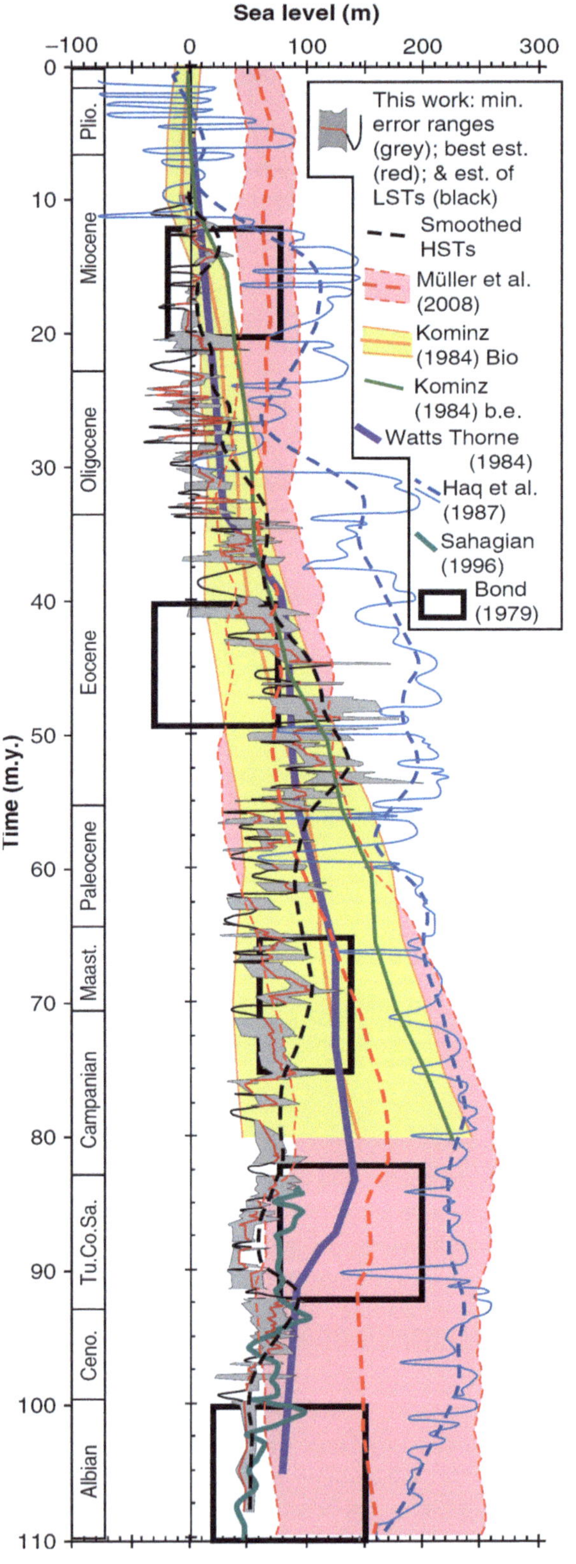



**Fig. 9.7** Comparisons of various estimates of eustatic sea-level change. Plio., Pliocene; Ceno., Cenomanian; Tu.Co.Sa, combined Turonian, Coniacian and Santonian; Maast., Maastrichtian (Kominz et al., 2008, Fig. 12)

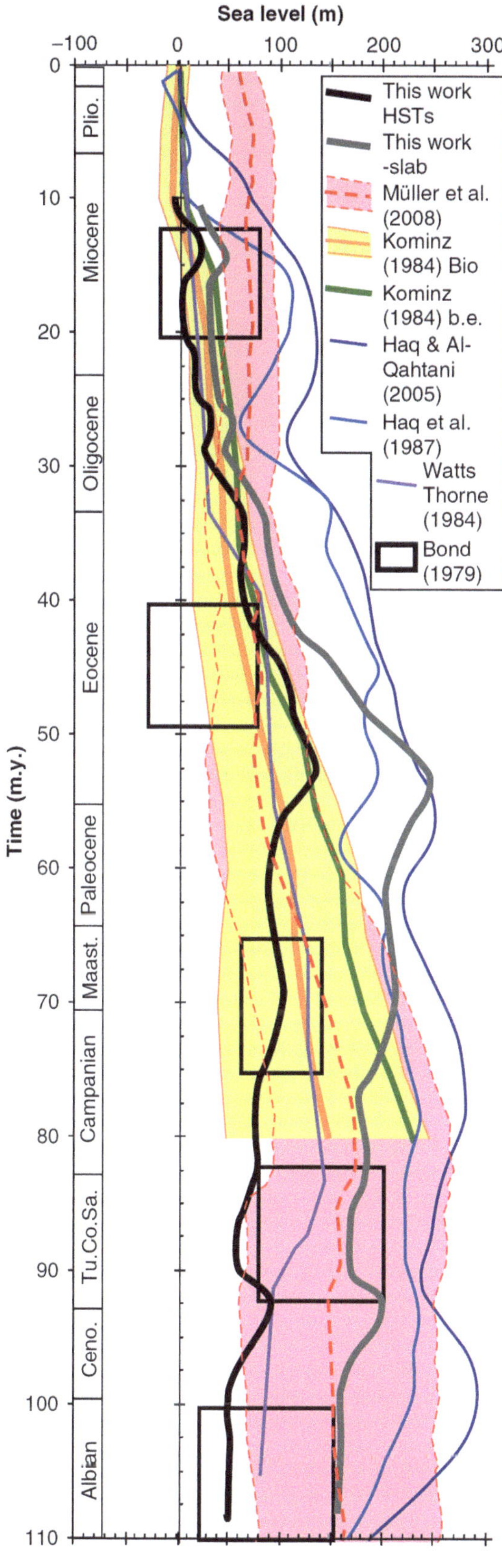



**Fig. 9.8** Comparisons of curves of long-term sea-level change. Ceno., Cenomanian; Tu.Co.Sa, combined Turonian, Coniacian and Santonian; Maast., Maastrichtian; Plio., Pliocene (Kominz et al., 2008, Fig. 13)

80 Ma (Kominz, 1984, b.e. curve). The Kominz et al. (2008) curve ("This work" in Fig. 9.7) is based on backstripping analysis of the subsurface stratigraphy of New Jersey, a study examined at some length in Sect. 14.6.1

Kominz et al. (2008) developed a smoothed version of their sea level curve by connecting highstands, and this is shown in Fig. 9.8, against other smoothed, long-term curves. At 80 Ma the discrepancy between the Müller et al. (2008) curve and that of Miller et al. (2005a) against which it was compared, implies a difference in sea level of 130 m. Müller et al. (2008) suggested that during the Late Cretaceous the dynamic load attributed to the subduction of the Farallon Plate (a component of the Panthalassa oceanic crust) beneath North America would have led to regional subsidence of the Atlantic margin of North America. Sea-level estimates based on stratigraphic studies in New Jersey would, therefore, underestimate global sea levels by that amount. Kominz et al. (2008) developed a correction to their new Jersey curve, which is shown in Fig. 9.8 as the curve labelled "This work—slab". The differences between the two Kominz et al. (2008) curves in Fig. 9.8 decrease from Eocene time on, as the dynamic topography effect of the downgoing Farallon Plate is thought to have decreased.

## 9.3.2 Dynamic Topography and Epeirogeny

Johnson (1971) was one of the first to emphasize the links between Sloss-type cycles of transgression-regression and regional orogeny. Burgess (2008) provided an updated evaluation of this relationship (Fig. 9.9). Johnson (1971) recognized that

> of the four major orogenies that occurred in North America during the Paleozoic and Mesozoic, three began, reached climactic stages, and went through waning stages during the time epicontinental seas were transgressing to their maximum extent and then regressing to form the great onlap-offlap cycles called sequences. ... The correspondence of orogenic events with onlap of the craton

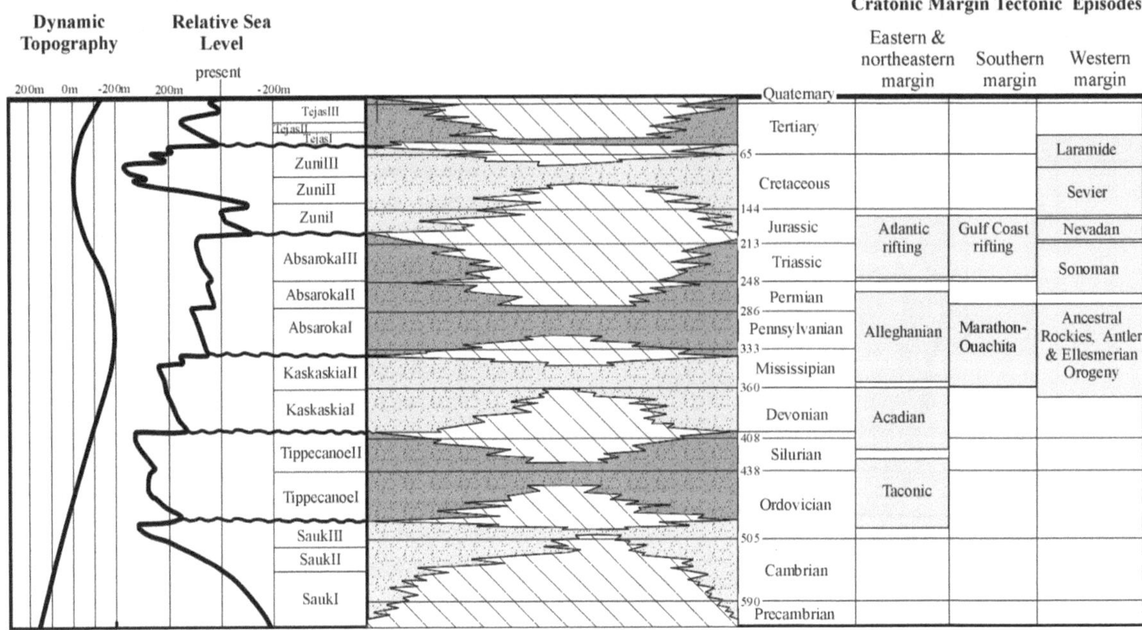



**Fig. 9.9** Correlation of the Sloss sequences with craton-margin tectonic episodes (Burgess, 2008)

is so consistent in general and even in detail, that it must reflect a fundamental relation.

In a series of papers Sloss (1972, 1979, 1982, 1984, 1988b; Sloss and Speed, 1974) elaborated the "Indian-name" sequences that he first established in 1963. He developed ever-more detailed isopach maps and discussed the effects of eustasy and tectonism in the development of the sequences. In one of the earlier of these papers (Sloss and Speed, 1974; see also Sloss, 1984) he attempted to subdivided the six sequences into two broad types. The first type, termed "submergent", and exemplified by the Sauk, Tippecanoe and Kaskaskia sequences,

> ... is dominated by flexure of the cratonic and interior margins. These sequences exhibit slow regional transgression and onlap, ... gentle subsidence of interior basins separated by less subsident domes and arches, and a lack of widespread brittle deformation manifested by faulting. The emergent episodes preceding each of these flexure-dominated times of deposition are the occasions for developing of the sequence-bounding unconformities and slow progressive transgression of the craton. (Sloss, 1984, p. 5)

Figure 5.15 illustrates the typical anatomy of the Sauk sequence. The second type of sequence Sloss termed "oscillatory", and

> ... is characterized by abrupt termination of a submergent episode through rapid cratonic uplift accompanied by high-angle faulting of basement crystalline rocks and, commonly, by faults that propagate from the basement to fracture and displace the overlying sedimentary cover. The Absaroka Sequence represents a typical oscillatory episode. (Sloss, 1984, p. 5)

Examples of these two types of sequence are illustrated by the contrasting subsidence maps of Figs. 5.9 and 5.10. The importance of broad, regional tilting of the craton in the formation of these sequences is apparent from regional cross-sections, such as that shown in Fig. 5.8.

In his later papers, Sloss considered epeirogeny and in-plane intraplate stresses and their effects on sequence architecture. For example, the late Mississippian sub-Absaroka unconformity is a time of pronounced change in cratonic subsidence and uplift patterns within the North American interior. The Canadian Shield emerged as an important source of sediment, and the southwestern Midcontinent (Texas, Oklahoma, New Mexico, Colorado) underwent pronounced submergence. Sloss (1988b) attributed this major change in continental configuration to intraplate stresses associated with plate convergence along the southern margin of the continent. There is now much additional support for these ideas, which are discussed at length here and in the next chapter.

Stratigraphers specializing in the study of continental interiors (e.g., Sloss and Speed, 1974) have for a long time appealed to a process that was termed *epeirogeny* by Gilbert (1890). The modern definition of epeirogeny (Bates and Jackson, 1987) defines it as "a form of diastrophism that has produced the larger features of the continents and oceans, for example plateaux and basins, in contrast to the more localized process of orogeny, which has produced mountain chains." The definition goes on to emphasize vertical motions of the earth's crust.

Modern studies of the thermal evolution of the mantle, supported by numerical modeling experiments, have provided a mechanism that explains the long term uplift, subsidence and tilting of continental areas, especially large cratonic interiors beyond the reach of the flexural effects of plate-margin extension or loading (Gurnis, 1988, 1990, 1992; Burgess and Gurnis, 1995; Burgess, 2008). These studies have shown that the earth's surface is maintained in the condition known as dynamic topography, reflecting the expansion and contraction of the crust resulting from thermal changes in the underlying lithosphere and mantle (Fig. 9.1). Much work remains to be done to test and apply these ideas by developing detailed numerical models of specific basinal stratigraphic histories, although it is already clear that dynamic topography is affected by both upwelling and downwelling currents on several scales. The following paragraphs describe a range of recent studies.

Bond (1978) provided some of the first important insights into epeirogenic processes by demonstrating that the earth's continents have had different histories of uplift and subsidence since Cretaceous time (Sect. 3.4). It is now possible to explain these differences using the concepts of dynamic topography. One of the most striking anomalies revealed by the hypsometric work is the elevation of Australia. "The interior of Australia became flooded by nearly 50% between 125 and 115 Ma and then became progressively exposed between 100 and 70 Ma at a time when nearly all other continents reached their maximum Cretaceous flooding" (Gurnis, 1992). Applying backstripping techniques to detailed isopach maps Russell and Gurnis (1994) estimated that although global sea level was about 180 m above the present level near the end of the Cretaceous, a smaller fraction of the Australian

continent was flooded than at the present day. Raising the continent an average of 235 m accounts for the end-Cretaceous paleogeography superimposed on a 180-m-high sea level. This result is related to cessation of subduction on the northeast margin of Australia at about 95 Ma. Subduction had generated a dynamic load by the presence of a cold crustal slab at depth, and the cessation of subduction allowed uplift. Then the northward migration of the continent as it split from Antarctica moved Australia off a dynamic topographic high and geoid low, toward a lower dynamic topography, resulting on continental lowering. Russell and Gurnis (1994) also discussed the broad tilting and basin development of the Australian interior. Burgess and Gurnis (1995) developed preliminary models for Sloss-type cratonic sequences invoking combinations of eustatic sea-level change and dynamic topography.

The thermal consequences of secondary mantle convection above a subducting slab have been invoked as a cause of enhanced subsidence in retroarc foreland basins. Cross (1986) showed how the changing pattern of regional isopachs in the Western Interior foreland basin of the Rocky Mountain region could be explained as a product of the generation of accommodation by crustal flexure caused by the imposition of the supracrustal load of the fold-thrust belt, or a result of regional subsidence over a cool, dense mantle. Mitrovica et al. (1989) developed this idea as an explanation for the anomalously broad extent of the Western Canada Sedimentary Basin during the Cretaceous, and a similar idea has been proposed for the Mesozoic basins of eastern Australia (Gallagher et al., 1994). Although this is clearly a plate-margin effect, the same concepts of dynamic topography apply as in the case of the broader epeirogenic processes discussed above.

Heat energy is mostly lost from the mantle via convection and consequent volcanism at oceanic spreading centers. Formation of supercontinents that persisted for tens of millions of years, covering a wide region of mantle, prevent this method of heat loss, trapping mantle heat produced by radioactive decay in the core. Such mantle insulation may produce a rise in temperature of ~20 K throughout the mantle (Gurnis and Torsvik, 1994). Thermal expansion due to this temperature increase produces stress on the base of the lithosphere, creating dynamic topography (Anderson, 1982) with an amplitude of ~150 m [Fig. 9.1]. As a consequence of this mechanism, continents should experience uplift during supercontinent formation and persistence, followed by subsidence as the supercontinent breaks up and the fragments drift off hot mantle onto adjacent, relatively cool mantle (Anderson, 1982; Gurnis, 1988). Similar effects can be produced by large descending plumes interacting with internal mantle viscosity boundaries (Pysklywec and Mitrovica, 1998) (quote from Burgess, 2008, p. 40).

The major change in cratonic subsidence patterns that occurred during the late Paleozoic (pre-Absaroka), as noted above, occurred at the time the North American continent was impacted by orogenic episodes on three of its four sides, the Alleghanian in the east, the Ouachita-Marathon in the south, and the Ellesmerian in the north (Fig. 9.5). These were three of the important orogenies that culminated in the construction of Pangea.

Burgess (2008, p. 40) pointed out that the two longest-duration lacunae in North America are the base-Sauk and the base-Zuni unconformities. Both formed during periods when North America was part of a supercontinent, Rodinia in the Late Precambrian, and Pangea in the late Paleozoic and early Mesozoic. Cratonic erosion and non-deposition would have been accentuated by dynamic topographic highs creating an emergent craton (Sloss and Speed, 1974). Conversely, the three Paleozoic unconformities are of shorter duration, and formed during a period when North America was one of several dispersed continents, overlying cool mantle, and thus having relatively low or "submergent" (Sloss and Speed, 1974) elevation. Subsidence analysis identifies an anomalously large subsidence event in Late Devonian to Mississippian time, explained by Kominz and Bond (1991) as due to the final stages of assembly of Pangea over a dynamic topographic low (Fig. 9.3).

The mechanism of subduction-induced mantle flow leading to widespread subsidence of foreland basins has been applied to the Carboniferous-Triassic subsidence of the Karoo Basin in southern Africa (Pysklywec and Mitrovica, 1998) and to the Silurian foreland basin adjacent to the Caledonian orogen, now underlying the Baltic Sea (Daradich et al., 2002).

Another importance aspect of dynamic topography is the large-scale continental tilts that can result from plate-margin collisions. Burgess (2008, p. 40) pointed out the absence of Mesozoic cratonic strata in eastern North America, compared with extensive Jurassic and Cretaceous deposition in the west. This may be in part due to slab-related dynamic topography and in part to continent-scale tilting up-to-the-east. Tilting would have resulted from increasing amounts of uplift

eastwards across North America towards the hottest mantle situated beneath the center of Pangea.

On the basis of coal moisture measurements and other measures of organic metamorphism, it is known that massive unroofing of the Appalachian and Ouachitan orogens occurred during the late Paleozoic-early Mesozoic. Removal of 4 km of sediment occurred within the undeformed Appalachian foreland basin; there was 7–13 km of cumulative erosion in the central Appalachians, and more than 13 km in the Ouachitas (Beaumont et al., 1988). Bally (1989, p. 430) said: "These amounts of erosion are stunning, and raise the question of the fate of the eroded material and how much of the unloading was due to as yet unrecognised tectonic unroofing." Dickinson (1988) had suggested that much of the thick accumulations of upper Paleozoic and Mesozoic fluvial and eolian strata in the southwestern United States had been derived from Appalachian sources, and this was confirmed by the zircon studies of Dickinson and Gehrels (2003). Neither the location nor scale of specific river systems could be identified by this research, but given the distance of transport and the volume of sediment displaying these provenance characteristics that is now located on the western continental margin, it seems highly probable that large-scale river systems were involved. A westerly tilt of the craton during the late Paleozoic and early Mesozoic could, in part, be attributed to continental-scale thermal doming preceding the rifting of Pangea, followed by the more localized heating and uplift of the rift shoulder of the newly formed Atlantic rift system. Ettensohn (2008) suggested that some of the detritus was also shed eastward, resulting in the accumulation of more than 5 km of sediment in southern New England in Early Jurassic time. The major uplift and erosional unroofing took place during the Late Permian to Early Triassic, and by Late Triassic time "parts of the old orogen in New England had been exhumed to nearly the present erosional level" (Ettensohn, 2008, p. 155; citing isotopic studies of Dallmeyer, 1989).

The extended duration of the sub-Tejas unconformity in eastern North America, and the current elevated topography of North America and the resultant predominantly erosional regime suggest that the dynamic topographic high is somehow persistent or that elevation is being maintained by some other mechanism.

The concepts of dynamic topography have also been invoked to explain cratonic basin formation and subsidence. As noted by Hartley and Allen (1994), there have been at least two major periods in earth history when suites of interior basins formed within large cratons. Both periods are associated with the breakup of supercontinents. The first of these was the early Paleozoic, when the Williston, Hudson Bay, Illinois, Michigan and other basins formed in the cratonic interior of North America. The second period was the Mesozoic breakup of Pangea, when a series of similar basins formed within the continent of Africa. Some of the African basins are undoubtedly related to plate-margin processes, and there has been much debate regarding the importance of reactivation of inherited crustal weaknesses as a cause of the North American basins (Quinlan, 1987). However, Hartley and Allen (1994) suggested that small-scale convective downwelling, decoupled from the large-scale motion, may be a significant factor in basin formation. They found strong evidence for this process in the formation of the Congo Basin. The stratigraphic histories of these suites of cratonic basins are similar but not identical (Quinlan, 1987), and as with the other examples discussed in this section, the processes that maintain dynamic topography are not thought to generate globally simultaneous (eustatic) changes in sea level.

Other examples of the large-scale flux of sediment across the interior of the North American continent, and their possible relationship to dynamic topography, were discussed by Miall (2008).

### 9.3.3 *The Origin of Sloss Sequences*

To conclude the discussion in this section: modern ideas about the $10^7$-year sequences (the "Sloss sequences": Fig. 5.5) are that they were generated by a combination of two main processes, eustasy and epeirogeny. Cycles of eustatic sea-level change are caused by variations in sea-floor spreading rate superimposed on the supercontinent cycle. These variations are a reflection of continental-scale adjustments to spreading patterns in response to plate rifting and collision events. They occur over time scales in the tens-of-millions of years range, and this can be confirmed by the detailed documentation of spreading histories from the magnetic striping of existing ocean floors, which reveal changes in spreading rate and trajectory over time scales of this magnitude.

Superimposed on the eustatic cycle are continental-scale episodes of warping and tilting driven by mantle thermal processes—the so-called dynamic topography process (Fig. 9.1). Complicating and overlaying both these processes are the regional mechanisms of crustal extension and crustal loading (Sect. 10.2) caused by plate-margin tectonism. In many places, either because the global processes are dominant, or because they are synchronous with regional tectonic episodicity, Sloss's sequences can easily be recognized within the overall stratigraphy. For example, a regional section through western Canada (Fig. 5.4) shows a clear subdivision into Sloss sequences and subsequences (capital letters up the left side of the diagram). Elsewhere, as in the Canadian Arctic Islands, the regional unconformities do not coincide with those defined by Sloss (Fig. 5.7), presumably because they were over-printed by the effects of regional tectonic episodes.

## 9.4 Main Conclusions

1. A long-term cycle of sea-level change ($10^8$-year time scale), termed the supercontinent cycle, results from the assembly and breakup of supercontinents on the earth's surface. Eustatic sea-level changes are driven by global changes in sea-floor spreading rate, variations in the average age of the oceanic crust, and variations in continental volumes caused by plate extension and collision.
2. Eustatic sea-level changes on a time scale of tens of millions of years are caused by variations in ocean-basin volume generated by episodic spreading, and by the variations in total length and age of the sea-floor spreading centres as supercontinents assemble, disassemble and disperse.
3. Many other processes have smaller effects on global sea levels through their effects on the volume of the ocean basins. These include oceanic volcanism, sedimentation, ocean temperature changes, and the desiccation of small ocean basins.
4. Epeirogenic effects are "dynamic topography" resulting from thermal effects of mantle convection associated with the supercontinent cycle. This

cycle has long-term consequences, including the generation of persistent geoid anomalies. Vertical continental movements are related to the thermal properties of large- and small-scale convection cells, and can involve continent-wide uplift, subsidence, and tilts, and cratonic basin formation. These movements do not correlate in sign or magnitude from continent to continent.

5. Comparisons of various curves of long-term sea-level change constructed using modern data sets reveal few similarities between curves. It is clear that sea-levels were in the range of at least 200 m higher during the Late Cretaceous than at present, and have slowly fallen since then, but the details of sea-level history even on a $10^{6-7}$-year time scale, remain unresolved.

# Chapter 10

# Tectonic Mechanisms

## Contents



## 10.1 Introduction

Sea-level changes within a basin may be caused by movements of the basement rather than changes in global sea level (eustasy). In Chap. 9 we examine long-term mechanisms of basement movement driven by deep-seated thermal processes, including the generation of dynamic topography, and we also discuss eustatic sea-level changes driven by changes in ocean-basin volume. These processes act over time periods of tens to hundreds of millions of years and are continental to global in scope. In this chapter, we discuss relative sea-level changes caused by continental tectonism that are regional to local in scope, and act over time periods of tens of millions to tens of thousands of years (possibly less). Tectonism of this type is driven primarily by crustal responses to plate-tectonic processes. Extensional and contractional movements accompanying the relative motions of plates cause crustal thinning and thickening and changes in regional thermal regimes, and this leads to regional uplift and subsidence. These stresses and strains are generated primarily at plate margins but, because of the rigidity of plates, they may be transmitted into plate interiors and affect entire continents.

By their very nature, these mechanisms of sea-level change are regional, possibly even continental in extent, but cannot be global, because they are driven by processes occurring within or beneath a single plate or by interaction between two plates. They affect the elevation of the plate itself, rather than the volume of the ocean basins or the water within them. This category of sea-level change is therefore not eustatic. However, as described in this chapter, they can simulate eustatic effects through the full range of geological episodicities over wide areas, and it is now thought by many researchers that tectonic mechanisms were responsible for many of the events that were used to define the global cycle charts of Vail et al. (1977), Haq et al. (1987, 1988a) and Graciansky et al. (1998). An important additional point is that because Earth is finite, regional plate-tectonic events, such as ridge reordering or adjustments in rotation vectors, may result in simultaneous kinematic changes elsewhere. It is possible,


A.D. Miall, *The Geology of Stratigraphic Sequences,* 2nd ed.,
DOI 10.1007/978-3-642-05027-5_10, © Springer-Verlag Berlin Heidelberg 2010

therefore, for tectonic episodes to be hemispheric or even global in scope. However, such episodes would take a different form (uplift, subsidence, extension, contraction, tilting, translation) within different plates and even within different regions of a given plate. Events of relative sea-level change could occur simultaneously over wide areas but would vary in magnitude and direction from one location to another. Possible examples identified by Embry (1993), Hiroki (1994) and Nielsen et al. (2007) are described in Sect. 10.4.3.

The Exxon school holds that tectonism is not responsible for the generation of sequence-bounding unconformities. As recently as 1991 Vail et al. (1991, p. 619) stated:

> ... even in ... tectonically active areas, the ages of sequence boundaries when dated at the minimum hiatus at their correlative unconformities match the age of the global eustatic falls and not the plate tectonic event causing the tectonism. Therefore, tectonism may enhance or subdue sequence and systems tract boundaries, but does not create them.

Vail and his coworkers placed primary emphasis on dating of sequence boundaries and their global correlatability, a methodological approach that has guided all their research. Problems of chronostratigraphic dating which bring this approach into serious question are discussed in Part IV. More important, Vail's statement is technically incorrect, as discussed below.

Embry (1990, p. 497) proposed a set of guidelines for identifying tectonism as a major control of relative sea-level change. His proposal was based on his examination of the Mesozoic stratigraphic record of Sverdrup Basin, Arctic Canada, a succession up to 9 km thick, which he subdivided into 30 stratigraphic sequences in the million-year frequency range. He suggested, however, that the guidelines have general applicability. They are as follows:

> (1) the sediment source area often varies greatly from one sequence to the next; (2) the sedimentary regime of the basin commonly changed drastically and abruptly across a sequence boundary; (3) faults terminate at sequence boundaries; (4) significant changes in subsidence and uplift patterns within the basin occurred across sequence boundaries; and, (5) there were significant differences in the magnitude and the extent of some of the subaerial unconformities recognized on the slowly subsiding margins of the Sverdrup Basin and time equivalent ones recognized by Vail et al. (1977, 1984) in areas of high subsidence.

To these points could be added: (6) sequence architecture can be genetically related to the developments of structures within a basin; (7) truncation of entire sequences beneath sequence boundaries indicates tectonic influence (e.g., Yoshida et al., 1996), and (8) a sequence architecture consisting of thick clastic wedges cannot be generated by passive sea-level changes (Galloway, 1989b).

Some types of widespread unconformity are now widely acknowledged as having tectonic origins, such as the *breakup unconformity* (Falvey, 1974), generated at the transition from rift to drift immediately prior to commencement of the flexural subsidence phase of an extensional continental margin (Sect. 10.2) and the *forebulge unconformity* (Beaumont, 1981; Sinclair et al., 1991) developed over the forebulge of a foreland basin (Sect. 10.3.2). Blair and Bilodeau (1988) used the term *tectonic cyclothem* for wedges of coarse, clastic sediment formed as a result of basin-margin tectonic activity. Howell and van der Pluijm (1999, p. 975) described what they termed *structural sequences,* which are "stratal sequences ... defined by significant changes in basin-subsidence patterns." The subdivided the Upper Cambrian to Lower Carboniferous fill of Michigan Basin into seven $10^7$-year structural sequences based on isopach patterns, indicating basin-centred subsidence, tilting, etc. These sequences do not necessarily coincide with the Sloss sequences, and do not necessarily have unconformable bounding surfaces. They reflect changes in the regional tectonic stress pattern. Ettensohn (1994, 2008) categorized large-scale stratigraphic packages of the Appalachian basin into *tectophases,* which were controlled by the basinal response to collisional orogeny and tectonic episodicity on a $10^7$–$10^8$-year time scale along the Appalachian orogen. Many other workers have used such terms as *tectonostratigraphic sequences* for sequences that can be genetically related to tectonic events or episodes.

One of the key areas in the construction of the original Vail curves was the North Sea Basin, and outcrop sections in areas adjacent to the North Sea were used extensively in the revisions of the curves published by Haq et al. (1987, 1988a). It is, therefore, of considerable significance that the Mesozoic-Cenozoic geology of the region has now been shown to have been strongly affected by almost continuous extensional tectonism throughout this period (Hardman and Brooks,

1990), the effects of which appear not to have been accounted for in the Exxon work. Vail and Todd (1981, pp. 216–217) stated:

> Tilting of beds as a result of fault-block rotation and differential subsidence occurred almost continuously throughout the Jurassic. Unconformity recognition is enhanced by periodic truncation of these tilted strata by lowstand erosion and/or onlap which developed during the subsequent rise. We find no evidence that the regional unconformities are caused by tectonic events within the northern North Sea basins, only that unconformity recognition is enhanced by the truncation and onlap patterns created by tectonic activity.

Most workers today would not agree with this view. Regional seismic data from this area, and its implications for sequence stratigraphy, are discussed in Sect. 10.2.2.

It is simplicity itself to demonstrate that tectonism may be an important factor in the timing of sequence boundaries. A modification of one of the standard Exxon diagrams can be used to illustrate this point (Fig. 10.1). Figure 2.2 shows how rates of subsidence and eustatic sea-level change may be integrated to develop a curve of relative changes of sea level. Sequence boundaries correspond to lowstands of relative sea level. The graphs comprising Fig. 10.1 have been constructed by integrating curves for extensional subsidence with curves of eustatic sea-level change. It is easy to see from these figures that changing the slope or position of the subsidence curve will change the position of lows and highs on the accommodation curve.

Catuneanu et al. (1998) and Allen and Allen (2005, Sect. 8.1) provided graphical and numerical solutions to the issue of integrating rates of subsidence, sea-level change and sediment supply. For example, Fig. 10.1b shows how the timing of sea-level highs and lows is affected when a curve of sea-level change is offset by varying subsidence rates. Allen and Allen (2005, p. 271) stated, with reference to Fig. 10.1b:

> Clearly when correlating different locations in a basin with spatially variable tectonic subsidence rate, we should expect significant diachroneity of the onset of erosion due to relative sea-level fall and of flooding during relative sea-level rise. The delay in the onset of erosion is a quarter of a eustatic wavelength or $\lambda/4$. The difference in the timing of flooding is also $\lambda/4$. Since diachroncity of key stratigraphic surfaces is to be expected from a globally synchronous eustatic change, it is curious that the inference of stratigraphic

synchroneity is seen as an acid test for a eustatic control.

Given that tectonism is not necessarily the slow, steady, background effect described in the Exxon work, but varies in rate and duration markedly from location to location, within and among basins, it seems unlikely that relative sea-level lows and highs would ever tend to occur at the same times in adjacent basins, let alone on different continents. In fact, it can be stated that if rates of eustatic sea-level change and tectonic subsidence are comparable and of the same order of magnitude *there can be no synchronous record of eustasy in the preserved stratigraphic record.* For example, it can be seen from Fig. 10.1a that if an Atlantic-type margin undergoes progressive rifting along strike ("unzipping"), $10^7$-year eustasy will not generate synchronous sea-level highs and lows. Parkinson and Summerhayes (1985) made a similar point.

The rates and magnitudes of various tectonic processes leading to uplift and subsidence are listed in Table 8.2 and are discussed in this chapter. Long-term and short-term rates, and magnitudes of total change, are in fact comparable to the rates and magnitudes of long-term and short-term eustatic sea-level change, which are also shown in Table 8.2. Therefore, tectonism cannot be relegated to a background effect of secondary importance. Eustasy and tectonism may be considered as independent curves of uplift and subsidence that go in and out of phase in a largely unpredictable manner. As discussed by Fortuin and de Smet (1991), this means that a careful tectonic analysis must be performed in an area of interest, including the calculation of subsidence histories by the methods of geohistory analysis or backstripping, before any conclusions can be drawn regarding the importance of eustasy. Vail et al. (1991) also referred to the need for detailed "tectono-stratigraphic analysis" and went to considerable lengths to describe how this should be done; but they did not take the next step, which is to show how the results should be tested against and integrated with the predictions made from the eustatic model of sequence stratigraphy.

Space considerations prohibit a complete review of the dependence of accommodation changes on tectonism. In this chapter I focus on extensional margins (rift basins and Atlantic-type continental margins), arc related basins, and foreland basins. Transform margins are not discussed.

**Fig. 10.1** Illustration of the shifts in peaks and troughs of accommodation space induced by changes in the rate of subsidence. (**a**) A simple asymptotic subsidence curve, with rates of subsidence derived from the values given in Table 8.2, is shown in two positions, T1 and T2, shifted 20 million years relative to each other. A *curve* simulating irregular eustasy with a $10^7$-year frequency is shown at *centre*, and at *top* the results of integrating this curve with the subsidence curve are shown. Comparison of the two accommodation curves A1 and A2 shows that if extensional subsidence is delayed by 20 million years but then follows the same pattern, peaks in the accommodation curve occur several million years later, whereas troughs tend to occur somewhat earlier (Miall, 1997); (**b**) Variations in the timing of relative sea-level highs and lows determined by integrating a curve of sea-level change with varying rates of tectonic subsidence. $\gamma$, $h_0$ = frequency and amplitude of the sea-level cycle, $\psi$ is a dimensionless parameter which varies from 0.2 to $2\pi$. Linear tectonic subsidence rates are indicated at the right. As the subsidence rate is increased, the duration of episodes of sea-level fall decreases, until it reaches a minimum at a subsidence rate of 3.142 mm/year (Allen and Allen, 2005, Fig. 8.2)

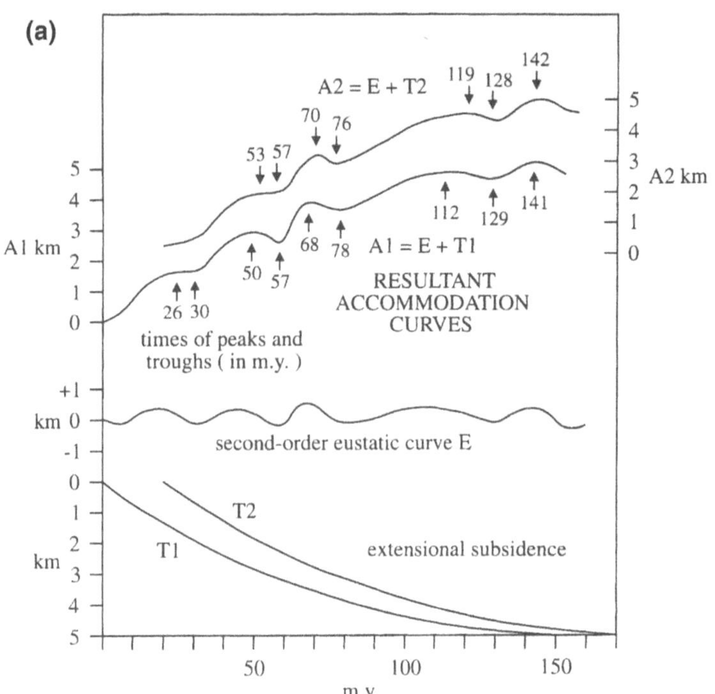


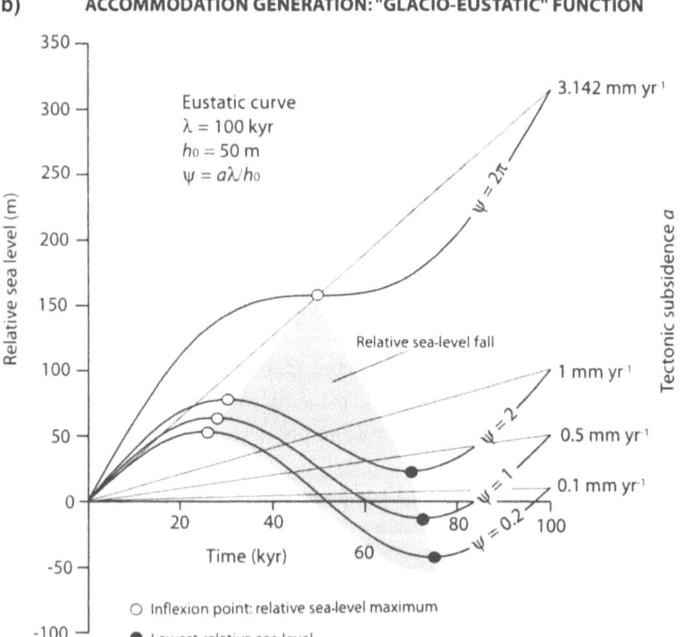

## 10.2 Rifting and Thermal Evolution of Divergent Plate Margins

### 10.2.1 Basic Geophysical Models and Their Implications for Sea-Level Change

The evolution of extensional continental margins is shown diagrammatically in Figs. 10.2 and 10.3. There is an initial rapid phase of continental stretching, which may be completed within a few million years, in the simplest case, although some basins, such as the North Sea, and the North Atlantic margins, undergo protracted or repeated rifting events over tens of million of years. This is the rift phase shown in Fig. 10.2. Deformation typically is brittle at the surface, taking the form of extensional faults. These may consist of repeated graben or half-graben faults, and may be listric, allowing for considerable extension of the crust. Beneath the brittle crust the lithosphere is plastic and extends by stretching or by the development of crustal

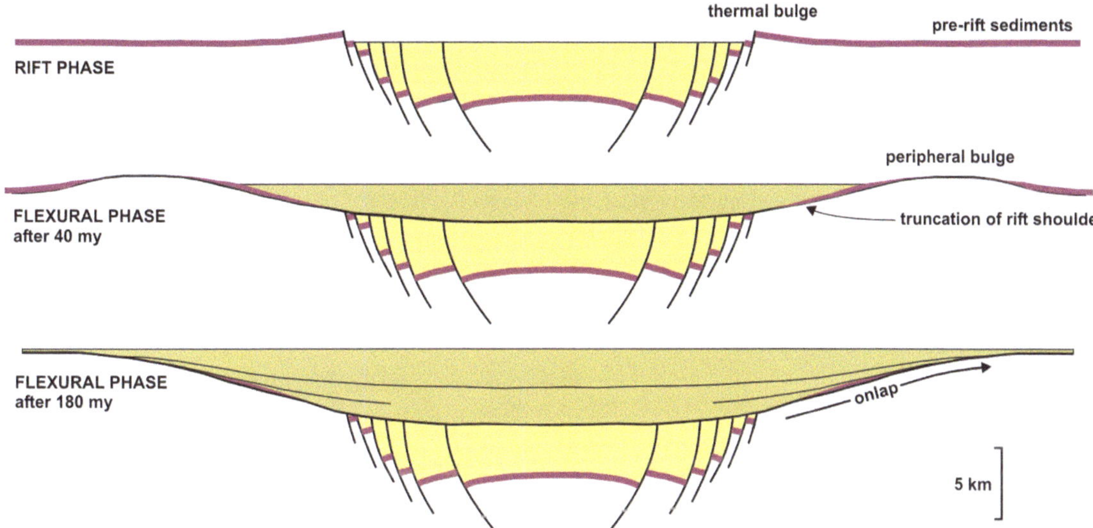


**Fig. 10.2** The *steer's head* or *Texas longhorn* model of the geometry of rift basins that develop during the stretching and dismemberment of a continental plate. This model shows the situation (exemplified by the North Sea) in which sea-floor spreading ceases during the flexural subsidence phase. An initial rift phase is generated by rapid crustal stretching, resulting in listric normal faulting and ductile flow of the lower lithosphere, with Airy-type isostatic subsidence. The thermal anomaly generated during this phase causes an uplifted thermal bulge, which typically is eroded subaerially, forming an unconformity. The thermal anomaly then decays, resulting in relatively slower subsidence. The flexural strength of the lithosphere causes the sediment load to be spread over a wider area than during the first phase of basin subsidence, creating the horns of the model (Dewey, 1982)

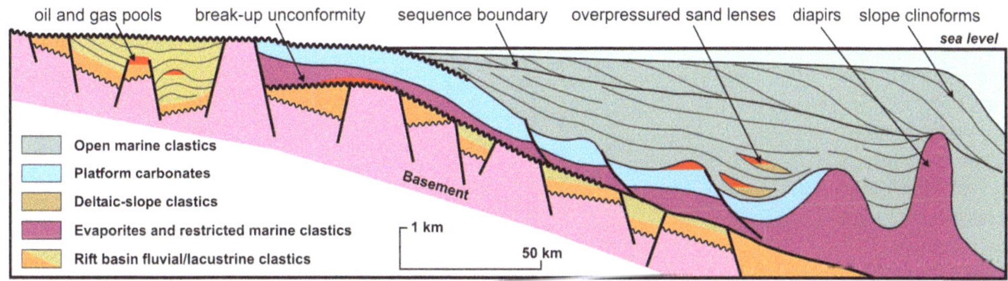


**Fig. 10.3** A simplified and generalized cross-section through the continental margin of southeastern Brazil. This margin exemplifies many of the common features of the classic "Atlantic-type" continental margin (based on Ponte et al., 1980)

detachment faults that descend to the base of the crust. This causes hot asthenospheric material to rise closer to the surface, resulting in heating and uplifting of the rift margins, as shown in Fig. 10.2. The uplift is equivalent to a relative fall in sea level, the rate and magnitude of which has been modeled by Watts et al. (1982), and is given in Table 8.2. The end of the rift phase is typically marked by a widespread unconformity, which is then followed by a lengthy cooling and flexural-subsidence phase, lasting for tens of millions of years. Thermally-driven subsidence, accompanied by water and sediment loading, leads to downward flexure of the continental margin. The flexural phase involves a much broader area of the continental margin than the rift phase, which accounts for the classic *steer's head* cross-section of extensional-margin basins (Fig. 10.2). Cooling and subsidence are rapid at first, but rates decreases asymptotically. The result is a subsidence curve that is concave-up (Fig. 10.4).

Continuation of sea-floor spreading at or near the centre of the rift results in breakup, and separation of the rifted margins by a widening ocean. Flexural subsidence of the flanks of this ocean generates architectures such as that shown in Fig. 10.3. The post-rift unconformity corresponds to the commencement of this phase, and is termed the *breakup unconformity* (Falvey, 1974). Flexural subsidence continues whether or not sea-floor spreading continues, or switches to another location (as happened, for example, in the case of the North Sea basin). In the latter case, subsidence and sedimentation result in a fully developed *steer's head basin*, as shown in the lowermost panel in Fig. 10.2. These have been called *failed rifts*, but the term carries anthropomorphic overtones, and should be avoided.

The basic physical model illustrated in Figs. 10.2 and 10.3 evolved from the work of Walcott, Sleep, Watts, and McKenzie, and is described in more detail in textbooks, such as Miall (1999) and Allen and Allen (2005). The mechanics of the model (heat flow, effects of water and sediment loading, crustal strength and elasticity, flexure) are well known, and have been simulated using numerical modeling. Stratigraphic and other data may be backstripped to reveal tectonic

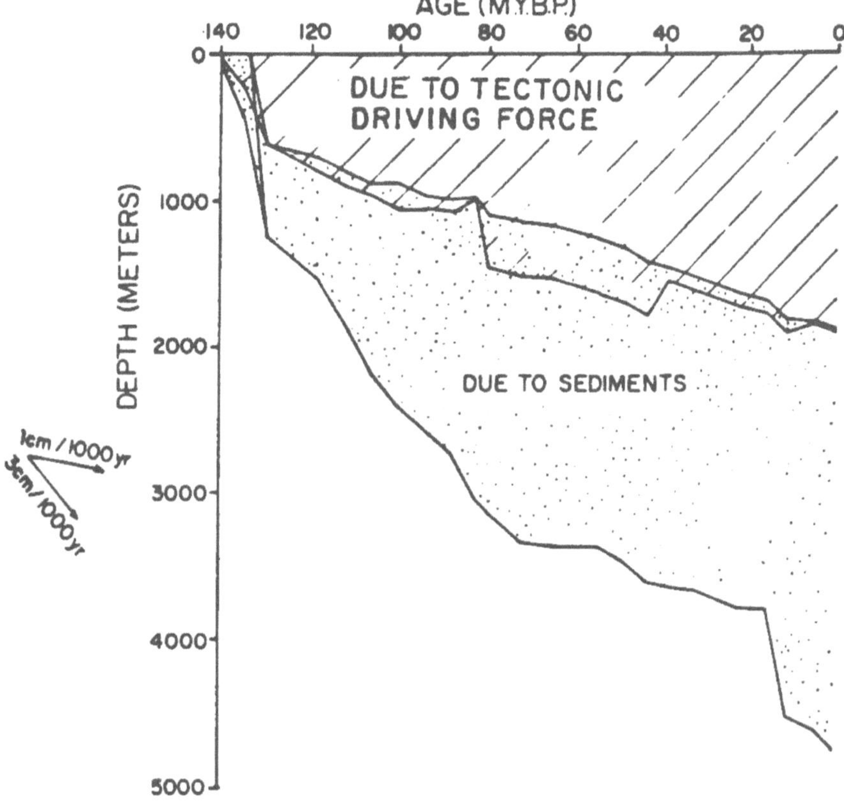



**Fig. 10.4** Typical subsidence *curve* for extensional continental margin. Note the *concave-up* shape. This *curve* was developed by backstripping procedures for the Atlantic margin of the United States off New York. Subsidence has been divided into that due to thermal subsidence (the tectonic driving mechanism), and that due to sediment loading (Steckler and Watts, 1978)

driving mechanisms, and these may then be fitted to equations of crustal behavior involving subtle modifications of the flexural model to incorporate possible geographical and temporal variations in flexural rigidity, viscous versus elastic behaviour, internal heat sources, such as radioactivity, and thermal-blanketing effects of sediments, which have low heat conductivity (Watts, 1981, 1989; Watts et al., 1982).

In the original models, it was assumed that one of the most important attributes of the thermal behavior of the lithosphere is that as it cools and thickens, its flexural rigidity increases. The incremental depression of the crust that accompanies each additional load of water and sediment, therefore, becomes progressively wider but shallower (Fig. 10.2). Watts (1981) illustrated two simple models of a continental margin that assume different stratigraphic behavior under similar conditions of gradually increasing flexural rigidity (Fig. 10.5). The upbuilding model is typical of carbonate-dominated environments, whereas the outbuilding model shows the characteristic

clinoform progradation of a clastic-dominated environment. In the upbuilding model, the load is added in the same place, and a virtually horizontal shelf develops. Deeper (outboard) sediments become backtilted toward the shoreline. In the outbuilding model, the load is progressively shifted seaward, and the backtilting effect is less pronounced or does not occur. Figure 10.6 is an enlargement of the landward edge of one of these models, showing how the increase in flexural rigidity of the lithosphere with age leads to the spreading of the sediment and water load over progressively wider areas of the crust, causing coastal onlap. Coastal onlap is also a feature of the basic steer's-head model shown in Fig. 10.2. A comparison of this model with a cross section through an actual continental margin (Fig. 10.7) shows that the model successfully mimics real stratigraphic architecture. This has some very important implications. Coastal onlap was one of the key indicators used in seismic stratigraphic models of sea-level change to postulate a rise in sea-level (Vail et al., 1977). But for sea-level changes on extensional

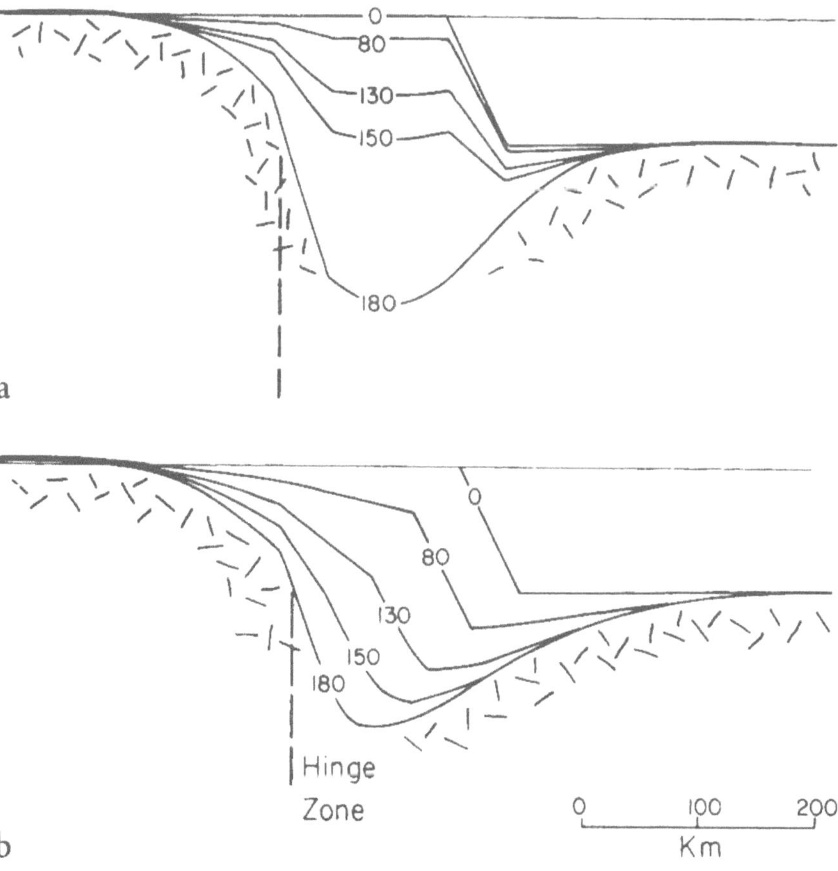



**Fig. 10.5** Two simple models for the development of a continental margin, based on (**a**) upbuilding, and (**b**) outbuilding (Watts, 1981). AAPG © 1981. Reprinted by permission of the AAPG whose permission is required for further use

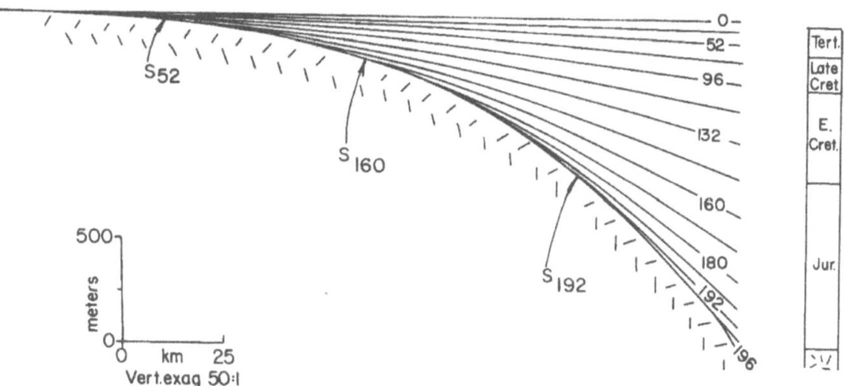



**Fig. 10.6** Enlargement of the landward edge of the stratigraphy generated in flexural models, such as those illustrated in Fig. 10.5. Note the progressive onlap of younger sediments onto the coastal plain (Watts, 1981). AAPG © 1981. Reprinted by permission of the AAPG whose permission is required for further use

continental margins occurring over a $10^7$-year time scale, we see that there is no need to invoke eustasy (Watts et al., 1982). This is a critical point with respect to Exxon's global cycle charts.

Later work by Watts (1989) suggested some revisions to the original steer's-head model. Continental crust may not show a significant increase in rigidity with time, in part because the sediment load inhibits heat loss and therefore slows the process of cooling. Cooling and subsidence of the underlying mantle contribute to development of flexural subsidence and coastal onlap, but a more significant effect is the flexural load of the sediment wedge itself. Considerable accommodation space is available for sediment to be deposited by lateral progradation of the continental margin, and Watts (1989) showed that this load can account for much of the onlap pattern observed on the Atlantic continental margin and elsewhere.

An example of a fully developed extensional continental margin is shown in Fig. 10.3, based on studies of the Atlantic margin of southeastern Brazil (Ponte et al., 1980). Many of the typical features of Atlantic-type margins are exemplified by this margin. Note the presence of a widespread breakup unconformity, which defines a distinct change in structural style between the two phases of margin development. Evaporites and restricted-marine deposits occur at the base of the flexural wedge, immediately above the breakup unconformity. These deposits record the first, incomplete marine connections of the incipient ocean formed immediately after breakup. Such deposits are common on Atlantic-type margins, including the Gulf of Mexico and the Atlantic margins of Nova Scotia and tropical west Africa.

One of the first ancient "passive", or Atlantic-type extensional continental margins to be identified in the

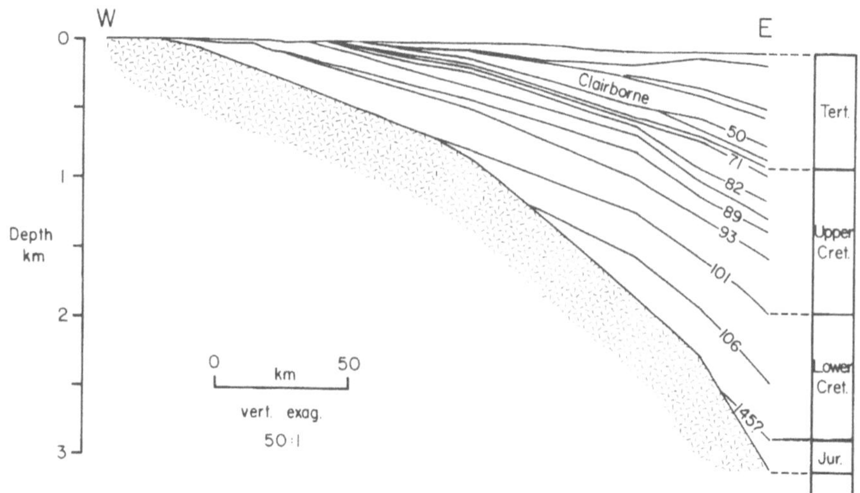



**Fig. 10.7** Observed stratigraphy of the Atlantic coastal plain of North Carolina. Note the coastal onlap of Cretaceous strata and compare this with the synthetic model of Fig. 10.6. The significance of this is discussed in the text (Watts, 1981). AAPG © 1981. Reprinted by permission of the AAPG whose permission is required for further use

geological record was the Lower Paleozoic margin of western North America (Stewart, 1972; Burchfiel and Davis, 1972), which began to develop late in the Precambrian as part of the rifting and dispersal of the Rodinia supercontinent. Bond and Kominz (1984) and Bond et al. (1985) applied backstripping methods to the ancient continental margin of western Canada and confirmed the appropriateness of the model of crustal stretching and thermal subsidence. They introduced an important modification to the backstripping method, which was required by the major difference in rock type between the Jurassic to modern sediments of the Atlantic margin, and the Precambrian-Cambrian rocks of British Columbia. On the Atlantic margin, the succession is primarily clastic. The sediments there had high initial porosities, which were gradually reduced by burial compaction. This compaction results in substantial thickness reductions during burial, which can

be compensated for in the subsidence calculations by using the local empirical relationship between porosity and depth. However, most of the miogeoclinal succession on the ancient British Columbia margin consists of carbonate sediments which typically undergo cementation and lithification early. Bond and Kominz (1984) constructed a different porosity-depth curve for these rocks based on studies of early carbonate diagenesis and tested against data from a well drilled off the Florida margin through a predominantly carbonate section. This allowed them to more accurately "delithify" the strata as a first step in the backstripping procedure.

Approximately 6 km of Lower Paleozoic miogeoclinal strata lie above a suite of Neoproterozoic strata that are thought to represent the rift phase of development of the continental margin (Fig. 10.8). The Paleozoic rocks are spectacularly exposed in the Rocky

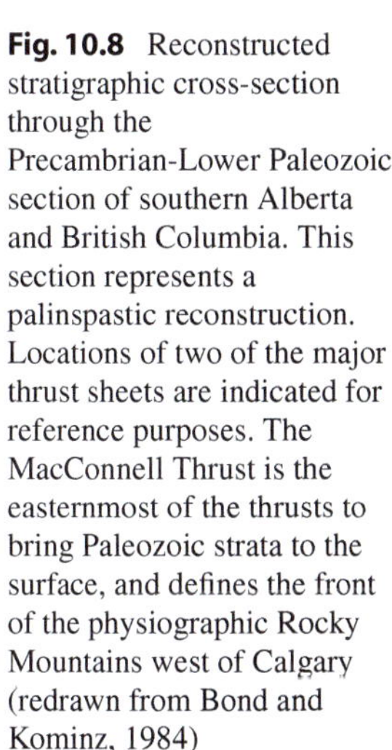

Fig. 10.8  Reconstructed stratigraphic cross-section through the Precambrian-Lower Paleozoic section of southern Alberta and British Columbia. This section represents a palinspastic reconstruction. Locations of two of the major thrust sheets are indicated for reference purposes. The MacConnell Thrust is the easternmost of the thrusts to bring Paleozoic strata to the surface, and defines the front of the physiographic Rocky Mountains west of Calgary (redrawn from Bond and Kominz, 1984)

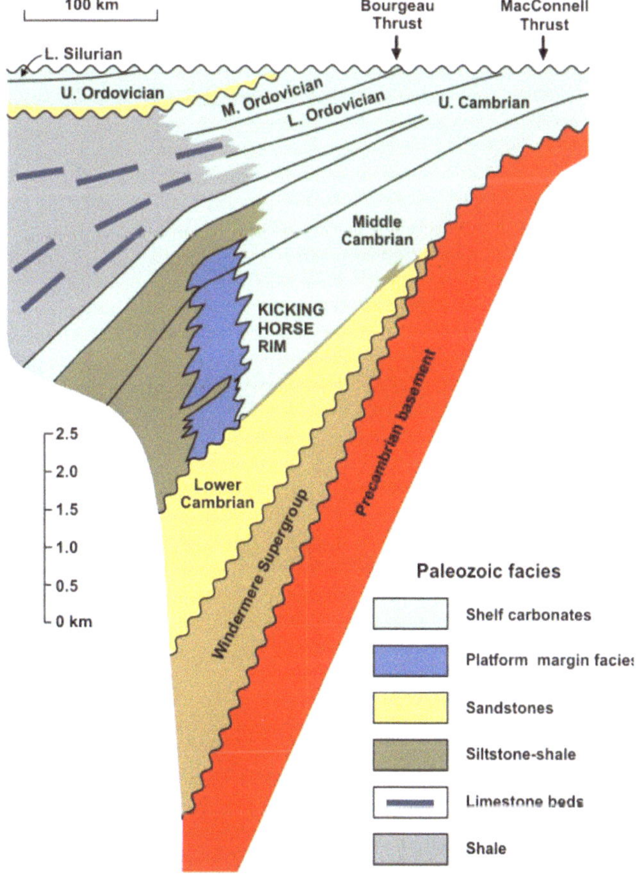

Mountains of Banff, Jasper and Yoho national parks. A succession of shallow-water, mainly carbonate sediments, extends from the craton westward to near the Alberta-British Columbia border, where a major facies change takes place over the "Kicking Horse Rim" (named after the famous mountain pass of the same name). The Paleozoic section thickens and changes facies westward into deeper water facies. The major sequence-bounding unconformities of Sloss (1963) can be recognized within the succession (Fig. 5.4). A major angular unconformity occurs near the middle of this succession; rocks of the Tippecanoe II sequence, corresponding to most of the Silurian System, are largely absent from the southern Canadian Rocky Mountains, owing to elevation of the West Alberta Arch (this is attributed to intraplate stresses: see Sect. 10.4). As the continental margin aged, and increased in flexural rigidity, and as the sediment load on the margin increased, the flexural hinge at the edge of the continental crust migrated cratonward, resulting in gradual onlap (Fig. 10.9).

Bond and Kominz (1984) compared a modeled two-dimensional cross-section of the continental margin with an actual, restored cross section (Fig. 3.14). The restored section exhibits a much greater than predicted thickness of the marginal flexural wedge. This probably indicates that both thermal subsidence and crustal thinning were underestimated in the back-stripping calculations. The thin wedge of Middle to Upper Cambrian strata extending across the craton is probably the result of the episode (or episodes) of eustatic sea-level rise (or dynamic-topography subsidence) that were responsible for development of the Sauk II and Sauk III sequences. Following this episode of high sea level, there was a widespread regression. Rocks of Middle Ordovician to Early Devonian age are almost entirely absent from western and central Alberta, owing to long-lived uplift, or to repeated episodes of sedimentation and erosion of the West Alberta Arch (see Sect. 10.4).

Summerhayes (1986) pointed out that the Exxon global cycle charts are biased by data from the North

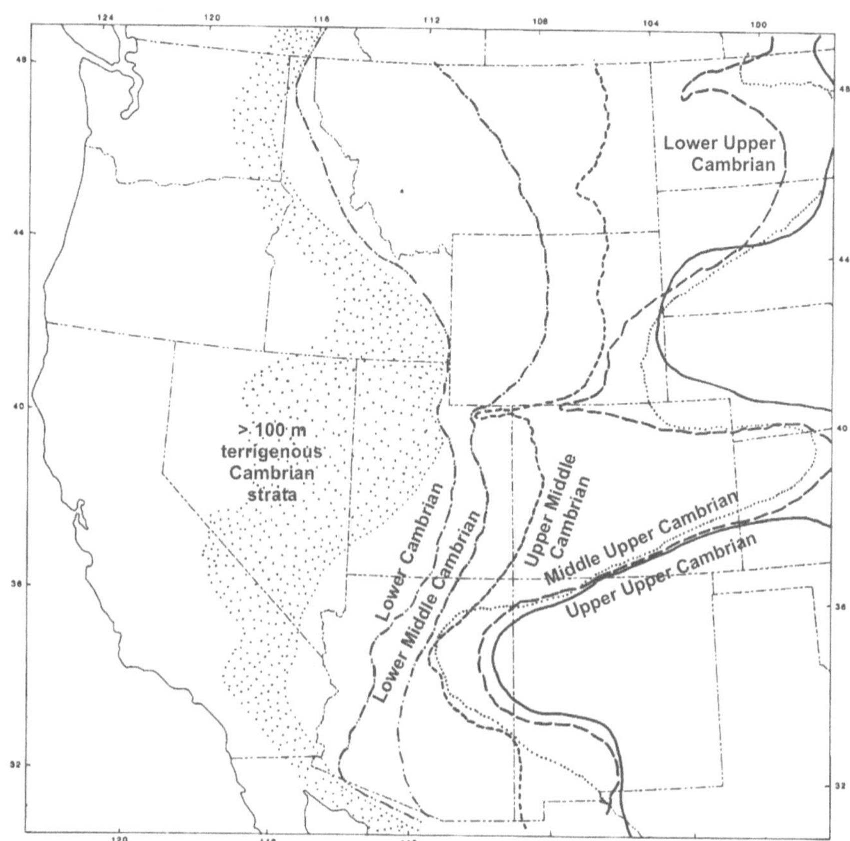


**Fig. 10.9** The extent of pre-drift sediments on the western margin of Laurentia in the United States (*stippled area*) and the onlap pattern of Cambrian strata (adapted from Stewart and Suczek, 1977)

Atlantic and from the North Sea and the Gulf of Mexico—both areas bordering the Atlantic Ocean. The question arises whether the results reflect not eustatic sea-level change but regionally synchronous tectonic effects resulting from widespread rifting and flexural subsidence. Summerhayes (1986) noted that many of the Cretaceous-Tertiary cycles on the Vail curve are not present in the Australian stratigraphic record, which appears to confirm this point. As noted above, Watts (1981, 1989) and Watts et al. (1982) demonstrated that coastal onlap on the $10^7$-year time scale could be caused by increases in flexural rigidity or sediment load with time on an extensional continental margin (compare Figs. 10.6 and 10.7), whereas times of low relative sea-level may represent times of marginal thermal uplift accompanying rifting events. Some other examples of apparent sea-level events reflecting local or regional tectonic episodes are discussed in Sect. 10.2.2.

The problem of what ultimately causes synchronous episodes of coastal onlap is a tricky one. During the breakup of supercontinents, rifting events along the margins of major new oceans are followed by flexural subsidence and consequent local transgression, but are also followed by a period of new-ocean formation, which causes a rise in sea-level because of the lengthening and thermal expansion of spreading centers. The disparity between the Vail and Watts models may therefore be more apparent than real, because both mechanisms for generating coastal onlap tie back to the same original causes, a point made by May and Warme (1987). However, Pitman and Golovchenko (1988) and Parkinson and Summerhayes (1985) examined the dependence of local relative rises and falls of sea-level (actual transgressions and regressions) on the balance between eustatic change and local tectonic (thermal) subsidence. They showed that if basins are in different stages of subsidence, having rifted at different times, eustatic effects may be masked to the extent that local sea-level changes may be completely different in each basin. This point was made in the introduction to this chapter, with reference to Fig. 10.1.

The breakup of the supercontinent Pangea, which initiated a long-term cycle of sea-level change, was accompanied by the development of extensional margins bordering the future Atlantic, Indian and Southern Oceans (Figs. 9.5 and 9.6). Segments of extensional margin were initiated at intervals of tens of millions of years between 250 and 5 Ma, and each underwent

a history of flexural subsidence. A complex pattern of tectonically driven (Watts-type) onlap and offlap was therefore superimposed on the long-term eustatic effects caused by this history of widening oceans. This combination of events may explain the development of Sloss-type sequences, which have $10^7$-year episodicity and demonstrate approximate correlations among continents (Fig. 5.14).


## 10.2.2  The Origins of Some Tectonostratigraphic Sequences

A survey of the documentation of the tectonic history of various divergent continental margins (Sect. 5.3.2) indicates little global, or even regional, correlation among the successions of events with $10^7$-year episodicities. Most unconformities relate to the regional history of rifting and flexural subsidence. In many cases the regional stratigraphy has been divided into tectonostratigraphic sequences of $10^7$-year duration and correlated with specific plate-tectonic events in the adjacent ocean. This type of correlation is particularly well illustrated in the case of the Grand Banks of Newfoundland, the development of which reflects a long-continued history of successive rifting and thermal events as the Atlantic Ocean gradually "unzipped" northward, around the promontory now comprising the Grand Banks continental margin (Fig. 10.10; Tankard and Welsink, 1987; Welsink and Tankard, 1988).

The North Atlantic margins and the North Sea basin constituted the birthplace of much of the Vail curve (e.g., Vail and Todd, 1981) and the controversy surrounding it (Miall, 1986). Extensional continental margins are commonly termed "passive" margins, to contrast them with "active" convergent margins. However, this may be misleading. For example, during the Jurassic, the northern North Sea was affected by extensional faults, wrench faults, diapiric movement of Zechstein (Permian) salt, volcanic activity, and regional isostatic subsidence (Hallam, 1988; Glennie, 1998). To separate the stratigraphic effects of this activity from those due to eustatic sea-level change requires meticulous documentation of local structure and stratigraphic architecture. This information was not available during the initial phase of exploration that was the basis for the Vail and Todd (1981) synthesis.

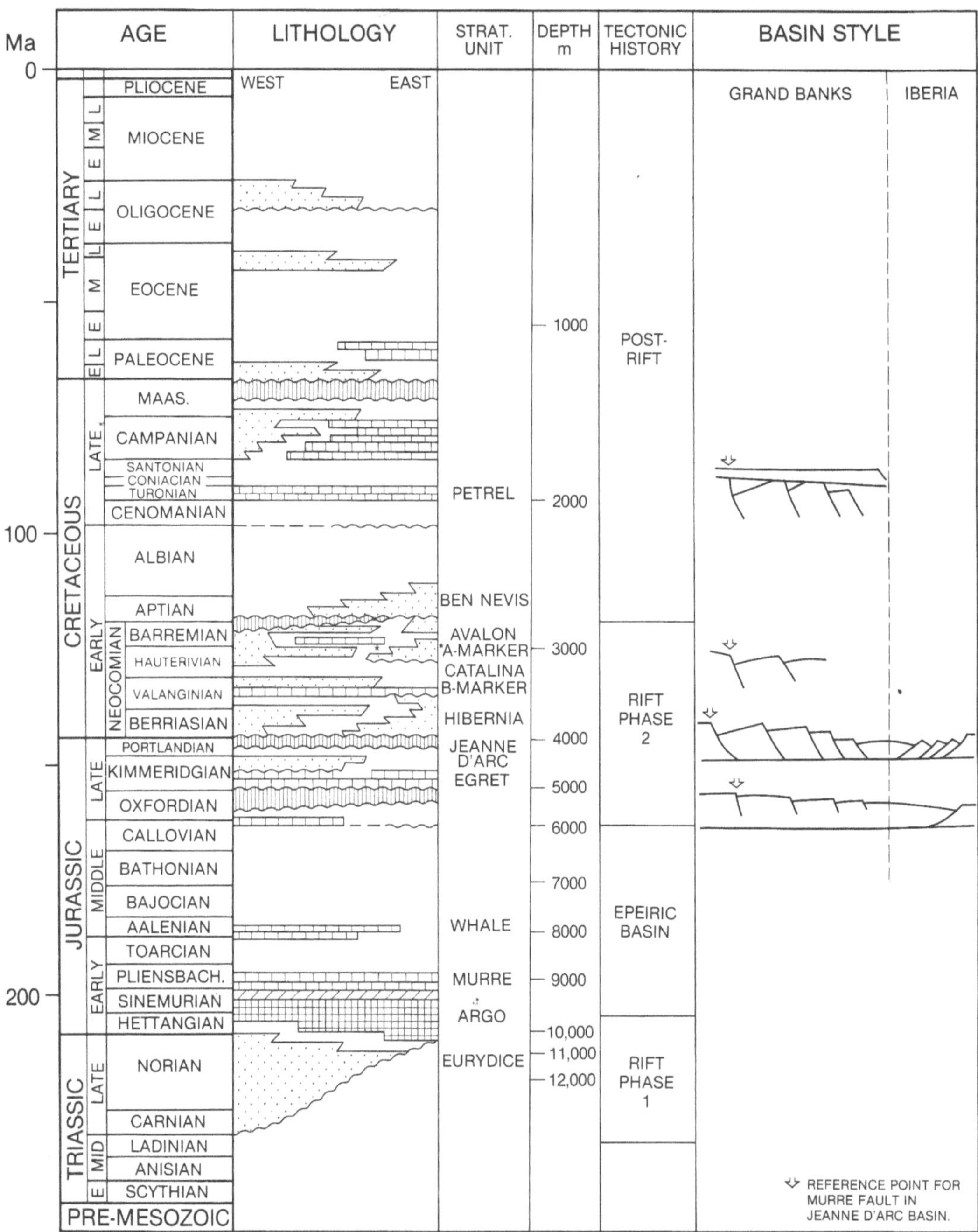



**Fig. 10.10** Tectonic sequence analysis of the Jeanne d'Arc Basin, Grand Banks of Newfoundland, showing the five major unconformity-bounded tectono-stratigraphic sequences interpreted on the basis of seismic-stratigraphic analysis, with the Exxon global cycle chart at *right*, for comparison (Tankard et al., 1989). AAPG © 1989. Reprinted by permission of the AAPG whose permission is required for further use

Subsequently, Badley et al. (1988) used a large seismic data set to examine the structural evolution of the North Viking Graben, and documented the ongoing nature of extensional tectonism during the Mesozoic. Hiscott et al. (1990) and Sinclair et al. (1994) have examined the Jurassic-Cretaceous rifting history of the basins of Maritime Canada, Iberia, and the continental margin west of the British Isles. Underhill (1991) reexamined the seismic data base of the Moray Firth, North Sea, on which much of the Jurassic part of the Vail curves was based. Underhill and Partington (1993a, b) demonstrated the occurrence of a significant thermal heating and uplift event in the mid-Jurassic. A book edited by Hardman and Brooks (1990) on the tectonic evolution of the British continental margins provides a wealth of data on this topic. The consensus is that vertical movements accompanying rift faulting and thermal effects were ubiquitous. Badley et al. (1988, p. 460) stated:

> Numerous authors have discussed unconformities and their causes in the northern North Sea (e.g. Vail & Todd 1981; Rawson & Riley 1982; Ziegler 1982; Miall 1986). The structural evolution of the northern Viking Graben presented below shows extension-related tectonics to have been the primary causal mechanism in the development of unconformities in the area. Sea-level changes appear to have played only a secondary role in unconformity development. Two types of tectonically-induced unconformity are commonly recognized in extensional basins: regionally-developed, angular stratal relationships related to differential block tilting, and local stratigraphic breaks on the crests of fault blocks elevated by footwall uplift, often involving erosion.

Badley et al. (1988) documented numerous examples of unconformities related to faulting, and wedge-shaped stratigraphic units developed by onlap during post-rift thermal subsidence. They recognized rift-thermal subsidence phases during the Triassic-Early Jurassic, Bathonian-Kimmeridgian, Kimmeridgian-Ryazanian, Albian, Turonian-Cenomanian and Late Cretaceous-Early Tertiary. They pointed out that most of these phases of movement were local, and should not be correlated with regional tectonic pulses, such as the Cimmerian, sub-Hercynian, Austrian or Laramide movements, as had been suggested by some writers (e.g., Ziegler, 1982).

The documentation of repeated regional rifting and thermal-subsidence events based on extensive analysis of seismic-reflection data (e.g., Badley et al., 1988), would seem to leave little cause to invoke eustasy as a mechanism for generating the observed regional

unconformities. Indeed, it seems that, fortuitously or not, the Exxon seismic line in the Moray Firth, off northeast Scotland, one of two used in the major analysis by Vail and Todd (1981) and Vail et al. (1984), did not cross the major faults whose movement, it is now thought, was responsible for developing the unconformities that they observed. Underhill (1991) showed that the major faults are oriented northeast-southwest in this area, and this is the orientation of the Exxon line! He demonstrated that rotation of fault blocks is a primary cause of unconformable onlap in this area (Fig. 10.11). When seen perpendicular to the fault trend this is revealed by a distinctive fanning of stratigraphic units, caused by block rotation during fault-induced subsidence and sedimentary onlap (Fig. 10.11). However, this is obscure in fault-parallel sections.

Another aspect of stratigraphic architecture that may be of tectonic or eustatic origin or may be unrelated to either is the downlapping pattern developed by the progradation of submarine fans. As pointed out by Miall (1986) onlapping by itself is an unreliable indicator of a relative rise of sea level, unless it can be demonstrated that the facies showing the onlap are coastal in origin, which is difficult to accomplish from seismic-facies data alone. However, most of the strata analyzed by Vail and Todd (1981) are deep-marine fan deposits. Fan progradation may lead to onlap of tilted basement surfaces, and could be interpreted as a product of fan growth during sea-level falling stage to lowstand (Fig. 2.8). However, fan rejuvenation may be triggered by fault movement, and fan switching can occur entirely as a result of autogenic channel-switching processes. A quantitative model of half-graben subsidence that explored the variables of sediment input, block rotation and isostatic response was offered by Roberts et al. (1993), who showed how the model could be applied to specific aspects of the Viking Graben. Prosser (1993) examined the paleogeography of evolving rift basins and the resulting seismic facies and architecture, and proposed a series of models of tectonic systems tracts linked to stages in rift evolution. White and Lovell (1997) demonstrated that the major submarine fans in the North Sea and Shetland-Faeroes Basin were developed during phases of magmatic underplating of the British Isles area, which caused uplift and erosion, and increased sediment delivery to the adjacent marine basins (Fig. 10.12).

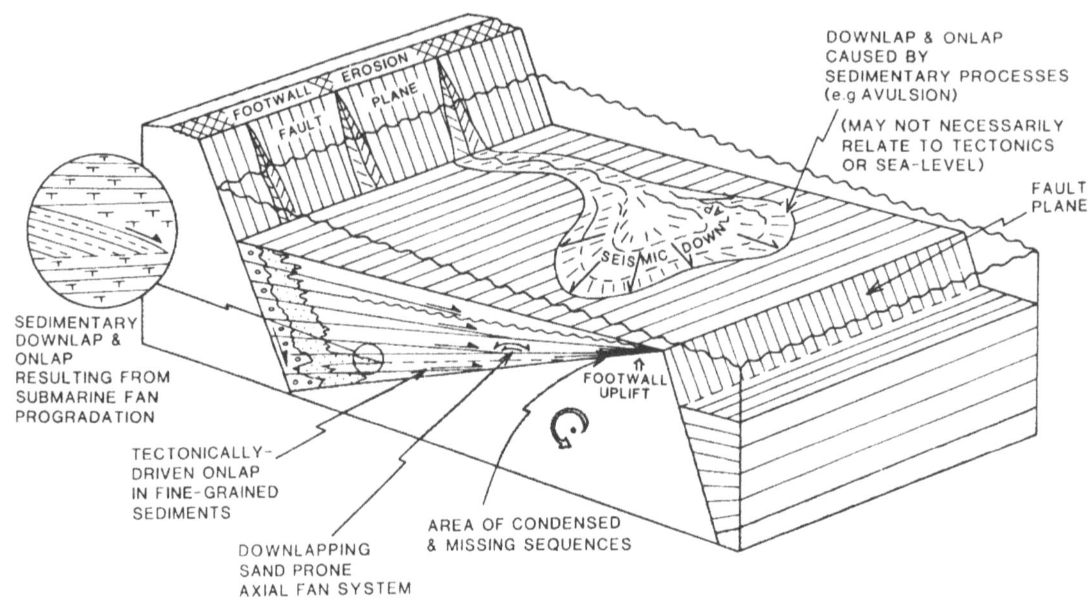



**Fig. 10.11** Diagram to illustrate how local sedimentary and tectonic processes can develop stratigraphic architectures that may be interpreted as sequence boundaries. Based on analysis of seismic data from the Inner Moray Firth, off northeast Scotland (Underhill, 1991)

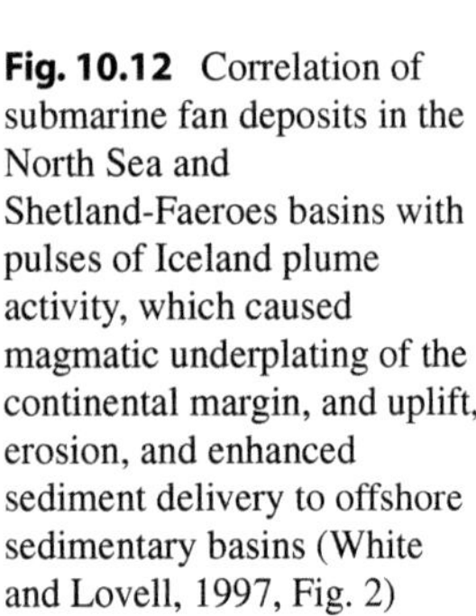

**Fig. 10.12** Correlation of submarine fan deposits in the North Sea and Shetland-Faeroes basins with pulses of Iceland plume activity, which caused magmatic underplating of the continental margin, and uplift, erosion, and enhanced sediment delivery to offshore sedimentary basins (White and Lovell, 1997, Fig. 2)

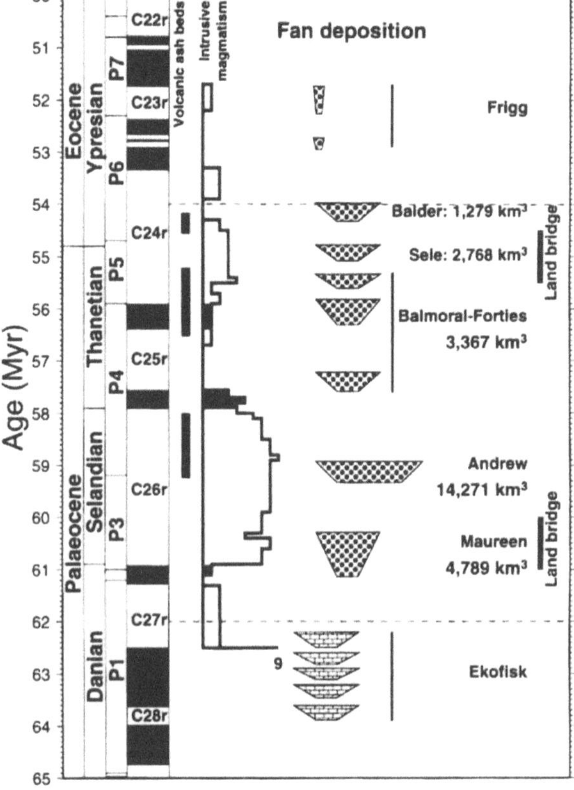

Based on a more complete understanding of regional tectonics and stratigraphic architecture than was available to Vail and Todd (1981), Underhill (1991) argued that the Inner Moray Firth sequence data could not be used to verify a global sea-level curve, thus undermining one of the original foundations of the Late Jurassic portion of the Vail et al. (1977) curves. Boldy and Brealy (1990) developed a similar view of the ubiquity of extensional tectonism in the Outer Moray Firth, and, together with the synthesis by Badley et al. (1988) dealing with the nearby Viking Graben, the consensus is clearly that the entire northern North Sea area was dominated by the effects of extensional tectonism.

Underhill and Partington (1993a, b) demonstrated that the intra-Aalenian "mid-Cimmerian" unconformity in the North Sea basin, which corresponds to the "first-order" 177-Ma sequence boundary in the Exxon global cycle chart of Haq et al. (1987, 1988a), is also a regional event. Underhill and Partington (1993a, b) pointed out that this event in the Exxon chart was based exclusively on stratigraphic sections within the North Sea Basin, and must therefore be questioned as a "global" event. Careful regional mapping of marker beds, particularly maximum-flooding surfaces, demonstrated that the unconformity developed as a result of subaerial erosion following thermal doming centred on a triple-point junction in the central North Sea (Fig. 10.13). Uplift resulted in removal of Jurassic strata and exposure of Triassic or older deposits at the centre of the uplift, with erosion to shallower levels extending out over a distance of at least 500 km (Fig. 10.13). Underhill and Partington (1993a, b) concluded that a transient plume developed at the point from which now radiates the Viking and Central Graben and the Moray Firth Basin. The uplift rapidly decayed during the later Aalenian, leading to progressive onlap of the unconformity.

In retrospect, it is now clear that the North Sea basin was a poor region to have chosen as a basis for the investigation of eustasy.

A comparative analysis of the rift basins flanking the North Atlantic region (Maritime Canada, Iberia, western British continental shelf) by Hiscott et al. (1990) demonstrated, again, the importance of local tectonism. The authors attempted to relate their sequence record to the global cycle chart of Haq et al. (1987, 1988a), but the documented record of sequence boundaries and varying subsidence rates showed few inter-basin correlations, and the authors appealed to various mechanisms of rift faulting, thermal doming, post-rift thermal subsidence, and transtensional faulting to explain the stratigraphy of each basin. Coupled with this was the suggestion that intraplate stress (Sect. 10.4) would have led to the transmission of many of these stress events to adjacent basins.

Sinclair et al. (1994) developed a general model for the tectonism and sedimentation in Atlantic-margin rifts (Fig. 10.14). Four phases of Late Cimmerian tectonism are shown. Generally, the left side of the diagram represents the Jeanne d'Arc Basin of the Grand Banks, while the right side represents the Porcupine Basin of the outer continental margin off Ireland, and the Outer Moray Firth area of the North Sea basin. However, this range of responses can occur within any one broad rift basin area. "Onset warp" represents a phase of regional subsidence variation without significant faulting. "Early synrift" represents the phase of initial brittle fracture of the crust, "mid synrift" represents a phase of maximum tectonic subsidence, while "late synrift" represents decreasing subsidence during the last phase of a rift episode. The same broad phases can be recognized throughout the North Atlantic region, but with variations from basin to basin in the timing and rate of subsidence, depositional conditions, etc.

A detailed analysis of the sequence stratigraphy and of the tectonic eustatic control of the Jurassic-Cretaceous succession in East Greenland was reported by Surlyk (1990, 1991). Rift basins there underwent a similar type of history of repeated rifting and thermal-flexural subsidence to that in the North Sea Basin, and Surlyk reached similar conclusions regarding the origin of the sequence of relative sea-level changes there. Regarding the Wollaston Forland Embayment he stated (p. 80):

A major phase of rotational block-faulting was initiated in the region in the Middle Volgian, lasted throughout the Late Volgian-Ryazanian, and gradually faded out in Valanginian time ... The tectonic overprint is so strong for this period, that it is a major challenge to separate the tectonic, eustatic, and regional sea-level signals. The Middle Volgian-Valanginian time interval is characterized by major fluctuations in the eustatic sea-level curve of Haq et al (1987). This curve is, however, strongly biased towards the North Sea where the same tectonic phase occurred; the apparent sea-level changes are probably controlled mainly by regional tectonics (cf. Hallam, 1988).

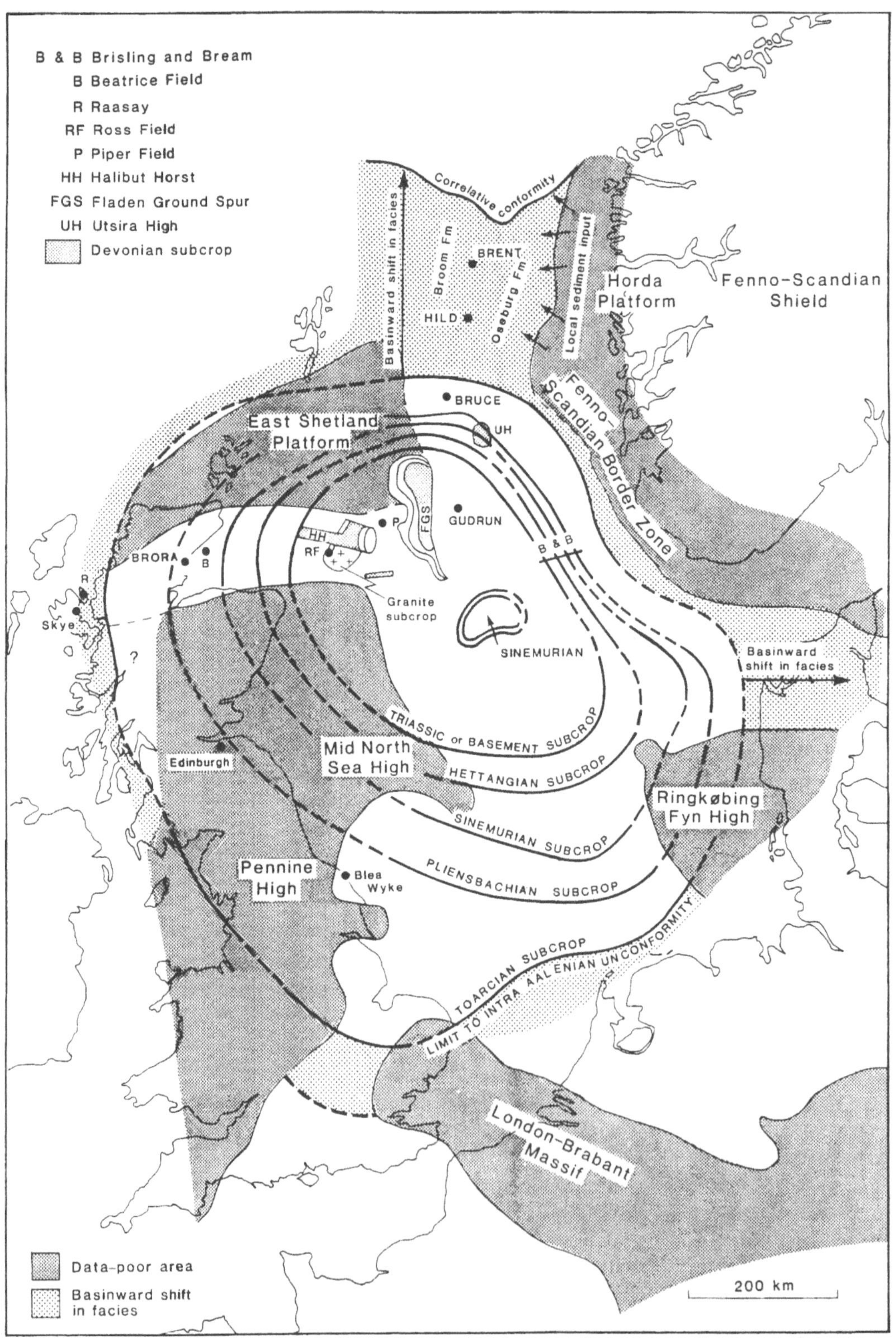


**Fig. 10.13** Subcrop map of the strata beneath the mid-Cimmerian unconformity of the North Sea basin. Maximum removal of strata below this mid-Aalenian surface occurs at the junction of the Viking and Central Grabens and the Moray Firth Basin. Uplift radiated outward over a distance of about 500 km. Beyond this, facies underwent a significant basinward shift (Underhill and Partington, 1993b). AAPG © 1993. Reprinted by permission of the AAPG whose permission is required for further use

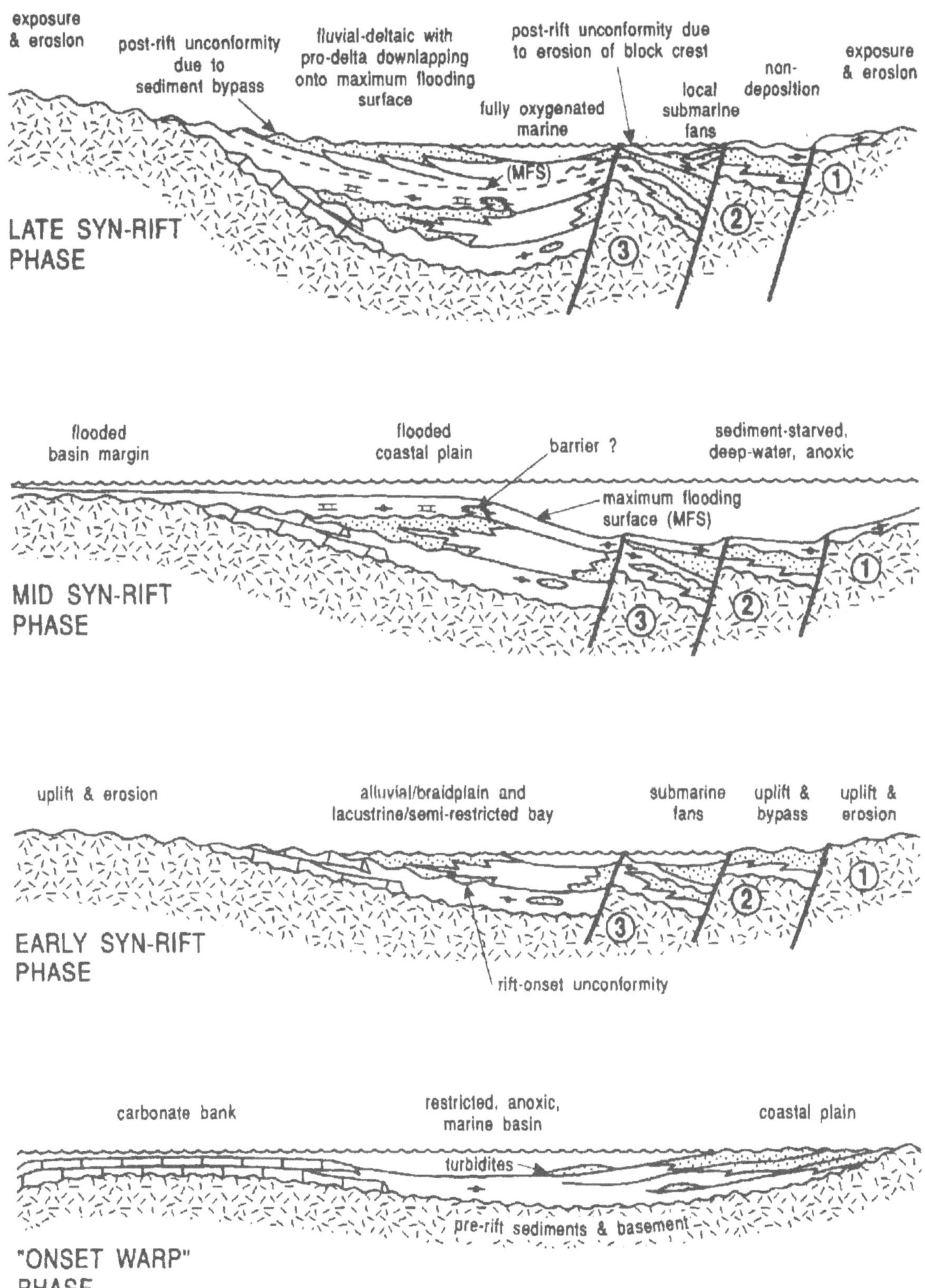


**Fig. 10.14**  A general model for the tectonism and sedimentation of rift basins marginal to the North Atlantic Ocean (Sinclair et al., 1994, Fig. 15). Stratigraphic documentation for selected basins is provided in Fig. 5.18

Elsewhere Surlyk stated:

> Substituting tectonism with sea level change as the domi-
> nating controlling factor on development of the Wollaston
> Forland Group is too simplistic. The basic aspects of the
> system in terms of geometry, environment, and processes
> were undoubtedly controlled by tectonism. The large-
> scale cyclicity probably is controlled to a large extent
> by sea level rises. However, I suggest that the sea level
> fluctuations were controlled by regional tectonism and
> that each major episode of faulting is associated with
> a rapid relative sea level fall followed by a longer sea
> level rise, reflecting increasing subsidence after rifting.
> This pattern would explain the presence and apparent
> contemporaneity of the numerous high-amplitude, short-
> wavelength fluctuations in relative sea level in the North
> Atlantic region during the tectonically very active inter-
> val spanning the Jurassic-Cretaceous boundary. (Surlyk,
> 1991, pp. 1483–5)

Rifting episodes can be linked directly to synchronous subsidence events and stratigraphic
developments on presently separate continents. The
opening of the present North Atlantic Ocean spanned
a period of well over 100 million years, commencing
during the Early or mid-Jurassic with breakup and
initiation of sea-floor spreading between the African
and North American continental plates, while the last
opening event occurred during Eocene time in the
Greenland/Norwegian sea area (Pitman and Talwani,
1972; Ziegler, 1988). "Rifting was neither continuous
nor random in time and place during this long period
of Pangaean disintegration. Splitting of the Pangaean
supercontinent did not occur as a continuous migration like the smooth opening of a zipper, but rather
occurred as discrete events, with the lateral movement
of diverging plates accommodated by large transfer
shear zones" (Sinclair et al., 1994, p. 212; see also
Fig. 9.6). As rifting preceded each breakup event
(Falvey, 1974), the structural and stratigraphic
responses at the end of each widely recognized rifting
event can be related to the initiation of sea-floor
spreading along a particular segment of the North
Atlantic.

Janssen et al. (1995) showed by backstripping analysis of over 200 well sections that subsidence patterns
in rifted basins within and on the flanks of Africa
can be related to changing intraplate stress patterns as
the oceans surrounding Africa opened between Late
Jurassic time and the present.

Loup and Wildi (1994) reported on a subsidence
analysis of the Paris Basin and a comparison with other
northwest European basins. None show the simple
concave-up subsidence-versus-time curve that would
be expected from a simple extensional basin (e.g.,
Fig. 10.4). Most are characterized by short concave
segments, with periods of slow or no subsidence, intervals of accelerated subsidence, and intervals of uplift.
Few of the events are synchronous over wide areas
of northwest Europe. These results are interpreted as
the product of rifting, transtension, and basin inversion resulting from regional plate-tectonic movements,
and subcrustal processes, including mantle plumes,
together with intraplate stresses resulting from all these
processes.

Submarine erosion of continental margins reduces
the sediment load and allows for flexural rebound.
Stratigraphically the result may look like a fall in
relative sea level. McGinnis et al. (1993) examined
the supposed Late Eocene-Early Oligocene global fall
in sea level, which has long been attributed to the
development of an Antarctic ice sheet. They suggested
that one important result of a global climatic cooling (which undoubtedly occurred in the mid-Cenozoic;
see Fig. 4.10) would have been enhancement of latitudinal thermal gradients, leading to strengthening of
bottom currents and increased global submarine erosion. The amount of eustatic fall associated with this
cooling event therefore may have been significantly
exaggerated.

## 10.3 Tectonism on Convergent Plate Margins and in Collision Zones

### 10.3.1 Magmatic Arcs and Subduction

It is a truism to state that areas of convergent tectonism
are areas of active tectonics. Several distinct processes
lead to local and regional relative changes in sea level,
which may be recorded as transgressive and regressive
sedimentary cycles in the major basins overlying and
flanking zones of convergence.

A study of an Upper Cretaceous volcaniclastic
forearc-basin succession in Baja California led Fulford
and Busby (1993) to conclude that tectonic basin subsidence and source-area uplift were by far the most
important controls on the architecture of the basin
fill. The overall succession indicates retrogradation,
suggesting rapid basin subsidence along bounding
faults. The succession includes a coarsening-upward

sequence that contains a significant change in sandstone composition relative to older parts of the succession, and this was interpreted as indicating source-area uplift and unroofing, coupled with an interpreted reduction in the fault-controlled subsidence of the basin.

The long-term subsidence (and sea-level) history of forearc basins depends on the evolution of the subduction zone (Dickinson, 1995). Subduction angles change in response to changes in the age of the subducting oceanic crust, and this leads to changes among extensional and contractional arc types (Dewey, 1980). For example, Moxon and Graham (1987) showed how in the Great Valley forearc basin of California areas distant from the arc underwent uplift as the subduction angle decreased in the Late Cretaceous. The arc itself retreated eastward, away from the subduction zone, and the arc flank, underlying the inner edge of the forearc basin, subsided asymptotically as the arc magma cooled. The subsidence pattern that developed in the overlying sedimentary basin is similar to that of extensional basins (concave up, cf. Fig. 10.4). The rate and magnitude of the regional, tectonically-driven relative sea-level changes in this basin are comparable to those of eustatic changes of $10^7$-year type.

Forearc regions are also affected by the nature of the material undergoing subduction. The entry of seamounts or aseismic ridges into subduction zones results in local uplift. Forearc basins may, therefore, be characterized by cycles of uplift followed by extensional subsidence. This process characterizes much of the west coast of South America. Tide-gauge records over the last 30 years record rapid neotectonic movements that, if continued for geologically significant periods of time, could readily account for the development of high-frequency sequences over $10^4$–$10^5$-year periods (Flint et al., 1991; Fig. 10.15). A network of extensional faults is present along much of the coast, and accommodates extensional movements during the process of subduction of an irregular oceanic plate. Flint et al. (1991, p. 101) stated:

> The fore-arc uplift—collapse process could result in a coupled lowering/raising of relative sea level over relatively short periods (c. 100,000 years). Furthermore, as the above studies both show that negligible accretion/obduction occur during seamount chain/ridge subduction, the inherent morphological/structural irregularities in these structures could result in cycles of minor fore-arc uplift/subsidence at such 100,000–200,00 year frequencies. This tectonic mechanism could thus produce

relative sea-level changes at a frequency equal to glacio-eustatic processes. These tectonically driven cycles will be restricted to arc-related sedimentary basins: they will not be of global extent nor globally synchronous.

The development of areally extensive tectonic sequences in southern Central America is discussed briefly in Sects. 5.3.2 and 6.3. The formation of regional unconformities that can be traced for distances of 900 km was attributed by Seyfried et al. (1991) to intraplate stresses (Sect. 10.4) reflecting tectonic adjustments during subduction.

Some authors claim to have observed or demonstrated a correlation between episodes of magmatic activity and periods of rapid sea-floor spreading and subduction. Winsemann and Seyfried (1991), who studied forearc sedimentation on the west coast of

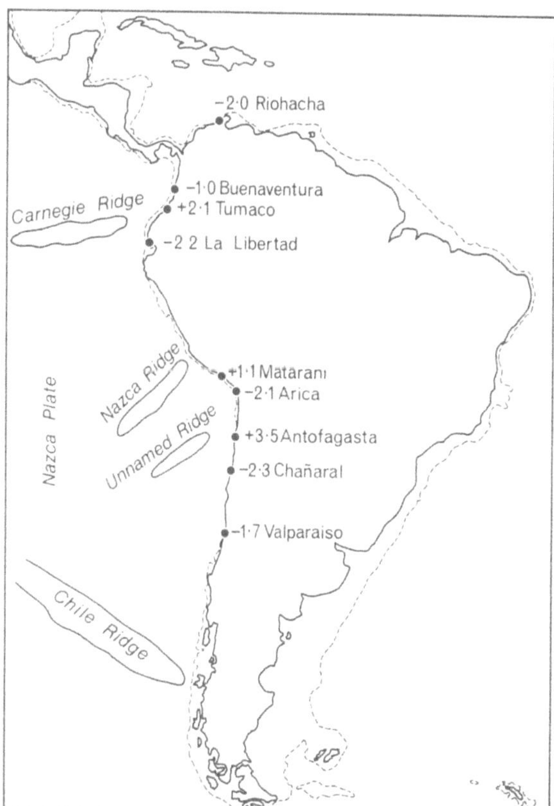


**Fig. 10.15**  Mean changes of relative sea level (in mm/year) during the last 30 years along the west coast of South America, based on tide-gauge records. The Andean margin is mostly undergoing subsidence (negative values), except where aseismic ridges have entered the subduction zone. Buoyancy of the subducting ridge is leading to relative uplift of the adjacent coastline (Flint et al., 1991)

Nicaragua, stated "the most striking feature is the almost simultaneity of tectonic/volcanic activity and second order cycles of eustatic sea-level fluctuations during Late Cretaceous, Paleocene, and mid-late Eocene times." They went on to say:

> Late Cretaceous to late Eocene drift rates calculated from the North Atlantic Ocean correlate closely with global orogenic phases which in turn coincide with crustal movements in Central America. These episodes of major tectonic/volcanic activity occurred at 80–75, 63, 55–53, and 42–38 Ma and agree with major tectonic and volcanic pulses in Central America … This corroborates that global tectonic processes play an important role in generating major sea-level fluctuations and hence in the formation of depositional sequences.

In and adjacent to arcs, magmatic intrusion results in regional doming over areas a few tens of kilometres across. Eruption is followed by deflation. The result is a lowering of relative sea level during the filling of the magma chamber, followed by a rise accompanying deflation of the chamber. Sloan and Williams (1991) documented transgressive-regressive cycles in the Devonian of western Ireland, which they interpreted as the product of this process. Regressive offshore-barrier sequences a few tens-of-metres thick developed during inflation of magma chambers and consequent uplift. The sequences are capped by volcanic intervals 1–190 m thick, recording eruptions, and the overlying sedimentary units are deeper-water offshore deposits indicating a rise in relative sea level following deflation and collapse of the magma chamber. The ages and durations of these sequences are not clear, but a $10^4$–$10^5$-year episodicity seems probable.

## 10.3.2 Rates of Uplift and Subsidence on Convergent Margins

One of the most impressive features of basins that develop in regions of convergent tectonism, including arc-related basins and those within and adjacent to suture zones, is the extremely rapid rates of tectonic and sedimentary processes that occur in these basins. Figure 10.16 shows the history of subsidence and uplift of various locations along the Banda Arc of Indonesia, which is an area of convergence and collision between Eurasia and the Australian and Pacific plates. At the bottom of this diagram is shown the global cycle chart of Haq et al. (1987, 1988a) at the same scale. Clearly the fluctuations indicated in this chart would be swamped by tectonic effects in an area undergoing a tectonic evolution comparable to that of the Banda Arc.

Rapid rates of uplift and subsidence have been calculated from many regions of active tectonism (rates are summarized in Table 8.2). Pedley and Grasso (1991) documented uplift rates in a subduction-accretion zone in southern Sicily, indicating localized Quaternary uplift of up to 1 km. Bishop (1991) studied raised terraces in the New Zealand Alps, a region characterized by transpressional uplift. Calculated uplift rates range from 5 to nearly 8 m/ka over periods of about 10 ka, with longer-term rates an order of magnitude smaller (1 m/ka for 100 ka). Data on rates of movement along the San Andreas transform system were summarized by May and Warme (1987), who noted measured modern subsidence rates of 1.5 m/ka in the Imperial Valley, and up to 2 m/ka in the Eocene Ventura Basin. Uplift rates in the Banda Arc are illustrated in Fig. 10.23, and range from 0.5 to 10 m/ka over periods of less than 2 million years (Fortuin and de Smet, 1991). Subsidence rates are not as rapid. In the Banda Arc, they range from 0.08 to 1 m/ka for periods of up to 15 million years, resulting in a total recorded subsidence of up to 4 km. In the Himalayan proforeland basin of Pakistan, Johnson et al. (1986) and Burbank et al. (1986) demonstrated subsidence (actually sediment-accumulation) rates of up to 0.6 m/ka persisting for 2–4 million years. The same overall rate of sediment accumulation, 0.6 m/ka averaged over 2.5 million years, was obtained for an Upper Cretaceous forearc-basin succession in Baja California (excluding possible but unknown eustatic effects) by Fulford and Busby (1993), and Eocene forearc basins in California subsided at rates between 0.1 and 0.7 m/ka for periods measured over millions of years. Hiroki (1994) determined vertical changes in relative sea level during the Quaternary in a forearc setting in coastal Japan and differentiated glacioeustatic from tectonic changes. He showed changes between subsidence and uplift over a $10^4$–$10^5$-year interval, with considerable local variation in rates, probably reflecting local structural controls. The maximum uplift rate, measured at one locality, was 0.29 m/ka, and the maximum subsidence rate was 0.45 m/ka.

The evidence indicates that uplift and subsidence are spasmodic and localized, with movement rates and

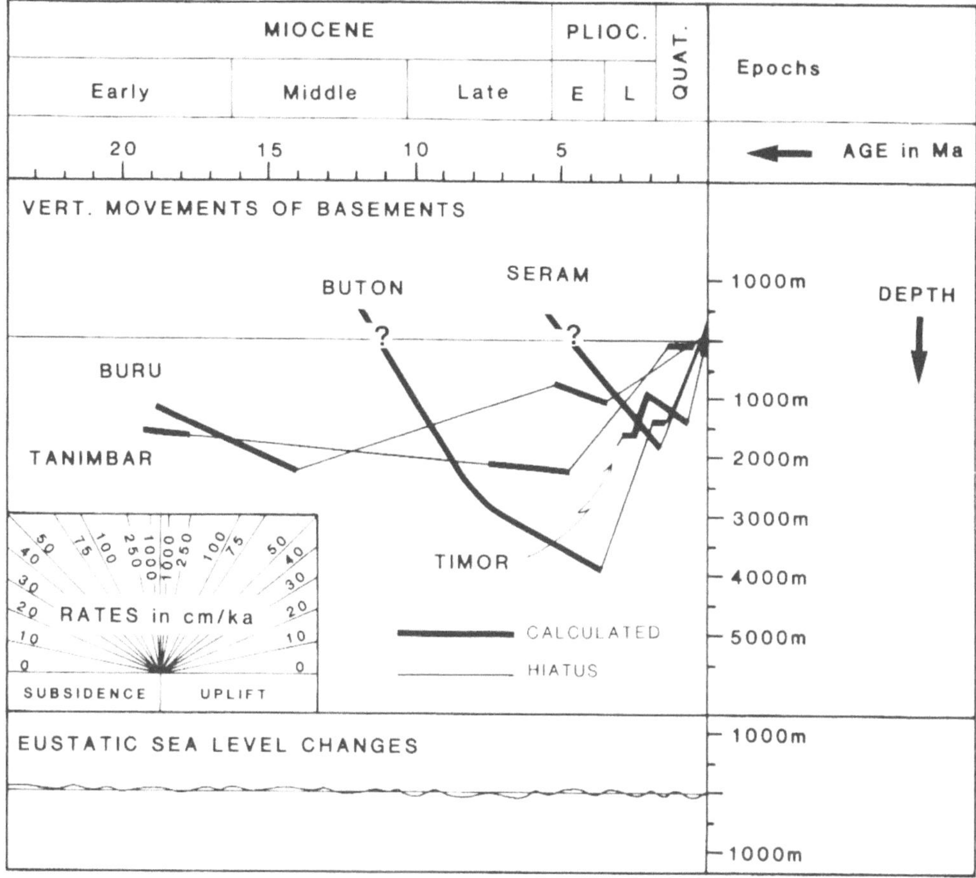


**Fig. 10.16**  Selection of age-versus-depth plots for various locations along the Banda Arc of Indonesia. *Thick lines* indicate plots derived from geohistory analysis. These are linked by *thin lines*, representing the average of vertical motions during erosional intervals. In each case uplift must have occurred to bring about erosion, but vertical amplitude is difficult or impossible to ascertain. At the base of the diagram, is plotted the global cycle chart of Haq et al. (1987, 1988) at the same vertical and horizontal scales (Fortuin and de Smet, 1991)

the total amount of movement varying over distances of only a few kilometres. Uplift on Timor is described as "regional arching over distances of 10 km to more than 100 km" (Fortuin and de Smet, 1991), and this is probably one of the more extensive areas to be affected simultaneously by a single episode of convergent tectonism. Based on measurements along a 40-km stretch of coastline Hiroki (1994) deduced that crustal bending had a wavelength of 50–100 km.

When examined in a regional context it would seem unlikely that tectonism could ever be confused with eustasy in tectonically active basins such as forcland or forearc basins. However, the paucity of stratigraphic data in many basins may tempt the regional geologist to attempt comparisons with the global cycle chart. Sengör (1992) argued that convergent margins are characterized by the successive development of numerous local tectonic unconformities. Poor dating control and limited outcrop may tempt the mapper to assign these to a single regional event, and even to correlate it with one of the events on the global cycle chart. In the case of old and no-longer active basins much of the evidence of active tectonism may have been lost to erosion. The difficulties are likely to be acute where only one or two exploration holes with limited biostratigraphic data are available for analysis. Even where careful structural and stratigraphic mapping clearly indicates tectonic control of sequence development, it has been a common practice in the past to discuss correlations of the sequence record with the global cycle chart (e.g., Deramond et al., 1993). Fortuin and de Smet (1991, p. 87) offered this critical appraisal of such correlations based on their observations in Timor:

> ... up to four sequence boundaries might be preserved according to the Haq et al. (1987) sea-level chart, notably at 3.0, 2.4, 1.6, and 0.8 Ma. We can indicate two distinct sequence boundaries, one around 2.1 Ma, when pelagic sedimentation was suddenly replaced by channel and fan deposits, and one around 0.2 Ma, which is the second uplift unconformity separating the marine sediments from overlying terrestrial conglomerates. ... The 2.1 Ma level has been interpreted to be of tectonic origin ... Around the 2.4 Ma eustatic low, bathyal pelagic chalks were being deposited, far from clastic sources, so that this and the 3.0 Ma lowstand period probably cannot be reflected in the sedimentary record. After the first pulse of uplift waned, quiet deposition, dominantly of marls, followed and the newly emerged hinterland of proto-Timor cannot have been very far away. The timing around 2.0 Ma of this facies change might well indicate control by a eustatic rise. In that case one would also expect to find a sequence boundary around 1.6 Ma, due to the next eustatic fall, but no facies change was observed there. Apart from possible age correlation errors these results seem an illustration of our conclusion that prediction of chronostratigraphic relations using the cycle chart (Haq et al., 1987) in basinal settings in tectonically active areas is uncertain, at least before some information concerning the ages of the main (tectonic) unconformities is available.

### 10.3.3 Tectonism Versus Eustasy in Foreland Basins

Foreland basins represent the "moat" formed by the depression of basement beneath the supracrustal load emplaced during contractional plate-margin tectonism (Price, 1973). A type of basin termed a *retroarc* or *retroforeland* basin is generated on the overriding plate behind a marginal orogen (Jordan, 1995), and a similar type of basin, termed a *peripheral* or *proforeland basin* is developed on the downgoing plate adjacent to a continental suture (Miall, 1995). Quantitative models for subsidence and its relation to crustal properties and to sediment loading were developed by Jordan (1981) and Beaumont (1981). DeCelles and Giles (1996) provided an essential overview of the architecture and evolution of foreland basins. A dynamic model of the coupled relationship between a fold-thrust belt and the evolving foreland basin was developed by Simpson (2006). Naylor and Sinclair (2007) developed a numerical model to explore the punctuated nature of fold-thrust-belt development, a process which has a direct effect on the episodic nature of subsidence and sedimentation in foreland basins.

Foreland basins may typically be subdivided into four distinct components (Fig. 10.17). The major controls on the tectonics and sedimentation of foreland basins are shown in Fig. 10.18. The *wedge-top system* represents a basin, or more typically, several basins, separated by thrust-fault ridges, located on top of the developing orogenic wedge. These have also been called *satellite* or *piggyback basins*. In many orogens, including the Cordilleran orogen in Canada, the wedge-top has been largely removed by erosion. The main depocentre is the *foredeep*. Outboard from that is the *forebulge*, a low-amplitude crustal upwarp representing the zone from which the basement develops the outboard downward flexure. The forebulge may be expressed as a linear inlier of older foredeep or basement strata, or it may be blanketed by basin-fill sediments, the latter occurring if there is a component of dynamic loading in the basin (Fig. 10.18). Beyond the forebulge is the *back-bulge basin*, a shallow "echo" of the foredeep, that tapers out gradually onto the craton.

The episodic nature of deformation is explained by *critical wedge theory* (Davis et al., 1983; Dahlen, 1984; Dahlen and Suppe, 1984). It can be demonstrated that as material deforms and piles up internally

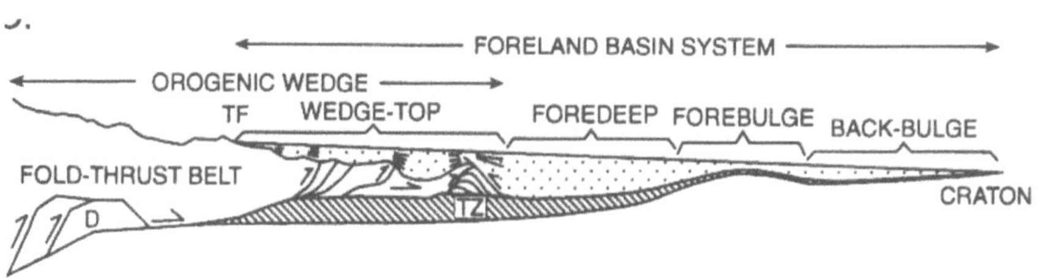


**Fig. 10.17** The subdivisions of a foreland basin (DeCelles and Giles, 1996)

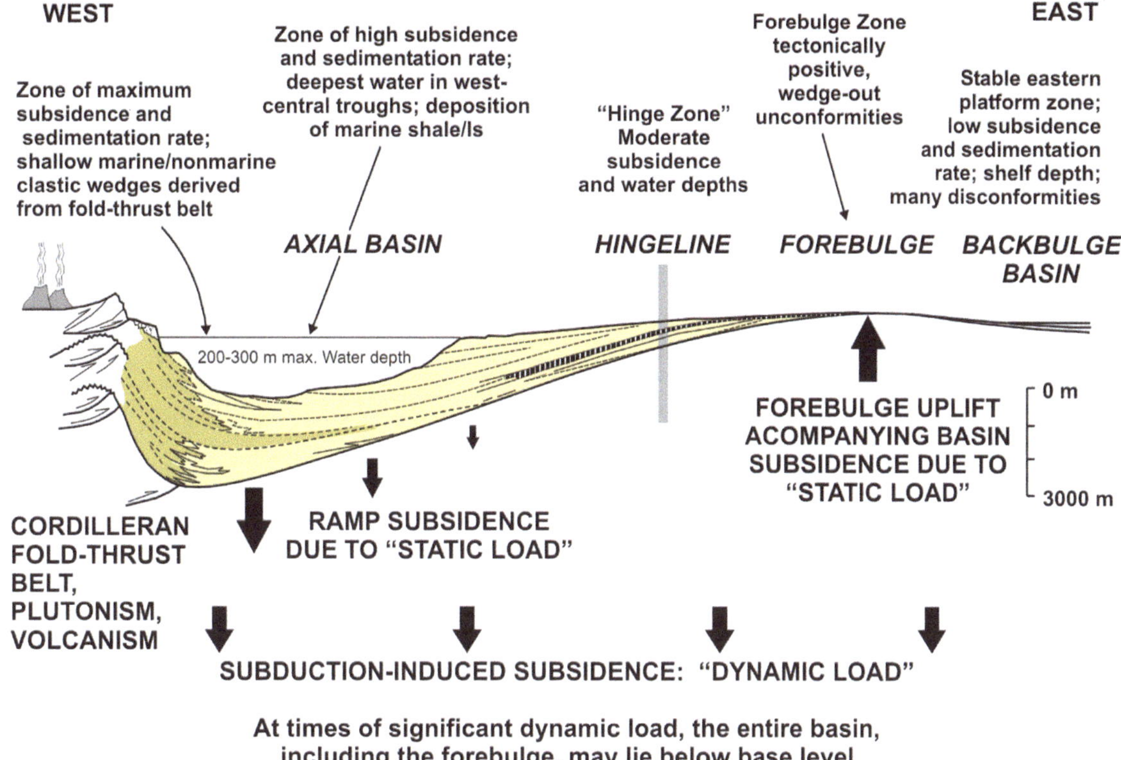


**Fig. 10.18**   The controls on subsidence and sedimentation in the Western Interior basin of North America (Miall et al., 2008)

in front of a thrust sheet, it will form a wedge shape. When the front surface slope reaches a critical angle, the wedge begins to slide along a basal detachment and begins accreting material at its front edge. This is analogous to the piling up of snow in front of a snow plow. The coherence and rigidity of a thrust sheet, and consequently its manner of deformation, depend on its composition. Episodic deformation reflects a range of controls, including the rate of regional crustal shortening, the magnitude of basal friction (which in turn depends on the nature of the detachment surface and the presence of lubricating pore fluids), and the rate at which material from the uplifting orogen is removed by erosion and sediment transfer.

> The maximum duration of activity on a thrust plane during frontal accretion is determined by the relative offset on the structure, divided by the convergence rate. The relative offset is approximated by the length of the accreted thrust sheet (Platt, 1988). This length is controlled by the thickness of accreted material. (i.e., the depth of detachment), the internal strength of the material, and the strength of the basal detachment (Naylor and Sinclair, 2007, p. 561).

Naylor and Sinclair (2007) demonstrated that the episodicity of thrust-belt activity ranges between 0.1 and 5 million years. This is of key importance from the point of view of the changing accommodation and rate of sediment delivery in the adjacent foreland basin.

The large-scale architecture of the foredeep depends primarily on the flexural strength of the underlying basement (Fig. 10.19). The Western Interior basin is an example of a foreland basin developed above ancient, rigid crust—the Canadian Shield. This basin was likely never more than 1 km deep, as indicated by modeling studies (Beaumont, 1981) and by the sedimentary facies of the basin fill. Younger crust is characterized by lower flexural rigidity. A given sedimentary load is therefore accommodated by a pattern of more localized subsidence, and the result is a narrower but deeper foredeep. The Taconic (Ordovician) basin of Tennessee is a good example. The basin fill is characterized by hundreds of metres of deep-marine submarine fan deposits (Tankard, 1986; Ettensohn, 2008).

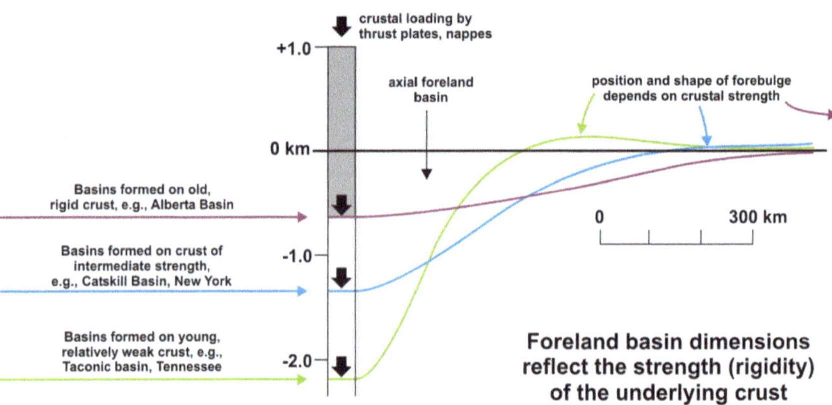



**Fig. 10.19** The cross-sectional shape of a foreland basin depends largely on the flexural rigidity of the underlying basement. Three contrasting conditions are shown (based on Beaumont, 1981)

Amongst the distinctive features of the flexural subsidence of a foreland basin is the "lozenge" shape of the basin-fill isopach (Fig. 10.20), which contrasts with the broader pattern of subsidence caused by dynamic topography (Sect. 9.3.2), as discussed below. The subsidence pattern that develop under flexural loading is distinctive (Allen et al., 1986; Cross, 1986). Subsidence rates tend to increase with time, and the curves commonly are characterized by one or more sharp inflexion points where the rate of subsidence rapidly increases as a result of a specific thrust-loading event (Fig. 10.21). The pattern of subsidence induced by flexural loading is quite different from that generated by other processes, such as the concave-up pattern that develops in extensional basins (Fig. 10.4). Figure 10.20 compares the linear, lozenge-shaped isopach pattern generated by flexural subsidence, with the broad, regional pattern of subsidence created by dynamic topography (Cross, 1986; Catuneanu et al., 1997). Subsidence rates may also vary over a $10^6$–$10^7$-year time scale because of heterogeneities in the underlying basement. The rate of

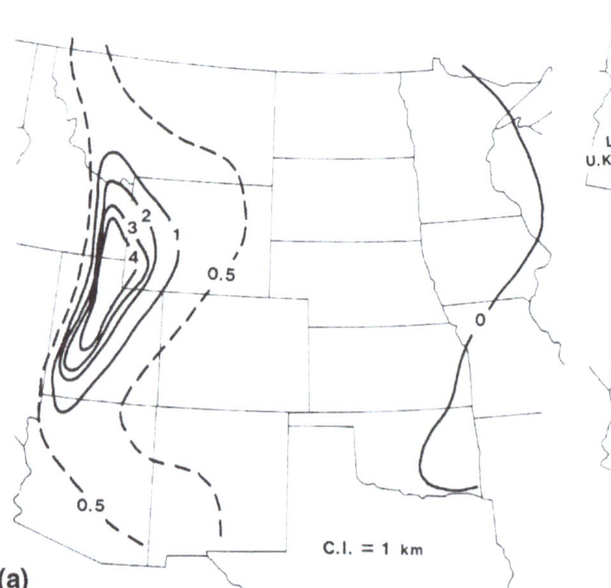



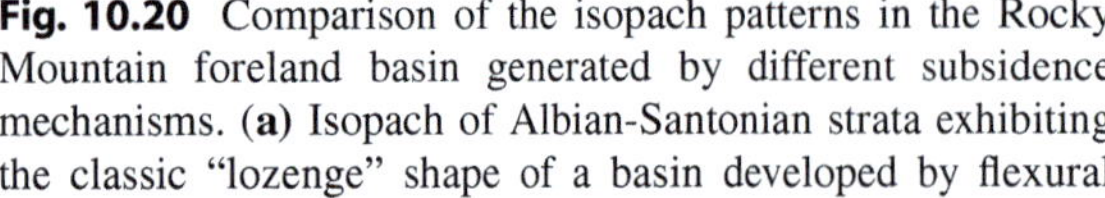

(a)

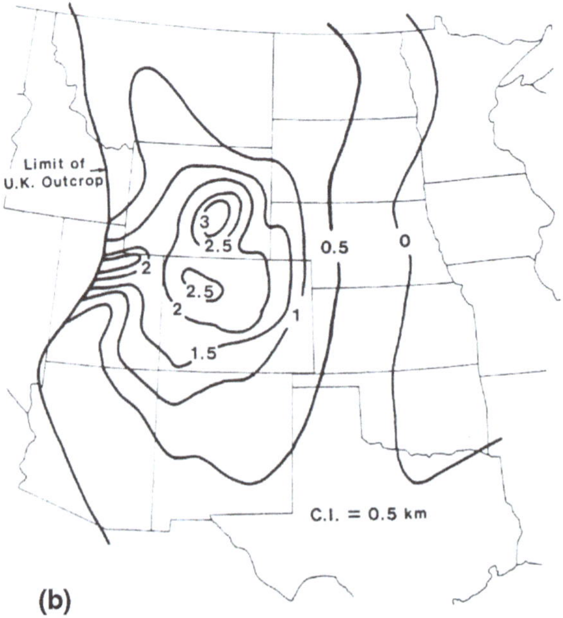



(b)

**Fig. 10.20** Comparison of the isopach patterns in the Rocky Mountain foreland basin generated by different subsidence mechanisms. (**a**) Isopach of Albian-Santonian strata exhibiting the classic "lozenge" shape of a basin developed by flexural loading. (**b**) Isopach of Campanian-Maastrichtian strata in the same area, exhibiting a more diffuse subsidence pattern. This is attributed to a regional loading effect related to dynamic topography (Cross, 1986, Fig. 4)

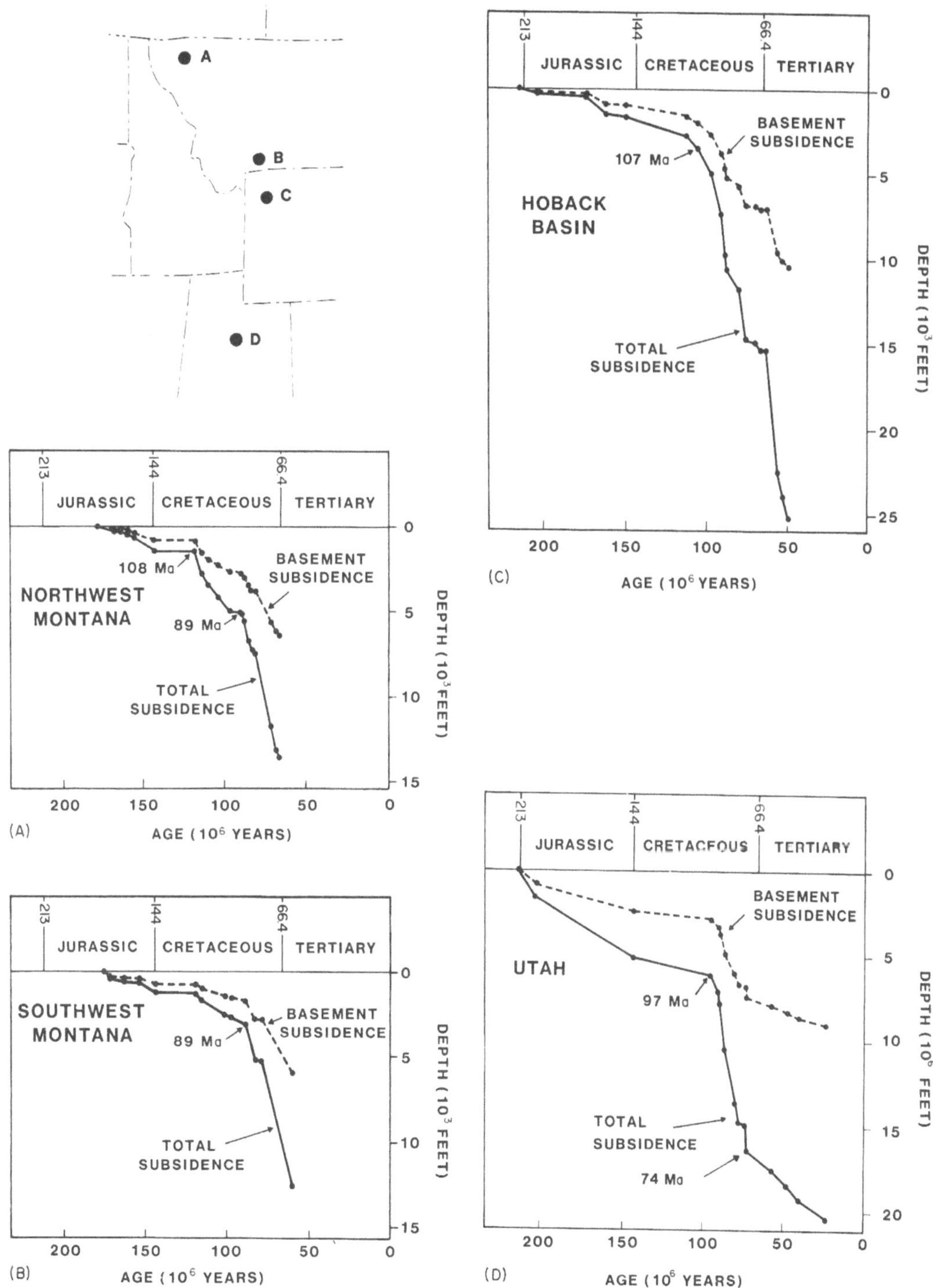


**Fig. 10.21** Examples of subsidence curves for the Rocky Mountain foreland basin, showing the characteristic *convex-up* shape, with inflexion points corresponding to thrust-loading events. This shape contrasts with the concave-up shape of subsidence curves derived from extensional basins, such as is shown in Fig. 10.4 (Cross, 1986)

flexural response to loading is governed largely by the strength and elasticity of the underlying crust, and if this varies laterally, subsidence rates will vary as different parts of the crust are loaded during contractional movements, including fold-thrust development and subduction (Waschbusch and Royden, 1992).

Simpson (2006) demonstrated that the rate at which sediment is removed from the uplifting fold-thrust belt by erosion and sediment transport has an important effect on the overall architecture of the system. With a slow rate of sediment transfer, as would occur, for example, in an arid system, sediment remains in wedge-top basins, and the foredeep remains deep and underfilled. The fold-thrust belt propagates rapidly. Where sediment transport rates are higher, the fold-thrust belt remains at a constant width, and the foredeep is rapidly filled. The imposition of the sediment load increases the crustal flexure and the basin widens, as a result. The gradual uplift of the fold-thrust belt may lead to feedback effects because of the modification to the local climate as the orogen increases in elevation. Typically, erosion and sediment-transport rates increase, resulting in accelerating sedimentation and subsidence.

Posamentier and Allen (1993) developed a useful model for the sequence stratigraphy of foreland ramp-type basins, in which the relationship between eustatic sea-level change and flexural subsidence was explored (Fig. 10.22). They suggested that in many foreland basins there may be a proximal region close to the fold-thrust belt (zone A) where subsidence is always faster than the rate of eustatic sea-level change. In this area eustasy does not lead to the generation of subaerial, erosional unconformities. Subsidence and sedimentation are more or less continuous, but may vary in rate. In more distal regions, the rate of flexural subsidence is less, and in this area, termed zone-B, eustatic fall may periodically outpace flexural subsidence, and erosional unconformities will be result. The distal margin of the basin is marked by a tectonic hinge, beyond which vertical motion of the forebulge is in the opposite direction to that of the basin. The location within the basin where the rate of tectonic subsidence and the rate of eustatic fall are equal is termed the equilibrium point. It moves basinward of the tectonic hinge during times of falling sea level, and toward the forebulge during times of rising sea level. The boundary between zone A and zone B is defined as the most proximal position reached by the equilibrium point, which it will attain at the inflection point during the falling leg of the sea-level curve. An application of this model to the Western Interior basin of North America is described in the next section.

It has long been known that foreland-basin strata are characterized by the intertonguing of marine and nonmarine strata, indicating episodic regression and transgression. For example, Weimer (1960) recognized four regional regressive-transgressive cycles in his classic work on the Upper Cretaceous of the Western Interior of the United States (Fig. 6.21), plus many minor events that he did not, at that time, attempt to correlate. More recent work on these cycles is described and illustrated in Sects. 6.2 and 7.6. Many nonmarine (molasse) successions consist of several separate wedges of coarse detritus that formed by episodic progradation into the basin (Miall, 1978; Van Houten, 1981; Blair and Bilodeau, 1988). Three scale of sequence cyclicity are exhibited by the Western Interior clastic wedge (Figs. 6.20, 6.22, 6.23, and 6.24).

It has commonly been assumed that wedges of coarse sediment prograding from a basin margin are syntectonic (Fig. 10.23a), and in many earlier studies, the dating of such wedges has been used to infer the timing of major orogenic episodes (see summary and references in Miall, 1981; Rust and Koster, 1984). It can now been seen that this interpretation is simplistic. In fact, the relationship between tectonism and sedimentation is complex, and depends on the balance between a range of controls, including sediment supply and sediment type, and the configuration and rigidity of the flexed basement. Some recent studies have demonstrated that tectonism does not necessarily coincide with the progradation of wedges of coarse sediment, but may precede such progradation by a significant period of time (Blair and Bilodeau, 1988; Jordan et al., 1988; Heller et al., 1988). In other cases, sediment input and progradation rates are adequate to keep pace with thrust-sheet loading (Sinclair et al., 1991). Tectonism, in the form of basin-margin fault movement, leads to increased rates of subsidence, particularly at the edge of the basin. Foreland basins, in particular, are developed as a result of crustal loading by overriding thrust sheets, and are characterized by syntectonic subsidence of the proximal regions of the basin. The immediate basinal response may be marine or lacustrine incursions, with ponding of coarse debris against the basin margin. Tectonism eventually leads to

**Fig. 10.22** The foreland ramp model of Posamentier and Allen (1993a). The rate of flexural subsidence decreases away from the thrust sheet. The equilibrium point is the point where the rate of flexural subsidence and the rate of sea-level change are equal. It moves across the basin as sea level rises and falls, as shown by the position of points T1 to T13. The location of the most proximal position of the equilibrium point defines the boundary between zones A and B

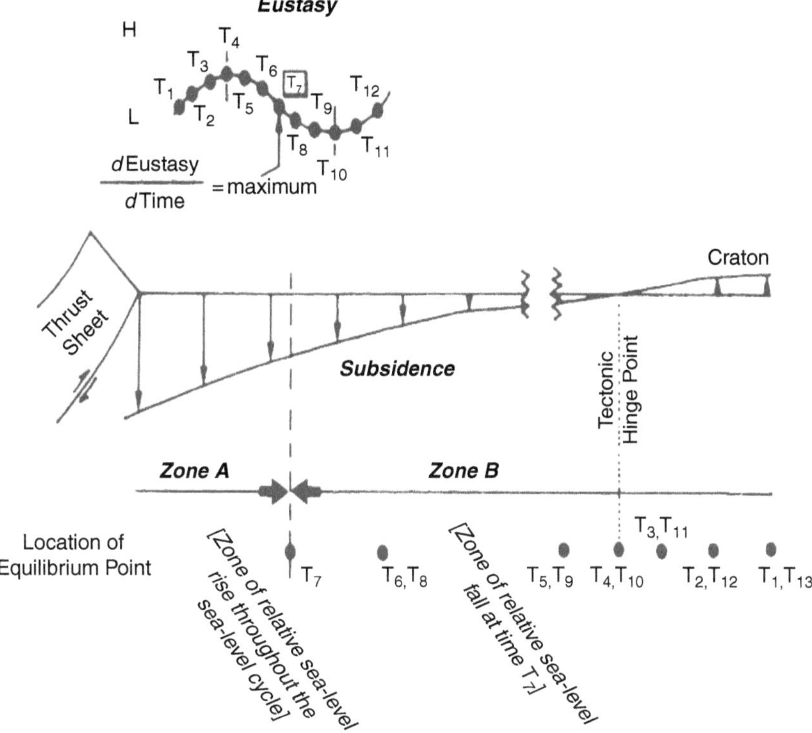



increased basin-margin relief and hence to a rejuvenation of the supply of coarse sediment, but this process takes time, and it has been suggested that clastic-wedge progradation is largely a post-tectonic phenomenon (Blair and Bilodeau, 1988; Heller et al., 1988). In this case, areally extensive coarse fluvial deposits may not be deposited until post-tectonic uplift of the basin takes place, driven by isostatic rebound following erosional unroofing of the fold-thrust belt. Following this scenario, increased rates of subsidence accompanying tectonism would be recorded by increased rates of sedimentation of fine-grained deposits, particularly shallow-marine and lacustrine sediments, at the basin margins, and at this time river systems may actually flow towards the fold belt (Burbank et al., 1988). Only in the most fault-proximal regions is tectonic activity likely to be recorded by the rapid development of a clastic wedge. Heller and Paola (1992) and Heller et al. (1988) referred to this model of coarse sedimentation as *antitectonic*, in contrast to the traditional *syntectonic* model described above (Fig. 10.23). There has been considerable debate in the literature regarding the validity of the syntectonic and antitectonic models with

respect to particular foreland basins (see summary in Miall, 1996, Sect. 10.3.6).

Some retroarc foreland basins are wider and deeper than can be accounted for by the flexural-loading model. Supracrustal loading typically can account for a basin up to about 400 km wide, whereas some basins are double this width. The Alberta basin, for example, is more than 1,000 km wide in its central part (from the fold-thrust belt to the edge of the Canadian Shield), occupying most of Alberta, plus the southern parts of the adjacent provinces of Saskatchewan and Manitoba (see Fig. 10.20, and discussion in next section). Beaumont (1982) suggested that a regional basinward tilting of the crust occurred at the time of the flexural loading, and proposed that the tilting was a response of the lithosphere to convective mantle flow coupled to a subduction zone. A pattern of secondary mantle flow beneath the overriding lithospheric plate was suggested by Toksöz and Bird (1977). Flow takes place toward the subduction zone and is drawn down parallel to the cold, descending oceanic plate. The crust is tilted toward the subduction zone by the mechanical drag effects of the downgoing current. When

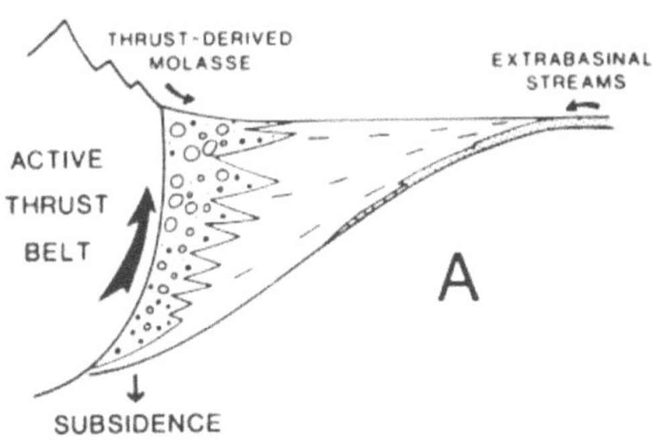



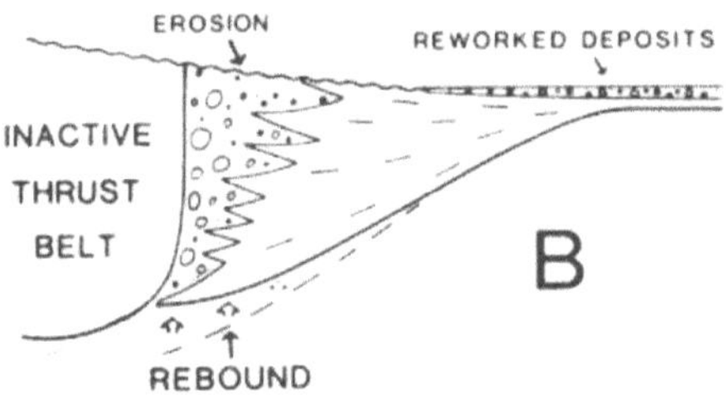



**Fig. 10.23** Two contrasting models for the development of pulses of coarse, clastic sedimentation in a foreland basin (Heller et al., 1988)

subduction ceases, buoyancy forces, coupled with erosional unroofing of the accreted terrane, combine to reverse the tilting process, leading to uplift of the basin. Mitrovica and Jarvis (1985) and Mitrovica et al. (1989) modelled this process, and showed that the width of the crust affected by the tilting increases as the subduction angle decreases. At subduction angles of 20°, the tilt effect extends for more than 1,500 km from the thrust front. The degree of tilting will also be affected by the flexural rigidity of the overriding lithosphere. The cessation of crustal shortening accompanies the termination of subduction and its associated mantle convection currents, so that the mechanical down-drag effect ceases, and the basin then tends to rebound (Mitrovica et al., 1989). The entire cycle takes a few tens of million years to complete, and would generate a cycle of relative sea-level change on a regional scale. This dependence of basin architecture and subsidence history on mantle thermal processes is an example of dynamic topography, a topic discussed in Sect. 9.3.2.

The question of causality of clastic large-scale cycles is particularly difficult to resolve in foreland basins, which, by their very nature, owe their origins to regional tectonic activity. The processes of continental collision, terrane accretion, nappe migration, thrust movement, and imbricate fault propagation, thicken and flexurally load the crust while generating tectonic highlands. Mantle thermal processes and regional flexural loading generate basement movements over a time scale of $10^6$–$10^7$ years, interspersed with episodes of isostatic uplift and erosion during orogenic quiescence. The local crustal response to regional contractional movements may generate flexural loading and vertical movements of the basin over much shorter time scales. Individual earthquakes, which may cause vertical movements of several metres, may have recurrence intervals of $10^1$–$10^2$ years, and longer term cycles of movement, over the $10^3$–$10^5$-year time band, may be related to the loading and uplift of large detached crustal sheets (terranes, nappes, thrust complexes) and their individual components, including imbricate slices (Fig. 10.24). Intraplate (in-plane) stress may transmit the effects of such tectonism throughout the basin and beyond (Sect. 10.4). These processes lead to episodic cycles of relative changes of sea-level on time scales of $10^4$–$10^7$ years. It has long been accepted that these cycles are responsible for large-scale molasse pulses (Miall, 1978; Van Houten, 1981). The numerical models of Jordan (1981) and Beaumont (1981), and the syntectonic and antitectonic models of Heller et al. (1988) provide the necessary theoretical background for explaining subsidence and sedimentation in terms of tectonism. The relationship between the time scale of sequences and possible tectonic mechanisms was examined in Pyrenean foreland basins by Deramond et al. (1993), from whose paper Table 10.1 is adapted.

Some studies appear to be specifically undertaken to explore the relationship between sedimentation and active basin-margin tectonism, and succeed in developing tectonic models for sequence development, but then offer correlations with the global cycle chart and go on to discuss eustatic control. The research by Deramond et al. (1993) in the tectonically highly active Pyrenean foreland basins is a good example of this. They referred to "tectonically enhanced" unconformities and stated "The apparent correlation between the two groups of independent phenomena is an artefact of the method which calibrates the tectonic evolution by comparison with eustatic fluctuations." This form of "calibration" is, of course, circular reasoning.

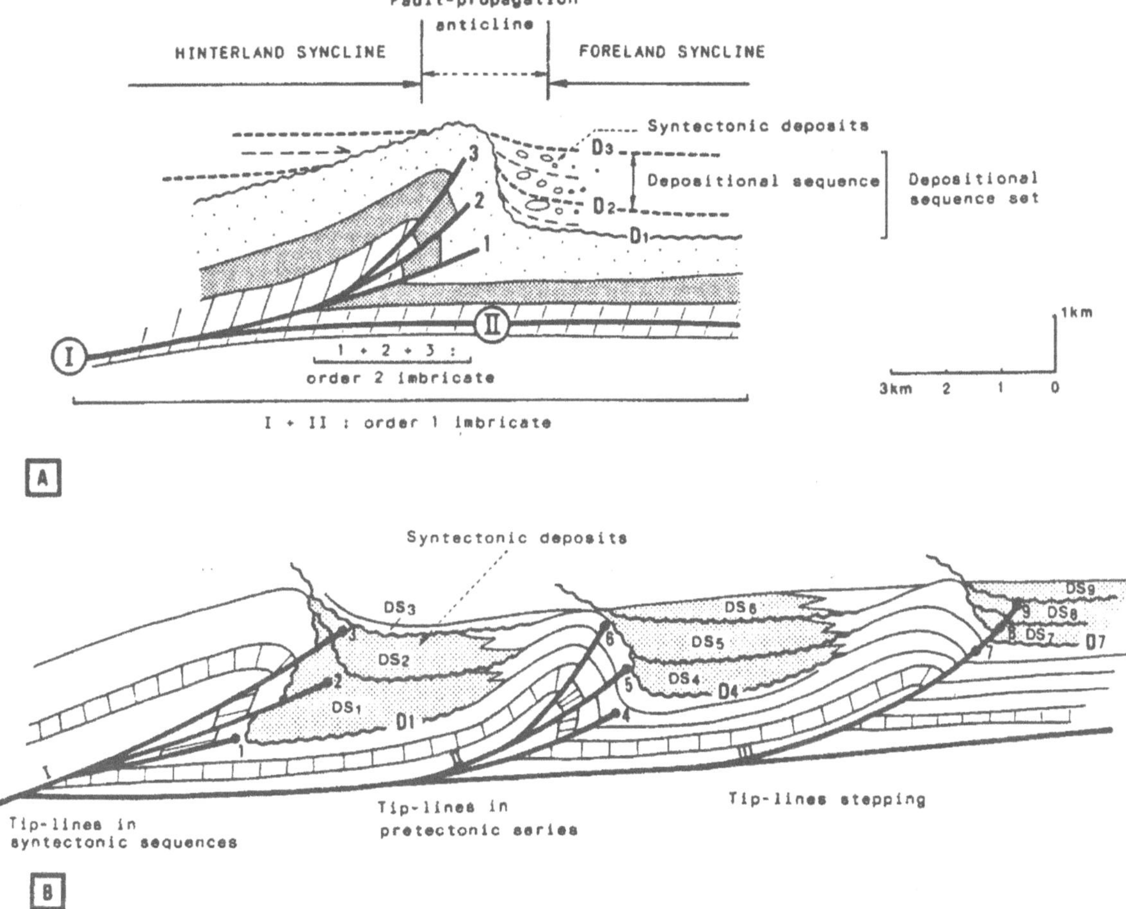



**Fig. 10.24** The relationship between the development of a fold-thrust belt and the stratigraphy of the adjacent foreland basin. Unconformities, numbered D1–D9, develop as imbricate thrust slices develop. Uplift of each slice is recorded by a corresponding numbered unconformity in the basin (Deramond et al., 1993)

Many workers have emphasized eustasy as a major control of Appalachian Basin stratigraphy (e.g., Dennison and Head, 1975). However, as Ettensohn (2008, p. 114) stated: "if the Appalachian Basin is in large part a composite foreland basin generated by deformational loading during four adjacent and nearly continuous, craton-margin orogenies, tectonism in the form of flexural subsidence and uplift and in the reactivation of various basement structures must have been an influence so overwhelming at times as to muffle the effects of coeval eustasy." Ettensohn (2008) suggested that stratigraphic recurrence interval might be a key interpretive clue, with cycles of durations of $10^6$–$10^8$ years likely of tectonic origin and those of $10^4$–$10^5$ years caused by glacioeustasy. However, high-resolution stratigraphic studies discussed in this section have shown that sequences with frequencies comparable to those of glacioeustatic fluctuations are widespread. As Ettensohn (2008, p. 114) noted, "movements on individual structures may induce comparable, small-scale changes at similar rates, but may be distinguished because of their very local influence."

It is becoming possible to suggest approximate correlations of individual clastic pulses with specific tectonic events, such as times of terrane accretion (e.g., Cant and Stockmal, 1989; Stockmal et al., 1992), or thrust faulting (e.g., Burbank and Raynolds, 1988; Deramond et al., 1993). More difficult to resolve is the question of causality for intertonguing fine-grained marine and nonmarine clastic successions far from sediment sources, where the influence of basin-margin

**Table 10.1**  The relationship between tectonic processes and stratigraphic signatures in foreland basins, at different time scales

| Duration (million years) | Scale | Tectonic process | Stratigraphic signature |
| --- | --- | --- | --- |
| >50 | Entire tectonic belt | Regional flexural loading, imbricate stacking | Regional foredeep basin |
| 10–50 | Regional | Terrane docking and accretion | Multiple "molasse" pulses |
| 10–50 | Regional | Effects of basement heterogeneities during crustal shortening | Local variations in subsidence rate; may lead to local transgressions/regressions |
| >5 | Regional | Fault-propagation anticline and foreland syncline | Sub-basin filled by sequence sets bounded by major enhanced unconformities |
| 5–0.5 | Local | Thrust overstep branches developing inside fault-propagation anticline | Enhanced sequence boundaries; structural truncation and rotation; decreasing upward dips; sharp onlaps; thick lowstands, syntectonic facies |
| <0.5 | Local | Movement of individual thrust plates, normal listric faults, minor folds | Depositional systems and bedsets geometrically controlled by tectonism and bounded by unconformable bedding-plane surfaces. Maximum flooding surfaces superimposed on growth-fault scarps. Shelf-perched lowstand deposits |

This table was adapted mainly from Deramond et al. (1993), with additional data from Waschbusch and Royden (1992), Stockmal et al. (1992).

tectonism is not obvious. The relative importance of tectonism and eustasy in generating these sequences is by no means clear, particularly where the occurrence of contemporaneous eustasy cannot be independently demonstrated. DeCelles (2004, pp. 150–151) said, regarding the question of tectonism versus eustasy, in the Rocky-Mountain foreland basin:

> … it is worthwhile to point out that thrusting events in the Sevier belt were mere increments of the larger deformation field that formed the Cordilleran orogenic belt, that regional trust loading was probably little affected by events in the frontal Sevier belt alone, and that loading was more or less continuous in time, rather than sporadic. … Thus, it may be unwise to look toward the Sevier thrust belt for individual thrusting and erosional events that could have controlled the development of individual depositional sequences, while excluding the potential effects of changing eustatic sea level, paleoclimate, sediment flux, and intrinsic processes.

These problems can be exemplified by a brief review of two North American basins where the stratigraphic record is particularly well know, the Cretaceous cyclicity of the North American Western Interior Seaway, and the late Paleozoic succession of the Appalachian foreland basin. Pyrenean basins and the Himalayan foredeep are also discussed briefly.

### 10.3.3.1  The North American Western Interior Basin

The sequence stratigraphy of this basin is described and illustrated in Sect. 6.2.1 (Figs. 6.19, 6.20, 6.21, 6.22, 6.23, 6.24, 6.25, 6.26, 6.27, 6.28, and 6.29), and Sect. 7.6 (Figs. 7.53, 7.54, 7.55, 7.56, 7.57, 7.58, and 7.59) (see also Miall et al., 2008). Ten major transgressive-regressive cycles have been recognized in the Rocky Mountain foreland basin (Fig. 6.21; Weimer, 1960, 1986; Kauffman, 1984). Transgressions are characterized by the development of thick and areally extensive mudstone units (e.g., Mancos Shale), and by fine-grained limestones (including chalk) in areas distant from sediment sources (e.g., Niobrara and Greenhorn Formations). Regressions gave rise to extensive clastic wedges, in which nonmarine sandstones and conglomerates pass basinward into shoreline and shelf sandbodies. The Indianola and Mesaverde groups of Utah and Colorado are good examples of major regressive stratigraphic packages, which contain numerous minor transgressive-regressive cycles nested within them (Fig. 6.20; Weimer, 1960; Molenaar, 1983; Fouch et al., 1983; Lawton, 1986a, b; Swift et al., 1987; Van Wagoner et al., 1990, 1991; Van Wagoner and Bertram, 1995; Olsen et al., 1995; Yoshida et al., 1996).

Are the broad patterns of sedimentation in the Western Interior Basin the result of regional flexural subsidence or eustatic sea-level change? Figures 10.25 and 10.26 provide some of details of the broad,

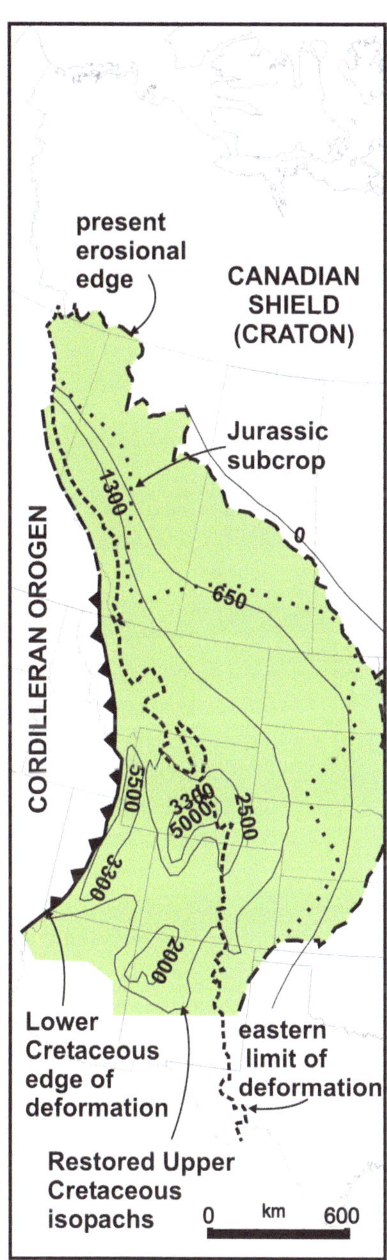


**Fig. 10.25** The Western Interior basin of North America. Note the position of the subcrop of Jurassic strata beneath the Cretaceous. The widening of the basin during the Cretaceous is attributed to subsidence caused by dynamic topography (Miall et al., 2008)

regional setting. The tectonic episodes that generated the clastic pulses are loosely correlated to the terrane-accretion events on the western continental margin (Fig. 10.26). The accretion process caused lithospheric delamination, which resulted in the terranes being emplaced onto the continental margin above a major detachment surface. This increased their flexural reach, and led to the generation of multiple pulses of contractional thrusting and flexural subsidence in the foreland basin, some hundreds of kilometres inboard from the continental margin.

The basin widened substantially during the Late Cretaceous, as a result of a major change in dynamic topography (Fig. 10.25: see also Sect. 9.3.2 and Fig. 10.17). Shifts in the primary locus of flexural loading of the continental margin, and a gradual cratonward migration of the fold-thrust belt can be tracked by changes in the position of the forebulge (Fig. 10.27). The importance of these tectonic processes in the generation of the stratigraphy is illustrated by the reconstruction of the stratigraphic architecture of a forebulge shown in Fig. 10.28.

Recent work by Pang and Nummedal (1995) has demonstrated the importance of basement heterogeneity as a control on subsidence patterns, and thereby confirmed the importance of flexural subsidence as a primary control of basin architecture. Pang and Nummedal (1995) constructed flexurally back-stripped subsidence profiles for six transects through Upper Cretaceous strata in the Western Interior Basin between Montana and New Mexico (Figs. 10.29 and 10.30). Variations in subsidence patterns reveal two important tectonic controls: regional differences in basement rigidity, the effects of which were discussed by Waschbusch and Royden (1992; see previous section), and the presence of "buttresses" in the basement that slowed or prevented crustal shortening during thrust movements, resulting in smaller flexural effects. The second point is exemplified by profiles (a) in Montana, (b) in Wyoming, and (f) in Utah-Colorado (Fig. 10.30), along which the rate of subsidence during the Late Cretaceous did not vary much, indicating very weak flexural effects. Reactivated basement structures and the presence of rigid basement blocks are cited as the reasons. Note also the effect of the Douglas Creek Arch (DCA in Fig. 10.30, profile e), near the Utah-Colorado border.

Some of the cycles with $10^6$-year periodicities in the Western Interior Basin may be global in origin,

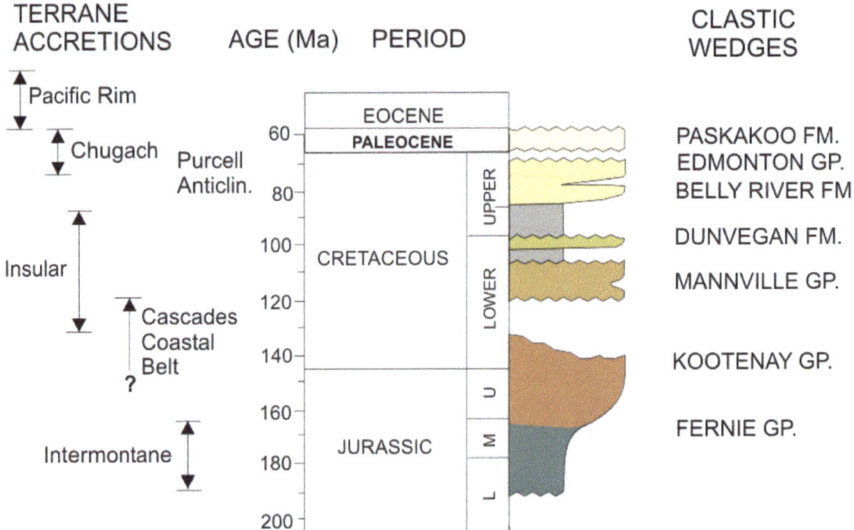



**Fig. 10.26** The major clastic pulses in the Alberta basin, plotted against the estimated times of terrane accretion on the Pacific margin of North America (adapted from Stockmal et al., 1992). See also Fig. 6.28, for additional details of these large-scale clastic sequences

and related to eustatic changes in sea level. Partial correlation between the sea-level curve for the western United States and that for northern Europe has been claimed (Hancock and Kauffman, 1979; Weimer, 1986), although correlations are bedeviled by the difficulties in reconciling various chronostratigraphic time scales. Arguments regarding the validity of this type of correlation, and the precision of the dating on which such correlation depends, are considered in Part IV. Kauffman (1984) claimed that there is a consistent correlation among transgression, thrusting (in Wyoming and Utah), and volcanism in the Western Interior. He also claimed that his $10^6$-year cycles are correlatable between North America and Europe (Hancock and Kauffman, 1979). The regional tectonic associations suggest that the cycles of sea-level change are regional in origin, related to episodic loading in the foreland fold-thrust belt and consequent basin subsidence effects. Yet, if the cycles can be correlated into Europe this would imply eustasy as the main control. All the tectonic events in the Cordillera are presumably tied to convergent plate movements between the paleo-Pacific Ocean and the North American continent. Yet Kauffman argued that changes in rates of seafloor spreading, which caused episodic tectonism, were also widespread enough to generate cycles of eustatic sea-level change, as demonstrated by the intercontinental correlation of his cycles in the Western Interior. One possibility is that $10^6$-year cyclicity represents the effects of regional plate rifting and convergence superimposed on $10^7$-year cycles of ridge length

and volume changes. However, the initial correlations between stratigraphic cyclicity, volcanism and thrust faulting on which these arguments are based seem weak.

There is no doubt that some of the transgressions are contemporaneous with thrust-faulting episodes within the Sevier orogen of Utah, possibly including episodes in the Albian, Santonian and Maastrichtian (compare Kauffman, 1984; Lawton, 1986a, b; Weimer, 1986). Other thrusting events within the Sevier orogen are not clearly correlated with regional changes in sea level in the basin, but are correlated with the development of major clastic wedges that prograded across the basin margins. Villien and Kligfield (1986), Lawton (1986a, b), and DeCelles et al. (1995) indicated a link between thrusting and clastic-wedge formation in Utah between the mid-Albian and the late Eocene, although it is not possible to provide the tight correlation between individual tectonic and stratigraphic events that is now available for other foreland basins, such as parts of the Himalayan foredeep and Pyrenean basins (Sect. 10.3.3.3), and on these grounds we cannot yet distinguish between the syntectonic and antitectonic model for all the units in this clastic wedge. An important exception to this generalization is the work of Liu et al. (2005), discussed below.

Are individual clastic tongues within major wedges (those discussed in Sect. 7.6) tectonic or eustatic in origin? There is no doubting an overriding tectonic control of sedimentation for many of these successions. Paleocurrent and petrographic evidence indicate

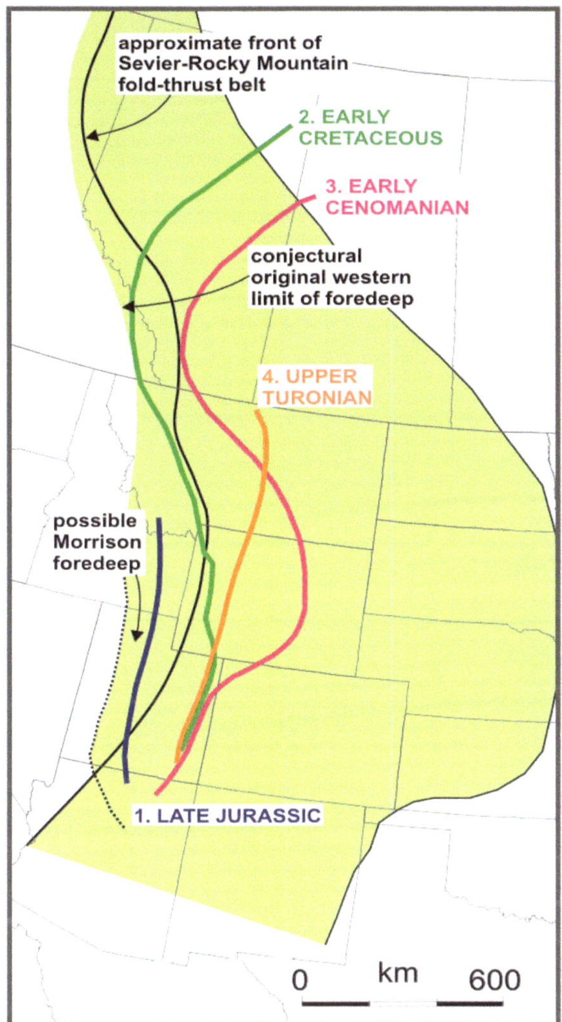



**Fig. 10.27** The shifting position of the forebulge in the Western Interior basin. Data for the US from DeCelles (2004); data for Canada from Yang and Miall (2008, 2009)

shifting sediment sources and changes in regional paleoslope during deposition of the Mesaverde Group and associated units in Utah. Lawton (1986a, b) documented unroofing of intrabasin uplifts from which early basin-fill sediments were cannibalized, and changes in dispersal patterns related to basin tilting. The volume of sediment within the wedge is too great to have been controlled by passive changes in sea level (see also Galloway, 1989b). But the question remains whether tectonically induced sediment input was modulated by sea-level control, leading to high-frequency ($10^4$–$10^5$-year) cyclicity along the fringes of the clastic wedge (such as in the Book Cliffs of

Utah: see Swift et al., 1987; Van Wagoner et al., 1990). Posamentier and Vail (1988) argued that fluvial coastal-plain progradation is switched on during initial sea-level fall after a time of highstand, as a result of the lateral shift in stream profiles and the creation of sedimentary accommodation space. However, Miall (1991a) argued that this concept is faulty on several grounds (e.g., base-level fall normally leads to incision), and maintained that considerations of sediment supply, driven by tectonism, are the major causes of coastal-plain progradation (see a more extended discussion of this problem in Miall, 1996; Holbrook et al., 2006).

The sequence architecture of the Castlegate Sandstone and equivalent units in the Book Cliffs of Utah (Fig. 10.31) was interpreted by Yoshida et al. (1996) and Miall and Arush (2001a) in terms of the simultaneous action of two types of tectonic control acting over different time scales. The single Castlegate sequence originally mapped at Price Canyon (Olsen et al. 1995) has been shown to consist of two amalgamated sequences (Willis, 2000; Yoshida, 2000), and represents one of a series of such cycles that are interpreted as the product of cyclic variations in the rate of subsidence on an approximately 5 million years time scale. Tectonic control is clearly indicated by the regional shifts in paleoslope from one sequence to the next (Fig. 10.32). During periods of slow subsidence, basinward sediment transport was facilitated, following the antitectonic model of Heller et al. (1988), leading to the development of extensive, sheet-like depositional units. The lower Castlegate Sandstone and the Bluecastle Sandstone are examples. These deposits represent tectonically-generated nonmarine analogs of lowstand systems tracts. They extend more than 150 km down depositional dip from the Price Canyon area. The sequence boundaries at their base may represent intervals of considerable erosion. The sequence boundary that truncates the marine shale unit constituting the Buck Tongue truncates tens of metres of strata in an updip (westerly) direction, and is a good example of what Miall and Arush (2001b) termed a *cryptic sequence boundary*, because it superimposes similar facies across a contact that is no different on the outcrop scale from a typical channel scour surface. An increase in the rate of subsidence during the deposition of the Upper Castlegate Sandstone (and equivalents to the east) led to higher rates of generation of accommodation

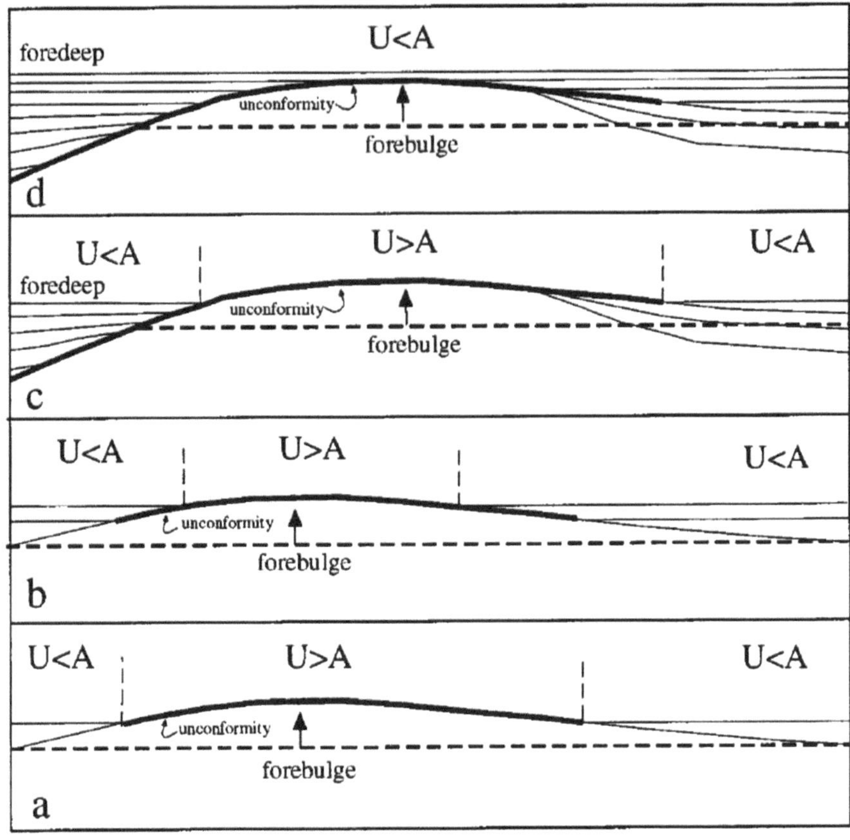



**Fig. 10.28** The evolution of a forebulge, showing the development of an unconformity as a result of uplift and cratonward migration, and the onlap of foredeep and back-bulge-basin strata (Currie, 1997)

space, with greater preservation potential for floodplain deposits, looser channel stacking patterns (Wright and Marriott, 1993), and possibly subtle changes in fluvial style (Schumm, 1993). The upper Castlegate Sandstone, which contains evidence of tidal invasion, is an example of this phase, corresponding to a transgressive systems tract.

This interpretation represents a modification of the model of Posamentier and Allen (1993), in which Yoshida et al. (1996) suggested that the long-term rate of flexural subsidence underwent cyclic variations. The amalgamated channel sandstones of the lower Castlegate Sandstone, the Bluecastle Tongue, and the other "lowstand" sandstones of Olsen et al. (1995), are attributed to slower rates of long-term subsidence in zone A, not to a geomorphic adjustment by the rivers to a short-term fall in base level in zone B, as in the original model of Posamentier and Allen (1993). Studies of the timing of thrusting in the Sevier orogen (DeCelles et al. 1995) indicate a pause in thrusting during the mid-Campanian, which may correspond to one or more of these times of uplift. In fact it is possible that considerable uplift and erosion took place throughout the project area, generating the erosional sequence boundaries that bracket the Castlegate sequence. This may in part explain why the sequence does not thicken westward toward the proximal part of the basin.

Other researchers have speculated about the possible tectonic control of high-frequency sequence architecture. Leithold (1994) pointed out that the Upper Cretaceous succession of southern Utah contains sequences with a frequency in the Milankovitch band and speculated about possible glacioeustatic mechanisms, but also discussed episodic thrust loading. Similarly, Kamola and Huntoon (1995) documented the pattern of repetitive parasequence progradation in the Blackhawk Formation, the unit below the Castlegate Sandstone in the Book Cliffs, and suggested episodic movement of imbricate slices in thrust sheets as a cause of loading and stratigraphic cyclicity.

A particularly clear relationship between sedimentation and tectonism (*contra* DeCelles, 2004, see quote

**Fig. 10.29** Location of transects a–f across the Western Interior Basin of the Rocky Mountain states, shown in Fig. 10.30 (Pang and Nummedal, 1995)

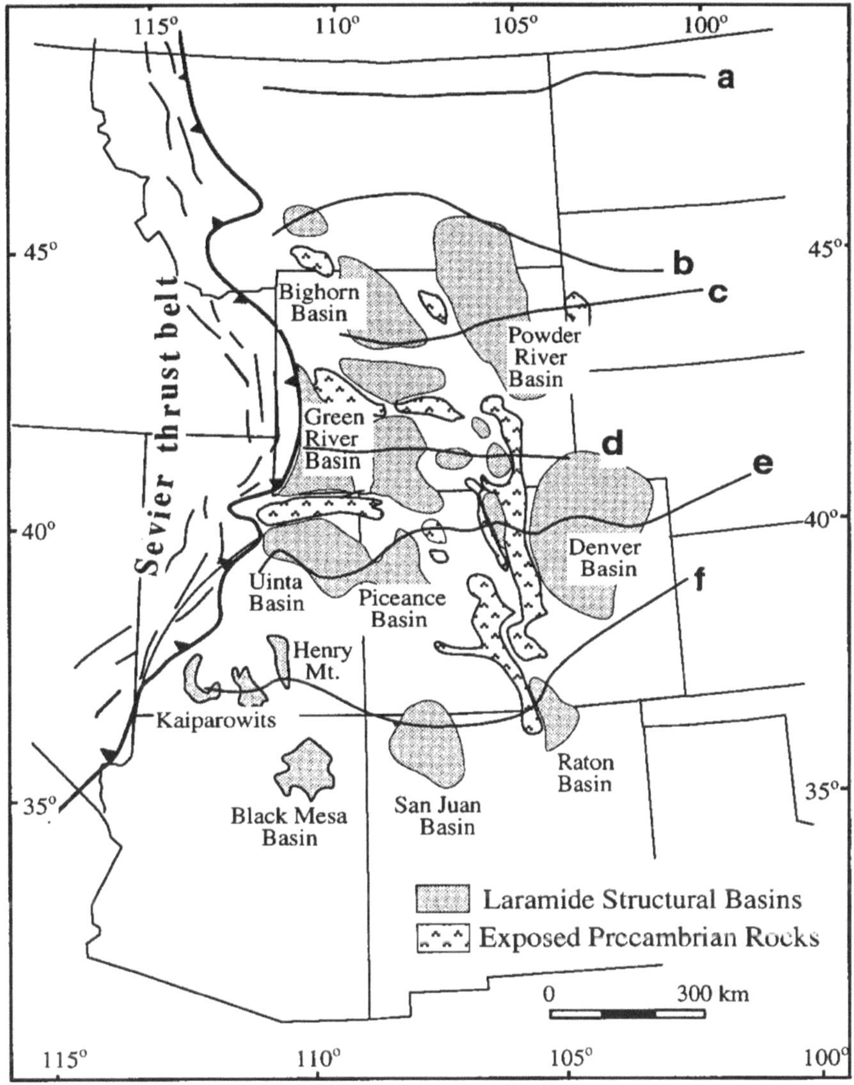


above) was developed by Liu et al. (2005) for the Upper Cretaceous stratigraphy of southern Wyoming (Location and structural cross-section in Figs. 10.33 and 10.34). Basic lithostratigraphic and chronostratigraphic documentation of these rocks is provided in Figs. 6.25, 6.26, and 6.27. Conglomerate units are interpreted as syntectonic products of uplift movement on specific major thrust-fault plates along the Sevier orogen (Fig. 6.25). The evidence for this is that the conglomerates are cut by and rest on the thrust faults. Sevier thrusting progressed gradually eastward, beginning with the Paris-Willard thrust from 140 to 112 Ma, and ending with the movement of the Absaroka thrust from 84 to 78.5 Ma.

The stratigraphy may be subdivided into five megasequences, the average duration of which is 6.24 million years (Fig. 6.27). Subsidence analysis using backstripping methods identified differential rates of subsidence along the line of section. Not surprisingly, the most rapid rates were in the west, decreasing eastward. A locus of minimum subsidence was, however, identified for each megasequence, and is interpreted as indicating the position of the forebulge. This was located near well-section #5 during the deposition of megasequence #1, moving eastward to near well #7 during the deposition of megasequence 4. There is no evidence of erosional episodes associated specifically with the forebulge, and Liu et al. (2005) suggested that this

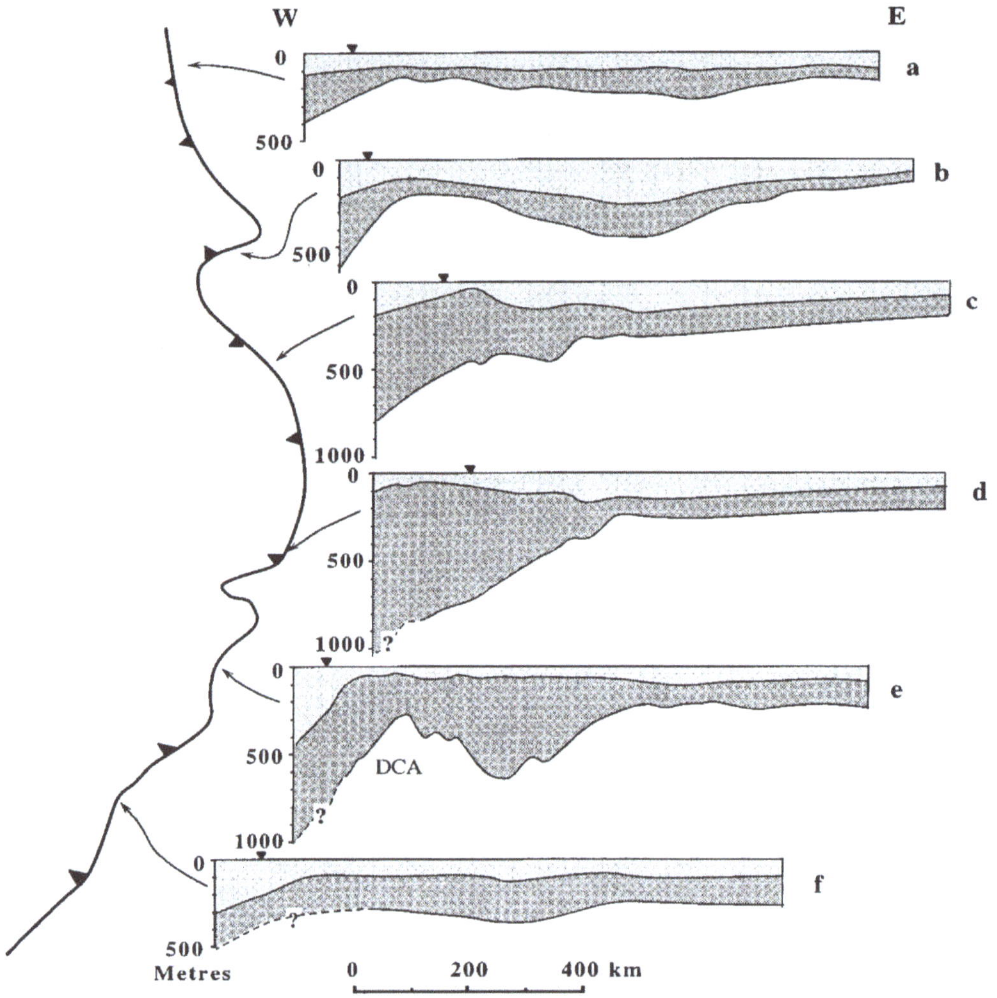

**Fig. 10.30**  Flexurally backstripped subsidence profiles along the transects shown in Fig. 10.29. *Light shading* indicates subsidence during the 97 (94 on e and f) to 90 Ma interval, *dark shading* shows subsidence during 90–80 Ma (Pang and Nummedal, 1995)

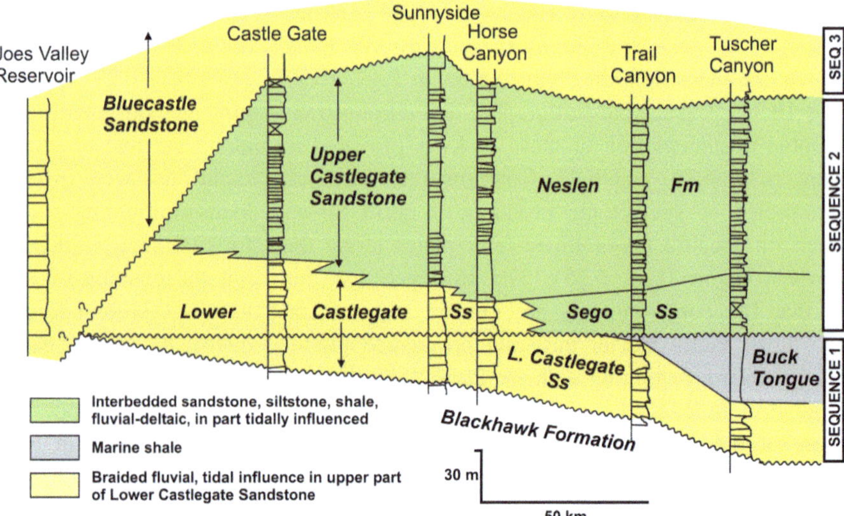


**Fig. 10.31**  Sequence stratigraphy of the Castlegate Sandstone and associated units, Book Cliffs, Utah (Miall and Arush, 2001a)

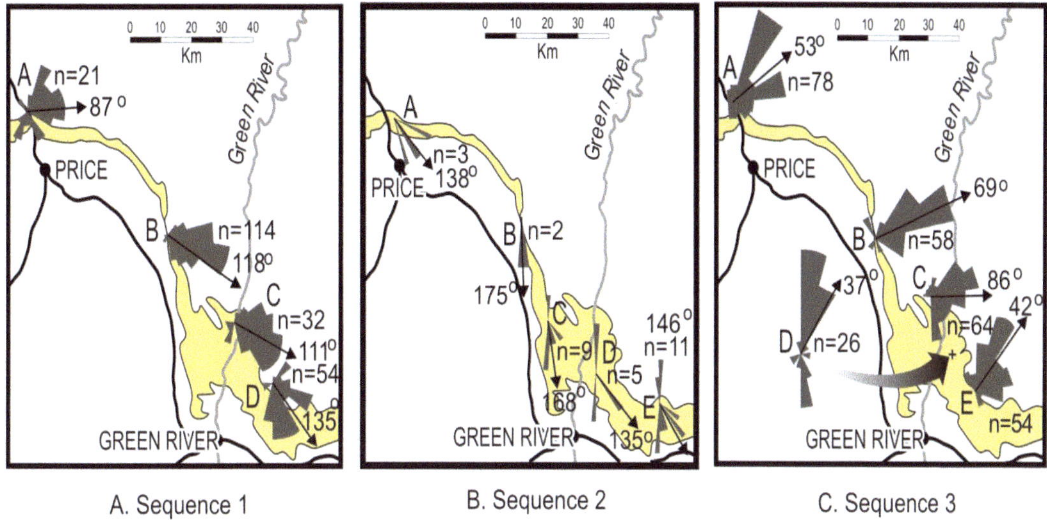


**Fig. 10.32** Regional paleocurrent patterns in the three sequences of the Castlegate-Bluecastle succession (sequence stratigraphy shown in Fig. 10.31) (Willis, 2000)

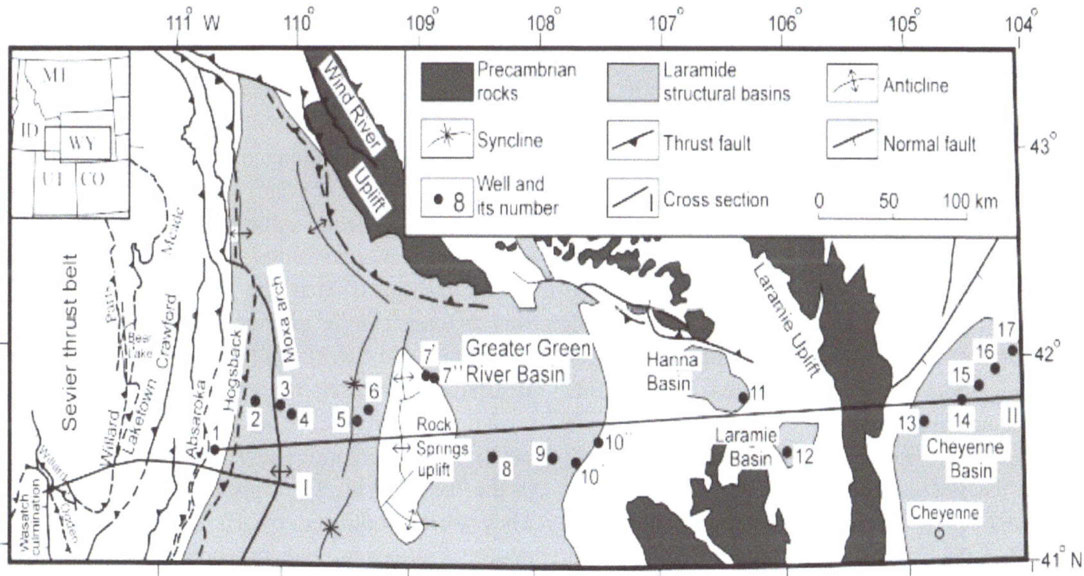


**Fig. 10.33** The Sevier fold-thrust belt and foreland basin of southwestern Wyoming and adjacent areas. Section I is a structural cross-section (Fig. 10.34) and Section II is a stratigraphic synthesis of the Cenomanian-Maastrichtian stratigraphy (Liu et al., Fig. 1)

may indicate sediment-induced subsidence, or "other subsidence not related to thrust loading" such as the dynamic loading discussed earlier (see Fig. 10.18).

The Greenhorn and Niobrara limestones are typically interpreted as the product of periods of eustatic sea-level highs (e.g., Kauffman, 1984), and it is unclear from the synthesis of Liu et al. (2005) how the phases of tectonism identified in Wyoming might have affected these deposits. They are more widespread than the Sevier tectonism, and are probably, therefore, genetically unrelated. This does not necessarily confirm a eustatic origin for these limestones, however,

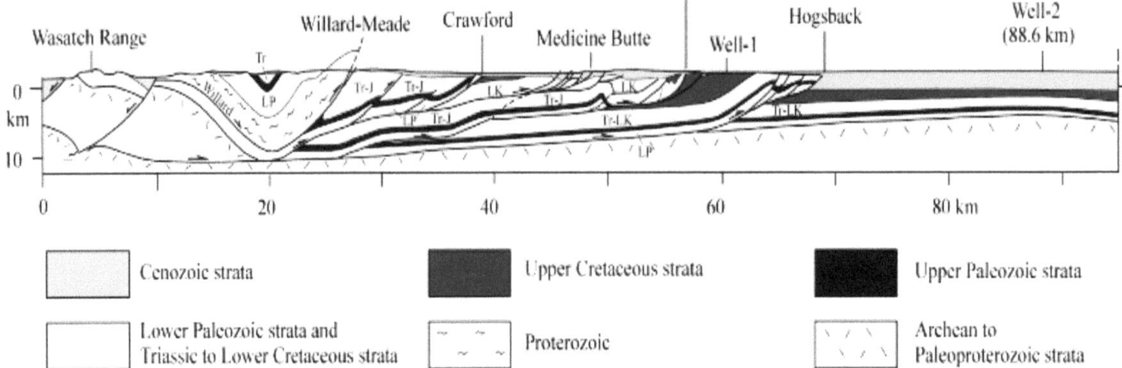


**Fig. 10.34** Structural cross-section through the Sevier fold-thrust belt, northeast Utah and southwest Wyoming. Location is shown by Section I in Fig. 10.32 (Liu et al., 2005, Fig. 2)

as other mechanisms (in-plane stress, dynamic topography) might be implicated. Liu et al. (2005) suggested that the widespread nature of the unconformity that caps megasequence #1 cannot be explained by foreland-basin processes. It extends from the foredeep to the back-bulge area and appears to truncate a comparable thickness of strata throughout the study area, except for the most proximal part of the foredeep. This feature of the stratigraphy is likewise also interpreted as a result of regional processes, such as a dynamic topography uplift or a eustatic sea-level low.

Figure 10.35 provides a depositional model for the Upper Cretaceous stratigraphy of southern Wyoming. Progradation of coarse clastics is associated with episodes of uplift along the thrust faults. The thickness of megasequence 3 is interpreted as a product of dynamic loading superimposed on foreland-basin subsidence. Uplift and erosion following this phase generated the unconformity that caps and truncates megasequence 3 towards the west, including, it is interpreted, a thick wedge of proximal Rock Springs sandstones. These eroded materials were transported eastward to form the relatively thin lens of deposits constituting megasequence #4.

Another detailed study of Upper Cretaceous strata in Wyoming revealed a suite of tectonically generated unconformities much more closely spaced in time. Vakarelov et al. (2006) utilized detailed bentonite correlations through the Cenomanian Frontier Formation to map four angular unconformities within strata spanning a 2.2 million years interval (Figs. 4.12 and 10.36). The unconformity surfaces were not generated by channel scour but by transgressive erosion, as indicated by the presence of the firm-ground *Glossifungites* ichnofacies below the unconformities, and pebble lags, sharks teeth and offshore marine faunas overlying the surfaces. One of these unconformities has been subsequently interpreted as the product of tectonic enhancement of an eustatic event (Gale et al., 2008); but see the discussion of this in Sect. 14.8.

Catuneanu et al. (1997, 1999, 2000) examined the sequence stratigraphy of the Bearpaw Formation (Campanian-Maastrichtian) across the Western Interior Basin from Alberta to Manitoba, a predominantly marine unit that represents the last marine transgression in the basin. Their examination was extended into younger (Paleocene) strata by Catuneanu and Sweet (1999) (Fig. 10.37a). The main Bearpaw transgression occurred in response to long-term flexural subsidence, and is one of several episodes of $10^6$-year oscillations that affected accommodation in the basin during the Upper Cretaceous-Paleogene. They were modulated by $10^5$-year cycles of subsidence. Catuneanu and his colleagues documented a succession of cycles with frequencies of $10^4$–$10^6$ years. Marine lithofacies, as indicated by wireline logs, reveal a pattern of "reciprocal" changes in water depth between the foredeep and the forebulge of the basin (Fig. 10.37b). During orogenic episodes, the foredeep deepens, generating a transgressive systems tract, and the forebulge is uplifted, forming a regressive systems tract (Fig. 10.38A). The switch between these modes takes place at a hingeline at the inner edge of the forebulge. During tectonically quiescent periods, the movements of these two components of the basin reverse (Fig. 10.38B). The "lozenge"

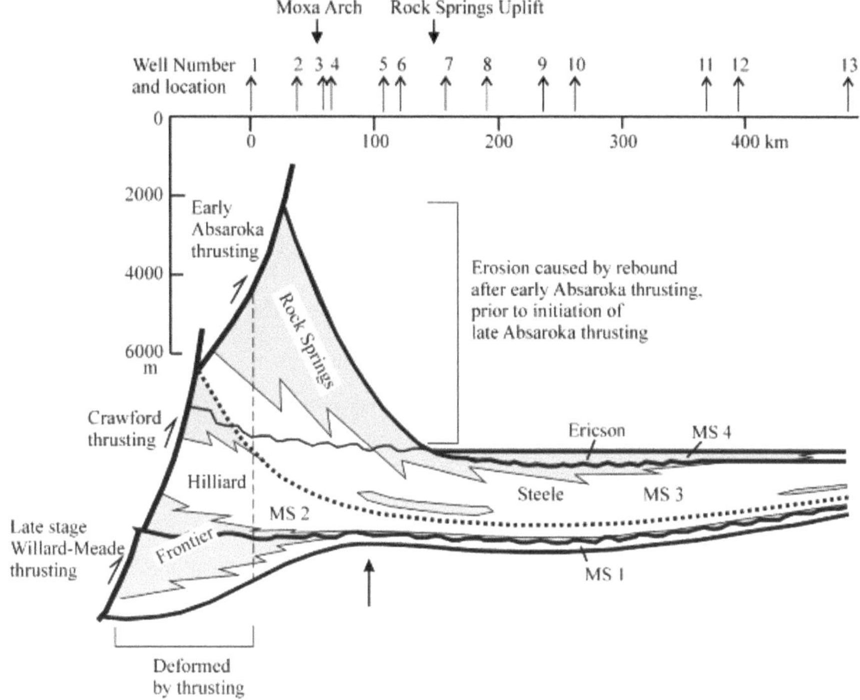


**Fig. 10.35** Depositional model for the Upper Cretaceous stratigraphy of southern Wyoming. *Grey shade* = sandstone, *white* = shale and limestone (Liu et al., 2005, Fig. 8)

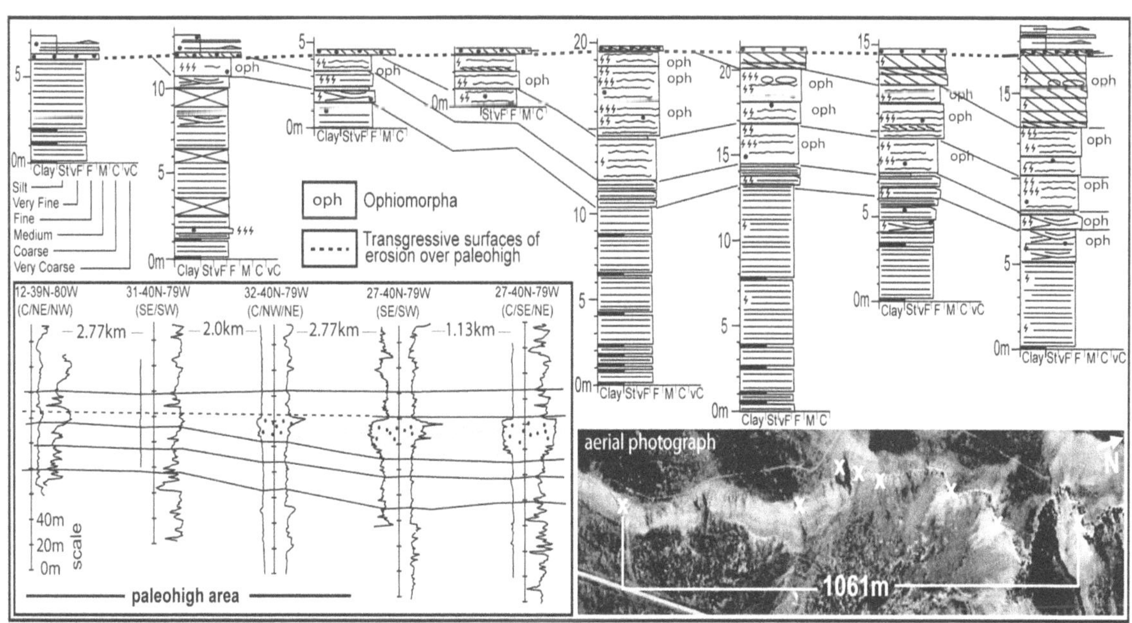


**Fig. 10.36** Examples of tectonically driven erosion of Second Frontier Sandstone in outcrop and subsurface. *Top*: Eight measured sections along continuous outcrop show increased removal of high-energy facies to left. *Bottom right*: Locations of measured sections on aerial photograph in same order from left to right. *Bottom left*: Expression of same erosional event in resistivity well logs. (Vakarelov et al., 2006, Fig. 2)

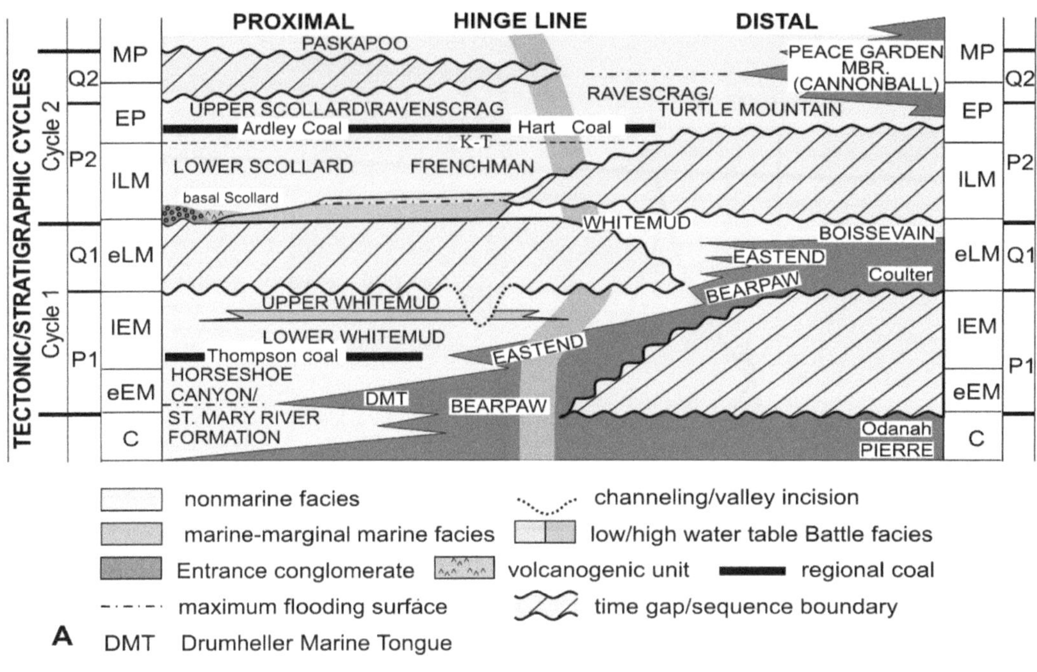



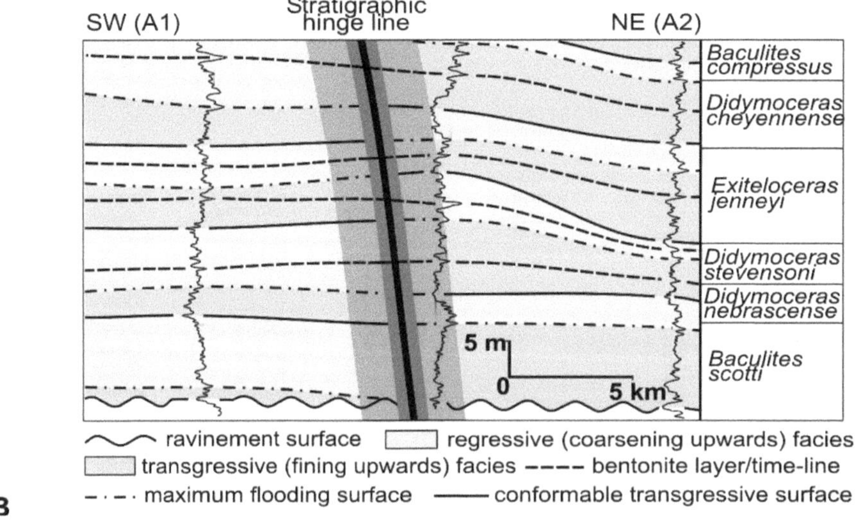



**Fig. 10.37** (a) Reciprocal stratigraphy of the Late Cretaceous-Paleocene strata of the Alberta Basin. Note that proximal unconformities (within the foredeep) correlate to continuous section distally (over the forebulge), and vice versa. (b) Example of the detailed biostratigraphic and bentonites correlations that demonstrate the high-frequency oscillation between transgressive and regressive phases within the foredeep, and its reciprocal relationship to facies changes on the forebulge (Catuneanu et al., 1997, 2000)

pattern of isopachs in a foredeep typically indicates a relatively localized flexural load (Fig. 10.20a). Mapping of the hingeline between the foredeep and the forebulge demonstrates a similar pattern and, moreover, it can indicate shifting of the major locus of flexural loading through time (Fig. 10.39). In the Western Interior Basin, the results of the mapping by Catuneanu and his colleagues indicate that the locus of load moved gradually along strike to the north during the Campanian-Maastrichtian (Fig. 10.39).

High-frequency tectonic forcing of allostratigraphy is suggested by the architecture of the Cenomanian

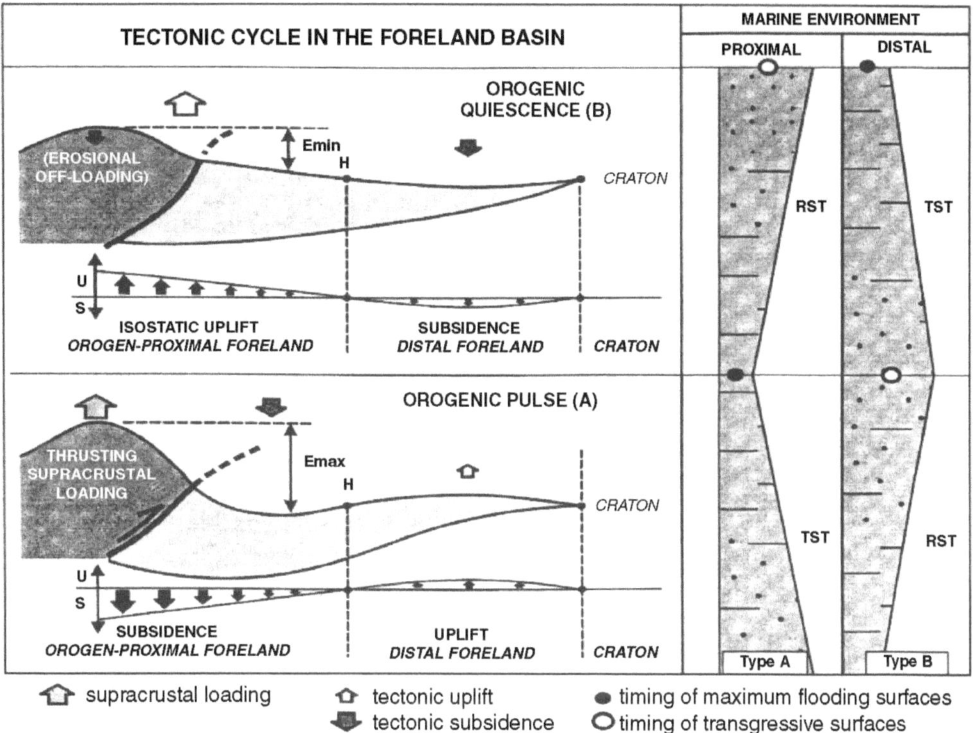



**Fig. 10.38**  Model of reciprocal sedimentation in foreland basins (Catuneanu et al., 1997)

to Coniacian stratigraphy of northern Alberta (Figs. 7.53 and 7.54). As discussed in Sect. 7.6, the stratigraphy contains a cyclicity with a frequency averaging 125 ka, which Varban and Plint (2008) tentatively attributed to high-frequency eustasy (discussed in Sect. 11.3.3). Twenty-seven of the cycles in the Kaskapau and overlying Cardium formations can be grouped into five "onlap cycles" that each represent, on average, 0.6 million years. The first seven of the allomembers demonstrate a pattern of southwesterly offlap of some 200 km relative to the forebulge of the basin. The remaining Kaskapau allomembers and allomembers C1 to C9 of the Cardium Formation then show a pulsed advance (onlap of the forebulge totaling some 350 km. The initial offlap probably indicates rapid subsidence of the foredeep, accompanied by uplift of the forebulge (as in Fig. 10.40a). The subsequent onlap of the forebulge is too large to simply represent the basinward shift of the foredeep due to the advance of the fold-thrust belt, which Varban and Plint (2008) estimated advanced only about 30 km during deposition of the Kaskapau formation.

They suggested, instead, that the onlap represents the enhanced subsidence caused by the addition of the sedimentary load to the flexural load of the fold-thrust belt (Fig. 10.40b), possibly enhanced by a long-term ($10^6$-year scale) eustatic rise in sea level (Fig. 10.40c).

### 10.3.3.2  The Appalachian Foreland Basin

This is another well-developed, classic example of a foreland basin, which was the subject of one of the early model studies of foreland-basin sedimentation and tectonism (Quinlan and Beaumont, 1984; Beaumont et al., 1988). Several major regional unconformities recognized within this basin contributed to the early sequence studies of L. L. Sloss (e.g., the Early to Middle Ordovician "Knox unconformity"), but modern work has indicated that repeated forebulge uplift and migration have played a significant role in developing the large-scale stratigraphic architecture of the basin. The Mississippian-Permian stratigraphy is a particularly interesting case, because this was a period

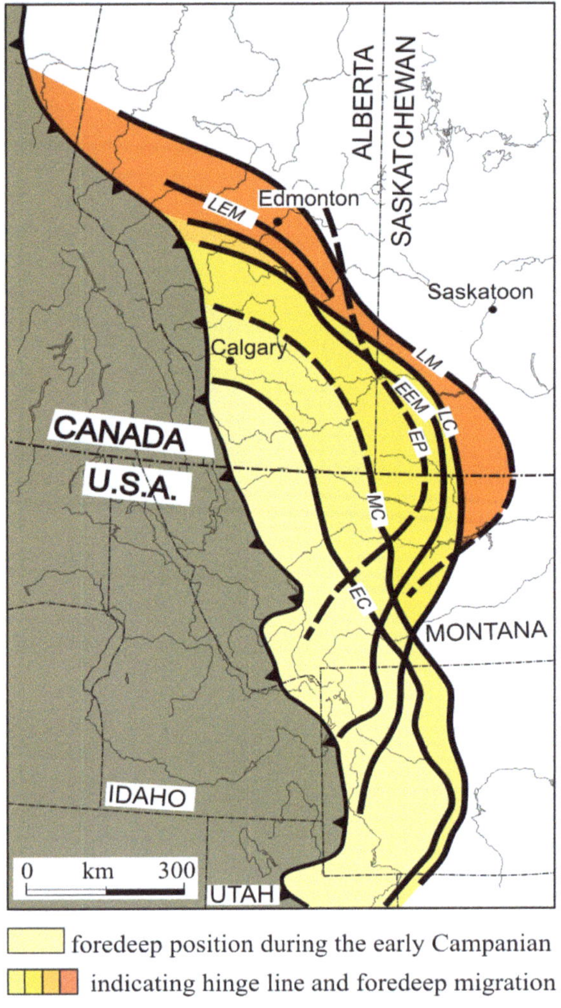



foredeep position during the early Campanian

indicating hinge line and foredeep migration

orogenic belt        ━ ━ ━ ·  inferred hinge line

━━━━━━ hinge line position based on data

**Fig. 10.39**  Migration of the hingeline between the foredeep and the forebulge reveals a gradual northward migration of the locus of flexural load between the Early Campanian (EC) and Late Maastrichtian (LM) (Catuneanu et al., 2000)

during which cycles of undisputed glacioeustatic origin, the cyclothems, were deposited in a basin that was tectonically active. Unravelling the complexities of tectonism and eustasy in this basin has been a major challenge, and many questions and disagreements between different specialists still remain. A major contribution to this debate is a set of papers edited by Dennison and Ettensohn (1994) dealing with the Upper Paleozoic cyclothems, in which tectonism and eustasy have been approached from different points of view by a wide variety of authors. The

following discussion of this basin is based mainly on papers from this book and from the review by Ettensohn (2008).

Ettensohn (1994) reviewed what used to be termed the *geosynclinal cycle*, in which stages of basin development are related to the evolving structural geology, including unconformities, and the sedimentary evolution, which tends to show a repetition of the succession from deep-water shale up through turbidite sandstone (*flysch*) to shallow marine-nonmarine sandstone (*molasse*). This model may be compared with that for the Alberta Basin by Cant and Stockmal (1989). Ettensohn (1994, p. 237) went on to provide a detailed review of the Paleozoic unconformities in the Appalachian basin, based on the cycle of flexural loading and unloading and the vertical motion and migration of the forebulge. He noted:

> The recurrence of unconformities, many accompanied by similar overlying sedimentary sequences, strongly suggests some type of cyclicity, even though recurrence intervals are irregular. Ten of the [thirteen interregional] unconformities . . . are interpreted to be primarily tectonic in origin. Interpretation of tectonic origin is based on the presence of a distinctive, overlying, flexural stratigraphic sequence, the coincidence of unconformity formation with the inception of established orogenies or tectophases therein, and the distribution of unconformities relative to probable loci of tectonism. In contrast, the absence of an overlying flexural sequence and no coincidence with orogeny suggests that unconformities were predominantly eustatic in origin. Only the unconformity at the Ordovician-Silurian boundary seems to be of this type.

Four of Ettensohn's (1994, 2008) *tectophases* are illustrated in Fig. 6.33, and Fig. 10.41 is a tectonic model that explains the development of a repeated succession of sedimentary facies during cycles of loading and unloading. The cycle begins with the development of a deep-water, anoxic basin during the initial cycle of flexural loading (Fig. 10.41a). The basin deepens as loading continues, causing the forebulge to elevate and to migrate towards the orogen. This movement of the forebulge generates a regional unconformity that sweeps towards the foredeep, contemporaneously with the emergence of orogenic uplands and the shedding of a major clastic wedge into the basin (Fig. 10.41b). There is therefore an upward coarsening of sedimentary facies, as exhibited by the Devonian tectophases shown in Fig. 6.33. During subsequent orogenic quiescence and isostatic rebounding, the clastic sediment flux decreases, and a carbonate platform develops on the craton

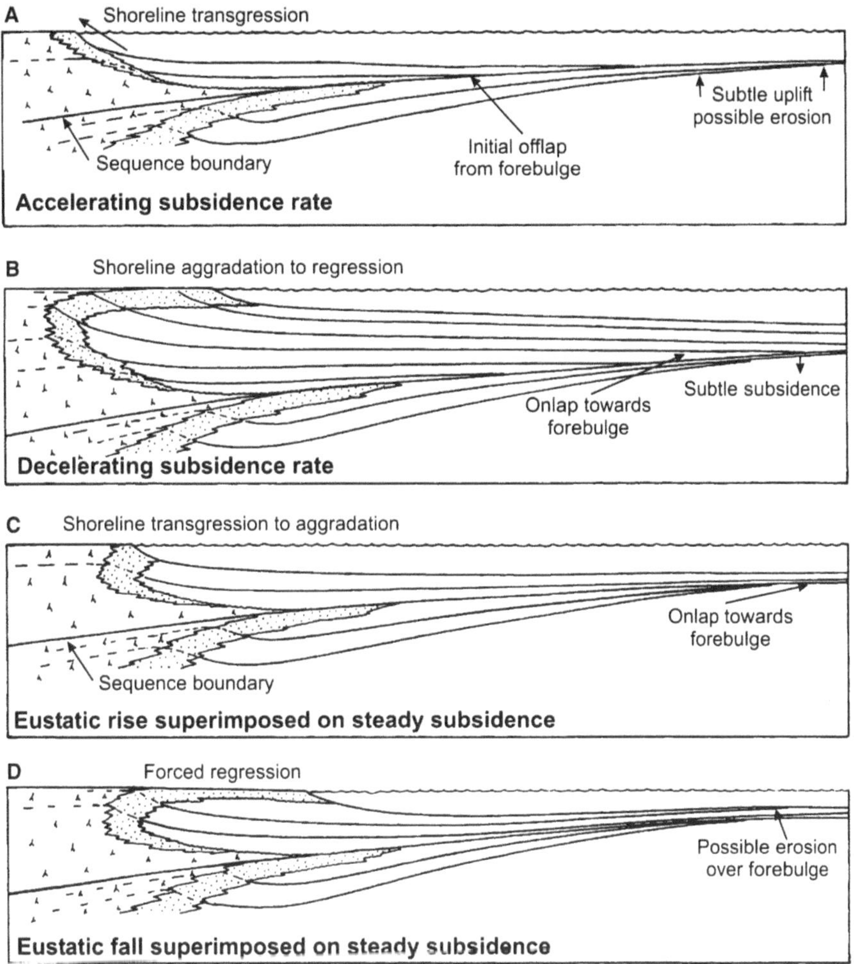



**Fig. 10.40** Cartoon summarizing possible sedimentary responses to tectonic (**a** and **b**) and eustatic (**c** and **d**) forcing across a low-gradient ramp, based on Jordan and Flemings (1991) and Sinclair et al. (1991) and other model studies. In (**a**), renewed thrusting causes an acceleration of proximal subsidence rate, resulting in shoreline back-step and then aggradation; the forebulge is also subtly elevated, promoting displacement of the point of sedimentary onlap towards the foredeep. In (**b**) the rate of flexural subsidence decelerates, promoting progradation of the shoreface; distally the forebulge tends to subside, permitting sediments to onlap towards the crest. In (**c**), eustatic rise creates new accommodation over the forebulge, leading to onlap, whereas the proximal shoreline tends to back-step or aggrade. In (**d**), eustatic fall decreases accommodation over the forebulge, shifting the point of onlap towards the foredeep, and simultaneously promotes forced regression of the proximal shoreline (Varban and Plint, 2008, Fig. 7)

(Fig. 10.41c). Post-orogenic subsidence of the forebulge provides accommodation for the basinward progradation of the carbonate platform (Fig. 10.41d).

Washington and Chisick (1994) examined the Taconic (Early-Middle Ordovician) unconformities between Newfoundland and Virginia in detail and claimed that the pattern of repeated sedimentary breaks makes more sense in terms of a eustatic model. However, they did not provide detailed evidence of correlation between the various unconformities

along strike, so their case remains unproven. A very widespread unconformity marking the end of the Ordovician may be the product of a short-lived glacial episode.

An alternative explanation of Middle Ordovician (Taconic) unconformities and sequence stratigraphy was offered by Joy et al. (2000), and is similar to that for the Devonian-Pennsylvanian tectophases discussed above. Flexural loading of the craton initiated extensional faulting within the craton margin,

**Fig. 10.41** Sequential tectonic cartoons illustrating the developing architecture of the Appalachian basin and its sedimentary fill during the Mississippian-Pennsylvanian cycle of flexural loading and unloading (Ettensohn, 1994, Fig. 15)

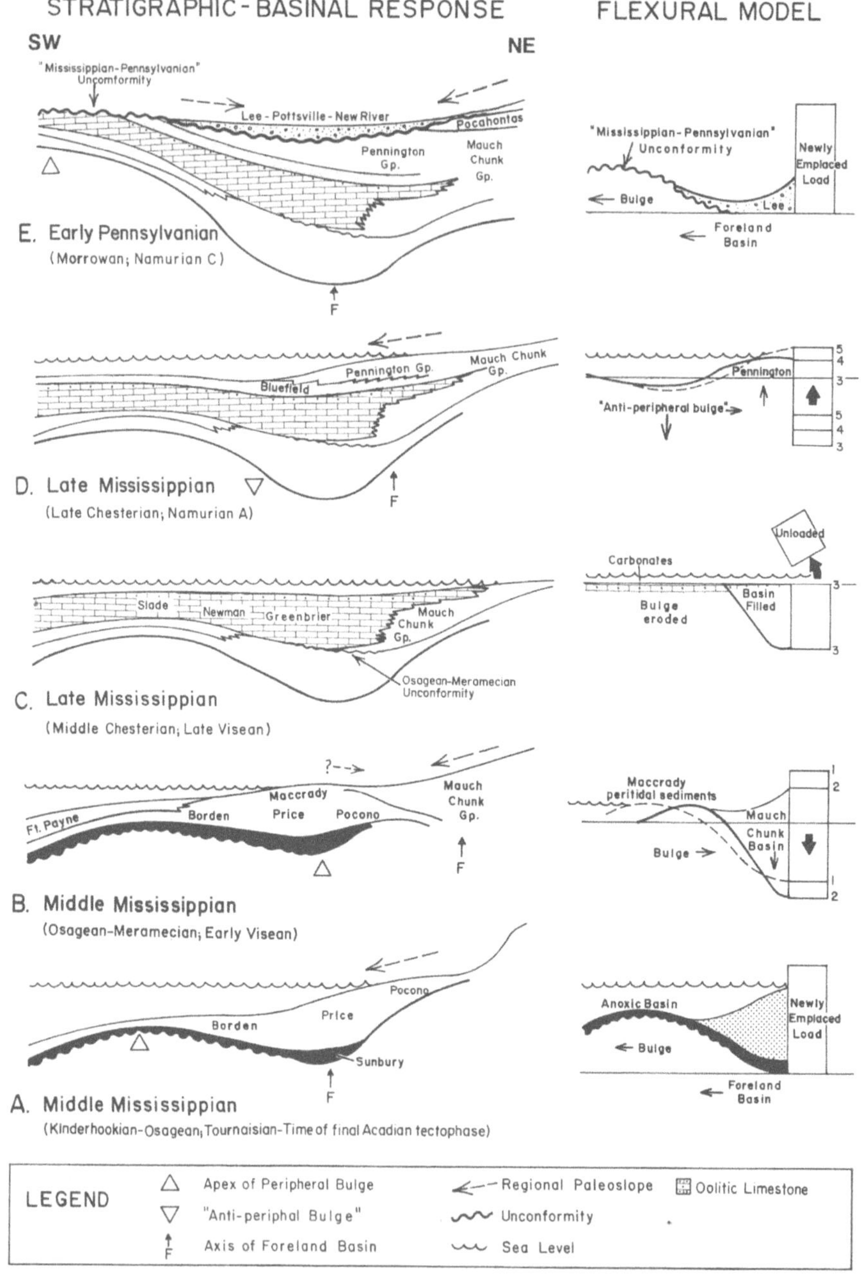


which dipped eastward into a deep-water foredeep, the latter receiving clastic input from the rising orogen to the east (Fig. 10.42). Platform carbonates simultaneously developed on the craton and across the forebulge, but clastics from the orogen interfingered with and eventually overstepped the carbonate units at the peak of orogenic or post-orogenic uplift. This sequence of events was repeated four times on a $10^6$ -year time scale during the Mid-Ordovician. As Joy et al. (2000, p. 730) stated: "Differential subsidence produced a disparity between maximum flooding in the basin and synchronous deposition of shallow-water carbonate sands on the shelf. These relationships provide the clearest evidence known to us for a tectonic forcing mechanism in the development of the Taconic sequences."

**Fig. 10.42** Development of the Taconic (Middle Ordovician) fill of the Appalachian basin in the Mohawk Valley, New York (Joy et al., 2000, Fig. 5)

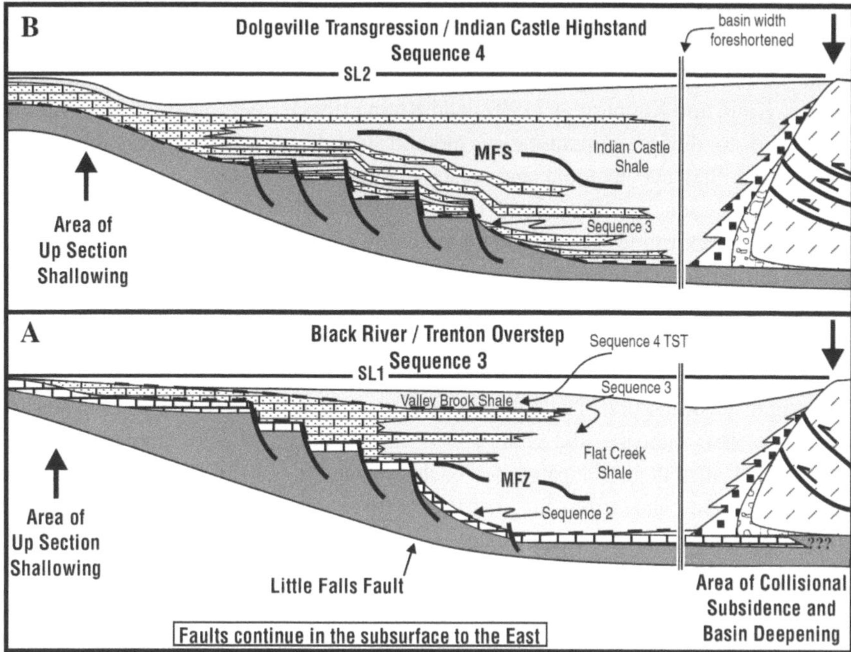



The glacioeustatic control of the Pennsylvanian cyclothems in the Appalachian basin is now well established, as discussed in Sect. 11.3.4. However, various lines of evidence suggest that contemporaneous tectonism was also important. Chesnut (1994) documented three scales of cycle in the Pennsylvanian of the central Appalachians: 1. a large-scale coarsening-upward succession corresponding to the Breathitt Group, which represents 9–34 million years, depending on which estimate of Pennsylvian time is used; 2. Within this are eight major-transgression cycles which average 1.1–4.3 million years (again, depending on time scale); 3. Within these are coal-clastic cycles averaging 0.2–0.7 million years. The large-scale cycle of the Breathitt Group represents long-term regional-scale flexural subsidence of the basin. The second scale of cyclicity is interpreted as tectonic, and in fact probably corresponds to a $10^6$-year cycle of tectonism, comparable to some of the tectonostratigraphic sequences described from the Western Interior basin, above. The smallest scale of cyclicity ($10^5$-year frequency) almost certainly represents Milankovitch glacioeustatic control.

Pashin (1994) constructed isopach maps for coal-bearing cyclothems of the Lower Pennsylvanian Pottsville Formation in the Black Warrior Basin of Alabama (Fig. 7.62). Each is estimated to represent 0.2–0.5 million years. He demonstrated that these cycles each show different isopach patterns, and stated (p. 103) that this demonstrates

> that major changes in basin geometry occurred as deformation loading of the Alabama promontory proceeded. Hence, significant spatial and temporal variations of subsidence rate occurred in the same time frame as deposition of each coarsening- and coaling-upward cycle. . . . Thus, flexure of the lithosphere below the Black Warrior Basin represents multiple events that occurred as elements of different tectonic terranes collided with and were thrust onto the Alabama promontory.

However, Pashin (1994, p. 103) also stated "although flexural subsidence may have amplified marine transgression, no tectonic causes of regional marine regression were identified that operated at the time scale of deposition of a single Pottsville cycle. For this reason, glacial eustasy is considered to have been the dominant cause of cyclicity in the study interval." In another study, Beuthin (1994) documented a fluvial paleovalley trunk-tributary system that appears to map out the slope of the Appalachian foreland basin and its flanks, indicating control by tectonic slope. However, a eustatic fall in sea level may have occurred to expose the deeper parts of the basin to fluvial erosion.

Both the studies noted in the previous paragraph emphasized glacioeustasy and failed to make use of the concept of high-frequency tectonic control that is now being proposed for some stratigraphic successions in the Western Interior Basin and the Pyrenean basins. Klein and Kupperman (1992) and Klein (1994) attempted to devise a quantitative method for distinguishing between tectonic and eustatic causes of sedimentary cyclicity. They used backstripping calculations to determine the amount of subsidence that can be attributed to tectonism during the accumulation of each cyclothem, and they used sedimentological methods, such as depths at which typical facies are deposited, to determine water-depth changes during cyclothem accumulation. The difference between these estimates then indicates the changes in water depth that can be attributed to eustasy for each cyclothem. Although these calculations are instructive, there are many sources of error in the estimates, and the results should probably be regarded as qualitative rather than precise in their indications. Undoubtedly tectonism was important in generating clastic sediment supply to the Appalachian Basin, and in increasing the rate of subsidence relative to the cratonic interior but, if Heckel (1994) is correct, high-frequency glacioeustasy can be detected in the Appalachian Basin. Careful biostratigraphic correlation between the Midcontinent, the Illinois Basin, and the Appalachian Basin suggests that individual cyclothems can be traced between the three regions. However, a bundling of the cyclothems is detectable in the Appalachian Basin. Bundles of cyclothems dominated by marine units are interbedded with bundles dominated by terrestrial deposits, and containing major paleosol intervals. The cyclicity responsible for the bundling is attributed by Heckel (1994) to tectonic flexure and unloading with a periodicity of about 3 million years.

### 10.3.3.3 Pyrenean and Himalayan Basins

In these basins the use of magnetostratigraphic data to supplement biostratigraphy has provided tight constraints on models of tectonism and sequence development. The technique may permit a precision of dating to within the nearest $10^5$ years. For example, Burbank and Raynolds (1984, 1988) used magnetostratigraphic techniques to date the fluvial deposits of the Himalayan foredeep of Pakistan. These results showed that, in the Himalayan foredeep, thrusting and uplift episodes were rapid and spasmodic, and that they did not

occur in sequence into the basin, as the classic model (e.g., Dahlstrom, 1970) would predict (Fig. 10.43). In one case, Burbank and Raynolds (1988) were able to demonstrate the uplift and removal of 3 km of sediment over the crest of an anticline within the basin over a period of 200,000 years, an average uplift and erosion rate of 1.5 cm/year. In most basins, the precision of dating obtained by these authors cannot be attempted, and this, therefore, suggests that caution should be used in assessing published reconstructions of the rates of convergence, crustal shortening, and subsidence, except where these are offered as long-term (>1 million years) averages. These results confirm the importance of high-frequency tectonism in the development of the fold-thrust belt and the potential for corresponding high-frequency sequence development in the adjacent deposits (Fig. 10.24).

Burbank et al. (1992) used magnetostratigraphic control to document timing of thickness and facies changes in the eastern Pyrenean foreland basin. They demonstrated that in most cases coarse-grained progradation occurs during times of slow subsidence, and rapid subsidence is generally reflected by thick fine-grained deposits, thus supporting the Heller et al. (1988) antitectonic model for foreland-basin development. Structural relationships, including faulted cross-cutting of sequences, enabled the history of deformation to be reconstructed and related to that of sequence development. The results (Fig. 10.44) show a complex pattern of thrusting, including the development of imbricates, and out-of-sequence thrusts. The sequences and their intervening unconformities therefore developed in response to a shifting pattern of flexural loading, with continual changes in the location of sediment sources and the rate of sediment input, and changes in areas of erosional unroofing and unloading.

Some of the detailed stratigraphic work in the Pyrenean basins is illustrated in Figs. 6.30, 6.31, and 6.32. Nijman (1998) demonstrated how the particulars of the stratigraphy and paleogeographic evolution of the Tremp-Ager basin could be explained with reference to a detailed tectonic model. The initial formation of the Tremp-Ager basin resulted from the southward movement of the Central-South Pyrenean Thrust sheet in the early Eocene (Ypresian). The subsequent evolution of the basin occurred in the form of one, or a combination of three different modes (Fig. 10.45). Deformation took the form of blind thrusting beneath the basin (Mode A: Fig. 10.45a), which is therefore a

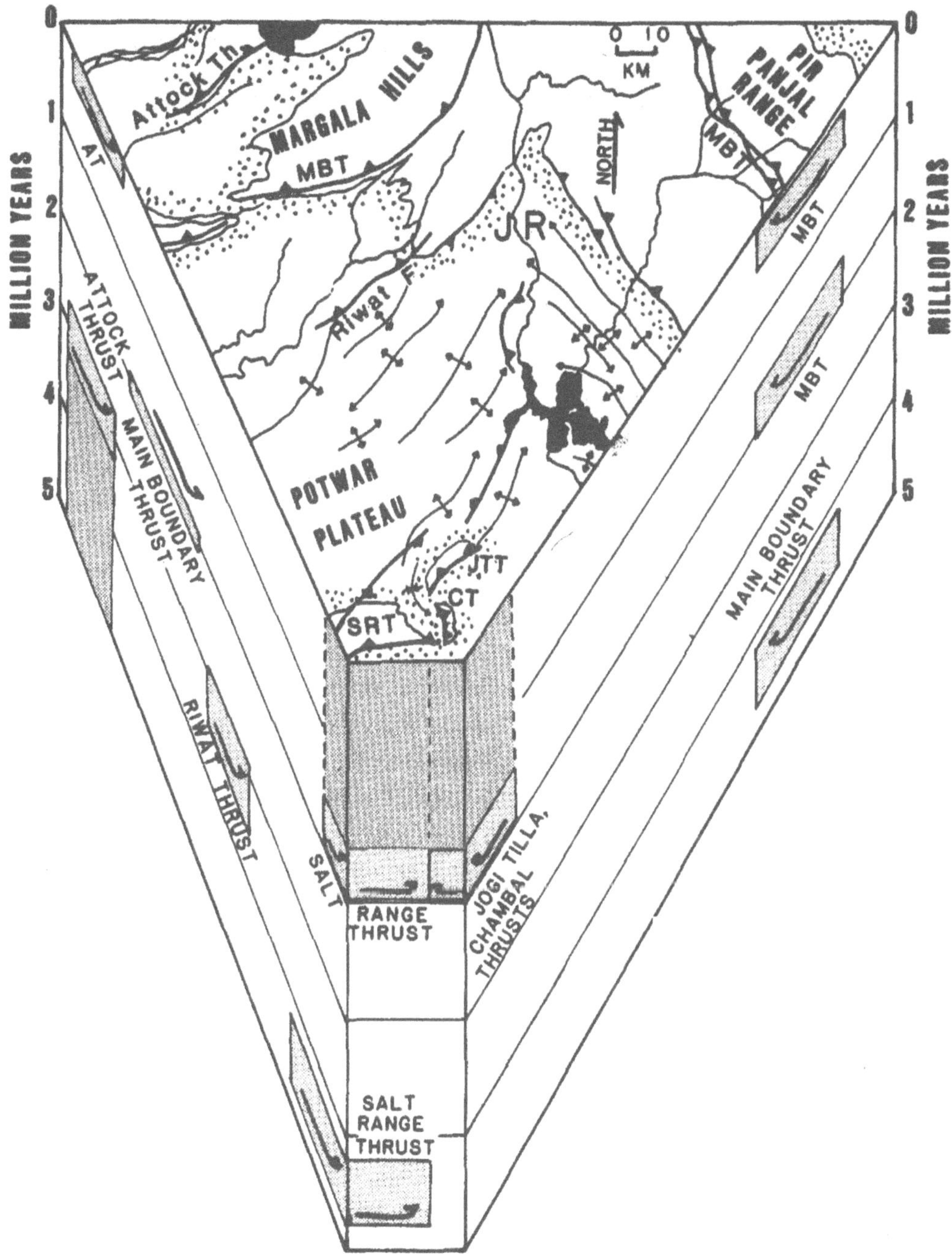


**Fig. 10.43** Chronology of fault motions within part of the Himalayan foredeep of Pakistan. The *shaded boxes* indicate the time and space domain over which each thrust is interpreted to have been active (Burbank and Raynolds, 1988)

wedge-top basin, in the terminology of DeCelles and Giles (1996). Episodic movement of the thrust sheets affected the surface paleogeography, and is regarded as the main control underlying the formation of the major tectono-stratigraphic cycles in the basin. Backthrusting led to uplift and erosion of the proximal part of the basin, which increased the detrital sediment supply from this northern margin (Mode B: Fig. 10.45b).

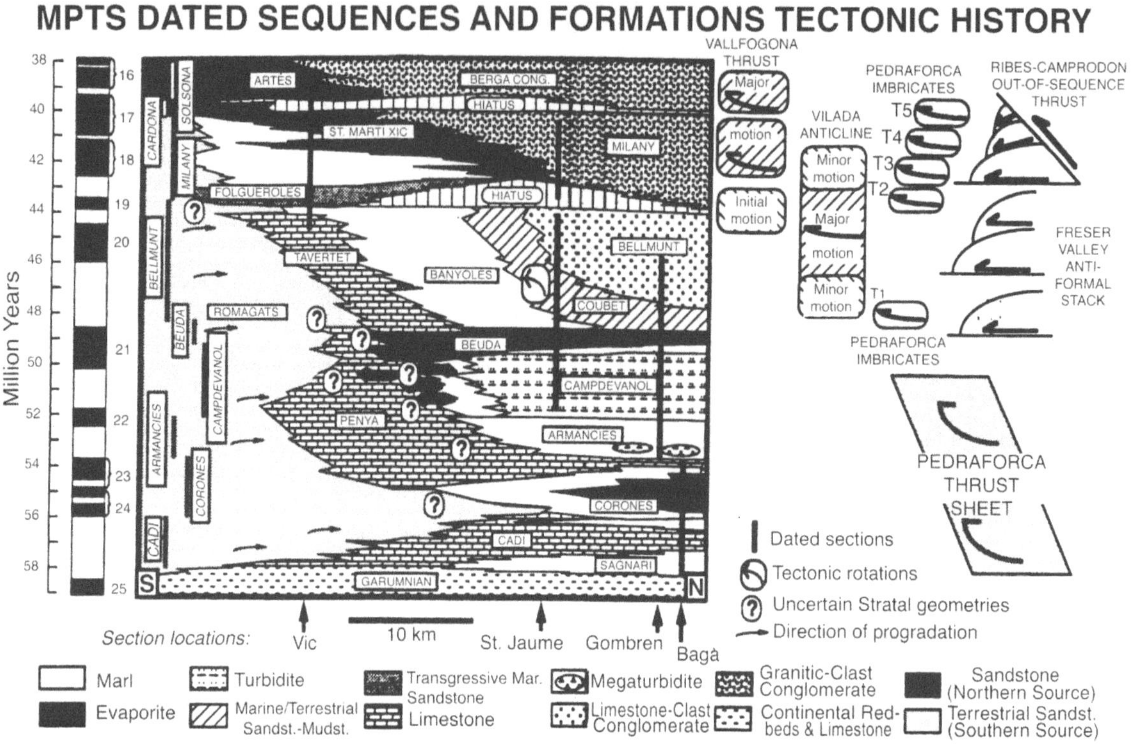

## MPTS DATED SEQUENCES AND FORMATIONS TECTONIC HISTORY

**Fig. 10.44** Schematic Wheeler diagram showing age, sedimentary facies, sequence development, and chronology of thrusting in the eastern Pyrenean foreland basin. The sequences named in the *vertical boxes* at left correspond to the sequences shown in Fig. 7.30 (Burbank et al., 1992)

In this mode, structural deformation of the basin and depositional progradation combined to shift the basin depocentre gradually southward. A third style of basin evolution involved southward gravity sliding of the frontal thrust stack (Mode C: Fig. 10.45c). This uplifted the southern margin of the basin, and enhanced the erosional supply of sediment from that side of the basin, resulting in northward fluvial progradation and movement of the basin depocentre in the same direction. The interfingering of the main sedimentary facies, and the back and forth movement of the depocentre, summarized in Fig. 6.32, are explained with reference to the three deformation modes in Fig. 10.46. Nijman (1998, p. 157) stated:

> Modes of structural control … determined the megasequence boundaries and architecture to a large extent and in combination. The zig-zag trace of the interdeltaic basin axis in cross-section of the Montanyana deposystem [Figs. 6.32, 10.46] correlates with the megasequence order and most of the megasequence boundaries are related to unconformities in one or both of the basin flanks and are marked by onlaps.

## 10.4 Intraplate Stress

### 10.4.1 The Pattern of Global Stress

"Ridge-push" and "slab-pull" are informal terms that have been used for some time in the plate-tectonics literature to refer to the horizontal (in-plane) forces associated with, respectively, the horizontal compressional effects resulting from the elevation differences between a spreading centre and the deep ocean basin, and the tensional effects on an oceanic plate generated by the downward movement under gravity into a subduction zone of a cold slab of oceanic crust. In-plane (horizontal) stress is a central theme of the paper by

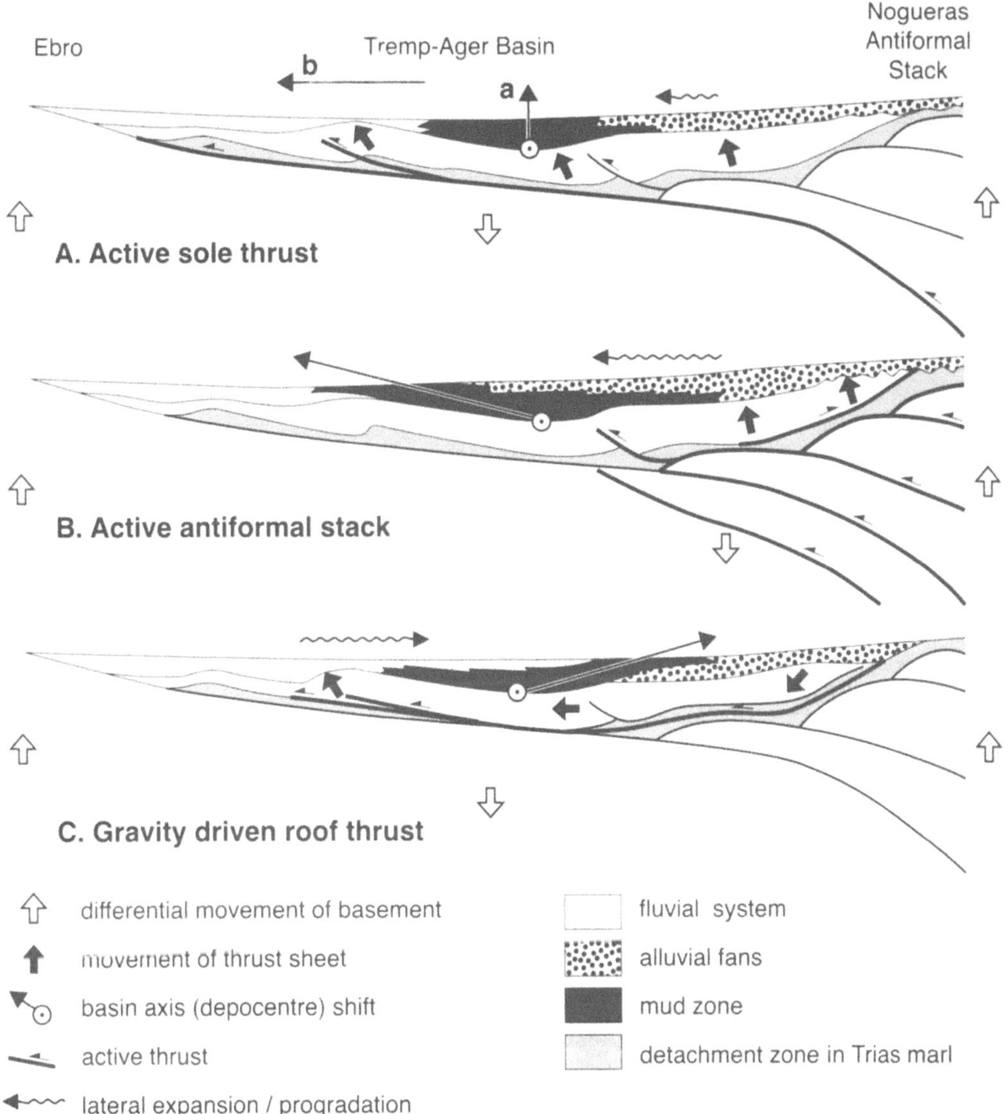


**Fig. 10.45**  Modes of tectonic evolution of the Tremp-Ager basin, Spain. The structural deformation took the form of three different styles, with correspondingly different effects on accommodation and basin paleogeography (Nijman, 1998, Fig. 15)

Molnar and Tapponnier (1975) describing a model of Himalayan collision, in which it was demonstrated that most of the Cenozoic structural geology of west China, extending for 3,000 km north of the Indus suture to the edge of the Siberian craton, could be explained as the product of deformation resulting from intraplate stresses transmitted into the continental interior from the India Asia collision zone. It came to be recognized that tectonic plates can store and transmit horizontal forces many thousands of kilometres from zones of plate-margin stress. The interpretation of Asian geology developed by Molnar and Tapponnier (1975) has subsequently received a remarkably detailed confirmation by studies of modern, real-time plate deformation using data obtained from the Global Positioning System (Zhang et al., 2004). The presence of resid ual stress fields in continental interiors has long been known from such evidence as the development of

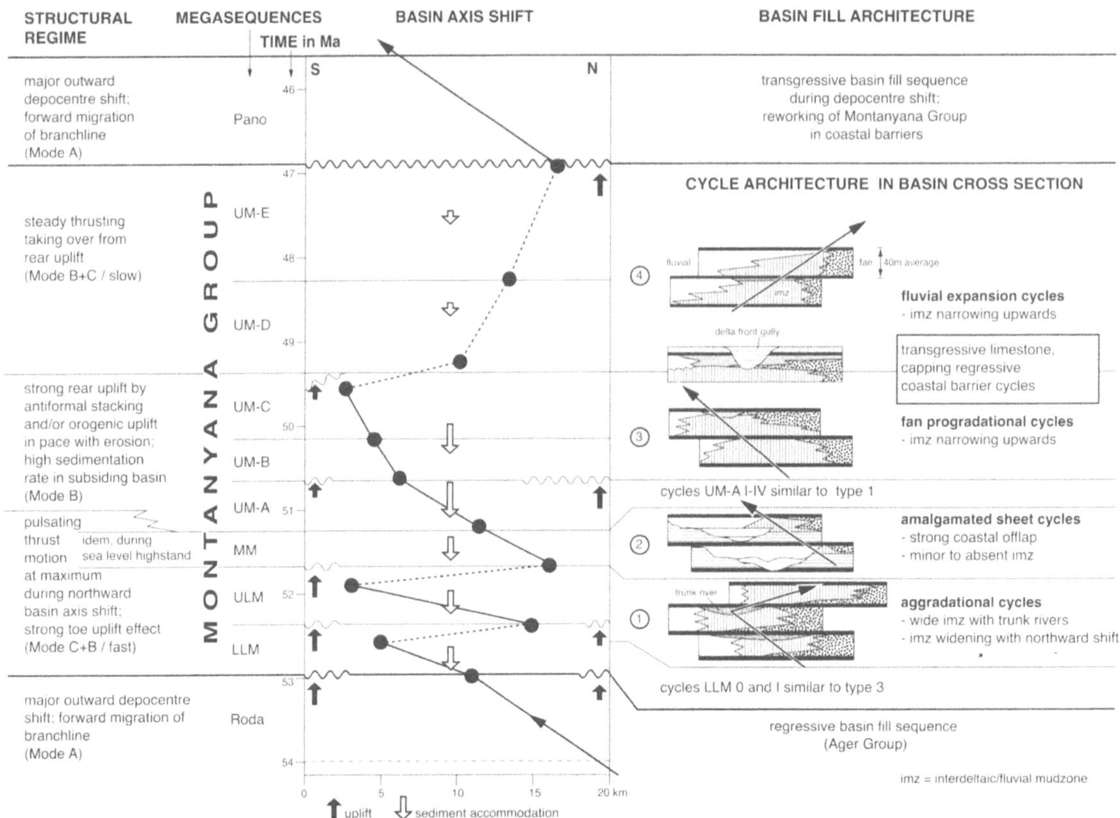


**Fig. 10.46** The evolving relationship between structural regime of Tremp-Ager Basin (the deformation modes of Fig. 10.45) and the architecture of megasequences and their component sedimentary cycles (Nijman, 1998, Fig. 16)

active joints ("break-outs") in exploration holes (summary in Cloetingh, 1988). Stress data for northwest Europe are shown in Fig. 10.47. Cloetingh (1988) provided similar data for the India-Australia plate, which revealed high compressive stresses, particularly in the northeast Indian Ocean, associated with the northward collision of this plate with Eurasia. There the seafloor has been deformed into broad flexural folds as a result of intraplate compressive stress. Cloetingh (1988, p. 206) suggested that:

the observed modern stress orientations show a remarkably consistent pattern [in northwest Europe], especially considering the heterogeneity in lithospheric structure in this area. These stress-orientation data indicate a propagation of stresses away from the Alpine collision front over large distances in the platform region.

In the late 1980s the study of in-plane stress evolved into the World Stress Map Project, under the auspices of the International Lithosphere Program.

Zoback (1992) provided a report on this project, as one of a series of papers discussing intraplate stress. She compiled over 7,300 in-situ stress orientation measurements, and provided a series of maps documenting the results. Zoback (1992) and Richardson (1992) confirmed the observation that stress orientations are remarkably consistent over large continental areas, including areas that are characterized by considerable crustal heterogeneity. Richardson (1992) pointed out that on a global scale orientation measurements are most readily interpreted with respect to the location and orientation of active spreading centres, and suggested that ridge-push forces are the most important in determining in-plane stress. Most stresses are compressional, with extensional stresses having been recorded mainly in areas of high topography. Stresses associated with plate collision are locally important, but the patterns of stress indicate a complex relationship between intraplate stress and tectonic deformation.

**Fig. 10.47** Compilation of observed maximum horizontal stress directions in the northwest European platform. 1 = in-situ measurements, 2 = horizontal stresses equal in all directions, as determined from in-situ measurements, 3 = determinations from earthquake focal-mechanism studies, 4 = well break-outs, 5 = location of Alpine fold-belt (Cloetingh, 1988)

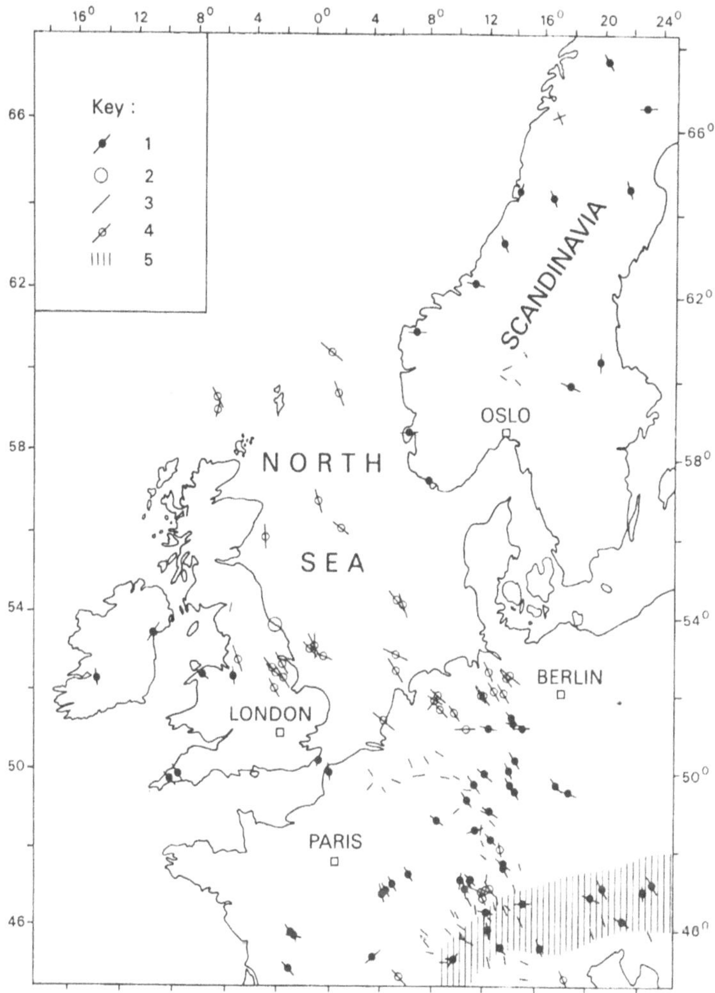



## 10.4.2 In-Plane Stress as a Control of Sequence Architecture

Cloetingh et al. (1985) were the first to recognize the significance of in-plane stress as a control on basin architecture. They argued

> that variations in regional stress fields acting within inhomogeneous lithospheric plates are capable of producing vertical movements of the Earth's surface or the apparent sealevel changes ... of a magnitude equal to those deduced from the stratigraphic record.

The important contribution which Cloetingh et al. (1985) made was to demonstrate by numerical modeling that horizontal stresses modify the effects of existing, known, vertical stresses on sedimentary basins (thermal and flexural subsidence, sediment loading), enlarging or reducing the amplitude of the resulting flexural deformation. The principles are illustrated in Fig. 10.48. They demonstrated that a horizontal stress of 1–2 kbar, well within the range of calculated and observed stresses resulting from plate motions, may result in a local uplift or subsidence of up to 100 m, at a rate of up to 0.1 m/ka (Table 8.2). Compressional stresses generate uplift of the flanks of a sedimentary basin and increased subsidence at the centre. Extensional stresses have the reverse effect (Fig. 10.48). The magnitude of the effect varies with the flexural age (rigidity) of the crust, as well as the magnitude of the stress itself.

The stratigraphic results of this process are extremely important. Figure 10.49 models the effects

**Fig. 10.48** Intraplate stress: principles of the geophysical model. At *top left* is shown a simple shelf-slope-rise sedimentary wedge at a continental margin. The thickness and age of this wedge are shown at *top*, *centre* and *right*. The resulting flexural deformation attributable to a horizontal stress of 1 kbar acting on this continental margin is shown in diagram. Compressional and tensional stresses yield equal but opposite effects (Cloetingh, 1988)

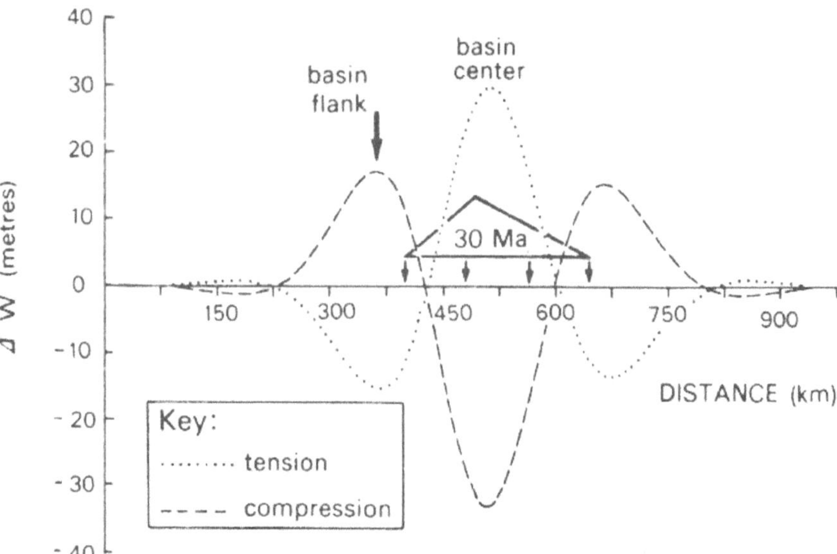



of imposing a horizontal stress of 500 bars on a continental margin undergoing long-term thermally-induced subsidence. Cloetingh (1988, p. 214) stated, with reference to this model:

> When horizontal compression occurs, the peripheral bulge is magnified while simultaneously migrating in a seaward direction, uplift of the basement takes place, an offlap develops, and an apparent fall in sea level results, possibly exposing the sediments to produce an erosional or weathering horizon. Simultaneously, the basin centre undergoes deepening [Fig. 10.49b], resulting in a steeper basin slope. For a horizontal tensional intraplate stress field, the flanks of the basin subside with its landward migration producing an apparent rise in sea level so that renewed deposition, with a corresponding facies change, is possible. In this case the centre of the basin shallows [Fig. 10.49c], and the basin slope is reduced.

These architectural features are locally identical to the onlap and offlap patterns from which Vail and his coworkers have derived their global cycle charts, although on a basinal scale sequence architectures show important differences. For example, build-up of compressional stresses over an extensional margin results in uplift of the continental shelf and offlap, and may lead to the generation of an erosional unconformity, at the same time as the basin centre undergoes deepening. Increased tilting of the continental margin may increase the tendency for slope failure and mass wasting, with increased potential for large-scale submarine-fan deposition at the base of the slope. Conversely, an increase in tensional stress may result

in flooding of the continental shelf and uplift and enhanced erosion of the basin floor, with the potential for the development of submarine unconformities. Changes of stress regime result in the typical indicators of relative sea-level change that are used in sequence analysis (unconformities, transgressions, progradation, retrogradation, etc.) being out of phase by a half cycle between the basin margin and the basin centre (although this may be impossible to demonstrate from the limited chronostratigraphic evidence that is commonly available). The kind of local, outcrop-scale analysis that is used as the basis for many sequence analyses (e.g., Van Wagoner et al., 1990) could potentially, therefore, give very misleading result. As Karner (1986) pointed out,

> basin margin and interior regions should experience opposite baselevel movements. The nature and distribution of sediment cyclicity across either a passive margin or intracratonic basin therefore offers an excellent opportunity to test the concept and importance of lateral stress-induced baselevel variations.

The elevation changes produced within any given sedimentary basin are small in areal extent, limited by the flexural wavelength of the basement underlying the basin. However, the point they missed themselves is that the actual intraplate stresses are much more widespread—they may be transmitted for thousands of kilometres, and will simultaneously affect all basins within that plate.

**Fig. 10.49** Idealized stratigraphy at a basin margin underlain by 40-Ma lithosphere. Diagram (**a**) shows the result of continuous onlap as a result of long-term cooling and flexural loading in the absence of intraplate stress (cf. Fig. 10.6). In (**b**) imposition of a 500-bar compressive intraplate stress at 30 Ma induces short-lived uplift, offlap, and the development of an unconformity on the basin flank. In (**c**) extensional stress generates enhanced subsidence and an increase in the rate of onlap. The stratigraphy in (**b**) and (**c**) indicate a relative fall and rise in sea-level, respectively (Cloetingh, 1988)

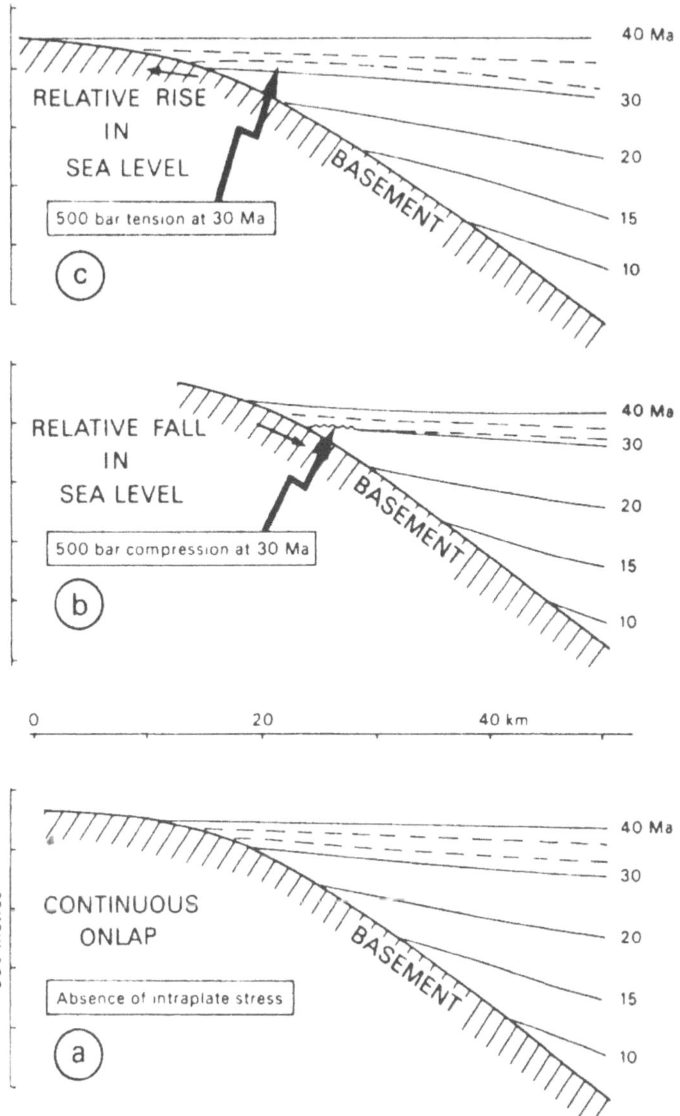


The initial hypothesis of Cloetingh et al. (1985) has been elaborated by Karner (1986), Cloetingh (1986, 1988), Lambeck et al. (1987), Cloetingh et al. (1990) and Cloetingh and Kooi (1990). Karner (1986), in particular, noted that "the simple relationship between in-plane stress and plate rigidity is likely to be complicated by the real rheological properties of the lithosphere", and that "on the application of an in-plane stress, the rigidity of the lithosphere changes and in so doing will modify any preexisting deformation." Karner (1986) carried out analysis and modeling of several aspects of this complex interrelationship of tectonic forces and processes. For example, he studied the change from in-plane extension to compression that characterizes the basin history in many complex orogens, such as the Tertiary basins of western Europe. Many of these basins are characterized by inversion, leading to uplift. The Wessex Basin is often quoted as the typical example of this process. Karner (1986) demonstrated that fault reactivation will lead to uplift and inversion, whereas unfaulted basins, or those whose basin-margin faults are locked, will experience enhanced subsidence under compressive in-plane stresses.

The importance of tectonism as a control of sea-level change, as evaluated by those who have studied it in detail, has led to a suggestion that Vail's work be virtually turned on its head. Karner (1986) stated in the conclusions to his work:

> Knowing the origin of the Vail curve puts its correlative powers onto a strong theoretical basis and helps to define where and when the curve is of use in seismic stratigraphy. Quite independently however, if this in-plane stress mechanism is right, it implies that the Vail curve can be used to measure the absolute value of paleo in-plane stress variations and also as a pointer to the timing of past plate boundary reconfigurations.

Elsewhere Karner (1986) stated:

> Because lateral force variations will be globally balanced (that is, an increase in relative compressive force at one plate boundary must be balanced by a relative increase in tensile forces at another), then in-plane stress-induced unconformities cannot be global. If there exists a causal (but undoubtedly complex) relationship between plate interactions and in-plane stress variations, then transgressive/regressive sequences from a variety of basins will be correlatable if they share the same in-plane stress system, which potentially may span a number of plates. In essence the Vail et al. (1977) coastal onlap curve can be correctly applied to widely spaced basins only when the above criteria are established.

Cloetingh and his colleagues have now carried out a considerable number of modeling experiments, in which they have applied the concepts of intraplate stress to extensional (Kooi and Cloetingh, 1992a, b) and contractional (Peper, 1994; Peper et al., 1992; Peper and Cloetingh, 1995; Heller et al., 1993) tectonic settings.

Kooi and Cloetingh (1992b) examined the effect of the depth at which crustal attenuation ("necking") occurs during crustal extension, and its implications for sequence architecture, and they discussed the importance of basinal tilting induced by flexural effects (Fig. 10.48). They showed that the stratigraphic effects of changes in intraplate stress regime are particularly pronounced and different from simple models of subsidence-with-eustasy in the case of extensional margins that undergo crustal thinning and "necking" at relatively shallow levels. Modeling experiments demonstrated that where necking occurs at deep levels, the resulting stratigraphic architecture is more similar to that of the eustatic model. Kooi and Cloetingh (1992b) showed that the angular unconformities that

result from the warping movements driven by changes in stress regime are very subtle, typically 0.5° or less, and therefore difficult to detect in most data sets. They confirmed that the observations suggested by Embry (1990) for the detection of tectonic influences in sequence generation (Sect. 10.1) are all consistent with tectonism driven by changes in the regional in-plane stress regime.

The response of foreland basins to flexural loading is discussed in Sect. 10.3.3. Most recent work has been concerned with tectonism on a $10^6$-year time scale. Recent studies of intraplate stress, aided by numerical modeling, have suggested that the crust may respond on much shorter time scales, possibly less than $10^4$ years (Cloetingh, 1988; Peper et al., 1992; Peper and Cloetingh, 1995; Heller et al., 1993). Peper et al. (1992) modeled responses of periodic tectonic movement at time scales down to that corresponding to the spacing of individual earthquakes ($10^0$–$10^1$ years), and made a convincing case for tectonic cyclicity that could explain cycles with frequencies of $10^5$–$10^4$ years. Heller et al. (1993) considered the effects of compressional tectonism in areas containing crustal heterogeneities, including active faults and inherited basement structures. They concluded that vertical motions of metres to a few tens of metres extending over distances of tens of kilometres could be attributed to such stresses. This is quite enough to overprint the effects of modest eustatic sea-level changes, and Heller et al. (1993) pointed to several subtle structural and stratigraphic features in the Rocky Mountain states that they suggested could be attributed to in-plane stress occurring as part of compressional tectonism during the Jurassic to mid-Cretaceous.

### 10.4.3 In-Plane Stress and Regional Histories of Sea-Level Change

Lambeck et al. (1987), and Cloetingh (1988) compiled stratigraphic and tectonic data for the North Sea Basin as a starting point for the evaluation of the importance of intraplate stress in this area, and Cloetingh et al. (1990) expanded the data-base to the entire North Atlantic region. They chose this region in part because of its importance in the original establishment of the global cycle chart by Vail in the 1970s. Many workers (e.g., Glennie, 1998; Miall, 1986; Hallam, 1988) had independently noted the importance of regional tectonism in controlling the stratigraphy of this area. Hallam (1988), in particular, noted "that a significant number of Jurassic unconformities are confined to the flanks of North Sea Basins, consistent with the predictions of [intraplate stress]." (Cloetingh, 1988). Such compilations are difficult to evaluate because of the uncertainties inherent in the positioning and relative importance of each event. The analysis which Cloetingh (1988) and Lambeck et al. (1987) offered of their own compilation is little more than "permissive", in the sense that they show that all relative sea-level events in Europe and the North Atlantic could be explained by changes in intraplate stress, but are unable to offer quantitative proof, in the form of detailed numerical models.

In another attempt to model paleostress Cloetingh and Kooi (1990) provided a curve of estimated changes in the paleostress field of the US Atlantic margins, based on a modification of the approach of Watts and his colleagues (which is discussed in Sect. 10.2), in which they superimposed short-term intraplate stress on the long-term flexural subsidence of the continental margin Changes in the stress regime of the US Atlantic margin occurred because of the continual change in the plate kinematics of the Atlantic region as the ocean opened and the spreading centre extended and underwent various changes in configuration. Important events, such as the initiation of spreading in Labrador Sea, the extension of spreading between Greenland and Europe, and various jumps in ridge position, brought about significant changes in plate trajectories, which would have been accompanied by changes in the direction and intensity of intraplate stresses.

Sea-floor spreading patterns in the north and central Atlantic Ocean (Srivastava and Tapscott, 1986; Klitgord and Schouten, 1986) provide a detailed history of plate motions. The reconstruction of a continuous succession of matched plate margins as the Atlantic Ocean opened indicates that the rotation poles of the North American plate relative to Europe and Africa changed at intervals of about 2–16 million years. This episodicity is comparable in magnitude to the duration of $10^6$-year stratigraphic cycles, a fact that is highly suggestive. Furthermore, some of the changes in rotation pole can be correlated in time with changes in the position of depocentres in the Mesozoic-Cenozoic stratigraphic record of the US

Atlantic continental margin. Poag and Sevon (1989) compiled isopach maps of the post-rift sedimentary record, and interpreted these in terms of the shifting pattern of denudation in and sediment transport from the Appalachian and Adirondack Mountains. It seems likely that many of the major existing rivers that presently carry sediment into this region have been in existence since the Mesozoic. Poag and Sevon (1989) were able to show many changes with time in the relative importance of these various sediment sources, and it is suggestive that several of these changes occurred at times of change in Atlantic plate kinematics. The sea-floor spreading record reveals at least fourteen changes in plate configuration in the central Atlantic since the Mid-Jurassic (Klitgord and Schouten, 1986). Seven of these changes, at 2.5, 10, 17, 50, 59, 67 and 150 Ma, correspond to times when the major sediment dispersal routes from the Appalachian Mountains to the continental shelf underwent a major shift (data from Poag and Sevon, 1989).

Tectonism is diachronous and, in the case of the intraplate stresses discussed by Cloetingh (1986, 1988), the effects vary across the plate. Adjacent plates may be expected to have dissimilar tectonic histories, except where major extensional or collisional events affect adjoining plates. However, there is no mechanism for generating globally simultaneous tectonic events of similar style and magnitude. If, as suggested by Cloetingh (1986, 1988), many $10^6$-year cycles are tectonic in origin, it should not be possible to construct a global cycle chart (but see below). The fact that such a chart has been constructed suggests that we should examine the basis on which it has been made. This examination forms the basis for Part IV of this book.

The fact that the earth's crust consists of a finite series of interacting plates suggests the possibility for widespread, even global tectonic episodes that are synchronous but of different style and magnitude in different areas. A possible example of this was noted by Hiroki (1994), who documented tectonic and glacioeustatic movements in a forearc basin on the east coast of Japan during the Quaternary. He noted that a change from subsidence to uplift between 331 and 122 ka B.P. occurred at about the same time as an increase in uplift rates in New Zealand, and a decrease in uplift rates in New Guinea. He stated that "the synchronous change in vertical crustal movement may be explained by a change in the regional horizontal stress field due to rotation of the Pacific plate."

Likewise, Embry (1993), in an analysis of Jurassic sea-level changes in the Canadian Arctic Islands, documented several that are clearly at least in part tectonic in origin, as they occurred at times of significant changes in thickness, facies or sediment dispersal patterns; yet they correlate remarkably well with the "global" events in Hallam's (1988) synthesis. Embry (1993) suggested that these events may be tectono-eustatic in origin. Careful documentation and correlation of deformation and sea-level histories in adjacent but separate regions may reveal more such genetically related episodes of tectonism in widely separated parts of the earth's surface.

In a later study, Embry (1997) examined Triassic sequence stratigraphy at six locations around the northern hemisphere, and compared the ages of sequence boundaries with those in the Triassic succession in western Canada. He identified twelve widespread unconformities within the Triassic system, with an average spacing of 4.3 million years. Most of the unconformities are angular, at least at basin margins, indicating that tectonism was involved in their generation (e.g., Fig. 10.50). An interregional comparison suggests widespread correlation of the sequence boundaries (Fig. 10.51), although the nature of the surface (amount of missing section, degree of angularity) varies from region to region. Embry (1997) noted that the conventional explanation of such widespread correlation would invoke eustatic sea-level change as the driving mechanism. However, he stated (p. 429):

> At various localities the data are sufficient for a given boundary to demonstrate it was formed by uplift and erosion (tectonic tilt test). Other features of these boundaries which indicate a tectonic influence on their formation are: 1) the sediment source area often varies from one sequence to the next, 2) the sedimentary regime of the basin commonly changed drastically and abruptly across the boundaries, 3) significant changes in subsidence and uplift patterns within a basin occurred across sequence boundaries. Also, for these localities, the sequence boundaries are the only horizons at which tectonism appears to have occurred. Thus, it seems inescapable that tectonics plays a significant role in the origin of global sequence boundaries. To account for both a eustatic and tectonic influence on global sequence boundaries, a plate tectonic explanation is appealed to. This appears to be the simplest means to account for these observations.

Embry (1997, p. 429) suggested that widespread contemporaneous sequence boundaries are a product

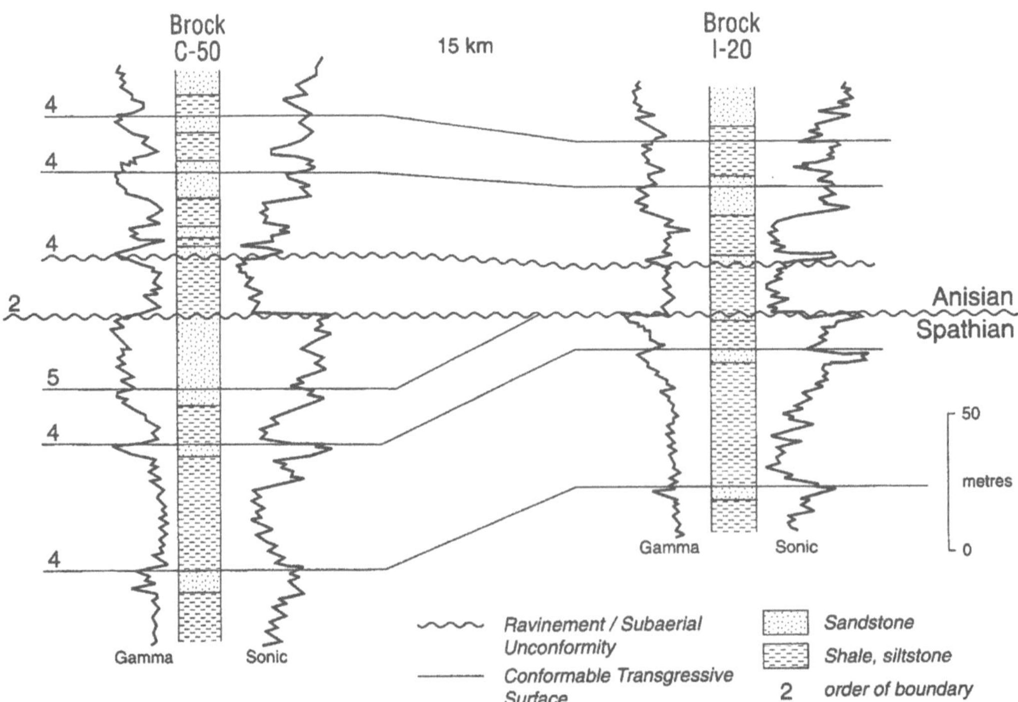



**Fig. 10.50** Application of the "tectonic tilt test": An example of an angular unconformity at a sequence boundary, Sverdrup Basin, Canada (Embry, 1997, Fig. 6)

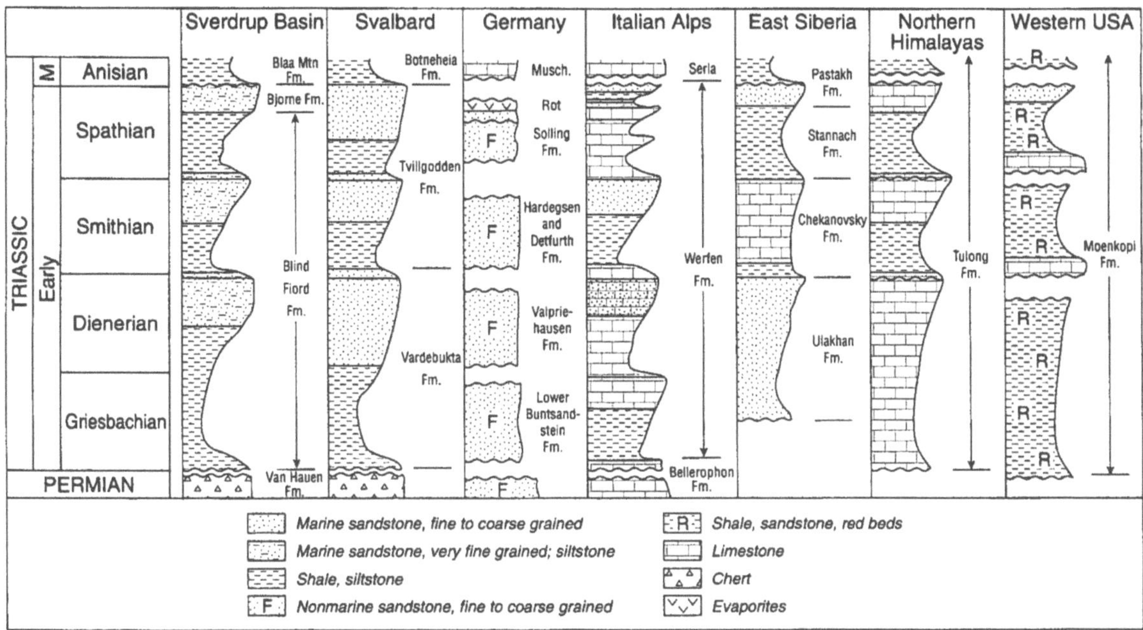



**Fig. 10.51** Comparison of the Lower Triassic stratigraphy at seven locations across the northern hemisphere (Embry, 1997, Fig. 5)

of episodes of global plate tectonic reorganization (Fig. 10.52). "At these times, there were significant changes in the rates and/or directions of spreading between one or more plates. This would result in a significant change in the horizontal stress regime of all the plates. With an increase of stress emanating from spreading ridges and subduction zones, both the oceanic and continental portions of each plate would be affected." Compressive stress, such as an increase in ridge-push, would lead to depression of the ocean basin and uplift of the continental margins, causing both an actual fall in eustatic sea level, because of an increase in ocean-basin volumes, and a tectonically-generated unconformity on continental margins. Relaxational movements would have the reverse effect. This process can therefore potentially explain the development of globally correlatable sequences and sequence boundaries, and can potentially also explain "tectonically-enhanced" unconformities, the term originally used by the Exxon school of sequence stratigraphy (see Sects. 5.3.2, 10.1, 12.5).

An excellent example of the simultaneous occurrence of a range of tectonic events over a large area was provided by Nielsen et al. (2007). They demonstrated how a change in the rate of convergence between Africa and Europe during the mid-Paleocene could be correlated temporally and genetically to an episode of rifting and strike-slip motion in the North Atlantic region, and to episodes of basin initiation and the relaxation of basin inversion in different regions of northwest Europe. Although the Icelandic and other mantle plumes were active at this time (~62 Ma), a flexural stress model indicated that the uplift caused by a major plume could not explain the observed geology, whereas a decrease in the north–south compressional stress caused by a pause in the convergence of Africa and Europe could modify the intraplate stresses enough to generate the observed effects. These include initiation of the separation of Greenland and Eurasia by relaxation along the major Hornsund fault of Svalbard, across which they were joined at that time, left-lateral displacement along a fault that became the Gakkel spreading ridge across the Arctic Ocean, left-lateral slip on the rift faults separating Greenland from the European margin (including the Rockall, Faeroes, Shetland basins), basin initiation in Svalbard, acceleration of the anticlockwise rotation of Greenland away from Labrador and against the Canadian Arctic Islands, and the relaxation of inversion structures from southern Britain to Denmark.

The interpretation of regional structural and stratigraphic events or episodes as the product of changes in the intraplate stress regime depends, in the first instance, on a demonstration of their contemporaneity

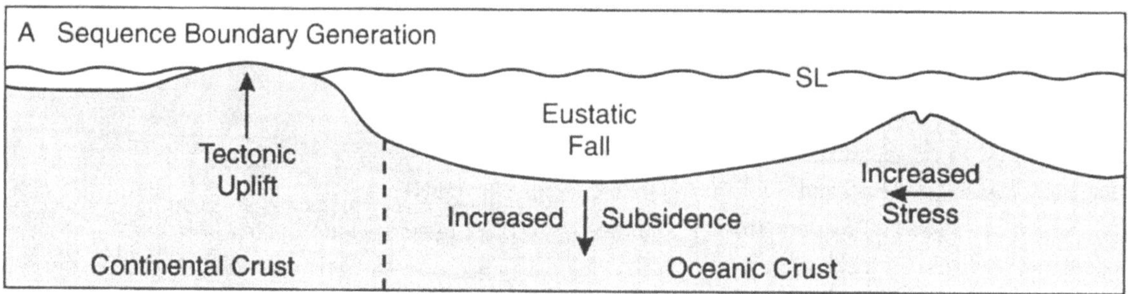



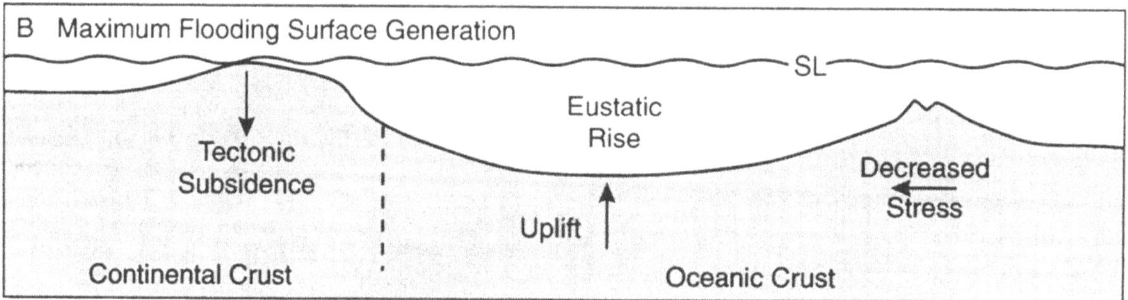



**Fig. 10.52** A speculative model for the roles of eustasy and tectonism in the formation of global sequence boundaries. Changes in horizontal stress regimes in plates due to changes in spreading rates and directions result in both eustatic and tectonic effects. These effects result in globally synchronous sequence boundaries which are in part tectonic in origin (Embry, 1997, Fig. 17)

or synchronicity. However, this raises the issue of the accuracy and precision of correlations, exactly the same question as arose over the hypothesis of global eustasy (Chap. 12). Given the limitations that still exist in our ability to test for global eustasy (Sect. 14.6), this question of causality must remain open for the time being. However, tectonic scenarios can be developed to explain synchronous regional unconformities, episodes of basin deepening or inversion, fault movement, etc., in terms of changes in the intraplate stress regime (e.g., Nielsen et al., 2007, as discussed above). For example, in the case of foreland basins, the contrasts in the deformation patterns due to flexural loading and to changes in intraplate stress are quite clear and unambiguous in terms of magnitude (amplitude), wavelength, and sign (Fig. 10.53).

## 10.5 Basement Control

The rate and magnitude of all the tectonic processes discussed in this chapter depend to a considerable extent on the nature of the basement underlying a sedimentary basin. The importance of ancient lines of weakness in Precambrian basement has long been discussed as a possible cause of the localization of cratonic basins (Quinlan, 1987; Miall, 1999, Sect. 9.3.6.1), and many authors have documented the system of uplifts and lineaments within large cratonic areas that are an expression of dynamic topography (e.g., Sanford et al., 1985). However, it is only relatively recently that sequence-stratigraphic analyses and numerical models of basin development have begun to taken into account the importance of local and regional variations in elasticity, and the effects of basement heterogeneities, such as deep-seated faults and buried sutures between tectonically dissimilar terranes (with different crustal thickness, strengths, and anisotropies). Within regions of the crust that have undergone a long history of repeated plate movements, with the overprinting of many separate tectonic episodes, the effects of basement heterogeneity on vertical movements of the crust are likely to be very complex, whether the major driving forces are flexural loading, crustal stretching, or mantle thermal effects (dynamic topography). Such regions of the earth would include most of northwest Europe, which has been

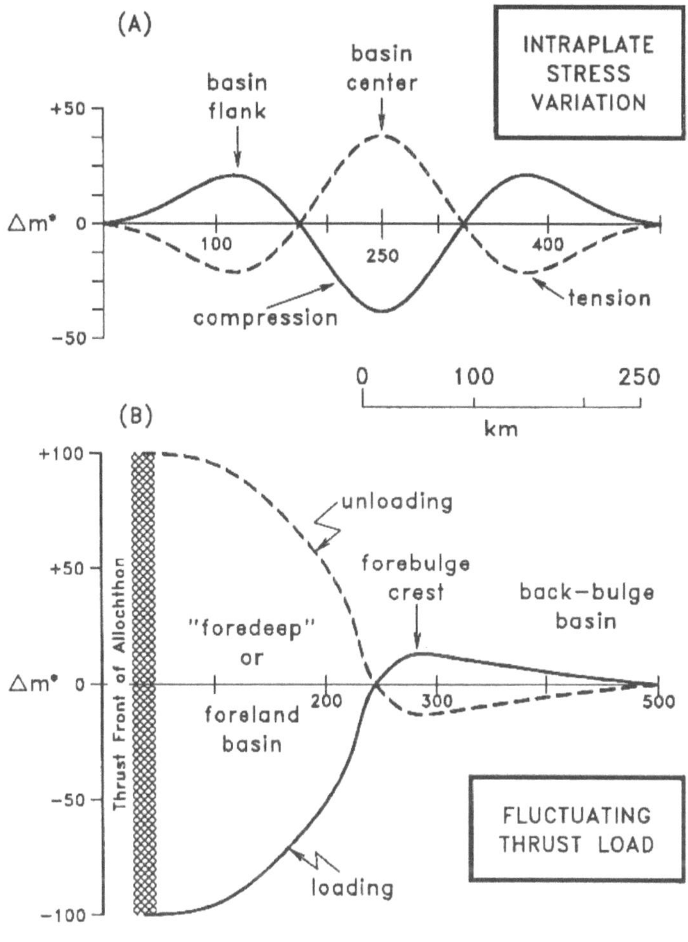



**Fig. 10.53** Comparison of the scale and wavelength of the crustal deflection caused by changes in the intraplate stress regime (*top*) and flexural loading of a continental margin (*bottom*) (Dickinson et al., 1994)

located astride major plate junctions for most of the Phanerozoic, and the southwest United States, which has been affected by repeated collisional and extensional events on its south and west margins throughout the Paleozoic and Mesozoic. Even the cratonic sedimentary cover above Precambrian basement may be affected by subtle basement heterogeneities during epeirogenic warping or during changes in the in-plane stress regime.

Several examples of this important type of local tectonic control have been referred to above. Most cited examples are located in foreland basins, and include the importance of buried faults in the Alberta Basin (Hart and Plint, 1993), and the influence of regional heterogeneities and basement "buttresses"

during flexural loading of the western continental margin of the United States (Pang and Nummedal, 1995). Yoshida et al. (1996) suggested that regional variations in sequence styles in the Upper Cretaceous Mesaverde Group of the Book Cliffs, Utah, could be explained in part as the differential response of Paleozoic-Mesozoic basement cover to flexural loading during the Sevier orogeny. Part of the study area is located above the Paradox Basin, an area of thinned crust overlain by upper Paleozoic evaporites and cut by numerous faults. Overlying this area the sequence stratigraphy of one interval, equivalent to the Castlegate Sandstone, is characterized by high-frequency sequences with well-developed trangressive tidal deposits and highstand marine units, whereas beyond the edge of the Paradox

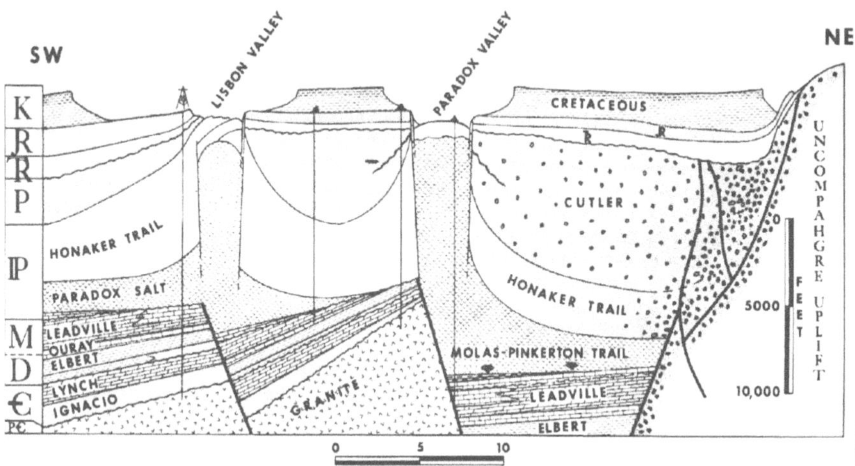



**Fig. 10.54** The Paradox Basin, Utah. Precambrian lineaments were reactivated repeatedly during the late Paleozoic and Mesozoic, as discussed in the text (Stevenson and Baars, 1986)

Basin these sequences appear to merge into a single sequence spanning about 5 million years. The relative changes in sea-level that generated this stratigraphy are attributed to vertical basement movements driven by in-plane stress transmitted from the Sevier orogen—the process discussed by Heller et al. (1993; see Sect. 10.4.2), and it is suggested that the area of the Paradox Basin had a lesser structural strength and integrity during the imposition of these stresses, and therefore demonstrated a more sensitive response to changes in the stress pattern. Similar basement tectonic control on older Mesozoic units overlying the Paradox Basin were described by Baars and Stevenson (1982) and Baars and Watney (1991).

The Paleozoic Paradox Basin of Utah and Colorado is a good example of what can happen when basement structures are repeatedly reactivated. The Basin was formed during a phase of Pennsylvanian tectonism in the southwestern United States, triggered by the collision of Laurasia and Gondwana (Blakey, 2008). This intraplate movement reactivated NW-SE-trending Precambrian structures and created the Ancestral Rockies, including the Paradox Basin and the adjacent Uncompahgre Uplift (Fig. 10.54). The basin filled with evaporites and carbonates, the latter primarily along the flanks of the basin. In the Middle Pennsylvanian, the Uncompahgre Fault, along the southwest flank of the uplift, accumulated up to 6 km of displacement as an equivalent thickness of coarse clastic debris was shed from the uplift south and west into the basin to form the Cutler Group (Stevenson and Baars, 1986). The sediment load on the salt initiated flowage, and caused the formation of giant salt anticlines along NW-SE trends, guided by movement on reactivated Precambrian faults. Movement on these anticlines was largely completed by Triassic time, but continued gentle upward movement of the salt tilted overlying Triassic sandstones, forming intraformational unconformities in the Chinle Formation (Hazel, 1994), which can be traced laterally into "cryptic sequence boundaries" that display no angularity (Miall and Arush, 2001b). Jurassic fluvial systems show major changes in paleocurrent patterns as a result of underlying diapiric salt movement (Bromley, 1991), and even in the Cretaceous strata that cap the succession there is evidence that transport directions in fluvial deposits, and the orientation of small erosional valleys was controlled by the structural grain of the underlying Paradox Basin structures (Yoshida et al., 1996).

## 10.6 Sediment Supply and the Importance of Big Rivers

Sediment supply is controlled primarily by tectonics and climate. In geologically simple areas, where the basin is fed directly from the adjacent margins and source-area uplift is related to basin subsidence, supply considerations are likely to be directly correlated to basin subsidence and eustasy as the major controls of basin architecture. Such is the case where subsidence is yoked to peripheral upwarps, or in proximal regions of foreland basins adjacent to fold-thrust belts. However, where the basin is supplied by long-distance fluvial transportation, complications are likely to arise.

Where the rate of sediment supply is high, it may overwhelm other influences to become a dominant control on sequence architecture.

Many sedimentary basins were filled by river systems whose drainage area has been subsequently remodeled by tectonism, and it may take considerable geological investigation to reconstruct their possible past positions. For example, stratigraphic successions may occur that cannot be related to the evolution of adjacent orogens. In North America, dynamic topographic processes have generated regional uplifts and continental tilts that have resulted in deep erosion and large-scale continental fluxes of detrital sediment (Miall, 2008, Sect. 2.2). For example, Rainbird (1992), Rainbird et al. (1992, 1997) and Hoffman and Grotzinger (1993) suggested that much of the detritus derived by uplift and erosion of the Grenville orogen of eastern North America during the late Precambrian may have ended up contributing to the thick Neoproterozoic sedimentary wedges on the western continental margin. Detailed study of detrital zircons from sedimentary rocks of this age in the western Canadian Arctic indicated that 50% of them are of Grenville age. Rainbird and his colleagues proposed that a major west-flowing river system was established during the late Proterozoic which transported this detritus some 3,000 km across the continental interior. Dickinson (1988) suggested that much of the thick accumulations of late Paleozoic and Mesozoic fluvial and eolian strata in the southwestern United States had been derived from Appalachian sources, and this was confirmed by the detrital-zircon studies of Dickinson and Gehrels (2003). McMillan (1973) envisaged a Tertiary river system draining from the continental interior of North America into Hudson Bay, ultimately delivering sediment to the Labrador Shelf. This has been supported by the studies of Cenozoic landforms and sediments by Duk-Rodkin and Hughes (1994).

Potter (1978) pointed out that major river systems may cross major tectonic boundaries, feeding sediment of a petrographic type unrelated to the receiving basin, into the basin at a rate unconnected in any way with the subsidence history of the basin itself. The modern Amazon river is a good example. It derives from the Andean Mountains, flows across and between, and is fed from several Precambrian shields, and debouches onto a major extensional continental margin. From the point of view of sequence stratigraphy, the important point is that large sediment supplies delivered to a shoreline may overwhelm the stratigraphic effects of variations in sea level. A region undergoing a relative or eustatic rise in sea level may still experience stratigraphic regression if large delta complexes are being built by major sediment-laden rivers. Holbrook et al. (2006) provided a useful discussion of the effects of upstream controls on the development of fluvial graded profiles, fluvial style and the development of nonmarine sequences downstream (Sect. 2.2).

Upstream controls may also be significant in the case of deep-marine deposits. Major episodes of submarine-fan sedimentation in the North Sea and Shetland-Faeroes basins correlate with pulses of Iceland plume activity, which caused magmatic underplating of the continental margin, and uplift, erosion, and enhanced sediment delivery to offshore sedimentary basins (Fig. 10.12: White and Lovell, 1997).

A significant example of this long-distance sedimentary control is the Cenozoic stratigraphic evolution of the Texas-Louisiana coast of the Gulf of Mexico (Worrall and Snelson, 1989). This continental margin is fed with sediment by rivers that have occupied essentially the same position since the early Tertiary (Fisher and McGowen, 1967). The rivers feed into the Gulf Coast from huge drainage basins occupying large areas of the North American Interior (Fig. 10.55). Progradation has extended the continental margin of the Gulf by up to 350 km. This has taken place episodically in both time and space, developing a series of major clastic wedges, some hundreds of metres in thickness (Figs. 10.56 and 10.57). According to the Exxon sequence-stratigraphy models these clastic wedges would be interpreted as highstand deposits, but Galloway (1989b, 2008) showed that their age distribution shows few correlations with the global cycle chart (Fig. 10.58). The major changes along strike of the thickness of these clastic wedges (Fig. 10.57) is also evidence against a control by passive sea-level change. Highly suggestive are the correlations with the tectonic events of the North American Interior; for example, the timing of the Lower and Upper Wilcox Group wedges relative to the timing of the Laramide orogenic pulses along the Cordillera (Fig. 10.58). It seems likely that sediment supply, driven by source-area tectonism, is the major control on the location, timing and thickness of the Gulf Coast clastic wedges. A secondary control is the nature of local tectonism on the continental margin itself, including growth faulting, evaporite

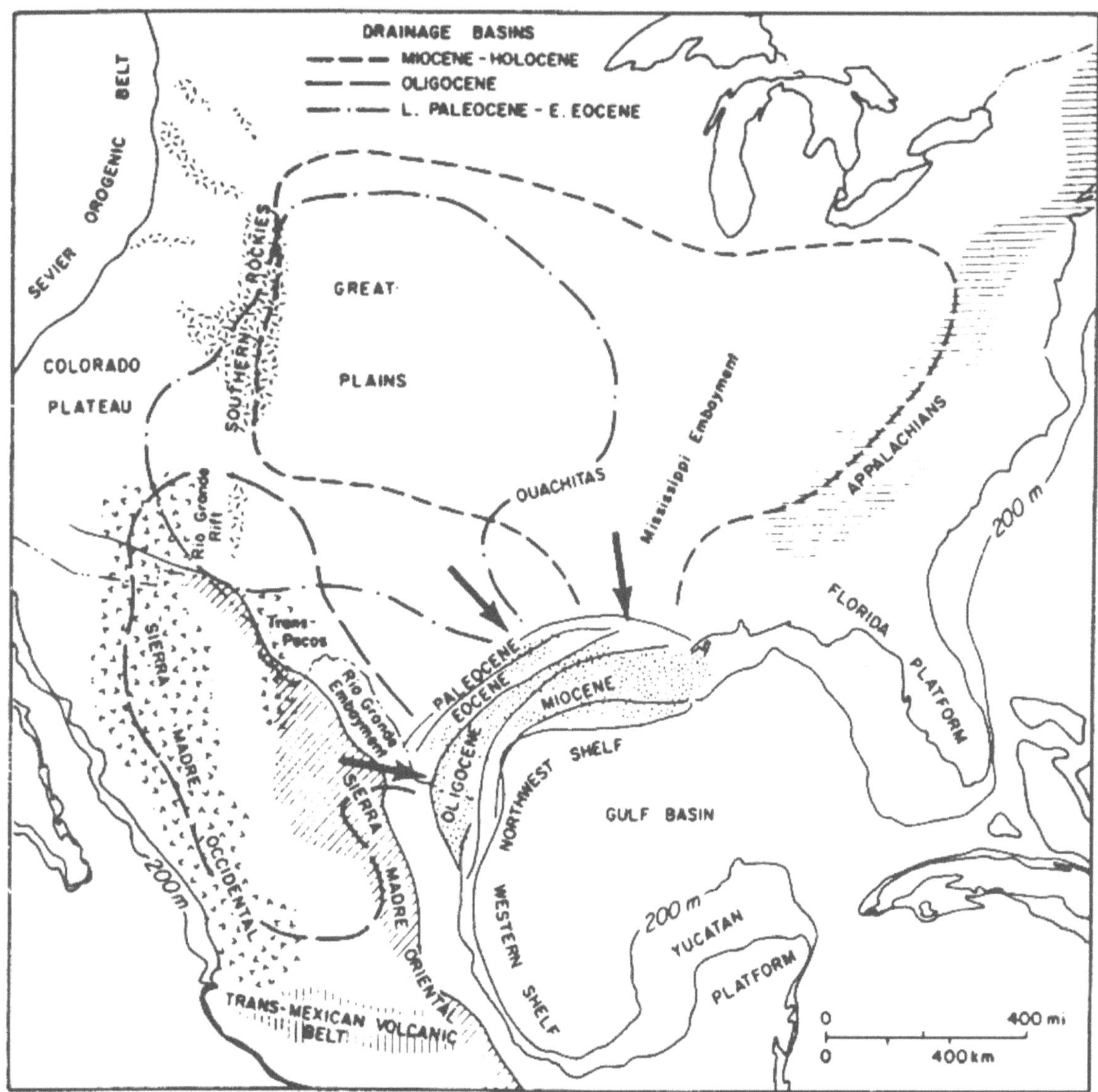



**Fig. 10.55** The major drainage basins which fed delta complexes on the Gulf Coast during the Cenozoic. *Arrows* indicate the position of three long-lasting "embayments" through which rivers entered the coastal region (Galloway, 1989b). AAPG © 1989. Reprinted by permission of the AAPG whose permission is required for further use

diapirism and gravity sliding. As shown by Shaub et al. (1984) and as reemphasized by Schlager (1993), variations in deep-marine sediment dispersal in the Gulf of Mexico show very similar patterns to the coastal and fluvial variations illustrated by Galloway (1989b, 2008). Large-scale submarine-fan systems are therefore dependent, also, on considerations of long-term sediment supply variation, which may be controlled by plate-margin tectonism, in-plane stress regime and dynamic topography.

In arc-related basins volcanic control of the sediment supply may overprint the effects of sea-level change. For example, Winsemann and Seyfried (1991) stated:

Sediment supply and tectonic activity overprinted the eustatic effects and enhanced or lessened them. If large supplies of clastics or uplift overcame the eustatic effects, deep marine sands were also deposited during highstand of sea level, whereas under conditions of low sediment input, thin-bedded turbidites were deposited even during lowstands of sea level.

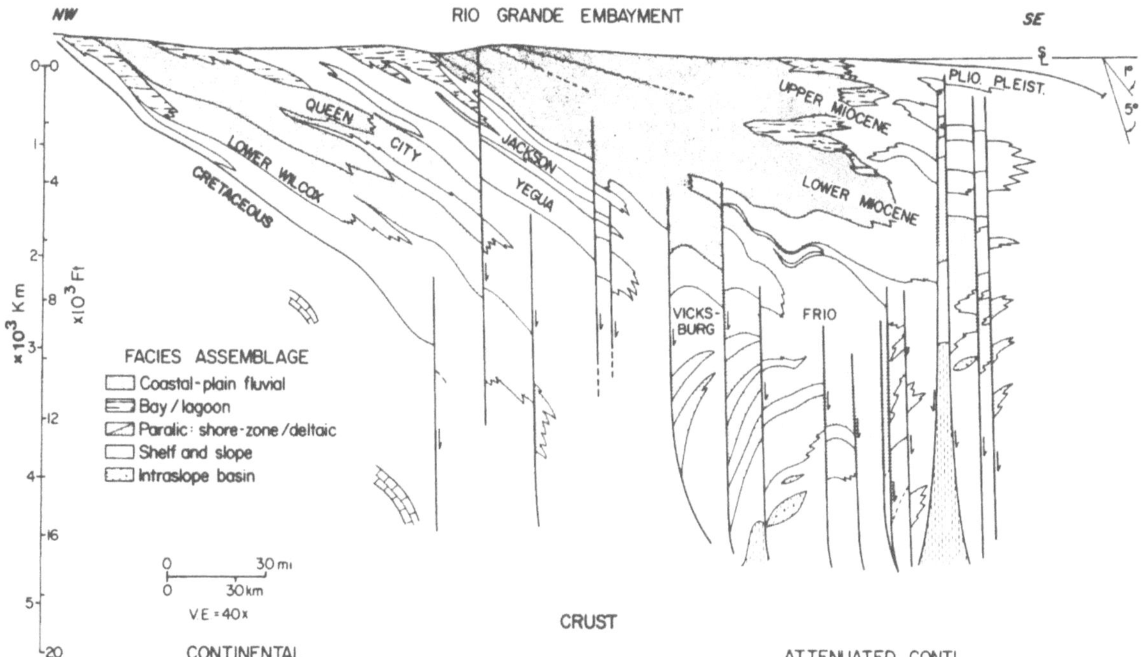



**Fig. 10.56** Generalized dip-oriented stratigraphic cross-section through the Rio Grande depocentre, in the northwest Gulf Coast (location shown in Fig. 10.33), indicating principal Cenozoic clastic wedges. Many of these thicken southward across contemporary growth faults (Galloway, 1989b). AAPG © 1989. Reprinted by permission of the AAPG whose permission is required for further use

**Fig. 10.57** Progradation of the Gulf Coast continental margin by the development of clastic wedges during the Cenozoic. Locations of the three principal embayments through which rivers entered the coastal plain are also shown (Galloway, 1989b). AAPG © 1989. Reprinted by permission of the AAPG whose permission is required for further use

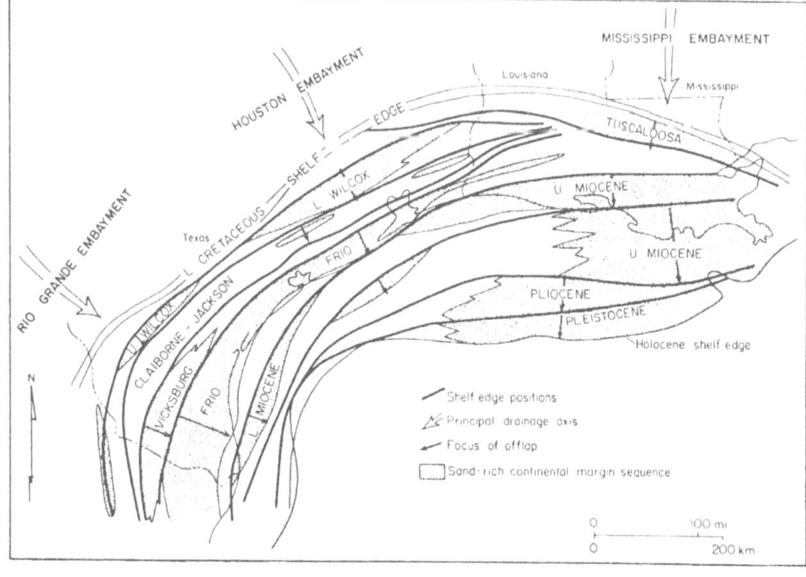



Other examples of the tectonic control of major sedimentary units are provided by the basins within and adjacent to the Alpine and Himalayan orogens. Sediments shed by the rising mountains drain into foreland basins, remnant ocean basins, strike-slip basins, and other internal basins. But the sediment supply is controlled entirely by uplift and by the tectonic control of dispersal routes. For example, Van Houten (1981) showed how the Oligocene Molasse of the Swiss proforeland basin was deposited by rivers

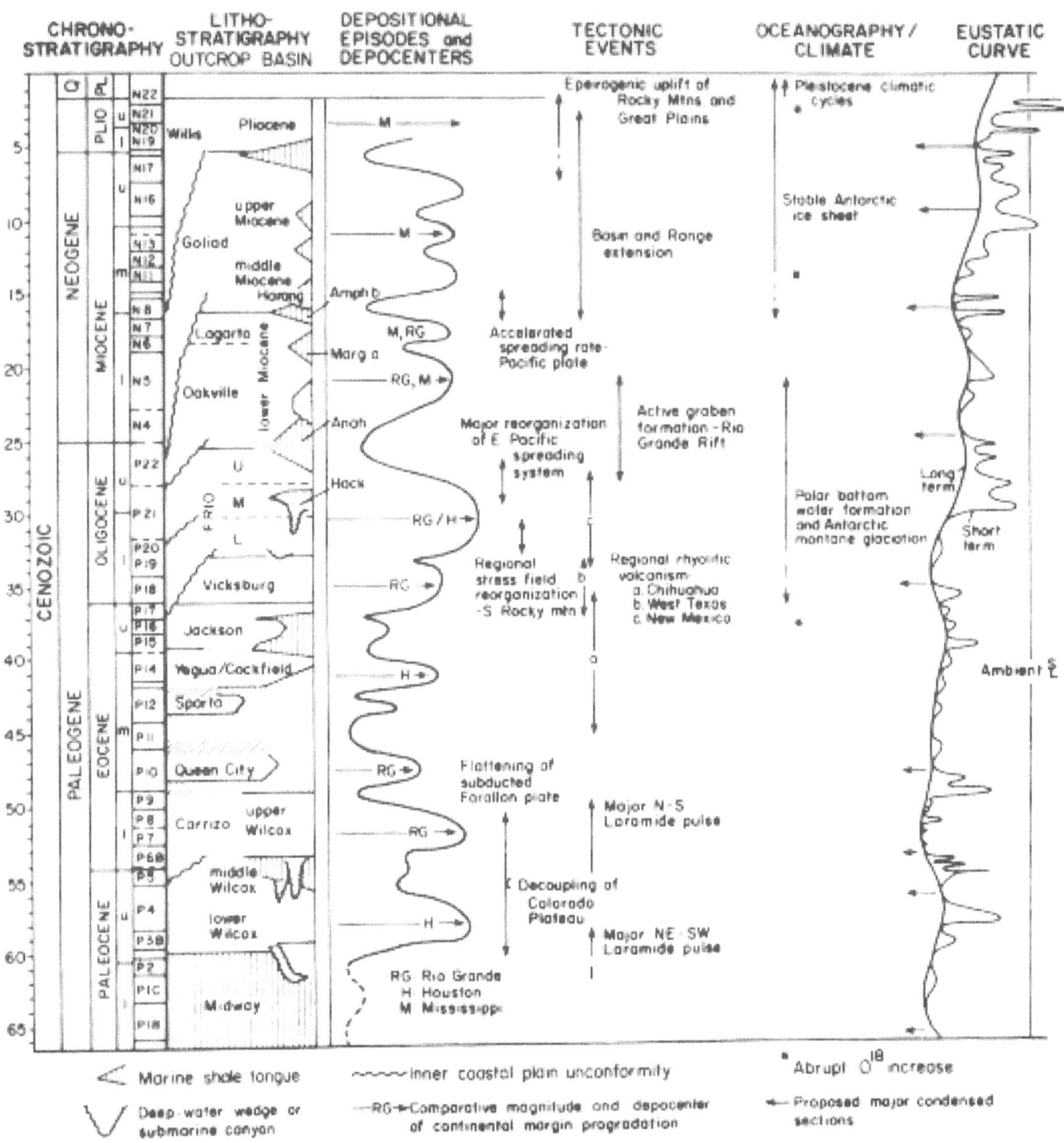



**Fig. 10.58** Age of the major clastic wedges along the Gulf Coast, compared with the age range of tectonic events in the North American Interior. The global cycle chart of Haq et al. (1987, 1988) is shown for comparative purposes at the right (Galloway, 1989b). AAPG © 1989. Reprinted by permission of the AAPG whose permission is required for further use

flowing axially along the basin, and that these underwent reversal in transport directions as a result of changes in the configuration of the basin and the collision zone during orogenesis. Brookfield (1992, 1993) discussed the shifting of dispersal routes through basins and fault valleys within the Himalayan orogen of central and southeast Asia. Some of the major rivers in the area (Tsangpo, Salween, Mekong) are known to have entirely switched to different basins during the evolution of the orogen. Much work remains to be done to relate the details of the stratigraphy in these various basins to the different controls of tectonic subsidence, tectonic control of sediment supply, and eustatic sea-level changes.

## 10.7 Environmental Change

Some sequence boundaries in carbonate sediments are not due to sea-level changes, but to environmental changes that ultimately are related to the tectonic evolution of the area. Two main processes have been documented. The first of these results in what Schlager (1989) termed drowning unconformities. Changes in nutrient supply, water chemistry, or the clastic content of the sea may cause a termination of carbonate-producing biogenic growth. Secondly, surfaces having the same character as sequence boundaries can be generated erosively by submarine currents, such as the Gulf Stream. Schlager (1992a) provided several examples of both these types of process, which are discussed in more detail and illustrated in Sect. 2.3.3 of this book. He argued that several events in the Exxon global cycle chart were generated by these non-eustatic processes.

## 10.8 Main Conclusions

Regarding extensional basins and continental margins:

1. The stratigraphic architectures that have been used by Vail and his coworkers to interpret sea-level changes on a $10^7$-year time scale may be generated by other processes. Coastal onlap (relative rises of sea level) may be caused by flexural downwarp of extensional continental margins. Relative falls in sea level may be caused by thermal doming accompanying rifting.
2. The processes which lead to long-term eustatic sea-level changes, such as changing rates of sea-floor spreading (Chap. 9) also lead to tectonic adjustments of the continents (e.g., initiation of the rift-thermal subsidence cycle), so that relative changes in sea-level may have multiple causes which are not necessarily simply interpretable in terms of eustasy.
3. The North Sea Basin, one of the type areas for the Vail curves, was affected by almost continuous tectonism during the Mesozoic and Cenozoic, and many of the sequence-boundary events and stratigraphic architectures there can now be interpreted in terms of specific local or regional episodes of rift faulting and thermal uplift and subsidence.

Regarding convergent plate margins and collision zones, and their associated foreland basins:

4. Arcs are characterized by active regional tectonism as a result of variations in subduction history, arc migration, and the filling and deflation of magma chambers. The rates, magnitudes and episodicities of the relative changes in sea level caused by this tectonism that have been determined by detailed mapping in modern arcs and other regions of convergent tectonism are consistent with sea-level changes interpreted to have generated stratigraphic sequences with $10^4$–$10^7$-year frequencies that have been mapped in arc-related basins.
5. The various elements of convergent tectonism that generate foreland basins by flexural loading (continental collision, terrane accretion, nappe migration, thrust movement, imbricate fault propagation) can cause relative changes of sea-level on a $10^4$–$10^7$-year time scale.
6. Stratigraphic sequences of $10^6$–$10^7$-year duration in foreland basins, such as the Western Interior of North America and Pyrenean basins, may commonly be correlated with specific tectonic episodes in the adjacent orogen, and there is increasing evidence that high-frequency sequences, of $10^4$–$10^5$-year duration, may be correlated with local tectonic events. The correlation of any of these sequences with a global cycle chart may, therefore, be fortuitous.

Regarding intraplate stress:

7. Stresses generated by plate-margin extension or compression, and transmitted horizontally, have important modifying effects on flexural behaviour of sedimentary basins, acting to amplify or subdue flexural deformation caused by regional tectonics. Changes in paleostress fields change these upwarps and downwarps, resulting in offlap and onlap patterns comparable in stratigraphic duration and magnitude to the "third-order" ($10^6$-year) stratigraphic cyclic changes of the Exxon global cycle chart.

8. High-frequency sequences (of $10^4$–$10^5$-year duration), of limited, regional extent, may be generated by the transmission of stresses induced by flexural loading of individual structures and their structural components, as thrust faults and their imbricates are propagated during contractional tectonism.

9. Stratigraphic patterns induced by intraplate stress are of opposite sign in basin margins and basin centres, and may correlate with tectonic episodes (such as fault movement, basin inversion) in adjacent regions The proposition of intraplate stress change as a widespread control of synchronous basinal events is, therefore, a testable hypothesis, requiring detailed correlation of unconformities and their correlative conformities and other structures across and between basins.

Regarding the importance of sediment supply:

10. The sediment supply to a basin may not be governed by local tectonics but by tectonic events in distant, structurally unrelated areas, with sediment transported across cratons and through orogens by major rivers. In such cases, the distribution and age of fluvial, coastal and submarine clastic wedges does not relate to the sea-level cycle in the basin, but to the tectonic evolution of the sediment source area.

Regarding the importance of environmental change:

11. Carbonate sedimentation may be interrupted by changes in nutrients, water chemistry or clastic content, which inhibit carbonate production. Carbonate platforms are also susceptible to erosion by shifting submarine currents. The result, in all these cases, may be breaks in sedimentation that appear similar to but are unrelated to sequence boundaries produced by sea-level fall.

General conclusions:

12. If tectonic subsidence varies in rate within and between basins, as is typically the case, but is comparable to the rate of eustatic sea-level change, sequence boundaries and flooding surfaces will be markedly diachronous. No synchronous eustatic signal can be generated in such cases.

13. The availability of tectonic mechanisms to explain stratigraphic cyclicity of all types and at all geological time scales removes the need for global eustasy as a primary mechanism for the generation of stratigraphic architectures. This being the case, the onus is on the supporters of eustasy to prove their case by quantitative documentation of eustatic processes, and by rigorous global correlation of supposed eustatic events.

14. Given the finite size of Earth, major plate-tectonic events, (e.g., collisions, ridge re-ordering events, changes in rotation vectors), and their stratigraphic responses, may be globally synchronous. The potential, therefore, exists for global stratigraphic correlation. However, the structural and stratigraphic signature would vary from region to region, such that, for example, a relative sea-level rise in one location may be genetically related to a fall elsewhere.

15. The invocation of tectonic mechanisms to explain stratigraphic architecture has led to the suggestion that Vail's charts be "inverted", to provide a source of data from which tectonic episodicity could be extracted. While this is an interesting idea, there remains the problem that the global accuracy and precision of the Vail curves are in question, and tectonic arguments are in as much danger as eustatic arguments of falling into the trap of false correlation and circular reasoning.

# Chapter 11

# Orbital Forcing

## Contents



## 11.1 Introduction

By the end of the nineteenth century it was recognized that during the Pleistocene the earth had undergone at least four major ice ages, and a search was underway for a mechanism that would alter climate so dramatically (useful historical summaries are given by Berger, 1988; Imbrie, 1985; Huggett, 1991, Sect. 2.3; Dott, 1992a, b; de Boer and Smith, 1994b). Continued stratigraphic work during the twentieth century produced evidence for many more cycles of ice formation, advance and melting, and it is now known that there were more than twenty such cycles, indicating successive major fluctuations in global climate.

The basis of the idea of orbital forcing can be traced back to the seventeenth century (Huggett, 1991, p. 24), but it was an amateur geologist and newspaper publisher, Charles MacLaren, who in 1842 first realized the implications of continental glaciation for major changes in sea level (Dott, 1992b). The rhythmicity in glaciation is now attributed to astronomical forcing. The Scotsman James Croll (1864) and the American G. K. Gilbert (1895) were the first to realize that variations in the earth's orbital behavior may affect the distribution of solar radiation received at the surface, by latitude and by season, and could be the cause of major climatic variations, but the ideas were not taken seriously for many years after this because of a lack of a quantifiable theory and supporting data. Gilbert invoked the idea to explain oscillations in carbonate content in some Cretaceous hemipelagic beds in Colorado (Fischer, 1986). Later, Bradley (1929) "recognized precessional cycles in oil shale-dolomite sequences of the Green River Formation in Colorado, Wyoming and Utah, using varves as an unusually precise measure of sedimentation rates" (de Boer and Smith, 1994b). Theoretical work on the distribution of insolation was carried out by the Serbian mathematician Milankovitch (1930, 1941), who showed how orbital oscillations could affect the distribution of solar radiation over the earth's surface. However, it was not for some years that the necessary data from the sedimentary record was obtained to support his model. Emiliani (1955) was the first to discover periodicities in the Pleistocene marine isotopic record, and the work by Hays et al. (1976) is regarded by many (e.g., de Boer and Smith, 1994b) as the definitive study that marked the beginning of a more widespread acceptance of so-called *Milankovitch processes* as the cause of stratigraphic cyclicity on a $10^4$–$10^5$-year frequency—what is now termed the *Milankovitch band* (Sect. 4.1). The model is now firmly established, particularly since accurate


A.D. Miall, *The Geology of Stratigraphic Sequences,* 2nd ed.,
DOI 10.1007/978-3-642-05027-5_11, © Springer-Verlag Berlin Heidelberg 2010

chronostratigraphic dating of marine sediments has led to the documentation of the record of faunal variations and temperature changes in numerous upper Cenozoic sections (Sect. 7.2 and 11.3). These show remarkably close agreement with the predictions made from astronomical observations.

There is an obvious link between climate and sea level. There is no doubting the efficacy of glacial advance and retreat as a mechanism for changing sea-level; we have the evidence of the Pleistocene glaciation at hand throughout the Northern Hemisphere. It has been calculated that during the period of maximum Pleistocene ice advance the sea was lowered by about 100 m (Donovan and Jones, 1979). Melting of the remaining ice caps would raise the present sea-level by about 40–50 m (Donovan and Jones, 1979). Recent history thus demonstrates a mechanism of changing sea-level by at least 150 m at a rate of about 0.01 m/year. This is fast enough to account for eustatic cycles with frequencies in the range of tens of thousands of years. Milankovitch processes are now widely accepted as the source of much cyclicity with cyclicities of $10^4$–$10^5$ years in the Late Paleozoic and Late Cenozoic sedimentary record—times when continental glaciation is known to have been widespread (Sects. 11.3.2 and Sect. 11.3.4). Other, more subtle (non-glacial) climatic variations may explain other forms of cyclicity of similar periodicity at times when there is no evidence for widespread continental ice sheets (Sect. 11.2.5). However, there is also increasing evidence for cyclicity of a similar time periodicity generated by tectonic mechanisms, especially in foreland basins (Sect. 10.3.3), and it is not safe to assume that sequences of $10^4$–$10^5$-year duration are necessarily generated by orbital forcing. This problem is examined further in Sect. 11.4.

There has been a considerable increase in interest in Milankovitch processes in recent years, particularly since the suggestion was made by House (1985) that the regularity of the orbital forcing process could be exploited for the construction of a new, highly accurate type of time scale. This has given rise to a vigorous new subdiscipline termed *cyclostratigraphy*. A number of books have appeared that treat the subject from various points of view. The astronomical processes that underpin Earth's climatic cyclicity are dealt with by Schwarzacher (1993, 2000), and Hinnov (2000), and by A. L. Berger, in a number of books, including Berger et al. (1984) and Berger

(1988). Collections of case studies, some with papers explaining orbital mechanics or climatic theory, have been edited by Fischer and Bottjer (1991), Berger et al. (1984), de Boer and Smith (1994a), House and Gale (1995) and D'Argenio et al. (2004a). Much additional information is contained in the books edited by Franseen et al. (1991), Dennison and Ettensohn (1994), and Weedon (2003). A major symposium on the cyclostratigraphic record was held by the Royal Society of London in 1998 (Shackleton et al., 1999). This symposium, and the development of the *cylostratigraphic* or *astronochronologic time scale* from the ancient sedimentary record are critically evaluated in Sect. 14.7.

## 11.2 The Nature of Milankovitch Processes

### 11.2.1 Components of Orbital Forcing

There are several separate components of orbital variation (Fig. 11.1). The present orbital behavior of the earth includes the following cyclic changes (Schwarzacher, 1993):

1. Variations in orbital *eccentricity* (the shape of the Earth's orbit around the sun). Several "wobbles", which have periods of 2,035.4, 412.8, 128.2, 99.5, 94.9, and 54 ka. The major periods are those at around 413 and 100 ka.
2. Changes of up to 3° in the *obliquity* of the ecliptic, with a major period of 41 ka, and minor periods of 53.6 and 39.7 ka.
3. *Precession* of the equinoxes. The Earth's orbit rotates like a spinning top, with a major period of 23.7 ka. This affects the timing of the perihelion (the position of closest approach of the earth to the sun on an elliptical orbit), which changes with a period of 19 ka.

Imbrie (1985, p. 423) explained the effects of these variables as follows:

Variations in obliquity alter the income side of the radiation budget in two fundamental ways: they modulate the intensity of the seasonal cycle, and they alter the annually integrated pole-to-equator insolation gradient on which

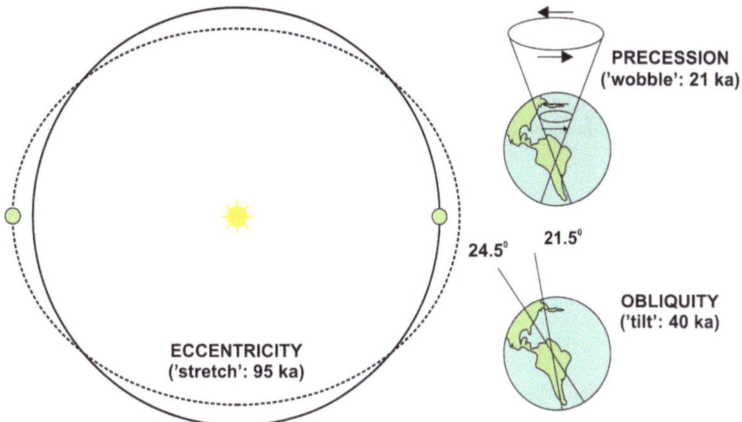

**Fig. 11.1** Perturbations in the orbital behaviour of the earth, showing the causes of Milankovitch cyclicity. Adapted from Imbrie and Imbrie (1979)



the intensity of the atmospheric and oceanic circulations largely depend. (Low values of obliquity correspond to lower seasonality and steeper insolation gradients.) Variations in precession, on the other hand, alter the structure of the seasonal cycle by moving the perihelion point along the orbit. The effect of this motion is to change the earth-sun distance at every season, and thereby change the intensity of incoming radiation at every season.

Each of these components is capable of causing significant climatic fluctuations given an adequate degree of global sensitivity to climate forcing. For example, when obliquity is low (rotation axis nearly normal to the ecliptic), more energy is delivered to the equator and less to the poles, giving rise to a steeper latitudinal temperature gradient and lower seasonality. Variations in precession alter the structure of the seasonal cycle, by moving the perihelion point along the orbit. This changes the earth-sun distance at every season, thus changing the intensity of insolation at each season. "For a given latitude and season typical departures from modern values are on the order of $\sim$5%" (Imbrie, 1985, p. 423). Because the forcing effects have different periods they go in and out of phase. One of the major contributions of Milankovitch was to demonstrate these phase relationships on the basis of laborious time-series calculations. These can now, of course, be readily carried out by computer (Fig. 11.2). The success of modern stratigraphic work has been to demonstrate the existence of curves of temperature change and other variables in the Cenozoic record that can be correlated directly with the curves of Fig. 11.2. For this purpose sophisticated time-series spectral analysis is performed on various

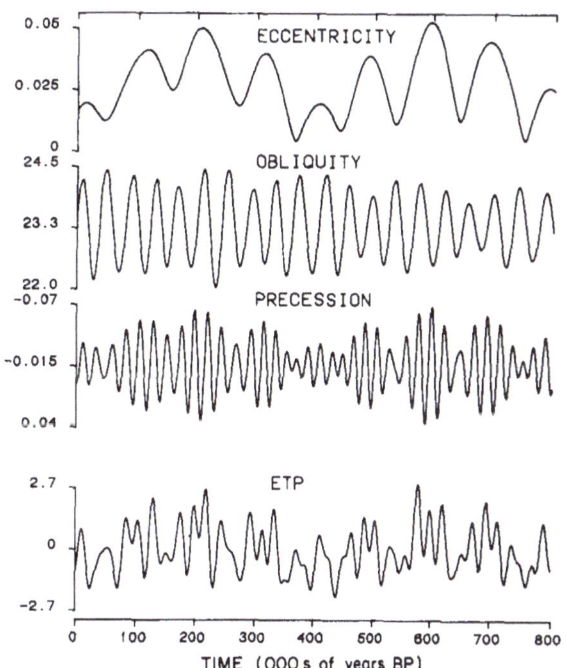



**Fig. 11.2** The three major orbital-forcing parameters, showing their combined effect in the eccentricity-tilt-precession (ETP) curve at *bottom*. Absolute eccentricity values are shown. Obliquity is measured in degrees. Precession is shown by a precession index. The ETP scale is in standard deviation units (Imbrie, 1985)

measured parameters, such as oxygen-isotope content (e.g., Fig. 11.3) or cycle thickness. This approach has led to the development of a special type of quantitative analysis termed *cyclostratigraphy* (House, 1985).

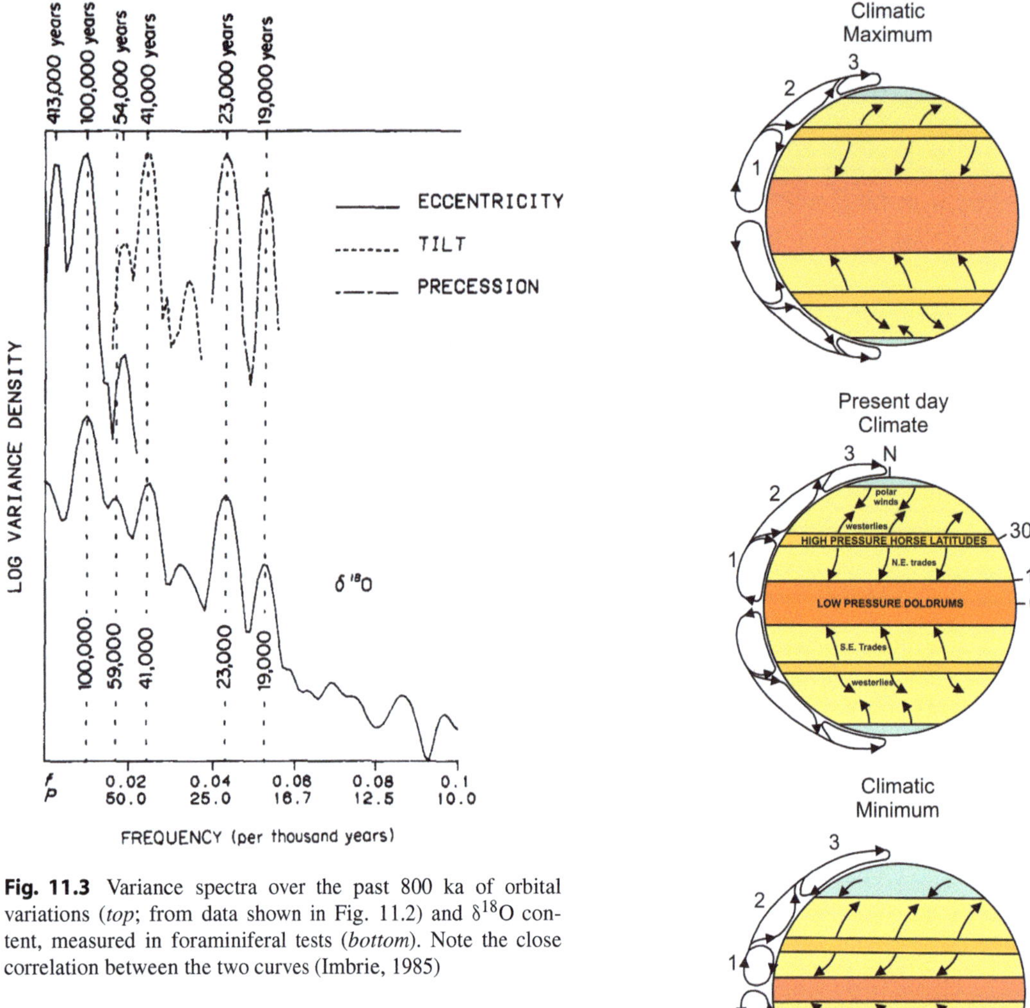



**Fig. 11.3** Variance spectra over the past 800 ka of orbital variations (*top*; from data shown in Fig. 11.2) and $\delta^{18}O$ content, measured in foraminiferal tests (*bottom*). Note the close correlation between the two curves (Imbrie, 1985)

**Fig. 11.4** Atmospheric circulation patterns for three climatic conditions. The present climate (*centre*) is intermediate between true glacial and interglacial conditions (adapted from Perlmutter and Matthews, 1990)

## 11.2.2 Basic Climatology

The major control of global climate is the coupled ocean-atmospheric circulation, which, in turn, controls humidity, rainfall and temperature. Figure 11.4 presents simple models of circulation for three climatic conditions. The earth is encircled by three cells, the Hadley, Ferrel and Polar cells. Consideration of this broad atmospheric structure and its effects on land and sea leads to a series of useful generalizations regarding regional climate. Where the circulation established by the three cells passes across the Earth's surface it sets up persistent wind patterns such as the trade winds. Atmospheric upwelling regions are characterized by greater humidity than downwelling regions, because of *adiabatic* effects which control air saturation. Upwelling along the equatorial belt is the cause

of the high rainfall and inconsistent wind patterns at low latitudes. Downwelling at the convergence of the Hadley and Ferrell cells delivers dry air to mid-latitudes and is the origin of the belt of high-pressure weather systems there. The location of equatorial rain forests and the major mid-latitude deserts (Sahara, Arabian, Namib, Simpson) are explained by these broad circulation patterns. Land and sea have different heat capacities, and therefore heat and cool at different rates, and this can set up regional atmospheric circulation effects. The cells change position with the seasons, leading to seasonal changes in circulation, such as the monsoonal changes in prevailing wind directions. Onshore winds carry moisture, which is released as rainfall when forced to rise over high relief (the adiabatic process). Windward slopes therefore tend to be wetter than leeward slopes, and the climate on the east and west sides of continents may be quite different.

During climatic minimums the downwelling region between the Hadley and Ferrel cells is located at about 15–35° latitude. Monsoon zones show little latitudinal shifting with the seasons, which also limits the amount of moisture transported to this zone. As conditions change to the climatic maximum, the Hadley-Ferrel downwelling zone shifts poleward, and the belt of deserts moves to 35–40°. Hadley and monsoonal circulation increase in efficiency, and the 10–35° zone becomes more humid. The upwelling arm of the Ferrel-Polar cell moves to about 70° latitude. The importance of these climatic shifts in the generation of non-glacial cyclic processes is discussed in Sect. 11.2.5.

de Boer and Smith (1994b, p. 5) summarized the orbital effects on climate as follows:

> At low latitudes, close to the equator, the influence of the cycle of precession, modulated by the varying eccentricity of the Earth's orbit, is dominant and causes latitudinal shifts of the caloric equator …. In turn, this causes significant shifts of the boundaries between adjacent climate zones. At mid-latitudes (20–40°) the orbital variations affect the relative length of the seasons and the contrast between summer and winter, and hence of monsoon intensity …. Toward higher latitudes (>40°) the effect of the varying obliquity becomes more prominent.

In detail, the response of the climate to astronomical forcing at any point on the earth's surface is extremely complex, reflecting various sensitivity factors, such as the relative positions and sizes of sea and land masses. Their latitude affects air and water circulation patterns and their relative position affects humidity (and hence rainfall), monsoon effects, and so on. These complexities are exemplified by Barron's (1983) discussion of the difficulties inherent in attempting to reconstruct the warm, equable climates of the Cretaceous. In general there is a lag effect between the astronomical force and the system's response, although on the kiloyear scale under consideration in this chapter the lag effect is probably relatively unimportant. An

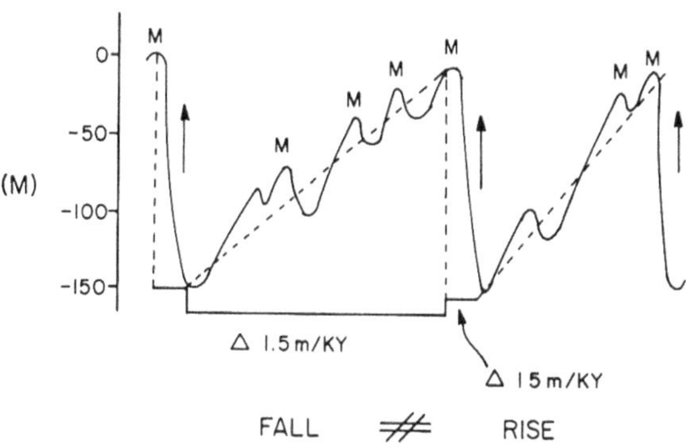


**Fig. 11.5** Schematic representation of the asymmetry of glacioeustatic rises and falls. Time moves from *right to left*. Two types of cycle are shown, with 100- and 40-ka (=KY) periodicities. Melting events are indicated by "M" (Williams, 1988)

exception is the build-up and melting of major ice sheets. In general, glacioeustatic transgressions (caused by ice melting) are much more rapid than regressions (which are caused by continental ice formation; Fig. 11.5). For Pleistocene 100-ka cycles, Hays et al. (1976) found that the sea-level fall occupied 85–90% of the total cycle period.

The rate of sea-level change brought about by the formation and melting of major ice sheets is conventionally estimated at 10 m/ka (e.g., Donovan and Jones, 1979). However, it has long been known from stratigraphic records that the Holocene has been characterized by short intervals of much more rapid sea-level rise. This has been attributed to the breakup of floating ice sheets (Anderson and Thomas, 1991). Melting of continental ice sheets results in a eustatic rise that eventually lifts off ice sheets grounded on the continental shelf. Once decoupled, these massive ice sheets are unstable and quickly break up, leading to pulses of very rapid sea-level rise, up to 30–50 m/ka for intervals of up to a few hundred years.

The correlation between Milankovitch periodicities and oxygen isotope content is shown in Fig. 11.3. The linkage between these parameters is as follows. $^{16}O$ is the lighter of the two main oxygen isotopes, and is therefore preferentially evaporated from seawater. During times of ice-free global climate this light oxygen is recycled to the oceans and no change in isotopic ratios occurs. However, continental ice buildups will be preferentially enriched in $^{16}O$, while the oceans are depleted in this isotope, with the result that the $\delta^{18}O$ content of the oceans is increased. Numerous studies since that of Emiliani (1955) have shown that the $^{16}O/^{18}O$ ratio is a sensitive indicator of global ocean temperatures and can therefore be used as an analog recorder of ice volumes (e.g., Matthews, 1984, 1988). The measurements are made on the carbonate in foraminiferal tests. Matthews (1984) suggested a calibration value of $\delta^{18}O$ variation of about 0.011‰ per metre of sea-level change. Precise calibrations are impossible because of uncertainties regarding the volume and content of floating pack ice, and diagenesis of foraminiferal tests. It is also important to take into account the lag between the maximum eustatic high and the highest surface ocean temperatures, which occur some $10^3$ years later.

### 11.2.3 Variations with Time in Orbital Periodicities

It is not known how the orbital periodicities determined for the Holocene, and which seem to have remained stable at least through the Neogene, might have varied in the distant geological past. Some studies have suggested that small perturbations in the motions of the planets during geologic time may have had significant effects on the orbital behaviour of the earth (Laskar, 1989). Plint et al. (1992) stated that "over longer periods of time, it is postulated that nonlinear amplifications of even very small perturbations in planetary orbits, as well as changes in the diameter of the core, and of the earth as a whole, will lead to chaotic, unpredictable planetary motions." Laskar (1999) stated that calculation of planetary motions cannot accurately retrodict Earth's orbital behaviour before about 35 Ma, because of the long-term chaotic behaviour of the planets and because of drag effects relating to the dynamics of the Earth's interior. Others (e.g., Berger et al., 1989, 1992; Berger and Loutre, 1994) reached different conclusions, suggesting that orbital periodicities should have remained fairly stable throughout the Phanerozoic. Some studies argue that as the "eccentricity periods are the product of interplanetary gravitational forces, they have probably remained stable through at least the last 600 million years" (Algeo and Wilkinson, 1988; after Walker and Zahnle, 1986), whereas the precession and obliquity periods have changed due to continued evolution of

the earth-moon system (Lambeck, 1980; Berger and Loutre, 1994).

Transfer of angular momentum to the moon has resulted in a decrease in the earth's rotational velocity and an increase in the moon's orbital velocity. The consequent recession of the moon has resulted in attenuation of the periods of the earth's precession and obliquity cycles. Approximation of orbital paleoperiods (Walker and Zahnle, 1986) indicate that the period of precession was about 17,000 yrs and that of obliquity about 28,000 yrs at the beginning of the Phanerozoic. (Algeo and Wilkinson, 1988)

Calculations by Berger and Loutre (1994) suggested that the precession periods 19 and 23 ka, would have been about 11.3 and 12.7 ka at 2.5 Ga, with little change in amplitude. By contrast, they suggested that the obliquity period would have had considerably greater amplitude at 2.5 Ga, but they predicted no change in period.

Bond et al. (1993) discussed the occurrence of Milankovitch cycles in the distant geological past and faced the problem that we have little hard data regarding Milankovitch periodicities in earlier geological time. In their opinion, astronomical calculations are too beset by assumptions and difficulties, and the best results may come from geological data. The gamma method (described in Sect. 11.2.6) may be the ideal method for examining this problem, and results from Cambrian and Cretaceous successions reported in their paper indicated periodicities consistent with those predicted by Berger et al. (1989, 1992).

Hinnov and Ogg (2007, p. 240) stated:

Three sources of uncertainty affect the ATS [astrochronological time scale]. First, lack of knowledge about Earth's past tidal dissipation and its effect on the precession translates into an accumulating bias in the timing of obliquity and precession cycles back in time (Berger et al. 1992). This effect is noted in [Fig. 14.22e] as a 'tidal error' in terms of potential deficit of years in the current La2004 precession model (Lourens et al. 2001, 2004). A second uncertainty source lies in chaotic diffusion in the solar system (Laskar 1990; Laskar et al. 1993, 2004). Earth's orbital eccentricity is likely stable throughout most of the Cenozoic; between 50–100 Ma, however, Earth-Mars orbital resonance is thought to have undergone a transition (Laskar et al. 2004). In particular, the 2.4 m.y. amplitude modulation of the ~100-kyr terms of Earth's orbital eccentricity may have been affected. The precise timing of this latest transition is not known; prior to the transition, orbital behavior cannot be modeled accurately. Fortunately, the 405-kyr orbital eccentricity term, from gravitational interaction between Jupiter and Venus, g2-g5, is thought to have remained very stable as well as dominant (due to the great mass of Jupiter over several hundreds of millions of years, with an estimate(uncertainty reaching only 500 ka at 250 Ma [see 'maximum error', Fig. 14.22e]).

There is clearly considerable disagreement and uncertainty in this area, and skepticism should be exercised in attempts to relate cycle periodicities in pre-Pleistocene rocks to specific orbital parameters, as has been done in some studies (see Sect. 14.7). House (1985) and House and Gale (1995) have even suggested using calculated periodicities to calibrate time scales, and discussed the use of interpreted Milankovitch cycles in the rock record to refine the geological time scale determined by biostratigraphic and radiometric means. At present this seems premature, because of the uncertainties regarding orbital changes in the past. In addition, as discussed in Sect. 14.7, there remains considerable uncertainty regarding the precision of the geological time scales within which Milankovitch refinements are to be calculated.

### 11.2.4 Isostasy and Geoid Changes

An additional complication concerning glacioeustasy is that isostatic and geoidal effects must be taken into consideration before changes in ice volume can be translated into changes in sea level. The geoid is defined as a sea-level surface encompassing the globe as if it extended continuously through the continents. Fjeldskaar (1989) stated that "under the continents the geoid can be thought of as the surface defined by the water level in narrow canals cut to sea level through the land masses." The geoid reflects gravity anomalies, and is therefore not spherical. Major continental ice sheets are several kilometres in thickness, and have a significant isostatic loading effect on the crust. Upon melting, the crust rebounds much more slowly than the rise in sea level brought about by melting. Locations close to the ice cap (within a few hundred kilometres) therefore undergo an initial rapid rise in sea level, followed by a slow fall, over periods of tens of thousands of years, a fall which is not recorded in areas beyond the isostatic reach of the ice cap. The ice mass also has a gravitational attraction that affects the shape of the geoid. Lambeck et al. (1987) modeled these two forces and produced the map shown in Fig. 11.6. The geoidal effect alone is shown in Fig. 11.7. This map shows the depression of the geoid (sea level) that would occur if the 20-ka ice cap underwent instant melting. The conclusion to be drawn from the studies of geoidal and isostatic effects is that the sea-level signature in the

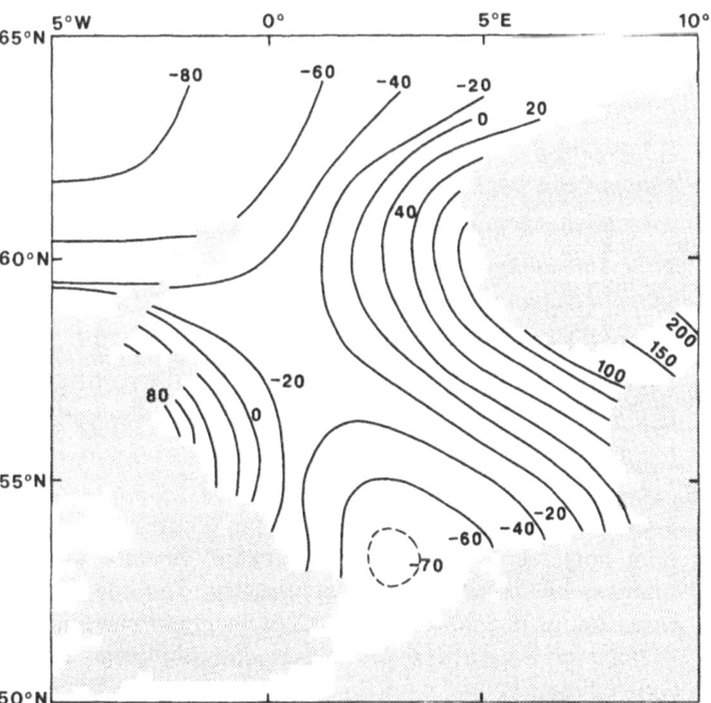


**Fig. 11.6**  Sea levels (in metres) relative to present-day sea levels in the North Sea, 20 ka ago, prior to the melting of the British and Scandinavian ice caps (Lambeck et al., 1987)

**Fig. 11.7** Theoretical
deflection of the geoid over
northwest Europe caused by
instantaneous deglaciation of
the 20-ka ice sheet. Contour
interval is 10 m (Fjeldskaar,
1989)

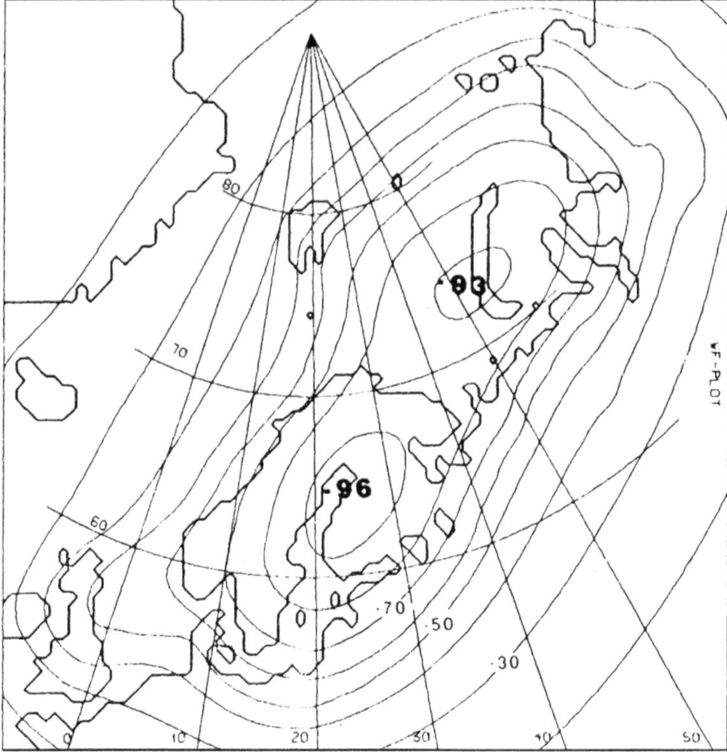



stratigraphic record close to major ice caps cannot be subjected to simple, straightforward interpretations of eustatic sea-level change. However, at distances of a few thousand kilometres from the edge of the ice cap these effects will be small enough to be negligible.

### 11.2.5 Nonglacial Milankovitch Cyclicity

Orbital variations may be capable of causing climatic change and consequent sedimentary cyclicity, even when the global cooling effects are inadequate to result in the formation of continental ice caps. For example, because of the interaction between the eccentricity and precession effects, there are phases when one hemisphere experiences short, hot summers and long, cold winters, while the other hemisphere undergoes long, cool summers and short, mild winters. These effects alternate between the hemispheres over periods of $10^4$ years, and the intensity of the effect waxes and wanes as the orbital variations go in and out of phase (Fischer, 1986). Such changes undoubtedly have dramatic effects on surface air- and water-temperature distributions, oceanic circulation and wind patterns.

There is, therefore, considerable potential for subtle sedimentological effects, without any change in sea-level.

Cecil (1990) provided a general model of the response of depositional systems to climate change. As climate belts shift as a result of orbital forcing, or the continents drift through climate belts as a result of plate motions, climate changes lead to changes in clastic sediment yield and patterns of chemical sedimentation. Vegetation patterns change in both elevation and latitude, as the global climate cycles through the changes in seasonality and total energy flux. For example, Fig. 11.8 illustrates the changes in the vegetation zones on the flanks of the Andean mountains, between warm and cool periods. Arid climates are times of low sediment yield and are accompanied by formation of pedogenic carbonates and evaporites. Sediment yield increases with increasing precipitation, leading to a predominance of clastic sedimentation. Maximum yields occur under temperate, seasonal wet/dry climatic conditions. Very humid climates are characterized by thick vegetation cover, which results in a reduction in sediment yield and an increasing importance of peat/coal formation.

**Fig. 11.8**  The vegetation
zones differentiated by
altitude on the flanks of the
Andes. These move up and
down the mountain slopes
according to changes in
climate from warm (*top*) to
cool (*bottom*)

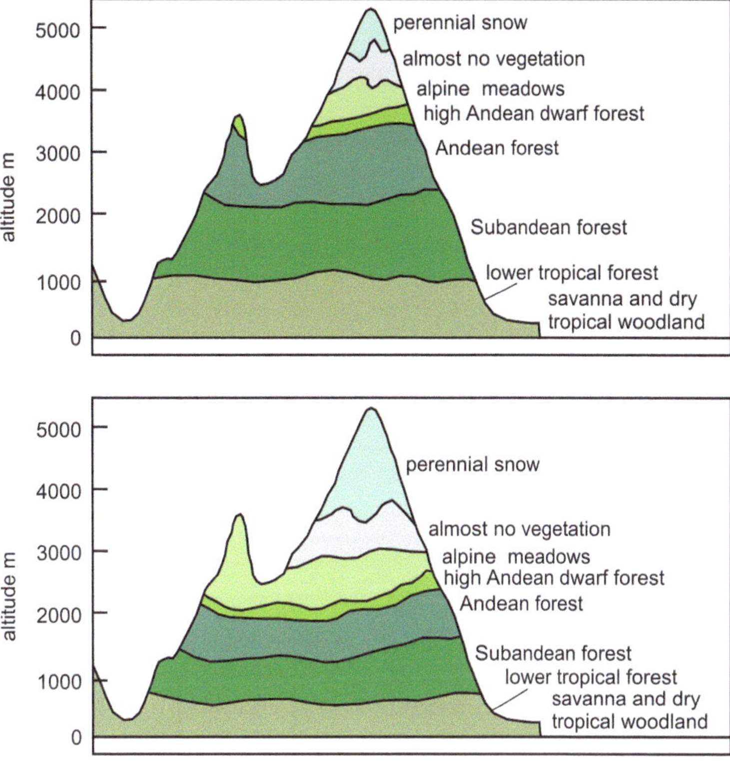


**Fig. 11.9**  The climate zones
of the Niger River watershed.
Movement of these belts as a
result of climate change can
be expected to exert a major
control on the amount and
type of sediment delivered to
the Niger delta (after Van der
Zwan, 2002)

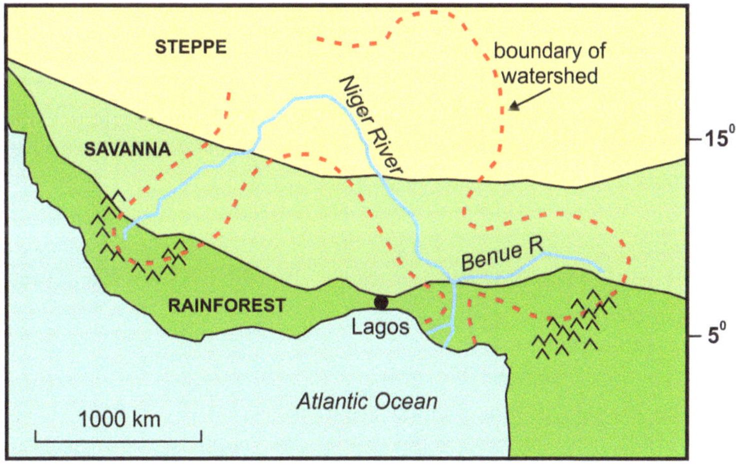


The Niger delta offers an interesting example of
how changes in climate might be directly reflected
in deltaic sedimentation (Fig. 11.9). The Niger river,
which is the major sediment source for the delta,
flows through three major climate belts that cross
tropical west Africa. In the north, the Sahel is an
area of steppe climate—arid, with very limited vegetation cover. Rainfall is sparse and flashy, leading to
erratic yield of coarse clastic detritus. The savanna is
an area of grassland, and the rainforest belt a zone
within which weathering and erosion processes will
tend to generate more suspended chemical sediment
load than clastic bedload. Cooler global climates, such
as those that characterize glacial episodes, will tend to
shift these climate belts southward, meaning that more
of the Niger watershed lies within the steppe zone,

with consequent increased levels of clastic sediment delivery to the delta. Conversely, during periods of warmer climate, the rainforest belt would be expected to expand northward, with a resulting reduction in the clastic sediment load. These variations might be expected to be present in the form of cyclic variations in sediment calibre or sand-bed thickness in the Niger delta, and it was just such an effect that Van der Zwan (2002) predicted when he set out to perform a time-series analysis of wireline-log records from the deltaic sediments. However, as agued in Sect. 14.7.1, there are too many other irregular, random and other processes, including those of autogenic origin, that could be expected to affect the accumulation of a major delta, and his results are regarded here as suspect.

Variations in organic productivity, in the depth of carbonate compensation (CCD), and in the degree of oxidation of marine waters are the kinds of effects that are controlled by these climatic variables and have direct consequences for sedimentation patterns (Fischer, 1986). For example, the rate of skeletal supply relates to organic productivity. Erba et al. (1992) documented rhythmic variations in species abundances in a Cretaceous clay using Fourier analysis, and interpreted these as "fertility cycles" driven by changes in oceanic temperature and circulation in response to variations in seasonality. Variations in the supply of clay due to variations in source area weathering and marine circulation affect the proportion of biogenic components in the resulting sediments. The carbonate fraction varies if the seafloor is close to the CCD. Various examples of laminated pelagic sediments in the geologic record were discussed by Fischer (1986) as illustrations of these processes (Fig. 11.10). Weltje and de Boer (1993) showed that the thicknesses and mineralogy in some rhythmic Pliocene turbidites in Greece can be attributed to precession-induced changes in precipitation and runoff feeding deltaic sediment sources. Redox cycles are another type of cycle resulting from varying sediment fluxes and oxygenation levels (de Boer and Smith, 1994b).

Many processes show a significant lag between changes in the forcing function and the sedimentary response. This is a well-known characteristic of the "carbonate factory", but may be a factor in other environments. For example, de Boer and Smith (1994b) cited examples of Pliocene carbonate-marl cycles and anoxic-aerobic cycles in the Mediterranean region that suggest a lag in the response of monsoonal systems to changes in insolation. Similarly, peat and methane production may not reach a peak for thousands of years after the related insolation peak has passed. Complex inter-relationships and feedback loops were summarized by Mörner (1994).

A particularly interesting study of Milankovitch cycles is that by Elder et al. (1994; see Fig. 7.59), who demonstrated a correlation between $10^5$-year clastic cycles on the margin of the Western Interior Seaway in Utah with carbonate cycles in the basin centre in Kansas, over a distance of about 1,500 km. They developed a model based on small-scale sea-level changes (10–15 m) driven by orbitally-forced climate changes. Their model included the formation and melting of small glaciers, but this component of the model does not seem to be necessary to this writer. Elder et al. (1994) suggested that during warm, dry phases thermal expansion of surface waters, glacial melting, or transfer of stored groundwater to oceans, could have contributed to transgression, flooding, and enhancement of carbonate productivity in the basin centre. During cool, wet phases sediment supply from the Sevier highlands to the west would have increased, leading to shoreface progradation, and a clay-rich phase in the basin centre (Fig. 11.11). There is no evidence of diachroneity of the sequence boundaries, as in some comparable sequences from this basin in Alberta (Sect. 10.3.3.1), therefore, for example, the reciprocal stratigraphy model developed there cannot be invoked in this case. However, there is some evidence of thickness changes at faults, and strandlines parallel the faults, so some tectonic influences were active during sedimentation.

The Greenhorn example suggests a general model for "non-glacial" Milanokovitch cyclicity, which is presented in Fig. 11.12. Idealized clastic and carbonate rhythms are shown correlating to an idealized oxygen isotope curve. We return to this area in Sect. 14.7.2 and 14.8 because the rocks have become a focus for some work on cyclostratigraphy and chemostratigraphy.

Hallam (1986) urged caution in that some cyclicity may relate to diagenetic effects and may not be a reflection of primary sea-level or climate controls. He discussed the case of limestone-shale cycles, such as the decimeter-scale couplets in the Blue Lias of southern England. House (1985), Weedon (1986), and others have studied these rhythms using refined biostratigraphic analysis to ascertain time spans and have performed spectral analysis on the cycle thicknesses in

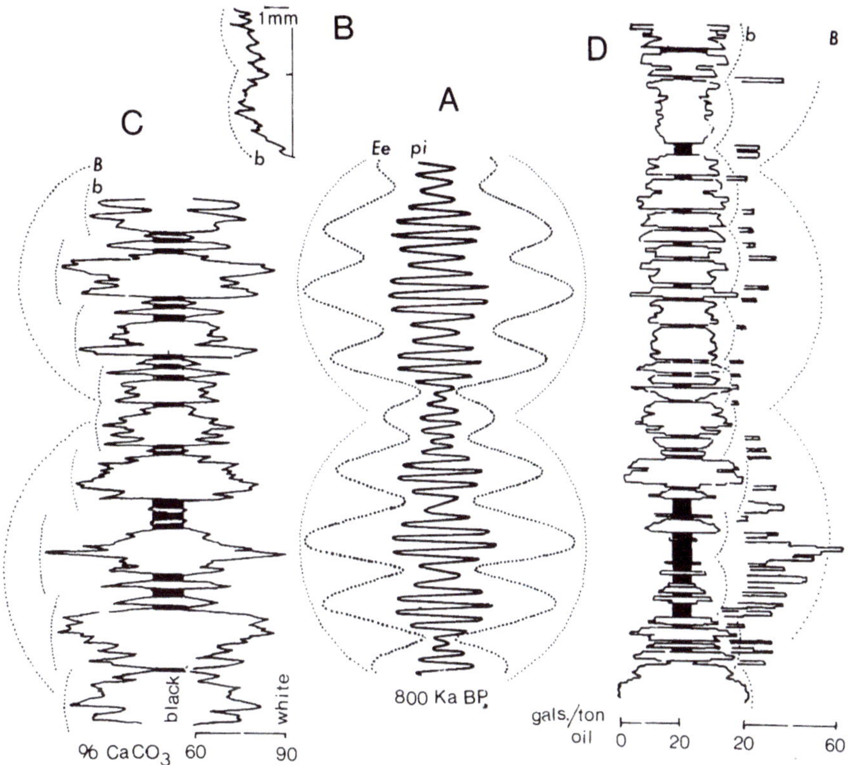



**Fig. 11.10** Comparison of Milankovitch cycles plotted at similar time scales. **a**, Milankovitch parameters plotted for the last 800 ka; pi, precession; Ee, long and short cycles of orbital eccentricity. **b**, Variations in thickness of evaporite varves, Castile Formation (Permian), New Mexico. **c**, Rythmicity in Fucoid Marl (Cretaceous), Italy. Decimeter-scale limestone-marl couplets are shown by fine detail in carbonate variation and are driven by precession. These are grouped into 50-cm bundles (b) by a redox cycle, in which marls are periodically *black* and sapropelic as a result of bottom stagnation, related to an eccentricity cycle. Superbundles (B) may reflect a long-term orbital eccentricity cycle. **d**, Variations in oil yield in the Parachute Creek Member, Green River Formation (Eocene), Colorado. Kerogen-rich layers formed during humid lacustrine phases, whereas kerogen-poor intervals reflect arid playa phases. Bundling in response to various orbital controls is shown by *dotted lines* (Fischer, 1986)

search of rhythmicity. Weedon (1986) claimed success in this endeavour and suggested a correlation with precession and obliquity effects. However, Hallam (1986) showed that in at least some of the cyclic sequences rhythm thickness is constant regardless of the overall sedimentation rate. He suggested that in such cases the rhythms may have been formed by diagenetic unmixing of calcium carbonate during diagenesis.

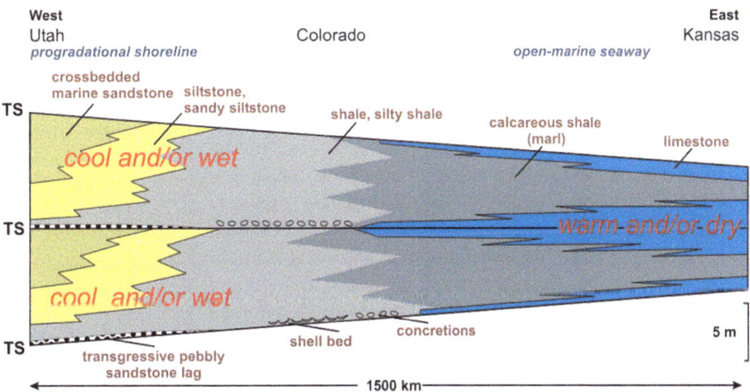



**Fig. 11.11** A model of the Greenhorn climate rhythms observed along a transect from Utah to Kansas (based on Elder et al., 1994)

**Fig. 11.12** A general model of "non-glacial" Milankovitch cyclicity, based on cycles observed in the Western Interior Seaway. These are correlated against an idealized oxygen isotope curve, showing the relationship between sea level, the alternation of warm and cool global temperatures, and the resulting sedimentary facies

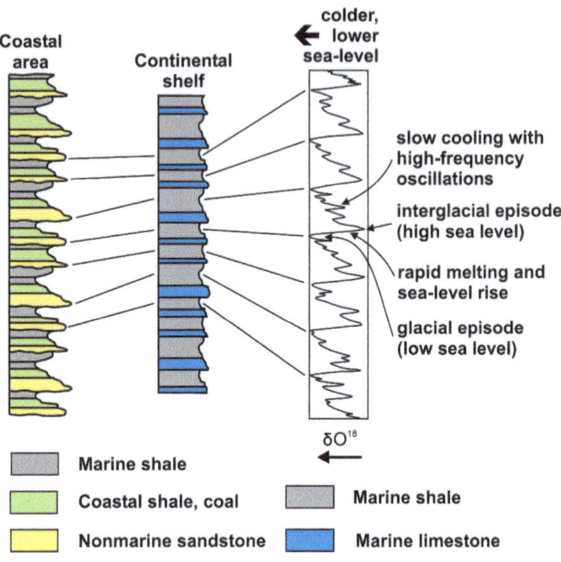



## 11.2.6 The Nature of the Cyclostratigraphic Data Base

Milankovitch cycles have been identified in the rock record primarily on the basis of observed rhythmicity in the rock record, particularly the occurrence of metre-scale or thinner (varved) cycles, typically in rocks of carbonate shelves, and fine-grained pelagic and lacustrine sediments. Some clastic or mixed carbonate-clastic successions, such as the upper Paleozoic cyclothems, are also attributed to Milankovitch processes. However, the presence of rhythmicity and an estimation of cycle durations of the appropriate magnitude does not prove orbital forcing as the cycle-generating mechanism. As stated by Algeo and Wilkinson (1988):

> Despite an often-claimed correspondence between cycle and Milankovitch orbital periods, factors independent of orbital modulation that affect cycle thickness and sedimentation rate may be responsible for such coincidence. For example, nearly all common processes of sediment transport and dispersal give rise to ordered depositional lithofacies sequences that span a relatively narrow range of thicknesses ... Further, long-term sediment accumulation rates are generally limited by long-term subsidence rates, which converge to a narrow range of values for very different sedimentary and tectonic environments (Sadler, 1981). In essence, the spectra of real-world cycle thicknesses and subsidence rates are relatively limited, and this in turn constrains the range of commonly-determined cycle periods. For many cyclic sequences, calculation of a Milankovitch-range period may be a virtual certainty,

regardless of the actual generic mechanism of cycle formation.

Algeo and Wilkinson (1988) studied cyclicity in more than 200 stratigraphic units, determining periodicities from cycle thickness, sedimentation rate, depositional environment and age range. They found that calculated periodicities are randomly distributed relative to all the Milankovitch periods except the 413-ka eccentricity cycle. They concluded that "if, in fact, a calculated average period within the broad range of Milankovitch periodicities is not a sufficient test of orbital modulation of sedimentary cycles, demonstration of such control becomes significantly more difficult than hitherto appreciated." Peper and Cloetingh (1995) demonstrated by numerical modeling the significance of the reverse problem, that certain non-Milankovitch processes, such as slope failure and intraplate stress changes, can generate perturbations that distort or obscure the stratigraphic response to Milankovitch processes.

Demonstration of a hierarchy of periodicities may be a useful indicator of orbital control. For example, where there are two orders of cycles with recurrence ratios of 5:1 a precession (20 ka) and eccentricity (100 ka) combination may be indicated. Fischer (1986) discussed examples of this, which he termed "cycle bundling" (Fig. 11.10).

Van Tassell (1994) provided a good example of the kind of problem described by Algeo and Wilkinson (1988). A wide spectrum of cycles is present in the

Devonian Catskill Delta of the Appalachian Basin, but proving they are of Milankovitch type (as claimed by the author) is problematic. Their thickness and duration fits that of Milankovitch and autogenic cycles, as Algeo and Wilkinson (1988) predicted would occur in these types of rocks. Demonstration of cycle duration is difficult, because of uncertainties about their age range. No Milankovitch-type bundling can be demonstrated, and the proof of correlation of cycles across major facies boundaries is very limited. A comparable, partly correlative suite of cycles was described by Filer (1994), based mainly on subsurface gamma-ray log correlations. His correlations seem unconvincing to this writer. No clear cyclic signatures can be traced from well to well. These date would benefit from application of the gamma technique, described below.

Drummond and Wilkinson (1993b) modeled carbonate cycles to study the effect of "lag time" (the time taken for the carbonate factory to reach full productivity rates when a platform is flooded). They showed that several discrete cycles could be produced during one smooth leg of sea level rise, as the factory overcomes the lag, races to deposit sediment, then is shut down as the platform is built to sea level. Therefore a one-for-one correspondence between sedimentary cycles and sea-level cycles cannot be expected, and cycle durations and hierarchies cannot be directly related to Milankovitch frequencies. However, this is just a numerical model, and they did not offer evidence of this from the rock record.

The requirements for the establishment of orbital forcing are twofold. Firstly it is necessary to demonstrate that the cycles are widespread, by examining their long-distance correlation. As noted by Pratt and James (1986) and Pratt et al. (1992), many shallow-marine cycles are generated by autogenic processes, and are laterally impersistent. Studies based on single vertical sections, however, detailed, are inadequate for the purpose of testing models of orbital forcing. Tracing and correlating sequences through major facies changes may also be indicative of orbital control. An example of this is illustrated in Fig. 7.59, and is discussed in Sect. 11.2.5. Then, given a demonstration of lateral persistence, it is necessary to demonstrate a persistent regularity of cycle periods, and cycle bundling. This may be done by precise measurement of facies thicknesses or parameters such as carbon or calcium content over lengthy sections

(see many examples in de Boer and Smith, 1994a). Spectral mapping methods may be particularly useful (Melnyk et al., 1994; Schwarzacher, 1993, 2000; Weedon, 2003). Cycle hierarchies and cycle stacking ratios have been discussed by Goldhammer et al. (1987, 1990) and Schwarzacher (1993). However, as pointed out in Sect. 14.7.2, in the absence of independent chronostratigraphic documentation of age, stacking ratios may be misleading. The use of Fischer plots for the documentation of metre-scale cycles is discussed in Sect. 3.7.

A method for independently testing the periodicity of metre-scale cycles, termed the *gamma method*, was devised by Kominz and Bond (1990). Cycles are subdivided into lithofacies according to a predetermined classification, and the thickness of each facies unit is measured in each cycle. It is assumed that each facies has an approximately constant accumulation rate, reflecting constant depositional conditions. A value for gamma for each facies is given by gamma = elapsed time/thickness. If it can be assumed that each cycle has the same duration (the absolute value of which does not need to be known) a set of equations can be written that determines the proportion of elapsed time represented by each facies in each cycle. If these values are reasonably consistent, then periodicity has been demonstrated. If a method of determining absolute age is available, for example by radiometric dating of bentonites that bracket the section under study, then the absolute duration of the cycle period can be determined, and if appropriate, this can be related to orbital frequencies. Given that orbital frequencies may have changed over time (Sect. 11.2.3), and that other sedimentary processes may generate cycles of comparable duration, these quantitative methods of cycle documentation are of considerable value.

## 11.3 The Geologic Record

### 11.3.1 The Sensitivity of the Earth to Glaciation

The modern literature on paleoclimatology (e.g., Barron, 1983; Barron and Thompson, 1990) indicates a considerable complexity and uncertainty in reconstructions of past climates. Although the control of

Milankovitch mechanisms on glacial cyclicity is now widely accepted, it is still not entirely clear why major glaciations start in the first place. Factors involved in sensitizing the earth to climatic change include the current climate, the plate configuration of the earth, atmospheric content (volcanic dust, $O_2$, $CO_2$), vegetation cover, and the nature of oceanic and atmospheric circulation (Barron, 1983; Barron and Thompson, 1990). Plate-tectonics has been a major control. As Eyles (2008, p. 109) noted: "Cooling after 55 Ma is increasingly being related to the tectonic and bathymetric evolution of the oceans as Pangea continued to disintegrate."

Low levels of carbon dioxide and correspondingly high oxygen levels are thought to be characteristic of times of low sea-level during supercontinent assembly phases, and tend to favour icehouse climates because of the reduced atmospheric greenhouse effect (Worsley et al., 1986, 1991; Worsley and Nance, 1989; see Sect. 5.1). Glaciation is favoured under such conditions, especially where large continental landmasses are located over polar regions, as in the case of the late Paleozoic glaciation (Sect. 11.3.4). Eyles (1993, 2008) made a strong case for the importance of tectonic control in the triggering of widespread continental glaciation. Uplift accompanying continental collision, and the widespread uplift of rift flanks during supercontinent dispersal, both lead to the development of broad, high-altitude areas where major Alpine ice-caps can easily form. Once formed, feedback effects, such as increased albedo, can lead to continued cooling. The geological record is replete with examples that can be interpreted in this way. For example, Eyles (1993, 2008) attributed the widespread late Proterozoic glaciation to tectonic uplift of rift flanks during the dispersal of the supercontinent of which Laurentia was the centre (see Hoffman, 1991). The assembly of Pangea was accompanied by the uplift of many orogenic highlands, and was followed by the long-lasting Gondwana glaciation (Sect. 11.3.4). Orogenic highlands formed in the Miocene, such as Tibet, may also have been instrumental in triggering of the Late Cenozoic glaciation (Ruddiman, 2008). Another major contributing factor in the case of Cenozoic glaciation was undoubtedly the plate-tectonic separation of Australia from Antarctica, which opened the Drake Passage and led to the development of circumpolar currents, which essentially isolated Antarctic weather systems from the rest of the southern hemisphere (Kennett, 1977).

Given these broad, long-term ($10^6$–$10^7$-year) regional to continental controls, Milankovitch processes then govern the amount of climate change and its periodicity within the Milankovitch time band. Considerable variation can be detected at this level. Imbrie (1985) carried out a spectral analysis of the $\delta^{18}O$ variations in deep sea cores for the last 782 ka. It shows a change in sensitivity at about 400 ka ago. The oxygen isotope record for the 782–400-ka record contains a major peak corresponding to a 100-ka period, and a lesser peak for a 23-ka period. The younger part of the record shows somewhat more variance, with stronger peaks in the longer-term periods, including a strongly represented 100-ka period, and an additional peak corresponding to a 41-ka period. Moore et al. (1982) studied the variation in carbonate concentrations in Upper Miocene and Quaternary deep-sea sediments. The variations are attributed to the degree of preservation of calcite microfossils, which, in turn, is related to carbonate solubility and the corrosiveness of bottom waters. Both factors are climate dependent. Moore et al. (1982) found that periods of 41, 100, and 400 ka were dominant in their spectral analyses. However, the 400-ka period is far more strongly recorded in the Miocene data, with the 100-ka period the dominant one in the Quaternary record. Imbrie (1985) interpreted the strong 100-ka peak as a combination of the two distinct eccentricity peaks at 95 and 123 ka.

Matthews and Frohlich (1991) extended Imbrie's (1985) analysis back to 2.5 Ma, and forward to a hypothetical 2.5 million years in the future. They demonstrated that the 100-ka eccentricity and 19-ka precession components combined to produce a 2-million years modulation, although the amplitudes of the various peaks varied according to input conditions. This result suggests that glacioeustasy may be capable of explaining longer-term cyclicity with million-year periodicities (equivalent to Vail's "third-order" cycles). To this analysis Matthews and Frohlich (1991) then added the effects of the presence of warm, saline waters in mid-latitude positions, such as would have occurred in the Tethyan Ocean before this water body was eliminated by subduction of its sea floor. "Warm water upwelling adjacent to a cold continent should create a very efficient mechanism for producing ice on the continent, thereby lowering sea level. Air descending over the cold continent flows out over warm water and picks up water vapor. Return flow aloft delivers this moisture to the interior of the continent as snow" (Matthews and Frohlich, 1991). It is postulated that thermohaline currents would have delivered this water to Antarctica, and could have accounted for large build-ups of snow there, with a calculated episodicity of 1.6–2.4 million years. Qualitatively, therefore, a glacioeustatic mechanism exists to explain million-year ("third-order" or "mesothemic") cyclicity during times when the earth was sensitive to glaciation. However, the lack of evidence for major continental ice caps throughout much of geologic time means that this cannot be regarded as a universal explanation.

Variations with time in the stratigraphic signal amplitude of the various Milankovitch periods may relate to the varying character of glacial regimes. The current view (e.g., Miller et al., 2005b) is that until the Miocene, only the Antarctic ice cap existed. This is mainly a land-based ice cap and is entirely surrounded by polar waters. Northern hemisphere ice commenced a significant buildup at about 14 Ma, with the major "ice ages" commencing at about 2.6 Ma. The Northern Hemisphere ice consisted of several continental-scale and smaller centers, some bordered to the north or south (or both) by water masses, and included numerous Alpine-type ice caps. It seems likely that the sensitivity of the ice caps to climate forcing would have been quite different in these different settings, with the large, stable Antarctic ice cap tending to respond much more slowly than the Northern Hemisphere ice centers.

Another example of the varying styles of glaciation affecting the styles of cyclicity was discussed by Wright (1992). He classified platform-carbonate cycles into two broad types. Type 1 consist of metre-scale "keep-up" cycles dominated by peritidal carbonates. They are common in the Cambro-Ordovician, early Devonian and mid-Triassic to Cretaceous sedimentary record. Type 2 cycles are 5–10 m thick. They consist mainly of shelf and deeper-water carbonates, with common paleokarstic and paleosol horizons (vadose caps). They are common in the Carboniferous sedimentary record. Wright (1992) suggested that the differences in cycle type may have been caused by differences in the type of glacial cycle, reflecting differences in the size of ice caps and their consequent sensitivity to change. He postulated that type-1 cycles represent small-scale ice caps, such as those formed during "greenhouse" climatic phases (Fig. 4.8),

When continental configuration or other factors did not facilitate the generation of major ice sheets. During such periods the amplitudes of these climatically induced sea-level changes were small, only slower third-order changes influenced the types of small cycles developed. In the absence of ice-buildups no rapid decametre-scale falls or rises in sea-level occurred. The falls in sea-level triggered by changes in insolation were small in amplitude and slow enough to be outpaced by subsidence, creating minor accommodation space, readily filled by peritidal cyclothems. ... The subsequent rises were slow enough to be outpaced by carbonate production to produce more peritidal cyclothems. (Wright, 1992, p. 2)

As noted elsewhere (Sect. 11.3.3) there is, however, doubt regarding the existence of significant ice caps for much of Phanerozoic time, and this mechanism for type-1 cyclothems remains unproven. Wright (1992) suggested that type-2 cyclothems developed in response to the growth of large continental ice caps during "icehouse" climatic phases, such as during the Carboniferous Gondwana glaciation, and the late Cenozoic glaciation. At such times high-amplitude sea-level falls lead to exposure of platform tops and to extensive meteoric diagenesis. Rapid rises lead to drowning in which only reefal organisms are capable of keeping pace with sea-level rise. Platform interiors are drowned, becoming areas of subtidal sedimentation, in which peritidal and reef sedimentation are restricted to nucleation sites on earlier topographic highs. This is the pattern of so-called "catch-up" sedimentation.

## 11.3.2 The Cenozoic Record

Many successions of Late Cenozoic age have now been correlated with the oxygen-isotope record, demonstrating a glacioeustatic control of the sequence stratigraphy. From numerous DSDP core records (Fig. 11.13), a standard isotopic record has been constructed, with numbered chrons providing a standard time scale for the Pliocene to recent (Fig. 11.14). Examples of the use of this system in chronostratigraphic dating of sequence records are shown in Figs. 7.3, 7.17, 7.22, 14.24 and 14.25.

A widespread sea-level fall of mid-Oligocene age has been suggested in many stratigraphic records. It appears on the Exxon global cycle charts, and also in many stratigraphic records that have been independently compiled for specific basins and

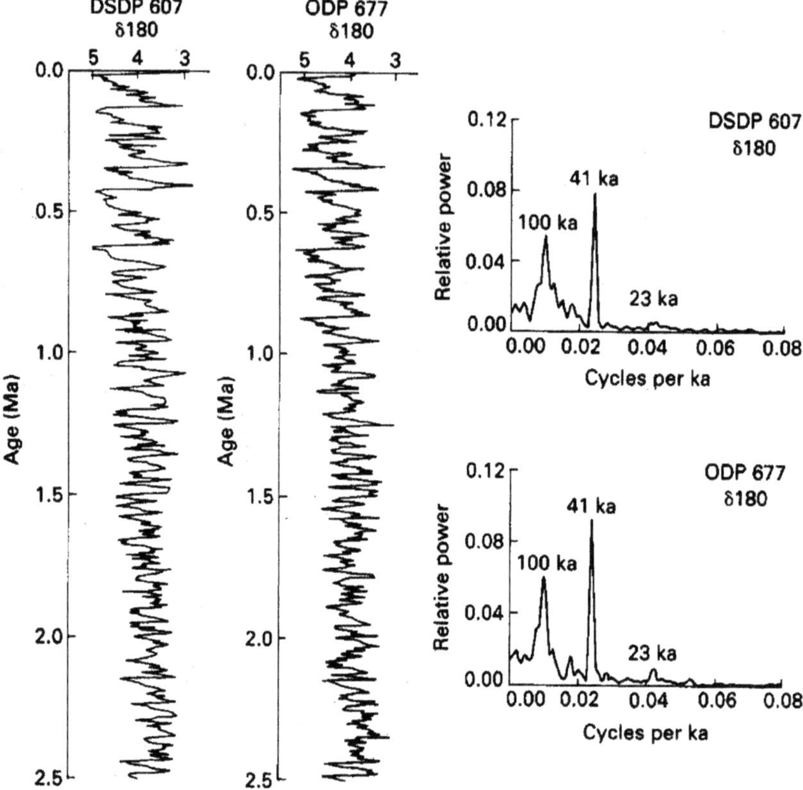



**Fig. 11.13** Examples of oxygen isotope records from two DSDP drill cores. Note the well-developed orbital signature, as shown by the spectral analyses at *right*, the gradual increase in the amplitude of the glacial to interglacial fluctuations with time, and the gradual overall cooling of the climate from 2.4 Ma, which is shortly after the time at which widespread northern hemisphere glaciation is believed to have commenced (Weedon, 1993)

continental margins. However, this age of inception of full-scale Antarctic glaciation has been gradually pushed back in time. Eyles (1993) stated: "the earliest input of glacial sediment to the Ross Sea margin of Antarctica occurred at about 36 Ma [latest Eocene] when temperate wet-based glaciers reached the Victoria Land rift basin. …. A continental ice sheet was probably in existence at this time … but a large continental ice mass did not become a permanent fixture of Antarctica until about 14 Ma." Isotopic evidence, referred to below, suggests that the major sea-level falls occurred at these two times (Fig. 11.15). In a later review, Eyles (2008, p. 110) said: "It has been thought that glacierization occurred early in the Antarctic by c. 44 Ma followed by ice growth around the Arctic at 14 Ma (the unipolar ice sheet model of Perlmutter and Plotnick, 2003) but recent work continues to push back the onset of glaciation in circum-Arctic regions ranging from 45 Ma (Moran et al., 2006) to c. 38 to 30 Ma."

The relationships between tectonic events, global cooling, and the development of southern and northern hemisphere ice is made clear in Fig. 11.16, from Eyles (2008), which summarizes the collation of a large body of research in the timing and controls of global climate changes. Regional uplift due to continental collisions, magmatic underplating, and the uplift of rift shoulders, all had the effect of elevating the continents and bringing about regional cooling. The cumulative effect of this was the development of major ice caps, first in Antarctica, and then in the northern hemisphere. Milankovitch forcing seems to have become particularly prominent in generating high-frequency glacial-interglacial fluctuation at about 3.5 Ma as the build-up of large-scale northern hemisphere ice was about to commence.

Between about 2.4 and 0.8 Ma, the 41 ka obliquity cycle was the dominant frequency, with the 100 ka eccentricity cycle becoming dominant during the last 800,000 years of earth history (Ruddiman and Raymo, 1988). Prior to 2.4 Ma, the 23 and 19 ka precessional frequencies were prominent. The switch coincides with the onset of major northern hemisphere glaciation.

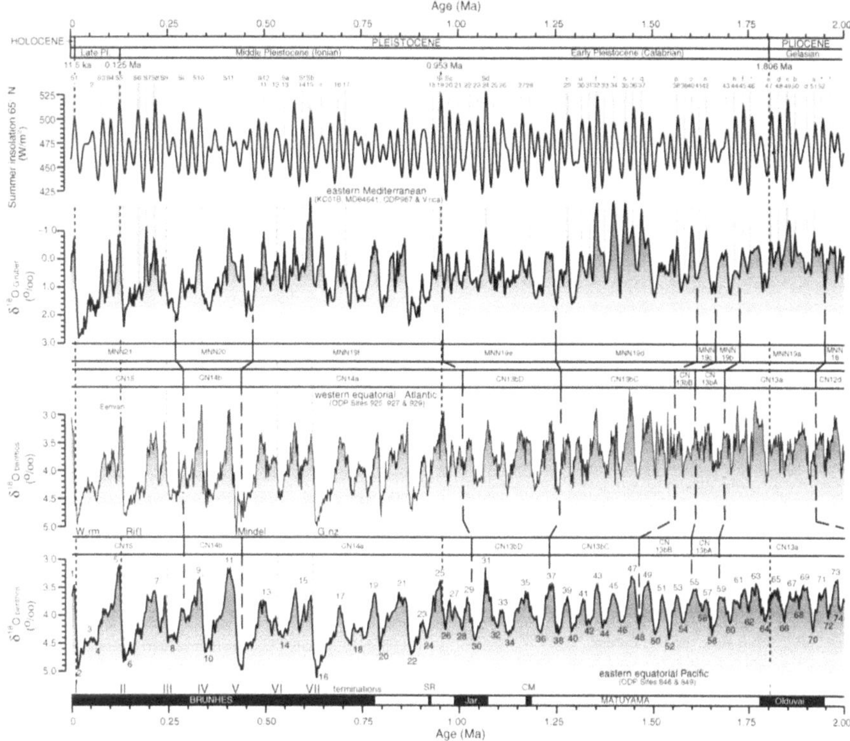

**Fig. 11.14** Comparison between the oxygen isotope chronologies of marine sequences in the Mediterranean, western equatorial Atlantic and eastern equatorial Pacific during the late Pliocene and Pleistocene (Gradstein et al., 2004, Fig. 21.4)

### 11.3.3 Glacioeustasy in the Mesozoic?

One major school of thought holds that all global high-frequency cycles with periodicities of $10^6$ years and less, are glacioeustatic in origin. Vail et al. (1991) made this assertion with regard to the "short-term eustatic curve" of Haq et al. (1987, 1988a), despite the lack of convincing evidence for continental glaciation for much of geologic time (Frakes, 1979, 1986; Hambrey and Harland, 1981; Barron, 1983; Francis and Frakes, 1993). As argued elsewhere in this book there is, in fact, doubt regarding the very existence of a global framework of cycles in the $10^6$-year range that would require such explanation (Chaps. 11 and 13), and non-glacial Milankovitch mechanisms can be invoked to explain many types of high-order cycle (Sect. 11.2.5). Nevertheless, glacioeustasy remains a popular hypothesis for all types of high-order cycle.

Various authors (e.g., Barron, 1983; Frakes, 1986) have examined the question of glaciation, or other evidence of cold climates during the Mesozoic, when most available evidence suggests that the earth was in a greenhouse climatic state (Fig. 4.8). Stable-isotope data indicate that ocean-water temperatures were higher than at present during the Cretaceous (Fig. 11.15), with a smaller equatorial-polar temperature gradient.

Frakes and Francis (1988) and Francis and Frakes (1993) provided summaries of the evidence for cold polar climates in the southern hemisphere during the Cretaceous. They noted that there is ample evidence for locally cold temperatures, perhaps of seasonal character, in many Lower Cretaceous high-latitude deposits. Many ice-rafted deposits of this age have also been recorded from such as areas as central Australia, northern Canada and northern Russia, which would have been located in high latitudes at the time. At the very least this indicates the presence of significant pack-ice fields. Frakes and Francis (1988) stated:

Given that the Earth is a sphere which receives solar insolation unequally over its surface and that heat transport in both atmosphere and oceans has its physical limits, it is difficult to explain how the polar zones could ever have

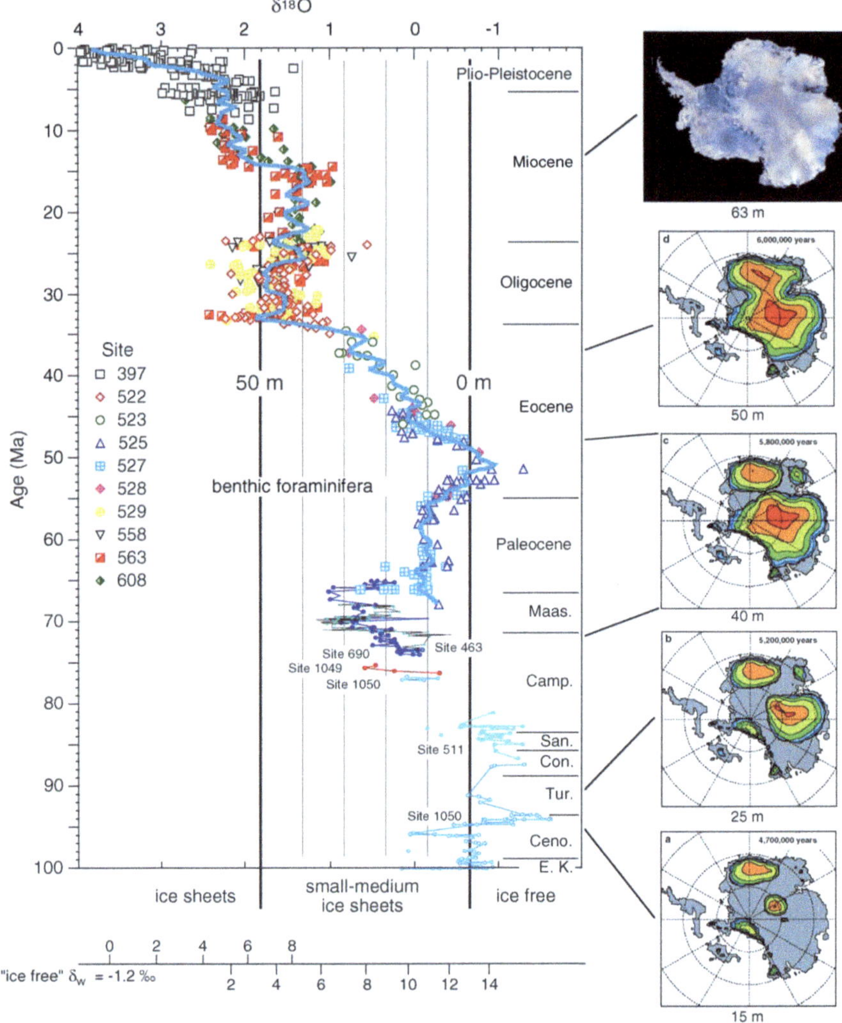

**Fig. 11.15** Compilation of $\delta^{18}O$ data for the Late Cretaceous to present, showing the magnitude of Antarctic ice caps and the maximum sea-level fall from a Cretaceous ice-free world that can be calculated from the isotope data. The maps are from DeConto and Pollard (2003). Illustration from Miller et al. (2005b, Fig. 11.1)



been warm enough to melt all traces of ice and snow there. The problem would be further exacerbated if 'normal' topographic relief to 1 to 2 km elevation existed on polar continents, producing much colder temperatures at high elevations. These factors suggest that at least seasonal ice might be expected to form in high latitudes throughout most of the Phanerozoic.

Frakes and Francis (1988) reported that "the geological literature reveals that reports of ice-rafted deposits exist for every period of the Phanerozoic Era except the Triassic." However, Marwick and Rowley (1998) reviewed the evidence for glaciation during the Triassic to present day, focusing on the record of supposed ice-rafted deposits, and concluded that the evidence was very ambiguous. They argued that most of the supposed glacially-related deposits could

be explained by other processes, such as the rafting of debris in tree roots. Rowley and Markwick (1992) tackled the problem of glacioeustasy during greenhouse climatic phases in a different way. They calculated the changes in oceanic water volume that would be required to yield changes in sea-level comparable to those indicated by the "third-order" cycles that comprise the main basis for the Exxon global cycle chart. Assuming glacioeustatic mechanisms as the dominant process, they calculated that it would require an Antarctic-sized ice sheet to evolve and dissipate about every 1–2 million years to create the magnitude of sea-level changes contained in this chart, but could find no evidence for this. Changes in oxygen isotope measurements from Jurassic through mid-Cenozoic

**Fig. 11.16** The cooling of the oceans since the mid-Cretaceous, and its relationship to global tectonic events. (**a**) The Late Cenozoic glacio-epoch after 55 Ma. The breakup of Pangea moved large landmasses into higher latitudes, isolated Antarctica and changed the configuration and bathymetry of ocean basins. The first ice appears at c. 40 Ma in both northern and southern hemispheres but Milankovitch-forced Northern hemisphere ice sheets only developed after 3 Ma. Milankovitch-forced continental ice sheets in the northern hemisphere were the culmination of some 50 million years of tectonically-influenced cooling. This may provide a model for the generation of glacio-epochs during earlier episodes of supercontinent breakup. (**b**) Geometry of continental extension during the breakup of Pangea. The most extensive uplifts were created where crust is old, thick and thus flexurally rigid (creating a high effective elastic plate thickness). The resulting isostatic rebound and uplift is distributed across a large area. Additionally, mantle convection below the broken margin, and magmatic underplating contributes to uplift of passive margins and was instrumental in creating uplifted plateau on which late Cenozoic ice sheets grew on the margins of the North Atlantic Ocean after 3 Ma (e.g., Laurentide, European ice sheets). Diagrams and captions (edited) from Eyles (2008, Figs. 11.7 and 11.8)

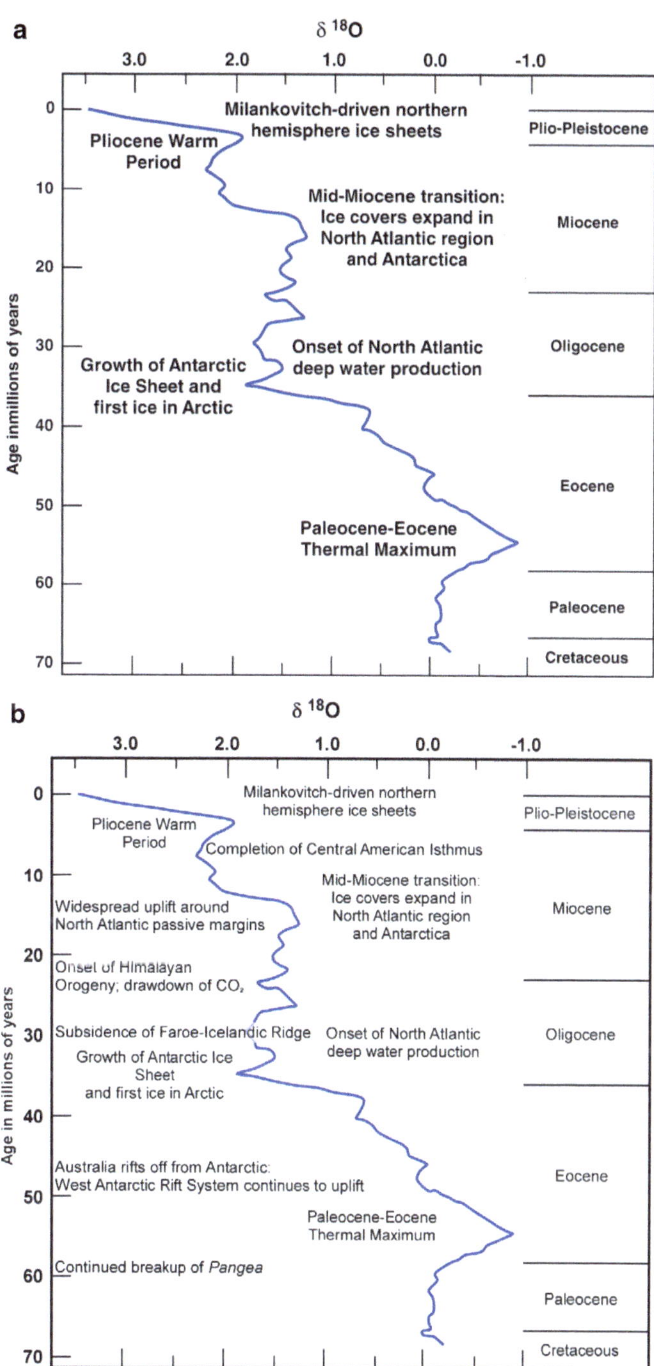



sediments in DSDP cores show quite different patterns to those predicted from a glacioeustatic mechanism. Rowley and Markwick (1992) were unable to identify generative causes for such changes during non-glacial times, such as during the Mesozoic and early Cenozoic. However, this research was predicated on the frequency and magnitude of the "third-order" cycles in the Haq et al. (1987) global cycle chart, which (it is argued elsewhere in this book) should no longer be regarded as a reliable indicator of the timing or magnitude of sea-level history. Miller et al. (2004, p. 389) concluded that the estimates of the magnitudes of sea-level fluctuations in the updated global cycle chart of Graciansky et al. (1998) were mostly much too high, and that the chart should be abandoned as a standard for sea-level changes.

Although there is little support, as yet, for the concept of continuous, ongoing glacioeustasy during the Mesozoic and early Cenozoic, there is an increasing body of evidence for periodic "cold snaps", to use Miller's useful phrase (Miller et al., 2005b, p. 226). Although the Cretaceous was generally a period dominated by a global greenhouse climate, there is an increasing body of oxygen isotope data for short-lived spikes that are best explained as the product of short-term episodes of cooling and continental-ice formation. Stoll and Schrag (1996) examined the oxygen and strontium isotope record in the early Cretaceous section in several DSDP cores and argued that excursions in these data indicated high-frequency sea-level changes that could only have been caused by glacioeustasy. Miller et al. (1999) examined the Maastrichtian record in coastal sediments in New Jersey. Dating of a sequence boundary there suggested a correlation with a $\delta^{18}O$ excursion between about 69.1 and 71.3 Ma (Fig. 14.35), suggesting a sea-level fall of 20–40 m, which would be equivalent to the formation and melting of 25–40% of the volume of the present Antarctic ice cap. Another study of the $\delta^{18}O$ record, by Bornemann et al. (2008) suggests a significant cooling episode during the Turonian. They evaluated and compared the oxygen isotope signature in planktic and benthic foraminifera and suggested a cool phase lasting about 200 ka at about 92 Ma, preceded and followed by several million years of "greenhouse" warmth. As discussed in Sect. 14.6.4, based on this kind of evidence, a good case can be made for glacioeustasy during at least some parts of the Cretaceous period, driven by the development and melting of small, temporary ice caps, probably located in Antarctica.

An increasing number of researchers are reporting cyclicity or rhythmicity in the Mesozoic record, and concluding that the evidence supports the existence of an orbital forcing mechanism, possibly including glacioeustasy (Plint, 1991; Elder et al., 1993; Sageman et al., 1997, 1998; Plint and Kreitner, 2007; Varban and Plint, 2008). For example, Varban and Plint (2008) pointed to the presence of repeated, widespread regional transgressive bounding surfaces in the Upper Cretaceous stratigraphy of northern

Alberta, and estimated "on geometric grounds" that glacioeustatic sea-level changes with amplitudes of around 10 m seemed to be suggested.

### 11.3.4 Late Paleozoic Cyclothems

It has long been widely accepted that the upper Paleozoic cyclothems are the product of high frequency eustatic sea-level fluctuations induced by repeated glaciations with $10^4$–$10^5$-year periodicities. This is an old idea (Wanless and Shepard, 1936; Wanless, 1950, 1972) that was recently been revived by Crowell (1978), Heckel (1986), and Veevers and Powell (1987). Crowell (1978) and Veevers and Powell (1987) showed that the Late Devonian-Permian glaciation of Gondwana had the same time span as the cyclothem-bearing sequences of the Northern Hemisphere.

A synthesis of the chronology and extent of glacigenic deposits in Gondwana by Caputo and Crowell (1985) provided an explanation for the widespread glaciations of Late Ordovician and Late Devonian to Permian age. They related the distribution of glacial deposits to the polar wandering paths established by paleomagnetism. Glacial episodes occurred when the South Pole lay over a major continental area. Between the Middle Silurian and the Middle Devonian, the supercontinent drifted away from the pole, which lay over the paleo-Pacific Ocean (Panthalassa). There was, therefore, no major continental glaciation at this time. Similarly, the continent drifted away from the pole in the Triassic, and glaciation ended, not to return to the Southern Hemisphere until the major cooling episode of the mid-Tertiary, when the Antarctic ice cap developed. Eyles (2008) emphasized the critical importance of tectonism in generating the large areas of elevated landmass which he argued are a necessary prerequisite for long-term global cooling. He stated (p. 109): "A tectonic control on the onset of glaciation is particularly clear for the Gondwanan glacio-epoch where ice growth accompanies continental collisions and the growth of Pangea. Nonetheless, a glacial record was not preserved in many areas until the onset of extensional basin formation and subsidence." His maps of drifting continents and ice extent demonstrate the diachroneity of glacial activity as continental plates drifted across the southern polar regions (Fig. 11.17).

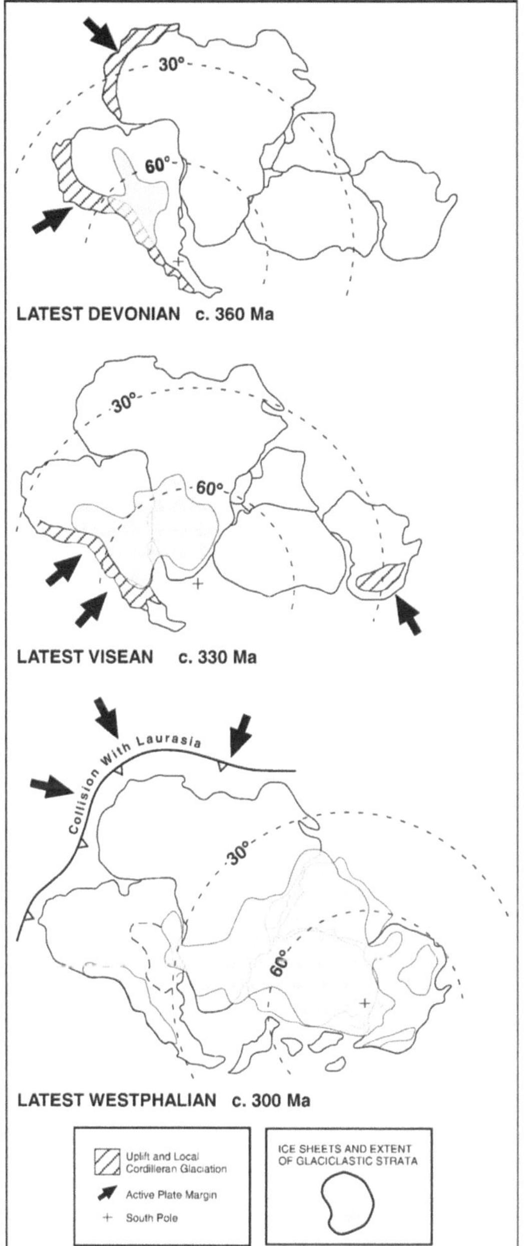



**Fig. 11.17** Ice growth phases during the Carboniferous-Permian Gondwanan glaciation. Glaciation began with a Devonian nucleus located over the high topography along the active South American plate margin. This expanded into southern Africa and then eastward across India, Australia and Antarctica concomitant with the drift of Gondwana across the south polar latitudes and the shift in the pole from North Africa to Antarctica and Australia by the Permian. The thickest rock record of Gondwanan ice covers occurs in marine intracratonic basins in Brazil, Oman and Southern Africa and in rifted margin basins of western Australia (Eyles, 2008, Fig. 11.6)

Studies of the changing sensitivity of the earth to orbital forcing in the Cenozoic (Sect. 11.3.1) suggest useful analogies for the interpretation of the Gondwana glacial record. The high-frequency cyclothemic periodicity determined by Heckel (1986) and others appear to fit closely the Cenozoic results discussed above. Polar wandering resulted in a progressive shifting of glacial centers across Gondwana (Fig. 11.17), which would have led to continuous variations in the sensitivity of the ice masses to climate forcing. The continuous variations in cycle length shown in the compilation by Veevers and Powell (1987) are, therefore, readily explained.

Crowley and Baum (1992) estimated the magnitude of sea-level changes that might have occurred during the Gondwana glaciation. They developed three scenarios for ice cover, ranging from a minimum to a maximum, based on basin analysis of the glacial deposits and tectonic considerations. Calculations of ice volumes from relationships developed for the Cenozoic glaciation, and conversion of these estimates into volumes of water, indicated possible total glacioeustatic fluctuations of between 45–75 (minimum ice cover) and 150–190 m, (maximum ice cover) allowing for isostatic adjustments. These are consistent with estimates derived from stratigraphic studies of the cyclothems, as shown by Klein (1992).

Klein (1992) also demonstrated that long-term climate change and tectonism also contributed to the total base-level changes that were responsible for developing the late Paleozoic cyclothems of the North American Midcontinent. It is a considerable challenge to sort out these various complexities. Cecil (1990) developed a model (discussed in Sect. 11.2.5) to explain variations in the importance of coal through the late Paleozoic cyclothems of North America. Both long- and short-term variations in climate can be detected. The long-term changes occur over periods of tens of millions of years, and are therefore related to the drift of the continent through climatic belts. Latest Devonian-earliest Mississipian cyclothems tend to contain thick clastics, suggesting a wet climate. An arid interval occurred in the late Osagean and early Meramecian, characterized by restricted clastic input and deposition of evaporites. North America then drifted into the tropical rain belt in the Early Pennsylvanian, resulting in an important phase of coal development.

**Fig. 11.18** Sea-level curve for part of the Middle-Upper Pennsylvanian sequence of the North American midcontinent, which ranges from 260 m thick in Iowa to 550 m thick in Kansas. The lateral extent of erosion surfaces at the base of the cycles is shown by diagonal hatching. Fluvial deltaic complexes are shown by a *series of dots*, and conodont-bearing shales by *lines with dots* (Heckel, 1986)

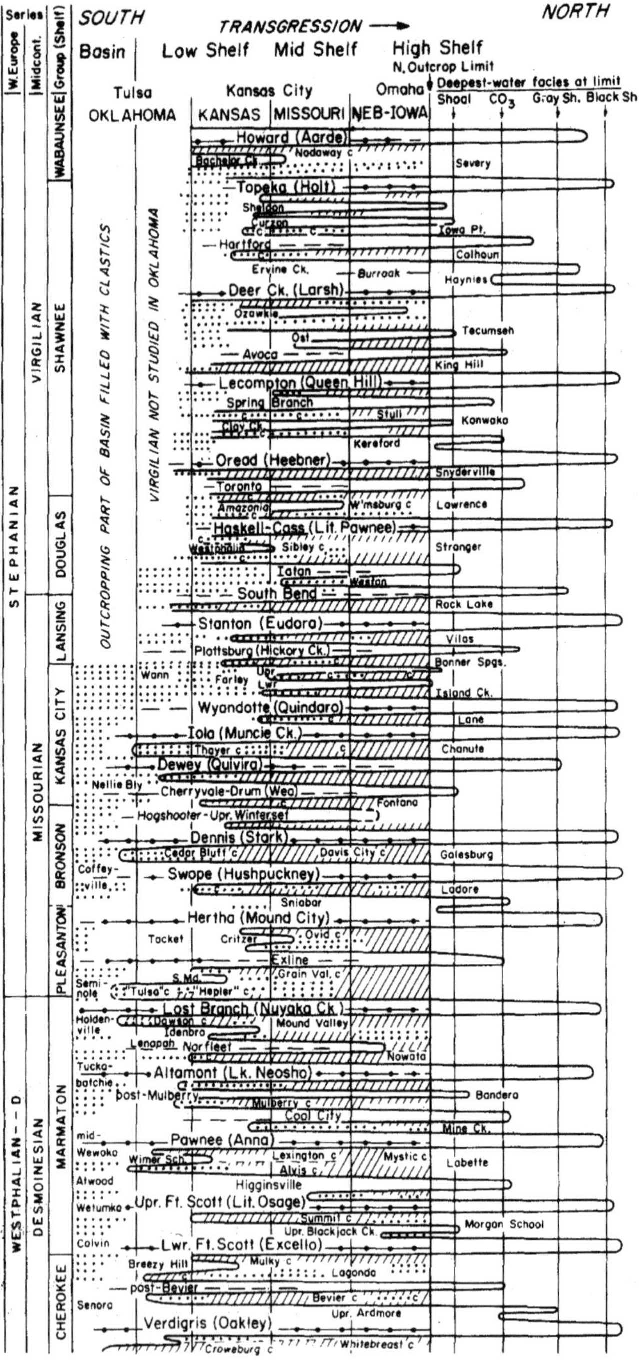


The paleogeographic interpretation of the cyclothems developed by Moore (1964; see Figs. 7.38) is practically identical to the classical sequence model of Posamentier and Vail (1988). The lower part of each cyclothem was formed during a transgression, which drowned clastic coastal-plain complexes and developed a sediment-starved shelf on which carbonates were deposited. This phase represents the melting of continental ice caps. With renewed cooling and ice formation, regression began, with the initiation of rapid deltaic progradation. The erosion surface at the base of the nonmarine sandstone may represent local deltaic channeling or widespread subaerial erosion. In Nova Scotia, calcareous paleosols occur at cyclothem sequence boundaries, formed at times of low stands of sea level, indicating seasonally dry climates during glacial episodes in the southern hemisphere. This contrasts with the humid (interglacial) climates indicated by coal-bearing deposits of the highstand systems tracts (Tandon and Gibling, 1994). Another study of paleosols that revealed fluctuations in climate related to the southern hemisphere glacial-interglacial cycles was reported by Miller et al. (1996). Variations in thickness and composition of the cyclic sequences were caused by local tectonic adjustments (epeirogenic warping, movement on fault blocks, etc.) and by proximity to clastic sources (Figs. 7.39 and 7.40). A similar model of transgression and regression was developed for the Kansas cyclothems by Heckel (1986). He distinguished major and minor cycles and estimated that the major cycles spanned 235–400 ka, while the minor cycles had durations of 40–120 ka. Detailed study of outcrops and cores enabled Heckel (1986, 1990) to erect a curve showing regional sea-level changes in the US Midcontinent (Fig. 11.18). However, the duration of these cycles may be questioned because of uncertainty regarding the periodicity of orbital parameters in the distant geological past, as discussed in Sect. 11.2.3.

Heckel's (1986) sea level curve was correlated with stratigraphic successions in Texas by Boardman and Heckel (1989), and tentative correlations have been extended into southern Arizona, northern Mexico, and the Paradox Basin, Utah, by Dickinson et al. (1994). As pointed out by Dickinson et al. (1994), that the cycles retain similar character and thickness throughout the entire North American interior, across cratonic areas and basins characterized by very different tectonic styles and subsidence histories, is convincing evidence against a tectonic origin for the cycles, and therefore supports the glacioeustatic model. We return to this point at the end of the chapter.

Van Veen and Simonsen (1991) attempted to extract a longer-term (million-year periodicity) signature from Heckel's curve, and used this to suggest correlations into Russia based on simple sequence matching. Boardman and Heckel (1991), in their comment on this proposal, noted the incompleteness of the data base, and expressed scepticism of the ad hoc way the correlations had been done, in the absence of biostratigraphic control. It left out many other factors that could have affected cycle development on longer term time scales, especially tectonism.

Major regressive events during the Devonian to Permian may correspond to times of exceptionally widespread Gondwanan glaciation. Veevers and Powell (1987) suggested that, at these times, regional episodes of tectonism may have uplifted broad areas of Gondwana, so that the cooling effects of elevated altitude were added to the climatic effects of the high latitudinal position of the supercontinent. The closure of the paleo-Tethyan Ocean as Gondwana collided with Laurentia, and the uplift of western South America and eastern Australia would have been particularly significant triggering events (Eyles, 1993, p. 129).

## 11.4 Distinguishing Between Orbital Forcing and Tectonic Driving Mechanisms

In this section we use the term *cyclothem* to refer to sequences generated by orbital forcing and the term *cyclothemic* to refer to cycles that may be similar to cyclothems in facies, thickness, and time span, but are not necessarily generated by orbital forcing. There is an overlap in frequency between cyclothems, as thus defined, and those attributed to high-frequency tectonism (Fig. 8.2). Dickinson et al. (1994) noted that there are four types of processes that can potentially generate cyclothemic-type deposits: (1) orbital forcing, (2) autogenic processes, such as delta-lobe switching, (3) movement of faults and folds and (4) flexural loading of individual thrust plates. Criteria for distinguishing between these various processes are clearly essential. A summary of the comparisons and contrasts between cyclothems and cycles of tectonic origin is presented in Fig. 11.19.

As discussed in Sect. 2.2.2, small-scale shoaling-upward facies successions, which have been called *parasequences*, may be of autogenic origin, and may potentially be confused with true cyclothemic deposits on the basis of vertical profile characteristics. However, autogenic deposits are limited in areal extent to the depositional system which they represent. The largest delta systems (e.g., Mississippi, Niger, Nile) are on the order of about 100 km across. Therefore, the lateral extent of a cyclic unit is clearly an important criterion

**Fig. 11.19** Table summarizing the key differences between sequences generated by tectonic mechanisms, and cyclothems generated by orbital forcing

| Tectonic cycles | Milankovitch cycles |
|---|---|
| May constitute an hierarchy of cycles spanning $10^3$-$10^7$-year durations | May constitute an hierarchy of cycles spanning $10^4$-$10^5$-year durations |
| Cycles correlate with tectonic episodes | No correlation with tectonism |
| No correlation with climate change | Cycles may be climatic or glacioeustatic. Cycles may contain rhythmic facies variations related to temperature/redox/productivity cycles |
| May contain internal angular unconformities | Internal angular unconformities not present |
| Clastic facies cycles correlate with cycle boundaries (f-u and c-u trends consistent with tectonic control) | Clastic and chemical cycles in different parts of a basin may correlate with each other |
| May show reciprocal vertical facies trends across subsidence/uplift hingelines | No facies relationships to structural trends |
| Confined to a particular basin or orogen | Cycles could potentially correlate across continent, or globally |

by which to distinguish autogenic facies successions from regional cycles caused by allogenic mechanisms.

The two major allogenic driving forces that develop cyclothemic deposits on a regional scale are orbital forcing and flexural loading. The immediate effects of flexural processes may be to generate accommodation close to the point of load, giving rise to the classic "lozenge-shaped" isopach distribution (Fig. 10.17a). However, because the crust has flexural strength and may transmit stress "in-plane" through the crust, flexural effects may be generated over wide areas of a continent. In fact, as discussed in Sect. 10.4, changes in intraplate stress may be capable of generating tectonic events and rapid changes in accommodation on a continental, hemispheric, even, potentially, on a global scale. However, high-frequency sequences that are deposited as a result of the changes in accommodation and sediment supply caused by tectonism will have characteristics that clearly distinguish them from true cyclothems. Cycle thickness and facies will show clear relationships to structural features within a basin. There may be changes in clastic grain-size that can be related to structural features. For example, coarse conglomerates may be cut by and rest on the thrust faults along which movement has generated the accommodation for a tectonic cyclothem. Examples of this are described in Sect. 6.2. Cycles generated by tectonism will typically vary in thickness and facies along trends that parallel tectonic grain, and may contain internal architectural features, such as onlap patterns and angular truncations that indicate the syndepositional nature of tectonism (e.g., Figs. 4.12, 10.24, 10.28,

10.36). The reciprocal-stratigraphy process described by Catuneanu et al. (1997b, 1999, 2000) is a particularly clear example of the way in which tectonism may leave an unmistakable imprint on stratigraphic architecture (Sect. 10.3.3.1).

Conversely, cyclothems and other sequences generated by orbital forcing may be expected to extend across tectonic elements, such as forebulges, without major change in thickness, but may change in facies. If eustatic sea-level change, or continental-scale climate change, are the major driving forces in sequence generation, then although sequences may show major internal facies changes, they may, nevertheless, extend across tectonic boundaries and be correlatable across and between sedimentary basins that are the product of a range of tectonic environments. This is the basis for the erection of the Greenhorn cycles of the Western Interior Basin (Figs. 7.59, 11.11) and the general model of non-glacial Milankovitch cycles shown in Fig. 11.12. A cyclic change in climate is the only mechanism that could generate sequences that change facies assemblage entirely from location to location. Likewise, cycles that consist entirely of changes in chemical sedimentation, such as marl-limestone rhythms, can only be the product of climatic forcing of changes in water chemistry or organic productivity.

Dickinson et al. (1994) used these ideas in the construction of some simple but elegant diagrams that clarified the differences between the effects of the two major allogenic driving mechanisms. Fig. 11.20 compares the successions in two contemporaneous basins,

**Fig. 11.20** Comparison of cycles deposited in two contemporaneous basins in the (Late Paleozoic) Ancestral Rockies region of the SW United States. On the *left*, cycles of alternating marine carbonate and terrestrial eolianite in the Paradox Basin. On the *right*, shoaling upward carbonate cycles of the Pedregosa Basin. The column style indicates upward decreases in water depth and/or increased subaerial relief, from *left to right*, in each column (Dickinson et al., 1994, Fig. 11.6)

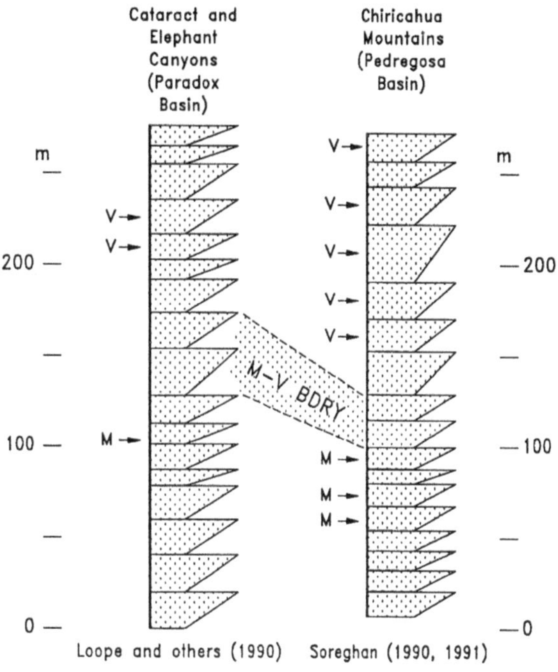


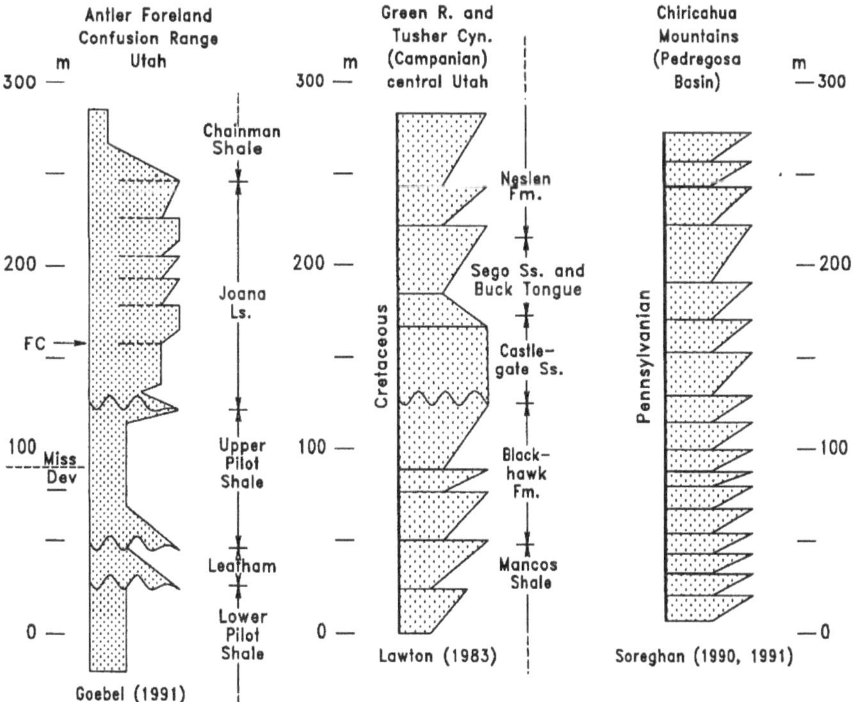


**Fig. 11.21** Comparison of the Pedregosa basin succession of cyclothems (*right*; from Fig. 11.20) with two different foreland-basin successions. On the *left*, the Devonian-Mississippian succession of the Antler foreland basin in Nevada; *at centre*, the Upper Cretaceous Sevier foreland basin succession of Utah (Dickinson et al., 1994, Figs. 11.8 and 11.9)

the Paradox basin in Utah, and the Pedregosa Basin in the Ouachita-Marathon foreland of southeast Arizona. Each column shows 17 cycles. The scale of the cycles is comparable, and it is concluded that they likely correlate, even though the absence of diagnostic fusulinids precludes a definitive correlation. As they stated (p. 30): "the very existence of persistent and widespread stratigraphic cycles may well afford the means to achieve stratigraphic correlation throughout the continent at a scale difficult to attempt with confidence using stratigraphic criteria alone."

Dickinson et al. (1994) went on to argue that the style of cycle generated by high-frequency tectonism as a result of flexural loading would be quite different. He compared two successions from foreland basin settings with the Pedregosa basin cyclothems (Fig. 11.21). That in the Antler foreland involve "alternations of oolitic to peloidal grainstones with intertidal to supratidal laminites. They reflect modest variations in water depth, from wave-washed shoals to peritidal lagoonal environments, with the carbonate platform that developed on the flexural forebulge" of the Antler foreland basin (Dickinson et al., 1994, p. 31). The succession is not consistent in facies or thickness across the basin. Contacts between the cycles become gradational as the cycles thicken basinward. The other foreland-basin succession is from the Cretaceous Sevier foreland of the Green River-Tusher Canyon area of central Utah. The deposits consist of interbedded shelf, prodelta, shoreface and delta plain deposits. Again, the cycle frequency and the facies are variable. At least in vertical section, these two foreland basin do not appear to display the regularity that would be expected from deposition under the control of an astronomically precise and regular climatic beat.

## 11.5 Main Conclusions

1. The evidence for Milankovitch control of climate and sedimentation in the late Cenozoic is now overwhelming. Stratigraphic cyclicities in many upper Cenozoic sections have been correlated with the oxygen-isotope record that tracks ocean-water temperature changes, and it is now generally agreed that the $\delta^{18}O$ record can be used as an analog recorder of eustatic sea-level, bearing in mind the lag between the eustatic high and the attainment of the highest temperatures. This body of knowledge permits very precise chronostratigraphic studies of Neogene sequence stratigraphy. The methods of *cyclostratigraphy*, as it is now termed, include sophisticated time-series analysis of cyclic parameters, such as cycle thickness and carbon content.

2. Earth's sensitivity to climate change depends on a wide range of parameters and feedback mechanisms, which make reconstruction of past climate change and of Milankovitch controls very difficult. Of particular importance, as demonstrated by the initiation of Gondwana glaciation and early Cenozoic glaciation of Antarctica, is the plate-tectonic control of large continental masses relative to the poles, and the resulting oceanic and atmospheric circulation patterns. Major glacial episodes may be initiated by regional uplift caused by collisional uplift of plateaus and mountain ranges, and by thermal uplift during supercontinent fragmentation.

3. Controversy remains regarding the periodicity of Milankovitch parameters in the geological past. Major changes may have occurred during the Phanerozoic because of changes in the orbital behaviour of the earth. In addition, sedimentological studies indicate that various autogenic mechanisms can generate apparent periodicities that simulate Milankovitch effects, some tectonic processes can also generate sequences with similar periodicities, and there is imprecision in biostratigraphic dating of most sequences. For these reasons, it is unwise to attempt to correlate suites of cycles in ancient "hanging" sections to specific periodicities or to invoke Milankovitch mechanisms purely on the basis of interpreted $10^4$–$10^5$-year cycle frequencies.

4. Glacioeustasy is markedly affected by isostatic and geoidal changes within hundreds to a few thousands of kilometres of the edge of a major continental ice cap. The timing, direction and magnitude of sea-level changes within this region may be entirely different from the signatures elsewhere.

5. The postulation of Milankovitch controls for any given cyclic sequence should only be attempted following rigorous spectral analysis of one or more cyclic parameters within the succession, because other autogenic and allogenic mechanisms can generate cycles with similar thickness and repetitiveness. Demonstration of the presence of a hierarchy

of cycle types with characteristic periodic ratios, such as the 1:5 precession:eccentricity combination, is additional evidence of Milankovitch controls, although the cautions noted in paragraph 3, above, need to be borne in mind.

6. There is evidence for at least localized glaciation for most periods during the Phanerozoic, but the evidence for major continental glaciation and consequent significant glacioeustasy outside the periods occupied by the late Paleozoic Gondwana glaciation and the Neogene glaciation is at present weak. There is increasing evidence for small, short-lived continental ice caps at several different intervals during the Cretaceous.

7. Glacioeustasy is not the only cycle-forming mechanism driven by orbital forcing. Variations in climate and oceanic circulation have significant effects on such parameters as sediment yield and organic productivity, which have generated many successions of metre-scale cycles throughout the geologic past. These are especially prominent in the deposits of many carbonate shelves and in fine-grained clastic deposits formed in lakes and deep seas.

8. Given the overlapping range of frequencies exhibited by the two major classes of high-frequency cycles, those generated by orbital forcing and those formed in response to tectonic processes, a range of tests needs to be performed to distinguish between them. Cyclothems generated by climate change extend regionally, potentially globally, but may change in facies from environment to environment. Cycles generated by tectonism are likely to be local to regional in extent and will contain internal evidence of tectonic control, including onlap and angular truncation relationships, coarse clastic facies cut by unconformities, and onlapping faults, etc.

**Part IV**
# Chronostratigraphy and Correlation: An Assessment of the Current Status of "Global Eustasy"

The purpose of this part of the book is to examine the nature of the chronostratigraphic record of cycle correlation, to examine the potential for error, and to review the various sea-level curves that have been prepared by Exxon and by other workers, in order to demonstrate the present level of uncertainty.

Are sequences in a given basin regional or global in distribution? Or are both types present? If so, how do we distinguish them? The basic premise of the Exxon approach is that there exists a globally correlatable suite of eustatic cycles, and that all stratigraphic data may be interpreted in keeping with this concept. However, this basic premise remains unproven, and it sidesteps the issue of regional tectonic control, of the type demonstrated in Parts II and III of this book. Unfortunately, the global cycle chart has commonly been presented as its own proof, with many researchers using the chart as a guide to correlation, and then citing the result as successful test of the chart. This practice is still in evidence, despite more than a decade of skeptical work on the issue, and is discussed in Chap. 12.

One of the critical tests of global cycle charts is to demonstrate that successions of cycles of the same age do indeed exist in many tectonically independent basins around the world. The chronostratigraphic accuracy and precision of the chart and of the field sections on which it is based are, therefore, of critical importance. This requires independent studies of sequence stratigraphies in tectonically unrelated basins. We examine this problem in Chap. 14.

The concept of the global cycle chart, as first formulated by Vail et al. (1977), has been accepted by many as a standard of geologic time. Various versions of the Exxon chart have appeared as part of many key synthesis charts, despite the absence of a rigorous, published examination of the supporting data. Such a casual approach to an apparently important new tool for correlation contrasts dramatically with the enormous effort that has been ongoing since the work of William Smith in the late eighteenth century to refine and calibrate biostratigraphic, radiometric, and magnetostratigraphic times scales. Revisions of these scales appear frequently (the latest in 2008), whereas the updated version of the Exxon global cycle chart published in 1987 continues in many quarters to be accepted as a finished product and as a standard for cross-comparison. Miall and Miall (2002) developed a set of explanations for this methodological approach and called it the "Exxon Factor". A newer (1998) "global" scale , based on European research, is now being used in the same way as the original chart by many workers, as discussed in Chap. 12.

- A sea-level curve may be compared with other regional data by plotting it on a chronostratigraphic correlation chart for the region … Such a combination shows the relations of sea-level changes to geologic age, distribution of depositional sequences, unconformities, facies and environment, and other information. (Vail et al., 1977, p. 77)
- It is … important to obtain precise ages of the sequence boundaries and rate and magnitude of apparent sea-level change to establish their global or regional nature. In this context, critical stratigraphic examination of seismic sections across the inner edge of passive margins to the seaward part is of particular importance. (Cloetingh, 1988, p. 219)
- Most refinement in correlation derives from a fuller understanding of the properties of the characters employed. There is always room for improvement, however there are several reasons why an apparent age will differ more or less from the true age. Most obvious is non-availability of appropriate characters and gaps in the record which lead to indeterminate results. In other cases human skill or error affect the uncertainty as in observational and experimental errors. These can in some degree be measured as standard errors, or as systematic errors due to mistaken assumptions (e.g. decay constant) or due to inadequate understanding of the characters used, taxonomic lumping, splitting, or mistakes. (Harland, 1978, p. 18)
- Biostratigraphy is not without its problem areas … Variations in the inferred relationship between absolute time and biostratigraphic zones comprise a major area of controversy. (Kauffman and Hazel, 1977b, p. iv)
- Without a precise time control the depositional mechanisms forming beds and sequences cannot be sufficiently understood. … timing has remained an elusive problem. Too many inaccuracies are involved in resolving stratigraphic durations, including a large range of error in radiometric age determinations, poor biostratigraphic as well as magnetostratigraphic resolution, and an incompleteness of sedimentary sections. As a result, time estimates are commonly imprecise, and the range of error is often larger than the actual time span considered … (Ricken, 1991, p. 773).
- … we believe that the Haq et al. (1987) curve is a "noisy" accumulation of a wide range of sea-level signals, and should not be used as a global benchmark. It should never be used as a chronostratigraphic tool by assuming a priori that a certain stratigraphic boundary has a globally synchronous and precise age, which it is therefore safe to extrapolate into a basin with poor age control (Allen and Allen, 2005, pp. 279–280).
- Global sequence stratigraphic/cycle charts … were always a fudge, and they aren't going to help much in chronostratigraphy at the resolution that is now needed (Christie-Blick et al., 2007, p. 223).

# Chapter 12

# The Concept of the Global Cycle Chart

## Contents



## 12.1  From Vail to Haq

The concept of eustasy dates back to the late nineteenth century. The term "eustatic" was coined by Suess (1885–1909) and, as discussed in Sect. 1.4, there was considerable discussion and theorizing about global processes, such as eustatic sea-level change, through the twentieth century. The topic received considerable importance with the development of seismic stratigraphy. One of the major theses of the Vail et al. (1977) work was that cycles of sea-level change can be correlated around the world, signifying that sequences are not a response to local tectonic events but the result of global or eustatic sea-level changes (Fig. 12.1). As noted in Chap. 1, in the early days of seismic stratigraphy, in the late 1970s and early 1980s, the global cycle chart was considered an inseparable part of the new methodology.

Vail et al. (1977, p. 96) had set out this paradigm clearly in their first publication, and this has remained their key concept:

One of the greatest potential applications of the global cycle chart is its use as an instrument of geochronology. Global cycles are geochronologic units defined by a single criterion-the global change in the relative position of sea level through time. Determination of these cycles is dependent on a synthesis of data from many branches of geology. As seen on the Phanerozoic chart …, the boundaries of the global cycles in several cases do not match the standard epoch and period boundaries, but several of the standard boundaries have been placed arbitrarily and remain controversial. Using global cycles with their natural and significant boundaries, an international system of geochronology can be developed on a rational basis. If geologists combine their efforts to prepare more accurate charts of regional cycles, and use them to improve the global chart, it can become a more accurate and meaningful standard for Phanerozoic time. (Vail et al., 1977, p. 96).

It was stated that much evidence had been amassed to demonstrate this idea, but the only evidence actually offered was a correlation between four Cretaceous to Recent basins in a chart reproduced here as Fig. 12.2. An important critique of this figure is that it is rare in geology for the evidence to permit such precise correlation of events that they can be indicated by the straight lines encompassing the globe (Miall, 1986). There is still much disagreement over the exact age of magnetic anomalies, and the absolute (numerical) age of biostratigraphic zones. Even in the case of Cenozoic stratigraphy, the determination of chronostratigraphic age is characterized by significant observational and experimental error, and this increases with the age of the sediments (these problems are discussed in Chap. 14). Yet there is no indication of these possible errors in Fig. 12.2. If such data are not provided, it is difficult to allow for revisions and refinements.

Global correlations such as those in Fig. 12.2 permitted Vail et al. (1977) to construct charts showing relative sea-level changes that, because of their


A.D. Miall, *The Geology of Stratigraphic Sequences,* 2nd ed.,
DOI 10.1007/978-3-642-05027-5_12, © Springer-Verlag Berlin Heidelberg 2010

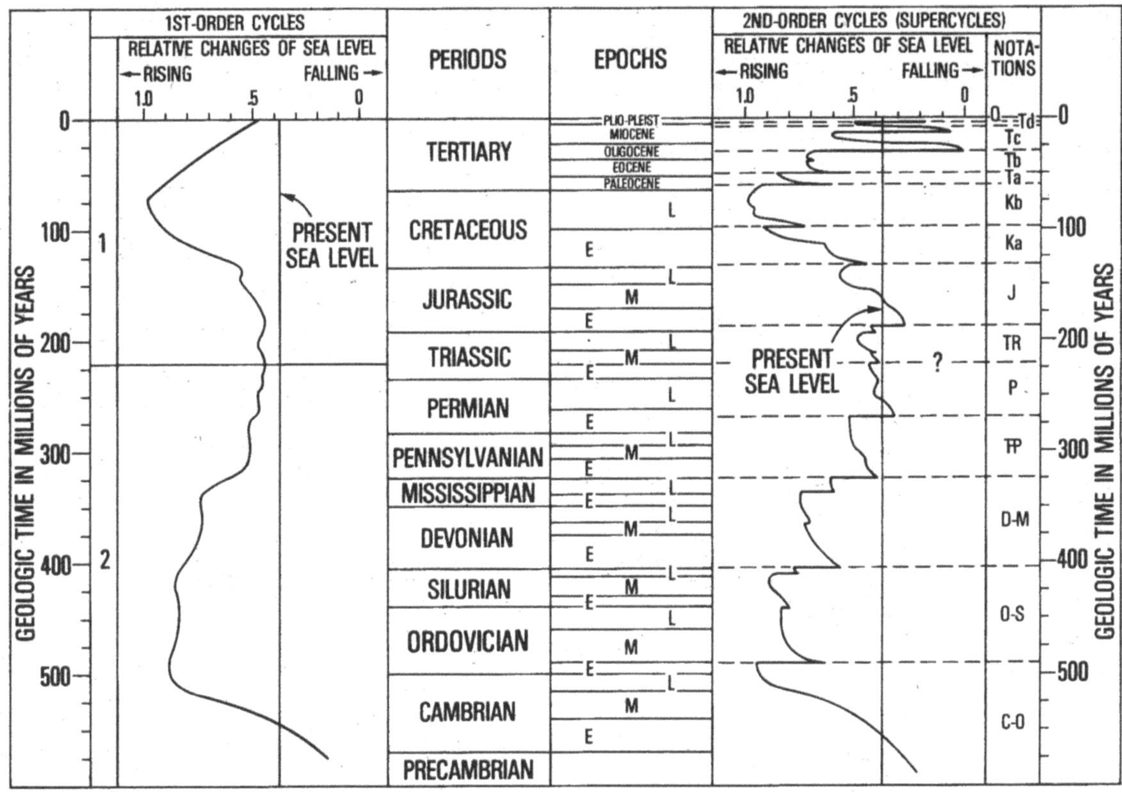


**Fig. 12.1**  The first published version of Vail's chart of long-term eustatic changes in sea level (Vail et al., 1977)

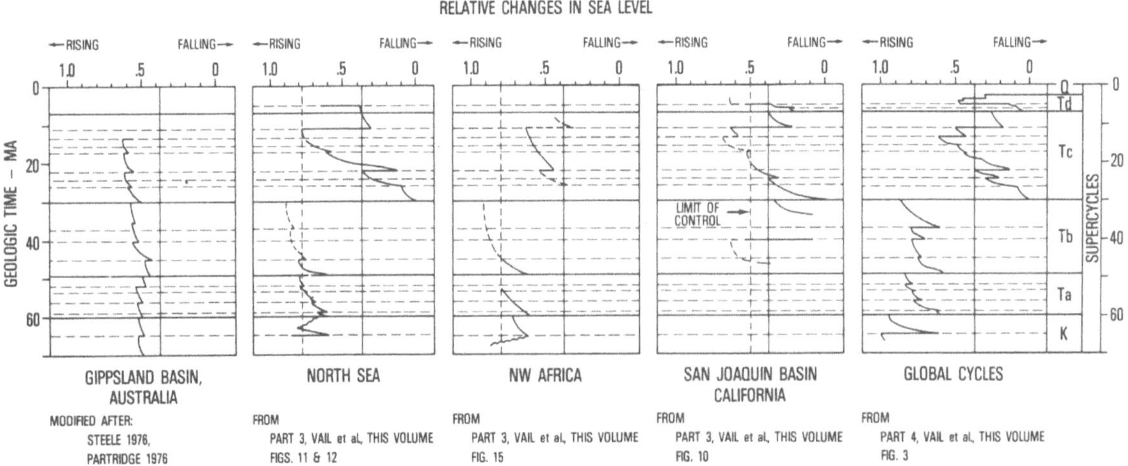


**Fig. 12.2**  Correlation of cycles of relative change of coastal onlap from four continents, showing how these have been averaged to produce the global cycle chart (Vail et al., 1977)

global correlatability, were interpreted as true, eustatic changes. A detailed chart showing sea-level change relative to magnetic reversal and biostratigraphic events was constructed for the Cenozoic, a less detailed chart for the Late Triassic to Recent, and a generalized chart for the entire Phanerozoic. In these charts, the rises and falls for individual regions were averaged to produce a global curve, and this was given in terms of a relative change, because it differs in amount from region to region depending on local tectonic events. The method of global averaging is obscure. These diagrams became among the most widely reproduced illustrations in the history of geology, because they purported to provide a key as important as that of plate tectonics for understanding worldwide stratigraphic patterns.

The pre-Jurassic cycles were constructed mainly from North American data (Vail et al., 1977, p. 88). For rocks of this age, correlations with oceanic events are not available (the oldest oceanic crust is probably Middle Jurassic), and so Vail and his co-workers presumably used much the same data as did Sloss, namely, subsurface information from the continental interior. It is, therefore, hardly surprising that the first four of Sloss's (1963) sequences (Fig. 1.12) are almost identical to the corresponding supercycles of Vail et al. (1977). Younger supercycles differ considerably from those of Sloss because of the availability of a wholly new and more detailed data base, offshore marine seismic and well records.

However, there is a fundamental contradiction inherent in this paradigm in that the global cycle chart was itself initially based on the existing global time scale, with all its imperfections (sequences were dated using fossils, radiometric dates, etc.), yet when the data were found to be inadequate, or conflicts arose, biostratigraphic and other independent data bearing on sequence age were subordinated to the pre-existing sequence framework. This approach was used throughout the early Exxon work. For example, with reference to the Jurassic of the North Sea, Vail and Todd (1981, p. 217) stated: "several unconformities cannot be dated precisely; in these cases their ages are based on our global cycle chart, with age assignment made on the basis of a best fit with the data." The assumption was made that important sea-level events in any given stratigraphic section represent eustatic events. From the paradigm of global eustasy it then follows that a

comparable pattern of sea-level events in other sections is, by definition, correlated with the original event, even when the chronostratigraphic data may not support such correlations. How can such correlations be subjected to independent testing if the data upon on which to base such a test (e.g. biostratigraphic zonation) are declared in advance to be untrustworthy? The dangers of circular reasoning and self-fulfilling hypotheses should have been obvious to scientists at large, but this was not the case.

One of the most serious problems that emerged as a result of the introduction of the Exxon method was the tendency to regard the Vail curves as some kind of approved global chronostratigraphic standard (Miall, 1986). Vail and his co-workers encouraged this trend and even favored seismic correlation over biostratigraphic correlation when the latter did not appear to fit their models. The global cycle chart became the primary reference frame against which new data are judged. The danger of circular reasoning in this method are self-evident, and the approach clearly does not lend itself to independent tests of the cycle chart, nor to a search for local or regional anomalies that might have major implications for local tectonic or climatic episodes (Miall, 1986). Examples of this approach abound in the literature from the 1980s. For example, Vail and Todd (1981, p. 230) stated "the late Pliensbachian hiatus described by Linsley and others (1979) fits the basal early Pliensbachian sequence boundary on our global cycle chart." In other words, the age assignment of the earlier workers is subordinated to the sequence framework. The Pliensbachian stage is now estimated to span approximately 7 million years, which provides an indication of the magnitude of the revision Vail and Todd (1981) were willing to make based on their sequence analysis. Vail et al. (1984, p. 143) stated:

> Interpretations [of stratigraphic sequences] based on lithofacies and biostratigraphy could be misleading unless they are placed within a context of detailed stratal chronostratigraphic correlations.

The context of the word "chronostratigraphic" in this reference implies correlation by tracing seismic reflections.

Baum and Vail (1988, p. 322) stated that

> sequence stratigraphy offers a unifying concept to divide the rock record into chronostratigraphic units, avoids the weaknesses and incorporates the strengths of other

methodologies, and provides a global framework for geochemical, geochronological, paleontological, and facies analyses.

Baum and Vail (1988) commented on the inconsistent placement of stage boundaries in the Cenozoic section of the Gulf Coast. Some are at sequence boundaries, others are at major surfaces within sequences, such as transgressive surfaces. They recommended the use of a sequence framework for redefining the stages, and defining the stage boundaries at the correlative conformities of the sequences. This approach indicates a misunderstanding of the modern concept of a chronostratigraphic time scale, in which boundaries are defined within continuous sections independent of tectonic, eustatic or other "events" that may or may not be global in scope.

An example of a standard Exxon-type sequence analysis was provided by Mitchum and Uliana (1988). Their correlation of a carbonate basin-margin section in a backarc setting with the global cycle chart was done on the basis of a general positioning of the stratigraphy within the Tithonian-Valanginian interval, by comparison (not detailed correlation) of the subsurface with nearby outcrops, where ammonite zonation had been carried out. No faunal data were available from the wells used to correlate the seismic section! However, the pattern of seismic sequence boundaries was said to match the global pattern for this interval.

In one of his later overview papers Vail et al. (1991) stated that sequences "can be used as chronostratigraphic units if the bounding unconformities are traced to the minimal hiatus at their conformable position and age dated with biostratigraphy." (p. 622) and "Sequence cycles provide the means to subdivide sedimentary strata into genetic chronostratigraphic intervals ... Sequences, systems tract, and parasequence surfaces provide a framework for correlation and mapping" (p. 659).

As is made clear by these quotes, the Exxon approach subordinated biostratigraphic and other data to the sequence framework, where conflicts arose. It failed to recognize the fundamentally independent nature of these data. The Exxon method is essentially that summarized in Fig. 12.3b. The assumption is made that an important sea-level event in any given stratigraphic section represents a eustatic event. From the paradigm of global eustasy it then follows that a comparable pattern of sea-level events in other sections

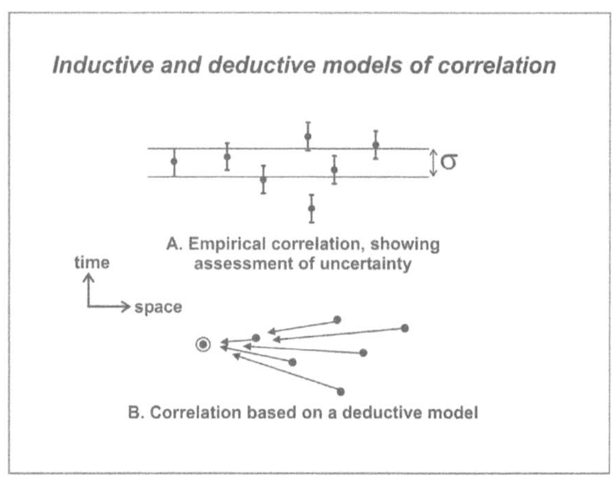


**Fig. 12.3** Two approaches to the correlation of the same set of stratigraphic events (*dots*). (**a**) age, with error bar, is determined independently for each location, with the degree of correlation determined by statistical means. (**b**) A sea-level event in one section is assumed to represent global eustasy, and it is assumed that all other nearby events correlate with it

is by definition correlated with the original event, even when the chronostratigraphic data may not support such correlations. Figure 12.3a illustrates schematically the quantitative approach to correlation, which is to attach error bars to assigned ages, based on numerical estimates of age, calculations from sedimentation rates, etc. A correlation band may then be erected based on standard error expressions, such as standard deviations. The accuracy and precision of correlations can readily be assessed from such diagrams.

Updated versions of the Vail curve were published in several books and papers (Haq et al., 1987, 1988a, b; Haq, 1991). These newer curves (e.g., Fig. 12.4) purported to incorporate a large volume of outcrop data, including sections in "western Europe and the trans-Tethys region, the USA Gulf and Atlantic coasts and the Western Interior Seaway, New Zealand and Australia, Pakistan, and Arctic islands of Bjørnoya and Svalbard" (as claimed by Haq et al., 1988a) reflecting work carried out by Exxon workers and their colleagues in academic and government institutions during the 1980s (Haq, 1991). Lists of these sections have been published (Haq et al., 1988a), but not the data derived from them, and so the criticism of the lack of published documentation remains. A few location-specific studies have now been made available by the Exxon group (Donovan et al., 1988; Baum and Vail,

**Fig. 12.4**  The revised global cycle chart of Haq et al. (1987, 1988a). This part shows the record of sequences established for the Cenozoic. Its first published appearance was accompanied by a detailed chart of biostratigraphic zonations

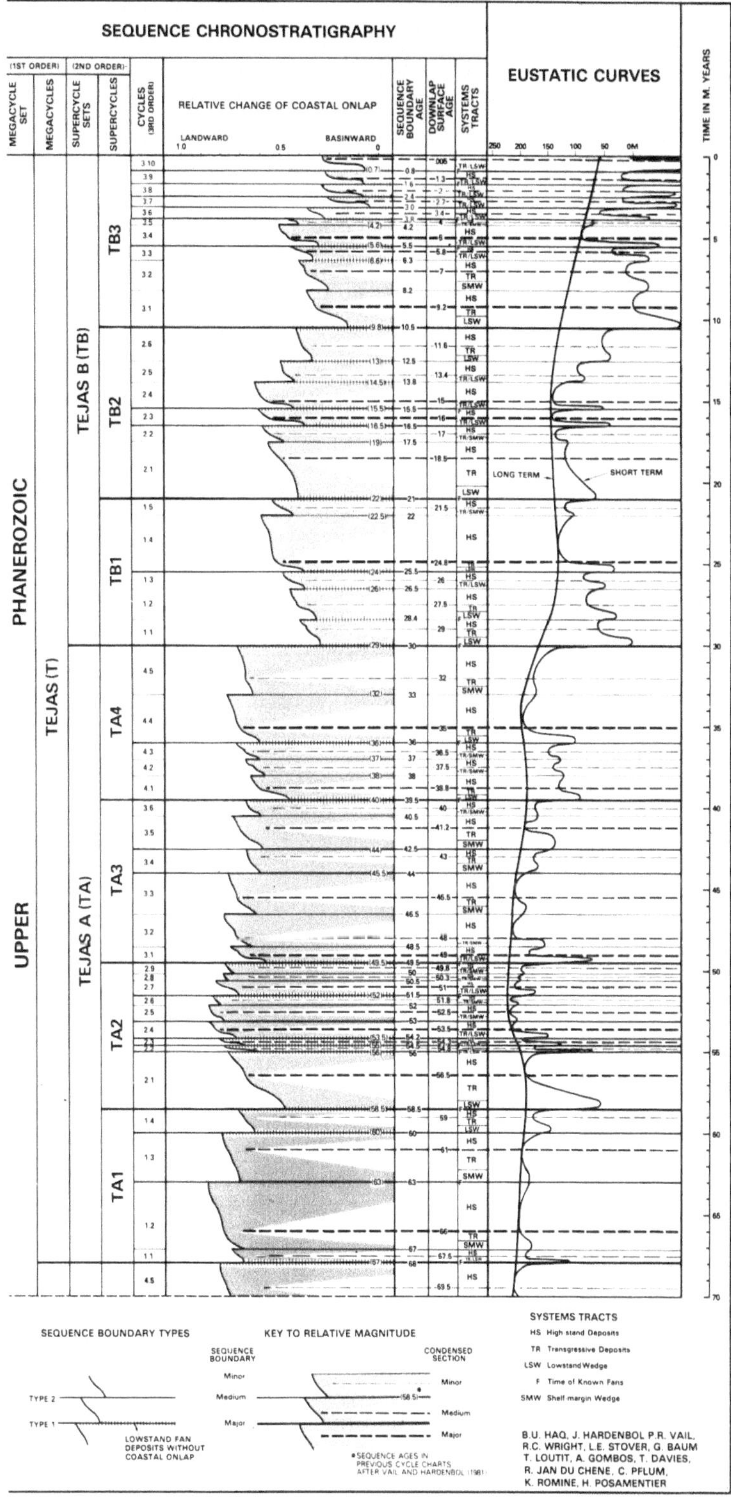

1988), but these barely begin to tackle the problem of inadequate documentation. Haq et al. (1988a) discussed in some detail their methods for developing a numerical time scale for the global cycle chart, pointing out possible sources of error in radiometric, magnetostratigraphic and biostratigraphic methods. But they did not incorporate these sources of error into the chart by providing error bands for the sea-level curve. The implications of this omission are discussed in Chap. 14.

Miller and Kent (1987, p. 53), while arguing for the need for careful chronostratigraphic correlation, pointed out that "the durations of the third-order cycles are at the limit of biostratigraphic resolution." They went on to state:

> We agree that in order to test the validity of the third-order cycles it is not necessary to establish that *every* [their emphasis] third-order cycle is precisely the same age on different margins. Haq and others (1987) utilized a sequence approach to recognize third-order events above known datum levels. Assuming that they observed the same patterns on different margins, their observation of the same ordinal hierarchy of events within a given time window on different margins argues against a local cause and points to eustatic control. ... However, the simple matching of third-order cycles between locations is complicated by gaps in the records, uncertainties in establishing datum planes, and the ability to discriminate between these cycles at the outcrop level.

Miller and Kent (1987), Christie-Blick et al. (1988), and Miall (1991a) have pointed out some of the problems and imprecisions in chronostratigraphic correlation. Ricken's (1991) comment regarding chronostratigraphic imprecision and error is quoted in the introduction to Part IV of this book.

The dangers inherent in simple pattern recognition, of the type alluded to by Miller and Kent (1987) are that, given the density of stratigraphic events present in the Vail curve, there is literally an "event for every occasion" (Miall, 1992). Practically any stratigraphic succession can be made to correlate with the Vail curve, even synthetic sections constructed from tables of random numbers (Miall, 1992; See Fig. 14.7 and discussion thereof). Dickinson (1993), in a discussion of Miall (1992), demonstrated that the average duration of the main ("third-order") cycles in the Exxon chart increases with age, and suggested that this reflects a decrease in the quality of the sequence data in older sections—in other words, the event spacing is at least in part an artefact of the data quality and the analytical

method used to construct the chart. Haq et al. (1988b) referred to a procedure of "rigorous pattern matching" of sequences and systems tracts, but have nowhere described their method or attempted to quantify the degree of "rigor".

It is important to remember that where tectonic subsidence and eustatic sea-level change have comparable rates, a synchronous eustatic signal may not be preserved in the geological record (Chap. 10). Therefore, even where eustasy was an important process, it may be difficult to demonstrate this from the stratigraphic record. Conversely, the absence of a synchronous record does not disprove the occurrence of eustatic sea-level change.

It does not help that few of the supporting data have been published. This is a criticism of the Exxon work that has been made several times (e.g., Miall, 1986), but Sloss (1988a), amongst others, leaped to the defence of his former students (p. 1664):

> A common complaint concerned the lack of supporting data and, indeed, a large measure of faith was required of the reader. Complaining parties failed to recognize that the 1977 curves represent two decades of analysis of data, much of it proprietary, including thousands of kilometres of seismic lines, hundreds of subsurface records, and untold man-hours of biostratigraphic work. Further curves showing a higher level of detail in the Cretaceous were not released for publication, and this omission added to the malaise of an ungrateful segment of the public. Many of the deficiencies of the 1977 product have now been corrected in subsequent publications; progress continues with activity now spread over a broad spectrum of industrial, academic, and governmental agencies.

Sloss (1988a) himself, though supporting the Exxon work and accepting most of its results, disputed the placement of several of the major sequence boundaries in the Exxon global cycle charts. Part of the problem reflects the difficulty in assessing how the relative importance of the sequence boundaries has been determined. The method of averaging the local curves to produce the global chart has not been explained in detail. In some cases, an ad-hoc approach seems to have been employed. For example, Haq (1991) stated "the long-term changes in relative magnitude are estimated using the method described in Hardenbol et al. (1981) with the Turonian high value adopted from Harrison (1986)" [Harrison's paper actually appeared in 1990 following a long delay in publication]. Without a rigorous, quantitative method of curve conflation, the relative magnitudes of the sea-level excursions cannot be assigned much significance, and therefore it follows that the assignment of certain sequence boundaries in the Exxon curves as "supercycle" boundaries carries little weight.

As noted by Miall (1992), the basic premise of the Exxon work is that there exists a globally correlatable suite of eustatic cycles, and that all field stratigraphic data may be interpreted in keeping with this concept. However, the basic premise remains unproven because of the lack of published documentation. All so-called tests of the Vail curve are suspect unless they provide a rigorous treatment of potential error and a discussion of alternative interpretations. In fact this is almost never done. Some examples of tests of the Vail curve are described in detail in Chap. 14.

Despite these critiques, there remains a serious conceptual disagreement about the appropriate approach to stratigraphic data. As recently as 2005, McGowran (2005, pp. 176–179), citing Loutit et al. (1988), referred to "the provision of a new physical-stratigraphic framework within which 'sample-based disciplines', such as biostratigraphy and geochemistry, can be evaluated." By this he meant the "powerful unifying force" of sequence stratigraphy, with its system of correlatable bounding unconformities: Referring to "three roots to the development of sequence stratigraphy," he suggested (p. 176) that the first root was the rise of sequence stratigraphy.

> As second root was to take unconformity-bounded stratigraphic bodies (allostratigraphic units) seriously as something more than a frustratingly imperfect stratigraphic record. Instead of unconformities being apologized for as the main manifestation of an all-too-imperfect geological record, they are information-rich and to be cherished, especially since Sloss et al. (1949) and Sloss (1963) proposed sequences and demonstrated cratonic sequences at the Phanerozoic scale separated by continent-wide breaks.

The third root was the concept of the sedimentary cycle.

McGowran (2005) acknowledged the controversies that have arisen around the "global curve" but said (p. 180) "it is important to affirm that inferred eustatic configurations are not the core of sequence stratigraphy which will survive any amount of eustatic/isostatic controversy." I refer McGowran and his readers to Carter et al. (1991), whose useful conceptual distinction provides the basis for the ensuing discussion of the controversies.

## 12.2 The Two-Paradigm Problem

Carter et al. (1991, pp. 42, 60) pointed out that sequence stratigraphy embraces two quite distinct concepts:

> Two different conceptual models underlie the application of sequence stratigraphy by Vail and his coworkers: one model relates to presumed [global] sea-level behaviour through time; the other model relates to the stratigraphic record produced during a single sea-level cycle. Though the two models are interrelated they are logically distinct, and we believe that it is important to test them separately. … Our studies lead us to have considerable confidence in the correctness and power of the Exxon sequence-stratigraphic model as applied to sea-level controlled, cyclothemic sequences. … At the same time, we suspect that the Exxon "Global" sea-level curve, in general, represents a patchwork through time of many different local relative sea-level curves.

In his essay on the structure of scientific revolutions, Kuhn (1962, p. 65) has noted that anomalies (data that do not "fit") appear only against the background provided by a paradigm. In this regard, he has argued that "if an anomaly is to evoke a crisis, it must usually be more than just an anomaly…. Sometimes an anomaly will clearly call into question explicit and fundamental generalizations of the paradigm" (Kuhn, 1962, p. 82). As Kuhn (1962, p. 76) has also observed, however, "philosophers of science have repeatedly demonstrated that more than one theoretical construction can always be placed upon a given collection of data." Miall and Miall (2001) demonstrated that views about the global-eustasy model have evolved into a broad spectrum of approaches, and it is useful to consider them as two, coexisting, end-member paradigms—the "global-eustasy paradigm" (the original concepts set out in Vail et al., 1977), and what Miall and Miall (2001) labeled the "complexity paradigm." Sequence stratigraphy, itself, including studies of sequence architecture, could be said to correspond now to Gutting's (1984) definition of a supertheory: a "wide-ranging set of fundamental beliefs about the nature of some domain of reality and about the proper methods by which to study that domain."

### 12.2.1 The Global-Eustasy Paradigm

Eustatic sea-level changes are global by definition, but the word "global" is employed here in the sense of "all-encompassing" or "universal", because this is the sense in which the model of eustatic sea-level change is employed by Vail and his colleagues. This group of workers attributes all sea-level change to eustatic processes, including changes in ocean basin volumes for the low frequency cycles, and glacioeustasy for those of high frequency, although it was conceded that apart from specific cases, such as the Gondwana glaciation of the late Paleozoic, there is little convincing evidence for glaciation prior to the Oligocene (Vail et al., 1977, pp. 92–94) (we return to this point in Sect. 14.6.4). Two quotes from Vail's first major publication states the case succinctly. The first is at the beginning of this chapter, the second is below:

> Glaciation and deglaciation are the only well-understood causal mechanisms that occur at the relatively rapid rate of third-order cycles (Vail et al., 1977, p. 94).

As noted above, Vail and his colleagues thought that they had developed an entirely new standard of geologic time and a new basis for geologic correlation, superior to that based on conventional chronostratigraphic data (biostratigraphy, magnetostratigraphy, radiometric dating)(see summary in Miall, 1997, pp. 282–284).

### 12.2.2 The Complexity Paradigm

This contrasting paradigm represents a body of ideas focusing on the hypothesis that sea-level change is affected by multiple processes operating simultaneously at different rates and over different ranges of time and space, possibly including eustatic sea-level change. No simple signal can therefore result, and the interpretation of each event in the stratigraphic record can only follow very careful, detailed local work, followed by meticulous regional and global correlations. An additional, and important argument is that because of limitations in our techniques for testing the reality of global eustasy, no conclusions can yet be drawn regarding the vast majority of the supposed eustatic events that have been proposed (e.g., Miall, 1997; Dewey and Pitman, 1998).

Underlying both paradigms are the principles and methods encompassed by the sequence-architecture model first introduced by Vail et al. (1977), and further elaborated by Haq et al. (1987, 1988a), Posamentier and Vail (1988), Posamentier et al. (1988), and Van

Wagoner et al. (1990), although details of definition and interpretation of the model remain to be resolved. The global-eustasy model and its global cycle chart could be said to represent the paradigm against which anomalies have arisen, and which have coalesced around the complexity paradigm.

As others have already argued, geology is characterized by complexity. The concept of the multiple working hypothesis originated with geologists (Gilbert, 1886; Chamberlin, 1897) and, as noted by Law (1980, p. 16), it is not uncommon for geologists to simultaneously consider contrasting or even conflicting methods and hypotheses, a condition he refers to as *conceptual pluralism*. Therefore, the growth of two conflicting views about global eustasy and their mutual survival for more than 15 years (since the first articles critical of the model appeared in the early 1980s) should not surprise historians or sociologists of science (or geologists). As Kuhn (1962, p. 149) observed, "the proponents of competing paradigms practice their trades in different worlds ... practicing in different worlds, the two groups of scientists see different things when they look from the same point of view in the same direction."

Kuhn (1996, pp. 181–187) also clarified the concept of the scientific paradigm by showing how it is constructed of a set of integrated parts. First, a paradigm *incorporates a system of symbolic generalizations or laws*. Second, it involves *commitments or beliefs in particular models*. Third, there are *ideas about values*, such as acceptable levels of quantitative rigor. Fourthly, *predictions are implied* by the paradigm. Fifth, as already discussed, there are the *exemplars*, or solved problems—"ways of seeing." These components of the two paradigms under consideration here are shown in Table 12.1. In the next section, we discuss how the assumptions of these paradigms guide the observations geologists make in "reality," and impact on the conclusions they draw and the anomalies they do or do not confront.

## 12.3 Defining and Deconstructing Global Eustasy and Complexity Texts

This section discusses the use of terms and language in sequence stratigraphy and how this use may influence perceptions and analysis. This is not to deny the

**Table 12.1**  The Components of the two paradigms[a]

| Component | Global-eustasy paradigm | Complexity paradigm |
|---|---|---|
| Symbolic generalizations or laws | (1) Seismic reflections are chronostratigraphic surfaces—they define time<br>(2) Tectonism may "enhance" an unconformity but does not affect its timing | (1) There are few global "surfaces" beyond those caused by very rare catastrophic events (e.g., terminal Cretaceous event).<br>(2) Seismic reflections, when defined using the most advanced processing techniques, do not yield simple globally correlatable surfaces<br>(3) Tectonism, eustasy, and varying sediment supply integrate to generate highly diachronous sequence boundaries |
| Commitments or beliefs in particular models | (1) Sequences are chronostratigraphic units<br>(2) Global eustasy is the dominant mechanisms for driving relative sea-level change | (1) No single technique can be used to precisely define geological time<br>(2) There are multiple causes of sea-level change, of varying frequency and areal extent |
| Values | (1) The conventional time scale is arbitrary in defining boundaries within continuous successions.<br>(2) Sequence chronostratigraphy is superior to conventional methods for dating and correlation | (1) Only by defining boundaries within continuous successions can we be sure we are defining a record of continuous time.<br>(2) Only a compilation and reconciliation of all age data can begin to approach an accurate and precise time scale |
| Values imply prediction | Global eustasy permits worldwide correlation | (1) Where precise chronostratigraphic methods can be used they will generally show that precisely dated sea-level events in different parts of the world do NOT correlate with each other<br>(2) Given the imprecision of chrono-stratigraphic methods, most data strings will be shown to be capable of correlation with the global cycle chart, even synthetic sections compiled from tables of random numbers. |
| Exemplars ("solved problems" or "ways of seeing") | (1) X out of Y unconformities in a given data set match the global cycle chart<br>(2) A previously known series of events can be shown to "match" or "correlate with" the chart (e.g., Alberta Basin molasse pulses, Pyrenean thrust-fault pulses) | Sequences in a given succession commonly can be correlated to local/regional tectonic/climatic/autogenic processes |

[a]The components are those defined by Kuhn (1996, pp. 181–187).

existence of the objective "real" world nor the role it may have in the production of scientific knowledge. Rather, this is an attempt to direct attention to how socially derived meanings and processes can shape how science, and in this instance, sequence stratigraphy is done (based on Miall and Miall, 2001).

Social theorists examining how humans interact in meaningful ways have argued that "definitions of situations" guide the process. Through the process of socialization and the learning of language, humans attribute meaning to the situations in which they find themselves whether these be processes of eating, socializing, or doing science. Specifically,

The crucial fact about a definition of a situation is that it is cognitive – it is our idea of our location in social time and space that constrains the way we act. When we have a definition of a situation, we cognitively configure acts, objects, and others in a way that makes sense to us as a basis for acting (Hewitt, 1997. p. 127).

These theorists have, as a central focus, the subjective standpoint of individual actors. "Unlike a more objectivist approach, which views the social world as a reality that exists independently of any individual's perception of it, phenomenology sees that reality as constituted by our view of it" (Hewitt, 1997, pp. 15–16).

Most interaction among humans takes place within routine situations with well-established definitions of the situation or subjective viewpoints that guide behaviour. According to McHugh (1968), there are three fundamental assumptions that influence routine interactions: (1) that the assumptions we hold about a situation are valid; (2) that others in the situation share our definition of the situation; and (3) that as long as our definition of the situation works, it will not be questioned or challenged. These theorists, therefore, emphasize the cognitive foundations of human conduct, manifested in language, and stress that what people know about a situation and what they do are interdependent (cf. Hewitt, 1997).

Miall and Miall (2001, p. 332) suggested that through the application of these principles to the study of science as a human activity, we can understand how different groups of scientists, with different convictions or definitions of the situation, could emphasize or de-emphasize particular types of data and hypotheses in favour of the desired goals of the research. In the global-eustasy model, for example, published supporting data are sparse, and the necessary supporting hypotheses are largely "internal" or "self-referential." Accordingly, the language of the global-eustasy school includes a number of key interpretive phrases, which serve to turn research results toward the desired model, to define the situation confronting the geologist in a routinized way that directs his or her analytic behaviour. Notably, these terms are not used by, or hold different meaning for, the nonmembers of this group, for example, those adhering to the complexity paradigm, who have their own set of definitions of the situation. Some of these terms that hold special meaning for the global-eustasy school, but are used selectively or not at all by other sequence stratigraphers, are presented in Table 12.2. They express

two of the three elements of the hermeneutic circle—the preconceptions and the presumed goals of the research.

In order to further understand and illustrate the consequences of the use of particular terms in the practice of geology, it is useful to draw on a form of literary criticism called *Deconstructionism*. As proposed by Jacques Derrida, who founded the approach, deconstructionism takes apart the logic of language in which authors make their arguments. According to Denzin (1992, p. 32), "it is a process which explores how a text is constructed and given meaning by its author or producer." It rejects the assumption that texts have logical meanings and argues for "demystifying texts instead of deciphering them" (Vogt, 1999, p. 74). To use an example, objectivist studies of history "...assume the facticity of objects of historical analysis as constituted prior to the observer's study of them" (Hall, 1990, p. 26). Deconstructionists challenge,

the artificial coherence of historical accounts, by showing how to locate 'transcendent' staging devices in historical discourses. Situated outside history, such devices render historical accounts plausible to readers by providing 'history with continuity and discourse with meaning' thematized by 'aboutness.' A 'history' of Nixon's Watergate crisis, for example, can only be narrated by telescoping events into a coherent story (Cohen, 1986:74–76). In this light, any notion that the historical object is simply 'out there,' waiting for the historian to discover and describe it, seems a self-serving conceit. (Hall, 1990, p. 26).

As Hall (1990, p. 35) further observed, "In an age when deconstructionists are busy assaulting texts as internally ordered assemblages, historical narrative has become suspect as a special kind of storytelling."

Miall and Miall (2001) did not subscribe to the extreme relativism of postmodernism and deconstructionism which are characterized by "an extreme or complete skepticism of, or disbelief in, the authenticity of human knowledge and practice" (Dawson and Prus, 1995, p. 107). Rather, the tenets of deconstructionism were used as an heuristic device to illustrate the role of language in scientific interpretations of the "real" world.

We can illustrate how phenomenological and deconstructionist approaches inform the scientific process of stratigraphic interpretation with a simple diagram (Fig. 12.3). Two interpretations of a stratigraphic "text" are shown. The text consists of a suite of seven data points in a time-space universe, such as sea-level lowstands within a tectonostratigraphic province, dated

**Table 12.2**  The Language of the Global Eustasy School. Key words and phrases that hold special meaning for the members of this school, but are used selectively or not at all by other sequence stratigraphers

| Term | Application in the context of the global eustasy model | Usage by followers of the complexity paradigm |
| --- | --- | --- |
| Eustasy | The universal control of sequence boundaries | A hypothesis for which the evidence is sparse and questionable |
| Global cycle chart | Universal stratigraphic template, applicable worldwide | A compilation of local and regional events incautiously labeled as "global" |
| Glacioeustasy | The mechanism for all high-frequency cycles | Operated only during Earth's major glacial periods, and must always be tested and calibrated against local and regional tectonic and other mechanisms |
| Sequence chronozone | Sequence interpreted as a primary chronostratigraphic unit | A hypothetical concept of no practical use (c.f. Hedberg's (1976) chronostratigraphic units) |
| Sechron | Same as above | Not used |
| 1st- to 5th-order cycles | The sequence hierarchy, based on an assumed grouping of cycle frequencies | An arbitrary classification not reflecting the documented variability in stratigraphic unit thickness or duration |
| Tectonically enhanced unconformity | An unconformity of eustatic origin, enhanced by uplift | Term not used |
| Tectonic overprint | Eustatic unconformity modified by local tectonics | Tectonic overprint on any other process, such as climate change, variations in sediment supply, or eustasy |
| Local tectonics | The reason why a sea-level event in the global cycle chart is absent from a particular stratigraphic section | Just that: local tectonics, with no specific generalized meaning |
| Number of sequences in a section | Used as an attribute of correlation | An incidental result of local processes with no global significance |
| Pattern matching | Technique of recognizing similar successions of stacked sequences to those in the GCC as an attribute of correlation with the GCC | A technique of questionable value given the several mechanisms for sequence generation that could be overprinted in any given succession |

GCC: global cycle chart

according to the best available methods. In Fig. 12.3b, we see the application of the global-eustasy "language" to the text by the first group, those who accept the primacy of the global-eustasy model. One of the data points, chosen, perhaps, because it represents an important type section, is interpreted as representing a global sea-level event, and, based on the assumptions of the global-eustasy language, all nearby points are referred to it. This is suggested by the arrows, which imply correlation, not movement of the data points, although implicit in the global-eustasy model is the idea that an eustatic event, as observed at a given location, can serve to define its age everywhere. Such statements as "13 out of 15 events in my data correlate with sequence boundaries in the global cycle chart" (e.g., see Kerr, 1980, 1984) are essentially applications of the language illustrated in

Fig. 12.3b. A powerful and entirely self-referential hermeneutic circle is at work in the construction of this type of model. Anomalies do not arise because the research is directed toward defining how observations fit.

By contrast, Fig. 12.3a shows the same data interpreted employing the language of the second group. Each data point is accompanied by a bar showing the "standard error" associated with the dating method, and the "language" of interpretation also includes an error band extending the standard error of the first data point across all seven points. The users of the data are then free to decide, based on their assessment of error, whether or not some of the data points fall outside the band they would consider to define an acceptable level of correlation with the first point. In this way, the light each new point throws on the question of sea-level change in this area can be assessed quantitatively, point by point, based on assessments of the various generation mechanisms. Here the definition of the situation (or hermeneutic circle) allows for greater consideration of anomaly in observations. Consequently, no quick conclusions are drawn by geologists using this approach and anomalies are more likely to be acknowledged. As W. I. Thomas (1931) has observed, if you define a situation as real, it is real in its consequences. For the protagonists of the two paradigms, each interpretation of the stratigraphic texts is "real" within the terms of their paradigm. For the adherents of the global-eustasy model the "consequence" is that discussion of the potential error in, or revision of, the age of a given sequence boundary is typically not undertaken, whereas error and revision are an integral part of the complexity paradigm.

In the next section, we discuss the interaction and degree of coexistence that the two paradigms exhibit, through an examination of the patterns of cross citation that exist between them.

## 12.4   Invisible Colleges and the Advancement of Knowledge

The study of citation patterns has long been an issue in social research on scientific knowledge and how it advances. Price (1961, 1963), for example, observed that some disciplines continually reference their classic or foundational works. Other disciplines cite only the most recent papers in a rapidly advancing research front. Citations also reveal clusters of interrelated researchers and underlying social networks, what Price referred to as "invisible colleges." According to Crane (1972, pp. 138–39),

> Analysis of the social organization of research areas in science has shown that social circles have invisible colleges that help to unify areas and provide coherence and direction to their fields. These central figures and some of their associates are closely linked by direct ties and develop a kind of solidarity that is useful in building morale and maintaining motivation among members.

However, she goes on to argue, in order for science to advance, "the exchange of ideas is important in generating new lines of inquiry and in producing some integration of the findings from diverse areas" (Crane, 1972, p. 114). Indeed, she concludes, scientific

communities may become completely subjective and dogmatic if they are unable to assimilate knowledge from other research areas. The preceding discussion demonstrates how the two paradigms under review have been constructed to yield different kinds of observations. We shall now consider the extent to which each addresses the anomalies arising from the research of the other, and the consequences this may have for the advancement of scientific knowledge (from Miall and Miall, 2001, pp. 334–340).

As a generalization, workers in the complexity paradigm cite the global-eustasy model as one of several possible scenarios for geologic interpretation, but the reverse is typically not the case. Thorne (1992), in a discussion of the assumptions underlying sequence stratigraphy, and which is generally supportive of the methodology, noted that "The stratigraphic literature since 1977, citing the Exxon sea-level cycle charts, is prodigious. (An ARCO [Atlantic Richfield Corporation] library author citation search of P. Vail contains 983 citations between 1977 and 1988). Various stratigraphic techniques have been used to test the applicability of the global sea-level cycle chart." Adherents of the global-eustasy paradigm may cite opposing work, but rarely make the connection between that work and its implications for their preferred model. Even publications by basin analysts from other large and prestigious petroleum companies, such as British Petroleum (e.g., Hubbard, 1988) have had little effect on the use of the model by its adherents. For example, Vail et al. (1991) discussed Hubbard's work on the tectonic origins of sequence boundaries, but made no connection between these results and their implications for the generality of his global-eustasy model.

An opportunity to examine the pattern of citation used by each paradigm has been offered by the appearance of two recent special publications of the Society for Sedimentary Geology (SEPM), both published in 1998. The first, a book entitled, *"Mesozoic and Cenozoic Sequence Stratigraphy of European Basins"* (Graciansky et al., 1998) of which Peter Vail is a co-editor and contributing author, takes what may be described as the classic global-eustasy approach to regional sequence studies, as clearly stated in the second paragraph of its Preface:

> Sequence stratigraphy applies the inherent premise that eustasy represents a global signal among the variables that play a role in shaping depositional sequences. This

global signal plays an essential role in shaping depositional sequences laid down in response to changes in relative sea level. Because of this global signal, bounding surfaces of depositional sequences (sequence boundaries at their correlative conformities) can be expected to be synchronous between basins. To demonstrate such synchroneity requires a very high stratigraphic resolution and a calibration of all stratigraphic disciplines.

Although the book consists of a series of regional studies, in which global correlation was not a stated objective, the assumption that eustatic control may be demonstrated by such studies is conveyed by this quotation, also from the Preface:

> The composite stratigraphic record of higher order eustatic sequences shows a significant increase in the number of sequences identified in the various European basins. Entries on the new charts include a composite stratigraphic record of 221 sequence boundaries in the Mesozoic and Cenozoic, compared to 119 sequences for the same interval identified by Haq et al. (1987, 1988a).

Published at about the same time as the Graciansky et al. (1998) text is the book *"Paleogeographic Evolution and Non-Glacial Eustasy, Northern South America,"* compiled and edited by Pindell and Drake (1998). Again, the Preface to this book sets out the philosophical approach adopted by the editors and contributing authors:

> Our ability to isolate eustasy in most settings is seriously hindered by the fact that relative sea-level history in any location is multi-variable, comprising the net effects of such components as eustasy, local and regional tectonism, local and global climatic variability and variable sediment supply.
>
> One of this book's prime conclusions … is that the eustatic component of short-term relative sea-level changes during non-glacial times cannot be confidently isolated in most settings. The inability to isolate short-term eustatic changes will preclude genetic correlation of short-term cycles and the use of short-term cycles on cycle charts as time scales of predictors of reservoir horizons in most basins.

In Table 12.3, we list twenty-six papers that may be regarded as the critical body of work exploring problems with the global-eustasy paradigm. In a few cases, some rows in this table contain more than one paper. This indicates that the relevant ideas have appeared in different form in several places. Omitted from this list are a few contemporary publications, such as Miller and Kent (1987) that, while making important points, were not published in journals normally read by petroleum geologists, and were, therefore, not widely noted or cited at the time of publication. Also not included is Pitman (1978), an important early work that discussed rates of sea-level change, particularly with respect to the rapidity of the sea-level falls implied by Vail's first "sawtooth" curve, but did not constitute a critique of the chart itself. The list begins with the work of A. B. Watts and his coworkers, who developed the flexural model for extensional continental margins. This model provides a powerful alternative model for patterns of coastal onlap of "second-order" type and thus challenges the assumptions of eustatic sea-level rise incorporated into the global cycle chart. Many of the authors listed in Table 12.3 also contributed to a series of Discussions of the Haq et al. (1987) paper that appeared in 1988, with Replies by the Exxon group (Christie-Blick et al., 1988).

Table 12.4 documents the pattern of citation of these critical articles by the contributing authors of the forty-five papers in the Graciansky et al. (1998) volume, less the one article that is published only in abstract form. The results confirm the assertion that data and arguments opposing the generality of the global-eustasy model generated by other specialists have not been widely responded to by adherents of the global-eustasy paradigm. Based on this citation analysis, the following observations may be made:

(1) Thirty-four of the 45 papers cite the global cycle chart of Haq et al. (1987, 1988a), mostly by showing how their sequences correlate to events in that chart.

(2) Twenty-eight of the papers cite none of the articles critical of the global-eustasy model, or any other critical articles not on our list. None cites more than five of those given in Table 12.3. Critical papers 2, 3, 5, 6, 8, 10, 18, 19, 20, and 24 from Table 12.3 are not cited at all. The Discussions and Replies of the 1987 version of the chart (Christie-Blick et al., 1988) are not cited anywhere in this book.

(3) Most articles acknowledge the importance of low-frequency ("second-order") tectonism in generating accommodation for sequences, but only one cites the early work in this area (Paper row #1 in Table 12.3). A few papers acknowledge the potential importance of in-plane stress (papers #4, 5, 11).

(4) Tectonic "enhancement" or "offsetting" of eustatic sequence boundaries is a common theme.

**Table 12.3** Key papers expressing doubt or opposition to parts or all of the Global-Eustasy Model

| # | Authors | Type | Brief summary of main points |
|---|---|---|---|
| 1 | Steckler and Watts (1978), Watts (1981, 1989), Watts et al. (1982) | M | Flexural model for onlap on extensional continental margins |
| 2 | Parkinson and Summerhayes (1985) | M | Integrating tectonism and eustasy does not generate clear eustatic signal |
| 3 | Summerhayes (1986) | S | Data for GCC largely from Atlantic and Gulf margins, so correlations there no proof of eustasy |
| 4 | Cloetingh et al. (1985), Cloetingh (1986) | M | First proposal for significance of in-plane stress as a mechanism for RSLC |
| 5 | Karner (1986) | M | Geophysical basis for in-plane stress |
| 6 | Miall (1986) | S | General critique of methodology of extracting sea-level history from seismic data |
| 7 | Burton et al. (1987), Kendall and Lerche (1988) | S | Too many unknowns to permit extraction of eustatic signal |
| 8 | Hubbard (1988) | M | Documentation of ages of sequence boundaries in several major basins that do not correlate to GCC (seismic data from a "rival" corporation: BP) |
| 9 | Blair and Bilodeau (1988) | M | Tectonic mechanisms for clastic-wedge generation |
| 10 | Algeo and Wilkinson (1988) | S | Cycle thicknesses are statistically not periodic and cannot be used to argue for Milankovitch control. |
| 11 | Cloetingh (1988) | M | Detailed application of in-plane stress model to basinal stratigraphy |
| 12 | Galloway (1989a) | M | Sequences may reflect variations in sediment supply caused by tectonic events in source area |
| 13 | Schlager (1989, 1991) | M | Some sequence boundaries are "drowning unconformities" |
| 14 | Embry (1990) | M | Methods for identifying tectonic influences in sequence generation |
| 15 | Christie-Blick et al. (1990), Christie-Blick (1991) | M, S | General critique of seismic methodology; emphasized importance of tectonism |
| 16 | Underhill (1991) | M | Tectonic origin for events in Moray Firth Basin that had been used in definition of the GCC |
| 17 | Miall (1991a) | C,M | Chronostratigraphic imprecision of sequence boundary ages. Tectonic origin of sequences, critique of onlap model |
| 18 | Aubry (1991, 1995) | C | Detailed tests of chronostratigraphic correlations of unconformities worldwide reveal no clear global signal |
| 19 | Fortuin and de Smet (1991) | M | High rates of tectonism can occur on convergent plate margins |
| 20 | Sloss (1991) | M | Tectonism must be considered as a primary mechanisms for sequence generation |
| 21 | Miall (1992) | C | GCC will correlate with any succession of events, including synthetic sections compiled from random numbers |
| 22 | Gurnis (1992) | M | Dynamic topography: a mechanism for sequence generation on cratons |
| 23 | Underhill and Partington (1993a, b) | M | Tectonic origin for major events in the central North Sea that had been used in definition of the GCC |
| 24 | Drummond and Wilkinson (1993a, 1996) | S | Cycle thickness reveals continuous distribution, therefore there is no cycle hierarchy |
| 25 | Miall (1994) | C | The inherent imprecision of chronostratigraphic methods—global correlations cannot yet be reliably tested |
| 26 | Christie-Blick and Driscoll (1995) | M | Review of sequence-generating mechanisms, including tectonism as a major factor |

Note: Abbreviations: GCC=global cycle chart, RSLC=relative sea-level change types of critique (column 3): C=discusses chronostratigraphic methods and the nature of detailed tests of the GCC, M=discusses alternative mechanisms for the generation of sea-level events, S=general critiques of methodology, circularity of correlations to GCC.

**Table 12.4** Citations of critical work of the Complexity Paradigm in Sepm Special Publication 60

| Author | GCC cited | Complexity citations[a] | Commentary |
| --- | --- | --- | --- |
| Hardenbol et al. | ● | none | No discussion of potential error in correlation |
| Jacquin and Graciansky | ● | 1, 4, 14, 22, 23 | Acknowledges tectonic influences but does not address the implications of integrating tectonism and eustasy |
| Jacquin and Graciansky | ● | 11, 14, 23 | |
| Duval et al. | ● | none | |
| Vecsei et al. | ● | 7, 13, 15, 17 | SBs as drowning unconformities, SBs "offset" by tectonism. Eustatic component isolated by correlation to GCC. The only paper to show SBs with error bars |
| Abreu et al. | ● | none | Cretaceous glacioeustasy |
| Vandenberghe and Hardenbol | ● | none | |
| Neal and Hardenbol | ● | none | Paleogene sequences increased from 26 to 36 |
| Michelsen et al. | ● | none | SBs "broadly comparable" to those in GCC |
| Vandenberghe et al. | ● | none | |
| Catalano et al. | ● | none | Tectonism enhances SBs but does not change their age |
| Vitale | | 9 | Tectonically-formed sequences correlate to GCC. High-frequency tectonism is only local |
| Flinch and Vail | | none | High-frequency tectonism purely local and does not affect timing of SBs |
| Vakarcs et al. | ● | 4 | Sequences are global because they correlate with GCC |
| Gnacolini et al. | ● | none | Sequences are global because they correlate with GCC |
| Abreu and Haddad | ● | 11, 25 | Chronostratigraphic problems acknowledged but not dealt with |
| Neal et al. | ● | none | |
| Geel et al. | ● | 4, 11, 14 | Applies Embry's criteria for tectonism but argues that this does not exclude eustasy as a control |
| Luterbacher | ● | none | Correlations with GCC may be artifact, notes danger of circular reasoning |
| Pujalte et al. | ● | none | Correlations with GCC may be because it was defined on data from same part of Europe |
| Hardenbol and Robaszynski | ● | none | "Number of sequences" and "sequence stacking" critical |
| Grafe and Wiedmann | ● | 13, 21, 26 | References cited but points not dealt with |
| Floquet | | none | Acknowledges importance of low-frequency tectonism but not its implications for eustatic correlations |
| Robaszynski et al. | ● | none | |
| Philip | | none | |
| Jacquin et al. | ● | none | |
| Ruffell and Wach | | none | |
| Hoedemaeker | ● | none | Uses French biostratigraphic system that does not always match GCC. "Number of sequences" significant |
| Jacquin et al. | ● | 23 | |
| Graciansky et al. | | 23 | Some successions of SBs cannot be correlated to GCC "with confidence" |
| Stephen and Davies | | 14, 16, 23 | Tectonism has not overprinted eustasy in North Sea, but "unsound" to use such a "local dataset" to invoke global mechanism |
| Leinfelder and Wilson | ● | none | |
| Gygi et al. | | none | |
| Van Buchem and Knox | ● | none | |
| Hesselbo and Jenkyns | ● | 12, 13, 17, 23 | References cited but points not dealt with |
| Graciansky et al. | | 23 | |
| Dumont | ● | 4, 11 | Poor correlation to GCC may indicate GCC "does not apply exactly to the European realm because global eustasy is not the single forcing mechanism" |

**Table 12.4** (continued)

| Author | GCC cited | Complexity citations[a] | Commentary |
|---|---|---|---|
| Gianolla and Jacquin | • | none | Tectonic modification of eustatic events |
| Skjold et al. | • | 4, 11 | Intraplate stress may have been important |
| Goggin and Jacquin | • | none | |
| Courel et al. | • | none | |
| Gaetanio et al. | • | 14 | Notes circularity of correlation to GCC that was mainly constructed from this project area. "Tectonism versus eustasy: an everlasting debate" |
| Gianolla et al. | • | none | |
| Ruffer and Bechstadt | | none | |

[a]The numbers in this column correspond to the numbered papers in Table 12.3.
Abbreviations: GCC=global cycle chart of Haq et al. (1987, 1988), SB=sequence boundary

(5) Several papers studied the influence of high-frequency tectonism in foreland basins and concluded that it is very local in effect and had no effect on the timing of sequence boundaries.

(6) A few papers cite the references discussing the problems with the imprecision of chronostratigraphic methods (papers #16, 21, 25), but none of them attempts to answer or deal with these problems.

(7) Only one paper (that by Vecsei et al.) shows error bars for the ages of sequence boundaries.

(8) The "number of sequences" in a given time interval is regarded as a significant point of comparison to the global cycle chart.

(9) Many papers claim that sequences must be global because they correlate with the global cycle chart, and use such correlations to extract or isolate the eustatic component.

(10) A few papers note the circularity of correlating with portions of the global cycle that were defined in the general vicinity of their project area.

By contrast, referencing by adherents of the complexity paradigm is much more inclusive of the other paradigm. For example, the paper by Dewey and Pitman (1998) in the Pindell and Drake (1998) volume, which examines "Sea-level changes: mechanisms, magnitudes and rates," cites the following articles from Table 12.3: papers 1, 2, 4, 5, 6, 17, 21, and 22. All of these articles are assessed in some detail in the text of the paper. This paper also cites the major works by Vail and his colleagues, beginning with the classic Vail et al. (1977) volume.

The paper by Dewey and Pitman (1998) concludes that there are multiple causes of sea-level change, most of them local to regional in scope. In this book, there is also an exhaustive review of the evidence for glaciation in the Mesozoic and Cenozoic, by Markwick and Rowley (1998), which concludes that there is no convincing evidence for widespread continental glaciation at any time during the Mesozoic (but see Sect. 14.6.4). Erikson and Pindell (1998), in this book, demonstrated that the relative sea-level changes which they can document in northern South America occur over time intervals longer than would be expected for glacioeustatic control (>1 million years), but that a few may be eustatic in origin. They compiled a total of six possible supraregional, potentially global (eustatic) events in the Cretaceous, at an average spacing of 13.3 million years. This number is close to what Miall (1992) estimated would be the case in the Cretaceous record, and is a marked contrast to the conclusions of the Graciansky et al. (1998) book.

The most recent expression of opinion by adherents of the complexity paradigm can be summed up in these quotations from Dewey and Pitman (1998, pp. 13–14):

We cannot conceive of a scenario that would allow the synchronous global eustatic frequency and amplitude implied by third-order cycles. ... An especial problem of using third-order sequence 'cycles' as indicators of global eustasy is the correlation problem. Miall (1992) has shown that 77% correlation with the Exxon Chart can be achieved between four random number-generated sequences. Whereas a 0.5 m.y. precision is needed for correlation of the most closely spaced third-order events, a precision of only 2 my can be achieved biostratigraphically. ... It is our contention that local or regional tectonic control of sea-level dominates eustatic effects, except for short-lived glacio-eustatic periods and, therefore, that most sequences and third-order cycles are controlled tectonically on a local and regional scale rather than being global-eustatically controlled. We suggest that eustatic sea-level changes are nowhere near as large as have been claimed and have been falsely correlated and constricted into a global eustatic time scale whereby a circular reasoning forces correlation and then uses that correlation as a standard into which all other sequences are squeezed.

Most recent textbooks on stratigraphy reach similar conclusions (e.g., see Hallam, 1998b, pp. 427–428; Nichols, 1999, p. 288).

Kuhn (1962, p. 93) has argued that "when paradigms enter, as they must, into a debate about paradigm choice, their role is necessarily circular. Each group uses its own paradigm to argue in that paradigm's defense." He has also stated that two scientific schools in disagreement will "talk through each other when debating the relative merits of their respective paradigms" (Kuhn, 1962, p. 108). While aspects of his argument hold true here, the examination of citation practices suggests that the complexity paradigm is more open to external influences in that it directly addresses the contributions and anomalies arising from the global-eustasy model. The global-eustasy model, however, adheres to a circular pattern of observation and justification, with little attempt to address anomalies arising from research outside the circle of researchers using it. The extent to which this practice will continue to elicit support for the paradigm is questionable. As Kuhn (1962, p. 93) has observed, "whatever its force, the status of the circular argument is only that of persuasion. It cannot be made logically or even probabilistically compelling for those who refuse to step into the circle."

In the next section, we address the tenets of the global-eustasy paradigm directly and consider its current status in the scientific community.

## 12.5 The Global-Eustasy Paradigm—A Revolution in Trouble?

According to Kuhn (1962, p. 93), "as in political revolutions, so in paradigm choice—there is no standard higher than the assent of the relevant community." Indeed, the most important component of a paradigm is the cognitive authority it is accorded, an authority based on the judgment of the scientific community that it deserves acceptance over other paradigms (cf. Gutting, 1984). We now examine the judgment of the scientific community as it pertains to the global-eustasy model itself, with particular reference to the recent book by Graciansky et al. (1998) discussed in the previous section.

The global-eustasy paradigm is built on the following assumptions, or hermeneutic prejudgments (Miall and Miall, 2001, p. 340):

1. That a global eustatic signal exists;
2. That eustatic control generates synchronous sequence boundaries;
3. That synchroneity can be demonstrated using current chronostratigraphic tools.

Those scientists skeptical of the global-eustasy model would argue that none of these three assumptions has been proven satisfactorily. Indeed, all three have been the subject of vigorous debate. For example, while the existence of a low-frequency ($10^7$–$10^9$-year cyclicity) eustatic signal is now generally acknowledged, there is no agreement on the existence of a high-frequency global signal during periods lacking widespread continental glaciation (e.g., during the Cretaceous). But it is not the purpose of this discussion to repeat the debate about these concepts and assumptions. Suffice it to note:

1. The existence of a suite of eustatic cycles of "third-order" magnitude, and global in extent—the basis for Peter Vail's first global cycle chart, has never been satisfactorily documented or proven (Miall, 1992, 1997), a crucial point largely ignored in the Graciansky et al. book, in which the Haq et al. (1987, 1988a) global cycle chart is taken as the starting point of their work.
2. It has been pointed out that where relative sea level is controlled by two or more processes of similar frequency and amplitude (e.g., thermal subsidence and second-order eustasy, or foreland basin tectonism and Milankovitch cyclicity), a synchronous regional (let alone global) signal cannot be generated (Parkinson and Summerhayes, 1985; Miall, 1997).
3. As noted earlier, conventional chronostratigraphic methods and results have been subordinated to the global cycle chart by the Exxon group, and independent efforts to chronostratigraphically calibrate the sequence record, such as those by A. Hallam (many papers, starting with Hallam, 1978), have been largely ignored by the Exxon group.

4. It has been argued that conventional chronostratigraphic techniques are inadequate to test the reality of high-frequency global synchroneity (Miall, 1994), except in the late Cenozoic, for which the techniques of cyclostratigraphy (calibration of the high-frequency sequence record by reference to the known frequencies generated by orbital forcing) are now providing a very refined time scale (House and Gale, 1995). Graphic correlation methods may change this, as noted in our Conclusion.

5. Several workers have demonstrated the inherent diachroneity of sequence boundaries, reflecting the integrated effects of varying local to regional subsidence rates and rates of sediment supply (Jordan and Flemings, 1991; Catuneanu et al., 1998).

Although the members of the global-eustasy school refer to "tectonic enhancement" of unconformities (Table 12.2), at the same time they effectively rule out consideration of the tectonic and other effects on sequence timing referred to in the previous paragraph. As Vail et al. (1991, p. 638) have stated:

> structure tends to enhance or subdue eustatically caused sequences and systems tract boundaries, but does not affect the age of the boundaries when dated at the minimum hiatus at their conformable position.

Even where studies appear to have been specifically undertaken to explore the relationship between sedimentation and tectonics, adherents of the global-eustasy school still regard correlation with the global cycle chart as significant. Thus Deramond et al. (1993; a study not part of the SEPM volume under examination in this section) explored the relationship between sedimentation and tectonism in tectonically active Pyrenean basins, and succeeded in developing tectonic models for sequence development. However, they then offered correlations of sequence boundaries with the global cycle chart and went on to discuss eustatic control. In referring to "tectonically enhanced" unconformities, they stated "The apparent correlation between the two groups of independent phenomena is an artifact of the method which calibrates the tectonic evolution by comparison with eustatic fluctuations." This form of "calibration" is, of course, circular reasoning.

Hardenbol et al. (1998) summarized the evidence used in the construction of a set of new chronostratigraphic cycle charts for the Mesozoic and Cenozoic that appeared in SEPM Special Publication 60. They noted that there are 221 sequence boundaries in this new synthesis, in contrast to the 119 that appeared in the chart published by Haq et al. (1987, 1988a). This is a boundary, on average, every 1.12 million years, in contrast to the average spacing of 2.08 million years in the earlier chart. A spacing of 2.08 million years in the earlier chart was already below chronostratigraphic resolution for most of the Mesozoic and Cenozoic. A potential error of ±0.5 million years is reasonable for the Cenozoic and parts of the Cretaceous, but potential errors of several millions of years must be factored into assigned ages for older parts of the Mesozoic record (Miall, 1994). Given a potential error of ±0.5 million years many events—those spaced at 1 million years or less—would overlap each other and would therefore be indistinguishable.

Even that earlier chart contains more putative sea-level events than any others published for the Mesozoic and Cenozoic, leading to the question: why does the Exxon global cycle chart contain so many more events than other sea-level curves? The question is even more apposite now. Miall (1997, p. 320) suggested that the paradigm of stratigraphic eustasy followed by the Exxon group allows them to interpret virtually all sea-level events as eustatic in origin. Hallam (1992a, p. 92) said, of the Vail and Todd (1981) article on the North Sea: "The very title of the Vail and Todd paper implies that, at that time at least, seismic sequence analysis of only one region was believed to be sufficient to obtain a global picture." Therefore, every new sea-level event that is discovered, can be added to the global chart.

This points to a different question: what does the Exxon chart actually represent? It undoubtedly constitutes a synthesis of real stratigraphic data from around the world. It has been suggest that most of the sequences are the product of regional events that originated as a result of plate-margin and intraplate tectonism, including basin loading and relaxation and in-plane stresses. These may have been well documented in one or more basins in areas of similar kinematic history, but their promotion to global events should be questioned. Some events may be duplicates of other events that have been miscorrelated because of chronostratigraphic error (Miall, 1992, p. 789).

Exactly the same assumptions and methodology appear to have been used in construction of the new charts in Graciansky et al. (1998). Sea-level "events" and corresponding sequence boundaries appear to have been added to the synthesis charts even when evidence from separate regional studies in this book indicates that many do not appear in each of the studied basins. Yet they are all labeled "global". This has been done despite the clearly stated problems of correlating between Boreal and Tethyan faunas within Europe, and the use of North American (Western Interior) chronostratigraphic standards where European data are inadequate, which potentially introduces still further sources of error and imprecision. Finally, because these sequences have been documented only in Europe, they cannot be defined in a blanket manner as having global (eustatic) significance.

A logical extension of this approach would be for the eustasy school to continue to carry out similar research in other parts of the world and to continue to add sequence boundaries to their "global" synthesis. We would predict a resulting chart containing many hundreds of events, and an average spacing in the $10^4$–$10^5$-year range. This may well be what this team has in mind, because, as noted below, they continue to hold to the mechanism of glacioeustasy as the main cause of most of their sea-level events, and adding still more events would quickly bring the average event spacing down to within the Milankovitch band, the normal glacioeustatic frequency.

A critical question that needs to be asked in this context, but has not been, concerns scientific methodology: how can the authors know that a "new" (as in newly observed) sea-level event is in fact a true "new" event and not a replication of another, previously known event, separated from it only by the margin of error inherent in the dating methods? While some new observations "prove" by "correlation" (the arrows in Fig. 12.3b) that a given sea-level event has been replicated, others are used to define new "events" in the global cycle chart. It is not clear how a distinction is made between these two alternatives. More importantly, how can the authors know that any of these events is eustatic, given that all their data are from Europe?

Correlations with the Haq et al. (1987, 1988a) curve are offered by most of the contributing authors in Graciansky et al. (1998) book (Table 12.4). Hardenbol et al. (1998) stated that this chart has been re-calibrated to the new time scale of Gradstein et al. (1995), but this has not been done in every case. For example, several

of the Haq et al. (1978, 1988a) sequence boundaries are given different ages in the summary chart of Hardenbol and Robaszynski (1998, Fig. 1) relative to that of Gräfe and Wiedmann (1998, Fig. 3). These errors provide an additional source of confusion.

Although this new European synthesis is offered as a major updating of the Haq et al. (1987) synthesis, the latter is still regarded as the basic reference point for a discussion of "global" standards in most of the papers in this new book. For example, Vecsei et al. (1998, pp. 69–70) discussed the matching of their sequence boundaries with those in the Haq et al. (1987, 1998a) chart. Their discussion contains statements such as:

> The brackets containing the ages of the mid-Cretaceous unconformity [and other sequence boundaries] each contain the age of a supersequence boundary as proposed by Haq et al. (1988a).
> The age of the boundary of SS 1/2 (Late Campanian) is the same as that of a sea-level lowstand ranked between the 2nd and 3rd orders by Haq et al. (1988a).

But then these authors also include statements like this:

> Four supersequence boundaries proposed by Haq et al. (1988a) to have occurred in Late Cretaceous to Miocene time could not be recognized on the Maiella platform margin.
> In the Maiella, 3rd-order sequences can only be recognized in part of the succession, and age resolution is not refined enough for their correlation with the onlap curve of Haq et al. (1988a).

Vecsei et al. (1998) employed the by-now well-known language of the eustasy school that the use of the Haq et al. (1987, 1988a) chart has helped to create, attributing mismatches with that chart to "regional tectonics" or "tectonic overprint." As noted by Miall (1997, Chap. 13), the presence of missing or extra sequences relative to the Exxon global cycle chart does not pose a problem for those using the chart. Such sequences are readily attributed to "local tectonics". To explain non-correlation, Vecsei et al. (1998) offered an appeal to chronostratigraphic imprecision and to the effects of local tectonics "offsetting" the global eustatic signal, but these authors did reasonably conclude "considering the age brackets in our data as well as the uncertainties in Haq et al. (1987, 1988a) curve, the supersequence boundaries can all be older or younger than the ages given by these authors." Our summary of this is that the results of this study can, by themselves, neither support nor negate the validity of the Haq et al. (1988) chart, and correlation or lack of correlation with that chart cannot be taken to mean anything very much. This paper is the only one in the book to show sequence-boundary ages with error bars, and is the paper most alive to the ideas encompassed by the complexity paradigm.

Even where sequence boundaries in an individual local study such as that of Vecsei et al. (1998) "match" or "correlate" with that chart, it cannot be claimed that this demonstrates the reality of the eustatic signal because this is only a local study, the global significance of which remains to be demonstrated. As noted by Miall (1997, p. 302), with regard to this type of *ad-hoc* argument: "to be rigorously objective, if sequence X is the result of tectonics, why not the identical sequence Y, which happens to correlate with the global cycle chart?" An independent test of the causes of sequence boundaries is required in order to validate the global-eustasy model.

In their summary of Upper Cretaceous sequence chronostratigraphy, Hardenbol and Robaszynski (1998) stated that some of the sequence boundaries in the local studies that follow "compare well" with boundaries in other local studies, while others "do not compare well". The total number of sequences recognized over a particular time interval is regarded as important information, and a table is offered listing the number of Upper Albian to Cenomanian and Turonian to basal Coniacian sequences identified in each of the four local studies that follow. These sequences are then added to each other to develop a "global" signal. This is very much the method that Miall (1997, Fig. 13.2) deduced from earlier Exxon work on the global cycle chart, and it helps to explain why the Exxon charts contain so many more events than those of any other worker in this field. It is not made clear in this or other work by members of the Exxon school, including Graciansky et al. (1998), what is the significance of the number of sequences in a particular interval, except that it suggests a similar frequency of cyclicity if different basins have roughly the same number (Table 12.2). This is, however, an integral part of the language of the eustasy model. Other scientists would claim that the number of sequences does not indicate correlation or global control because there are several mechanisms that can generate cycles of most known

frequencies. The validity of the "stacking" method is not explained.

Since the early work of Vail et al. (1977), glacioeustasy has been proposed by Vail and his coworkers as the main mechanism for generating high-frequency sea-level changes. As part of this new book on European basins, Abreu et al. (1998) compiled a new synthesis of oxygen isotope data and used this compilation as a basis for constructing curves showing interpreted low and high-frequency eustatic sea-level changes. The Cenozoic ice age is generally thought to have commenced in the Late Eocene, with the initial build-up of the Antarctic ice cap (Eyles, 1993). For the post-Late Eocene time period, high-frequency glacioeustatic fluctuations are well documented in the stratigraphic record, but the evidence for the pre-Late Eocene Cenozoic is equivocal. All compilations of oxygen isotope data, including that of Abreu et al. (1998), show generally lighter isotopic values in the pre-Eocene record, indicating generally warmer ocean temperatures than during the Neogene, and most based on the sedimentary and stratigraphic studies investigating the existence of continental ice during the Mesozoic have concluded that there was no widespread glaciation (Eyles, 1993; see Fig. 11.16). Abreu et al. (1998) noted that, "acceptance of continental ice during the Cretaceous and Paleogene is still highly controversial among stratigraphers," and that "Data supporting the current consensus in the scientific community on the absence of continental ice in the Cretaceous are not conclusive and may need to be challenged." We return to the issue of continental ice during the Mesozoic in Sect. 14.6.4.

Hardenbol et al. (1998, p. 7) made explicit their preferred model, that the record of high-frequency sequences is of glacioeustatic origin. If they are correct that glacioeustasy was the major controlling factor (and this is debatable; the occurrence of glacioeustasy in the pre-Neogene is discussed in Sects. 11.3.3 and 14.6.4) then their cycle charts contain far too *few* sequence boundaries. If only the long-term 405 ka eccentricity frequency is capable of being consistently recognized in the stratigraphic record for the more distant geological past, as suggested by Hinnov and Ogg (2007), the chart for the Cenozoic should contain about 160 regularly-spaced sequences, far more than the 58 shown in the Paleogene-Neogene charts published by Hardenbol et al. (1998). The irregular spacing between the events in these charts is not consistent with the rhythmicity expected of an astrochronological control, and no attempt was made in this work to seek continuous sections to which time-series analysis could be applied.

In conclusion: the differences between the two SEPM Special Publications are marked, in terms of the language of the interpretations, the referents of the contributing authors and editors, and the conclusions reached. Both books consist of compilations of new, detailed field data, compiled using the most modern field and laboratory methods. If the primary mechanism of sequence-boundary generation is eustasy, then the compilations of sea-level events in the two books should be virtually identical, subject to the normal margins of error to be expected from these types of stratigraphic data, and allowing for differences in sequence preservation in the two areas (proximal versus distal successions, variable influence of tectonic overprint, etc.) Even allowing for such differences, we suggest that the very different sea-level event spacings resulting from the two analyses (1.12 versus 13.3 million years), points to the completely different reference frames of the two sets of researchers. They represent values and exemplars of contrasting paradigms. Textual analysis of the works indicates that the stratigraphic texts—the field records—hold different meanings for the two groups of sequence stratigraphers. In the words of deconstructionism, the texts hold multiple conflicting interpretations. According to Kuhn (1962, p. 144),

> In the sciences, the testing situation never consists, as puzzle-solving does, simply in the comparison of a single paradigm with nature. Instead, testing occurs as part of the competition between two rival paradigms for the allegiance of the scientific community.

It remains for future analysis to determine whether the cognitive authority of the global-eustasy model will be maintained. However, there remains an active body of workers who continue to pursue this methodological approach. Haq and Schutter (2008) offered a new "Chronology of Paleozoic sea-level changes" built from the same approach used by Vail et al. (1977), Haq et al. (1987) and Graciansky et al. (1998). Successions of sequences from selected "Reference Districts" were assembled into a single "global" chart. Each sequence boundary was assigned an absolute age to the nearest 0.1 million years, but no assessments of potential error were provided for these ages. This chart contains

172 "discrete third-order events" averaging 1.7 million years per cycle. It needs to be pointed out that current estimates of potential error in the geological time scale range between $\pm 2$ and 3 million years, except for the Pennsylvanian and Permian, where the standard error drops to about $\pm 1$ million years (Hinnov and Ogg, 2007; see Fig. 14.22). The level of assumed accuracy and precision of this chart is therefore beyond what is technically possible. None of the laborious and rigorous procedures described in Sect. 14.5 were followed in the assembly of this synthesis. As noted by Ricken (1991, p. 773) with reference to earlier attempts to date and correlate short-term events, the "range of error is often larger than the actual time span considered." It is recommended that this "new" chart be ignored.

As discussed in Chap. 10, when there is more than one process driving accommodation change, the timing of the highs and lows is likely to vary from location to location as a result of variations in local rates of change. Tectonic processes, in particular, are not constant in rate. As Allen and Allen (2005, p. 271) noted, when integrating a curve of eustatic sea-level change with a variable rate of tectonic subsidence, the higher the subsidence rate the narrower the time band during which the integrated curve indicates periods of actual fall in relative sea level (Fig. 10.1b).

> Clearly, when correlating different locations in a basin with spatially variable tectonic subsidence rate, we should expect significant diachroneity of the onset of erosion due to relative sea-level fall and of flooding during relative sea-level rise. The delay in the onset of erosion is a quarter of a eustatic wavelength or $\lambda/4$. The difference in the timing of flooding is also $\lambda/4$. Since diachroneity of key stratigraphic surfaces is to be expected from a globally synchronous eustatic change, it is curious that the inference of stratigraphic synchroneity is seen as an acid test for a eustatic control. [$\lambda$=frequency of the sea-level curve. Quote from Allen and Allen, 2005, p. 271].

## 12.6 Conclusions

In this Chapter, we have examined the practice of geological science as it pertains to sequence stratigraphy— an example of what Dott (1998) referred to as an "historical" science. More specifically, we have examined the global-eustasy model and its global cycle chart. As part of our analysis, we have examined the nature of stratigraphic data and its importance in the process of validating this paradigm. Through the use of philosophical and sociological analyses of the nature of human activity, and in particular the work of Thomas Kuhn, we have also attempted to illustrate (1) that the preconceptions of geologists shape their observations in nature; (2) that the working environment can contribute to the consensus that develops around a theoretical approach with a concomitant disregard for anomalous data that may arise; (3) that a theoretical argument can be accepted by the geological community in the absence of "proofs" such as documentation and primary data; (4) that the definition of a situation and the use or non-use of geological language "texts" can direct geological interpretive processes in one direction or another; and (5) that citation patterns and clusters of interrelated "invisible colleges" of geologists can extend or thwart the advancement of geological knowledge.

We return to the question posed at the beginning of this book (Sect. 1.2.1): What is unique about geological reasoning? Miall and Miall (2001, p. 344) argued that there is nothing unique about geological reasoning per se. As has been demonstrated in this chapter, the processes of paradigm construction and testing reflect those advanced by Kuhn in his examination of how science proceeds. In this regard, geology is no different from the other natural sciences. Further, we have demonstrated that geological reasoning, like all human activity, is influenced by social processes that shape perceptions and direct discovery, influences that have not been commonly recognized by working scientists.

Where geological reasoning does differ is in the nature of the reality it attempts to explain. Stratigraphic data are, of necessity, historical and incomplete. This brings to the fore processes of interpretation that constitute "explanations that work" and that have a "narrative logic," which are consistent with all the available facts. Having said that, chronostratigraphic data that bear on the problem of sequence correlation are quantifiable, and should be treated the way they would be treated by physicists and chemists, with attention paid to experimental source and to both accuracy and precision. For example, the employment of graphical correlation methods in this work is to be encouraged (Mann and Lane, 1995). New research carried out with these principles in mind, testing the nature of regional

event correlations (e.g., Aubry, 1995; Immenhauser and Scott, 1999) is revealing a complex pattern of sea-level changes that would appear to us to be inconsistent with the global-eustasy paradigm.

Finally, we suggest that geologists must be cognizant of the influences that preconceptions and group processes can bring to bear on scientific theorizing. By becoming conscious of these influences in the daily work environment, geologists can minimize the bias in observation and interpretation that will inevitably result.

What if Vail and his colleagues are wrong? What if there is, in fact, no suite of global cycles dependent on and accurately reflecting a history of eustatic sea-level change? Vail's basic premise has not yet been proved. Part III of this book demonstrates in detail that many alternative mechanisms exist for the generation of regional suites of stratigraphic sequences, at comparable rates to the major eustatic processes, and producing stratigraphic results of identical character. There is, therefore, no reason for favouring Vail's interpretation of the cycle charts over, say, regional tectonic mechanisms, unless very accurate and precise global correlations among the cycles can be demonstrated. This is the only possible independent proof of Vail's paradigm. The demonstration of global correlation of sequence frameworks between distant and tectonically unrelated basins is the essential prerequisite before Vail's paradigm can be taken seriously as a testable geological hypothesis although, as noted above, there are limits to how precise we should expect such correlations to be. The Exxon workers have not even begun to address the serious problems that must be dealt with before such a proof can be developed.

Some years ago (Miall, 1992) I stated:

Stratigraphic correlation with the Exxon chart will almost always succeed, but it is questionable what this proves. The occurrence of numerous third-order cycles during the Mesozoic and Cenozoic (as implied by the Haq et al. [1988a] chart) remains to be demonstrated. The existing Exxon cycle chart should be abandoned—it is too flawed to be fixed. Its continued use drags us all further into the pit of conflicting correlations and circular reasoning. We should start again, by building a framework of independent sequence stratotypes, following the methods used so successfully in recent years by stratigraphers who have been perfecting the global time scale by the careful selection and meticulous documentation of boundary stratotypes, without preconceptions as to the results.

It is the purpose of Part IV of this book to demonstrate that existing methods are at present not capable of providing such precise correlations for cycles with periodicities of $10^6$ years and less, except in certain exceptional situations. We are many years away from even knowing whether it will ever be possible to construct a chart of global cycles in the million-year range.

We do not yet know that such a framework exists. This has not inhibited other workers from attempting to construct charts of sea-level change. For example, a large-scale research program on the continental shelf of New Jersey has devoted substantial resources to this objective. This and other comparable work are discussed in Chap. 14.

# Chapter 13

# Time in Sequence Stratigraphy

## Contents



## 13.1  Introduction

The purpose of this chapter is to demonstrate the existence of a hierarchy of geologic time extending over some fourteen orders of magnitude, and the importance of time breaks in the sedimentary record.

## 13.2  Hierarchies of Time and the Completeness of the Stratigraphic Record

It has become a geological truism that many sedimentary units accumulate as a result of short intervals of rapid sedimentation separated by long intervals of time when little or no sediment is deposited (Ager, 1981). Time is continuous, but the stratigraphic record of time is not; and as recorded in the stratigraphic record, time is discontinuous on several time scales (Fig. 13.1). Breaks in the record range from such trivial events as the nondeposition or erosion that takes place in front of an advancing bedform (a few seconds to minutes), to the nondeposition due to drying out at ebb tide (a few hours), to the summer dry periods following spring run-off events (several months), to the

surfaces of erosion corresponding to sequence boundaries (tens to at least hundreds of thousands of years), to the longer breaks caused by tectonism, up to the major regional unconformities generated by orogeny (millions of years). There is a similarly wide variation in actual rates of continuous accumulation, from the rapid sandflow or grainfall accumulation of a cross-bed foreset lamina (time measured in seconds), and the dumping of graded beds from a turbidity current (time measured in hours to days), to the slow pelagic fill of an oceanic abyssal plain (undisturbed for hundreds or thousands of years, or more) (Fig. 13.2).

It is now widely realized that rates of sedimentation measured in modern depositional environments or the ancient record vary in inverse proportion to the time scale over which they are measured. Sadler (1981) documented this in detail, using 25,000 records of accumulation rates (Fig. 13.3). His synthesis showed that measured sedimentation rates vary by eleven orders of magnitude, from $10^{-4}$ to $10^7$ m/ka. This huge range of values reflects the increasing number and length of intervals of nondeposition or erosion factored into the measurements as the length of the measured stratigraphic record increases. Miall (1991b) suggested that the sedimentary time scale constitutes a natural hierarchy corresponding to the natural hierarchy of temporal processes (diurnal, lunar, seasonal, geomorphic threshold, tectonic, etc.). Crowley (1984) determined by modeling experiments that as sedimentation rate decreases, the number of time lines preserved decreases exponentially, and the completeness of the record of depositional events decreases linearly. Low-magnitude depositional events are progressively eliminated from the record.

Many workers, including Berggren and Van Couvering (1978), Ager (1981), Sadler (1981),


A.D. Miall, *The Geology of Stratigraphic Sequences,* 2nd ed.,
DOI 10.1007/978-3-642-05027-5_13, © Springer-Verlag Berlin Heidelberg 2010

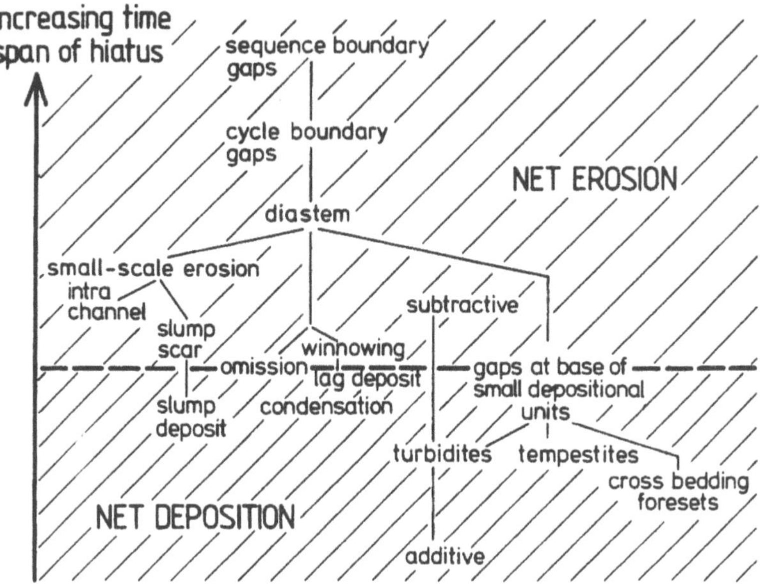



**Fig. 13.1** The types of stratigraphic gap, arranged in a hierarchy to illustrate the relationship to time spans (Ricken, 1991)

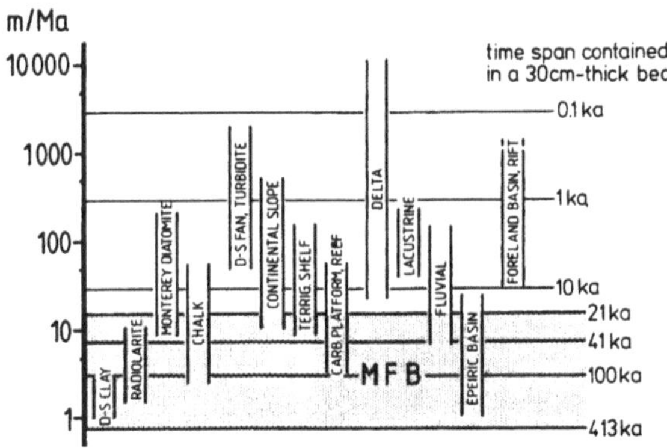



**Fig. 13.2** Long-term sedimentation rates for various types of deposit. Time spans at right indicate time required for accumulation of a 30-cm bed at the given sedimentation rate. Shaded area corresponds to time scales within the Milankovitch frequency band (MFB) (Ricken, 1991)

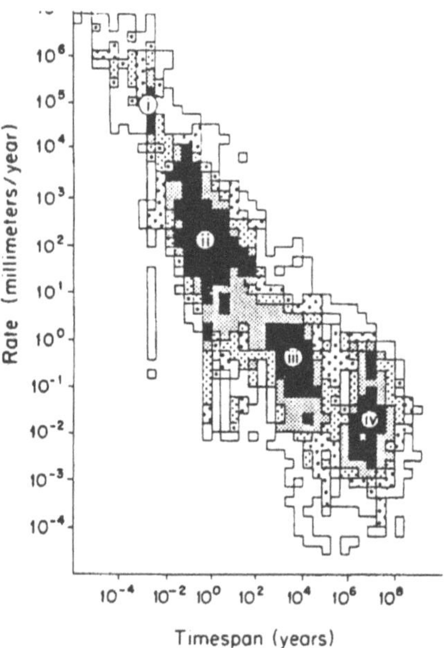



**Fig. 13.3** A plot of sedimentary accumulation rates against timespan, compiled by Sadler (1981). The data group into four clusters, labelled i to iv. These clusters Sadler (1981) explained as follows: i, continuous observation of a modern environment; ii, reoccupation of survey stations after a lapse of time; iii, data points determined by radiocarbon dating; and iv, data points determined by biostratigraphy, calibrated by radiometric dating, magnetostratigraphy, etc.

McShea and Raup (1986), Einsele et al. (1991b), and Ricken (1991) have been aware of the hierarchical nature of stratigraphic events, and the problem this poses for evaluating the correlation of events of very different time spans. Algeo (1993) proposed a method for estimating stratigraphic completeness based on preservation of magnetic reversal events. The method is most suitable for time intervals such as the Late Jurassic-Early Cretaceous, and the latest

Cretaceous-Present, during which time periods, reversal frequency was in the range 1–5 million years. At other times, reversal frequency was much lower, and the sensitivity of magnetostratigraphic methods correspondingly less.

Miall (1991b) developed a table which classifies depositional units into a hierarchy of physical scales, and compiled data on sedimentation rates for the units at the various scales (Table 13.1). It is important to note that sedimentation rates vary by up to an order of magnitude between one rank of this hierarchy and the next.

A simple illustration of the hierarchy of sedimentation rates, and the important consequences this has for correlation is shown in Fig. 13.4. Sedimentation rates used in this exercise and quoted below are based on the compilation of Miall (1991b). The total elapsed time for the succession of four sequences in this diagram is 1 million years, based on conventional geological dating methods, such as the use of biostratigraphy.

**Table 13.1** Hierarchies of architectural units in clastic deposits

| Group | Time scale of process (years) | Examples of processes | Instantaneous sed. rate (m/ka) | Fluvial, deltaic *Miall* | Eolian *Brookfield, Kocurek* | Coastal, estuarine *Allen[a] Nio and Yang[b]* | Shelf *Dott and Bourgeois,[c] Shurr[d]* | Submarine fan *Mutti & Normark* |
|---|---|---|---|---|---|---|---|---|
| 1. | $10^{-6}$ | Burst-sweep cycle | | Lamina | Grainflow grainfall | Lamina | Lamina | |
| 2. | $10^{-5}$–$10^{-4}$ | Bedform migration | $10^5$ | Ripple (microform) [1st-order surface] | Ripple | Ripple [E3 surface[a]] [A[b]] | [3-surface in HCS[c]] | |
| 3. | $10^{-3}$ | Diurnal tidal cycle | $10^5$ | Diurnal dune incr., react. surf. [1st-order surface] | Daily cycle [3rd-order surface | Tidal bundle [E2 surface[a]] [A[b]] | [2-surface in HCS[c]] | |
| 4. | $10^{-2}$–$10^{-1}$ | Neap-spring tidal cycle | $10^4$ | Dune (mesoform) [2nd-order surface] | Dune [3rd-order surface] | Neap-spring bundle, storm, layer [E2[a], B[b]] | HCS sequence [1-surface[c]] | |
| 5. | $10^0$–$10^1$ | Seasonal to 10 year flood | $10^{2-3}$ | Macroform growth increment [3rd-order surface] | Reactivation [2nd, 3rd-order] surfaces, annual cycle | Sand wave, major storm layer [E1[a], C[b]] | HCS sequence [1-surface[c]] | |
| 6. | $10^2$–$10^3$ | 100 year flood | $10^{2-3}$ | Macroform, e.g. point bar levee, splay [4th-order surface | Dune, draa [1st-, 2nd-order surfaces] | Sand wave field, washover fan [D[b]] | [facies pack-age (V)[d]] | Macroform [5] |
| 7. | $10^3$–$10^4$ | Long term geomorphic processes | $10^0$–$10^1$ | Channel, delta lobe [5th-order surface] | Draa, erg [1st-order, super surface] | Sand-ridge, barrier island, tidal channel [E[b]] | [elongate lens (IV)[d]] | Minor lobe, channel-levee [4] |
| 8. | $10^4$–$10^5$ | 5th-order (Milankovitch) cycles | $10^{-1}$ | Channel belt [6th-order surface] | Erg [super surface] | Sand-ridge field, c-u cycle [F[b]] | [regional lentil (III)[d]] | Major lobe [turb. stage: 3] |
| 9. | $10^5$–$10^6$ | 4th-order (Milankovitch) cycles | $10^{-1}$–$10^{-2}$ | Depo. system, alluvial fan, major delta | Erg [super surface] | c-u cycle [G[b]] | [ss sheet (II)[d]] | Depo. System [2] |
| 10. | $10^6$–$10^7$ | 3rd-order cycles | $10^{-1}$–$10^{-2}$ | Basin-fill complex | Basin-fill complex | Coastal-plain complex [H[b]] | [lithosome (I)[d]] | Fan complex [1] |

Hierarchical subdivisions of other authors are given in square brackets. Names of authors are at head of each column. [a,b,c,d]: terminology of authors indicated at head of columns. HCS: hummocky cross-stratification. c-u: coarsening-upward. Adapted from Miall (1991b).

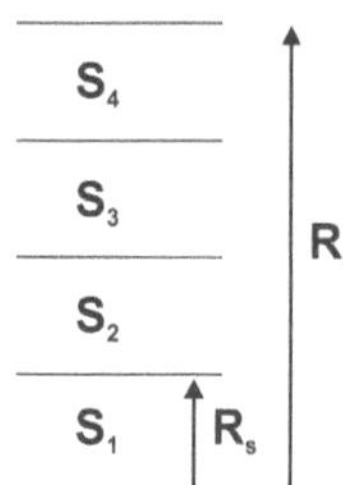



**Thickness of each sequence , S = 10 m**
**Total thickness of section = 40 m**
**If average sedimentation rate, R$_a$ = 0.04 m/ka**
**total elapsed time for the accumulation**
**of S$_1$ to S$_4$ = 1 m.y.**

**For individual sequences**
**R$_s$ = 0.1 m/ka**
**Elapsed time for one sequence = 100 ka**
**For all four sequences = 400 ka**

**Fig. 13.4** Comparison of sedimentation rates measured at two different scales. The typical range of sedimentation rates for long-term sedimentation at a geomorphic scale, comparable to that which can be measured for many fifth-order and some fourth-order stratigraphic sequences, is 0.1–0.01 m/ka (groups 8 and 9 of Table 13.1). Longer-term sedimentation rates, such as those estimated from geological, chronostratigraphic data, are in the order of 0.01 m/ka (group 10, 3rd-order cycles, Table 13.1). Calculations of total elapsed time using these two rates give results that are an order of magnitude different, indicating considerable "missing" time corresponding to nondeposition and erosion

However, in modern environments where similar sedimentary successions are accumulating, such as on prograding shorelines, short-term sedimentation rates typically are much higher. Elapsed time calculated on the basis of continuous sedimentation of an individual sequence amounts to considerably less (total of 400 ka for the four sequences), indicating a significant amount of "missing" time (600 ka). This time is represented by the sedimentary breaks between the sequences. Algeo and Wilkinson (1988) concluded, following a similar discussion of sedimentation rates, that in most stratigraphic sections, only about one thirtieth of elapsed time is represented by sediment. Even in the deep oceans it has been demonstrated that in some cases "there is almost as much geological time represented by unconformities as there is time represented by sediments"(Aubry, 1995).

As discussed below, a further analysis could take into account the rapid sedimentation of individual

subenvironments within the shoreline (tidal channels, beaches, washover fans, etc.), and this would demonstrate the presence of missing time at a smaller scale, within the 100 ka represented by each sequence (e.g., Swift and Thorne, 1991, Fig. 16, p. 23). Thus, facies successions ("parasequences", in the Exxon terminology; e.g., Van Wagoner et al., 1990) that constitute the components of sequences, such as delta lobes and regressive beaches, represent $10^3$–$10^4$ years and have short-term sedimentation rates up to an order of magnitude higher than high-frequency sequences, in the 1–10 m/ka range.

The important point to emerge from this simple exercise is that the sedimentary breaks between supposedly continuous successions may represent significant lengths of time, much longer than is suggested by calculations of long-term sedimentation rates. This opens the possibility that the sequences that were deposited during the time span between the breaks may not actually correlate in time at all. Physical tracing of sequences by use of marker horizons, mapping of erosion surfaces and sequence boundaries, etc., may confirm the existence of a regional sequence framework, but these sequences could, in principal, be markedly diachronous, and correlation among basins, where no such physical tracing is possible, should be viewed with extreme caution.

Figure 13.5 makes the same point regarding missing time in a different way. Many detailed chronostratigraphic compilations have shown that marine stratigraphic successions commonly consist of intervals of "continuous" section representing up to a few million years of sedimentation, separated by disconformities spanning a few-hundred-thousand years to more than one-million years (e.g., MacLeod and Keller, 1991, Fig. 15; Aubry, 1991, Fig. 6). The first column of Fig. 13.5, labelled MC (for cycles in the million-year range), illustrates an example of such a succession. Detailed studies of such cycles demonstrate that only a fraction of elapsed time is represented by sediment. For example, Crampton et al. (2006) demonstrated that on the million-year scale, an average of 24% of time is recorded in a suite of Upper Cretaceous sections in New Zealand, whereas in a suite of drill cores through the Lower Cretaceous to Miocene stratigraphic record of New Jersey, the plots of Browning et al. (2008) show that about 82% of elapsed time is represented by sediments, although some sections are more complete than others. Each

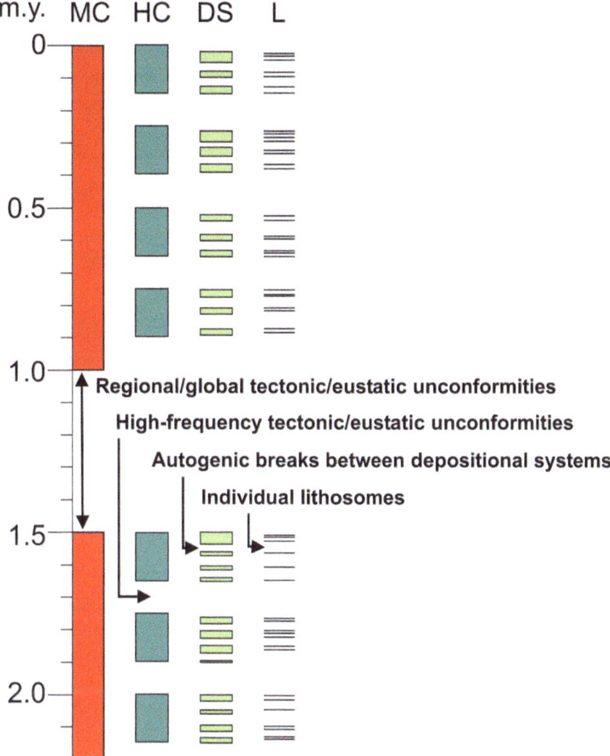


**Fig. 13.5** A demonstration of the predominance of missing time in the sedimentary record. Two cycles with frequencies in the million-year range are plotted on a chronostratigraphic scale (column MC), and successively broken down into components that reflect an increasingly fine scale of chronostratigraphic subdivision. The second column shows hundred-thousand-year cycles (HC), followed by depositional systems (DS) and individual lithosomes (L), such as channels, deltas, beaches, etc. At this scale chronostratigraphic subdivision is at the limit of line thickness, and is therefore generalized, but does not represent the limit of subdivision that should be indicated, based on the control of deposition by events of shorter duration and recurrence interval (e.g., infrequent hurricanes, seasonal dynamic events, etc.)

million-year cycle may be composed of a suite of high-frequency cycles, such as those in the hundred-thousand-year range, labelled HC in Fig. 13.5. Detailed chronostratigraphic compilations for many successions demonstrate that the hiatuses between the cycles represent as much or more missing time than is recorded by actual sediment (e.g., Ramsbottom, 1979; Heckel, 1986; Kamp and Turner, 1990; for example, see Figs. 6.5, 6.6, 6.37, 6.38, 7.17, 11.18). Sedimentation rates calculated for such sequences (Table 13.1) confirm this, and the second column of Fig. 13.5 indicates a possible chronostratigraphic breakdown of the third-order cycles into component Milankovitch-band cycles, which may similarly represent incomplete preservation. For example, a detailed chronostratigraphic correlation of the coastal Wanganui Basin sequences in new Zealand ("5th-order Milankovitch cycles of "group 8 in Table 13.1) shows that at the $10^5$-year time scale only 47% of elapsed time is represented by sediments (Kamp and Turner, 1990). Each Milankovitch cycle consists of superimposed depositional systems (column DS) such as delta or barrier-strandplain complexes, and each of these, in turn, is made up of individual lithosomes (column L), including fluvial and tidal channels, beaches, delta lobes, etc. According to the hierarchical breakdown of Table 13.1, the four columns in Fig. 13.5 correspond to sediment groups 10, 9, 8, and 7, in order from left to right. In each case, moving (from left to right) to a smaller scale of depositional unit focuses attention on a finer scale of depositional subdivision, including contained discontinuities. The evidence clearly confirms Ager's (1981) assertion that the sedimentary record consists of "more gap than record".

Devine's (1991) lithostratigraphic and chronostratigraphic model of a typical marginal-marine sequence

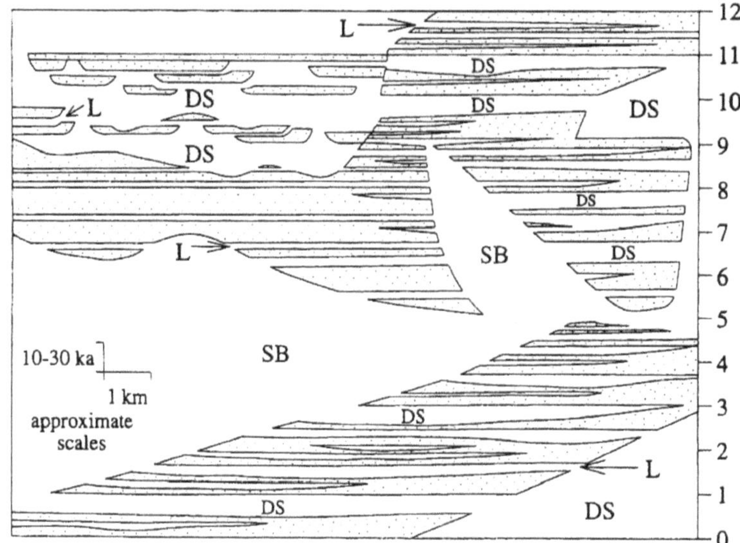



**Fig. 13.6** A redrawing of the sequence model of Devine (1991) to emphasize the gaps in the stratigraphic record. The most significant break in sedimentation, labelled SB, represents a minor local disconformity between parasequences in Devine's original work. In larger-scale systems this would be the position of the sequence boundary, and would correspond to the kind of gap shown in column HC of Fig. 13.5. Intermediate-scale gaps are those between individual depositional systems (DS), and correspond to some of the time-line "events" in Devine's original model. The smallest gaps (L) are those between individual lithosomes, caused by storm erosion, unusually low tides, etc. Only a few of these are labelled

demonstrates the importance of missing time at the sequence boundary (his subaerial hiatus). Shorter breaks in his model, such as the estuarine scours, correspond to breaks between depositional systems, but more are present in such a succession than Devine (1991) has indicated. His chronostratigraphic diagram is redrawn in Fig. 13.6 to emphasize sedimentary breaks, and numerous additional discontinuities have been indicated, corresponding to the types of breaks in the record introduced by switches in depositional systems, channel avulsions, storms and hurricanes, etc. Cartwright et al. (1993) made a similar point regarding the complexity of the preserved record, particularly in marginal-marine deposits, and commented on the difficulty of "forcing-through" meaningful stratigraphic correlations using seismic-reflection data.

The log-linear relationship between sedimentation rates and time span was addressed by Plotnick (1986), who demonstrated that it could be interpreted using the fractal "Cantor bar" model of Mandelbrot (1983). Using the process described by Mandlebrot (1983) as "curdling", Plotnick (1986) developed a Cantor bar for a hypothetical stratigraphic section by successively emplacing hiatuses within portions of the section, at ever increasing levels of detail. The result is illustrated in Fig. 13.7:

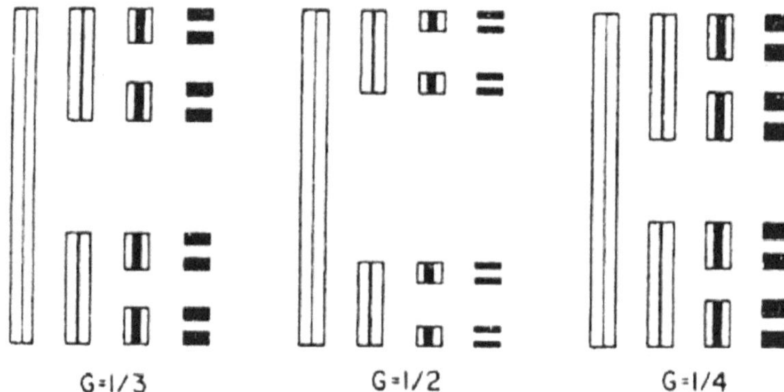



**Fig. 13.7** Cantor bars generated using three different gap sizes (G) (Plotnick, 1986, Fig. 1)

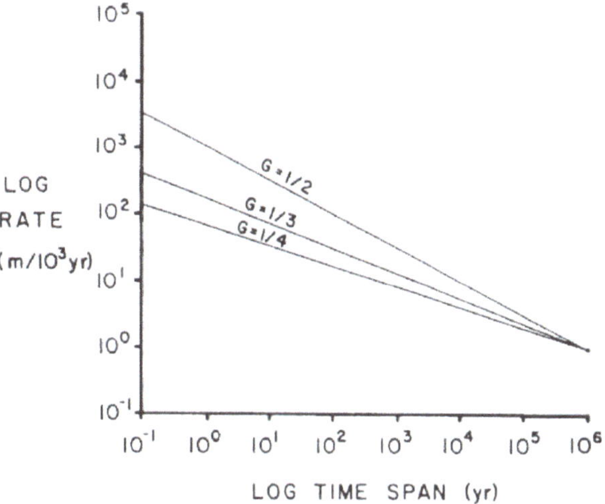

**Fig. 13.8** Relationship between sedimentation rate and time span of observation for different values of G, assuming that the rates are the same ($1 \text{ m}/10^3$ years) at a time span of $10^6$ years (Plotnick, 1986, Fig. 2)



Assume a sedimentary pile 1000 m thick, deposited over a total interval of 1,000,000 years. The measured sedimentation rate for the entire pile is, therefore, 1 m/1000 years. Now assume that a recognizable hiatus exists exactly in the middle of the section, corresponding to a third of the total time (i.e., 333,333 years). The subpiles above and below the hiatus each contain 500 m of sediment, each deposited over 333,333 years, so that the measured sedimentation rate for each subpile is 1.5 m/1000 years. We now repeat the process, introducing hiatuses of 111,111 years in each of the two subpiles. This produces 4 subpiles, each 250 m thick, each with a duration of 111,111 years. The measured sedimentation rate is now 2.25 m/1000 years. The process can be reiterated endlessly, producing subpiles representing progressively shorter periods of time with higher sedimentation rates (fig. la). Nevertheless, because the total sediment thickness is conserved at each step, the sedimentation rate of the entire pile remains 1 m/l000 years (Plotnick, 1986, p. 885).

The selection of one third as the length of the hiatus, or gap (G in Fig. 13.7) is arbitrary. Other gap lengths generate Cantor bars that differ only in detail. Figure 13.7 is remarkably similar to Fig. 13.5, which was constructed for the first edition of this book, based on the hierarchies of sedimentation rates compiled by Miall (1991b), but with no knowledge of fractals. Figure 13.8 illustrates the consequence of varying the value of G, and explains the clustering of the data points in the data set (Fig. 13.3) compiled by Sadler (1981).

The dependence of sedimentary accumulation on the availability of accommodation was understood by Barrell (1917; Fig. 1.3 of this book) and is the basis of modern sequence stratigraphy (Van Wagoner et al.,

1990; Fig. 2.3 of this book). The fractal model provides an elegant basis for integrating this knowledge with the data on varying sedimentation rates and varying scales of hiatuses discussed in the paragraphs above. Mandelbrot (1983) and Plotnick (1986) provided a version of an accumulation graph, called a Devil's staircase (Fig. 13.9). This shows how sediments accumulate as a series of clusters of varying lengths. Vertical increments of the graph correspond to intervals of sedimentation; horizontal plateaus represent periods of non-accumulation (or sedimentation removed by erosion). Sequence stratigraphy is

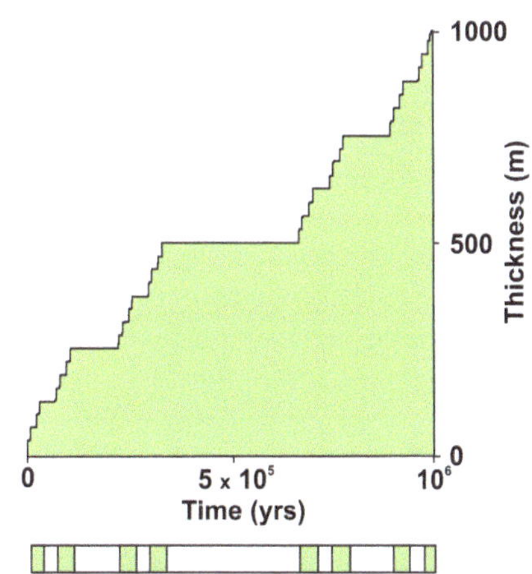



**Fig. 13.9** Stratigraphic thickness accumulation viewed as a Cantor function—what has been termed a Devils's staircase—constructed with $G = \frac{1}{2}$. The corresponding Cantor bar is shown at a lower resolution below the graph (adapted from Plotnick, 1986, Fig. 5)

essentially a study of the repetitive cycle of accumulation followed by gap, at various scales. The larger, more obvious gaps (the longer plateaus in Fig. 13.9) define for us the major sequences, over a range of time scales. The prominence of particular ranges of "accumulation+gap" length in the first data sets compiled by Vail et al. (1977) was what led to the establishment of the sequence hierarchy of first-, second-order, and so on. That this has now been shown to be an incomplete representation of nature (Fig. 4.3) does not alter the fact that there is a limited range of processes that control accumulation, and these have fairly well defined rates which, nevertheless, overlap in time to some extent.

As Plotnick (1986, Table 1) demonstrated, the incompleteness of the stratigraphic record depends on the scale at which that record is examined. Sections spanning several million years may only represent as little as 10% of elapsed time at the 1,000-year measurement scale, although this is below the resolution normally obtainable in geological data. Sequences, as we know them, are essentially clusters of the shorter "accumulation+gap" intervals separated by the longer gaps—those more readily recognizable from geological data.

How does the fragmentary nature of preserved time affect the value and meaning of geological time scales? A time scale that focuses on continuity is to be preferred over one that is built on unconformities. The modern method of refining the geological time scale, uses "continuous" sections for the definition of chronostratigraphic boundaries and encompasses a method for the incorporation of missing time by defining only the base of chronostratigraphic units, not their tops (Miall, 1999, Fig. 3.7.3). If missing time is subsequently documented in the stratigraphic record by careful chronostratigraphic interpretation it is assigned to the underlying chronostratigraphic unit, thereby avoiding the need for a redefinition of the unit (Ager, 1964; McLaren, 1970; Bassett, 1985).

In the Exxon work much use is made of the term "correlative conformities", as in their original definition of a sequence as "a relatively conformable succession of genetically related strata bounded at its top and base by unconformities or their correlative conformities" (Vail et al., 1977, p. 210). However, sequence boundaries are diachronous. The transgression and regression that constitute a sea-level cycle generate breaks in sedimentation at different times in different

parts of a basin margin. Christie-Blick et al. (2007, p. 222) noted that "at some scale, unconformities pass laterally not into correlative conformities, but into correlative intervals. Such considerations begin to be important as the resolution of the geological timescale improves at a global scale." Kidwell (1988) demonstrated that the major break in sedimentation on the open shelf occurs as a result of erosion during lowstand and sediment bypass or starvation during transgression, whereas in marginal-marine environments the major break occurs during regression. The sequence-boundary unconformities are, therefore, offset by as much as a half cycle between basin-margin and basin-centre locations. Kidwell (1988) referred to this process as reciprocal sedimentation. Variations in the balance among subsidence rate, sediment supply and sea-level change also generate substantial diachronism in sequence boundaries.

Given the diachronous nature of sequence boundaries, the accuracy of sequence correlation could be improved by dating of the correlative conformities in deep-marine settings, where breaks in sedimentation are likely to be at a minimum, but it is doubtful if this is often possible. It requires that sequences be physically traced from basin margins into deep-water environments, introducing problems of physical correlation in areas of limited data, and problems of chronostratigraphic correlation between different sedimentary environments in which zonal assemblages are likely to be of different type. As discussed in Chap. 12, problems of correlation across environmental and faunal-province boundaries are often significant. There is also a theoretical debate regarding the extension of subaerial erosion surfaces into the shallow-marine environment (Chap. 2). In view of the ubiquity of breaks in sedimentation in the stratigraphic record it is arguable whether, in fact, the concept of the correlative conformity is realistic.

The significance of missing section and the ambiguity surrounding the correlation of unconformities is discussed further in Chap. 14.

Correlation of sequence frameworks that cannot be related to each other by physical tracing must be based on methods that provide an adequate degree of accuracy and precision. What is required is correlation to a precision corresponding to a fraction of the smallest possible time span of an individual sequence, at the scale under consideration. Cycles of $10^7$–$10^8$-year duration require dating with an accuracy to within a few millions of years. This is not a difficult requirement, at least within marine rocks of Phanerozoic age, and a global framework of second-order cycles is within our grasp. In fact, Sloss began the process of establishing this framework many years ago, as described in Sect. 1.4. Whether the Exxon framework of "second-order" sequences is to be considered reliable as indicators of global control is another question, which is addressed in Sect. 9.3.3.

Cycles with million-year periodicities require a precision of $\pm0.5$ million years, or better for reliable correlation. As is demonstrated in Chap. 14, conventional geological techniques, such as biostratigraphy calibrated with radiometric, chemostratigraphic, and/or magnetostratigraphic data, can rarely attain this degree of precision (except in the youngest Cenozoic), and it is, therefore, highly questionable whether the data currently permit us to attempt the erection of an interbasinal framework of "third-order" cycles. In special cases, use of high-resolution correlation techniques, including graphic correlation, permit precision to within about $\pm100$ ka in rocks of early Cenozoic and Mesozoic age (Kauffman et al., 1991). In such cases, a framework of regional correlation can be built with some degree of confidence. However, it is doubtful if we are anywhere near ready to extend such frameworks to other basins on different continents, in different tectonic settings, and in different climatic zones.

High-frequency sequences (those with periodicities in the Milankovitch band) cannot yet be reliably correlated interregionally, except for the glacioeustatic sequences of the late Cenozoic, as documented in Sects. 11.3.2 and 14.6.3. For this most recent interval of geological time rapid changes in sea level have climatic causes which have left a reliable chronostratigraphic signature in the form of the oxygen-isotope record.

One of the major problems with the global-eustasy paradigm is circular reasoning in placing undue emphasis on significant "events" that are presumed to be synchronous in different areas. This presumption causes difficulties in performing meaningful tests of the global cycle chart, as shown in Chap. 14. Secondly, there is no reason why changes in sea level, which are the events defining stratigraphic sequences, should show any relationship to zone and stage boundaries that are based on biological evolution. Changes in sea level affect the ecology of marine environments, but the relationship of such changes to biotic evolution and extinction is complex. We should not, therefore, expect any relationship between sequence boundaries and biostratigraphically-based boundaries in the standard geological time scale.

## 13.3 Main Conclusions

1. Only a small fraction of elapsed time normally is represented by sediment preserved in the geological record.
2. The hierarchy of sequences represent time spans and sedimentation rates that differ from each other by orders of magnitude.
3. Accurate and precise estimates of time spans of the component sequences and hiatuses in a succession cannot be achieved by determining bracketing ages from stratigraphically widely spaced marker beds that enclose several or many sequences.
4. Correlation of sequences requires that they be dated with a precision, or potential error range, equal to a fraction of the range of the shortest sequence in the succession. For example, sequences with $10^6$-year periodicities (those comprising the main basis of the Exxon global cycle chart) require a correlation precision of $\pm0.5$ million years, or less.

# Chapter 14

# Chronostratigraphy, Correlation, and Modern Tests for Global Eustasy

## Contents



## 14.1 Introduction

As discussed in Chap. 12, sequence stratigraphy is currently characterized by two alternative, opposing paradigms, the *global-eustasy paradigm*, and the *complexity paradigm*. McGowran (2005, p. 365) has pointed out that these are conceptual end-members of a "broad spectrum of responses and approaches to the controversy over global eustasy." McGowran (2005, p. 366) agreed that such markers as sequence boundaries and maximum flooding surfaces "should not be intrinsic to geochronology," *contra* Vail, but he stated "I do not share the Miall-Wilson view that sequence-surfaces have no promise in correlation."

Here lies the crux of the problem. What do sequence surfaces signify? One of the major purposes of this book has been to demonstrate the complex, multifaceted nature of the sequence record. Sequences have multiple causes, only two of which, (1) changing rates of sea-floor spreading, affecting ocean-volumes (Chap. 9), and (2) glacioeustasy (Chap. 11), can be convincingly demonstrated to generate global effects on the sedimentary record that potentially may be correlated and become contributors to the global stratigraphic data base. Other sequences are strictly local to regional in extent. The problem is that sequence-generating processes are scale independent (e.g., see Catuneanu, 2006), and whether they are controlled by eustatic sea-level change or tectonic changes in accommodation, such locally observable attributes as facies and thickness do not necessarily provide any clues as to ultimate allogenic mechanisms. Even the key attribute of sequence thickness has been demonstrated to carry no useful signal of causation (Algeo and Wilkinson, 1988; Drummond and Wilkinson, 1996), and Schlager (2004) showed that distribution of sequence durations is fractal in character, indicating that the sequence hierarchy concept of Vail et al. (1977) cannot be supported. Schwarzacher (2000, p. 52) stated that

A.D. Miall, *The Geology of Stratigraphic Sequences*, 2nd ed.,
DOI 10.1007/978-3-642-05027-5_14, © Springer-Verlag Berlin Heidelberg 2010

This system of ordering cycles is widely used but has serious shortcomings. By pretending that there exists a hierarchy of cycles, repetitive patterns, which have nothing in common except their duration, are mixed together. Such a classification is just as meaningless as grouping elephants and fleas into an order based on their size.

The most important single test that can be performed on a suite of sequences is correlation. In the first place, the reality of a succession of sequences must be established by the establishment of local stratigraphic relationships, in order to distinguish facies successions generated by local autogenic causes, such as delta switching and abandonment, from those of true allogenic origin (this is the parasequence problem discussed in Sect. 2.2.2). Sequences that can be correlated beyond a particular basin into an adjacent basin, into a different tectonic province, or across an ocean to another tectonic plate, are clearly the product of a widespread process, and only eustatic sea-level change seems capable of generating global changes in accommodation that are synchronous and in the same direction, although, as illustrated by Fig. 10.1, and the discussion in Sect. 10.1, where a curve of eustatic sea-level change is integrated with varying rates of subsidence, the timing of sea-level highs and lows may be offset by as much as one quarter of the wavelength of the sea-level cycle (Allen and Allen, 2005, p. 271). As discussed in Sect. 10.4, intraplate stress changes may be capable of generating tectonic events that are synchronous but of locally different character, but this remains a partially tested hypothesis.

The test of global correlation is, therefore, fundamental to the issue of sequence mechanism. Until there are multiple, repeated demonstrations of global correlation, the ubiquity and central importance of eustasy as a stratigraphic control must remain a largely unproven hypothesis. But here the problem of the hermeneutic circle arises, and here lies the methodological divide between the inductive and deductive approaches to stratigraphy.

McGowran (2005, p. xvi), citing Popper, argued:

As Sir Karl Popper advocated tirelessly, the inductivist model of science, whereby we collect data and draw conclusions from it, is not very useful or stimulating or realistic. No: we always have theories and biases influencing our choice of observations and our perceptions of 'facts', and biostratigraphy is no different. We do not simply identify, range, zone and correlate in tedious induction – for we have stubborn, crazy, powerful and pet ideas about tectonism, climate change, or evolutionary relationships whose triumph or tragedy can depend utterly on the fragility of robustness of the chronology or correlation.

The problem is that pet ideas can introduce bias. This is what Vail and Todd (1981, p. 230) were about when they stated "the late Pliensbachian hiatus described by Linsley and others (1979) fits the basal early Pliensbachian sequence boundary on our global cycle chart." (see Sect. 12.1). To allow such adjustments to the biostratigraphic record is to render inoperable the concept of an independent test of the hypothesis. In Sect. 12.1 I point out the obvious dangers of circular reasoning involved in the deductivist approach to sequence stratigraphy. As McGowran (2005, p. 190) noted, it is a "forbiddingly difficult task ... [to] disentangle the triad of processes controlling the stratigraphic record on the continental margin: namely eustasy, tectonics (including thermal subsidence, isostasy, compaction, flexure), and sediment supply." The last thing that needs to be added as a component of this task is bias.

In the next section we illustrate this problem further with reference to two examples.

## 14.2 Chronostratigraphic Models and the Testing of Correlations

Adherents of the global-eustasy paradigm believe that sequence stratigraphy may be used as "an instrument of geochronology" (Vail et al., 1977, p. 96). The defining feature of the model is the belief that sequence boundaries are global chronostratigraphic indicators, and that their ages are not influenced by the tectonic behaviour of sedimentary basins. From this first principle follows the use of pattern recognition as a means of correlation. In correlation charts constructed with time as the ordinate, correlation lines will always be parallel ("railroad correlation lines")—this was one of the distinguishing characteristics of the first published chart to show the relationship of regional cycle charts to each other and to the derived global chart (Fig. 12.2). The following discussion of chronostratigraphic models is taken largely from Miall and Miall (2004).

Despite the considerable body of critical discussion about the global cycle chart that has appeared since the mid-1980s, adherents of the global-eustasy model continue to practice a science that emphasizes the predominant importance of this deductive model

(Sect. 12.2). A typical example is a recent study of the correlations between Cenomanian (Upper Cretaceous) sections in the Anglo-Paris Basin and in southeast India. This work specifically set out "to demonstrate that sea-level changes are globally synchronous and therefore must be eustatically controlled" (Gale et al., 2002, p. 291). The key data diagram in this short paper is a chart showing the relationships between sequence and systems tracts in the two basins (Fig. 14.1). The ordinate in this diagram is an arbitrary scale which assigns each sequence equal space. Sequence boundaries, and even systems tracts, can, therefore, be correlated between the two basins, which are on opposite sides of the world, using parallel "railroad-line" correlation lines. There is no indication that the sections in the two basins are of very different thicknesses (which can be deduced from a derived sea-level curve included in the paper), nor is there any suggestion in the diagram that the sequences might represent varying time intervals or that they may include significant disconformities. Indeed, it is claimed that the sequences correspond to the 400-ka Milankovitch-band eccentricity

cycles, identified in a different study of another basin by one of the authors, based on spectral analysis of grayscale reflectance data of chalk. The only actual empirical data provided in this paper are the ranges of key ammonites, shown relative to the sequence and systems-tract boundaries, and they reveal some discrepancies in the sequence correlations.

As Kuhn (1962) has observed, "Results which confirm already accepted theories are paid attention to, while disconfirming results are ignored. Knowing what results should be expected from research, scientists may be able to devise techniques that obtain them." In the authors' eyes, therefore, their correlation diagram (Gale et al., 2002, Fig. 3) serves the purpose of a "challenge successfully met" (Kuhn, 1996, p. 204), the challenge of correlating sequences of events in widely spaced basins having no tectonic relationship to each other. As Kuhn (1996, p. 205) also noted, "The demonstrated ability to set up and to solve puzzles presented by nature is, in case of value conflict, the dominant criterion for most members of a scientific group." The work of Gale et al. (2002) appears to present a "puzzle"

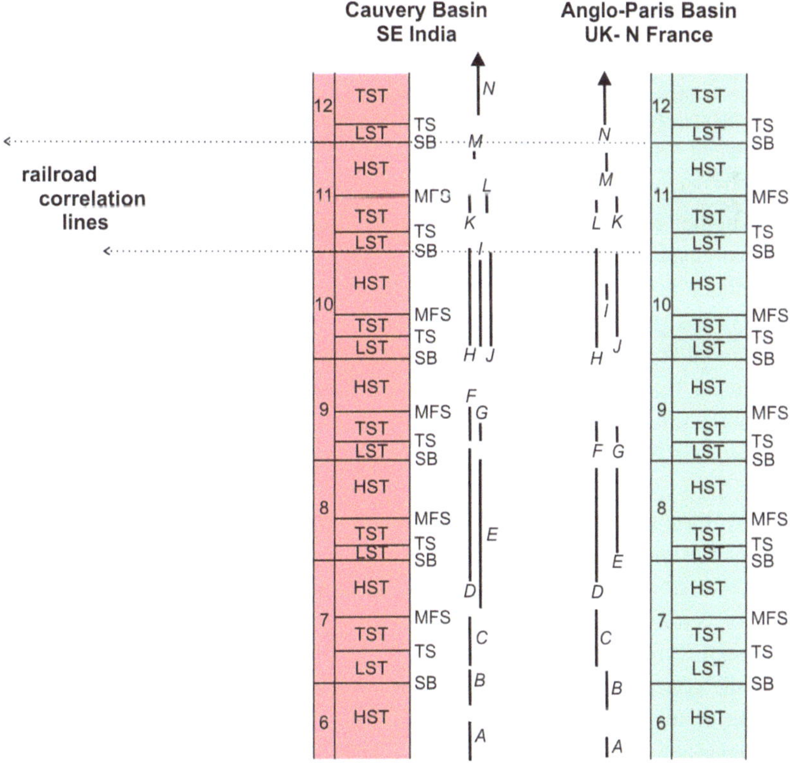


**Fig. 14.1** Correlations between Cenomanian sections in southeast India and the Anglo-Paris Basin, plotted on an arbitrary ordinate. Note the exact parallelism of all correlation lines between key surfaces (sequence and systems tract boundaries) throughout this diagram (adapted from Gale et al., 2002)

to those who doubt the reality of global eustasy. But does it? Perhaps it all lies in how the data are presented.

We cite here a different opinion, based on a different kind of test of sequence correlations: Prothero (2001) carried out a series of correlation tests on rocks of Paleogene age in California. Amongst his conclusions:

> Sequence stratigraphic methods are now routinely applied to the correlation of strata in a wide variety of depositional settings. In many cases the sequence boundaries are correlated to the global cycle chart of Haq et al., (1987, 1988) without further testing by biostratigraphy or other chronostratigraphic techniques. Emery and Myers (1996, p. 89) noted that "sequence stratigraphy has now largely superceded [sic] biostratigraphy as the primary correlative tool in subsurface basin analysis." The last decade of layoffs of biostratigraphers from most major oil companies also seems to indicate that some geologists think they can get along fine without biostratigraphic data. Where the biostratigraphic data are very low in resolution or highly facies-controlled, perhaps sequence stratigraphic correlations work better.
>
> But if there is any lesson that two centuries of geological investigation since the days of William Smith have taught us, it is that *biostratigraphy is the ultimate arbiter of chronostratigraphic correlations.* The literature is full of lithostratigraphic correlation schemes that have failed because of insufficient attention to biostratigraphy. ... . Blindly correlating stratigraphic events to the outdated onlap-offlap curve of Haq et al. (1987, 1988a) without determining whether biostratigraphic data support their correlations, continues the trend of poor science.

Another recent example of the application of the global-eustasy paradigm, in which every sequence boundary is assumed to be of eustatic origin (Sect. 12.2), was provided by Haq and Schutter (2008), as discussed at the conclusion of Sect. 12.5. The lack of technical rigour in the dating and correlation of the events in this chart stands in stark contrast to the modern work in New Jersey and New Zealand, discussed in Sect. 14.6.

We turn now to another study of high-resolution ammonite biostratigraphy, one that tells a completely different story. The Jurassic strata of western Europe are of central importance in the history of Geology. Williams Smith receives most of the credit for inventing the concept of the geological map with his early work in the Jurassic outcrops near Bath, England (Winchester, 2001). Gressly (1838) developed his ideas about facies working on the Jurassic strata of the Jura Mountains. Concepts of the stage, the zone and the hemera were all worked out from detailed biostratigraphic study of Jurassic strata in England,

France and Germany (Sect. 1.3). As a result, there is an immense body of knowledge available dealing with the biostratigraphy of these rocks. We refer here to a single study, one by Callomon (1995), who returned to the very detailed work of S. S. Buckman and R. Brinkmann during the first decades of the twentieth century on the Middle Jurassic Inferior Oolite formation of southern England (See Sect. 1.3.2).

Building on this early work, Callomon (1995) identified fifty-six faunal horizons based on ammonites in the Inferior Oolite, a shallow-marine limestone succession some 5 m in thickness, spanning the Aalenian and Bajocian stages. Callomon's calculations demonstrate that these horizons average 140 ka in duration, but without any implication that these horizons might be equally spaced. Sedimentological study shows the succession to contain numerous scour surfaces and erosion surfaces, but only the detailed biostratigraphic data could have revealed the complexity of the stratigraphic picture that Callomon presented. A detailed comparison among thirteen sections through this short interval, spaced out along an 80-km transect across Somerset and Dorset, in southern England, shows that each section is different in almost every detail (Fig. 14.2). Approximately half of the faunal horizons are missing in each section, and it is a different suite of missing events in each case. None of the horizons is present in all of the thirteen sections. Callomon (1995, Table 5) demonstrated that on average the sections are only 43% complete.

> The 'gaps' united by thin bands of 'deposit' are evident. The time durations that left no record ... or whose record has been destroyed ... are often greater than the time intervals ... between the biochrons of adjacent faunal horizons. What is less evident, however, is any coherent relationship between the lengths of the gaps and their positions, such as might be explicable by simple sequence stratigraphy—and this across a distance of only 80 km in a single basin (Callomon, 1995, p. 140).

Note that Fig. 14.2 also uses an arbitrary ordinate, but in this case an argument can be made that it probably represents an insignificant distortion of the original empirical data. Callomon's paper includes samples of three of the actual stratigraphic sections, plotted with a thickness ordinate, which show that the major biozone increments are indeed of comparable thickness (Callomon, 1995, Fig. 4). Callomon's paper is, in fact, rigorously empirical. The interpretation of continuous, if fragmentary, successions in Fig. 14.3 has been added for the purpose of this book to emphasize

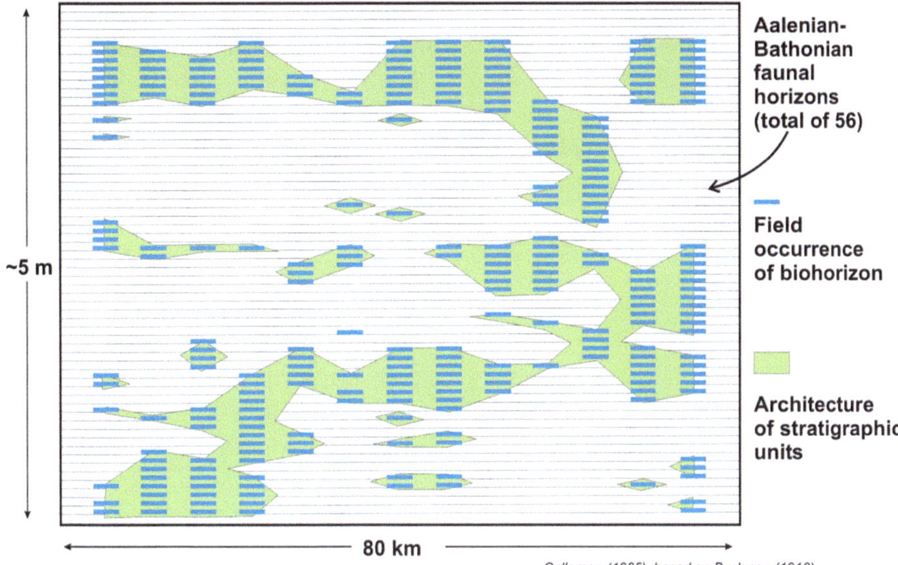



**Fig. 14.2** The fifty-six ammonite faunal horizons in the Inferior Oolite of southern England, simplified from Callomon (1995, Fig. 5), shown for simplicity as *horizontal lines*. The presence of each horizon in each of the sections is shown as a *rectangular box*, and the possible occurrence of conformable units within and between the sample sections is indicated by the *shaded-in areas* (my addition)

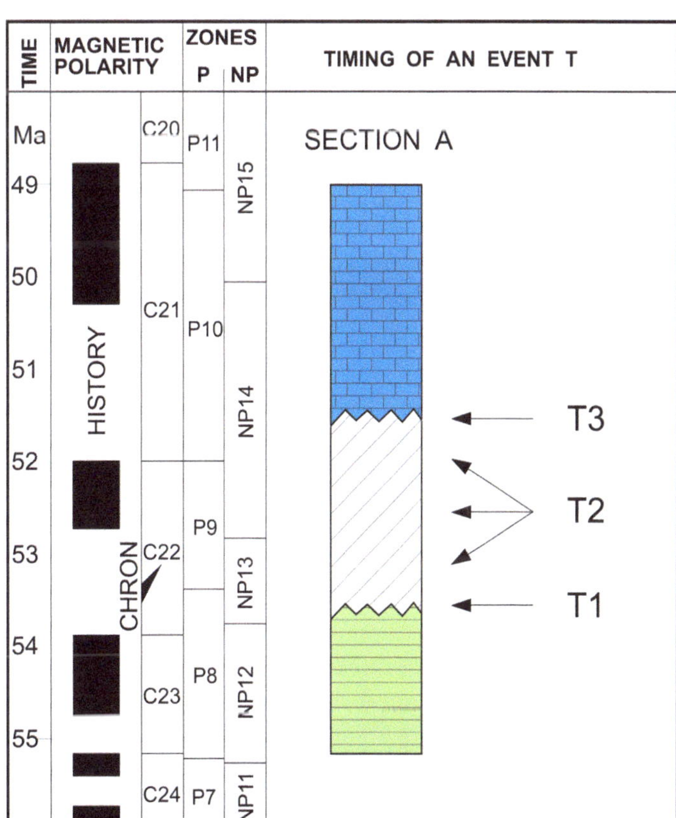



**Fig. 14.3** The interpretation of unconformities. The timing of the event could have been T1, T2, T3 or a longer episode which overlapped one or more of these times (Aubry, 1991)

the patchiness of the geological record, a point to which we return below. An interpretation of this stratigraphic pattern would seem to require calling on a variety of simultaneous, interacting processes, such as slight tectonic movements, variations in sediment supply and in the strength of marine currents, as well as possible changes in sea-level. In other words, this example of the stratigraphic record would seem to be a good example to bring forth in support of what we have called the *complexity paradigm* (Miall and Miall, 2001).

The differences between the Gale et al. (2002) study and that by Callomon (1995) cannot be attributed to differences in geology, such as differences in the depositional environments of the rocks. Undoubtedly, the Middle Jurassic succession of southern England is replete with erosion surfaces, but so are the sections studied by Gale et al. (2002). These authors based their sequence definitions on the recognition of erosion surfaces, which they classified as sequence boundaries, and on facies changes from shelf to non-marine deposits, upon which they based their definitions of systems tracts. Even successions that are entirely marine have typically been found to contain numerous breaks in the record, and complex patterns of non-correlating disconformities and diastems when subjected to detailed biostratigraphic study (e.g., Aubry, 1991, 1995). The presence of numerous diastems even in deep-marine deposits has long been known (Fischer and Arthur, 1977).

Clearly, the two studies can each be assigned to one of the two paradigms that we developed in Miall and Miall (2001): Gale et al. (2002) to the global-eustasy paradigm and Callomon (1995) to the complexity paradigm (Miall and Miall, 2004, p. 32). We suggested that the differences between the two studies have less to do with the geology described by the authors than the way the data have been presented, based on what Stewart (1986) has termed the "Interests perspective." Building on Kuhn (1962), Stewart (1986, p. 262) suggested that when there is competition between two or more scientific theories or paradigms "the key process determining choices between paradigms is persuasion based on widely shared values, such as quantitative predictions, accuracy of results, simplicity, and scope." The choice between paradigms "is not determined by which one can explain the most 'facts'—for what is accepted as a fact depends to a large degree on one's accepted paradigm." The choice between paradigms

can also represent "desires to protect the basis of one's previous intellectual contributions" (Stewart, 1986, p. 263). For example, one of the six authors of the Gale et al. paper, is also a co-author of several of the papers we classify in our earlier work (Miall and Miall, 2001) as representing the global-eustasy paradigm. As we noted elsewhere (Miall and Miall, 2002, p. 322), the authors who accepted and used the global-eustasy model and the global cycle chart as unproblematic premises for their own research, gave social support to these practices and thereby, contributed to "transforming them into 'facts of measurement and effect estimation'" (Fuchs, 1992, p. 50; Latour, 1987). Further, they contributed to the transformation of these "facts" into an unproblematic *black box* and "an unquestioned foundation for subsequent scientific work" (Fuchs, 1992, p. 48). As Golinski (1998, p. 140) has summarized, "when an instrument ... assumes the status of an accepted means of producing valid phenomena, then it can be said to have become a 'black box.'" We return to the concept of the black box below.

One of the key differences between the two studies examined thus far (Gale et al., 2002; Callomon, 1995) is the difference in the way the authors treat missing time. One of Callomon's main results was the demonstration of the extremely fragmentary nature of the preserved record of the Inferior Oolite in southern England. By contrast, the implication of the Gale et al. (2002) study is that sedimentation is virtually continuous.

Given the outcome of this exercise in comparative biostratigraphy it would seem clear that what is required is a rigorously defined, objective chronostratigraphic scale against which sequence correlations may be tested. One of the major purposes of this chapter is to explain how the modern chronostratigraphic time scale is being assembled and used to test stratigraphic concepts, and the strengths and limitations of this body of work as it affects the practice of sequence stratigraphy.

## 14.3 Chronostratigraphic Meaning of Unconformities

Sequence boundaries are unconformities, but assigning an age to an unconformity surface is not necessarily a simple matter, as illustrated in the useful theoretical discussion by Aubry (1991). An unconformity represents a finite time span at any one location; it may have a complex genesis, representing amalgamation of more than one event. It may also be markedly diachronous, because the transgressions and regressions that occur during the genesis of a stratigraphic sequence could span the entire duration of the sequence. As noted in Sect. 13.1, Kidwell (1988) demonstrated that this results in an offset in sequence-boundary unconformities by as much as one half of a cycle between basin centre and basin margin.

Unconformities represent amalgamations of two surfaces, the surface of truncation of older strata and the surface of transgression of younger strata. These two surfaces may vary in age considerably from location to location, as indicated by chronostratigraphic diagrams (e.g., Vail et al., 1977, Fig. 13, p. 78). Even if the two surfaces have been dated, this still does not provide an accurate estimate of the timing of the event or events that generated the unconformity. As shown in Fig. 14.3, a fall in relative sea level may have occurred at time T1, in which case no erosion or deposition took place prior to deposition of the overlying sequence. An alternative is that sedimentation continued until time T3, followed by a rapid erosional event and transgression. Or the sea-level event may have occurred at any time T2.

A given major unconformity may represent the combined effects of two or more unrelated sea-level events, of eustatic and tectonic origin (Fig. 14.4). Recognizing the occurrence of more than one event requires the location and dating of sections where the hiatus is short, as in the various possible sections B in Fig. 14.4. As explained by Aubry (1991, p. 6646);

> If an unconformity Y with a short hiatus on the shelves of basin A can be shown to be exactly correlative (=isochronous=synchronous) in the stratigraphic sense with an unconformity y with a short hiatus on the shelves on basin B (i.e., if the two hiatuses overlap almost exactly), it is probable, although not certain, that both unconformities Y and y are correlative, in the genetic sense, with a unique event T. Overlap between hiatuses of stratigraphically correlatable unconformities in two widely separated basins fulfills a condition required but insufficient to establish global eustasy. Unconformities Y and y will become a global eustatic signal if other correlative (=synchronous) unconformities ... with short hiatus can be recognized on as many shelves as possible of widely separated basins.

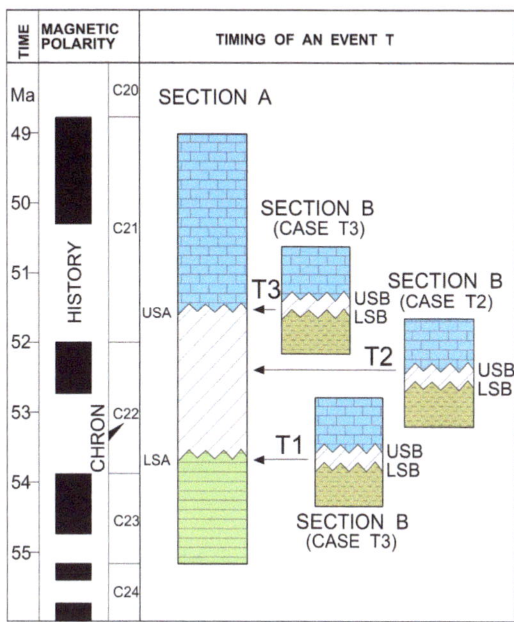


**Fig. 14.4** The generation of a single unconformity by the amalgamation of more than one event (Aubry, 1991)

Even this procedure begs the question of the interpreted duration of sea-level events. Given the rapid events that characterize glacioeustasy (frequency in the $10^4$–$10^5$-year range) this is not a trivial question. Aubry's figures (Figs. 14.3 and 14.4) were drawn to illustrate actual problems which arose in attempts to assign ages to unconformities in Eocene sections. The difference between times T1 and T3 in Fig. 14.4 is 2 million years, which represents a significant potential range of error. It could correspond to one entire million-year sequence.

Vail and his coworkers assert that sequence boundaries are the same age everywhere, despite the influence of tectonism, sediment supply or other factors (e.g., Vail et al., 1991). They believe that sediment-supply factors may affect shoreline position but not the timing of sequence boundaries, based on the models of Jervey (1988). However, this conclusion is not supported by later modeling work. Theoretical studies by Pitman and Golovchenko (1988) demonstrated a phase-lag between sea-level change and the stratigraphic response, particularly in areas of slow subsidence and sea-level change, and rapid sediment supply. This was confirmed by the computer modeling experiments of Jordan and Flemings (1991), who

**Fig. 14.5** The interpretation of the stratigraphy of part of the Lower-Middle Eocene interval at selected sites around the Atlantic basin (Aubry, 1991)

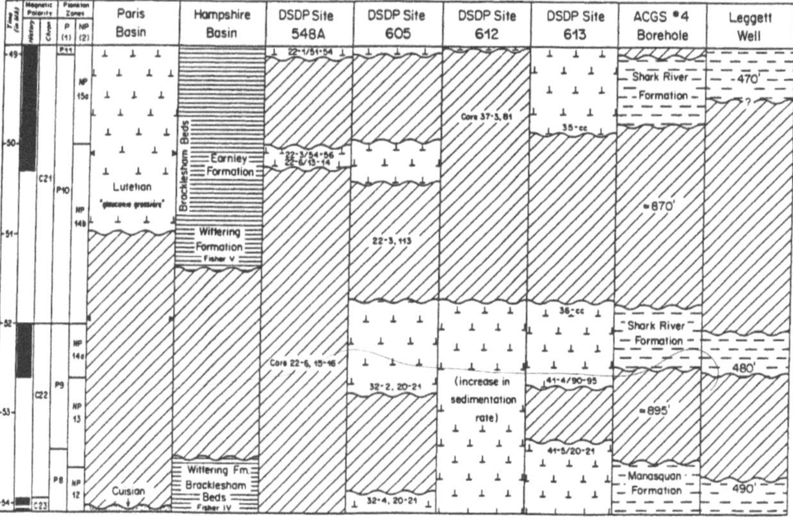

stated that "the sequence boundary for an identical sea level history could be of different ages and the ages could differ by as much as 1/4 cycle." Schlager (1993) and Martinsen and Helland-Hansen (1994) pointed out that the deltas of big rivers such as the Mississippi and the Rhone have prograded while the nearby coast is still retreating as a result of the post-glacial sea-level rise. In effect, this is equivalent to a 1/4-phase shift in the sequence cycle, a difference between "highstand" deltas and "transgressive" shore-lines. The lag may vary from basin to basin depend-ing on variations in sediment flux, subsidence rates, and the amplitude of the sea-level cycle. In the case of the $10^6$-year ("third-order") cycles considered in these experiments, this amounts to a variation of up to several millions of years. This work by Pitman, Jordan and their coworkers is important because it indi-cates that sequence boundaries are inherently impre-cise recorders of sea-level change. Practical exam-ples of this were illustrated by Leckie and Krystinik (1993).

Aubry (1991) provided a detailed discussion of a practical example of the problem of interpret-ing unconformities. The main purpose of her paper was to use the method of rigorous evaluation of unconformities to examine the framework of Lower-Middle Eocene sedimentation in Europe, North Africa, the Atlantic margins, and North America,

and including data from several DSDP sites. A summary of some of the main results is given in Fig. 14.5, and the paper contains a detailed dis-cussion of the biostratigraphic and magnetostrati-graphic evidence, including an evaluation of sev-eral zonal schemes and their relationships in various locations.

Amongst the results of Aubry's analysis was the discovery of more than one unconformity at the Lower-Middle Eocene boundary, within the CP12A and NP14 nannofossil zones, which Haq et al. (1987, 1988a) date as 49 Ma, but which Aubry (1991) assigned an age of 52 Ma (a time difference corresponding to four sequences in the Exxon global cycle chart). Aubry (1991, p. 6672) stated:

Assuming that sequence stratigraphy reflects global changes in sea level, hiatuses around the lower mid-dle Eocene boundary on the North Atlantic margins, in Cyrenaica and in California would be indicative of at least two global eustatic events.

The oldest of these unconformities would have occurred between 52.8 and 53.3 Ma or between 52.8 and 52.9 Ma, depending on two different interpretations.

Correlative unconformities would correspond to the TA2.9/TA3.1 (the 49.5 Ma) sequence boundary. A younger event would have occurred between 50.3 and 51.7 Ma as deduced from the stratigraphic record on the North Atlantic margins, and the correlative unconformities would represent the TA3.1/TA3.2 (the 48.5 Ma) sequence boundary. That hiatuses overlap is a condition required for their physical expressions, the unconformities, to be genetically related. It is not however a sufficient condition, unless the hiatuses are so short that the timing of the causative event can be precisely determined and correlated with a genetic mechanism. The hiatuses associated with the uppermost lower Eocene unconformities are short (less than 1 m.y.) ... However, these unconformities appear to have limited regional extent ... which suggests that they may be unrelated. In contrast the lower middle Eocene unconformities, which have a broader geographic extent ... are associated with long hiatuses (over 2 m.y.), and may be polygenetic.

The discussion continues in this vein for several additional paragraphs. The sense that emerges is that no single clear unconformable signal can be extracted for any level within this short pile of strata. Aubry (1991) developed four possible scenarios to explain the observed data (Fig. 14.6). The first alternative is that both unconformities result from a single eustatic event, within the NP14 zone, which would seem to correspond to the 48.5 Ma sequence boundary. However, this raises problems of different interpretations of Aubry (1991) and Haq et al. (1987, 1988a) regarding the relationships between the specific zones, stage boundaries, and sequence boundaries, a discussion of which is beyond the scope of this book. An interpretation involving two successive eustatic events represents Aubry's second possible interpretation (Fig. 14.6). However, without the framework of sequence stratigraphy to tie the events in the various locations together their differing character and time duration would suggest alternative mechanisms, including regional tectonism. For example, Aubry (1991) stated, "the formation of a nodular bed in Cyrenaica and the development of an unconformity on the New Jersey continental margin at apparently the same time (within the limits of

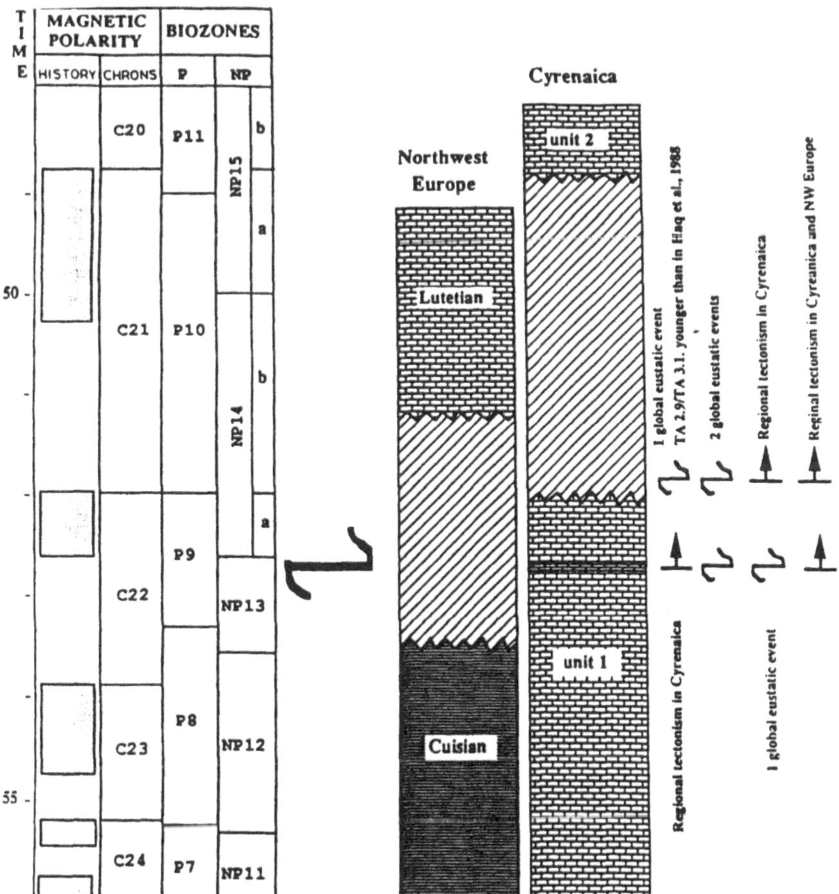


**Fig. 14.6** Correlation of *Lower-Middle* Eocene stratigraphy in northwest Europe and Cyrenaica, and four possible interpretations of the observed breaks in sedimentation. Explained in text (Aubry, 1991)

biochronologic resolution) may or may not be related." This type of argument accounts for the other two interpretations given in Fig. 14.6.

In her conclusions, Aubry (1991) argued that:

Since it cannot be demonstrated that lower middle Eocene unconformities (and probably no other Paleogene unconformities) formed simultaneously on the margins of a basin, there is no a priori reason to assume that they result from a single event, i.e., a global sea level fall. It may be that unconformities on the slope and rise ... form independently from one another at different times, the difference in timing between two erosional events at two different localities being so little (a few tens to a few hundreds of thousands of years perhaps) that the resulting unconformities are apparently synchronous.

In a later paper, Aubry (1995) extended her analysis of Lower and Middle Eocene unconformities to the entire central and southern Atlantic Ocean, examining over one-hundred deep-sea sites. She demonstrated the presence of multiple unconformities, and no clear pattern that would support the construction of a single summary section that could serve to summarize all the data.

The conclusions that may be drawn from this detailed examination of unconformities in general, and the Eocene record in particular, is that the attempt to identify and date sequence boundaries by the evidence of unconformities preserved in the stratigraphic record may prove a frustrating and ambiguous enterprise.


## 14.4  A Correlation Experiment

I conducted a simple experiment to illustrate the questionable value of the Exxon cycle chart (Miall, 1992). The experiment replicates the correlation exercise performed by basin analysts attempting to compare data from a new outcrop section or well to an existing regional or global standard. It is basic geologic reasoning that if such a correlation exercise results in the definition of numerous tie lines between the new data and the standard, two conclusions may be drawn: (1) The events documented in the regional or global standard also occurred in the area of the new section (2) The new data confirm and strengthen the validity of the standard. There is, of course, a danger of circular reasoning in such an analytical procedure.

Figure 14.7 shows the 40 Cretaceous sequence boundaries in the Exxon chart compared with the event boundaries recognized in four other sections. In these four sections the events have been classified into three kinds: (1) events dated to within ±0.5 million years of an event in the Exxon chart (2) events outside this range but falling within ±1 million years of an event in the Exxon chart (3) events that differ by >1 million years from any event in the Exxon chart. Table 14.1 lists the number of such events, and demonstrates a high degree of correlation of all four sections with the Exxon chart. In fact, the lowest correlation success rate is 77%, from Sect. 3. By normal geologic standards, considering the fuzziness and messiness of most geologic data, this would be regarded by most experienced geologists as indicating a high degree of correlation. This is especially the case when we bear in mind the normal types of imprecision inherent in chronostratigraphic data. An error range of ±1 million years is unrealistic in most cases. With an error range of ±2 million years the correlation success rate would be almost 100%. The data in Fig. 14.7 would therefore appear to provide a successful confirmation of the Exxon chart. The catch is that all four of the test sections were constructed by random-number generation!

Columns 1 and 2 were constructed by selecting 40 values from a table of two-digit random numbers ranging from 0 to 99. These were normalized to the range 66.5–131, this being the range (Ma) of the Cretaceous Period. The resulting numbers were then treated as stratigraphic events and were plotted to produce the columns seen in Fig. 14.7, their "ages" rounded to the nearest 0.5 million years. The symbol used to plot each "event" indicates the proximity in age to one of the events on the Exxon chart (see key for Fig. 14.7). Upon rounding, several of the events overlapped. Therefore, the number of events plotted in each case is less than 40. Columns 3 and 4 were constructed by a different method. Event spacing in the Exxon chart for the Cretaceous ranges from 0.5 to 4.5 million years. Suites of 40 random numbers were normalized to this range, rounded to the nearest 0.5 million years, and then used to construct a stratigraphic sequence. The first (oldest) event shown in columns 3 and 4 of Fig. 14.7 represents 131 minus the first normalized random-number value. The second event is the age of the first plotted event, minus the second normalized value, and so on. Again, some events overlapped, and are not shown, so the resulting columns contain fewer than 40 events.

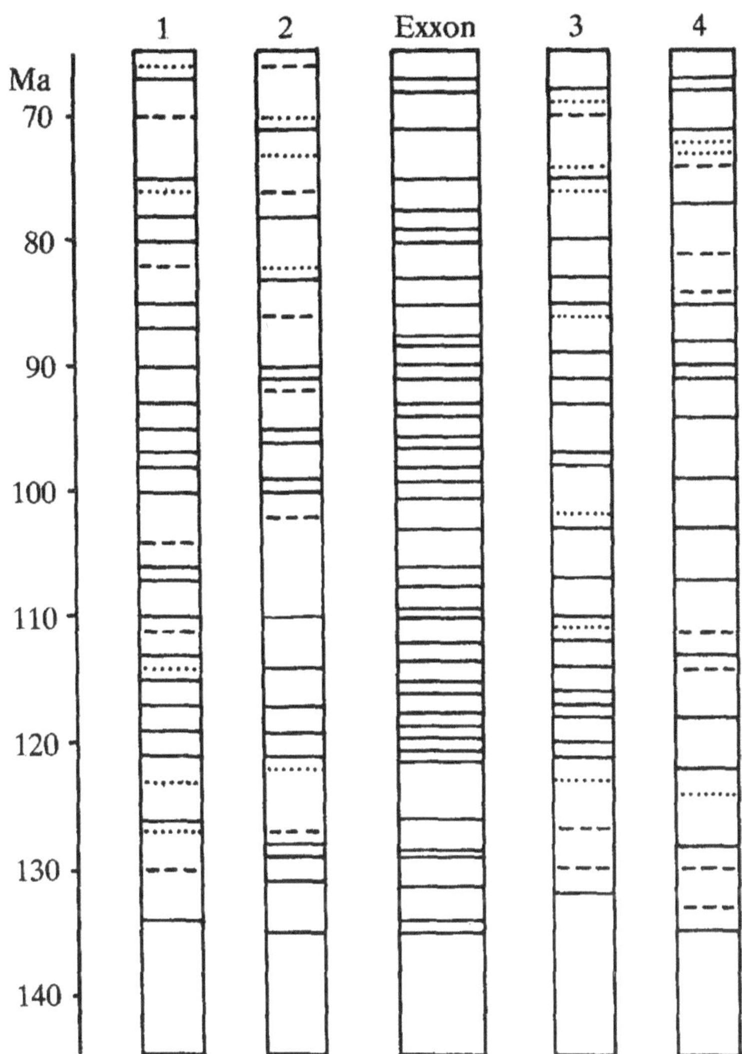

**Fig. 14.7** Event boundaries in the Exxon global cycle chart (total of 40, centre column). Columns 1–4 are independent data sets used in the correlation experiment (Miall, 1992)



**Table 14.1**   Results of correlation experiment (Miall, 1992)

| Section | No. of events | ±0.5 million years ties | | ±1 million years ties | Total ties | | Mismatches |
| | | No. | % Fit | No. | No. | % Fit | |
| --- | --- | --- | --- | --- | --- | --- | --- |
| 1 | 32 | 22 | 69 | 5 | 27 | 84 | 5 |
| 2 | 28 | 18 | 64 | 6 | 24 | 86 | 4 |
| 3 | 31 | 21 | 68 | 3 | 24 | 77 | 7 |
| 4 | 27 | 17 | 63 | 7 | 24 | 89 | 3 |

Convincing correlations have been achieved by both random positioning of stratigraphic events and random sequencing of events. Such a high degree of correlation must therefore be expected to exist, as a kind of background noise, in any actual field test of the Exxon chart. How can we be sure that any given event on the chart is not the product of local processes and compounded error, if random processes can produce such convincing correlations? The answer is that we cannot.

This exercise was offered by Miall (1992) in the same spirit as the classic paper "Cycles and psychology" by Zeller (1964). In that paper, Zeller succeeded in convincing his colleagues to construct stratigraphic correlations between sections whose rock types and thicknesses had been compiled entirely from random numbers. As Zeller stated, his colleagues "should have found it completely impossible to make any correlation … with plotted stratigraphic sections." But they did, thus demonstrating the training geologists receive in pattern-recognition techniques.

## 14.5 Testing for Eustasy: The Way Forward

### 14.5.1 Introduction

In an editorial in the journal *Paleoceanography*, Miller (1994) set out the nature of the challenge. As he stated, during the last 5 years (he meant the time leading up to 1994), "there has been a great deal of interest in global sea level (eustatic) changes and their relationship to sequence stratigraphy." He noted the difficulty in unraveling the triad of processes that control sedimentation, eustasy, tectonics (thermal subsidence, isostasy, compaction and flexure), and sediment supply, and the challenge of approaching this problem objectively. In order to test for the very existence of a eustatic control on the stratigraphic record, the approach he recommended was to undertake a very detailed study of Neogene continental margins, for which time period, there is already an existing, objective record of eustatic sea-level change in the form of the $\delta^{18}O$ record.

The Neogene would certainly be a good place to start, for several important reasons (1) for this most recent portion of the stratigraphic record, there are excellent, undisturbed successions on the

world's continental margins; (2) the chronostratigraphic record for the Neogene is exceptionally good because of the accuracy and precision of the combined magnetostratigraphic-biostratigraphic data-base, supplemented by the $\delta^{18}O$ record; and (3) the $\delta^{18}O$ record itself is particularly useful, precisely because it is a record of the repeated, large glacioeustatic fluctuations that have occurred on Earth over the last few million years. In other words, we know that there was a prominent eustatic signal during the Neogene, and it would be an important first step to test the nature of the stratigraphic record through that period of Earth history.

However, this is only a beginning, because the main question is whether the stratigraphic record is good enough to demonstrate unequivocally the record of eustatic sea-level change during periods for which we have no clear, independent evidence of an eustatic control. As discussed earlier (Chaps. 7 and 11) there is increasing, but still incomplete, evidence for modest, orbitally-forced eustasy during the so-called greenhouse climatic era of the Mesozoic, and the record of the Sloss cycles (Chap. 5) is by now well documented. However, the evidence of what used to be termed "third-order" eustasy ($10^6$-year periodicity) is patchy at best, the question of tectonic control remains very much alive, and high-resolution documentation of global correlations between pre-Neogene sequence records at all time scales is largely absent. Part IV of this book, leading up to this point, has attempted to demonstrate that we must begin from scratch in the endeavour to build a chart of global sea-level events. If we persist in using the Haq et al. (1987, 1988a) or the Graciansky et al. (1998) curves as starting points, we are prejudicing the results from the outset, because the body of work which those publications represent was essentially based on an inversion of the null hypothesis (no eustatic control) from its inception. Following Vail's original concept which, as noted elsewhere (Miall and Miall, 2002), he formulated as a graduate student in the 1950s, Vail, Haq, Graciansky and their colleagues, took it as a starting assumption that sequence boundaries are a record of eustatic sea-level change, and every boundary was interpreted in that way (Fig. 12.3b). Christie-Blick et al. (2007, p. 217), citing Jan Hardenbol, one of the Vail co-workers, stated:

As Jan Hardenbol writes in his abstract [Hardenbol et al., 1998], "A well-calibrated regional biochronostratigraphic framework is seen as an essential first step towards an eventual *demonstration* [Christie-Blick's italics] of sychroneity of sequence in basins with different tectonic histories." In other words, a fundamental assumption of the 1998 compilation, and inevitably the outcome of a new global synthesis, is that there is such a synchronous pattern to be discovered.

The test of global correlation rests on, and can only be as good as, the chronostratigraphic data base. There are two parts to such tests: (1) The reliability of a correlation tests rests, in the first instance, on the accuracy and precision of the global time scale; what McGowran (2005, p. 51) called the "integrated magne-tobiochronological timescale" or IMBTS; and (2) Each test section must be capable of being fitted to or corre-lated with the IMBTS with an accuracy and precision adequate to demonstrate the necessary degree of con-temporaneity between the test sequence boundaries, taking into account the frequency of the eustatic sig-nal under consideration. As demonstrated in Sect. 14.2, the significance of unconformities may be ambiguous. For example, a large gap in the record may hide the record of several high-frequency sea-level events. It is also shown in Sect. 14.3 that the reality of correlation depends on the relationship between dating error and event spacing. A quote from Ricken (1991, p. 773) at the beginning of part IV, makes the point succinctly: "the range of error is often larger than the actual time span considered."

The nature of the chronostratigraphic time scale is clearly critical. In the 12 years since the first edition of this book was published, a great deal has changed, and much of what was argued in Chap. 13 of that edition (based largely on a study completed in 1994) is now out of date, and is, therefore, not repeated in this new edition of the book. Integrated time scales have been under development since at least the 1960s (Sect. 1.3.3; see also Miall, 1999, Chap. 3). A substantial evolution in the quality of the scale is in evidence from the work of Berggren et al. (1995) and Gradstein et al. (2004). Nowadays, the IMBTS is managed by the International Commission on Stratigraphy (ICS), the largest scien-tific body within the International Union of Geological Sciences (IUGS), and the only organization concerned with stratigraphy on a global scale. The website of the ICS, at http://www.stratigraphy.org provides continu-ally updated references to all aspects of the developing time scale, and detailed documentation of every com-ponent of it on a regional and global basis. A new

journal, *Stratigraphy*, was founded in 2004, to serve as a platform for research publications in the field, and the IUGS journal *Episodes* continues to provide updates on noteworthy events.

In the next section, we discuss the general methods of correlation and dating currently in use, concluding with a review of the accuracy and precision with which sequences can be tied to the global time scale.

## 14.5.2 The Dating and Correlation of Stratigraphic Events: Potential Sources of Uncertainty

The dating and correlation of stratigraphic events between basins, where physical tracing-out of beds cannot be performed, involve the use of biostratigraphy and other chronostratigraphic methods. The process is a complex one, fraught with many possible sources of error. What follows is an attempt to break the process down into a series of discrete "steps", although in prac-tice, dating successions and erecting local and global time scales is an iterative process, and no individual stratigrapher follows the entire procedural order as set out here. The geologist is able to draw on the accumu-lated knowledge of the geological time scale that has (as noted earlier) been undergoing improvements for more than 200 years, but each new case study presents its own unique problems.

There are two related but distinct problems to be addressed in the construction of a global stratigraphic framework. On the one hand, we need to test whether similar stratigraphic events (such as sequence bound-aries) and successions of events (e.g., successions of sequences) have the same ages in different parts of the earth, in order to test models of global causality (e.g., eustasy). From this perspective, precise correlation is critical, but determination of absolute age is not. On the other hand, there is the ongoing effort to deter-mine the precise age of stratigraphic events, leading to continuing refinements in the global time scale. In either case, the potential error, or uncertainty, involved with one or more of the "steps" described above must be acknowledged and addressed as part of a determi-nation of the level of refinement to be expected in stratigraphic interpretation.

Dating and correlation methods are discussed in standard textbooks (e.g., Miall, 1999, Chap. 3; Boggs,

2005) and in advanced research texts (e.g., Harland et al., 1990; Doyle and Bennett, 1998; Gradstein et al., 2004).

Six main "steps" are involved in the dating and correlation of stratigraphic events (Miall, 1991a, 1994). Figure 14.8 summarizes these steps and provides generalized estimates of the magnitude of the uncertainty associated with each aspect of the correlation and dating of the stratigraphic record. Some of these errors may be cumulative, as discussed in the subsequent sections. The assignment of ages and of correlations with global frameworks is an iterative process that, in some areas, has been underway for many years. There is much feed-back and cross-checking from one step to another. What follows should be viewed, therefore, as an attempt to break down the practical business of dating and correlation into more readily understandable pieces, all of which may be employed at one time or another in the unraveling of regional and global stratigraphies. The main steps are as follows:

- Identification of sequence boundaries. Determining the position of the sequence boundary may or may not be a straightforward procedure. There are several potential sources of error and confusion (Sect. 14.5.2.1).

- Determining the chronostratigraphic significance of unconformities. Unconformities, such as sequence boundaries, represent finite time spans which vary in duration from place to place. In any given location this time span could encompass the time span represented by several different sedimentary breaks at other locations (Sects. 14.3 and 14.5.2.2).

- Determination of the biostratigraphic framework (Sect. 14.5.2.3). One or more fossil groups is used to assign the selected event to a biozone framework. Error and uncertainty may be introduced because of the incompleteness of the fossil record (Sect. 14.5.2.4).

- Assessment of relative biostratigraphic precision. The length of time represented by biozones depends on such factors as faunal diversity and rates of

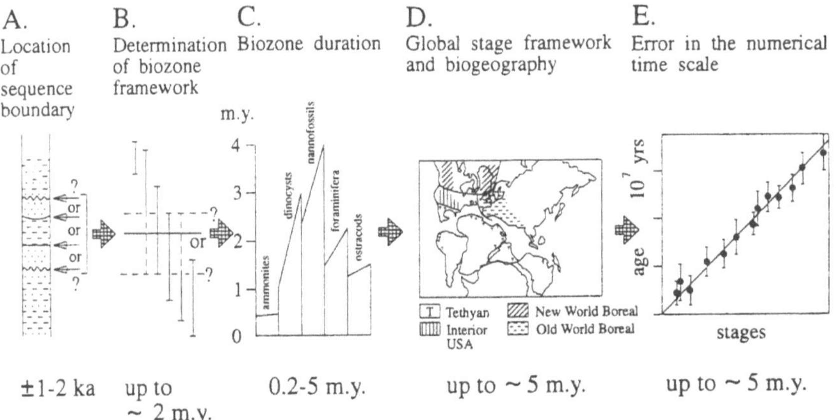


**Fig. 14.8** Steps in the correlation and dating of stratigraphic events. e = typical range of error associated with each step. (**a**) In the case of the sequence framework, location of sequence boundaries may not be a simple matter, but depends on interpretation of the rock record using sequence principles. (**b**) Assignment of the boundary event to the biozone framework. An incomplete record of preserved taxa (almost always the case) may lead to ambiguity in the placement of biozone boundaries. (**c**) The precision of biozone correlation depends on biozone duration. Shown here is a simplification of Cox's (1990) summary of the duration of zones in Jurassic sediments of the North Sea Basin. (**d**) The building of a global stage framework is fundamental to the development of a global time scale. However, global correlation is hampered by faunal provincialism. Shown here is a simplification of the faunal provinces of Cretaceous ammonites, shown on a mid-Cretaceous plate-tectonic reconstruction. Based on Kennedy and Cobban (1977) and Kauffman (1984). (**e**) The assignment of numerical ages to stage boundaries and other stratigraphic events contains inherent experimental error and also the error involved in the original correlation of the datable horizon(s) to the stratigraphic event in question. Diagrams of this type are a standard feature of any discussion of the global time scale (e.g., Haq et al., 1988a; Harland et al., 1990). The establishment of a global biostratigraphically-based sequence framework involves the accumulation of uncertainty over steps (**a–d**). Potential error may be reduced by the application of radiometric, magnetostratigraphic or chemostratigraphic techniques which, nonetheless, contain their own inherent uncertainties (step **e**)

evolution. It varies considerably through geological time and between different fossil groups (Sect. 14.5.4).

- Correlation of biozones with the global stage framework. The existing stage framework was, with notable exceptions, built from the study of macrofossils in European type sections. Correlation with this framework raises questions of environmental limitations on biozone extent, our ability to interrelate zonal schemes built from different fossil groups, and problems of global faunal and floral provinciality (Sect. 14.5.5).
- Assignment of absolute ages. The use of radiometric and magnetostratigraphic dating methods, plus the increasing use of chemostratigraphy (oxygen and strontium isotope concentrations) permits the assignment of absolute ages in years to the biostratigraphic framework. Such techniques also constitute methods of correlation in their own right, especially where fossils are sparse (Sect. 14.5.6). Carbon isotope data have also been used in a few specialized studies (Sect. 14.8).

### 14.5.2.1 Identification of Sequence Boundaries

The first step is that a well section, outcrop profile or seismic record is analyzed and the positions of sequence boundaries are determined from the vertical succession of lithofacies or from the architecture of the seismic reflections.

There is substantial residual disagreement about the definition of sequences, as discussed in Chap. 2 (see Fig. 2.22). For example, are the falling-stage deposits assigned to the top of one sequence or to the base of the overlying sequence? The difference in the placement, and therefore the age, of the sequence boundary changes significantly, depending on the sequence model used.

In outcrop and well data, other possible errors in this procedure arise from the potential for confusion between allogenic and autogenic causes for the breaks in the stratigraphic record. Autogenic causes include condensed sequences, ravinement surfaces, and channel scours. Various other processes also generate sedimentary breaks that may be confused with sequence boundaries, such as environmental change in carbonate settings (e.g., drowning unconformities). The subject is discussed at some length in Chap. 2. These errors

could lead to possible errors in boundary placement of at least several meters. Given a typical sedimentation rate of 0.1 m/ka, a 10 m error is equal to 100 ka.

In seismic sections, poor seismic resolution may result in the miscorrelation of thin units (e.g., Cartwright et al., 1993). Armentrout et al. (1993) used petrophysical logs tied to seismic cross-sections to develop a correlation grid in Paleogene deposits in the North Sea. In this basin, where wells are up to tens of kilometres apart, they determined that tracing markers around correlation loops could result in mismatches of up to 30 m. Errors of this magnitude might not be expected in mature areas, such as the Alberta Basin or the US Gulf Coast.

Vail et al. (1977) had proposed that sequence boundaries could be dated more accurately by tracing them into areas of continuous sedimentation, such as in deep-marine settings. The equivalent of the unconformable sequence boundary was to be termed the *correlative conformity*. However, in conformable, continuous sections there is not likely to be any potential for the development of recognizable reflections (changes in acoustic impedance), and so the value of the correlative conformity concept is hypothetical. Furthermore, given the problems of tracing surfaces on seismic sections through coastal successions, a region of substantial autogenic facies change and local scour, into the deep marine, the likelihood of being to locate precisely the correct horizon might be questioned. In any case, as Christie-Blick et al. (2007, p. 222) pointed out, the diachroneity of sequence boundaries, the fact that they represent spans of time, rather than instantaneous events, means that "at some scale, unconformities pass laterally not into correlative conformities, but into *correlative intervals* [my italics]. Such considerations begin to be important as the resolution of the geological timescale improves at a global scale"

### 14.5.2.2 Chronostratigraphic Meaning of Unconformities

Assigning an age to an unconformity surface is not necessarily a simple matter, as discussed in Sect. 14.3. An unconformity represents a finite time span; it may have a complex genesis, representing amalgamation of more than one event, more than one different type of event (e.g., a combination of tectonism and eustasy), or even a phase-amplification effect of the superimposition of two or more different eustatic signals. Unconformities may be markedly diachronous, because the transgressions and regressions that occur during the genesis of a stratigraphic sequence could span the entire duration of the sequence. As noted in Sect. 13.1, Kidwell (1988) demonstrated that this results in an offset in sequence boundary unconformities by as much as one half of a cycle between basin centre and basin margin.

Detailed chronostratigraphic analysis of some well dated sections (see summary in Sect. 13.1) has shown that as little as 24% to as much as 82% of elapsed time is stored in the sedimentary succession, with the remainder of the time span represented by the unconformities, in other words, the sequence boundaries. Ager (1981, 1993) argued for even greater extremes of missing time, but we concern ourselves here with the (hopefully) more continuous sections that form the basis for the modern definitions of key sequence profiles (e.g., those on the New Jersey continental margin, and the Wanganui Basin of New Zealand). Within tectonostratigraphic successions lasting tens of millions of years, such partial preservation of the sedimentary record could include unconformities representing many millions of years, which could mean that lengthy portions of any high-frequency sequence record are entirely missing. At the $10^{6-7}$-year time scale, chronostratigraphic precision should be adequate at least to reveal the scale of such gaps, but it is this problem of missing cycles buried within a lower-frequency record that underlies the problem of testing for a regional (let alone, global) record of high-frequency sequences. The problem is particularly acute in the case of cyclostratigraphy—the current research endeavour to extend the astrochronological time scale into deep geologic time, beyond the limits of "continuous" deep-sea cores and exposed pelagic sections (e.g., Hilgen, 1991) into the realm of the "floating" section (Miall and Miall, 2004). We return to this topic in Sect. 14.7.

### 14.5.2.3 Determination of the Biostratigraphic Framework

Sequence boundaries are dated and correlated by biostratigraphic, radiometric or magnetostratigraphic techniques. The potential for error here stems from the imperfection of the stratigraphic record, including its fossil content. Ranges of useful biostratigraphic indicators are commonly denoted by the acronyms FAD and LAD, which mean *first-* [lowest] and *last-* [highest] *appearance datums*. First and last taxon occurrences could vary in position by several meters to even a few tens of meters among sections of the same age. Errors in age assignment of complete biozones or half zones are not uncommon.

The biostratigraphic data base for the global time scale has experienced an orders-of-magnitude expansion since microfossils began to take on increasing importance in subsurface petroleum exploration in the post-war years, and following the commencement of the Deep Sea Drilling Project (DSDP) in 1968 (Berggren et al., 1995; McGowran, 2005). The main basis for the global time scale consists of the classical macrofossil assemblages (e.g., ammonites, graptolites) that have been in use, in many cases, since the nineteenth century. Supplementing macrofossil and microfossil collections from outcrops, drilling operations on land and beneath the oceans have contributed a vast biostratigraphic sample base over the last few decades, permitting the evaluation of a wide range of microfossils for biostratigraphic purposes in all the world's tectonic and climatic zones. The result has been a considerable improvement in the flexibility and precision of dating methods, and much incremental improvement in the global time scale. The foraminiferal and nannofossil record for the Cenozoic is now precise enough that meaningful tests of the global-eustasy paradigm can now be attempted, and this is discussed in Sect. 14.6.

As noted by Cope (1993), detailed, meticulous taxonomic work still holds the potential for much improved biostratigraphic resolution and flexibility at all levels of the geological column. In older Mesozoic and Paleozoic strata, such fossil groups as conodonts and graptolites have yielded very refined biostratigraphic zonal systems. However, this potential has yet to be fully tapped by sequence stratigraphers.

The various types of zones, and the methods for the erection of zonal schemes, are subjects dealt with in standard textbooks and reviews (e.g., Kauffman and Hazel, 1977a; Doyle and Bennett, 1998; Miall, 1999, Chap. 3; McGowran, 2005), and are not discussed here. The purpose of this section is to focus on two major problems that affect the accuracy and precision of biostratigraphic correlation.

### 14.5.2.4 The Problem of Incomplete Biostratigraphic Recovery

Because of incomplete preservation, poor recovery, or environmental factors, the rocks rarely yield a complete record through time of each biozone fauna or flora. Practical, measured biostratigraphic time, as indicated by the imperfect fossil record, may be different from hypothetical "real" time, which can rarely be perfectly defined (Murphy, 1977; Johnson, 1992a, b). This is illustrated in Fig. 14.9, in which the differences between observed lines of correlation and hypothetical but invisible and unmeasureable time lines is indicated by the grey-coloured gaps.

How important are the gaps? FADs and LADs should correspond to time planes in the rocks, but in real basins the situation is often like that shown in Fig. 14.9. There are several reasons why fossils would be found at younger horizons than predicted. The depositional environment may be unsuitable in some locations at the time the organism first evolved, or the fossils have been destroyed by diagenesis. Or perhaps sampling of the section was simply incomplete. This is an unlikely reason in the case of micro-organisms because they typically occur in very large numbers, but could be the case where outcrops are sampled for macrofossils. The occurrence of a key form at a horizon *lower* than predicted may simply indicate that the type section was inadequately sampled or that local environmental factors intervened there. In such cases it may be useful to revise downwards the record of the FAD in the stratotype, for future reference purposes.

In subsurface work, the positions of FADs and LADs may be distorted by cavings. For example, over much of the Canadian interior, Cretaceous rocks rest unconformably on a succession of early Paleozoic age. It is not uncommon to encounter Cretaceous microfossils, such as foraminifera, in well cuttings from Paleozoic horizons. The explanation is simply that upper levels of the drill hole yielded cavings into the mud stream while the well was being drilled into the underlying Paleozoic rocks. An intermixing of materials such as this is easy to detect, but would be much more difficult to deal with if caving is a persistent problem, occurring at several or many stratigraphic levels within a continuous section. In such cases it may be wise not to make use of FADs at all, because of their potential inaccuracy.

Many researchers have attempted to sidestep these problem, by treating fossil occurrences quantitatively, and applying statistical treatments to assessments of preservation and correlation (e.g., Riedel, 1981; McKinney, 1986; Agterberg, 1990; Guex, 1991). The method of graphic correlation (Sect. 14.5.4), which focuses attention on the incompleteness of the fossil record, allows us to examine this question with

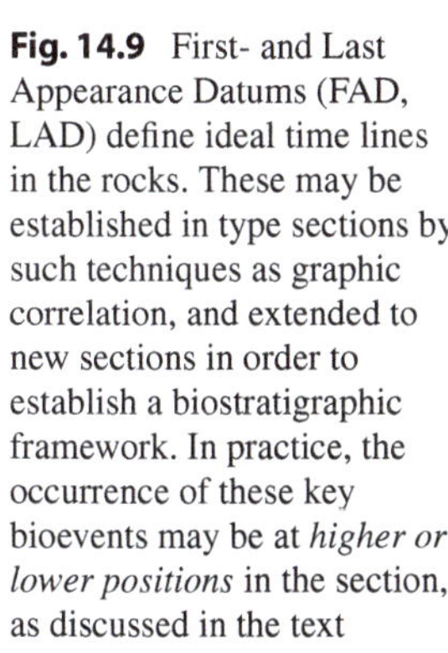

**Fig. 14.9** First- and Last Appearance Datums (FAD, LAD) define ideal time lines in the rocks. These may be established in type sections by such techniques as graphic correlation, and extended to new sections in order to establish a biostratigraphic framework. In practice, the occurrence of these key bioevents may be at *higher or lower positions* in the section, as discussed in the text

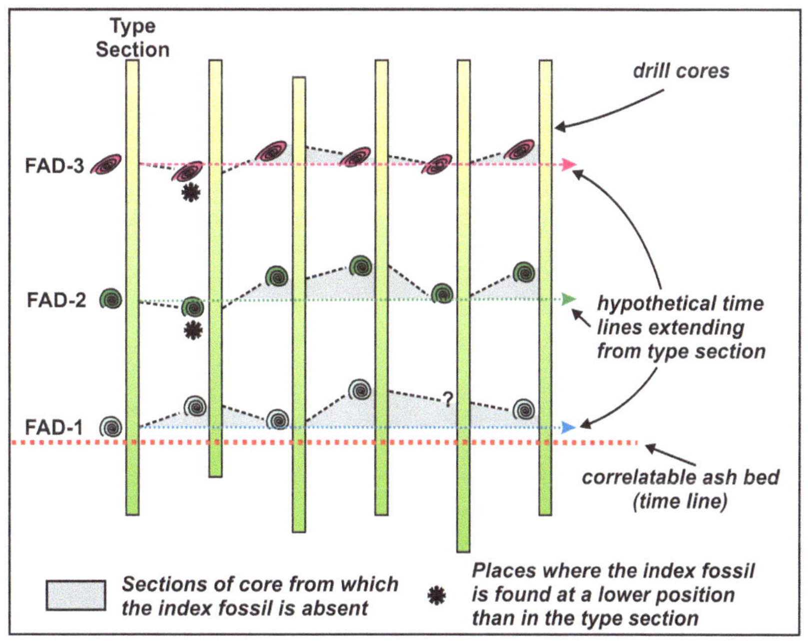

some precision. Results presented by Edwards (1989) indicate that differences of up to 10 m between biostratigraphic and hypothetical "real" time lines are not uncommon. Doyle (1977) illustrated in detail an attempt to correlate two wells using palynological data (Fig. 14.10). Ranges of correlation error up to 30 m are apparent from his data, and result from incomplete preservation or wide sample spacing. At a sedimentation rate of 0.1 m/ka 30 m of section is equivalent to 300 ka, more than enough to lead to miscorrelation of sequences in the Milankovitch band.

Conventional wisdom has it that pelagic organisms are the best biostratigraphic indicators because they are widely distributed by ocean currents and tend

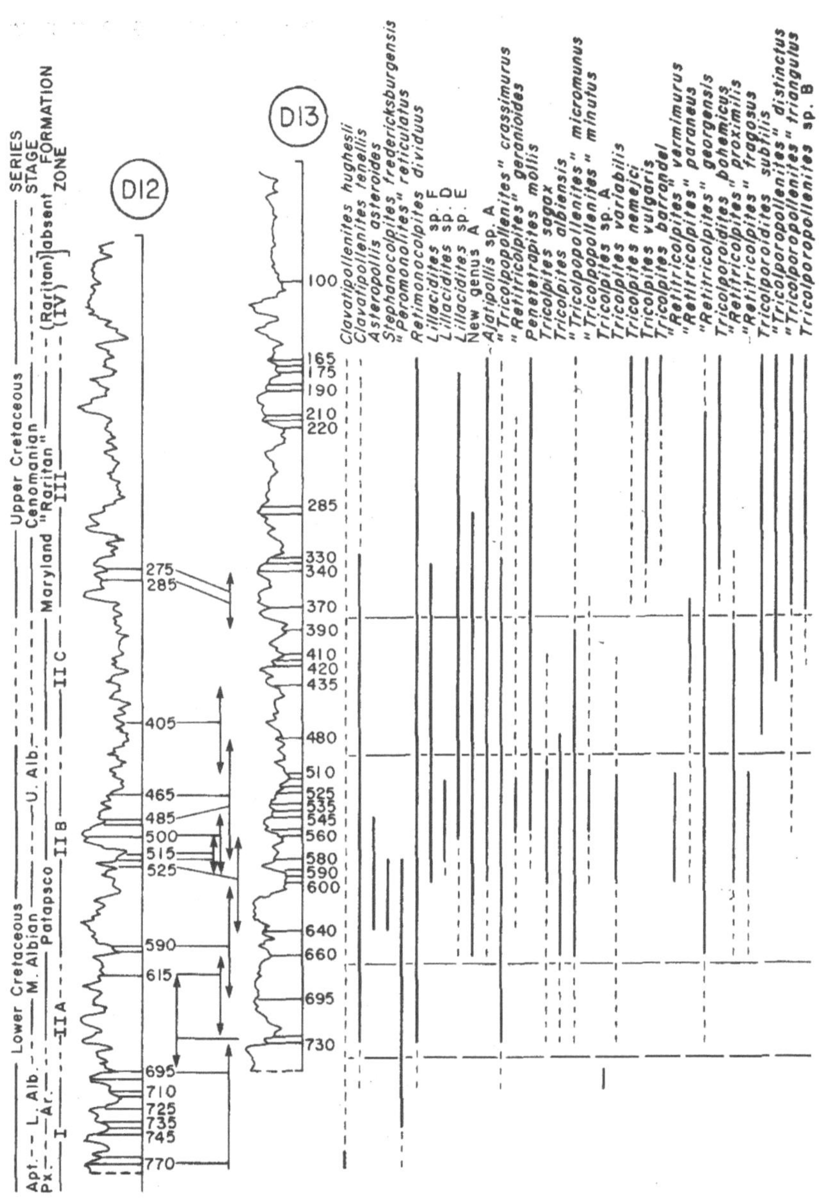

**Fig. 14.10** Correlation of two wells through Cretaceous strata in Delaware, using palynological data. Correlation brackets (*double-headed arrows*) terminate just above and below samples in well D13 that bracket the age of the indicated sample in well D12 (Doyle, 1977)

to be less environmentally sensitive. However, recent detailed studies of nektonic and planktonic forms have indicated a wide range of factors that lead to uneven distribution and preservation even of these more desirable forms. Thierstein (1981) and Roth and Bowdler (1981), in studies of Cretaceous nannoplankton, discussed mechanisms that affected distribution of these organisms. Among the most important environmental factors are changes in ocean currents, which affect water temperatures and the concentrations of oxygen and carbon dioxide and nutrients in the seas. Changes in temperature and sea level affect the position of the carbonate compensation depth in the oceans, and hence affect the rates of dissolution and preservability of calcareous forms in deep-sea sediments. Conversely, bottom dwelling faunas, which are specifically limited in their geographical distribution because of ecological factors, may exhibit a diversity and rapid evolutionary turn-over that makes them ideal as biostratigraphic indicators in specific stratigraphic settings. For example, many corals have been found to be of great biostratigraphic utility in the study of reef limestones of all ages.

### 14.5.2.5 Diachroneity of the Biostratigraphic Record

Another common item of conventional wisdom is that evolutionary changes in faunal assemblages are dispersed so rapidly that, on geological time scales, they can essentially be regarded as instantaneous. This argument is used, in particular, to justify the interpretation of FADs as time-stratigraphic events (setting aside the problems of preservation discussed above). However, this is not always the case. Some examples of detailed work have demonstrated considerable diachroneity in important pelagic fossil groups.

MacLeod and Keller (1991) explored the completeness of the stratigraphic sections that span the Cretaceous-Tertiary boundary, as a basis for an examination of the various hypotheses that have been proposed to explain the dramatic global extinction occurring at that time. They used graphic correlation methods, and were able to demonstrate that many foraminiferal FADs and LADs are diachronous. Maximum diachroneity at this time is indicated by the species *Subbotina pseudobulloides,* the FAD of

which may vary by up to 250 ka between Texas and North Africa. However, it is not clear how much of this apparent diachroneity is due to preservational factors.

An even more startling example of diachroneity is that reported by Jenkins and Gamson (1993). The FAD of the Neogene foraminifera *Globorotalia truncatulinoides* differs by 600 ka between the southeast Pacific Ocean and the North Atlantic Ocean, based on analysis of much DSDP material. This is interpreted as indicating the time taken for the organism to migrate northward from the South Pacific following its first evolutionary appearance there. As Jenkins and Gamson (1993) concluded

> The implications are that some of the well documented evolutionary lineages in the Cenozoic may show similar patterns of evolution being limited to discrete ocean water masses followed by later migration into other oceans … If this is true, then some of these so-called 'datum planes' are diachronous.

This conclusion is of considerable importance, because the result is derived from excellent data, and can, therefore, be regarded as highly reliable, and relates to one of the most universally preferred fossil groups for Mesozoic-Cenozoic biostratigraphic purposes, the foraminifera. It would appear to suggest a limit of up to about one half million years on the precision that can be expected of any biostratigraphic event.

The two cases reported here may or may not be a fair representation of the magnitude of diachroneity in general. After a great deal of study, experienced biostratigraphers commonly determine that some species are more reliable or consistent in their occurrence than others. Such forms may be termed *index fossils,* and receive a prominence reflecting their usefulness in stratigraphic studies. Studies may indicate that some groups are more reliable than others as biostratigraphic indicators. For example, Ziegler et al. (1968) demonstrated that brachiopod successions in the Welsh Paleozoic record were facies controlled and markedly diachronous, based on the use of the zonal scheme provided by graptolites as the primary indicator of relative time. Armentrout (1981) used diatom zones to demonstrate that molluscan stages are time transgressive in the Cenozoic rocks of the Northwest US. Wignall (1991) demonstrated the diachroneity of Jurassic ostracod zones.

### 14.5.3 The Value of Quantitative Biostratigraphic Methods

Much work has been carried out in attempts to apply quantitative, statistical methods to biostratigraphic data, in order to refine stratigraphic correlations and to permit these correlations to be evaluated in probabilistic terms. Excellent syntheses have been provided by Agterberg (1990) and Guex (1991). However, many problems remain, because the biostratigraphic data base does not necessarily meet some of the necessary assumptions required for statistical work. Guex (1991, p. 179), in discussing the use of multivariate methods, quoted Millendorf and Heffner (1978, p. 313), who stated:

> This approach ignores the effects of faunal gradation within an isochronous unit with respect to geographic position. Thus, if the faunal composition of such an isochronous unit changes across the study area, samples taken from distant points in the unit might be dissimilar enough not to cluster. Simply, the greater the lateral variation and the larger the distance between them, the less similar are the two isochronous samples.

Guex (1991, p. 180) himself stated:

> In one way or another, all methods based on global resemblance between fossil samples end in fixing the boundaries of statistical biofacies, and they do not make it possible to find, within a fossil assemblage, the species that are characteristic of the relative age of the deposits under study.

Biofacies distributions are determined in part by environment. The definition of "statistical biofacies" is therefore not a very useful contribution to problems of correlation. Not only biofacies, but sampling methods and questions of preservation also affect the distribution of fossils (e.g., Agterberg, 1990, Chap. 2; see Fig. 14.9). It is questionable, therefore, whether statistical methods can assist directly with solving the problem of assessing error in global correlation.

There is one important exception to this generalization, and that is a technique known as graphic correlation. The method was proposed by Shaw (1964), and has been developed by Miller (1977) and Edwards (1984, 1989). The procedure is summarized by Miall (1999, pp. 110–112). Mann and Lane (1995) edited a collection of papers describing modern methods and case studies. The data on foraminiferal diachroneity at the Cretaceous-Tertiary boundary quoted above from MacLeod and Keller (1991) were obtained using the graphic correlation method. The statistical methods that have evolved for ranking, scaling and correlation (Agterberg, 1990; Gradstein et al., 1990; Sadler, 2004) are a form of quantified graphic correlation.

As with conventional biostratigraphy, the graphic method relies on the careful field or laboratory recording of occurrence data, and focuses on the collation and interpretation of FADs and LADs. The objective is to define the local ranges for many taxa in at least three complete sections through the succession of interest. The more sections that are used, the more nearly these ranges will correspond to the total (true) ranges of the taxa. To compare the sections, a simple graphical method is used. One particularly complete and well-sampled section is chosen as a standard reference section. Eventually, data from several other good sections are amalgamated with it to produce a composite standard reference section. A particularly thorough paleontologic study should be carried out on the standard reference section, as this enables later sections, for example, those produced by exploration drilling, to be correlated with it rapidly and accurately.

The graphic technique is used both to amalgamate data for the production of the composite standard and for correlating the standard with new sections. Figure 14.11 shows a two-dimensional graph in which the thicknesses of two sections X and Y have been marked off on the corresponding axes. The FADs and LADs are marked on the sections by circles and crosses, respectively. If the taxon occurs in both sections, points can be drawn within the graph corresponding to FADs and LADs by tracing lines perpendicular to the X and Y axes until they intersect. For example, the plot for the top of fossil 7 is the coincidence of points X = 350 and Y = 355.

If all the taxa occur over their total range in both sections and if sedimentation rates are constant (but not necessarily the same) in both sections, the points on the graph fall on a straight line, called the line of correlation. In most cases, however, there will be a scatter of points. The X section is chosen as the standard reference section, with the expectation that ranges will be more complete there. The line of correlation is then drawn so that it falls below most of the FADs and above most of the LADs. FADs to the left of the line indicate late first appearance of the taxon in section Y. Those to the right of the line indicate late first appearance in section X. If X is the composite standard, it can

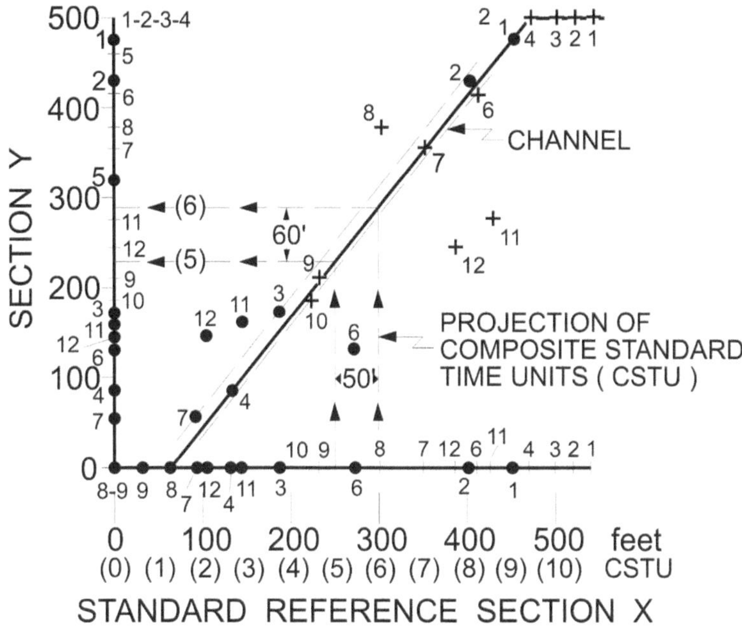



**Fig. 14.11** Typical data plot used in the graphic correlation method. Section X is chosen as the standard reference section, and section Y is any other section to be correlated with it. FADs are shown by circles and LADs by crosses, plotted along the axis of each section. Data points within the graph indicate correlations of FADs and LADs, and are the basis for defining the line of correlation (the *diagonal line*). Points off the line reflect incomplete sampling or absence as a result of ecological factors. Progressive correlations of other sections to the standard enables "true" ranges of each taxon to be refined in the standard section, resulting in a detailed basis for further correlation and the erection of standard time units (Miller, 1977)

be corrected by using the occurrence in section Y to determine where the taxon should have first appeared in the standard.

If the average, long-term rate of sedimentation changes in one or other of the sections, the line of correlation will bend. If there is a hiatus (or a fault) in the new, untested sections (sections Y), the line will show a horizontal terrace. Obviously, the standard reference section should be chosen so as to avoid these problems as far as possible. Harper and Crowley (1985) pointed out that sedimentation rates are in fact never constant and that stratigraphic sections are full of gaps of varying lengths (Sect. 13.1). For this reason, they questioned the value of the graphic correlation method. However, Edwards (1985) responded that when due regard is paid to the scale of intraformational stratigraphic gaps, versus the (usually) much coarser scale of biostratigraphic correlation, the presence of gaps is not of critical importance. Longer gaps, of the scale that can be detected in biostratigraphic data (e.g., missing biozones) will give rise to obvious hiatuses in the line of correlation, as noted previously.

The advantage of the graphic method is that once a reliable composite standard reference section has been drawn up it facilitates chronostratigraphic correlation between any point within it and the correct point on any comparison section. Correlation points may simply be read off the line of correlation. The range of error arising from such correlation depends on the accuracy with which the line of correlation can be drawn. Hay and Southam (1978) recommended using linear regression techniques to determine the correlation line, but this approach assigns equal weight to all data points instead of using one standard section as a basis for a continuing process of improvement. As Edwards (1984) noted, all data points do not necessarily have equal value; the judgment and experience of the biostratigrapher are essential in evaluating the input data. For this reason, statistical treatment of the data is inappropriate.

Figure 14.12 illustrates an example of the use of the graphic method in correlating an Upper Cretaceous succession in the Green River Basin, Wyoming, using palynological data (from Miller, 1977). The composite standard reference section has been converted from

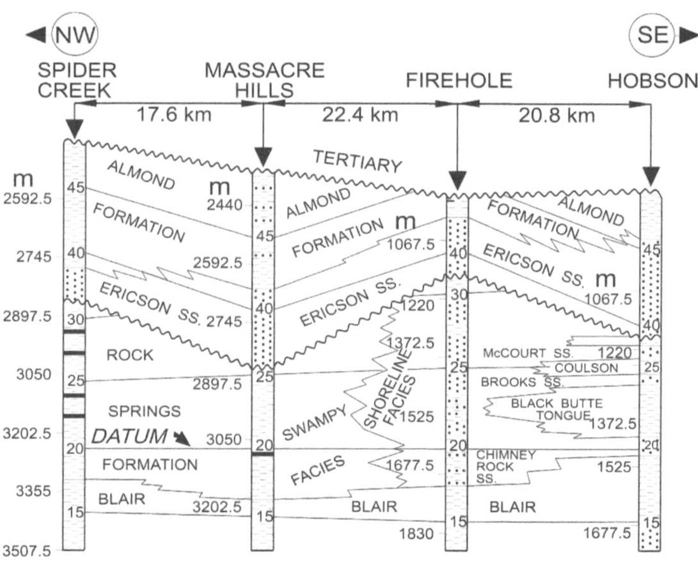



**Fig. 14.12** An example of the correlation of four wells through Cretaceous sections in Wyoming. Numbers within each log are composite standard time units derived from graphic correlation with the standard reference section. They can be used to generate correlations on an interval scale rather than the ordinal scale obtainable using conventional biostratigraphic zonations (Miller, 1977)

thickness into composite standard time units, by dividing it up arbitrarily into units of equal thickness. As long as the rate of sedimentation in the reference section is constant, these time units will be of constant duration, although we cannot determine by this method alone what their duration is in years. Isochrons may be drawn to connect stratigraphic sections at any selected level of the composite standard time scale. These isochrons assist in defining the architecture of the succession. For example, time unit 30 in Fig. 14.12 is truncated, indicating the presence of an unconformity.

An important difference between the graphical method and conventional zoning schemes is that zoning methods provide little more than an ordinal level of correlation (biozones, as expressed in the rock record, have a finite thickness which commonly cannot be further subdivided), whereas the graphic method provides interval data (the ability to make graduated subdivisions of relative time). Given appropriate ties to the global time frame, the composite standard time units can be correlated to absolute ages in years, and used to make interpolations of the age of any given horizon (such as a sequence boundary) between fossil occurrences and tie points. The precision of these estimates is limited solely by the accuracy and precision obtainable during the correlation to the global standard. MacLeod and Keller (1991) provided excellent examples of this procedure, and their results suggest an

obtainable precision of less than ±100 ka. Other examples of the use of graphic correlation are given by Scott et al. (1988), although no data plots are presented. In a later paper Scott et al. (1993) used graphic correlation methods in a study of core data to demonstrate diachroneity of some Cretaceous sequence boundaries of more than 0.5 million years. Additional examples of graphic correlation were given by Mann and Lane (1995).

The availability of large computers capable of storing and processing very large amounts of data has permitted the development of complex programs that can maximize the use of the information available in microfossil and palynomorph recoveries, which typically are very large. A Ranking and Scaling program is now available from www.stratigraphy.org, and a form of automated graphic correlation has been developed by P. M. Sadler (in Harries, 2003). This technique, called Constrained Optimization (CONOP), is designed to filter data to manage errors and inconsistencies in data sets, such as incomplete collecting, or misidentification of species, and to develop best-fit correlations. When taxa occurrences are ordered in this way, unconformities are identified by "clusters" of correlations, that is to say, an artifact of the program is that it identifies "anomalously high numbers of apparently coeval events at a single composite event level" (Crampton et al., 2006). As discussed below (Sect. 14.6) the application of this method can considerably facilitate the task of regional and interregional correlation, achieving precision in local, relative correlations in the $10^5$-year range.

### 14.5.4  Assessment of Relative Biostratigraphic Precision

The precision of biostratigraphic zonation is partly a reflection of the diversity and rate of evolution of the fossil group used to define the zonal scheme. This varies considerably over time and between different fossil groups. For example, Fig. 14.8c is a simplified version of a chart provided by Cox (1990) to illustrate the time resolution of various fossil groups used in the subsurface correlation of Jurassic strata in the North Sea Basin. The best time resolution is obtained from ammonites, which have been subdivided into biozones representing about 0.5 million years. Hallam (1992a) and Cope (1993) quoted detailed studies of British ammonites that yield a local resolution estimated at less than 200 ka, but such a level of accuracy cannot yet be extended globally for the purpose of testing the global cyclicity model, and ammonites are rarely obtainable in subsurface work. The microfossil groups that are typically used provide time resolution ranging from 1 to 4 million years. The sloping caps to each bar in Fig. 14.8c illustrate the varying length of biozones for each fossil group through the Jurassic. For example dinocyst zones each represent about 1 million years in the Late Jurassic but are about 3 million years long in the Early Jurassic.

Moore and Romine (1981), in a study of the contributions to stratigraphy of the DSDP project, examined the question of biostratigraphic resolution in detail. In 1975 (the latest data examined in this paper) the resolution of foraminiferal zonation, as expressed by the average duration of biozones, varied from 4 million years during much of the Cretaceous, to 1 million years during parts of the Neogene. Srinivasan and Kennett (1981) found that foraminiferal zones in the Neogene ranged from 0.4 to 2.0 million years. They suggested that

> Experience shows that this resolution seems to have reached its practical limit. This . . . is largely constrained by the evolution of important new species within distinctive and useful lineages. Further subdivision of the existing zones is of course possible when additional criteria are employed, but further subdivision of zones into shorter time-intervals does not guarantee a practical scheme for biostratigraphic subdivision; that is, such zones may not be widely applicable.

Combinations of foraminiferal zones with other microorganisms occurring in the same sediments, such as calcareous nannofossils and radiolaria, increase biostratigraphic precision (Moore and Romine, 1981; Srinivasan and Kennett, 1981; Cope, 1993), but not necessarily by a large amount. Figure 14.13 shows the biostratigraphic resolution that was achievable in 1975 based on combinations of all three fossil groups. The combination of three fossil groups does increases accuracy and precision (to a 0.3–2.0 million years range), but commonly zonal boundaries of more than one group coincide in time, so that no additional precision is provided. It was not thought likely that precision would increase by very much. In fact, the system of numbered microfossil zones that was established early during the DSDP project still formed the basis for the Exxon global cycle charts of the late 1980s. As noted earlier in this book, the analysis of different fossil groups from the same stratigraphic sections may also indicate that some groups are more facies controlled than others, and demonstrate diachronism. Cross-checking between these different groups may therefore be important in the reduction of biostratigraphic uncertainty (e.g., Ziegler et al., 1968; Armentrout, 1981).

McGowran (2005, p. 245) discussed progress since the review of Moore and Romine (1981). He noted that "zones are not only unequal in duration but are clustered, so that chunks of well-resolved time contrast with poorly resolved chunks." His update of the data for the Cenozoic is shown in Fig. 14.14, and shows that the duration of microfossil biozones ranges from a minimum of 200 ka for planktonic foraminifera in the Pliocene, to 2 million years for planktonic microfossils of all types in the Oligocene. All the evidence indicates wider spacings for sequence boundaries and bioevents in the Eocene, which may reflect unknown oceanographic or climatic causes.

It has been suggested that because sequence boundaries are dated primarily by biostratigraphic data, they should be referred to and correlated on this basis, without reference to the absolute time scale, in order to circumvent the imprecisions associated with this scale (as discussed in the next sections). However, even

**Fig. 14.13** Biostratigraphic resolution of marine strata based on combined zonation of foraminifera, radiolaria and calcareous nannofossils, expressed as the number of biozone boundaries per million years, averaged over individual epochs. Average resolution as it existed in 1969, at the beginning of the DSDP project, is shown by the blue, cross-hatched columns. Improvements up to 1975 are indicated by the *yellow top* to each column (Moore and Romine, 1981)

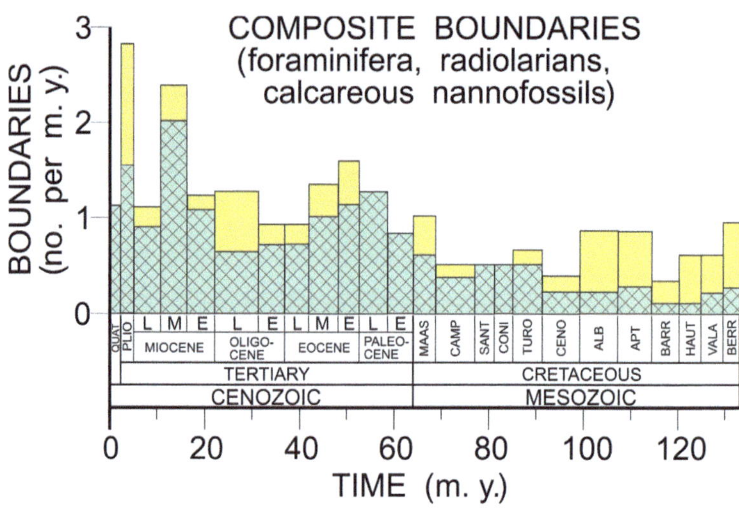



**Fig. 14.14** Histograms of boundaries per million years, averaged over each sub-epoch. The *bottom chart* shows two sets of zone-defining calcareous microfossil FAD's and LADs, based on Berggren et al. (1995). The *middle diagram* provides the same information for calcareous microfossils. The seemingly greater refinement in subdivisions may simply reflect more study of these taxa. The *top diagram* compares the density of sequence boundaries in the Haq et al. (1987, 1988a) global cycle chart with that of Hardenbol et al. (1998). All three diagrams reveal the results of progress since the start of the DSDP project, but the third diagram may simply reflect the application of the deductive model to the interpretation of the sequence record, as discussed in Chap. 12. Diagram from McGowran (2005)

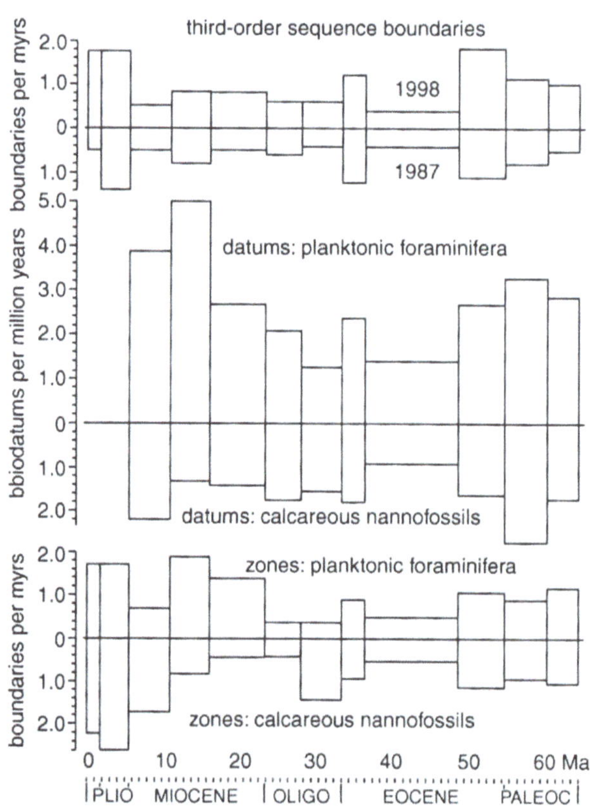



where no attempt is made to provide absolute ages for sequence boundaries, this discussion has shown that a built-in biostratigraphic imprecision of between about one half million years and (in the worst case) several million years must be accepted for the ages of sequence boundaries. Errors are larger for earlier parts of the Phanerozoic time scale. A potential error in this range, which is typical of most regional biostratigraphic frameworks, is already too great to permit the interregional correlation of sequences that are less than a few million years in duration (most of the "third-order" cycles that constitute the main basis for the global cycle chart).

As noted in the previous section, the use of graphic correlation and its automated equivalents can yield precision in relative correlations in the $10^5$-year range. However, until the biostratigraphic framework becomes more completely global in nature, and can be supplemented more extensively with locally-fixed radiometric or magnetostratigraphic tie points, this does not directly aid in the testing of *global* stratigraphic events.

### 14.5.5 Correlation of Biozones with the Global Stage Framework

If a succession of sequences is to be tested for a possible global eustatic imprint, it must be correlated with contemporaneous successions around the world. As discussed in Sects. 11.4 and 11.5, some researchers, commencing with Vail, have claimed that a eustatic signature is demonstrated if successions in different parts of the world spanning the same time interval contain the same number of sequences. But in the modern era of high-precision chronostratigraphy this is simply not good enough. Each sequence boundary

must be correlated precisely, independently of others, and assessed as a potential global event on that basis.

A significant problem with inter-regional correlation is that of faunal provincialism. For example, Berry (1987) reported that correlation of Early and Middle Ordovician graptolite faunas between Europe and North America was fraught with controversy because of faunal provincialism. At that time the proto-Atlantic (Iapteus) Ocean was at its widest. Similarly, in a classic study of global ammonite distributions, Kennedy and Cobban (1977) demonstrated the diversity of styles of distribution, including latitudinal and longitudinal restrictions imposed by oceanic barriers and climatic differences (Fig. 14.15). Correlation between boreal (northern) and tethyan faunas across the Tethyan Ocean was at that time difficult within the "Old World" of Europe and Africa, but was facilitated by the fact that many of the same genera occurred within the Western Interior Basin of North America, where intermingling of faunas was brought about by transgressions and regressions accompanying sea-level changes, and by shifts in climatic belts. As Hancock (1993a, p. 8) stated "Every wide-ranging stratigrapher working on the Mesozoic meets the difficulty of correlations between boreal and tethyan realms." Hancock (1993a, p. 8) also stated "It is seldom realized by geologists at

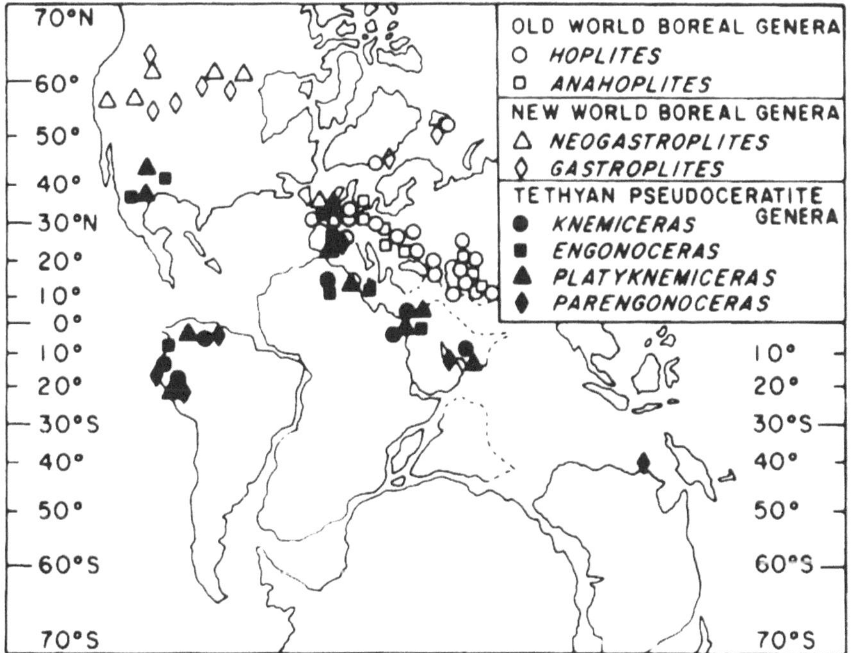


**Fig. 14.15** Examples of latitudinally-distributed ammonite genera (Kennedy and Cobban, 1977)

large how insecure most zonal schemes are, and how few are the regions in which any one scheme has been successfully tested." A synthesis with which he was involved (Birkelund et al., 1984; see also Hancock, 1993b) noted as many as nine different possible biostratigraphic standards that could be used to define specific stage boundaries in the Cretaceous. Many are partly in conflict with one another. Faunal provincialism is also a problem with microfossils, including most of those favoured by biostratigraphers for intercontinental correlation. Thierstein (1981) and Roth and Bowdler (1981) described latitudinally-controlled biogeographic distribution of Cretaceous nannoplankton. The latter authors also described neritic-oceanic biogeographic gradients.

Surlyk (1990, 1991) studied the sequence stratigraphy of the Jurassic section of East Greenland and derived a sea-level curve for this area, comparing it to the curves of Hallam (1988) and Haq et al. (1987, 1988a). In doing so he noted "that the correlation between the Boreal and Tethyan stages across the Jurassic-Cretaceous boundary is only precise within 1/2 stage, rendering the eustatic ... nature of sea-level curves rather meaningless for this time interval."

The problems of biogeography and faunal provincialism are substantially reduced by the incorporation of independent methods of correlation, including magnetostratigraphy, radiometric dating and chemostratigraphy, particularly (for the Cenozoic) the

$\delta^{18}O$ record, as discussed in the next section. However, much modern work also makes reference to the global cycle chart of Haq et al. (1987a, 1988a) or the more recent work of Hardenbol et al. (1998) or other chapters in Graciansky et al. (1998). As this part of the book should, by now, have made clear, this is highly problematic, because it raises the question of whether and by how much correlations are adjusted to fit the "requirement" to line up sequence boundaries against the boundaries appearing in one of those charts.

A good example of the problem is the case of the Cenozoic Paratethys. This an area now occupied by central Europe (Austria, Hungary, and the Czech and Slovak republics) which, during the Mesozoic was part of the Tethyan Ocean. With the gradual closure of Africa-Arabia against Asia, the ocean was transformed into a series of small oceans and intervening landmasses impeding or blocking marine circulation. Faunas became increasing provincial by the end of the Oligocene, and by the mid-Miocene the area had become almost completely isolated.

The recognition of Paratethys as a biogeographic entity with Neogene faunal assemblages different from the those of the Mediterranean dates back to the 1920s (Piller et al., 2007). McGowran (2005, pp. 312–316) and Piller et al. (2007) discussed the research, commencing in the 1960s, to integrate the regional biozone framework with the better-known Mediterranean chronostratigraphy. Figure 14.16 illustrates the state

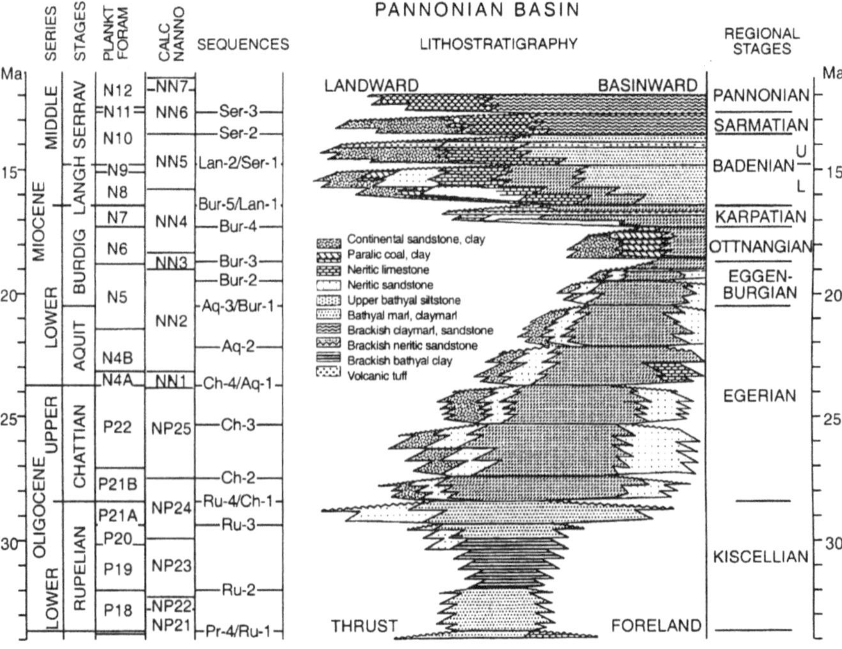


**Fig. 14.16** Lithostratigraphy and chronostratigraphy of the Paleogene Hungarian basin and the Early Neogene Pannonian basin in Hungary. The Geochronological correlations shown at *left* are from Berggren et al. (1995), the sequences from Hardenbol et al. (1998). Simplified by McGowran (2005) from Vakarcs et al. (1998)

of Paratethys correlations at the end of the 1990s. Figure 14.17 provides the latest version of the correlation framework. A careful comparison between the two charts reveals some revisions in the ages of the Paratethyan and standard stages. This is to be expected. However, what is unclear is the extent to which the "sea-level curve" has been used to calibrate or tune the chronostratigraphy. Both charts show the sequence framework of Hardenbol et al. (1998). Here is the discussion in Piller et al. (2007, p. 152) of the Egerian Stage, the regional Paratethyan stage that straddles the Oligocene-Miocene boundary:

**Correlation:** This stage straddles the Oligocene/Miocene boundary (Baldi and Senes, 1975) in comprising the upper part of the Chattian and the lower part of the Aquitanian (fig. 1). As pointed out by Baldi et al. (1999) the distribution of larger benthic foraminifers implies a correlation of its lower boundary with the lower boundary of the Shallow Benthic Zone SBZ 22,

that is calibrated in the Mediterranean and NE Atlantic with the base of the planktonic foraminiferal zone P22 (Cahuzac and Poignant. 1997). Moreover, the recalibration of 3rd order sea level sequences supported by the biostratigraphic results of Mandic and Steininger (2003), implies the position of the upper Egerian boundary in the mid-Aquitanian, and not at its top (fig. 1). Although suggested already by Hungarian stratigraphers (e.g., Baldi et al., 1999), this interpretation contrasts substantially with the current stratigraphic concept (e.g., Rogl et al., 1979; Rogl and Steininger, 1983; Steininger et al., 1985; Vakarcs et al., 1998; Rogl, 1998b; Mandic and Steininger, 2003). The Paleogene/Neogene boundary is difficult to detect in the Central Paratethys since the index fossil for the Aquitanian, *Paragloborotalia kugleri,* is absent. Correlations are usually based on calcareous nannofossils including uppermost NP 24 to NN 1/2 nannozones (Rogl, 1998b). In addition, *Miogypsina* species are very useful for biostratigraphic correlation. Whereas the lower Egerian deposits belong to SBZ 23 the upper Egerian limestones with *Miogypsina gunteri* found at Bretka (E Slovakia) (Baldi and Senes, 1975) belong to the lower part of SBZ 24 and thus

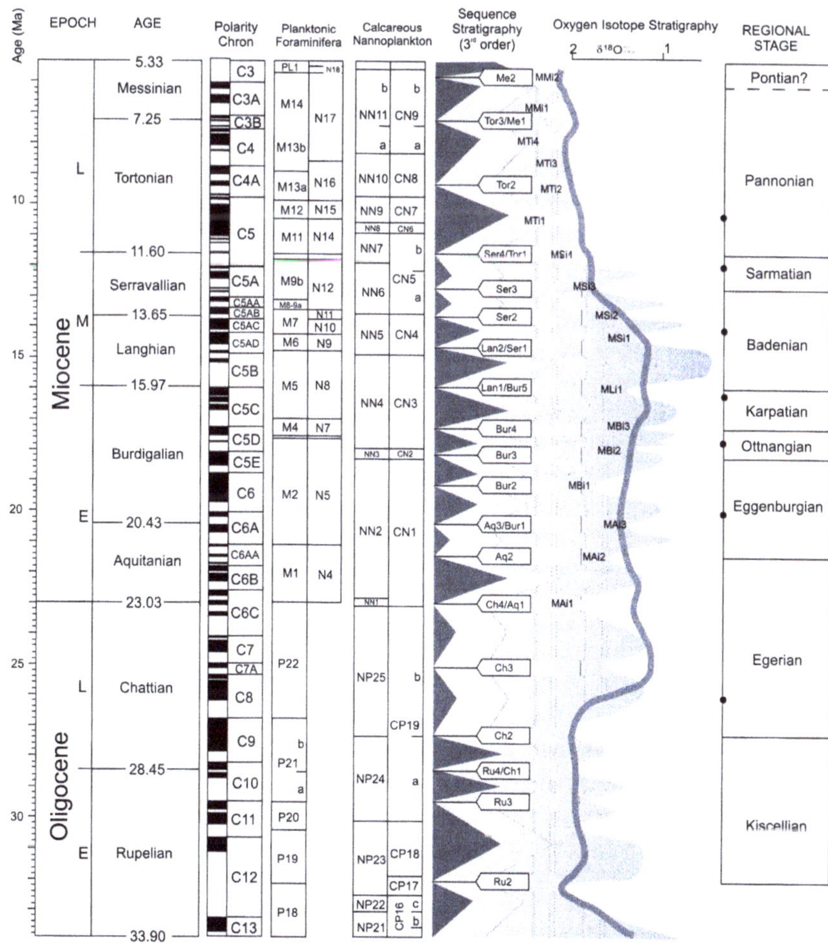

**Fig. 14.17**
Oligocene-Miocene geochronology for Paratethys, from Piller et al. (2007)

correspond to the lower Aquitanian (Cahuzac and Poignant, 1997). Consequently, in terms of sequence stratigraphy the Egerian/Eggenburgian boundary corresponds with the Aq 2 sea level lowstand of Hardenbol et al. (1998). The following 3rd order transgression-regression cycle already includes Eggenburgian deposits (see below). This interpretation is in accordance with the general regressive trend in the upper Egerian sediments and with erosional unconformities frequently forming their top. In continuous sections the sediments at the boundary were often deposited in very shallow water environments characterized by brackish water faunas. Continuous deep marine sections are only known from the strongly tectonised thrust sheets of the Outer West Carpathians and their equivalents (Krhovsky et al., 2001).

This is an excellent example of the type of detailed comparison between different biostratigraphic assemblages and biozones that comprises the basis for the modern geological time scale. But note the inclusion of such phrases as "the recalibration of 3rd order sea level sequences", "Consequently, in terms of sequence stratigraphy the Egerian/Eggenburgian boundary corresponds with the Aq 2 sea level lowstand", and "The following 3rd order transgression-regression cycle already includes Eggenburgian deposits." Which way is the correlation or correction going in this process? Is the biostratigraphy being adjusted against the sequence classification of Hardenbol et al. (1998), or is the independently-established chronostratigraphy being used as a test or confirmation of the Hardenbol chart? This is not made clear.

Over most of the Oligocene-Miocene interval, foraminiferal and nannoplankton biozones have durations of a million years or longer (Fig. 14.17), and over most intervals in this time span, these durations exceed those of the sequences comprising this portion of the Hardenbol chart. How important are additional calibrations provided by magnetostratigraphy and radiometric dating? Such cross-calibration is nowadays an integral and essential part of the process by which the global time scale has been brought to its present state of refinement (see next section). However, it is not clear whether and by how much such additional documentation has been used, to calibrate either the sections used by Hardenbol et al. (1998) or those used to construct the Paratethys regional timescale. Without such independent data, correlations between the Hardenbol chart and regional time scales, such as those shown in Figs. 14.16 and 14.17 need to be evaluated with caution because of the possible contamination by circular reasoning.

The closure of Tethys by the collision of the Indian and Arabian plates with Asia also separated the classical European-Mediterranean seas from those of the far east. As described by McGowran (2005, pp. 316–325) biostratigraphic correlation of Cenozoic marine strata in such areas as Indonesia (important in its own right for the purposes of petroleum exploration) has long been hampered by faunal provincialism. His own work on the larger foraminifera resulted in the chart reproduced here as Fig. 14.18. A system of zones identified by letters had been established in the 1920s; the current framework is shown in this figure. Note that all of these zones are now known to span several million years; some are more than 10 million years in duration. They are inadequate on their own, therefore, to provide the basis for tests of the $10^6$-year sequence framework of the Hardenbol chart.

The Indo-Pacific zonal scheme shown in Fig. 14.18 is of interest because it provides a particularly clear example of a latitudinal control on faunal distributions. Note that most of the genera used in the construction of the scheme are limited to a zone between about 16° north and south of the paleoequator, with some rare occurrences occurring as far north or south as latitude 50°.

### 14.5.6 Assignment of Absolute Ages and the Importance of the Modern Time Scale

The test for global eustasy is to match precisely dated sea-level events, such as sequence boundaries, in tectonically unrelated successions around the world. This requires that each succession be dated to a high level of accuracy and precision. The Hardenbol et al. (1998) system of Mesozoic-Cenozoic sequences is widely regarded as one of the best established succession of sequence boundaries, and largely because of its pedigree—it represents the expanded and updated version of the Haq et al. (1987, 1988a) chart which, itself, was updated from Vail et al. (1977)—it has achieved an informal, unofficial (but nonetheless widespread) status as some sort of global standard. However, it is no such thing. At best, it represents an amalgam of largely European stratigraphic events, but as discussed at length in Chap. 12, even this status could be challenged.

**Fig. 14.18** The foraminiferal letter zone scheme in use for the tropical Indo-Pacific region, as revised and updated by McGowran (2005). Most genera are confined to tropical latitudes, but occasional examples of the assemblages are found as far north or south as latitude 50°. This is thought to indicate that the forms in question were more pandemic or cosmopolitan in their ecological tolerance

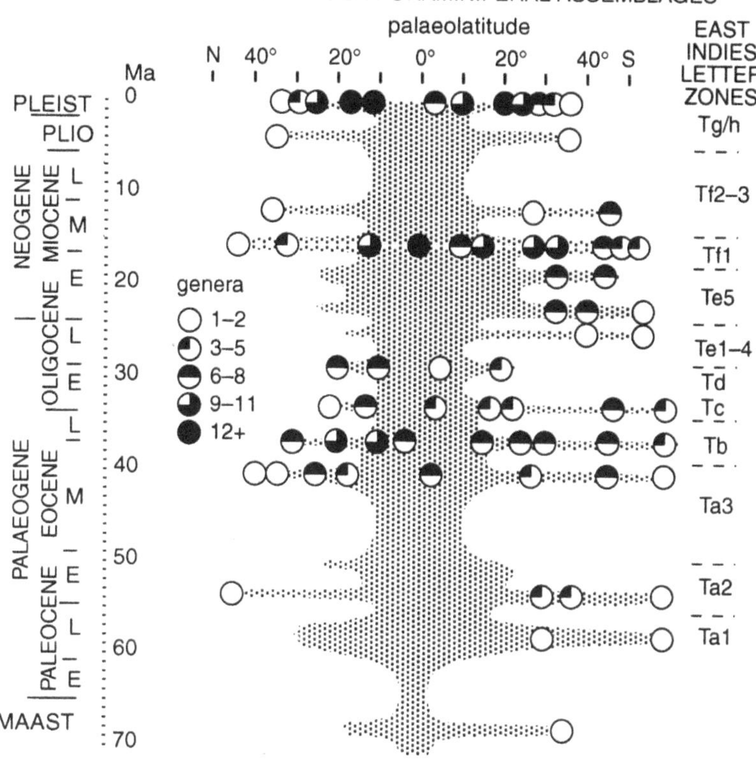



Given the discussion of regional time scales in the previous section it is noteworthy that the Hardenbol et al. (1998) chronostratigraphic framework is built from an amalgamation of scales from different locations. For example:

In the Upper Cretaceous (Coniacian through Campanian) sequence boundaries identified in boreal and tethyan basins could not be calibrated reliably to the temporal framework. Instead, for the Coniacian through lower Maastrichtian interval, a record of sequence boundaries from North America is included on the chart which could be calibrated to the North American ammonite zones included by Gradstein et al. (1994) in their Mesozoic time scale. Sequences of Haq et al. (1987) in the Coniacian through lower Maastrichtian interval were also based on the North American record and were tentatively calibrated to the standard stages (Hardenbol et al., 1998, p. 7).

The Hardenbol chart is therefore not even based entirely on European data, and cannot be considered a completely reliable basis for assessing European events. This appears to have been forgotten in the execution of the research project to document the sequence record on the New Jersey continental margin, (the work of K. G. Miller and his colleagues; e.g., Browning et al., 2008), in which one of the major objectives of the project and a central test of the eustatic origin of the succession of sequences was to compare the New Jersey record against that of Hardenbol et al. (1998)! It is a fair question, in evaluating the correlation of the New Jersey record with that of Hardenbol et al. (1998), to ask to what extent the New Jersey succession is, in part, being correlated with itself?

It is one thing to establish a detailed chronostratigraphic framework and template for regional correlation and dating, as Graciansky et al. (1998) set out to do as a basis for the new cycle chart. It is something else entirely to use that framework for the dating of local successions of sequences. The ages of sequence boundaries at each location are only as accurate and precise as the chronostratigraphic data that are available *at that location* to date the succession. These are important arguments, because the Hardenbol chart is widely used as a template against which other successions are compared and dated. This last step, to use the Hardenbol chart to calibrate or tune successions in other parts of the Earth, is the step that introduces the hazard of circular reasoning into stratigraphic synthesis, and makes the results of global tests particularly difficult to evaluate.

The accuracy and precision of time resolution varies with the chronostratigraphic methods used and the level of the stratigraphic column. Various authors have estimated the dating precision over various intervals of geological time, and have assessed the incremental improvements in the global time scale that have been built from local, regional, and inter-regional correlation programs. Kidd and Hailwood (1993) estimated the potential resolution for various time slices back to the Triassic (Table 14.2).

Kauffman et al. (1991) described a method of high-resolution correlation that utilizes biostratigraphy (graphic correlation methods), magnetostratigraphy, chemostratigraphy and the correlation of event beds. Numerous datable ash beds are present in his sections. A potential uncertainty of only ±100 ka is claimed for Cretaceous beds of the Western Interior of the US. However, this provides only a regional framework, and it is unlikely that it could be extended to other unrelated basins with the same degree of precision.

Miller (1990) indicated that chronostratigraphic resolution of Cenozoic sections could ideally attain accuracies of ±100 ka, based on the use of modern chronostratigraphic techniques and biostratigraphic data bases, but he conceded that uncertainties of 0.5–2.0 million years are common in many actual case studies. Aubry (1991), in a discussion of the early Eocene record of sea-level change, stated that under ideal conditions combinations of biostratigraphic (mainly microfossil) and magnetostratigraphic data should permit dating to within 0.2–0.3 million years. Yet commonly, according to her, the data from specific locations are inadequate to permit the use of

all available tools. This is another of the problems with the use of sequence frameworks such as the Hardenbol chart as the basis for regional or global correlation. Even if we could accept that they provided a valid basis for global correlation (which they do not), the potential error associated with the dating of each sequence boundary at each location is not stated.

Continuing revisions of the geological time scale result in repeated adjustments of assigned ages for major boundaries. Figure 14.19 illustrates the refinements in the Mesozoic time scale that had been achieved up to 2004. The shifts in assigned ages of up to at least 5 million years, in some cases, should be borne in mind in assessing the validity of correlation frameworks that compare data from widely separated sites constructed over different time periods and based on different chains of chronostratigraphic reasoning. The revisions are far from complete. There remain a significant number of GSSPs that have not been formally and officially designated, and residual disputes, relating to new field data, the recognition of potentially significant "event" markers, and debates about the importance of precedence and priority mean that some important chronostratigraphic boundaries remain in dispute. For example, Berggren (2007) and Fluegeman (2007) discussed residual disagreements concerning the establishment of stage/age boundaries in the Cenozoic. Three different methods of defining the top of Eocene-Oligocene boundary in the type Priabonian section, all based on good science or historical precedent, differ in practice by about 30 m of section, which translates into a time difference of about 0.4 million years.

There is no question that the work of the International Commission on Stratigraphy over the last 2 decades has brought about dramatic improvements in the accuracy, precision, and applicability of the global time scale. But the dating of each sequence boundary must go back to the first principles of stratigraphic dating until such time as we can convincingly demonstrate the origins of a given sequence framework. Only with that theoretical framework in place should we then be permitted to turn to a deductive model of sequence stratigraphy and use a given sequence chronostratigraphy as a template for regional or global correlation. As discussed in a later section (Sect. 14.7) the astrochronological time scale is beginning to permit this important interpretive step in certain limited situations.

**Table 14.2** Achievable resolution for integrated stratigraphy in marine successions

| Quaternary | <1–3 ka |
| --- | --- |
| Late Cenozoic | 5–10 ka |
| Early Cenozoic | 10 ka – 1 million years |
| Late Cretaceous | 100 ka – 1 million years |
| Early Cretaceous | ~10 million years |
| Jurassic | 50–150 ka |
| Triassic | 225 ka – 2 million years |

Simplified from Kidd and Hailwood (1993)

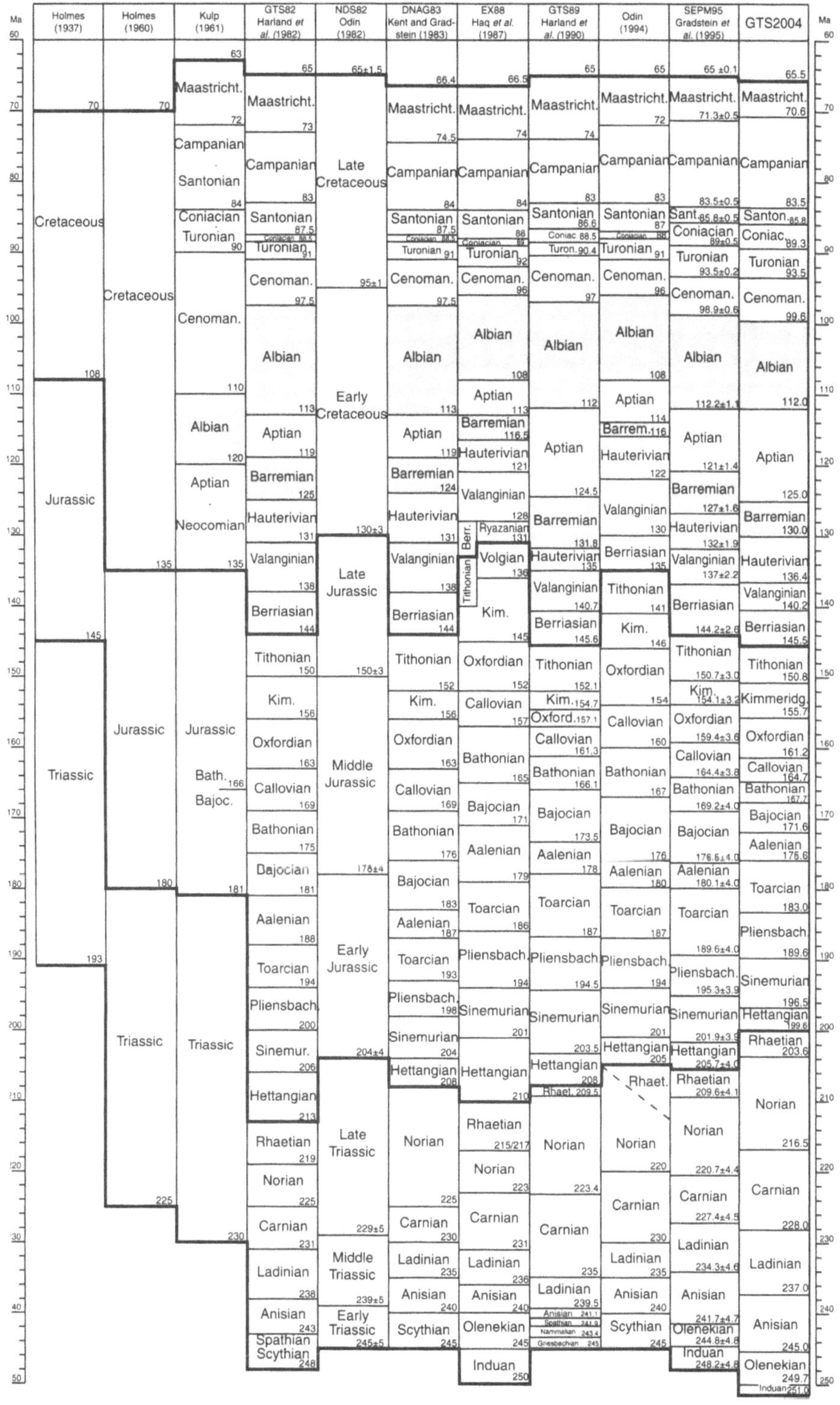


**Fig. 14.19**  A comparison of Mesozoic stage-boundary ages in selected time scales, including the first attempt at a quantitative time scale by Holmes (1937). GTS2004 refers to the source of the diagram, the time scale by Gradstein et al. (2004)

Until this ideal status is reached, the methods for stratigraphic dating and correlation need to follow rigorous empirical procedures and must take account of the basic problems illustrated in Figs. 14.20 and 14.21. Each sequence boundary needs to be assessed relative to whatever chronostratigraphic data are available in the immediate stratigraphic vicinity. Figure 14.20 illustrates a common situation: a stratigraphic event of interest—it could be a sequence boundary or an important biohorizon— is bracketed by volcanic ash beds that provide for an accurate estimate of absolute age. However, the calculation of this age (see

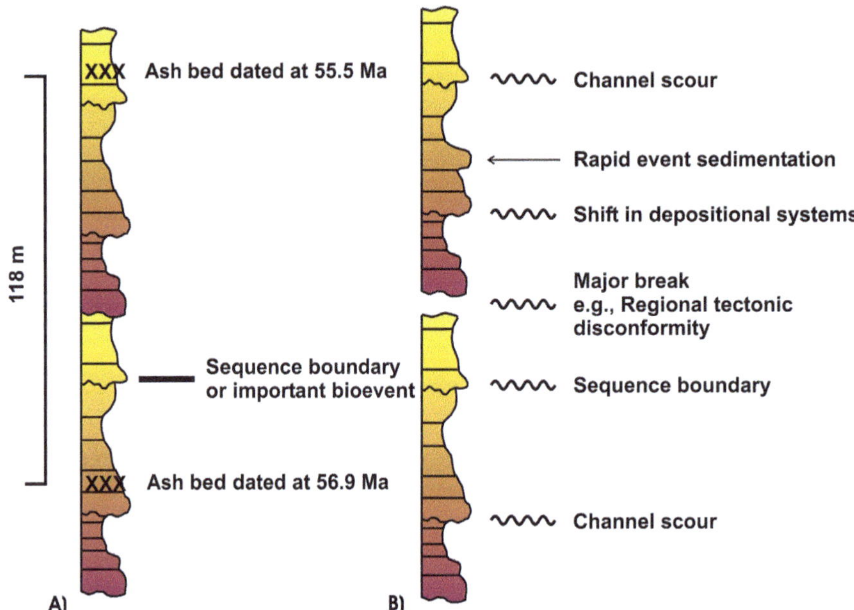



**Fig. 14.20** *Left:* The standard bracketing method of estimating ages for a bed that cannot be directly dated. In this hypothetical section, 118 m of beds accumulated in (56.9–55.5=) 1.4 million years. The average sedimentation rate is therefore 118/1.4 m/million years = 84 m/million years. The sequence boundary is 30.3 m above the *lower* ash bed. Assuming a constant sedimentation rate and no hiatuses, 30.3 m of beds accumulate in 30.3/84 = 0.36 million years. Therefore the age of this bed is 56.9–0.36 = 56.54 Ma. *Right:* the problem is that, in practice, real sections are replete with breaks in sedimentation and are normally characterized by variable sedimentation rates. The use of simple arithmetic proportion is therefore unlikely to yield very reliable results

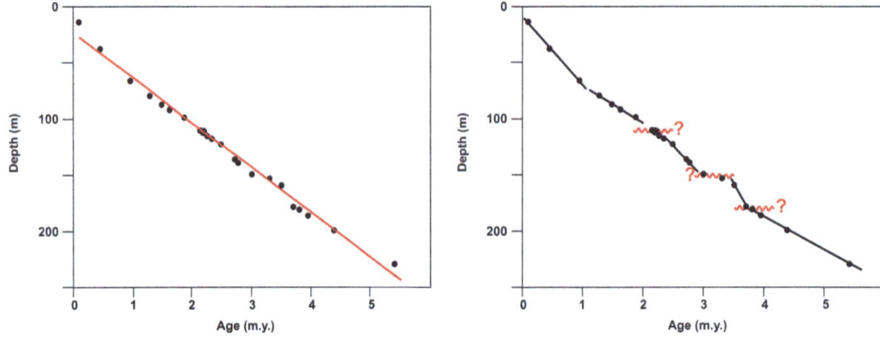



**Fig. 14.21** The problem of calibration. The data points in both graphs represent the same set of real data from a well in the Gulf Coast, off Florida (Roof et al., 1991). Each point is a biomarker that has been dated against the available time scale. The question then arises, does this data represent continuous sedimentation at a constant rate? The *straight line* of correlation in the *left-hand figure* suggests that this is the case, making allowance for small errors in the dating of the fossils. But it is possible to look at the data a different way, as seen in the *right-hand figure*. See text for further discussion

caption to the figure) requires that the section between the ash beds accumulated at a constant rate and contains no sedimentary breaks. This condition is commonly not satisfied, and it is often difficult to ascertain that this is the case. Inconspicuous breaks in sedimentation, the presence of subtle sour surfaces, surfaces of microkarst, beds indicating rapid event sedimentation, such as storm beds or turbidites, all complicate the application of simple arithmetic proportion methods for determining the age of a bed of interest. This problem is one that is almost universally encountered, because it is rare for a stratigraphic event of importance, such as a sequence boundary, to be associated with accurately datable chronostratigraphic material on the same horizon or in the same bed.

Figure 14.21 takes this problem one step further, to the averaging that may be attempted for an entire section. The data set is from an offshore drilling project that set out to examine the record of Milankovitch cyclic frequencies in a drill core from the Gulf of Mexico. The data points in Fig. 14.21 represent individual biohorizons occurring in the drill core, dated using the time scale of Berggren et al. (1985a, b). The same set of data points is used in both the graphs in this figure. On the left, the points have been interpreted in terms of a simple straight line intersecting the points as closely as possible. This correlation looks reasonable, and could be used to calculate an average sedimentation rate for the entire succession. However, this ignores the problems raised with respect to Fig. 14.20, the possibility of breaks in sedimentation and variable sedimentation rate. A straight-line correlation was not, in fact, suggested by Roof et al. (1991), who discussed such influence as influxes of sediment from the nearby Mississippi delta. In their graph they simply joined each point to its neighbours, but despite the irregularities this reveals, they proceeded to assume a constant sedimentation rate and used a linear transformation of core depth to time for the purpose of data display and calculation of cyclic frequencies.

Another way of looking at the data is indicated by the right-hand graph of Fig. 14.21. Some limited clusters of adjacent points appear to lie on short straight-line segments, suggesting changes from one constant sedimentation rate to another. Some pairs of points are situated nearly horizontally relative to each other, suggesting a jump in age over a limited vertical space. This could indicate the presence of disconformities, a possibility not discussed by Roof et al. (1991). How can we distinguish between all the possibilities for correlation and dating of these data points? The only possible answer is to return to the original cores and dissect them in great detail, with a focus on sedimentary textures and structures with the objective of constructing a more detailed sedimentary history.

The modern geological time scale (GTS2004: Gradstein et al., 2004, and the updated scale maintained at www.stratigraphy.org) takes into account all of the issues raised with respect to Figs. 14.20 and 14.21 by collating data from multiple sources. The construction of Composite Standard Reference Sections using graphic correlation methods (Sect. 14.5.4) is part of this process. Currently finalized Global stratotypes for systems, series and stages are identified in Gradstein et al. (2004) and on the website, with references to published documentation, most of which consists of reports in the journal *Episodes* by representatives from boundary working groups. Realistic error estimates are provided for Phanerozoic stages, and range from very small values ($10^4$-year range) for most of the Cenozoic, the time scale for which is increasingly linked to an astrochronological record, to as much as $\pm 4$ million years for several stages between the Middle Jurassic and Early Cretaceous (Fig. 14.22).

Correlation of sequence successions into the geological time scale cannot avoid the types of systematic problem discussed above with reference to Figs. 14.20 and 14.21. At the risk of being repetitive, it must be pointed out, again, that the average event spacing of 1.12 million years for the Hardenbol et al. (1998) Mesozoic-Cenozoic chart (Sect. 12.5) is less than the potential error for stage boundaries in the geological time scale over virtually all of the time span of the Mesozoic (the time scale as it existed in the late 1990s), although Hardenbol et al. (1998) did not provide error estimates. Error estimates for the Cenozoic are now much smaller because of the introduction of astrochronology, but this did not constitute part of the basis for the time scale or the field stratigraphic records used by Hardenbol et al. (1998). The Late Cretaceous record contains 28 sequences, according to Graciansky et al. (1998). The average spacing of these events is 1.2 million years. Potential error on stage boundary ages in the modern time scale for the Upper Cretaceous range from 0.6 to 1.0 million years, to which must be

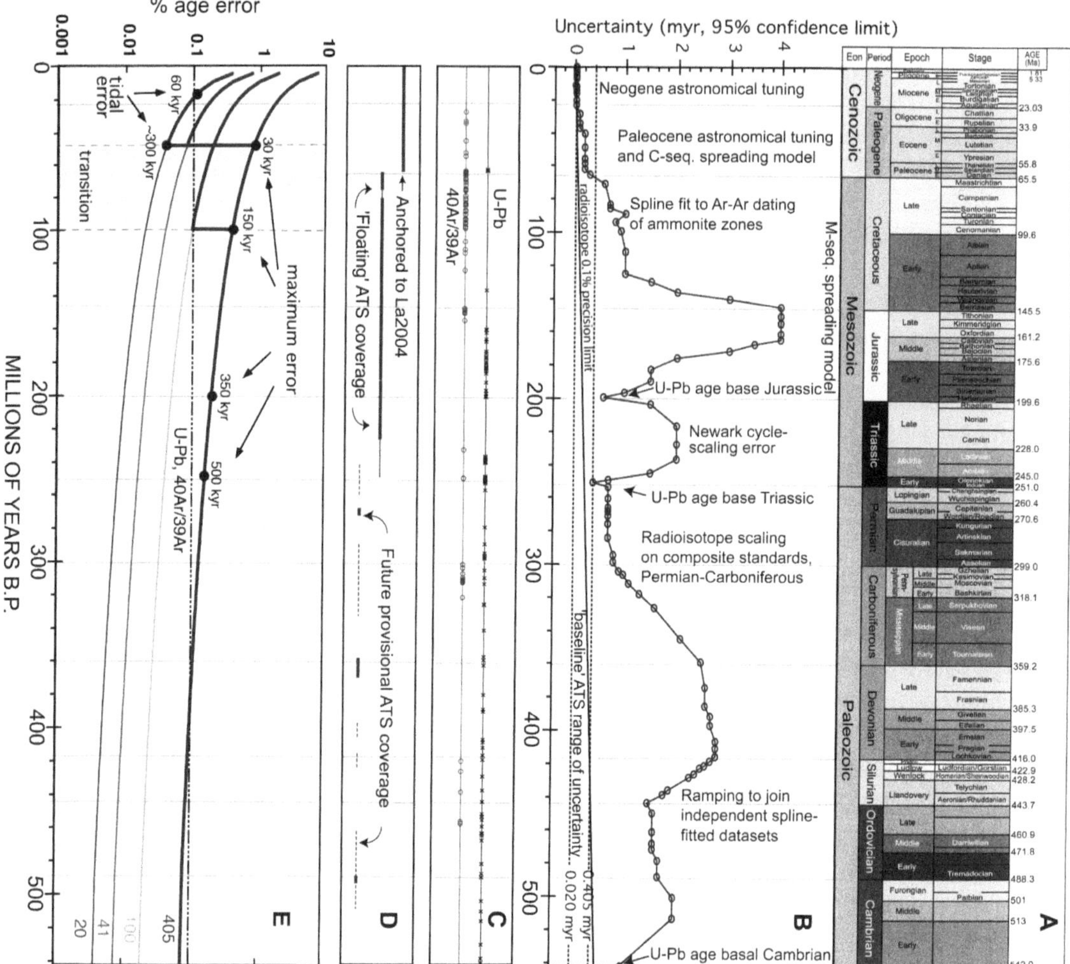


**Fig. 14.22** (**a**) The standard divisions of the Phanerozoic International Geologic Time Scale (Gradstein et al., 2004). (**b**) Estimated uncertainty (95% confidence level) in the ages of stage boundaries. (**c**) Distribution of radiometric ages used in the construction of the time scale. (**d**) Documented and poten-tial astrochronological time series. (**e**) Estimated error to be expected when the astrochronological time scale through the Phanerozoic is completed and integrated into the International Geologic Time Scale. This optimistic scenario is discussed in Sect. 14.7. Diagram from Hinnov and Ogg (2007)

added any error incurred in the process of correlating sequence successions from their type localities into the standard time scale.

The basis for modern chronostratigraphic methods should be rigorous empiricism. For example, the purpose of defining chronostratigraphic boundaries within continuous sections—where "nothing happened", to use McLaren's (1970, p. 802) felicitous phrase—is precisely in order to avoid having to interpret the meaning of unconformities or the significance of a prominent stratigraphic event that might otherwise make for a readily recognizable local marker. In other words deductive models are avoided at all costs. As Berggren et al. (1995) put it in their Introduction to a major compilation of new studies of geologic time: "The biostratigrapher deals not so much with falsification of rival hypotheses, the definitive mode of scientific reasoning described by Karl Popper, but rather with the progressive enhancement of what is already known."

That does not mean to say that other factors have not come into play in the selection of boundaries and of boundary stratotypes. Simmons et al. (1997) referred to "political and personal preferences and pressures" in the choice of one location over another. However, by and large the principle is to let the definition of boundaries evolve from the natural empirical data base that is built up world wide for each interval of geologic time. Such data bases are now complex and multi-faceted, involving not just multiple biostratigraphic records, but radiometric dating, magnetostratigraphy, and a growing body of chemostratigraphic evidence (e.g., see Gradstein et al., 1995, 2004). Notably, however, there remain serious local problems relating to the diachroneity of biohorizons, limitations in the use of magnetostratigraphy at low latitudes, imprecision of radiometric ages, and the effects of species variation on oxygen isotope stratigraphy (Kidd and Hailwood 1993; Smith, 1993). Several examples of problematic boundary definitions were discussed by Berggren (2007). Elsewhere (Miall, 2004) I discuss a recent shift from the emphasis on empirical data synthesis in the definition of chronostratigraphic units, with some workers favouring the use of prominent marker horizons as the basis for definition, even where these require revisions of long-established stage boundaries (see also Sect. 1.3.3).

The whole edifice of our geological time construction is in danger of losing its empirical purity if deductive models are brought in to play a major role in the perfection of the time scale. These range from the local problems alluded to above, any of which are susceptible to the effects of personal choice and bias, to much larger and potentially more fundamental problems. The dangers and contradictions involved in the use of models based on sequence stratigraphy have now been well aired (Chap. 12; Miall and Miall, 2001), and certainly include the concept of using sequence boundaries as empirical indicators of global sea-level events. Many of the same comments could be applied to the use of "event stratigraphy" as a basis for dating and correlation. While the most famous of geological "events" in the stratigraphic record, the Cretaceous-Tertiary boundary, now appears to be extremely well established worldwide as a universal geological marker, other events are not so clearly demarcated. Extinction events, major volcanic events, regional storm events and other large scale, sudden processes on Earth are commonly assumed to have left regional or global stratigraphic signatures. While many of these may turn out to be of global significance (e.g., the extinction event marking the end of the Permian), all such events are the product of deductive models that build an hypothesis of extent and significance from an originally small, localized field data

base. There exists a strong temptation to look for confirming evidence of preconceived concepts or models, rather than to rigorously test and attempt to falsify preliminary hypotheses. Stratigraphers should beware! These remarks are relevant to the application of the so-called *"hyper-pragmatic"* approach (Castradori, 2002) to the definition of chronostratigraphic boundaries (see also Miall, 2004, and Sect. 1.3.3).

As Aubry (2007, p. 133) wrote, in defence of the traditional empirical definition of stratotypes and boundary points:

> Chronostratigraphy is in danger of dogmatism when stratigraphic levels are selected on the claim that they yield the expression of temporally significant events, whether evolutionary, geophysical or geochemical. Such chronostratigraphic boundaries form an inflexible system in which the time-rock relationships are difficult to appraise. We must never forget that we do not date and correlate events, we date and correlate strata!

The greatest dangers, in my view, now involve the growth of the field of "cyclostratigraphy." House (1985, following a suggestion by G. K. Gilbert in 1895) was amongst the first to propose that the record of orbital frequencies preserved in the rock record could assist in the refinement of a geologic time scale by providing calibration of biostratigraphic data with a $10^4$–$10^5$-year precision. House and Gale (1995, Preface) remarked that "the reality of orbital forcing of climate was established as a fact" in the 1970s. Is this a reasonable statement? It is instructive to trace the development of this idea in the geological literature, which we do here in Sect. 14.7

## 14.6 Modern Tests of the Global-Eustasy Paradigm

In the first edition of this book, Chap. 14 was devoted to a comparison of early and later versions of parts of the Exxon global cycle charts with each other and with other independently constructed charts of changes in sea level through geologic time, focusing mainly on the Mesozoic record. The analysis revealed virtually no consistency within this body of work, in terms of the event spacing or actual ages of sea-level events through geologic time. Later work dealing with regional analyses of sea-level changes, including the two SEPM (Society for Sedimentary Geology) syntheses published in 1998, was analyzed by Miall and Miall (2001), a critique that forms part of Chap. 12 of this volume.

In the dozen years since this book first appeared, a considerable body of work has been carried out to examine local and regional sequence frameworks, in part to test concepts of global correlation and causation. The vast improvements in chronostratigraphic methods documented in this chapter, the improvements in seismic-reflection methods, and the devotion of several onshore and offshore drilling projects to these problems, have provide several important new data sets, which it is the purpose of this section to briefly review. The material discussed in Chap. 14 of the first edition is now sufficiently out of date, and irrelevant in terms of modern techniques and data sets that it is not repeated here.

### 14.6.1 Cretaceous-Paleogene Sequence Stratigraphy of New Jersey

The most important modern data set is that collected from the shallow subsurface of the New Jersey coastal plain and the immediately adjacent offshore. A leg of the Ocean Drilling Program that was actually drilled onshore (Leg 174AX), was devoted to these problems, supplementing earlier legs (Leg 150, 150X) that were drilled on the New Jersey continental shelf. The major

early results were described by Miller et al. (1998, 2004) and Browning et al. (2005), and updated and revised results were presented in a special issue of *Basin Research* (see, in particular, Kominz et al., 2008; Browning et al., 2008). Of particular interest from the perspective of this book is the attempt by Miller et al. (2005a, b) to correlate the New Jersey sequence record for the Upper Cretaceous to an emerging chronology of glacioeustasy postulated by Matthews and Frohlich (2002). A summary of the New Jersey sequences is provided in this book as Fig. 4.9. Details of Oligocene sequences are shown in Fig. 14.23. Comparisons with other emerging detailed stratigraphies are discussed in the next section.

This research project made full use of the armoury of dating methods now available for detailed stratigraphic studies. In order to assess the quality of the data, and its potential value as a body of tests for the concept of global eustasy, it is necessary to track the dating back to its sources, in order to evaluate the accuracy and precision of the chronostratigraphy. The authors clearly decided early in the project that the Exxon global cycle charts were not to be fully trusted as reliable indicators of sea-level change. They questioned the accuracy of some parts of the Graciansky et al. (1998) time scale, and concluded that the estimates of the magnitudes of sea-level fluctuations were mostly much too high. Miller et al. (2004,

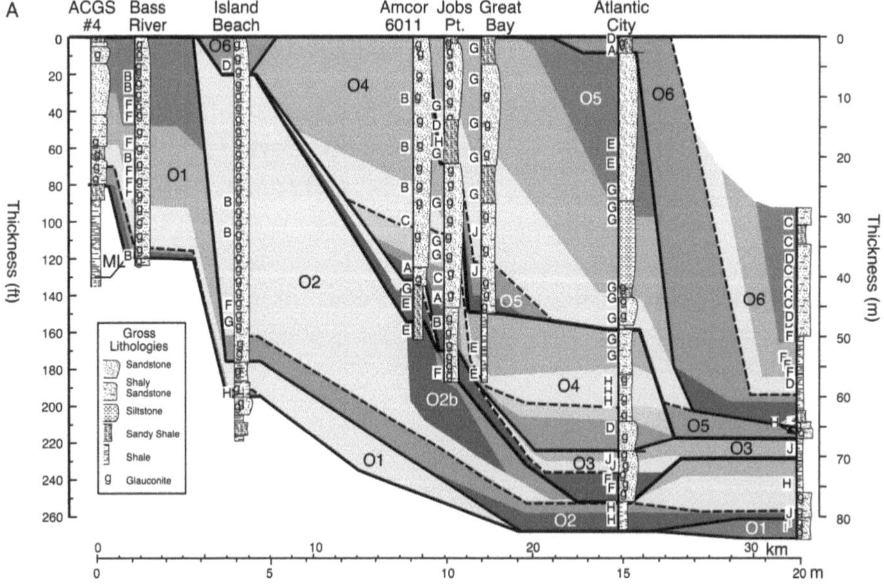



**Fig. 14.23** Details of the Oligocene sequences O1 to O6 forming a set of seaward-dipping clinoforms along a NW-SE transect across the New Jersey continental margin. Gross lithologies for each borehole are indicated. Upper-case letters to the left of each borehole indicate foraminiferal biofacies, identified by factor analysis (Kominz and Pekar, 2001)

p. 389) stated "We conclude that it is time to abandon the use of the EPR [Exxon Exploration Production Research] record for the late Cretaceous and suggest that the New Jersey and Russian Platform backstripped records provide the best substitute." Notwithstanding this remark, the Haq et al. (1987) chart is shown against their and other data for reference purposes in some of their most recent research (e.g., Kominz et al., 2008).

Figures 14.24 and 14.25 reproduce the correlation diagrams from Miller et al. (1998) for the Upper Oligocene to Middle/Upper Miocene record of two boreholes drilled on the continental slope. Figure 14.26 provides the correlation diagrams from Miller et al. (2004) for two of the onshore core holes that sampled the Upper Cretaceous record. The sequences, with their boundaries, were recognized on the basis of facies studies in the drill cores, correlated with reflections in seismic lines run through or close to the drill-hole locations. The primary chronostratigraphic correlation method used throughout this study was strontium-isotope chemostratigraphy. Strontium sample points

are indicated by black circles with error bars in each of these figures, positioned against the absolute time scale, which is arranged along the abscissa of each figure. Supplementary magnetostratigraphic data were used for the Cenozoic sequences, and key biostratigraphic indicators were also employed, all of which are indicated on the three figures.

The time scale derived from strontium-isotope stratigraphy forms a critical basis for the New Jersey project. Figure 14.27 illustrates the distribution of Upper Cretaceous samples and reference sources for the time scale used by Miller et al. (2004). A standard scale has evolved since the 1980s by the accumulation of increasing numbers of samples tied into the global time scale by biostratigraphy, magnetostratigraphy and, more recently, astrochronology. However, as McArthur and Howarth (2004, p. 100) stated, in their chapter in GTS2004 (Gradstein et al., 2004):

The difficulty of assigning numerical ages to sedimentary rocks by the first two methods is well known. Users of the calibration curve, and the equivalent look-up

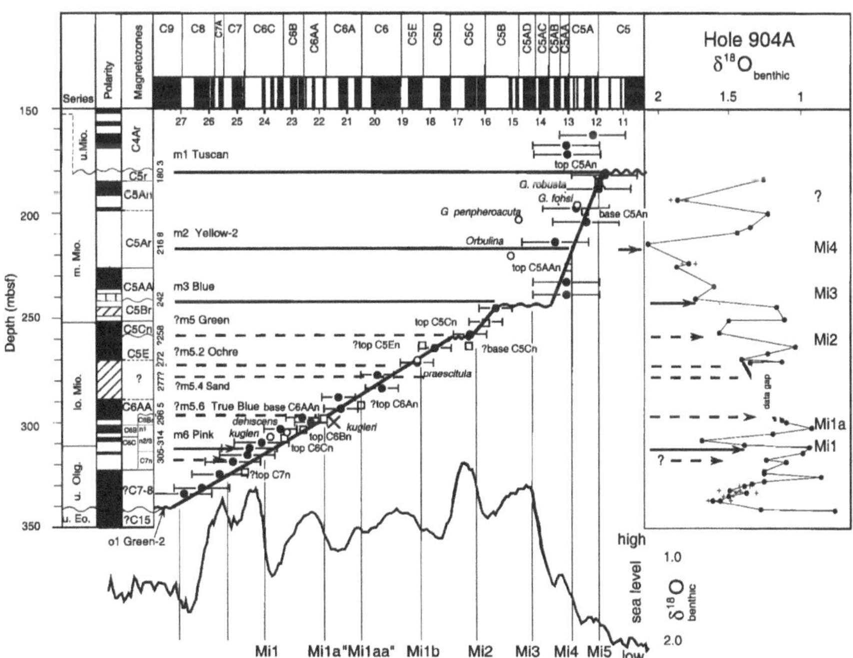


**Fig. 14.24** Age-depth diagram for Hole 904, New Jersey continental slope. The core record, with magnetozones, is indicated at left. Sequence boundaries, recognized from seismic and facies studies, are shown by the horizontal lines extending from the core log across to the right, to the sloping line of correlation. Sequences are named, from the top down, "m1 Tuscan", etc. The standard magnetostratigraphic time scale is shown across the top of the figure and the δ¹⁸O record is shown along the bottom.

Dated sample points are indicated by symbols along the line of correlation: strontium isotope values, with error bars black circles), planktonic foraminifer lowest occurrences (open circles) and highest occurrences (crosses), and magnetostratigraphic reversal boundaries (squares with chron number). δ¹⁸O values from hole 904A are indicated in the box at right. Unconformities are indicated by wavy lines in the line of correlation (Miller, et al., 1998)

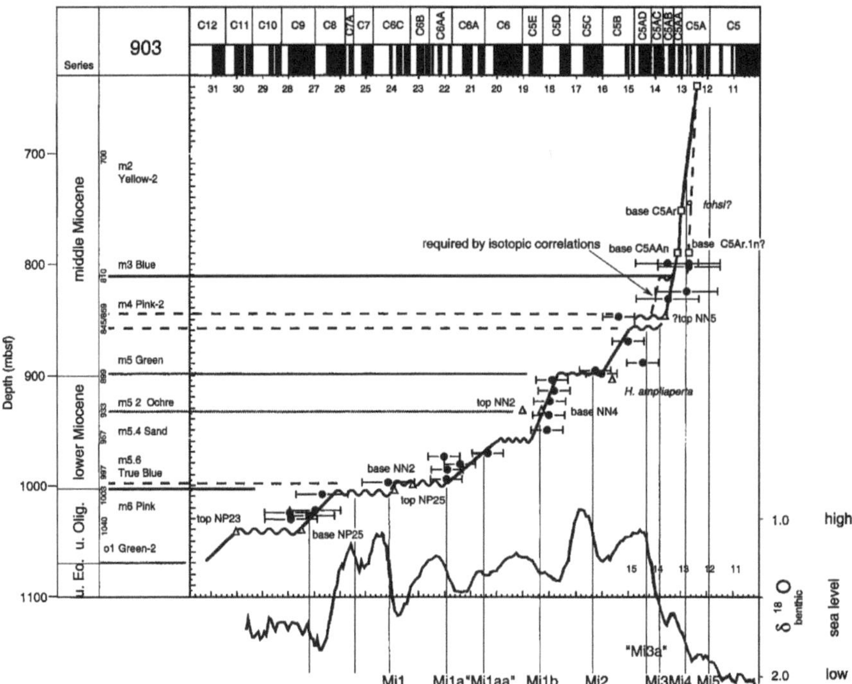



**Fig. 14.25** Age-depth diagram for hole 903, New Jersey continental slope. See Fig. 14.24 for explanation. Note the larger number of unconformities indicated in this section (see text for discussion) (Miller et al., 1998)

tables derived from it that enable rapid conversion of $^{87}Sr/^{86}Sr$ to age and vice versa …., must recognize that the original numerical ages on which the curve is based may include uncertainties derived from interpolation, extrapolation, and indirect stratigraphic correlations and may suffer from problems of boundary recognition (both bio- and magnetostratigraphic), diachroneity, and assumptions concerning sedimentation rate, all of which contribute uncertainty to the age models used to generate the calibration line. Further more, age models are ultimately based (mostly) on radiometric dates and are as accurate as those dates.

These are the problems discussed in the previous section with reference to Figs. 14.20 and 14.21. As an illustration of the type of data on which Miller et al.'s correlations in Fig. 14.27 were based, a figure from one of their primary reference sources is included here as Fig. 14.28. McArthur et al. (1993) compared and correlated samples from sections through the Upper Cretaceous in three locations in Germany, England and the US Western Interior. Their purpose was to contribute to the network of intersecting correlations among isotope data and biostratigraphic and magnetostratigraphic samples in order to refine mutual correlations and reduce error. In each of the three cases, positioning of the biozone and stage boundaries involves some interpolation and extrapolation, including the absence (at that time) of international agreement on the relative ages of all of the various age indicators in these different biogeographic locations.

**Fig. 14.26** Age-depth diagram for the Upper Cretaceous stratigraphic record in two holes drilled on the New Jersey continental shelf. Sequences, with their boundaries, recognized from detailed core studies, are shown by their assigned names down the left side of each diagram. Strontium isotope values, with error bars, are shown by the black circles. Open circles are diagenetically altered Sr values. Planktonic foraminiferal ages are indicated by "x", nannofossil zones by the open triangles, relative to the nannofossil zones shown along the abscissa at the bottom of the diagram. Average, smoothed, long-term sedimentation rates (ignoring unconformities) are indicated by the segments of straight black lines. The thicker, shorter, grey lines are lines of correlation based on tighter control by the chronostratigraphic data for each sequence. The horizontal offsets in these lines at the sequence boundaries provide estimates of the time missing at these surfaces. This information is summarized in the pattern of blocks at the base of the diagram. Grey areas indicate spans of time represented by continuous section at the $10^6$-year time scale (Miller et al., 2004)

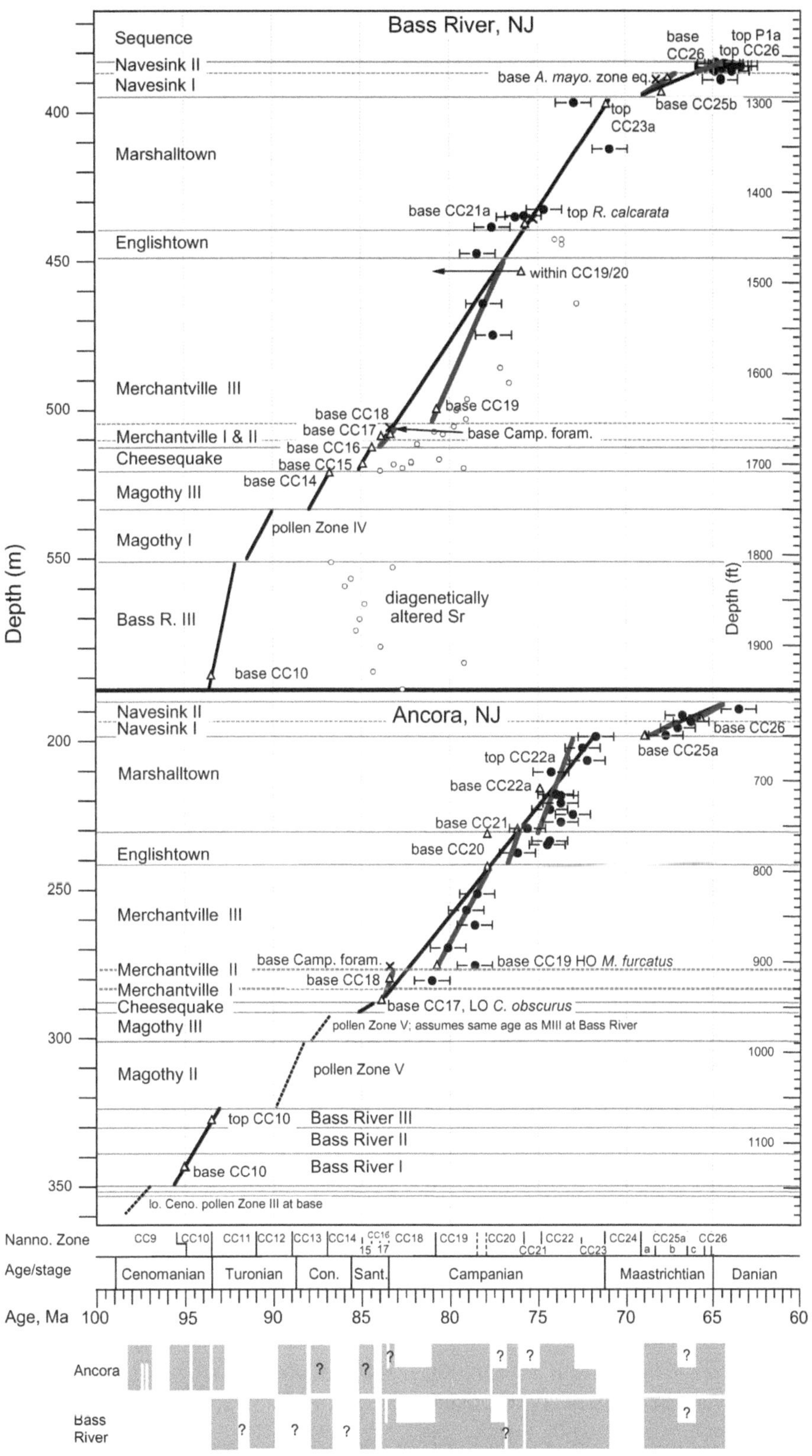
Bass River, NJ
Sequence
Navesink II
Navesink I
Marshalltown
Englishtown
Merchantville  III
Merchantville I & II
Cheesequake
Magothy III
Magothy I
Bass R. III
base CC26
top P1a
top CC26
base A. mayo. zone eq.
base CC25b
top CC23a
base CC21a
top R. calcarata
within CC19/20
base CC18
base CC19
base CC17
base Camp. foram.
base CC16
base CC15
base CC14
pollen Zone IV
diagenetically
altered Sr
base CC10
400
450
500
550
Depth (m)
1300
1400
1500
1600
1700
1800
1900
Depth (ft)
Ancora, NJ
Navesink II
Navesink I
Marshalltown
Englishtown
Merchantville  III
Merchantville  II
Merchantville  I
Cheesequake
Magothy III
Magothy II
Bass River III
Bass River II
Bass River I
base CC26
base CC25a
top CC22a
base CC22a
base CC21
base CC20
base Camp. foram.
base CC19 HO M. furcatus
base CC18
base CC17, LO C. obscurus
pollen Zone V; assumes same age as MIII at Bass River
pollen Zone V
top CC10
base CC10
lo. Ceno. pollen Zone III at base
200
250
300
350
700
800
900
1000
1100
Nanno. Zone
CC9
CC10
CC11
CC12
CC13
CC14
CC16
15  17
CC18
CC19
CC20
CC21
CC22
CC23
CC24
CC25a
a  b  c
CC26
Age/stage
Cenomanian
Turonian
Con.
Sant.
Campanian
Maastrichtian
Danian
Age, Ma
100
95
90
85
80
75
70
65
60
Ancora
Bass
River
?

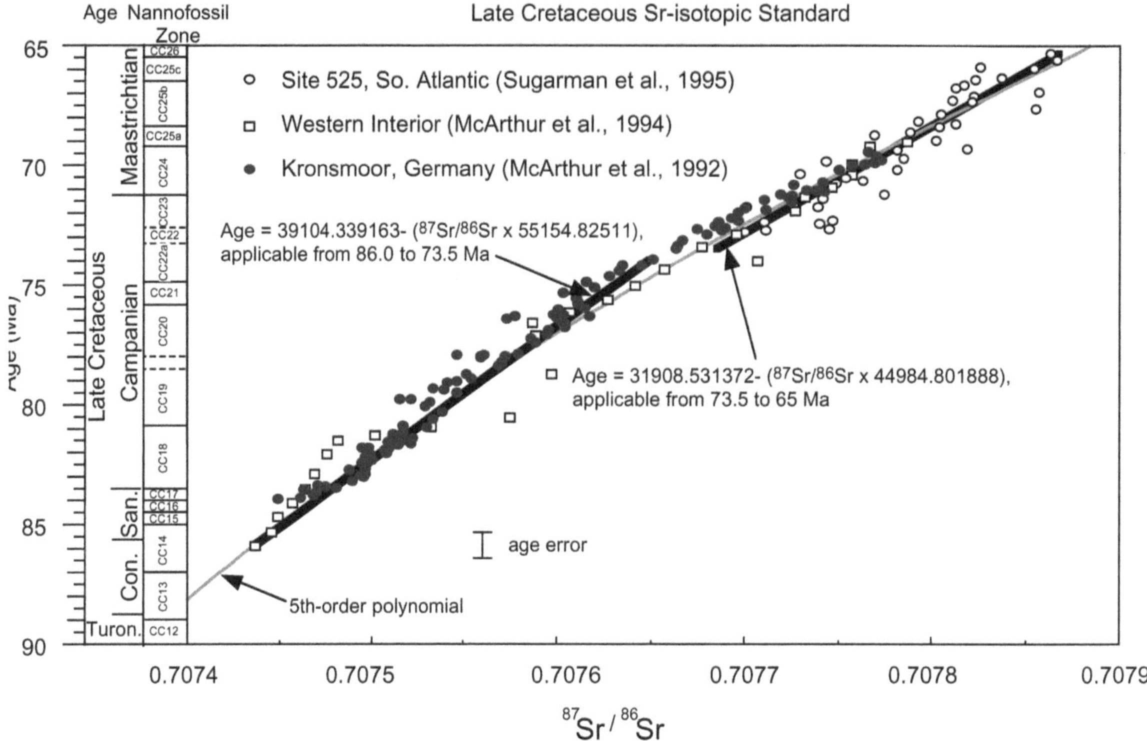



**Fig. 14.27**  Calibration of the strontium-isotope curve for the Late Cretaceous, as used by Miller et al. (2004)

Positioning errors of chronostratigraphic boundaries of a few metres in the sections translates into age uncertainties in the range of ±0.5 million years for a given point within a succession of strontium measurements. Reading across the graphs in Figs. 14.27 and 14.28 indicates that there are uncertainties in the age of individual measurements, if taken in isolation, of up to about 1.6 million years. This is the width of the spread of values relative to age in Fig. 14.27. The vertical range of about 25 m over which any given value could be obtained from the Lägerdorf, Germany, section in Fig. 14.28a corresponds to a time interval of about 1.6 million years, according to the raw data reported in McArthur et al. (1993).

To bring this back to the New Jersey study, this means that minimum error estimates of ±0.5 million years must be assumed for all sample points in Figs. 14.24, 14.25 and 14.26, and in the several cases where sequences have been dated by only one or two strontium samples, errors of up to about ±0.8 million years must be allowed for. The coarseness of the dating of parts of the succession is indicated by the fact that lines of correlation are drawn through several sequence

boundaries without deflection in Figs. 14.24, 14.25 and 14.26. To put this another way, the construction of a line of correlation through parts of the borehole sections does not reveal the likely presence of offsets at these sequence boundaries. We know that the sequence boundaries are there on the basis of facies and other studies of the core materials, but the derived age interpretations cannot in these cases, corroborate the existence of significant time gaps, even if they are actually present.

An examination of Figs. 14.24, 14.25 and 14.26 reveals many places where the line of correlation could have been drawn differently. This is not to criticize the research, which has clearly been carried out in a most meticulous fashion, but to ensure that the results are used appropriately as a basis for inter-regional correlation. The null hypothesis regarding the New Jersey data is that there is no clear eustatic signal, and it is therefore critically important, as the authors clearly intend, that the observed record be evaluated appropriately, taking due regard of its internal incompleteness and imprecision. The problem with the Exxon curves and their successors is that they always seem to be the tail

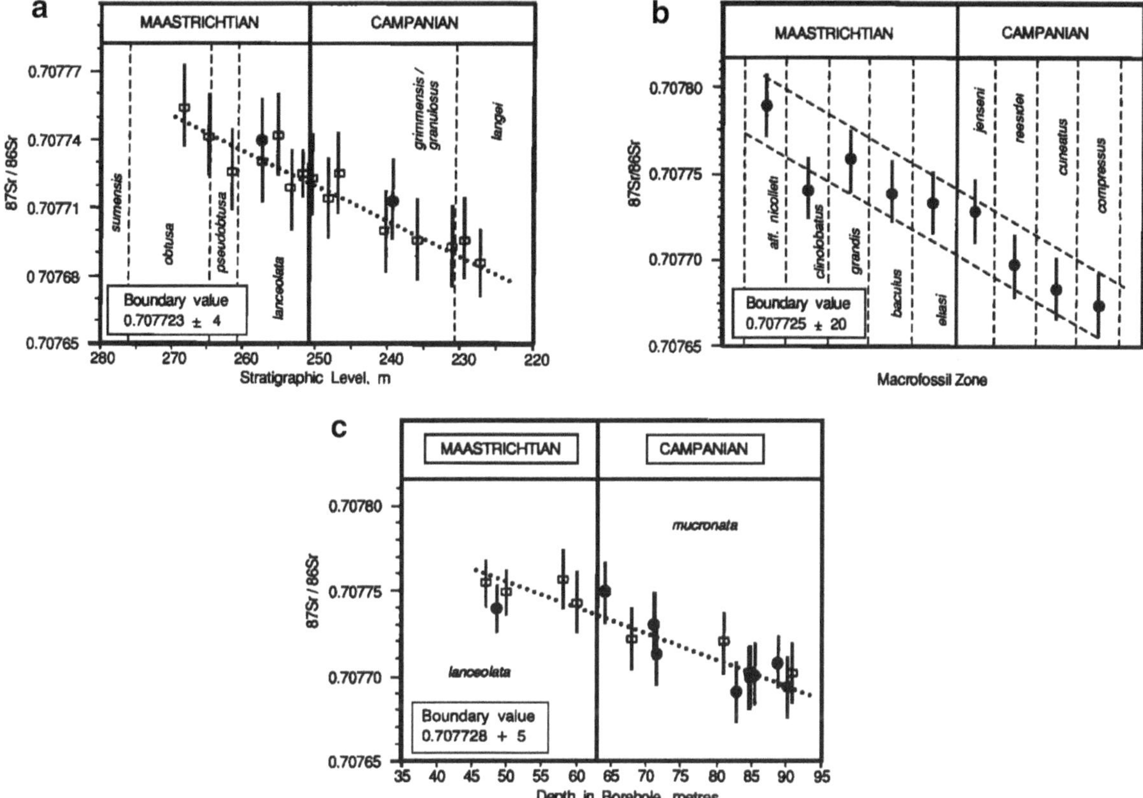

**Fig. 14.28** Trends in isotope ratios across the Campanian-Maastrichtian boundary in three locations in, (**a**) Germany, (**b**) US Western Interior, and (**c**) England (McArthur et al., 1993)

wagging the dog. One of the primary objectives of this book is to develop a discussion of the ways to avoid this trap. We return to his issue in the next section, in which global correlations with the New Jersey data are evaluated.

Given the sequence succession shown in Figs. 14.23, 14.24, 14.25 and 14.26, with the best estimates of sequence-boundary ages, the next step is to examine the construction of a local sea-level curve. The procedures were described by Kominz and Pekar (2001) and Kominz et al. (2008), who constructed curves for each drill core using backstripping methods (See Sect. 3.5). Water depths during sedimentation were estimated from sedimentary facies and biofacies, and at any one time these varied from location to location, reflecting the position of the core holes in different parts of the contemporaneous continental margin and in different parts of regional depositional systems. The downward flexure of the continental margin as a result of thermal cooling and sediment

loading was accounted for by a two-dimensional backstripping procedure. Unconformities in the succession correspond to sea-level lowstands, the elevations of which cannot, of course, be calculated because of erosion.

Results for the Oligocene portion of the sequence succession are shown in Fig. 14.29, and a complete set of sea-level curves for the New Jersey study is shown in Fig. 14.30. In the latter, estimates of lowstands have been added. Differences among the curves in Fig. 14.30 arise from adjustments in estimates of ages and additional data from newer boreholes available for the 2008 publication.

What does the curve in Fig. 14.30 (and also in Fig. 4.10) mean, in terms of sedimentary processes? The sequences which it summarizes were formed in environments ranging from nonmarine to outer neritic, at water depths up to about 145 m (Kominz and Pekar, 2001; Browning et al., 2008). Sequence boundary spacing indicates an episodicity in the $10^6$-year

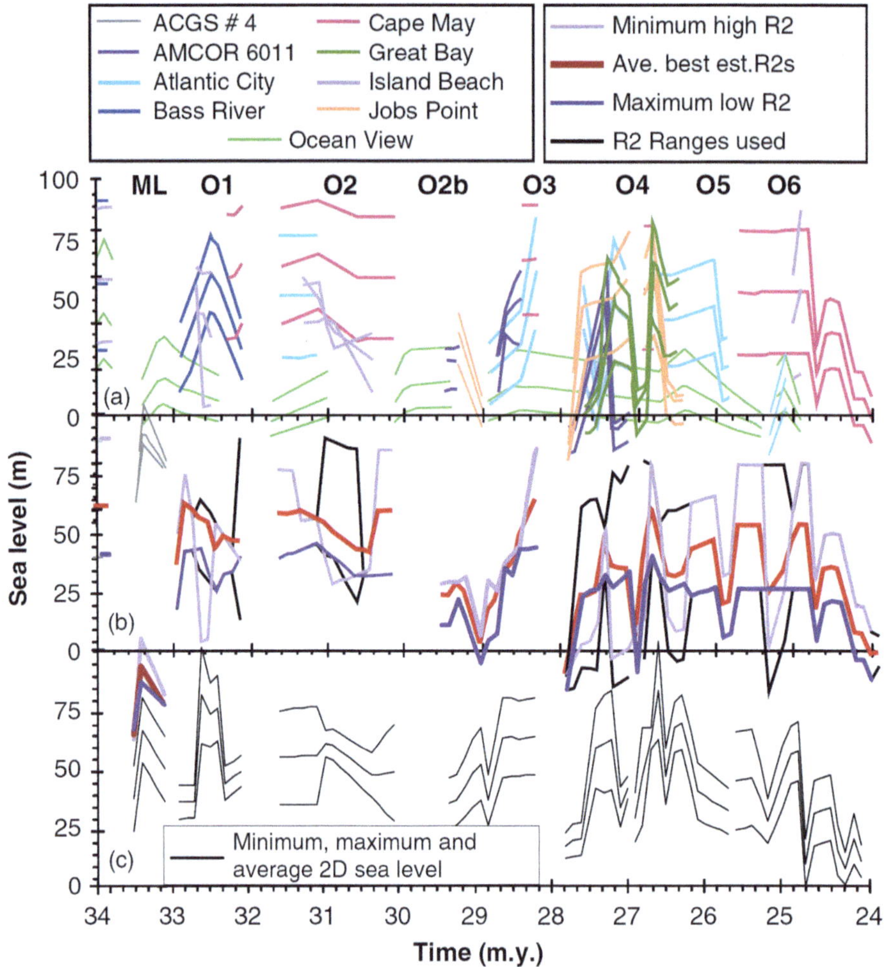


**Fig. 14.29** Estimates of sea-level change in the Oligocene sequences in eight New Jersey boreholes. Diagrams (**a**) and (**b**) relate to results published by Kominz et al. (2008); diagram (**c**) provides earlier results, from Kominz and Pekar (2001), for comparison (Kominz et al., 2008)

range. Long-term sea-level change evaluated (but not discussed here) by Kominz et al. (2008) indicates a high from late Paleocene to mid-Eocene time, but is the specific sequence record and the indication of long-term sea-level change a true indicator of eustatic sea-level change? Given the care taken to date the sections and to account for other influences through the application of backstripping methods, it is clear that the New Jersey data set seems far more likely to contain some version of an eustatic signal than any of the Exxon curves that preceded it. However, until it can be cross-checked against other curves constructed with similar care in other parts of the world, this must remain an untested hypothesis (see discussion of long-term sea-level changes in Sect. 9.3). Miller et al. (2004,

p. 389) discussed the possibility of in-plane stress changes as a cause of changes in accommodation, but cited internal evidence that seemed to rule this out. Dynamic topographic effects could also be contributing to the long-term changes in accommodation, and are discussed briefly in Sect. 9.3. It seems likely that during the Late Cretaceous the long-term average may have been as much as 100 m higher than that shown in Fig. 14.30. This does not affect the timing or magnitude of the relative changes in sea level on the $10^6$-year time scale that are the subject of the discussion here. In the next section we discuss the nature of global correlations, and the evidence, discussed in several papers by K. G. Miller and his colleagues, for the influence of glacioeustasy.

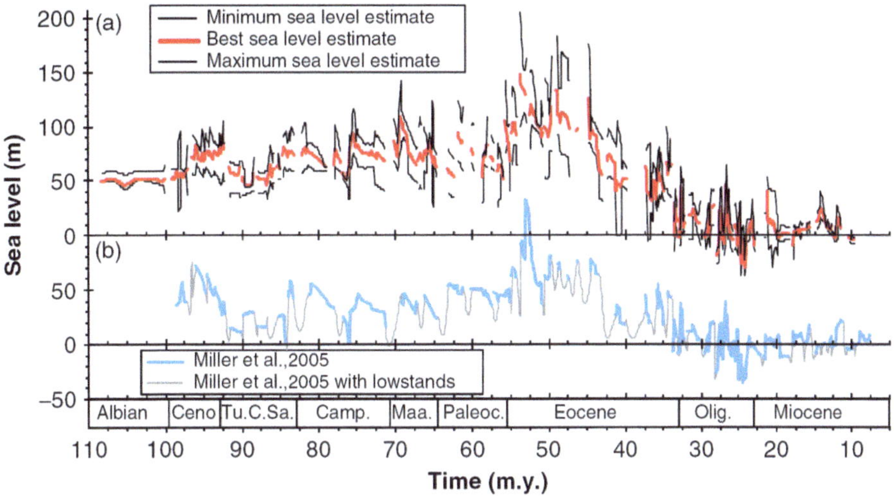


**Fig. 14.30** *Top:* the sea-level curve, with estimated error, derived from one-dimensional backstripping by Kominz et al. (2008), compared with *Bottom:* the *curve* developed by Miller et al. (2005 a, b)

## 14.6.2 Other Modern High-Resolution Studies of Cretaceous-Paleogene Sequence Stratigraphy

An Upper Cretaceous data set described by Crampton et al. (2006) provides an opportunity to attempt a test of the potential global template that has emerged from the New Jersey study. The quantitative biostratigraphic methods they employed are discussed briefly above, in Sect. 14.5.4. Fifteen outcrop sections were correlated to generate a composite section, the local use of which permits relative correlation with a resolution of 130 ka. However, they stated (p. 981) that "sadly, there are few suitable events in the Coniacian to Maastrichtian interval in New Zealand, and calibration to the local and international time scale is rather problematic." They identified seven dinoflagellate events (6 LAD's and one FAD) which can be tied with some degree of accuracy to the global time scale. The relationship of these events to their composite section is shown in Fig. 14.31. One event (#5) is eliminated as an outlier, and the remaining events, based on linear regression, permit the dating of the New Zealand sequences with a precision of ±2.5 million years. Crampton et al. (2006, p. 985) state that "we can identify seven out of the eight Santonian to Maastrichtian hiatuses described by Miller et al. (2004) in the New Jersey coastal plain."

The updated calibration of the New Jersey data presented by Kominz et al. (2008) led them to state (p. 221): "Eleven of the New Zealand unconformities correspond with sequence boundaries in our new sea-level curve . . .. As suggested by Crampton et al. (2006) the fact that 10–11 of the 15 New Zealand unconformities match New Jersey sequence boundaries and two are associated with sea level falls is a strong indication that this signal is eustatic in origin." These correlations are shown in Fig. 14.32, and are discussed below.

Another valuable independent study of sea-level change in the Mesozoic stratigraphic record was provided by Sahagian et al. (1996). These authors examined the outcrop and well records of the Middle Jurassic to Upper Cretaceous succession on the Russian platform, a succession that has received long study, and has been documented in detail by macrofossils (ammonites, bivalves), microfossils (foraminifera) and palynomorphs. The biostratigraphic framework was calibrated against the chronostratigraphic framework of Harland et al. (1990). Backstripping was performed on the data set, using a range of sedimentological criteria to estimate paleo-water depths (Fig. 14.33). Only the Upper Cretaceous part of their eustatic curve overlaps the New Jersey data set and can be compared directly to it. The interpreted lowstand events from this curve are shown in Fig. 14.32.

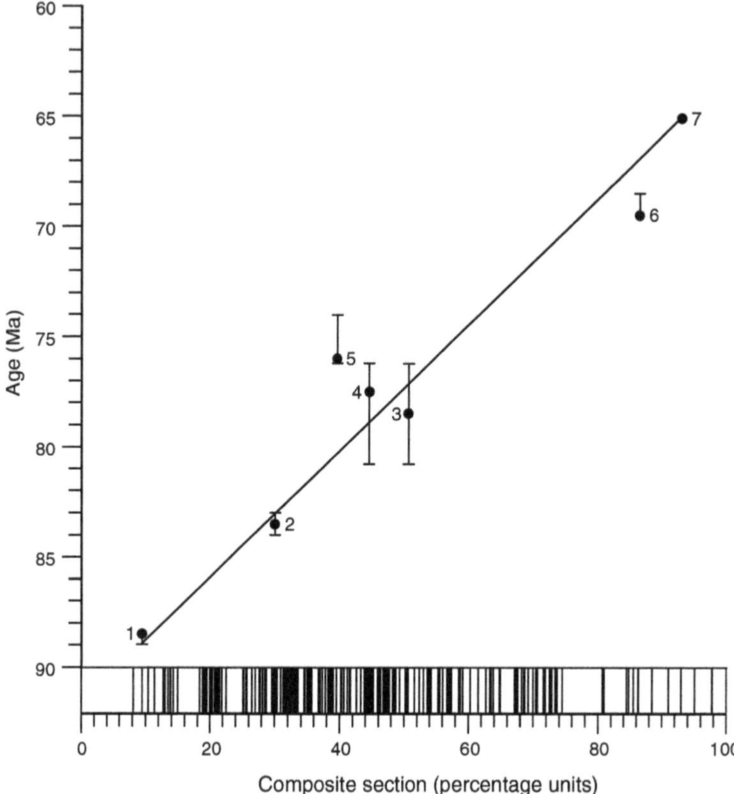



**Fig. 14.31** The age model used to calibrate the New Zealand composite section. Events located by relative dating are shown within the bar at the *bottom*. The seven globally datable dinoflagellate events are indicated, with estimated age errors, and are joined by a *straight line* which can be used to interpolate the absolute ages of the local events. Dinoflagellate event 5 is regarded as an outlier, and is not used in this age model (Crampton et al., 2006)

There are currently few other independent data sets with which to evaluate the Jurassic to Paleogene portion of the sequence record. We do not discuss any of the Vail/Haq charts in this context because of the questions, raised earlier, regarding the built-in circular reasoning involved in the development of this body of work. While the New Jersey data set establishes a very high standard for completeness and accuracy, it nevertheless only represents one location. Miller et al. (2008a) reported on a thorough re-evaluation of a quarry exposing Eocene-Oligocene strata in Alabama, and demonstrated a significant degree of similarity to the corresponding parts of the New Jersey sequence succession (see Sect. 6.1.1). Although Alabama and New Jersey are both located within the same Atlantic-margin tectonic province, the distance between them is such that some differences in subsidence and sedimentary history could be expected, and this comparison can be accepted as a meaningful test of eustatic control. They remarked (p. 50) that "The minor differences in the ages of preserved sequences between the regions reflect the influence of both tectonics and sedimentation on the two basins and minor discrepancies in the fine tuning of ages and correlating sequences." A complete test of global eustasy awaits the availability of similarly comprehensive data sets from tectonically unrelated parts of the world.

In the meantime, the limited comparisons that can be made between the New Jersey, Russian and New Zealand data are shown in Fig. 14.32. Column 4 indicates the position of sea-level lowstands predicted from a model of glacioeustasy proposed for the Cretaceous (and early Cenozoic, not shown here). The increasing evidence for the existence of continental ice during the so-called greenhouse climatic era of the Cretaceous is discussed in the next section. At this point it can be noted that, despite the claims of Crampton et al. (2006) and Kominz et al. (2008), there is very little, if any, similarity between the ages of sea-level lowstands, as indicated in this chart. Only one of the predicted glacioeustatic lowstands, at about 76 Ma, is recorded both in the New Jersey and the New Zealand sections. One event occurs in both the New Jersey and Russian successions, at about 91.5 Ma (this is below the stratigraphic limit of the New Zealand data set). Modest revisions of assigned ages of sequence-

**Fig. 14.32** Comparison of the ages of sequence-bounding unconformities and other data relating to sea-level changes in four independent data sets spanning the Upper Cretaceous sedimentary record. The *grey boxes* indicate the interpreted durations of unconformities. *Column 1:* sea-level lowstands derived from the curves of Sahagian et al. (1996), whose data set deals with Santonian to Bajocian data and therefore ends at about 84 Ma. *Column 2:* Unconformities on the New Jersey continental shelf (Browning et al., 2008). *Column 3:* Sea-level lowstand events in the New Zealand stratigraphic record (Crampton et al., 2006). Each event here has been arbitrarily assigned a time span of 1.5 million years. *Column 4:* Sea-level lowstands predicted from the model of Cretaceous glacioeustasy by Matthews and Frohlich (2002), as reproduced by Miller et al. (2005b). The 2.4 million years long-eccentricity cycle is estimated to have been the dominant frequency

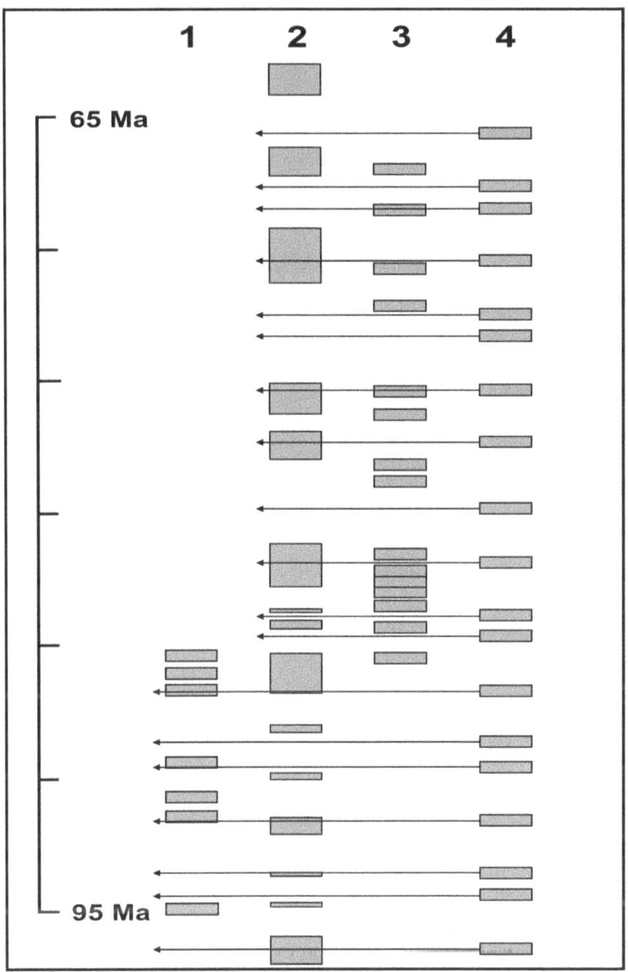


bounding unconformities would bring several other events into line as contemporaneous events, and some events may be unrepresented because of incomplete sampling, unrecognized tectonic overprinting, or local autogenic causes. It could be argued that the margins of error associated with the global correlations between New Jersey, Russia and New Zealand would allow for adjustments of assigned sequence boundary ages by as much as 2–3 million years, which would make a substantial difference to the picture that emerges from comparisons such as that shown in Fig. 14.32. However, this would be to return to the tail-wagging-dog circular reasoning that this book has tried to demonstrate should be avoided if the science is to make any real progress. As an examination of global eustasy, it must be concluded that the results of this synthesis, as it currently stands, are inconclusive.

### 14.6.3 Sequence Stratigraphy of the Neogene

In this section we examine Miocene-Pliocene sequence stratigraphy on the $10^6$-year scale as preserved on some modern continental margins. There are several excellent carbonate and clastic successions that illuminate the issue of eustatic sea-level control. Glacioeustasy is a given for this period, as discussed in the next section. However, we focus here on the low-frequency sequence record, the preservation of which is less regular than the Milankovitch-band cyclicity that is now being used to construct an astrochronological time scale (Sect. 14.7).

As discussed in Chap. 6, modern, high-resolution seismic data are providing a wealth of new information

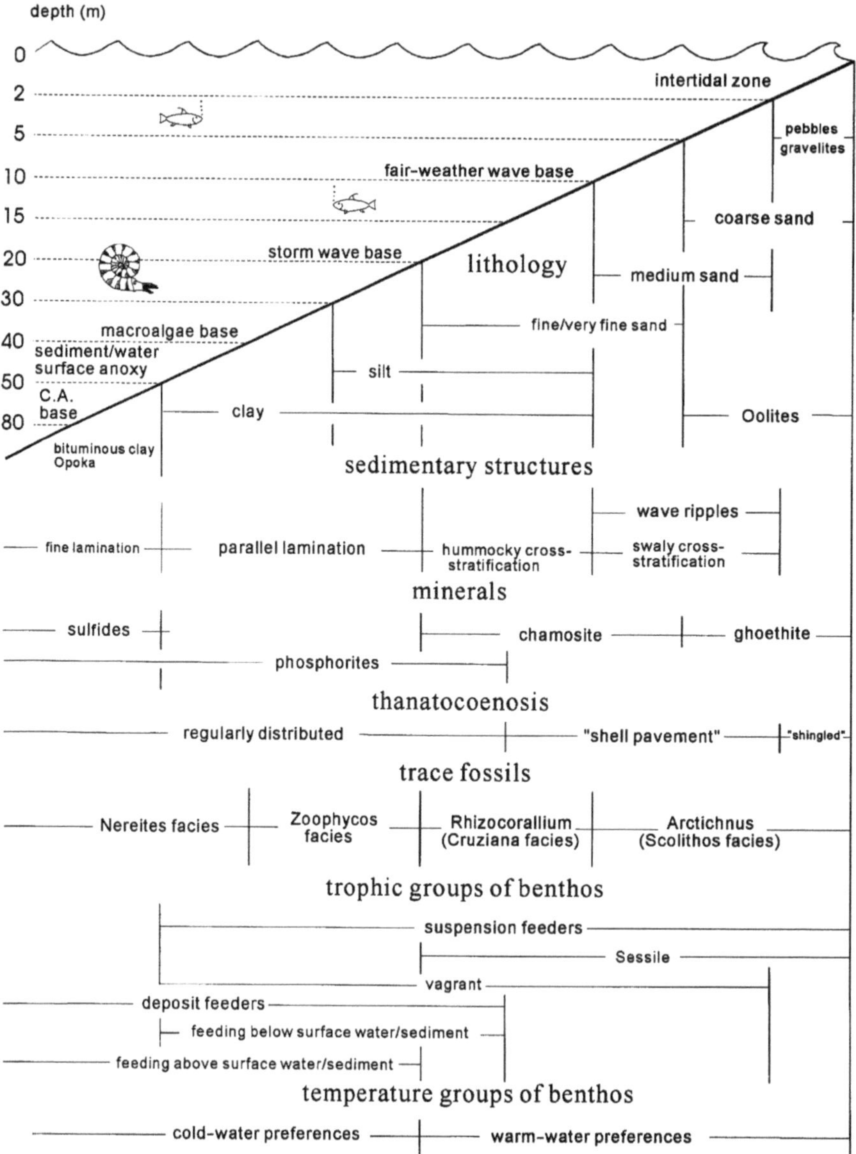


**Fig. 14.33** Criteria used to estimate water depths in the Middle Jurassic-Upper Cretaceous succession in the Russian Platform (Sahagian et al., 1996)

concerning the response of carbonate margins to changes in accommodation and sediment supply. Many of these carbonate margins have also been extensively drilled, and the resulting core data provide the necessary facies and biostratigraphic data to permit an examination of the question of global sea-level control. Figure 14.34 reproduces a correlation diagram developed by Betzler et al. (2000) comparing the Miocene-Pliocene $10^6$-year sequence record on the Bahamas Bank and the Queensland plateau, to which has been added the sequence boundaries from the same interval of the New Jersey continental margin succession (from Kominz et al., 2008). The Queensland-Bahamas study is based on correlations of calcareous nannoplankton and planktonic foraminifera. The analysis of the Bahamas data was provided by Eberli et al. (1997); the seismic section that formed the basis of this study is illustrated in Fig. 6.9. The density of the biostratigraphic data ranges from one to three biohorizons per million

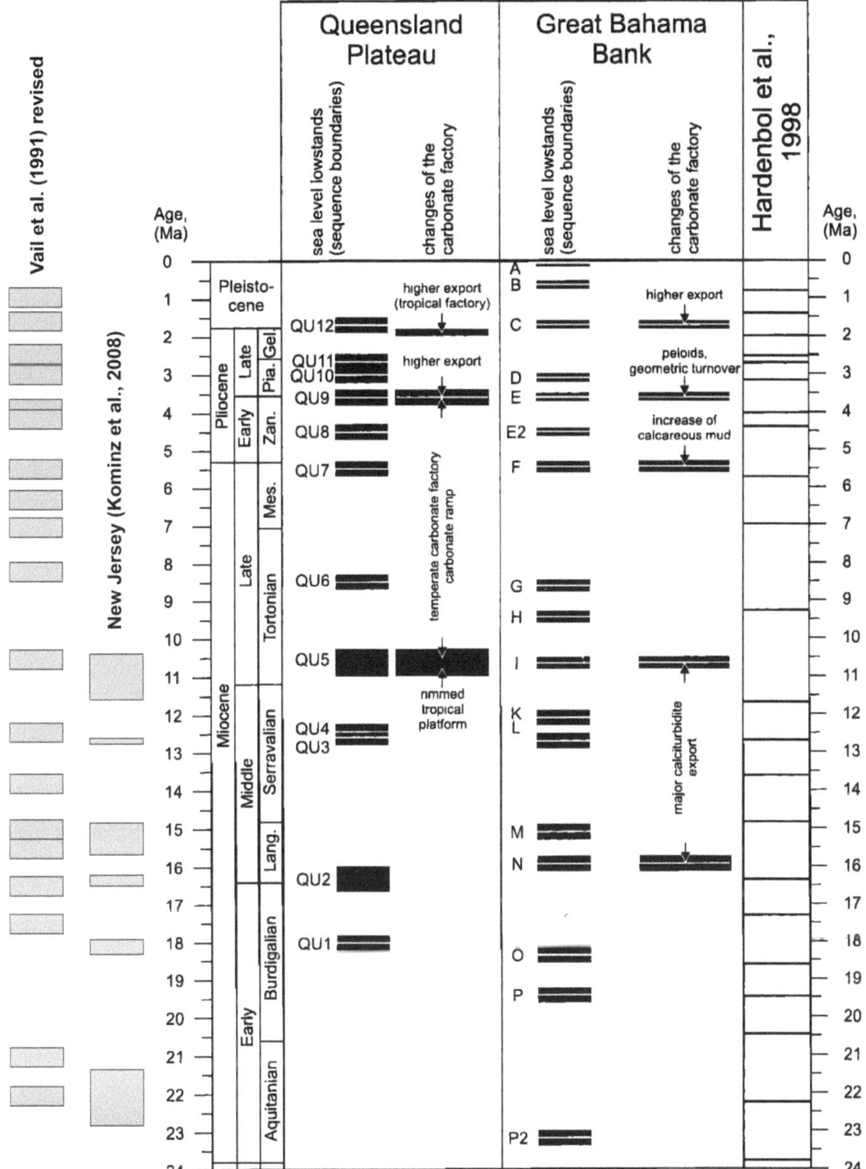



**Fig. 14.34** Comparison of Neogene sequence boundaries in the platform carbonate margins of Queensland and the Bahamas Bank. The sequence boundaries from the Hardenbol et al. (1998) scale are shown at *right*. From Betzler et al. (2000). Two additional sets of sequence boundaries have been added, at *left*, years, calibrated to the Berggren et al. (1995) time scale. Gradstein et al. (2004, Fig. 1.7) indicated that stage-boundary ages have been adjusted by between about 0.1 and 0.8 million years from the Berggren et al. (1995) scale, not enough to significantly affect the level of correlation with the more recently dated New Jersey section.

the sequence boundaries from the Miocene portion of the New Jersey record (from Kominz et al., 2008), and the Neogene record of the Antarctic (from Vail et al., 1991, with boundary ages revised by Hardenbol et al., 1998. Data from McGowran, 2005)

Figure 14.34 indicates a high degree of correlation between Queensland and the Bahamas. Betzler et al. (2000, p. 727) stated that throughout the Miocene-Pliocene:

The isochroneity of sea level lowstands in two tectonically unrelated carbonate platforms is strong evidence for eustatic sea level changes as the controlling factor on large. to medium-scale (1–5 Ma) stratigraphic packaging. The timing of most of the currently recognized eustatic sea level fluctuations in these independent carbonate systems does, however, not seem to match as clearly with the timing of global sea level lowstands as proposed by Haq et al. (1988) and Hardenbol et. al. (1998) in the Mesozoic and Cenozoic Sequence Chronostratigraphic Chart (MCSSC) (Figure 3)[Fig. 14.34 of this book]. This discrepancy may relate to the resolution of biostratigraphic dating. The accuracy of dating depends on a number of factors: (1) the number of biostratigraphic datum levels found, (2) the quality of the biostratigraphic events used. Factors such as abundance of the species play a role and the exact appearance or exit levels of a species should be quantified, which is hardly done owing to time constraints, and (3) consistency of sedimentation rates in between datum levels because the interpolation technique is used for the timing of a certain event.

Betzler et al. (2000) went on to note that the number of sequence boundaries in their data does not correspond to the number of boundaries in the Exxon chart—an issue that, as noted earlier (see Table 11.2), constitutes one of the criteria for correlation that has consistently misled and confused stratigraphers. We do not consider the lack of correlation with the Hardenbol chart further, for reasons that should by now be obvious to readers of this book. Betzler et al. (2000) suggested that the accuracy of their sequence boundary correlation might be improved by more intensive biostratigraphic analysis. However, for our purposes, it is much more significant to evaluate the correlations with the sequence boundary record from New Jersey (accurate sequence boundary ages for the succession younger than 10 Ma are not available from this area). This is a comparison between two independent studies and therefore fulfills one of the most important criteria for a meaningful test of the global-eustasy paradigm. In this case, the evidence indicates a strong degree of correlation between New Jersey, The Bahamas, and Queensland. Despite the small differences in the time scales used in these two different interpretations, the comparisons between the data sets is striking. The adjustments that would be required to recalibrate the Betzler et al. (2000) study to GTS2004 would not significantly change the pattern visible in Fig. 14.34. The case for eustatic control of Miocene stratigraphy is considerably strengthened by this comparison.

Also shown in Fig. 14.34, for reference purposes, are the sequence boundaries documented in the "great Neogene sedimentary wedge" (wording of McGowran, 2005, p. 190) reconstructed by Vail et al. (1991), based

in part on an analysis of the Antarctic margin (Bartek et al., 1991). The model of the Neogene wedge (see Fig. 6.4) claims to illustrate a global pattern of sea-level change and accompanying stratigraphic architecture that can be recognized worldwide. However, there is very little correspondence between these sequence boundaries and the New Jersey-Queensland-Bahamas events documented in Fig. 14.34, which calls into question the chronostratigraphic basis on which the wedge model was drawn. McGowran (2005, p. 191) indicated that the number and ages of the sequence boundaries has been modified by Hardenbol et al. (1998), but the boundaries and ages indicated in his redrawn version of the Vail et al. (1991, Fig. 12) diagram are the same as in that diagram, which were, in turn, based on the Haq et al. (1987) global cycle chart.

Additional data from other studies provide strong support for eustatic control during the Miocene, although it is also clear that this is not the only significant control on platform carbonate development. Figure 6.14 is a summary diagram from the detailed study of the Gulf of Papua carbonate platform by Tcherepanov et al. (2008). This diagram shows the Queensland-Bahamas sequence boundaries from Betzler et al. (2000), although the positions of these boundaries have changed significantly from those shown in the original source (Fig. 14.34), for reasons that are unclear. They are much greater adjustments than would be required to revise the correlations to GTS2004. The most important sea-level lowstand in the Gulf of Papua data is one at 11.0 Ma, which clearly correlates with lowstands in New Jersey, Queensland and the Bahamas (Fig. 14.34).

### 14.6.4 *The Growing Evidence for Glacioeustasy in the Mesozoic and Early Cenozoic*

As noted by Vail et al. (1977, p. 93), glacioeustasy is the only process known that can generate large-amplitude eustatic changes in sea level over geologically short periods of time, but they acknowledged that (at that time) there was no convincing evidence of continental glaciation in the Mesozoic-Cenozoic record prior to the Oligocene. At a time when oxygen-isotope data were very limited, Matthews and Poore (1980) and Matthews (1984) suggested that the world was ice free until about 100 Ma, after which ice build-up began, and proceeded in several significant steps until the late Cenozoic. But they acknowledged that, in general, evidence for glaciation in the Cretaceous was very inadequate. In fact it has been widely assumed that the Cretaceous was a period of continuous "greenhouse" climates (Frakes, 1979).

However, results of the Deep-Sea Drilling Project, including the gradual accumulation and calibration of a large data-base of oxygen-isotope data, are dramatically changing this world view. Frakes and Francis (1988) suggested that there may never have been an entirely ice-free world, on the basis of such evidence as the rare occurrence of erratics, dropstones and other indicators of floating ice, in the geological record. But many of these erratics, collected from a variety of Cretaceous marine units in Europe and Australia, may have been rafted by plants or animals (Markwick and Rowley, 1998), and cannot be taken as reliable evidence of glaciation. More convincing evidence for glacial climates preceding the late Cenozoic ice age came from ODP Leg 119 in the Southern Ocean. Ehrmann and Mackensen (1992) reported the retrieval of diamictites of middle-Late Eocene age on this leg, interpreted as ice-rafted deposits, and suggesting the existence of continental ice on Antarctica during the Eocene.

Stoll and Schrag (1996) examined the oxygen- and strontium-isotope record in the Lower Cretaceous section in several DSDP cores and argued that excursions in these data indicated high-frequency sea-level changes that could only have been caused by glacioeustasy.

Given the diametrically opposite points of view to the question of global eustasy taken by Pindell and Drake (1998) and Graciansky et al. (1998) (see discussion in Chap. 12), it is perhaps not surprising that contributors to these volumes would arrive at different conclusions regarding the likelihood of continental glaciation, and therefore the probability of glacioeustasy, during the Cretaceous-Paleogene. Abreu et al. (1998, p. 77), a contributor to the "global eustasy" volume, noted negative $\delta^{18}O$ events near the Cenomanian-Turonian boundary and in the Upper Turonian, and attempted to show how long-term temperature changes indicated by the sparse isotopic data paralleled the long-term trends in the Hardenbol global cycle chart. In contrast, Markwick and Rowley (1998), in the Pindell and Drake (1998) volume (which argues for complexity in stratigraphic controls), while also

noting the short-term spikes in oxygen isotope data from the Cenomanian and Turonian, concluded that for most of the Mesozoic and early Cenozoic the evidence for glaciation is "ambiguous at best". They devoted considerable space in their analysis to a demonstration of the impossibility of the very large eustatic changes in sea level required by the Haq et al. cycle chart, by calculating the volumes of continental ice that would have been required to account for the amounts of water from the oceans necessary to create the sea-level changes indicated in this chart. At this stage in the development of isotope geochemistry, many of the factors influencing the isotope signature in fossil organisms remained to be resolved satisfactorily, including the different chemistries of shell formation by planktonic and benthic organisms, the question of whether deep- or shallow-water ocean temperatures were being indicated, the effects of post-depositional diagenesis, and so on.

Largely because of the large volume of oxygen isotope data acquired during ODP cruises over the last decade, the use of this class of data to evaluate late Cenozoic glacioeustasy and oceanic temperature changes has become so much part of standard methodology that the $\delta^{18}O$ curve is now accepted as an independent measure of geologic time. We are not at that point for the Mesozoic-early Cenozoic yet, but having said that, oxygen isotope data are now providing many essential new insights on climatic and oceanographic changes through that period. Several recent publications review this data set and argue the case for periodic ice-cap development on the Antarctic continent through the so-called greenhouse period of the Mesozoic-early Cenozoic (e.g., Miller et al., 1999, 2005a, b, 2008b). Two detailed studies focused on specific parts of the Cretaceous record. Miller et al. (1999) examined the Maastrichtian record in coastal sediments in New Jersey, and Borneman et al. (2008) discussed the implications of a $\delta^{18}O$ excursion during the Turonian.

One of the more prominent sequence boundaries exposed in the New Jersey coastal plain is an early Maastrichtian surface containing clear evidence of subaerial exposure overlain and underlain by shallow-marine deposits. Lithofacies and biofacies data were interpreted by Miller et al. (1999) to suggest a sea-level change of about 40 m associated with this boundary. Dating of the sequence boundary suggested a correlation with a $\delta^{18}O$ excursion between about 69.1

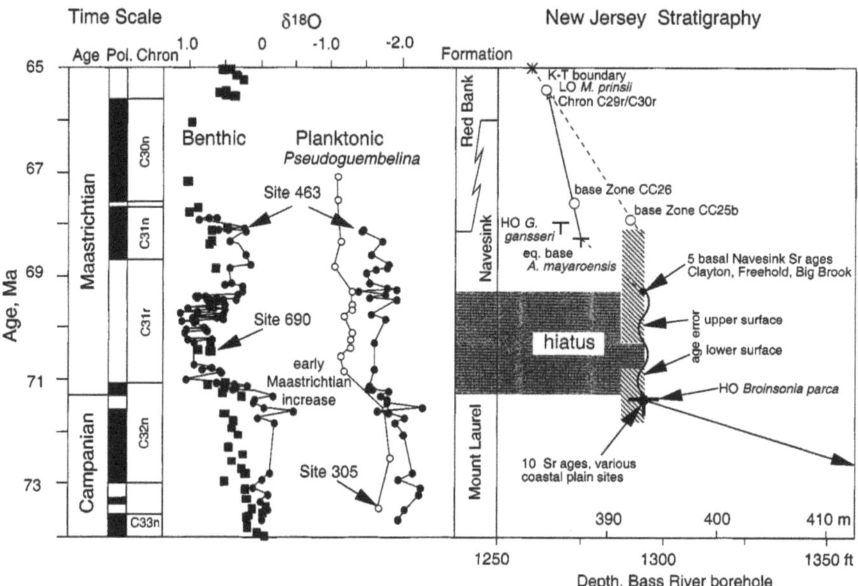


**Fig. 14.35** Comparison of the δ¹⁸O record from ODP data with outcrop data from the New Jersey coastal plain. Dating of the *upper and lower surface* of the hiatus was achieved by strontium isotope and biostratigraphic data and clearly suggests that the hiatus coincides with a positive δ¹⁸O excursion (Miller et al., 1999)

and 71.3 Ma (Fig. 14.35). When calibrated against the δ¹⁸O record, the estimated sea-level fall at this time is 20–40 m, which would be equivalent to the formation and melting of 25–40% of the volume of the present Antarctic ice cap. Assuming the minimum value of 20 m sea-level fall followed by a 20 m rise, during the estimated 2.2 million years time span of the hiatus, this requires an average rate of change of about 18 m/million years. This is two orders of magnitude slower that the sea-level changes associated with the late Cenozoic glaciation, and compares with rates of change associated with some regional tectonic mechanisms, such as intraplate stress changes and thermal subsidence (see Table 8.2). However, there is no tectonic reason why a δ¹⁸O excursion would correlate with a sequence boundary. In support of a glacioeustatic origin it may be pointed out that the time span of 2.2 million years is close to the 2.4 million years period of a long eccentricity cycle modeled by Matthews and Frohlich (2002).

Another study of the δ¹⁸O record, by Bornemann (2008) suggests a significant cooling episode during the Turonian. They evaluated and compared the oxygen isotope signature in planktic and benthic foraminifera and suggested a cool phase lasting about 200 ka at about 92 Ma, preceded and followed by several million years of "greenhouse" warmth.

Miller et al. (2005a, b) argued for a new interpretation of the Cretaceous to early Cenozoic climatic record. While acknowledging the sparseness of the isotopic data, and the susceptibility to diagenetic modification of samples from older, relatively deeply buried sections, they argued that enough evidence had now been assembled for a rethinking of the concept of a long period of greenhouse climatic conditions. They suggested that the climate was characterized by "cool snaps"—the development of small ice caps lasting in the order of 100 ka. It was postulated that these ice caps must have been located in Antarctica, which was, as it is now, located in a polar position. The volume of ice required for a given sea-level fall is a simple enough calculation. Given the estimates of sea-level change required by the facies and backstripping analysis of the New Jersey record, Miller et al. (2005b) used the ice-cap models of DeConto and Pollard (2003) to suggest an evolution of ice cover through the Late Cretaceous to Eocene (Fig. 11.15). In principal this model seems reasonable, but in detail it cannot yet be independently verified by the discovery of well-documented Cretaceous glacial deposits in Antarctica, or by a correlation with the insolation curve derived by astronomical prediction. Column 4 of Fig. 14.32 shows the predicted lowstands from the orbital-forcing model of Matthews and Frohlich (2002), as incorporated into the

New Jersey studies of Miller et al. (2005a). As can be seen in this figure, there is little correspondence between the predicted lowstands and the calculated ages of sequence boundaries in New Jersey and New Zealand (despite the claim to the contrary by Miller et al., 2005b, p. 228).

In conclusion: an increasing number of stratigraphic studies (documented in Chaps. 6 and Chap. 7) is providing strong evidence for high-frequency sequence-generating mechanisms during the Cretaceous and early Cenozoic. Some sequence records can be explained by climate change (without glacioeustasy), of probable Milankovitch origin, while other studies, some of them highlighted in this section, provide a strong case for glacioeustasy, driven by the development of small, temporary ice caps, probably located in Antarctica. The chronostratigraphic evidence for this process remains weak for the Cretaceous (Fig. 14.32), which could reflect residual errors and imprecisions in the global chronostratigraphic data base. The evidence is much more convincing for the Neogene (Fig. 14.34), which has long been accepted as a period characterized by major Antarctic ice build-up.

## 14.7 Cyclostratigraphy and Astrochronology

### 14.7.1 Historical Background of Cyclostratigraphy

The principles underlying the orbital forcing of stratigraphy are set out in Chap. 11. The purpose of this section is to examine the work that has been underway since the mid-1990s to develop a time scale based on the "pacemaker" of orbital forcing. This is relevant to Part IV of this book in that cyclostratigraphy and the astrochronological time scale are the "new eustasy"— the time scale is being developed under the assumption of an overriding control of the orbital "pacemaker" and the assumed regularity of the resulting stratigraphy, which includes an interpreted driving force for high-frequency sequence stratigraphy.

In 1995, in a landmark volume establishing new time scales and new methodologies for refining these scales, Herbert et al. (1995, p. 81) introduced

"Milankovitch cycles" with reference to the classic work of Hays et al. (1976) and others, and stated that:

> Stratigraphers may now be in a position to invert the usual argument; the "Milankovitch" hypothesis may be used *in appropriate settings* to improve standard time scales. (italics as in original)

Rather than an attempt to use empirical data to "constrain the predictions made from celestial mechanics", as recommended by Fischer and Schwarzacher (1984), rather than a careful attempt to evaluate possible differences in orbital parameters in the geological past, this approach set out to confirm the reality of the Milankovitch signal by demonstrating orbital parameters similar to those acting at the present day.

In 1998 the Royal Society of London held a symposium on the topic "Astronomical (Milankovitch) calibration of the geological time scale", which was attended by many of the leading researchers in this field. The results were published in the Philosophical Transactions of the Royal Society (Shackleton, et al., 1999; papers in this set are referred to by author below, but are not listed separately in the references at the end of this paper). The symposium led off with an astronomical study of orbital frequencies by Laskar. He concluded that calculation of planetary motions cannot accurately retrodict Earth's orbital behaviour before about 35 Ma, because of the long-term chaotic behaviour of the planets and because of drag effects relating to the dynamics of the Earth's interior. However, he suggested that "The uncertainty of the dissipative effects due to tidal dissipation, core-mantle interactions, and changes in dynamical ellipticity are real, but if the geological data are precise enough, this should not be a real problem for the orbital solution." His argument was that cyclostratigraphers could rely on accurate data from the geologic record in order to improve their astronomical calculations.

However, a review of the remaining papers, constituting the body of the symposium proceedings, does not reassure us that such data are becoming available. Most of the papers consist of detailed studies of cyclic geological data which have been subject to time series analysis, filtering and tuning in order to highlight cycle frequencies. Most of these studies result in the reconstruction of frequencies similar or identical to those known from the present day, despite the warnings of Laskar (1999), or earlier warnings of similar character (e.g., Berger and Loutre, 1994) that cycle frequencies could be significantly different in the more distant geological past. Having said this, these frequencies are not reported with any consistency. For example, a "long eccentricity" period is variously reported as having a period of 400 ka (Gale et al.), 404 ka (Olsen et al., Hinnov and Park, Herbert), 406 ka (Shackleton et al.), and 413 ka (Hilgen et al.), without reference to any such long-term variation in frequency.

All these individual studies represent analysis of "hanging" or "floating" sections, that is, sections that are not rigorously tied in to any existing cyclostratigraphic stratotype (because these do not exist for pre-Pliocene strata), but are dated according to conventional chronostratigraphic methods, that is, by use of biostratigraphy, with or without independent radiometric or magnetostratigraphic calibration. The problem with this is that even the best such chronostratigraphic calibration is associated with significant error, as much as $\pm 4$ Ma in the Jurassic (Weedon et al.). There is, therefore, no method to rigorously constrain tuning exercises. For example, a one-million-year error in age range would encompass 48 potential precessional cycles with a frequency of 21 ka. Calibrations may, therefore, be affected by very large errors. In addition, nearly all the individual studies report variations in sedimentary facies, and actual or suspected missing section, indicating difficulties in arriving at reliable estimates of sedimentation rate for the depth-time transformation. The interpretation of frequencies in many of these papers is constrained by the identification of a cycle "bundling", such as the 1:5 bundling that is said to characterize the combined effect of eccentricity and precessional cycles, but such bundling is rarely precise and, in any case, also ignores the warnings of Laskar and others that cycle periods may have been significantly different in the distant past. Individual cycles, like magnetic reversal events, are difficult to individually characterize. As Murphy and Salvador said of the latter (online version of the International Stratigraphic Guide by Michael A. Murphy and Amos Salvador; www.stratigraphy.org), reversals "have relatively little individuality, one reversal looks like another." Comparisons and correlations between cycle strings are therefore all too easy to accomplish, and there may be little about such successions to indicate the presence of missing section. Torrens (2002, p. 257) referred to what he called the "bar-code effect" of dealing with "basically repetitious, often binary" data of orbital cycles and magnetic reversals:

> The problem is that, if one bar-line is missed or remains unread, the bar-code becomes that, not of the next object, but that of a quite different object. The proximity of the next object, becomes no proximity at all.

Miall and Miall (2004, p. 39) suggested that attempts to develop a time scale with an accuracy and precision in the $10^4$-year range by calibrating it against conventional chronostratigraphic dates up to two orders of magnitude less precise represents a fundamentally flawed methodology. The best that can said about the Royal Society symposium is that the results are "permissive"—they point to a possible future potential, but one that is very far from being realized. Nonetheless, the editors of the symposium proceedings, state, in their Preface: "We believe that the calibration of at least the past 100 Ma is feasible over the next few years."

In another incident of the "tail wagging the dog", Van der Zwan (2002) generated orbital signals from subsurface digital gamma-ray data run on fluvial, deltaic and turbidite sediments from the Niger Delta. The potential for climate change to affect the sediments of the delta is discussed in Sect. 11.2.5. The delta plain and delta front are sedimentary environments characterized by frequent breaks in sedimentation and by very large fluctuations in sedimentation rate as a result of changes in river and tidal energy, storm activity, submarine landslides, etc.—not environments conducive to the preservation of rhythmic regularity generated by an external signal. Nevertheless, his time-series analysis generated a power spectrum and, with the aid of dates provided by biostratigraphy, he converted the cycles to a time frequency, claiming that these demonstrated the periodicity of orbital eccentricity. When questioned about this methodology, Van der Zwan responded (e-mail correspondence, 2003):

> With respect to the key question, the validity of the method, for me the ultimate test is whether it is possible to forward model the stratigraphy of the calibration well correctly using these climatic input signals. In this respect, the fact that Milankovitch cyclicity is recorded in both pollen spectra and gamma ray records convinced me of the presence of such climatic signals in the geological record.

And this:

> An ultimate final control is whether the input parameters used make geological sense.

An analysis of a Paleocene pelagic rhythmic succession in Italy (Polletti et al., 2004) provides a good example of the "inversion" of the Milankovitch argument. These researchers extracted cyclic frequencies from their data by spectral analysis, using a constant sedimentation rate calculated from the age range of the section (using the biostratigraphic time scale of Berggren et al., 1995). They commented:

> The apparent disagreement with eccentricity, obliquity and precession periods does not imply a stochastic distribution. Parts of the succession may actually be missing due to hiatuses not detectable by the traditional biostratigraphy and the calculated sedimentation rate may thus not reflect the real deposition history and account for the apparent disagreement.

Their next step was to adjust the sedimentation rate so that the resultant cyclic frequencies matched present-day orbital frequencies. The ratios of some of the cycle lengths in the adjusted cycle spectrum then matched the ratios between some of the Milankovitch frequencies, although the relative strengths of the cyclic signals varied considerably from those of the present day.

Here is one of the problems with Frodeman's "narrative logic" (Sect. 1.2.4). How do we tell when the logically elegant is, nonetheless, wrong? Experimentation that makes use of numerical models is commonly used in the Earth Sciences to verify or validate a quantitative model. However, Oreskes et al. (1994) have pointed out the logical fallacy involved in "affirming the consequent"—the demonstration by numerical modeling of a predicted relationship. This does not constitute confirmation of the model; it merely indicates a certain probability that one family of solutions to a problem is feasible. Denzin (1970, p. 9) defined the "fallacy of objectivism" as the researcher's belief that if "formulations are theoretically or methodologically sound they must have relevance in the empirical world."

To summarize this section, between the time of the initial, cautious work of Hays et al. (1976) and Berger et al. (1984) and the present day, the "Milankovitch theory" has been inverted. At first, researchers attempted to use geological data to explore the range of cyclic climatic periodicities in the

geological past, with a view to testing the evidence for the presence of an orbital-forcing signature. Now, the demonstration of cyclic periods falling somewhere within the "Milankovitch band" is enough for researchers to assert that the controlling mechanism was orbital forcing. Initial cautions based on an understanding of the incompleteness of the geological record have given way to a predisposition to respect the power of time series analysis to generate the expected signals. At least one astronomer has even suggested that geological data may be used to constrain the astronomical calculations (Laskar, 1999). For Laskar, geological data had become a "black box" of unquestioned veracity. Geologists experienced in the incompleteness and inconsistencies of field data and knowledgeable about the warnings associated with the use of time series analysis offered by signal theorists (Rial, 1999, 2004), would be very skeptical about this last approach. Rial (1999) warned that "chronologies based on orbital tuning cannot be used because orbital tuning subtly forces the astronomical signal into the data." As Dott (1992b, p. 13) pointed out, "the Milankovitch theory is very accommodating, for it provides a period to suit nearly every purpose."

### 14.7.2 The Building of a Time Scale

The tuning of stratigraphic records with the use of preserved orbital signals has become a fruitful area of research (House and Gale 1995; Hinnov, 2000; Weedon, 2003), but much remains to be done to clarify possible changes with time in orbital frequencies, and the broad framework of absolute ages within which refined cyclostratigraphic determinations can be carried out.

Cyclostratigraphic studies depend on a hierarchy of five theoretical assumptions (Miall and Miall, 2004, p. 40):

1. The section is continuous, or
2. (alternate): Discontinuities in the section can be recognized and accounted for in the subsequent analysis;
3. Sedimentation rate was constant, or events (such as turbidites) beds can all be recognized and discounted.

4. Orbital frequencies can be predicted for the distant geological past, based on independent age-bracketing of the section;
5. Thickness can be converted to time using a simple sedimentation-rate transformation;
6. The variabilities in stratigraphic preservation (facies changes, hiatuses) can be effectively managed by pattern-matching techniques.

Do we have reliable tests of any of these assumptions? At present there are still many questions. "Bundling" of three to six cycles into larger groupings has been suggested as one distinctive feature of orbital control, indicating nesting of obliquity or precessional cycles within the longer eccentricity cycles (Fischer, 1986; Cotillon, 1995), but given the natural variability in the record, and the tendency of geologists to "see" cycles in virtually any data string (Zeller, 1964), this approach needs to be used with caution.

Few researchers refer to the possibility of variations in orbital behaviour through geologic time; yet a considerable amount of work has been carried out on this problem by astrophysicists, and some of this is published in mainstream geological literature. For example, Laskar (1999) concluded that calculation of planetary motions cannot accurately retrodict Earth's orbital behaviour before about 35 Ma. Berger and Loutre (1994) suggested that the obliquity and precessional periods have steadily lengthened through geologic time. Taking only their calculations for the Late Cretaceous-Cenozoic, and omitting any consideration of chaotic behaviour, they indicated that the 19,000-year precessional period would have been 18,645 years at 72 Ma, and the 41,000-year obliquity period would have been 39,381 years. What this means is that at 72 Ma, the Earth, over a 10-million years time period, would have experienced ten more precessional and obliquity cycles than the present-day periods would have predicted. This represents a very significant difference, and one that cannot be ignored if cyclostratigraphic data are to be used to refine the geologic time scale. Yet no mention of this is made in current literature on this topic. And this may not be all. Murray and Holman (1999) demonstrated that the orbits of the giant outer planets are chaotic on a $10^7$-year time scale. The implications of this result for the inner planets, including Earth, have yet to be resolved, but Murray (pers. comm., 2003) suggested that Earth's orbit has undoubtedly been affected by similar gravitational effects on a comparable time scale.

A reliable cyclostratigraphic (astrochronologic) time scale was first established for the youngest Cenozoic strata, back to about 5 Ma (Hilgen, 1991; Berggren et al., 1995; Hilgen et al., 2006; see Figs. 14.36 and 14.37). At the time of writing, astronomically calibrated sections have been used to extend the astrochronological time scale back to 14.84 Ma, the base of the Serravallian stage, in the mid-Miocene (www.stratigraphy.org), and research is proceeding to extend the time scale not only to the base of the Cenozoic, but through at least the Mesozoic (Hinnov and Ogg, 2007). However, Westphal et al. (2008) offered a sharply critical review of the field data base on which part of the astrochronological time scale is based, pointing to problems of diagenesis and differential compaction of contrasting lithologies, that render direct one-for-one correlations between the critical field sections problematic. A careful comparison of the correlations between two critical field sections shows that the correlations of the astronomical cycles are not always supported by the correlations of bioevents. In some cases, there are different numbers of cycles between the occurrences of key bioevents (Fig. 14.38).

The problem, as we perceive it (Miall and Miall, 2004), is that a powerful model may, once again, as in the case of the Vail curves, be used to drive the development of correlations rather than the empirical data being used to test the model. Remarks about orbital control being a demonstrated "fact" tempt geologists to ignore the null hypothesis and to readily assume that cyclic successions were deposited under the influence of orbital forcing. Cyclostratigraphy can only work in continuous sections or where the existence of hiatuses has been carefully evaluated. As I have attempted to demonstrate (Sect. 14.3; see also, in particular, Aubry, 1991, 1995), evaluation of unconformities is one of the most difficult and most neglected aspects of stratigraphic study. One approach developed specifically for cyclostratigraphic successions has been offered by Meyers and Sageman (2004). It requires that the stratigraphy be expressed as a numerical parameter that can be subjected to time-series analysis, such as the optical densitometry gray-scale analysis of the Bridge Creek Limestone used as a test in this paper. The data string for the section are then sampled in a moving window and subjected to a harmonic analysis.

**Fig. 14.36** The Punta di Maiata section on Sicily. Punta di Maiata is the middle partial section of the Rossello Composite and part of the Zanclean unit stratotypes, which defines the base of the Pliocene (Van Couvering et al., 2000; Hilgen et al., 2006). Larger-scale eccentricity-related cycles are clearly visible in the weathering profile of the cape. Small-scale quadripartite cycles are precession-related; precession-obliquity interference patterns are present in particular in the older 400-kyr carbonate maximum indicated in *blue*. All cycles have been tuned in detail and the section has an excellent magnetostratigraphy, calcareous plankton biostratigraphy and stable isotope stratigraphy (Hilgen et al., 2006)

Continuous sections yield a constant harmonic signature reflecting orbital controls, but the harmonics are disrupted by hiatuses and this can be made very clear by a graphical display of the harmonic signature of each sample window. Some work can be done to eliminate anomalies in sedimentation rate by careful examination of sections to identify such units as storm deposits and turbidites that can be eliminated from the section for the purpose of time-series analysis. We return to the Bridge Creek Limestone at the end of this section.

A good example of the potential pitfalls in an otherwise interesting paper is the study of Miocene pelagic carbonates reported by Cleaveland et al. (2002). The objective of this paper was to use the preserved orbital signature in the rocks to "tune" ages derived from biostratigraphic and radiometric data, based on calibration against a theoretical orbital eccentricity curve. Their data set consisted of a set of $CaCO_3$ values obtained by analysis of a tightly sampled limestone section. Sample position in the section was first converted to age based on an average sedimentation rate derived from the ages obtained from radiometric dates on ashes at the top and bottom of the section. This requires two assumptions: constant sedimentation rate, and absence of hiatuses. This derived data set was then subjected to two separate smoothing procedures to enhance the visibility of the predicted cycle frequency (two more assumptions). The result is twenty cycles, the same number as in a theoretical orbital eccentricity curve for the geologic interval. The match between the massaged data set and the theoretical curve is visually excellent. The results, which now incorporate five assumptions (four from the data and the fifth being the theoretical curve), were then used to adjust the age of an important stage boundary, determined from a biohorizon that occurs within the section.

The radiometric ages on the ashes, $12.86 \pm 0.16$ and $11.48 \pm 0.13$ Ma, yield extreme possible age ranges for the studied section of between 1.09 and 1.67 million years. Using the ages within this range preferred by the authors, the cycle frequencies calculate to about 94 ka. This was equated to a supposed 100 ka eccentricity frequency by Cleaveland et al. (2002), although the present-day observed eccentricity frequencies (House, 1995, p. 12) are 54, 106 and 410 ka, not 100 ka, and

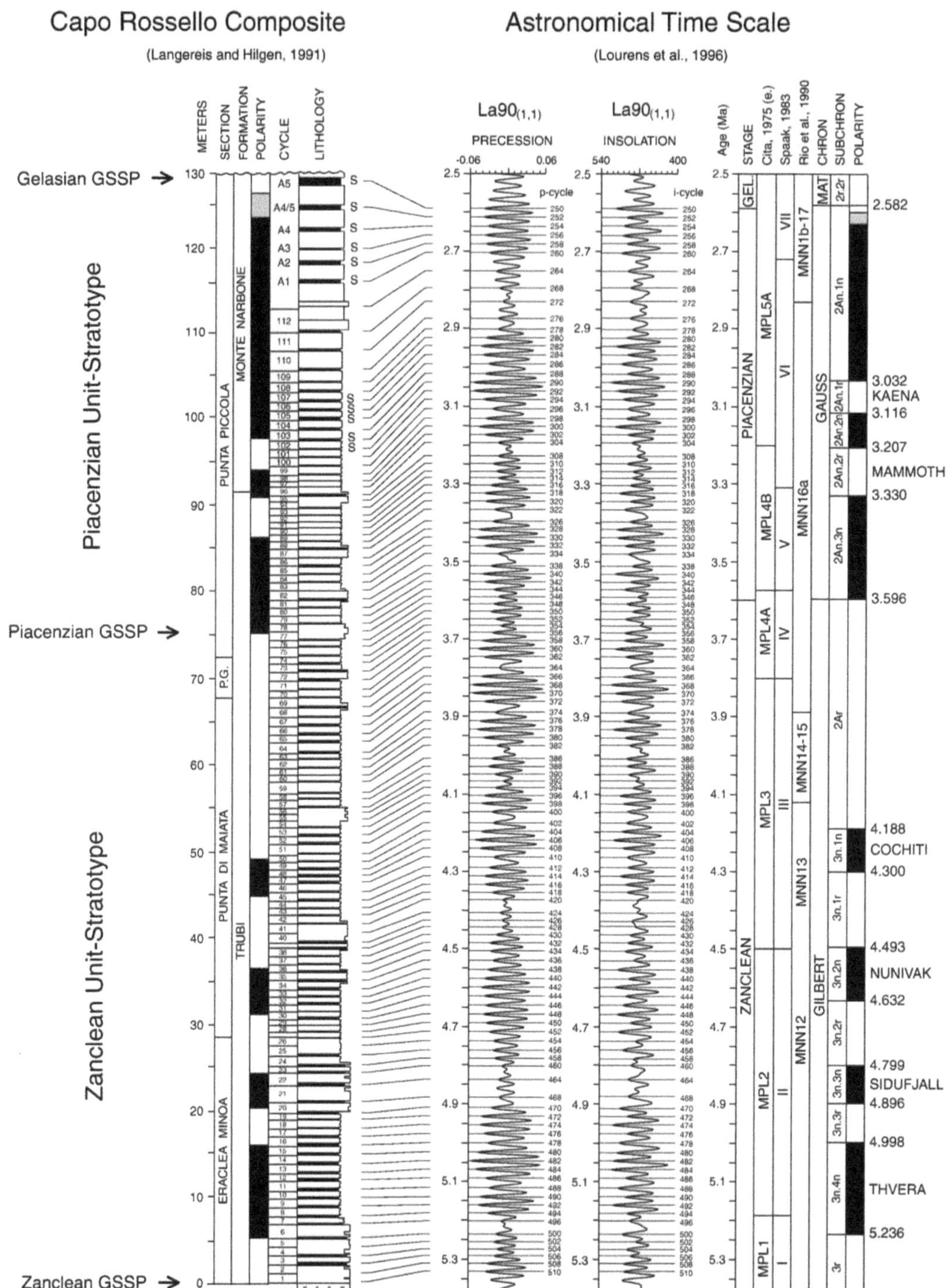


**Fig. 14.37** The Rossello Composite Section (RCS, Sicily, Italy,) the unit stratotypes for series spanning the base of the Pliocene, incorporating the orbital tuning of the basic precession-controlled sedimentary cycles and the resulting astronomical time scale with accurate and precise astronomical ages for sedimentary cycles, calcareous plankton events and magnetic reversal boundaries. The Zanclean and Piacenzian GSSPs are formally defined in the RCS while the level that time-stratigraphically correlates with the Gelasian GSSP is found in the topmost part of the section. The well tuned RCS lies at the base of the Early–Middle Pliocene part of the astrochronological time scale and the Global Standard Chronostratigraphic Scale and as such could serve as unit stratotype for both the Zanclean and Piacenzian Stage (Hilgen et al., 2006)

**Fig. 14.38** Comparison of the cycle record in two of the key sections in Italy used for the construction of the astrochronological time scale. Some of the bioevents markers enclose different numbers of cycles in the two sections, which calls into question the precision that is claimed for this methodology (Westphal et al., 2008, Fig. 10)

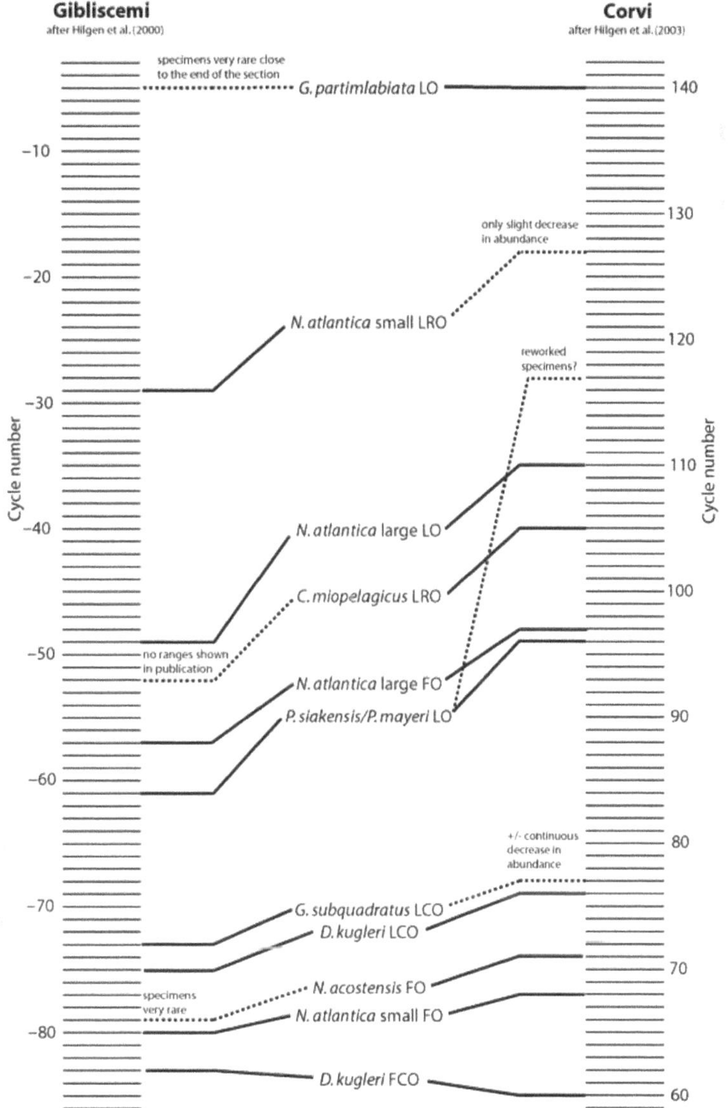



the theoretical eccentricity curve on which Cleaveland et al. (2002) based their comparison includes a frequency of 95 ka. The range of possibilities for cycle frequency in the Cleaveland et al. (2002) data that are yielded by the error limits of the radiometric ages extends from 73 to 111 ka, which just about encompasses the 106 ka frequency reported by House (1995). This, however, was not the number used by Cleaveland et al. (2002), and their use of a 100-ka value is not explained. The work on stratigraphic incompleteness by Smith (1993) and Aubry (1995) should serve as a warning about the potential for missing cycles and,

consequently, errors in correlation and rate calculations in studies of this type.

An example of the use of multiple tie points for age transformation is provided in another interesting study by Roof et al. (1991), which is the first paper in a special issue of *Journal of Sedimentary Research* devoted to "Orbital forcing and sedimentary sequences." A core some 235 m long yielded 23 biostratigraphic control points, in the form of planktonic foraminiferal recoveries. Their data are discussed in Sect. 14.5.6 (see Fig. 14.21), a discussion which can be read as an example of the problems associated with the testing of the assumptions listed at the beginning this section. Most ages are expressed in the Roof et al. (1991) paper (their Table 1) to the nearest 10,000 years, although no error ranges are indicated. A simple, linear arithmetic transformation is assumed throughout the rest of their paper, in that all core logs are plotted with time as the ordinate instead of depth, based on the calculated average. The data, once transformed, were taken as an empirical standard against which other measurements were plotted, and were then used in time series analysis to extract orbital frequencies. These authors chose not to tune their data, with the result that spectral analysis generated a wide range of frequencies for different intervals of the core. In only some of the core do these frequencies compare with those suspected to be the result of orbital forcing.

Based on the poor correlation with predicted Milankovitch frequencies, Roof et al. (1991) concluded that shifts in ocean currents and sediment delivery in their sample area (Gulf of Mexico), especially that delivered by the nearby Mississippi River, accounted for much of the variability in sedimentation rate, and that this overprinted climatic effects within much of the core.

There is nothing wrong with the science in these papers in so far as they are reports of specific field case studies. Problems may arise, however, when results of this type are used as confirming evidence for the general model of orbital control of sedimentation and astronomical calibration of the geological time scale. For example, the presentation by Cleaveland et al. (2002) of their correlations is followed by this statement:

> Our astronomical correlation is made based on a combination of pattern matching between the carbonate and eccentricity curves, radiometric age constraints provided by the two volcanic ashes in the section, and results from spectral analysis of the tuned data set. Of other possible correlations consistent with the radiometric age constraints, we found greatest enhancement in the obliquity and precessional bands when the correlation shown in Fig. 3 is applied, and so favor this correlation over other possibilities, despite some discrepancies between the amplitudes of corresponding carbonate and eccentricity peaks.

These authors have made use of the qualitative "pattern matching" method, which is easily susceptible to bias and error. However, they did note internal discrepancies and possible alternative correlations. The assumptions of the model, therefore, must be continuously borne in mind.

A practical problem with dating and correlation at the very fine scale required by cyclostratigraphy is the problem of disturbance and mixing by bioturbation, and diagenetic modification of the sedimentary record. Anderson (2001) and Zalasiewicz et al. (2007) discussed these problems. Gale et al. (2005) reported an obliquity and eccentricity signal in Eocene-Oligocene paleosols based on illite abundance, a product of diagenetic modification of smectite soils. Such alternation may (but may not) reflect the product of actual orbital climatic changes, but as Zalasiewicz et al. (2007, p. 142) remarked, the offset between deposition and diagenesis may be hundreds to thousands of years. Hallam (1986) suspected that many cyclic limestone-shale deposits, such as the Lower Jurassic Lias Formation of southern Britain, are largely, if not entirely, the product of diagenetic unmixing, and therefore contain no meaningful cyclostratigraphic signal.

McGowran (2005, pp. 65–84) and Hinnov and Ogg (2007) discussed the modern work underway to extend the astrochronological time scale back through deeper and deeper geologic time. As McGowran (2005, p. 70) noted: "The power of the Cenozoic IMBS [integrated magneto-biostratigraphic time scale] lies in the incessant triangulating between bio-, magneto- and radio-chronologies against the linearizer, seafloor spreading. Astrochronology now is an integral part of this procedure." As noted by Hinnov and Ogg (2007, p. 240) astrochronology can improve the precision of local correlation by an order of magnitude where continuous sections of orbitally-forced cycles can be independently fixed in time by radiometric or magnetostratigraphic dating. Radiometric dating can now achieve precision of $\pm 0.1\%$, although this is still inadequate to firmly fix the position of floating sections. A potential $\pm 0.1\%$ error at 100 Ma amounts to $\pm 100$ ka, which is enough to prevent accurate correlations of successions of precessional or obliquity cycles. Hinnov and Ogg (2007), citing recent astronomical work, suggest that the 405 ka eccentricity cycle might be the only reliable cyclic frequency to carry the time scale back into the Mesozoic.

Because of its use in improving local correlations by an order of magnitude (to a $10^4$-year scale), calibrating marine cyclic sections by astrochronology can "liberate" a great deal of biostratigraphic detail, such as abundance data and minor evolutionary changes that had previously been little more than noise in the data (McGowran, 2005, p. 78). This can potentially increase the accuracy and precision of *regional* biostratigraphic correlation where the other forms of high-resolution correlation and dating, such as magnetostratigraphy, are unavailable. However, given the problems with faunal provincialism discussed in Sect. 14.5.6, it is unlikely that such additional detail would be of use in intercontinental correlation.

Does astrochronology offer any potential for the testing of sequence models? In that the presence of $10^{4-5}$-year cyclicity in the rock record is itself a demonstration of the importance of orbital forcing as a driver of climate and, potentially, glacioeustasy (see discussion in previous section), the answer is yes. Three examples from Europe and one from the US are briefly discussed here to illustrate both the nature of the contribution astrochronology can make, and the pitfalls that can surround it. The first example exemplifies the way in which a refined astrochonological time scale may be employed to carry out refined tests of geological hypotheses. The second example illustrates the pitfalls of building cyclostratigraphic interpretations based on incomplete evidence. The third example discusses interpretive practices that this writer finds problematic. The fourth example demonstrates that the closer one looks and the more refined the data and methodology, the more irregular the rocks appear to be, which calls into question our ability to apply routinized statistical methods to the ancient record.

Hilgen et al. (2007) examined the age and correlation of the Messinian evaporites in the Mediterranean Sea in order to explore the several different hypotheses concerning the origin of the dramatic desiccation of this large marine basin. Alternative models for the isolation of the basin include glacioeustatic lowering of sea level below the Gibralter Strait sill, and tectonic closure of the marine connection through this strait. Astrochronological correlation of sections throughout the basin suggests that the commencement of evaporation of the sea began at the identical time throughout the basin. Careful correlation of Mediterranean sections with the oxygen isotope record from the North Atlantic Ocean indicates that initiation of evaporation coincided with an episode of glacioeustatic sea-level rise, not fall. Likewise, the termination of evaporation, as a result of re-flooding of the basin, does not correlate to a glacioeustatic sea-level rise, which, according to Hilgen et al (2007, p. 234) lends credibility to other hypotheses for the flooding of the basin, such as headward erosion of rivers incising the Gibralter Strait from the Mediterranean side.

A more problematic area of research concerns the well-known Middle-Upper Triassic carbonate cycles of the Italian Dolomites. These were first known by the work carried out on the so-called Lofer Cycles, named after the Dachstein Limestone in the Lofer district of the Northern Calcareous Alps—the Dolomites—in Italy (Fischer, 1964; Enos and Samankassou, 1998). The Middle Triassic Latemar carbonate buildup, nearby, is also a well known area of high-frequency cyclicity (Goldhammer et al., 1987; Hinnov and Goldhammer, 1991; Preto et al., 2001). The sedimentology and cyclicity of these cycles are described briefly in Sect. 7.3. The problematic nature of this area of research is that independent work to accurately date the rocks indicates a time span for the cyclic succession that is not consistent with Milankovitch frequencies. U-Pb dating of zircons from interbedded volcaniclastic rocks suggested a shorter time span than expected for the succession, making it difficult to fit the successions to Milankovitch frequencies. This problem triggered the so-called "Latemar controversy," which included suggestions that the cycles represent previously unknown sub-Milankovitch frequencies, or even that they are not allocyclic in origin (Zühlke et al., 2003).

Following the development of these concerns regarding the chronostratigraphy of the Latemar cycles, Preto et al. (2001) resampled the field sections, measuring a more complete (longer) section and employing a more detailed classification of the carbonate facies into four depth-related categories. Using available radiometric ages, they recalculated the age range of the disputed succession, and by tuning the sedimentation rate to a precessional index they found that three expected eccentricity frequencies and an expected (but weakly recorded) obliquity frequency were extracted by time series analysis. The 1:5 cycle stacking pattern, which it has been suggested is characteristic of orbitally forced successions (Chap. 11), was also recognized. However, Zühlke et al. (2003) insisted that their own even more detailed section measurements and correlations, coupled with high-precision U-Pb dating (Mundil et al., 2003) and integrated biostratigraphy yielded cycle periods of 2.20–2.89 ka (up to 6.4 ka when calculated using maximum uncertainties), which cannot be reconciled with the expected Triassic "standard" precessional periodicity of 18–21 ka. They then experimented with various mathematical models for tuning the data series, and arrived at a most-likely solution, consistent with the chronostratigraphic data, that if the basic cycle period is set at 4.2 ka, then the various higher frequencies and bundling ratios appear to fall into place. Spectral analysis with the basic cycle set at 4.2 ka yielded a set of higher frequencies that can be assigned to predicted Milankovitch frequencies, including a 1: 4.7 ratio with a frequency in the 18.1–21.5 ka range, assumed to be a precessional frequency. This result was later essentially confirmed by independent statistical testing by Meyers (2008). Sedimentation (carbonate production) rates for the Latemar buildup calculated from this revised chronostratigraphic model yield values comparable to "moderately prolific (sub)recent carbonate platforms" (Zühlke et al., 2003, p. 76). The time-scale of cyclicity raised speculations about possible comparisons to the millennial climatic fluctuations of the Quaternary era, which included such high-frequency events as the Dansgaard-Oeschger cycles (Broecker and Denton, 1989) and Heinrich events (Heinrich, 1988). However, it was recognized that under the assumed greenhouse climatic conditions of the Triassic, a quite different regimen of hemispheric and global climatic oscillations would have been expected to occur.

The implications of the Latemar controversy are, as Zühlke et al. (2003, p. 79) pointed out, that in the case of high-frequency cyclicity, simple cycle stacking ratios cannot automatically be assigned to eccentricity-obliquity or eccentricity-precessional bundling. In fact, this highlights the entire questionable issue of automatically assigning Milankovitch frequencies to cyclic successions deposited during the remote geological past simply on the basis of bracketing ages that indicate frequencies in the order of $10^4$–$10^5$ years. Enormous care will be required to extend the astrochronological time scale back through the Cenozoic, and this author remains very skeptical that it will ever be possible to accurately assign floating sections to an astrochronological time scale in the pre-Neogene.

The third example discussed here deals with the cyclostratigraphic analysis of some shallow-water Aptian-Albian carbonates in the southern Apennine, Italy, by D'Argenio et al. (2004b). Four sections of the carbonates were studied, consisting mainly of successions of subtidal to peritidal wackestones and grainstones. Cycle boundaries were recognized on the basis of microkarst or pedogenesis. The authors claim that statistical analysis of recurring sedimentary features generated ratios that match those of orbital frequencies with a high degree of statistical probability, However, their stratigraphic sections show cycles with highly variable thicknesses (Fig. 14.39). In one of the sections cycle thickness ranges from 0.3 to 5.55 m. Nevertheless, the analysis goes on to erect a system of "superbundles," each consisting of two or three cycles. Three to five of the superbundles are said to define transgressive-regressive sets. Correlations between the sections show that in some locations some of the cycles are missing, which is attributed to low accommodation. From these sections a regional correlation framework was constructed, using carbon isotope data as a supplementary means of correlation. The next step of the analysis was to construct a regional "orbital chronostratigraphy," although "in each section both location and time of missing intervals are variable." This is explained in terms of "accommodation space as well as bed thickness and facies of time-equivalent cycles are controlled predominantly by regional changes in subsidence and carbonate productivity. Therefore a minor number of missing bundles is recorded in successions formed on more open carbonate-platform environments (hence producing higher sediment volumes) and vice versa." (D'Argenio et al., 2004b, p. 115). In fact, in the composite chronostratigraphic column (Fig. 14.40) about one third of the bundles are missing. This composite section, with its carbon isotope curve, was then correlated to a standard isotope curve and tied into the eustatic "third-order" sequences of Jacquin et al. (in Graciansky et al., 1998).

The expected standard error for the Aptian-Albian is about $\pm 1$ million years (Fig. 14.22) and there seems to be good reasons to expect a more precise correlation of the composite section than this, given the matching of the carbon isotope curve to a standard curve for the Aptian-Albian (not shown here). However, the proposed correlations of the eustatic events with the orbital chronology seems highly unrealistic to this writer. No range of potential error is indicated for the "third-order" events in the Graciansky et al. (1998) cycle chart, including that portion of it shown in Fig. 14.22. Nor, given the numerous missing bundles in the composite section, should the correlation with this section be regarded as definitive. A potential error of $\pm 1$ million years in the cycle chart allows for miscorrelation by one or two superbundles. The pattern of missing bundles in the Apennines section is reminiscent of the pattern of missing faunal horizons in the Inferior Oolite of southern England (Fig. 14.2), and

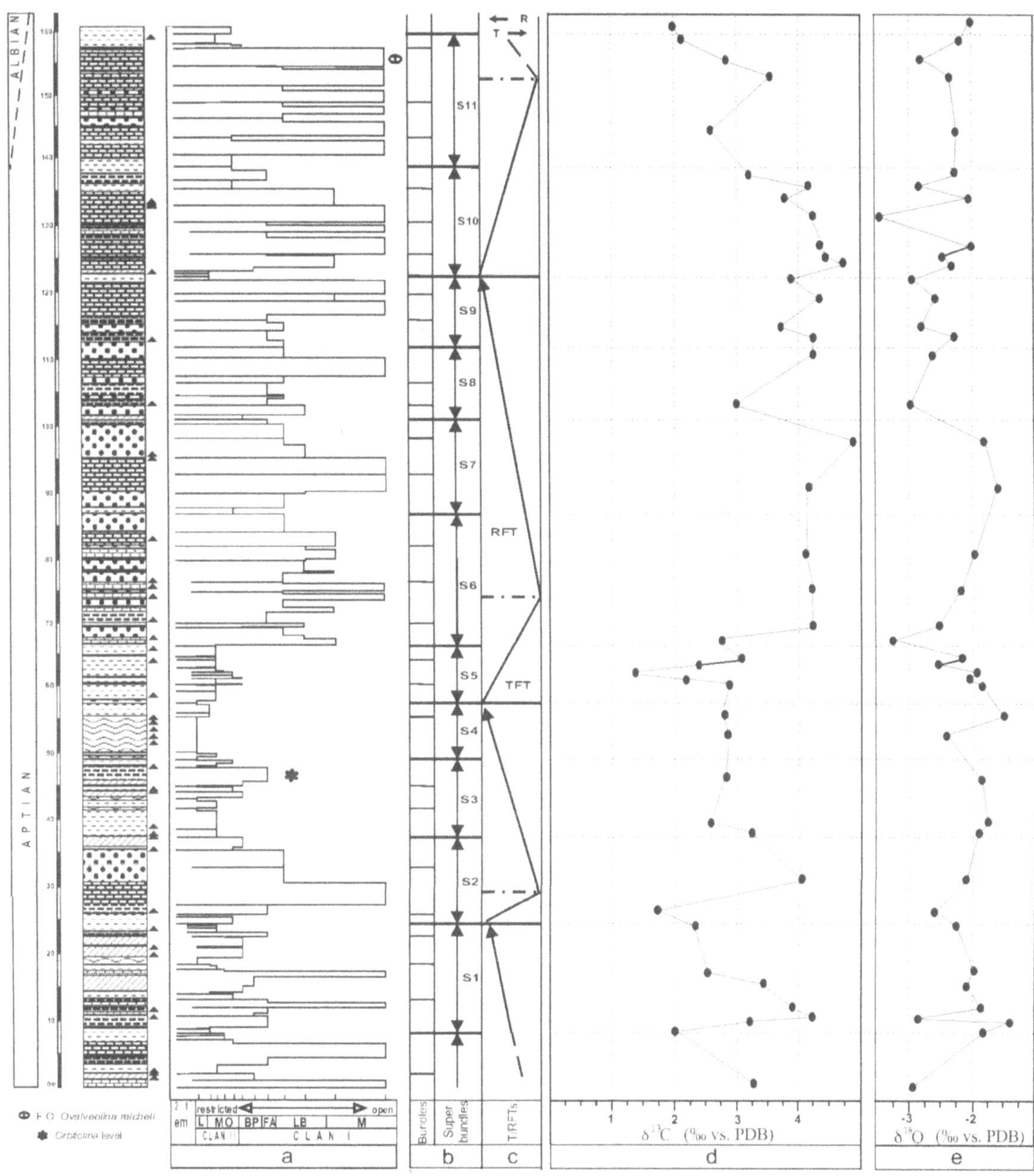

**Fig. 14.39** Lithology, cycles, and stable-isotope data for one of the sections studied by D'Argenio et al. (2004b, Fig. 3)

may represent a similar shallow-marine environment, in which sedimentation is impersistent and preservation is very patchy. Although the authors claim to have demonstrated orbital periodicities in their section data, the raw data have none of the regular, rhythmic characteristics of other more obviously rhythmic sections, such as those illustrated in Chaps. 7 and 11. That a few of the superbundle boundaries may or may not be said to correlate with the global cycle "third-order" events raises all kinds of questions. If the section really is controlled by orbital forcing, why are some of the bundles missing, and why do only three

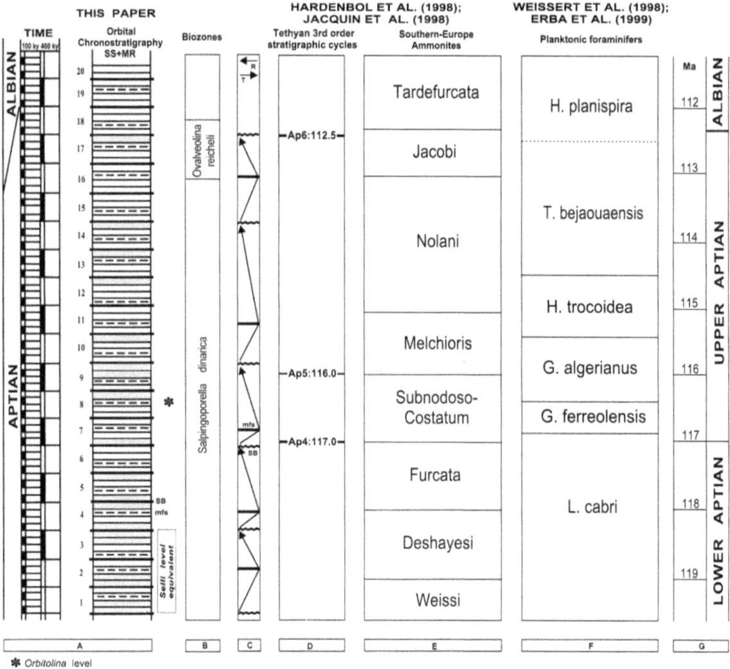



**Fig. 14.40** Orbital chronostratigraphy and correlation of a composite section of the Aptian-Albian section of southern Italy (D'Argenio et al., 2004b, Fig. 11)

of the superbundle boundaries correlate with sequence boundaries claimed to be present elsewhere in Europe? If the superbundle boundaries are the product of orbital forcing why are there not more of them in the European cycle chart? In conclusion, the procedure illustrated in Fig. 14.40 compares the results of two completely different methods of developing local chronostratigraphies, both of them, in the view of this writer, seriously flawed.

For the final example, we turn to Cenomanian-Turonian sections in Colorado, which have received intensive examination from several perspectives. In this section we discuss cyclostratigraphy. We return to the same sections in Sect. 14.8 to explore the record there of a major "oceanic anoxic event." and its significance for the themes of this chapter of global correlation and its application to sequence stratigraphy. Sageman et al. (1997, 1998) summarized the history of investigation of these rocks. Gilbert (1895) had hypothesized that the rhythmically bedded limestone/marl couplets of the Bridge Creek Limestone were generated by orbital forcing, and numerous workers since that time have explored this hypothesis. Sageman et al. (1997, 1998) were able to make use of a continuous core drilled for stratigraphic purposes, from which to describe a detailed sedimentological log

and carry out various chemical analyses, plus a optical grey-scale sampling of the varying carbonate-clay composition. This was followed up by the application of sophisticated harmonic analysis of the optical densitometry data by Meyers and Sageman (2004), the proposal of an orbital time scale for the section by Sageman et al. (2006), and an attempted detailed correlation of the beds to a contemporaneous section in Britain by Gale et al. (2008).

In the first, detailed, modern cyclostratigraphic analysis of the Bridge Creek core, Sageman et al. (1997, p. 286) noted the "somewhat chaotic record" of the succession. The simplified lithologic log shown in Fig. 14.41 reveals that limestones are quite erratically spaced through the section, and this is shown in more detail in the optical densitometer log provided by Meyers and Sageman (2004, Fig. 8). Visually, the Bridge Creek Limestone Member appears rhythmic, but with nothing like the almost mathematical regularity of some of the pelagic Neogene sections of southern Italy (e.g., Fig. 14.36), from which the cyclostratigraphic time scale has been constructed. Careful facies analysis revealed variations in limestone lithologies, suggesting variations in environment (and hence, sedimentation rate), and at least one hiatus in the section was identified. This would not seem to be a good

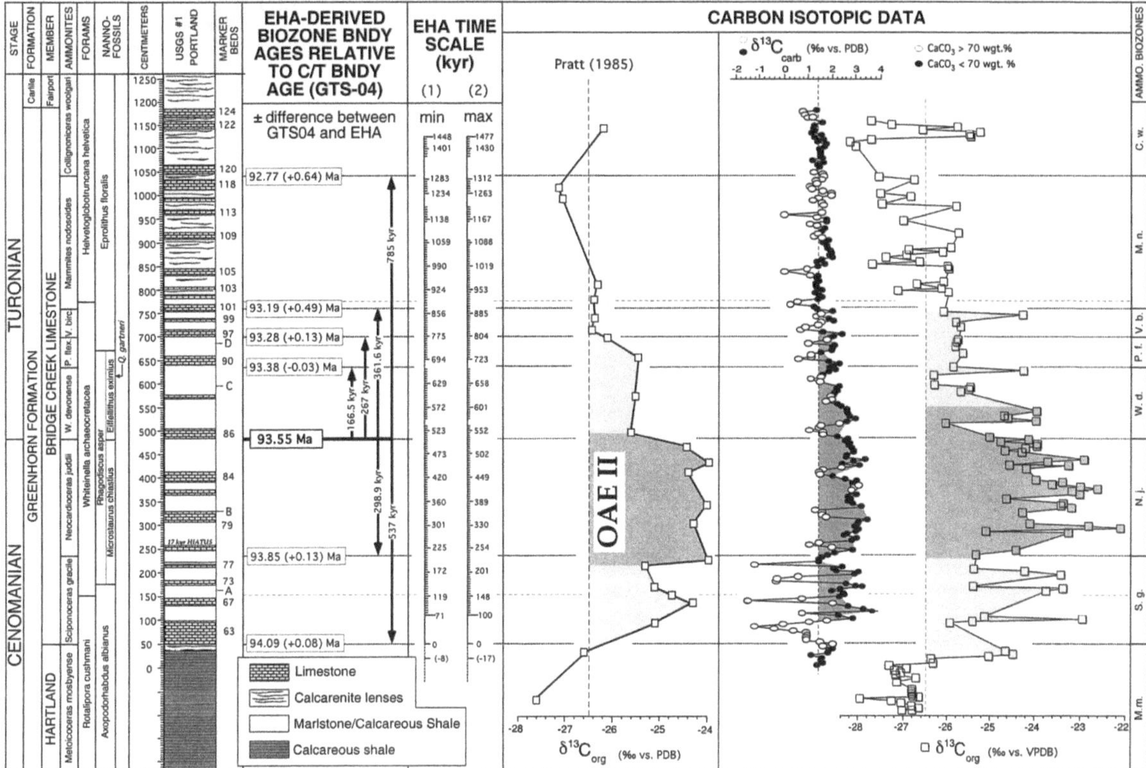

Fig. 14.41   Chronology and isotope data for the Cenomanian-Turonian stratotypes, as documented in a drill core from central Colorado (Sageman et al., 2006, Fig. 1)

candidate for cyclostratigraphic analysis. However, Meyers and Sageman (2004) employed a statistical method (Evolutive Harmonic Analysis, or EHA) that calculated amplitude spectra from a moving window through the densitometer core data. This method identifies hiatuses and variable amplitude frequencies, that may be tuned to predicted time frequencies. They concluded that the 95 ka eccentricity frequency was dominant, and when the record was tuned to this frequency, the spectra contained peaks that matched other predicted frequencies with small errors. Two different EHA time scales are shown in Fig. 14.41, which make use of two slightly different independent chronostratigraphic frameworks for the age range of the section. Examination of this time scale indicates that equal thickness increments in the core do not correspond to equal time increments. Note, also, the differences between the EHA scale and the chronostratigraphic ages derived from application of the Gradstein et al. (2004) time scale, differences which range from 30 to 640 ka. Calculations of sedimentation rate based on radiometric dating of bentonites indicate a value

of 0.84 cm/ka in the basal part of the section, and 2.76 cm/ka near the top (Sageman et al., 2006). The limestone beds are not spaced regularly, and therefore cannot represent any known orbital frequency. In detail, therefore, this "rhythmic" section is, in fact, quite irregular, and not a good candidate for establishing a cyclostratigraphic time scale. We return to this point at the end of the next section, because this section has been employed in a study of orbitally-forced eustatic sea-level change in the Cretaceous.

## 14.8 Testing Correlations with Carbon Isotope Chemostratigraphy

As discussed at length earlier in this chapter, chemostratigraphy has become an integral, indeed, essential component of the geological time scale (the IMBTS referred to by McGowran, 2005). Strontium isotope stratigraphy is now widely used for dating and correlation throughout the Phanerozoic (McArthur and Howarth, 2004; see discussion of its uses in Sect. 14.6.1). Oxygen-isotope chemostratigraphy is an essential underpinning of the Neogene chronostratigraphic scale (Gradstein et al., 2004), and is being interpreted to suggest the occurrence of glacioeustatic events during the Cretaceous and Paleogene (Sect. 14.6.4). These methods rely on the assumption, proven by multiple practical studies and tests, that isotopic levels in the world's marine waters become thoroughly mixed over geologically short time periods, and that therefore isotopic signatures may be reliable time markers. Correlations using these methods are typically achieved by pattern recognition—the establishment of standard curves for intervals of geologic time, based on detailed sampling (e.g., in DSDP cores; see Figs. 11.13 and 11.14). In the case of strontium isotopes, the absolute values of the $^{87}Sr/^{86}Sr$ ratio may in some cases be enough to provide a tight constraint on geological age, as these values have proven to be consistent worldwide.

A parallel body of work has explored the variation in carbon isotopic concentrations through geologic time, with a view to understanding the global carbon cycle and to provide a supplementary tool for global correlation (Scholle and Arthur, 1980; Veizer et al., 1999; Stoll and Schrag, 2000; Swart and Eberli, 2005; Katz et al., 2005). This is, however, a rather different story.

The $\delta^{13}C$ composition of what Katz et al. (2005, p. 323) termed the "ocean's mobile carbon reservoir" is controlled by biological productivity, volcanic and hydrothermal outgassing and weathering, and by oceanic mixing patterns. Different organisms show different patterns of carbon isotope fractionation. Stratigraphic $\delta^{13}C$ signatures are therefore typically determined from bulk sediment samples, rather than from specific microorganisms, as in the case of the oxygen isotope record. However, this signature may be affected by changes in the redox state of the oceans (which affects the distribution of oxidized carbonate and reduced organic matter), by differences in burial rate, and by diagenesis. It has also been demonstrated that samples of the same age from shallow- and deep-water water platform carbonates and from the open ocean may have dissimilar $\delta^{13}C$ values. Swart and Eberli (2005) published curves showing quite different $\delta^{13}C$ histories for Bahamas Bank platform, slope and open-ocean samples back through the last 25 Ma (Fig. 14.42). One reason for this may be the different

isotopic fractionation patterns generated by photosynthesis of shallow-water versus deep-water pelagic microorganisms.

The pattern of carbon $^{13}C$ variations illustrated in Fig. 14.42, and its explanation, might be thought to have discouraged any attempt to develop a carbon isotope chronostratigraphy. Not so. Given the continual changes in Earth's climate, changes in tectonic conditions, additions to atmospheric carbon reservoir from volcanic emissions, and so on, the carbon reservoir is in constant flux, and it can be shown that the changes are more or less well reflected by changes in the sedimentary record worldwide. An event stratigraphy based on the $\delta^{13}C$ record has been constructed for several intervals of the geological column, but as discussed below, many problems and questions remain. I focus here on the Cenomanian-Turonian interval of the Late Cretaceous, during which a major Oceanic Anoxic Event occurred (Schlanger and Jenkyns, 1976). Detailed studies of this event have been carried out in southeast England (Gale et al., 1993), Italy (Parente et al., 2008; Scopelli et al., 2008; Galeotti et al., 2009), the Western Interior Seaway (Sageman et al., 2006; Gale et al., 2008), New Jersey (Bowman and Bralower, 2005), and China (Li et al., 2006), and intercontinental comparisons have been presented by Tsikos et al. (2004), Gale et al. (2008), and Galeotti et al. (2009).

The Cenomanian-Turonian anoxic event, known as OAE2, is recorded by the "excess burial of organic carbon" (Tsikos et al., 2004, p. 711) and is well represented in black shales, such as the Livello Bonarelli of the Apennines in central Italy. The "event" spans approximately 2–15 m of section in each location, and has been recorded in a variety of rock types, including chalks, pelagic carbonates, shales and siltstones. The total carbon content varies over a considerable range between these various facies and locations, but a significant positive $\delta^{13}C$ excursion spanning nearly 1 million years in the upper Cenomanian and lowermost Turonian has been recorded in all the studied sections (Fig. 14.43). Although there is a broad match of high $^{13}C$ spanning the latter part of the Cenomanian and the Cenomanian-Turonian boundary, "the most organic-rich and/or carbonate-poor horizons may be significantly diachronous when considered against integrated chemo- and biostratigraphy ... probably reflecting the unique interaction between local palaeoceanographic and diagenetic conditions at each site and the global

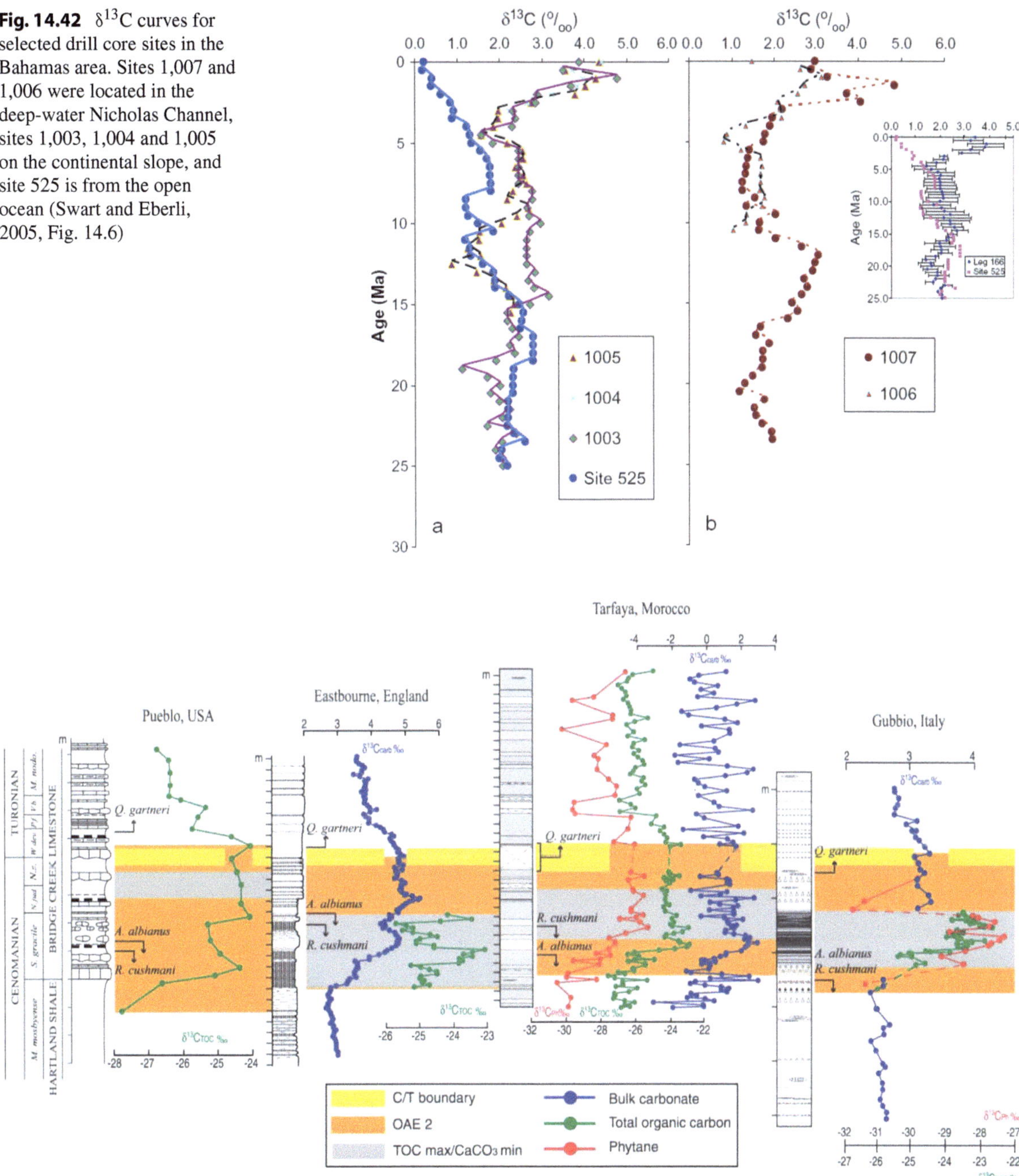

**Fig. 14.42** $\delta^{13}$C curves for selected drill core sites in the Bahamas area. Sites 1,007 and 1,006 were located in the deep-water Nicholas Channel, sites 1,003, 1,004 and 1,005 on the continental slope, and site 525 is from the open ocean (Swart and Eberli, 2005, Fig. 14.6)



**Fig. 14.43**   Comparison of $\delta^{13}$C records for four sections in North America, Europe and north Africa (Tsikos et al., 2004, Fig. 14.6)

event" (Tsikos et al., 2004, p. 716). Tsikos et al. (2004, p. 716) suggested that the data demonstrate that the high $^{13}$C-bearing units are "essentially coeval." In detail, however, the curves for the four locations show significant differences, which Tsikos et al. (2004) attributed to local variations in organic productivity, preservation, and diagenesis.

The exposure at Pueblo, Colorado, has been designated the GSSP (Global Boundary Stratotype Section and Point) for the base of the Turonian (see

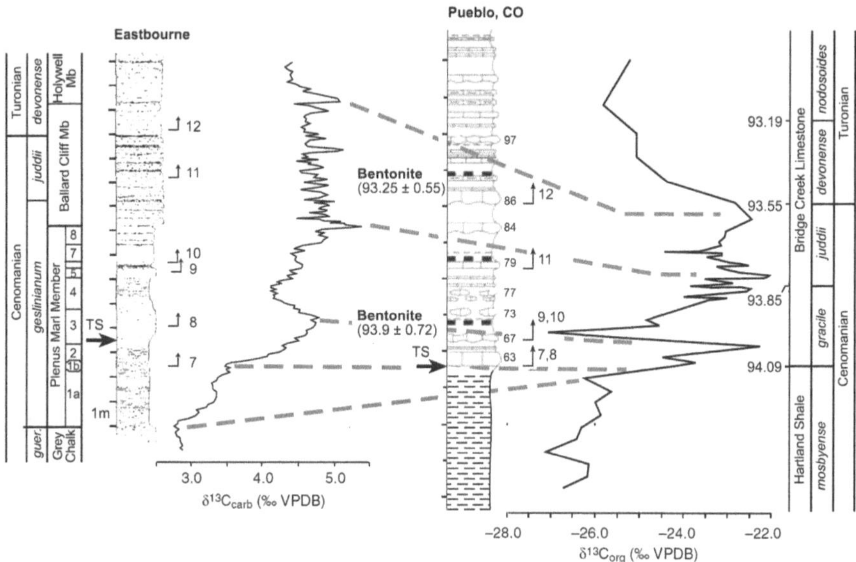

**Fig. 14.44** Correlation of the Cenomanian-Turonian boundary sections between Pueblo, Colorado (the GSSP for the base of the Turonian) and Eastbourne, UK (proposed reference section for this same boundary). Chronostratigraphic correlations slope between the sections because they are plotted on a thickness scale (Gale et al., 2008, Fig. 14.4)

www.stratigraphy.org) and the section at Eastbourne, UK, has been proposed as a reference section for this boundary (Paul et al., 1999). The chronostratigraphic resolution of these sections is therefore excellent, and accordingly they provide a good basis for an examination of how chronostratigraphically precise the $\delta^{13}$C curves are (we discussed the cyclostratigraphy of the Colorado succession in the Sect. 14.7.2). Figure 14.44 indicates detailed correlations between the Pueblo and Eastbourne sections. Dashed lines link distinctive inflection points in the $\delta^{13}$C curves. These suggest a very close comparison between $\delta^{13}$C values at a detailed level. However, a careful examination reveals that when cross comparisons are made between biostratigraphic markers (the base and top of the *juddii* zone), the $\delta^{13}$C curve inflection points, and interpreted sequence boundaries (the numbered events 7–12), there are differences of more than 1 m in position of these markers relative to each other in the two sections. This represents, on average, an error of about 130 ka, according to the detailed time scale generated for the Pueblo section by Sageman et al. (2006). Comparable differences in the ages of interpreted $\delta^{13}$C inflection points are indicated by a correlation of the Eastbourne section to three Italian sections and an ODP site (Fig. 14.45), and even larger differences in the patterns of $\delta^{13}$C variation is shown in Fig. 14.46. The correlations of peaks in this diagram suggests shifts in the timing of some peaks in the $\delta^{13}$C curves of as much as a million

years. This is not "correlation" as we have learned to understand it!

To return to the Pueblo and Eastbourne type and reference sections, Gale et al. (2008) proposed a high-frequency sequence correlation between the two locations, as a further demonstration of their hypothesis of orbitally forced sequence boundaries and eustatic sea-level events in the Late Cretaceous. They claimed that twelve sequence boundaries can be correlated between the two sections, and they further claimed that these tie into a framework of sequences defined in earlier work that are supposedly controlled by a 405-ka eccentricity cycle (this earlier paper, by Gale et al., 2002, was analyzed in Sect. 14.2). Their diagram showing six of the sequence boundaries is shown here as Fig. 14.44. The sequences are defined on the basis of comparisons between lithologic and facies observations in the sections at the two locations. They are not evenly spaced in either of the sections but, given the irregular sedimentation rate calculated for the Pueblo section, this is not surprising. But note that nowhere in the original analysis of the orbital timescale at Pueblo is a 405-ka periodicity discussed (Meyers and Sageman, 2004; Sageman et al., 2006). A careful comparison of these events with the relative orbital time scale provided by Sageman et al. (2006) (Fig. 14.41), provides the following data (Table 14.3) :

Shown in Fig. 14.44 but not discussed by Sageman et al. (2006) is the apparent coincidence of sequence boundaries 7 and 8 and boundaries 9 and 10, pre-

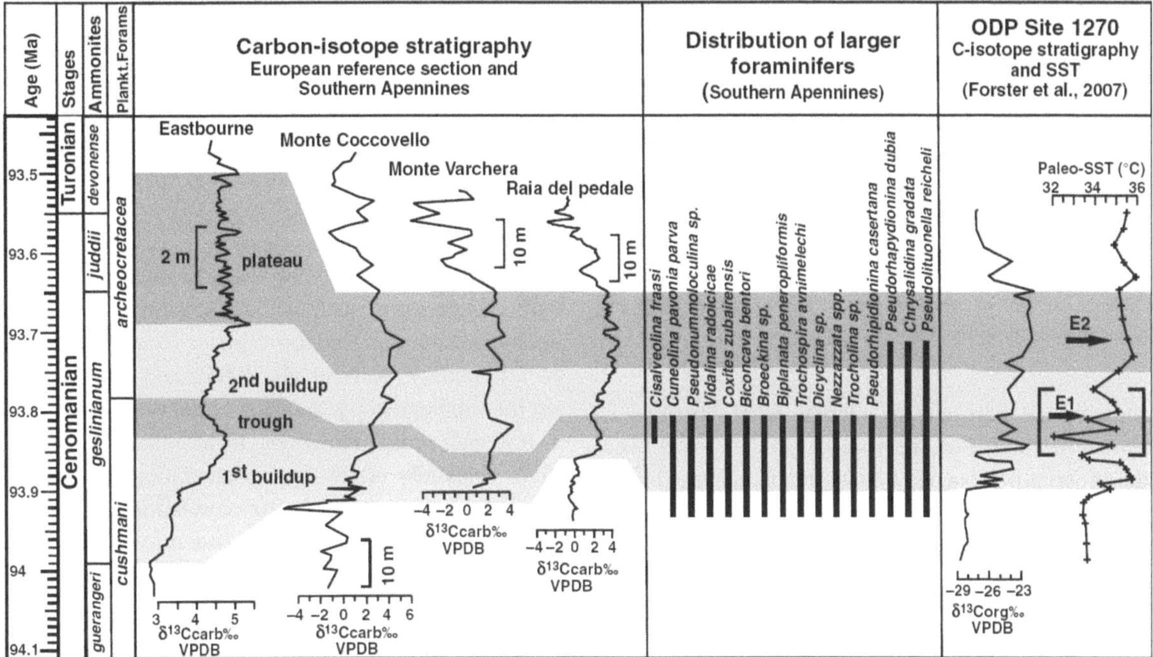



**Fig. 14.45** Comparison and correlation of the $\delta^{13}C$ record for the Eastbourne reference section for the base-Turonian boundary, with three other sections in Italy, and an ODP site. All these sections have been adjusted so that their thickness scale approximates to the interpreted age scale (Parente, et al., 2008, Fig. 14.1)

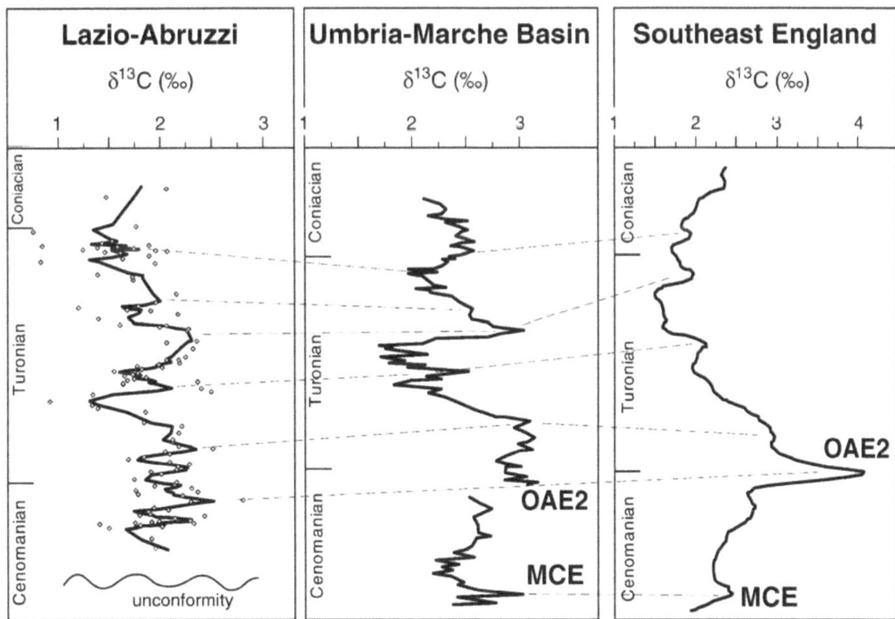



**Fig. 14.46** Comparison of $\delta^{13}C$ records for the Cenomanian-Coniacian interval for three locations in Europe (Galeotti et al., 2009, Fig. 14.9)

**Table 14.3**

| SB#[1] | Pueblo[2] bed # | Thickness[3] (m) | Relative age (ka)[4] |
|---|---|---|---|
| 12 | 86 | | 508 |
| 11 | 79 | 1.8 | 307 |
| 9,10 | 67 | 2.0 | 95 |
| 7,8 | 63 | 0.7 | 0 |

[1] Sequence boundary numbers from Gale et al. (2008)
[2] Bed numbers from Sageman et al. (2006)
[3] Sequence thickness measured from Fig. 14.44
[4] Ages of SB interpolated from EHA (1) time scale of Sageman et al. (2006, Fig. 14.1)

sumably as a result of internal local erosion. As shown in Table 14.3 the spacing (thickness) between the preserved sequence boundaries is not regular and does not accord to a 405 ka time scale. A regularity in thickness might have been expected in a section supposedly dominated by cyclostratigraphic processes. In the Eastbourne section (Fig. 14.44) sequence thicknesses vary widely, between 0.4 and 3.5 m.

Gale et al. (2008) correlated their sequence boundary 2 with one of the unconformities mapped by Vakarelov et al. (2006) in Wyoming, on the basis of the occurrence of the Soap Creek Bentonite a few metres above this surface in both locations. The Wyoming section is reproduced here as Fig. 4.12, where the bentonite is labeled SC(7). Gale et al. (2008, p. 861) suggested that this indicates an "overriding eustatic control to the erosional unconformity" and that this indicates that the unconformity is "tectonically enhanced." The latter term is a favourite of those practicing the global-eustasy paradigm (Sect. 12.3); it is one of a number of terms that have special meanings for this group (Table 12.2). In fact, there is no process that would genetically link tectonism and eustasy on the tight time scale being discussed here. The Vakarelov et al. (2006) study is briefly discussed in Sect. 10.3.3.1 as an example of the type of stratigraphic architecture generated by high-frequency tectonism. It seems likely to this writer that the unexplained coincidences of sequence boundaries (7–8 and 9–10) at Pueblo could also be the product of local high-frequency tectonism.

The Late Cretaceous is by no means the only interval of geologic time for which detailed $\delta^{13}C$ data are available. For example, the Early Jurassic is discussed by Mattioli et al. (2004), and Lower Cretaceous (Aptian-Albian) data are discussed by Herrle et al. (2004) and Wortmann et al. (2004). Good interregional

correlations of $\delta^{13}C$ excursions are shown in these studies, but much depends on the prior availability of good biostratigraphic control, because the shape of the $\delta^{13}C$ curve varies considerably from place to place, and without additional data on which to base correlations, many of the spikes and inflection points would be difficult to match.

To conclude this section, carbon isotope chemostratigraphy may be used to document major carbon excursions, such as those generated by oceanic anoxic events and sudden climate change (e.g., the K-T boundary: Katz et al., 2005), and may provide useful supplementary data for correlation on at least a regional scale. However, local influences of all kinds markedly affect local preservation of the $\delta^{13}C$ signature, and high-resolution correlation at a <1 million years scale is not to be regarded as reliable. The one detailed attempt to combine chemostratigraphy and cyclostratigraphy, in studies of the Cenomanian-Turonian boundary interval (discussed in this section and in Sect. 14.7.2), raises more questions than answers. Much may depend on the prior availability of a good chronostratigraphic data base, and the results may be markedly affected by sampling strategies.

## 14.9 Main Conclusions

1. Geological practice has taken two contrasting approaches to the issue of regional and global correlation. The *global-eustasy paradigm* has taken as a starting assumption that all sequence boundaries are driven by eustatic sea-level change, permitting the construction of a global time scale from sequence boundaries. Discussions presented here argue that this approach is fundamentally flawed. The alternative *complexity paradigm* assumes that multiple processes have affected the timing of sequence boundaries, requiring that rigorous, independent methods of chronostratigraphic dating and correlation must be employed to test concepts of sequence correlation and causation.

2. Methods of dating and correlation have improved dramatically in recent decades, but errors and inconsistencies remain, requiring great care and circumspection in the construction and testing of models that depend on accurate age information.

3. A few excellent modern data sets are now available that provide rigorous examination of subsidence and sedimentation histories calibrated against a refined chronostratigraphic record. However, there is still an inadequate data base from which to come to reliable conclusions regarding the prevalence or importance of eustasy during the Mesozoic and Paleogene. Correlations of sequence boundaries for the Neogene have provided an emerging picture of global sea-level control, almost certainly driven primarily by glacioeustasy.

4. Limited evidence is being assembled for episodic glaciation and glacioeustasy during the Cretaceous.

5. A global time scale is now being constructed from the record of orbital forcing. Termed *astrochronology* or *cyclostratigraphy*, the technique can only be regarded as reliable where it is based on continuous sections or composite sections that provide a complete record extending backwards from the present day. Currently a reliable scale extends back to the mid-Miocene. Many attempts to develop partial scales for earlier time periods, or to explain older stratigraphic successions in terms of orbital forcing, have involved questionable practices, and many of the results need to be evaluated with caution and skepticism.

6. Lessons learned from an examination of the global-eustasy paradigm, and from the more recent research trends in cyclostratigraphy, suggest that extreme caution needs to be employed in the application of the deductive approach to stratigraphic interpretation. While hypothesis building and testing is fundamental to the scientific method, it has been too commonly the case that the attractiveness of models has overwhelmed a lack of supporting data.

# Chapter 15

# Future Directions

## Contents



## 15.1 Research Methods

Whether the objective of research is to explore regional basin history and plate kinematics, or to focus on petroleum potential, the exploration of a new areas requires that logical procedures be followed, to ensure maximum efficiency. This is discussed in the concluding section of Chap. 3 (Sect. 3.8). Vail et al. (1991) had suggested a series of steps that commenced with a determination of the physical chronostratigraphic framework by interpreting sequences, systems tracts and parasequences and/or simple sequences on outcrops, well logs and seismic data and age date with high resolution biostratigraphy.

A problem with this approach is that it is built on the assumption that a chronostratigraphic framework can be constructed as the first step of the analysis. This, of course, arises directly from the assumptions of the *global eustasy paradigm*, amongst which is the central idea that all sequence boundaries are chronostratigraphic surfaces. Given this assumption, the ability to then immediately carry out a geohistory (backstripping) analysis would follow logically.

Following a discussion of the application of Exxon sequence analysis methods and concepts to two basin examples, Miall (1997, Sect. 17.3.3) suggested a modified set of research procedures:

1. Develop an allostratigraphic framework based on detailed lithostratigraphic well correlation and/or seismic facies analysis.
2. Develop a suite of possible stratigraphic models that conform with available stratigraphic and facies data.
3. Establish a regional sequence framework with the use of all available chronostratigraphic information.
4. Determine the relationships between sequence boundaries and tectonic events by tracing sequence boundaries into areas of structural deformation, and documenting the architecture of onlap/offlap relationships, fault offsets, unconformable discordancies, etc.
5. Establish the relationship between sequence boundaries and regional tectonic history, based on plate-kinematic reconstructions.
6. Refine the sequence-stratigraphic model. Subdivide sequences into depositional systems tracts and interpret facies.
7. Construct regional structural, isopach, and facies maps, interpret paleogeographic evolution, and develop plays and prospects based on this analysis.
8. Develop detailed subsidence and thermal-maturation history by backstripping/geohistory analysis.

It was suggested that subsidence and maturation analysis (Step 8) be carried out at the conclusion


A.D. Miall, *The Geology of Stratigraphic Sequences*, 2nd ed.,
DOI 10.1007/978-3-642-05027-5_15, © Springer-Verlag Berlin Heidelberg 2010

of the detailed analysis, rather than near the beginning, as proposed in the Exxon approach. The reason is that a complete, thorough analysis requires the input of a considerable amount of stratigraphic data. Corrections for changing water depths and for porosity/lithification characteristics, which are an integral part of such analysis, all require a detailed knowledge of the stratigraphic and paleogeographic evolution of the basin.

Catuneanu (2006, pp. 63–71) provided a detailed set of suggestions for completing a regional sequence analysis. His "workflow", which is summarized here (an expanded summary of his approach is included in Sect. 3.8), is based on "a general understanding [that] the larger-scale tectonic and depositional setting must be achieved first, before the smaller-scale details can be tackled in the most efficient way and in the right geological context." His workflow therefore proceeds from the large-scale through a decreasing scale of observation and an increasing level of detail. Figure 15.1 is his tabulation of the utility of various types of data set for the purpose of the analysis, and Fig. 15.2 tabulates the value of each type of data set for a regional sequence analysis. These are the basic components of the workflow:

1. Interpret the tectonic setting. This determines the type of sedimentary basin, and therefore controls the subsidence pattern and the general style of stratigraphic architecture. Ideally, the analysis should start from regional seismic lines.
2. Determine the broad regional paleographic setting, including the orientation of regional depositional dip and strike, from which broad predictions may be made about the position and orientation of coastlines, and regional facies architecture.
3. Determine paleodepositional environments, making use of well-log ("well-log motifs"), core data, and 3-D seismic data, as available.
4. From the regional concepts developed by Steps 1–3, depositional trends may be predicted, and this provides essential diagnostic clues for the interpretation of sequence architecture. The focus should now be on the recognition and mapping of seismic terminations and major bounding surfaces. For example, coastal onlap may indicate transgression (retrogradation) and downlap indicates regression (progradation). Only after the depositional trends are constrained, can the sequence stratigraphic surfaces that mark changes in such trends be mapped and labeled accordingly.
5. The last step of a sequence analysis is to the identify systems tracts.

Catuneanu (2006) was not primarily concerned in his book with driving mechanisms, and so his workflow ends with the construction of a detailed sequence stratigraphy. It is only once this has been completed that geohistory (backstripping) analysis can be carried out, a reconstruction of the history of relative sea-level change constructed, and a search for sequence-generating mechanisms be undertaken.

Modern sequence stratigraphic research is making the following essential points:

1. It clear how important seismic-reflection data are to a complete and reliable interpretation. Many of the examples described in this book help to make that

| Key: √√√ good<br>√√ fair<br>√ poor | Rock data | | | Geophysical data | | |
|---|---|---|---|---|---|---|
| | Outcrops | | | | Seismic data | |
| | Large-scale | Small-scale | Core | Well logs | 2D | 3D |
| Tectonic setting | √√ | √ | √ | √√ | √√√ | √√√ |
| Lithofacies | √√√ | √√√ | √√√ | √√ | √ | √√ |
| Depositional elements | √√√ | √√ | √√ | √√ | √ | √√√ |
| Depositional systems | √√√ | √√ | √√ | √√ | √ | √√√ |
| Depositional trends | √√√ | √√ | √√ | √√ | √√ | √√√ |
| Stratal terminations | √√√ | √ | √ | √ | √√√ | √√√ |
| Nature of contacts | √√√ | √√√ | √√√ | √√ | √√ | √√ |

**Fig. 15.1**  The utility of different data sets for constructing tectonic and stratigraphic interpretations (Catuneanu, 2006, Fig. 2.70)

| Data set | Main applications /contributions to sequence stratigraphic analysis |
| --- | --- |
| Seismic data | Continuous subsurface imaging; tectonic setting; structural styles; regional stratigraphic architecture; imaging of depositional elements; geomorphology |
| Well-log data | Vertical stacking patterns; grading trends; depositional systems; depositional elements; inferred lateral facies trends; calibration of seismic data |
| Core data | Lithology; textures and sedimentary structures; nature of stratigraphic contacts; physical rock properties; paleocurrents in oriented core; calibration of well-log and seismic data |
| Outcrop data | 3-D control on facies architecture; insights into process sedimentology; lithofacies; depositional elements; depositional systems; all other applications afforded by core data |
| Geochemical data | Depositional environment; depositional processes; diagenesis; absolute ages; paleoclimate |
| Paleontological data | Depositional environment; depositional processes; ecology; relative ages |

**Fig. 15.2** What various data sets contribute to a regional sequence-stratigraphic analysis (Catuneanu, 2006, Fig. 2.71)

point, and it is one that Catuneanu (2006) emphasizes in his discussion of "workflow." Seismic geomorphology (Davies et al., 2007), which exploits the ability of modern 3-D seismic to image ancient depositional systems in great detail, is becoming invaluable in the construction of the internal architecture of sequences (a topic not addressed in detail in this book).

2. A full comprehension of the internal architecture of sequences and their mode of construction is immeasurably facilitated by the use of numerical and graphical modeling. Catuneanu (2006, Chap. 7) addresses this issue in detail, and it is also usefully explored at the website established by C. G. St. C. (Chris) Kendall at http://strata.geol.sc.edu/

3. At a large scale of analysis (the regional to global), beyond the continuous coverage of a network of seismic lines or well logs, the exploration of sequence architectures and driving mechanisms requires that sequence stratigraphies be related to each other by indirect correlation. This places primary emphasis on the importance of chronostratigraphic methods, and the accuracy and precision of the correlations they provide. As discussed at length in Part IV of the book, much remains to be done in this area, and we are still not in a position to carry out definitive tests of the importance of the global eustasy paradigm.

## 15.2 Remaining Questions

### 15.2.1 Future Advances in Cyclostratigraphy?

There is a perceived danger involving the growth of the field of cyclostratigraphy, particularly as applied to the analysis of "hanging" sections from the distant geological past (pre-Neogene). As discussed in Sect. 14.7, between the time of the initial, cautious work of Hays et al. (1976) and Berger et al. (1984) and the present day, the "Milankovitch theory" has been inverted. At first, researchers attempted to use geological data to explore the range of cyclic climatic periodicities in the geological past, with a view to testing the evidence for the presence of an orbital-forcing signature. Now, the demonstration of cyclic periods falling somewhere within the "Milankovitch band" is enough for researchers to assert that the controlling mechanism was orbital forcing. Initial cautions based on an understanding of the incompleteness of the geological record have given way to a predisposition to respect the power of time series analysis to generate the expected signals. At least one astronomer has even suggested that geological data may be used to constrain the astronomical calculations (Laskar, 1999). For Laskar, geological data had become a "black box" of unquestioned veracity. House and Gale (1995, Preface) remarked that "the reality of orbital forcing of climate was established as a fact." Counting the number of cycles in a succession has become a criterion for correlation, a simplistic method reminiscent of that used in support of the global eustasy model (Table 12.2), and which may lead to over-simplifications regarding missing section (Westphal et al., 2008).

In Sect. 12.1, I point out the obvious dangers of circular reasoning involved in the deductivist approach to sequence stratigraphy. Miall and Miall (2004, p. 39) suggested that attempts to develop a time scale with an accuracy and precision in the $10^4$-year range by calibrating it against conventional chronostratigraphic dates up to two orders of magnitude less precise represents a fundamentally flawed methodology.

Geologists experienced in the incompleteness and inconsistencies of field data and knowledgeable about the warnings associated with the use of time series analysis offered by signal theorists (Rial, 1999, 2004), should be skeptical. Rial (1999) warned that "chronologies based on orbital tuning cannot be used because orbital tuning subtly forces the astronomical signal into the data." As Dott (1992b, p. 13) pointed out, "the Milankovitch theory is very accommodating, for it provides a period to suit nearly every purpose."

### 15.2.2 Tectonic Mechanisms of Sequence Generation

There is scope for much work in the investigation of tectonic mechanisms for specific sequence sets. In particular, the detailed stratigraphic work now being carried out in foreland basins demonstrates a considerable sensitivity of the depositional process to high-frequency tectonic control (Sect. 10.3.3), and an opportunity to test models of tectonic control against alternatives, particularly models based on orbital forcing (see below).

Another potential area of fruitful study is the model of regional to global intraplate stress change proposed by Embry (1997). He suggested in his study (p. 429) that widespread contemporaneous sequence boundaries are a product of episodes of global plate tectonic reorganization (Fig. 10.52). "At these times, there were significant changes in the rates and/or directions of spreading between one or more plates. This

would result in a significant change in the horizontal stress regime of all the plates. With an increase of stress emanating from spreading ridges and subduction zones, both the oceanic and continental portions of each plate would be affected." What would appear to be an excellent example of this mechanism is the simultaneous occurrence of a range of tectonic events over a large area of the North Atlantic region documented by Nielsen et al. (2007). They demonstrated how a change in the rate of convergence between Africa and Europe during the mid-Paleocene could be correlated temporally and genetically to an episode of rifting and strike-slip motion in the North Atlantic region, and to episodes of basin initiation and the relaxation of basin inversion in different regions of northwest Europe (Sect. 10.4.3).

### 15.2.3 Orbital Forcing

Despite the cautions expressed above in Sect. 15.2.1, it seems likely that much fruitful research can be carried out to explore the influence of orbital forcing on the geological record. There is increasing evidence from the stable isotope record for the occasional development of small-scale ice-caps during the Cretaceous (Sect. 14.6.4), and several workers compiling very detailed stratigraphic documentation of the high-frequency sequence record of the pre-Neogene, such as A. G. Plint, are finding increasingly convincing evidence for sea-level changes of small amplitude but widespread extent, that could best be explained as result of glacioeustasy (Sect. 11.3.3). Other workers (e.g., W. P. Elder, B. B. Sageman) are documenting marine cycles that can best be explained as the product of other processes of orbital forcing, including changes temperature and precipitation regimes that affect organic productivity and sediment delivery.

### 15.2.4 The Codification of Sequence Nomenclature

Catuneanu et al. (2009, p. 3) stated: "The process of standardization is hampered mainly because consensus needs to be reached between "schools" that promote rather different approaches (or models) with respect to how the sequence stratigraphic method should be applied to the rock record." However, as argued by Catuneanu (2006, p. 340) "standardizing sequence stratigraphy is in fact within reach." In the concluding chapter of his 2006 book he has sections entitled "Theory vs. Reality in Sequence Stratigraphy" and "Uses and Abuses in Sequence Stratigraphy" in which he suggests that "the greatest danger in sequence stratigraphy is dogma." The present book has deliberately not entered the debate about the definition of sequences, largely because this debate has been very effectively explored by Catuneanu (2006) and Catuneanu et al. (2009), but progress of the type that can be expected in the near future, as discussed in the preceding sections of this chapter, requires that communication between specialists improve, and this further requires the use of a common language.

# References


Abreu, V. S., Hardenbol, J., Haddad, G. A, Baum, G. R., Droxler, A. W., and Vail, P. R., 1998, Oxygen isotope synthesis: a cretaceous ice-house? in Graciansky, P.-C. de, Hardenbol, J., Jacquin, T., and Vail, P. R., eds., Mesozoic and Cenozoic sequence stratigraphy of European basins: Society for Sedimentary Geology (SEPM) Special Publication 60, pp. 75–80.

Ager, D. V., 1964, The British Mesozoic Committee: Nature, 203, p. 1059.

Ager, D. V., 1973, The nature of the stratigraphical record: John Wiley, New York, 114p.

Ager, D. V., 1981, The nature of the stratigraphical record, Second edition: John Wiley, New York, 122p.

Ager, D. V., 1993, The new catastrophism: Cambridge University Press, Cambridge, 231p.

Agterberg, F. P., 1990, Automated stratigraphic correlation: Elsevier, Amsterdam, 424p.

Aitken, J. D., 1966, Middle Cambrian to Middle Ordovician cyclic sedimentation, southern Rocky Mountains of Alberta: Bulletin of Canadian Petroleum Geology, 14, pp. 405–441.

Aitken, J. D., 1978, Revised models for depositional grand cycles, Cambrian of the southern Rocky Mountains, Canada: Bulletin of Canadian Petroleum Geology, 26, pp. 515–542.

Algeo, T. J., 1993, Quantifying stratigraphic completeness: a probabilistic approach using paleomagnetic data: Journal of Geology, 101, pp. 421–433.

Algeo, T. J., and Seslavinsky, K. B., 1995, The Paleozoic world: continental flooding, hypsometry, and sealevel: American Journal of Science, 295, pp. 787–822.

Algeo, T. J., and Wilkinson, B. H., 1988, Periodicity of mesoscale Phanerozoic sedimentary cycles and the role of Milankovitch orbital modulation: Journal of Geology, 96, pp. 313–322.

Allen, J. R. L., 1980, Sand waves: a model of origin and internal structure: Sedimentary Geology, 26, pp. 281–328.

Allen, P. A., and Allen, J. R., 2005, Basin analysis, principles and applications, Second edition: Wiley-Blackwell, Hoboken, New Jersey, 500p.

Allen, P. A., and Collinson, J. D., 1986, Lakes, in Reading, H. G., ed., Sedimentary environments and facies: Blackwell Scientific Publications, Oxford, pp. 63–94

Allen, P. A., and Etienne, J. L., 2008, Sedimentary challenge to Snowball Earth: Nature Geoscience, 1, pp. 817–825.

Allen, P. A., Homewood, P., and Williams, G. D., 1986, Foreland basins: an introduction, in Allen, P. A., and Homewood, P., eds., Foreland basins: International Association of Sedimentologists Special Publication 8, pp. 3–12.

Ambrose, W. A., Hentz, T. F., Bonaffé, F., Loucks, R. G., Brown, L. F., Jr., Wang, F. P., and Potter, E. C., 2009, Sequence-stratigraphic controls on complex reservoir architecture of highstand fluvial-dominated deltaic and lowstand valley-fill deposits in the Upper Cretaceous (Cenomanian) Woodbine Group, East Texas field: regional and local perspectives: American Association of Petroleum Geologists Bulletin, 93, pp. 231–269.

Anadón, P., Cabrera, L., Colombo, F., Marzo, M., and Riba, O., 1986, Syntectonic intraformational unconformities in alluvial fan deposits, eastern Ebro basin margins (NE Spain), in Allen, P. A., and Homewood, P., eds., Foreland basins: International Association of Sedimentologists Special Publication 8, pp. 259–271.

Anderson, D. L., 1982, Hotspots, polar wander, Mesozoic convection, and the geoid: Nature, 297, pp. 391–393.

Anderson, D. L., 1984, The earth as a planet: paradigms and paradoxes: Science, 223, pp. 347–355.

Anderson, D. L., 1989, Theory of the Earth: Blackwell Scientific Publications, Oxford, 366p.

Anderson, D. L., 1994, Superplumes or supercontinents? Geology, 22, pp. 39–42.

Anderson, D. M., 2001, Attenuation of millennial-scale events by bioturbation in marine sediments: Paleoceanography, 16, pp. 352–357.

Anderson, J. B., and Thomas, M. A., 1991, Marine ice-sheet decoupling as a mechanism for rapid, episodic sea-level change: the record of such events and their influence on sedimentation: Sedimentary Geology, 70, pp. 87–104.

Arkell, W. J., 1946, Standard of the European Jurassic: Geological Society of America Bulletin, 57, pp. 1–34.

Armentrout, J. M., 1981, Correlation and ages of Cenozoic chronostratigraphic units in Oregon and Washington: Geological Society of America Special Paper 184, pp. 137–148.

Armentrout, J. M., Malecek, S. J., Fearn, L. B., Sheppard, C. E., Naylor, P. H., Miles, A. W., Desmarais, R. J., and Dunay, R. E., 1993, Log-motif analysis of Paleogene depositional systems tracts, central and northern North Sea: defined



A.D. Miall, *The Geology of Stratigraphic Sequences,* 2nd ed.,
DOI 10.1007/978-3-642-05027-5, © Springer-Verlag Berlin Heidelberg 2010

by sequence stratigraphic analysis, in Parker, J. R., ed., Petroleum geology of northwest Europe: Proceedings of 4th Conference: Geological Society, London, pp. 45–57.

Armentrout, J. M., and Perkins, B. F., eds., 1991, Sequence stratigraphy as an exploration tool: Concepts and practices in the Gulf Coast: Eleventh Annual Research Conference, Society of Economic Paleontologists and Mineralogists, 417p.

Aubry, M.-P., 1991, Sequence stratigraphy: eustasy or tectonic imprint: Journal of Geophysical Research, 96B, pp. 6641–6679.

Aubry, M.-P., 1995, From chronology to stratigraphy: interpreting the Lower and Middle Eocene stratigraphic record in the Atlantic Ocean, in Berggren, W. A., Kent, D. V., Aubry, M.-P., and Hardenbol, J., eds., Geochronology, time scales and global stratigraphic correlation: Society for Sedimentary Geology Special Publication 54, pp. 213–274.

Aubry, M.-P., 2007, Chronostratigraphy beyond the GSSP: Stratigraphy, 4, pp. 127–137.

Aubry, M.-P., Van Couvering, J., Berggren, W. A., and Steininger, F., 1999, Problems in chronostratigraphy: stages, series, unit and boundary stratotype section and point and tarnished golden spikes: Earth-Science Reviews, 46, pp. 99–148.

Aubry, M.-P., Van Couvering, J., Berggren, W. A., and Steininger, F., 2000, Should the gold spike glitter: Episodes, 23, pp. 203–210.

Autin, W. J., 1992, Use of alloformations for definition of Holocene meander belts in the middle Amite River, southeastern Louisiana: Geological Society of America Bulletin, 104, pp. 233–241.

Baars, D. L., and Stevenson, G. M., 1982, Subtle stratigraphic traps in Paleozoic rocks of Paradox Basin, in Halbouty, M. T., ed., The deliberate search for the subtle trap: American Association of Petroleum Geologists Memoir 32, pp. 131–158.

Baars, D. L., and Watney, W. L., 1991, Paleotectonic control of reservoir facies, in Franseen, E. K., Watney, W. L., and Kendall, C. G. St.C., eds., Sedimentary modeling: computer simulations and methods for improved parameter definition: Kansas Geological Survey Bulletin 233, pp. 253–262.

Badley, M. E., Price, J. D., Dahl, C. R., and Agdestein, T., 1988, The structural evolution of the northern Viking Graben and its bearing upon extensional modes of basin formation: Journal of the Geological Society, London, 145, pp. 455–472.

Bally, A. W., ed., 1987, Atlas of seismic stratigraphy: American Association of Petroleum Geologists Studies in Geology 27, in 3 vols.

Bally, A. W., 1989, Phanerozoic basins of North America, in Bally, A. W., and Palmer, A. R., eds., The geology of North America—an overview: The geology of North America, Geological Society of America, A, pp. 397–446.

Bally, A. W., and Snelson, S., 1980, Realms of subsidence, in Miall, A. D., ed., Facts and principles of world petroleum occurrence: Canadian Society of Petroleum Geologists Memoir 6, pp. 9–94.

Barnes, B., Bloor, D., and Henry, J., 1996, Scientific knowledge: a sociological analysis: The University of Chicago press, Chicago, 230p.

Barrell, J., 1917, Rhythms and the measurement of geologic time: Geological Society of America Bulletin, 28, pp. 745–904.

Barron, E. J., 1983, A warm, equable Cretaceous: the nature of the problem: Earth Science Reviews, 19, pp. 305–338.

Barron, E. J., and Thompson, S. L., 1990, Sea level and climate change, in Revelle, R., ed., Sea-level change: National Research Council, Studies in Geophysics, National Academy Press, Washington pp. 185–192.

Bartek, L. R., Vail, P. R., Anderson, J. B., Emmet, P. A., and Wu, S., 1991, Effect of Cenozoic ice sheet fluctuations in Antarctica on the stratigraphic signature of the Neogene: Journal of Geophysical Research, 96B, pp. 6753–6778.

Bates, R. L., and Jackson, J. A., 1987, Glossary of geology, Third edition: American Geological Institute, Alexandria, 788p.

Bassett, M. G., 1985, Towards a "common language" in stratigraphy: Episodes, 8, pp. 86–92.

Baum, G. R., and Vail, P. R., 1988, Sequence stratigraphic concepts applied to Paleogene outcrops, Gulf and Atlantic basins, in Wilgus, C. K., Hastings, B. S., Kendall, C. G. St. C., Posamentier, H. W., Ross, C. A., and Van Wagoner, J. C., eds., Sea-level changes: an integrated approach: Society of Economic Paleontologists and Mineralogists Special Publication 42, pp. 309–327.

Beaumont, C., 1981, Foreland basins: Geophysical Journal of the Royal Astronomical Society, 65, pp. 291–329.

Beaumont, C., 1982, Platform sedimentation: International Association of Sedimentologists, 11th International Sedimentology Congress, Hamilton, Ontario, Abstracts, p. 132.

Beaumont, C., Quinlan, G., and Hamilton, J., 1988, Orogeny and stratigraphy: numerical models of the Paleozoic in the eastern interior of North America: Tectonics, 7, pp. 389–416.

Beaumont, C., Quinlan, G. M., and Stockmal, G. S., 1993, The evolution of the Western Interior Basin: Causes, consequences and unsolved problems, in Caldwell, W. G. E., and Kauffman, E. G., eds., Evolution of the Western Interior Basin: Geological Association of Canada Special Paper 39, pp. 97–117.

Belopolsky, A. V., and Droxler, A. W., 2004, Seismic expression of prograding carbonate bank margins: middle Miocene progradation in the Maldives, Indian Ocean, in Eberli, G. P., et al., ed., Seismic imaging of carbonate reservoirs and systems: American Association of Petroleum Geologists memoir 81, pp. 267–290.

Berger, A., 1988, Milankovitch theory and climate: Reviews of Geophysics, 26, pp. 624–657.

Berger, A. L., Imbrie, J., Hays, J., Kukla, G., and Saltzman, B., eds., 1984, Milankovitch and climate: NATO ASI Series, D: Reidel Publishing Company, Dordrecht, 2 vols., 895p.

Berger, A. L., and Loutre, M. F., 1994, Astronomical forcing through geological time, in de Boer, P. L., and Smith, D. G., eds., Orbital forcing and cyclic sequences: International Association of Sedimentologists Special Publication 19, pp. 15–24.

Berger, A. L., Loutre, M. F., and Dehant, V., 1989, Influence of the changing lunar orbit on the astronomical frequencies of pre-Quaternary insolation patterns: Paleoceanography, 4, pp. 555–564.

Berger, A. L., Loutre, M. F., and Laskar, J., 1992, Stability of the astronomical frequencies over the earth's history for paleoclimate studies: Science, 255, pp. 560–566.

Berggren, W. A., 1972, A Cenozoic time-scale-some implications for regional geology and paleobiology: Lethaia, 5, pp. 195–215.

Berggren, W. A., 1998, The Cenozoic era: Lyellian (chrono)stratigraphy and nomenclatural record at the millennium, in Blundell, D. J., and Scott, A. C., eds., Lyell: The past is the key to the present: Geological Society of London Special Publication 143, pp. 111–132.

Berggren, W. A., 2007, Status of the hierarchical subdivision of higher order marine Cenozoic chronostratigraphic units: Stratigraphy, 4, pp. 99–108.

Berggren, W. A., Kent, D. V., Flynn, J. J., and Van Couvering, J. A., 1985a, The Neogene, part 2. Neogene geochronology and chronostratigraphy, in Snelling, N. J., ed., The chronology of the Geologic record: Geological Society of America memoir, 10, pp. 211–260.

Berggren, W. A., Kent, D. V., Flynn, J. J., and Van Couvering, J. A., 1985b, Cenozoic geochronology: Geological Society of America Bulletin, 96, pp. 1407–1418.

Berggren, W. A., Kent, D. V., Aubry, M.-P., and Hardenbol, J., eds., 1995, Geochronology, time scales and global stratigraphic correlation: Society for Sedimentary Geology Special Publication 54, 386p.

Berggren, W. A., and Van Couvering, J. A., 1978, Biochronology, in Cohee, G. V., Glaessner, M. F., and Hedberg, H. D., eds., Contributions to the geologic time scale: American Association of Petroleum Geologists Studies in Geology 6, pp. 39–55.

Bergman, K. M., and Walker, R. G., 1987, The importance of sea level fluctuations in the formation of linear conglomerate bodies: Carrot Creek Member of the Cardium Formation, Cretaceous Western Interior Seaway, Alberta, Canada: Journal of Sedimentary Petrology, 57, pp. 651–665.

Berner, R. A., Lasaga, A. C., and Garrels, R. M., 1983, The carbonate-silicate geochemical cycle and its effect on atmospheric carbon dioxide over the past 100 million years: American Journal of Science, 283, pp. 641–683.

Berry, E. W., 1929, Shall we return to cataclysmical geology? American Journal of Science, 5th Series, 17, pp. 1–12.

Berry, W. B. N., 1987, Growth of prehistoric time scale based on organic evolution, Revised edition: Blackwell Science, Oxford, 202p.

Betzler, C., Kroon, D., and Reijmer, J. G. J., 2000, Sychroneity of major late Neogene sea level fluctuations and paleoceanographically controlled changes as recorded by two carbonate platforms: Paleoceanography, 15, pp. 722–730.

Betzler, C., Reijmer, J. J. G., Bernet, K., Eberli, G. P., and Anselmetti, F. S., 1999, Sedimentary patterns and geometries of the Bahamian outer carbonate ramp (Miocene-Lower Pliocene, Great Bahama Bank): Sedimentology, 46, pp. 1127–1143.

Beuthin, J. D., 1994, A sub-Pennyslvanian paleovalley system in the central Appalachian basin and its implications for tectonic and eustatic controls on the origin of the regional Mississippian-Pennyslvanian unconformity, in Dennison, J. M., and Ettensohn, F. R., eds., Tectonic and eustatic controls on sedimentary cycles: Society for Sedimentary Geology, Concepts in Sedimentology and Paleontology, 4, pp. 107–120.

Bhattacharya, J., 1988, Autocyclic and allocyclic sequences in river- and wave-dominated deltaic sediments of the Upper Cretaceous Dunvegan Formation, Alberta: core examples, in James, D. P., and Leckie, D. A., eds., Sequences, stratigraphy, sedimentology: surface and subsurface: Canadian Society of Petroleum Geologists Memoir 15, pp. 25–32.

Bhattacharya, J., 1991, Regional to sub-regional facies architecture of river-dominated deltas, Upper Cretaceous Dunvegan Formation, Alberta subsurface, in Miall, A. D., and Tyler, N., eds., The three-dimensional facies architecture of terrigenous clastic sediments and its implications for hydrocarbon discovery and recovery, Society of Economic Paleontologists and Mineralogists, Concepts in Sedimentology and Paleontology, 3, pp. 189–206.

Bhattacharya, J. P., 1993, The expression and interpretation of marine flooding surfaces and erosional surfaces in core; examples from the Upper Cretaceous Dunvegan Formation, Alberta foreland basin, Canada, in Posamentier, H. W., Summerhayes, C. P., Haq, B. U., and Allen, G. P., eds., Sequence stratigraphy and facies associations: International Association of Sedimentologists Special Publication 18, pp. 125–160.

Bhattacharya, J., and Walker, R. G., 1991, Allostratigraphic subdivision of the Upper Cretaceous Dunvegan, Shaftesbury, and Kaskapau formations in the northwestern Alberta subsurface: Bulletin of Canadian Petroleum Geology, 39, pp. 145–164.

Bhattacharya, J., and Walker, R. G., 1992, Deltas, in Walker, R. G., and James, N. P., eds., Facies models: response to sea level change: Geological Association of Canada, pp. 157–177.

Biddle, K. T., and Schlager, W., eds., 1991, The record of sea-level fluctuations: Sedimentary Geology, 70, No. 2/4 (special issue), pp. 85–270.

Birkelund, T., Hancock, J. M., Hart. M. B., Rawson, P. F., Remane, J., Robaszynski, F., Schmid, F., and Surlyk, F., 1984, Cretaceous stage boundaries-proposals: Bulletin of the Geological Society of Denmark, 33, pp. 3–20.

Bishop, D. G., 1991, High-level marine terraces in western and southern New Zealand: indicators of the tectonic tempo of an active continental margin, in Macdonald, D. I. M., ed., 1991, Sedimentation, tectonics and eustasy: sea-level changes at active margins: International Association of Sedimentologists Special Publication 12, p. 69–78.

Bjørlykke, D., Elvsborg, A., and Hoy, R., 1976, Late Precambrian sedimentation in the central Sparagmite basin of south Norway: Norsk Geologisk Tidskrift, 56, pp. 233–290.

Blackwelder, E., 1909, The valuation of unconformities: Journal of Geology, 17, pp. 289–299.

Blair, T. C., and Bilodeau, W. L., 1988, Development of tectonic cyclothems in rift, pull-apart, and foreland basins: sedimentary response to episodic tectonism: Geology, 16, pp. 517–520.

Blakey, R., 2008, Pennsylvanian-Jurassic sedimentary basins of the Colorado Plateau and Southern Rocky Mountains, in Miall, A. D., ed., The Phanerozoic sedimentary basins of the United States and Canada: Sedimentary Basins of the World, Elsevier, Amsterdam, vol. 5, pp. 245–296.

Blum, M. D., 1994, Genesis and architecture of incised valley fill sequences: a late Quaternary example from the Colorado River, Gulf Coastal Plain of Texas, in Weimer, P., and Posamentier, H. W., eds., Siliciclastic sequence stratigraphy: Recent developments and applications: American Association of Petroleum Geologists Memoir 58, p. 259–283.

Boardman, D. R., and Heckel, P. H., 1989, Glacial-eustatic sea-level curve for early Late Pennsylvanian sequence in north-central Texas and biostratigraphic correlation with curve for midcontinent North America: Geology, 17, pp. 802–805.

Boardman, D. H., and Heckel, P. H., 1991, Reply to comment on "glacial-eustatic sea-level curve for early late Pennsylvanian sequence in north-central Texas and biostratigraphic correlation with curve for midcontinent North America": Geology, 19, pp. 92–94.

Boggs, S., 2005, Principles of sedimentology and stratigraphy, Fourth Edition: Prentice-Hall, Englewood Cliffs, New Jersey, 662p.

Bohacs, K., and Suter, J., 1997, Sequence stratigraphic distribution of coaly rocks: fundamental controls and paralic examples: American Association of Petroleum Geologists Bulletin, 81, pp. 1612–1639.

Boldy, S. A. R., and Brealey, S., 1990, Timing, nature and sedimentary result of Jurassic tectonism in the Outer Moray Firth, in Hardman, R. F. P., and Brooks, J., eds., Tectonic events responsible for Britains's oil and gas reserves: Geological Society, London, Special Publication 55, pp. 259–279.

Bond, G., 1976, Evidence for continental subsidence in North America during the Late Cretaceous global submergence: Geology, 4, pp. 557–560.

Bond, G., 1978, Speculations on real sea-level changes and vertical motions of continents at selected times in the Cretaceous and Tertiary periods: Geology, 6, pp. 247–250.

Bond, G. C., Christie-Blick, N., Kominz, M. A., and Devlin, W. J., 1985, An Early Cambrian rift to post-rift transition in the Cordillera of western North America: Nature, 315, pp. 742–746.

Bond, G. C., Devlin, W. J., Kominz, M., Beavan, J., and McManus, J., 1993, Evidence of astronomical forcing of the Earth's climate in Cretaceous and Cambrian times: Tectonophysics, 222, pp. 295–315.

Bond, G. C., and Kominz, M. A., 1984, Construction of tectonic subsidence curves for the early Paleozoic miogeocline, southern Canadian Rocky Mountains: implications for subsidence mechanisms, age of breakup, and crustal thinning: Geological Society of America Bulletin, 95, pp. 155–173.

Bond, G. C., and Kominz, M. A., 1991a, Disentangling Middle Paleozoic sea level and tectonic events in cratonic margins and cratonic basins of North America: Journal of Geophysical Research, 96B, pp. 6619–6639.

Bond, G. C., and Kominz, M. A., 1991b, Some comments on the problem of using vertical facies changes to infer accommodation and eustatic sea-level histories with examples from Utah and the southern Canadian Rockies, in Franseen, E. K., Watney, W. L., and Kendall, C. G. St.C., eds., Sedimentary modeling: computer simulations and methods for improved parameter definition: Kansas Geological Survey Bulletin 233, pp. 273–291.

Bond, G. C., Kominz, M. A., Steckler, M. S., and Grotzinger, J. P., 1989, Role of thermal subsidence, flexure, and eustasy in the evolution of early Paleozoic passive margin carbonate platforms, in Crevello, P., Wilson, J., Sarg, R., and Read, F., eds., Controls on evolution of carbonate platforms and basin development: Society of Economic Paleontologists and Mineralogists Special Publication 44, pp. 39–61.

Boreen, T., and Walker, R. G., 1991, Definition of allomembers and their facies assemblages in the Viking Formation, Willesden Green area, Alberta: Bulletin of Canadian Petroleum Geology, 39, pp. 123–144.

Borer, J. M., and Harris, P. M., 1991, Lithofacies and cyclicity of the Yates Formation, Permian Basin: implications for reservoir heterogeneity: American Association of Petroleum Geologists Bulletin, 75, pp. 726–779.

Bornemann, A., Norris, R. J., Friedrich, O., Beckmann, B., Schouten, S., Damste, J. S, Vogel, J., Hofmann, P., and Wagner, T., 2008, Isotopic evidence for glaciation during the Cretaceous supergreenhouse: Science, 319, pp. 189–192.

Boss, S. K., and Rasmussen, K. A., 1995, Misuse of Fischer plots as sea-level curves: Geology, 23, pp. 221–224.

Bosellini, A., 1984, Progradation geometries of carbonate platforms: examples from the Triassic of the Dolomites, Northern Italy: Sedimentology, 31, pp. 1–24.

Bouma, A. H., Normark, W. R., and Barnes, N. E., eds., 1985, Submarine fans and related turbidite systems: Springer-Verlag Inc., Berlin and New York, 351p.

Bowman, A. R., and Bralower, T. J., 2005, Paleoceanographic significance of high-resolution carbon isotope records across the Cenomanian-Turonian bin the western inerior and New Jersey coastl plain, USA: Marine Geology, 217, pp. 305–321.

Boyd, R., and Penland, S., 1988, A geomorphic model for Mississippi delta evolution: Gulf Coast Association of Geological Societies, Transactions, 38, pp. 443–452.

Boyd, R., Suter, J., and Penland, S., 1989, Sequence stratigraphy of the Mississippi delta: Transactions of the Gulf Coast Association of Geological Societies, 39, pp. 331–340.

Bradley, W. B., 1929, The varves and climate of the Green River epoch: U. S. Geological Survey Professional Paper 158, pp. 87–110.

Broecker, W. S., and Denton, G. H., 1989, The role of ocean-atmosphere reorganization in glacial cycles: Geochimica Cosmochimica Acta, 53, pp. 2465–2501.

Bromley, M. H., 1991, Architectural features of the Kayenta Formation (Lower Jurassic), Colorado Plateau, USA: relationship to salt tectonics in the Paradox Basin: Sedimentary Geology, 73, pp. 77–99.

Bronn, H. G., 1858, On the laws of evolution of the organic world during the formation of the crust of the Earth: Annals and Magazine of Natural History, series 3, 4, pp. 89–90.

Brookfield, M. E., 1977, The origin of bounding surfaces in ancient aeolian sandstones: Sedimentology, 24, pp. 303–332.

Brookfield, M. E., 1992, The paleorivers of central Asia: the interrelationship of Cenozoic tectonism, erosion and sedimentation: 29th International Geological Congress, Kyoto, Japan, Abstracts, 2, p. 292.

Brookfield, M. E., 1993, The interrelations of post-collision tectonism and sedimentation in Central Asia, in Frostick, L. E., and Steel, R. J., eds., Tectonic controls and signatures in sedimentary successions: International Association of Sedimentologists Special Publication 20, pp. 13–35.

Brown, L. F., Jr., and Fisher, W. L., 1977, Seismic-stratigraphic interpretation of depositional systems: examples from Brazilian rift and pull-apart basins, in Payton, C. E., ed., Seismic stratigraphy ö applications to hydrocarbon exploration: American Association of Petroleum Geologists Memoir 26, pp. 213–248.

Browning, J. V., Miller, K. G., Sugarman, P. J., Kominz, M. A., McLaughlin, P. P. Kulpecz, A. A., and Feigenson, M. D., 2008, 100 Myr record of sequences, sedimentary facies and sea level changes from Ocean Drilling Program onshore coreholes, US Mid-Atlantic coastal plain: Basin Research, 20, pp. 227–248.

Browning, J. V., Miller, K. G., McLaughlin, P. P., Kominz, M. A., Sugarman, P. J., Monteverde, D. H., Feigenson, M. D., and Hernandez, J. C., 2005, Quantification of the effects of eustasy, subsidence, and sediment supply on Miocene sequences, mid-Atlantic margin of the United States: Geological Society of America Bulletin, 118, pp. 567–588.

Buchanan, R. C., and Maples, C. G., 1992, R. C. Moore and concepts of sea-level change in the midcontinent, in Dott, R. H., Jr., ed., Eustasy: the ups and downs of a major geological concept: Geological Society of America Memoir 180, pp. 73–81.

Buckman, S. S., 1893, On the Bajocian of the Sherbourne district: its relation to subjacent and superjacent strata: Geological Society of London, Quarterly Journal, 49, pp. 479–522.

Buckman, S. S., 1898, On the grouping of some divisions of so-called Jurassic time: Geological Society of London, Quarterly Journal, 54, pp. 442–462.

Buckman, S. S., 1910, Certain Jurassic ('Inferior Oolite') species of ammonites and brachiopoda: Geological Society of London, Quarterly Journal, 66, pp. 90–110.

Burbank, D. W., Puigdefábregas, C., and Muñoz, J. A., 1992, The chronology of the Eocene tectonic and stratigraphic development of the eastern Pyrenean foreland basin, northeast Spain: Geological Society of America Bulletin, 104, pp. 1101–1120.

Burbank, D. W., Beck, R. A., Raynolds, R. G. H., Hobbs, R., and Tahirkheli, R. A. K., 1988, Thrusting and gravel progradation in foreland basins: a test of post-thrusting gravel dispersal: Geology, 16, pp. 1143–1146.

Burbank, D. W., and Raynolds, R. G. H., 1984, Sequential late Cenozoic structural disruption of the northern Himalayan foredeep: Nature, 311, pp. 114–118.

Burbank, D. W., and Raynolds, R. G. H., 1988, Stratigraphic keys to the timing of thrusting in terrestrial foreland basins: applications to the northwestern Himalaya, in Kleinspehn, K. L., and Paola, C., eds., New perspectives in basin analysis: Springer-Verlag, New York, pp. 331–351.

Burbank, D. W., Raynolds, R. G. H., and Johnson, G. D., 1986, Late Cenozoic tectonics and sedimentation in the north-western Himalayan foredeep: II. Eastern limb of the Northwest Syntaxis and regional synthesis, in Allen, P. A., and Homewood, P., eds., Foreland basins: International Association of Sedimentologists Special Publication 8, pp. 293–306.

Burchfiel, B. C., and Davis, G. A., 1972, Structural framework and evolution of the southern part of the Cordilleran Orogen, western United States: American Journal of Science, 272, pp. 97–118.

Burgess, P. M., 2008, Phanerozoic evolution of the sedimentary cover of the North American craton, in Miall, A. D., ed., The Sedimentary Basins of the United States and Canada: Sedimentary basins of the World, vol. 5, K. J. Hsü, Series Editor, Elsevier Science, Amsterdam, pp. 31–63.

Burgess, P. M., and Gurnis, M., 1995, Mechanisms for the formation of cratonic stratigraphic sequences: Earth and Planetary Science Letters, 136, pp. 647–663.

Burgess, P. M., and Hovius, N., 1998, Rates of delta progradation during highstands; consequences for timing of deposition in deep-marine systems: Geological Society, London, Journal, 155, pp. 217–222.

Burton, R., Kendall, C. G. St. C., and Lerche, I., 1987, Out of our depth: on the impossibility of fathoming eustasy from the stratigraphic record: Earth Science Reviews, 24, pp. 237–277.

Busch, R. M., and Rollins, H. B., 1984, Correlation of Carboniferous strata using a hierarchy of transgressive-regressive units: Geology, 12, pp. 471–474.

Butterworth, P. J., 1991, The role of eustasy in the development of a regional shallowing event in a tectonically active basin: Fossil Bluff Group (Jurassic-Cretaceous), Alexander Island, Antarctica, in Macdonald, D. I. M., ed., 1991, Sedimentation, tectonics and eustasy: sea-level changes at active margins: International Association of Sedimentologists Special Publication 12, pp. 307–329.

Callomon, J. H., 1995, Time from fossils: S. S. Buckman and Jurassic high-resolution geochronology, in Le Bas, M. J., ed., Milestones in Geology: Geological Society of London Memoir 16, pp. 127–150.

Callomon, J. H., 2001, Fossils as geological clocks, in C. L. E. Lewis and S. J. Knell, eds., The age of the Earth: from 4004 BC to AD 2002: Geological Society of London Special Publication 190, pp. 237–252.

Campbell, C. V., 1967, Lamina, laminaset, bed and betset: Sedimentology, 8, pp. 7–26.

Cant, D. J., 1979, Storm-dominated shallow marine sediments of the Arisaig Group (Silurian-Devonian) of Nova Scotia: Canadian Journal of Earth Sciences, 17, pp. 120–131.

Cant, D. J., 1984, Development of shoreline-shelf sand bodies in a Cretaceous epeiric sea deposit: Journal of Sedimentary Petrology, 54, pp. 541–556.

Cant, D. J., 1992, Subsurface facies analysis, in Walker, R. G., and James, N. P., eds., Facies models: response to sea level change: Geological Association of Canada, pp. 27–45.

Cant, D. J., 1995, Sequence stratigraphic analysis of individual depositional successions: effects of marine/nonmarine sediment partitioning and longitudinal sediment transport, Mannville Group, Alberta foreland basin, Canada: American Association of Petroleum Geologists Bulletin, 79, pp. 749–762.

Cant, D. J., 1996, Sedimentological and stratigraphic organization of a foreland basin clastic wedge, Mannville Group, western Canada basin: Journal of Sedimentary Research, 66, pp. 1137–1147.

Cant, D. J., 1998, Sequence stratigraphy, subsidence rates, and alluvial facies, Mannville Group, Alberta foreland basin, in Shanley, K. W., and McCabe, P. J., eds., Relative role of eustasy, climate, and tectonism in continental rocks: Society for Sedimentary Geology (SEPM) Special Publication 59, pp. 49–63.

Cant, D. J., and Stockmal, G. S., 1989, The Alberta foreland basin: relationship between stratigraphy and terrane-accretion events: Canadian Journal of Earth Sciences, 26, pp. 1964–1975.

Caputo, M. V., and Crowell, J. C., 1985, Migration of glacial centers across Gondwana during Paleozoic era: Geological Society of America Bulletin, 96, pp. 1020–1036.

Carmichael, S. M. M., 1988, Linear estuarine conglomerate bodies formed during a mid-Albian marine transgression; "Upper Gates" Formation, Rocky Mountain Foothills of northeastern British Columbia, in James, D. P., and Leckie, D. A., eds., Sequences, stratigraphy, sedimentology: surface and subsurface: Canadian Society of Petroleum Geologists Memoir 15, pp. 49–62.

Carter, R. M., 2007, Stratigraphy into the 21st century: Stratigraphy, 4, pp. 187–193.

Carter, R. M., Abbott, S. T., Fulthorpe, C. S., Haywick, D. W., and Henderson, R. A., 1991, Application of global sea-level and sequence-stratigraphic models in southern hemisphere Neogene strata from New Zealand, in Macdonald, D. I. M., ed., 1991, Sedimentation, tectonics and eustasy: sea-level changes at active margins: International Association of Sedimentologists Special Publication 12, pp. 41–65.

Cartwright, J. A., Haddock, R. C., and Pinheiro, L. M., 1993, The lateral extent of sequence boundaries, in Williams, G. D., and Dobb, A., eds., Tectonics and seismic sequence stratigraphy: Geological Society, London, Special Publication 71, pp. 15–34.

Carvajal, C. R., and Steel, R. J., 2006, Thick turbidite successions from supply-dominated shelves during sea-level highstand: Geology, 34, pp. 665–668.

Castradori, D., 2002, A complete standard chronostratigraphic scale: how to turn a dream into reality? Episodes, 25, pp. 107–110.

Catuneanu, O., 2006, Principles of sequence stratigraphy: Elsevier, Amsterdam, 375p.

Catuneanu, O., Abreu, V., Bhattacharya, J. P., Blum, M. D., Dalrymple, R. W., Eriksson, P. G., Fielding, C. R., Fisher, W. L., Galloway, W. E., Gibling, M. R., Giles, K. A., Holbrook, J. M., Jordan, R., Kendall, C. G. St. C., Macurda, B., Martinsen, O. J., Miall, A. D., Neal, J. E., Nummedal, D., Pomar, L., Posamentier, H. W., Pratt, B. R., Sarg, J. F., Shanley, K. W., Steel, R. J., Strasser, A., Tucker, M. E., and Winker, C., 2009, Toward the Standardization of Sequence Stratigraphy: Earth Science Reviews, 92, pp. 1–33.

Catuneanu, O., Beaumont, C., and Waschbusch, P., 1997a, Interplay of static loads and subduction dynamics in foreland basins: reciprocal stratigraphies and the "missing" peripheral bulge: Geology, 25, pp. 1087–1090.

Catuneanu, O., and Eriksson, P. G., 1999, The sequence stratigraphic concept and the Precambrian rock record: an example from the 2.7-2.1 Ga Transvaal Supergroup, Kaapvaal craton: Precambrian Research, 97, pp. 215–251.

Catuneanu, O, and Sweet, A. R., 1999, Maastrichtian-Paleocene foreland basin stratigraphies, Western Canada: a reciprocal sequence architecture: Canadian Journal of Earth Sciences, 36: pp. 685–703.

Catuneanu, O., Sweet, A. R., and Miall, A. D., 1997b, Reciprocal architecture of Bearpaw T-R sequences, uppermost Cretaceous, Western Canada Sedimentary Basin: Bulletin of Canadian Petroleum Geology, 45, pp. 75–94.

Catuneanu, O., Sweet, A., and Miall, A. D., 1999, Concept and styles of reciprocal stratigraphies: Western Canada foreland system: Terra Nova, 11, pp. 1–8.

Catuneanu, O., Sweet, A., and Miall, A. D., 2000, Reciprocal stratigraphy of the Campanian-Paleocene Western Interior of North America: Sedimentary Geology, 134, pp. 235–255.

Catuneanu, O., Willis, A., and Miall, A. D., 1998, Temporal significance of sequence boundaries: Sedimentary Geology, 121, pp. 157–178.

Cecil, C. B., 1990, Paleoclimate controls on stratigraphic repetition of chemical and siliciclastic rocks: Geology, 18, pp. 533–536.

Célérier, B., 1988, Paleobathymetry and geodynamic models for subsidence: Palaios, 3, pp. 454–463.

Chamberlin, T. C., 1897, The method of multiple working hypotheses: Journal of Geology, 5, pp. 837–845.

Chamberlin, T. C., 1898, The ulterior basis of time divisions and the classification of geologic history: Journal of Geology, 6, pp. 449–462.

Chamberlin, T. C., 1909, Diastrophism as the ultimate basis of correlation: Journal of Geology, 17, pp. 685–693.

Chang, K. H., 1975, Unconformity-bounded stratigraphic units: Geological Society of America Bulletin, 86, pp. 1544–1552.

Chesnut, D. R., Jr., 1994, Eustatic and tectonic control of deposition of the Lower and Middle Pennsylvanian strata of the central Appalachian Basin, in Dennison, J. M., and Ettensohn, F. R., eds., Tectonic and eustatic controls on sedimentary cycles: Society for Sedimentary Geology, Concepts in Sedimentology and Paleontology, 4, pp. 51–64.

Chlupác, I., 1972, The Silurian-Devonian boundary in the Barrandian: Bulletin of Canadian Petroleum Geology, 20, pp. 104–174.

Christie-Blick, N., 1991, Onlap, offlap, and the origin of unconformity-bounded depositional sequences: Marine Geology, 97, pp. 35–56.

Christie-Blick, N., and Driscoll, N. W., 1995, Sequence stratigraphy: Annual Review of Earth and Planetary Sciences, 23, pp. 451–478.

Christie-Blick, N., Mountain, G. S., Miller, K. G., Matthews, R. K., Gradstein, F. M., Agterberg, F. P., Aubry, M.-P., Berggren, W. A., Flynn, J. J., Hewitt, R., Kent, D. V., Klitgord, K. D., Obradovitch, J., Ogg, J. G., Haq, B. U., Vail, P. R., Hardenbol, J., and Van Wagoner, J. C., 1988, Sea-level history (Discussions and replies): Science, 241, pp. 596–602.

Christie-Blick, N., Mountain, G. S., and Miller, K. G., 1990, Seismic stratigraphy: record of sea-level change, in Revelle, R., ed., Sea-level change: National Research Council, Studies in Geophysics: National Academy Press, Washington, pp. 116–140.

Christie-Blick, N., Pekar, S. F., and Madof, A. S., 2007, Is there a role for sequence stratigraphy in chronostratigraphy? Stratigraphy, 4, pp. 217–229.

Clarke, A., and Gerson, E., 1990, Symbolic interactionism in social studies of science, in Becker, H., and McCall, M., eds., Symbolic Interaction and Cultural Studies: University of Chicago Press, Chicago, Illinois, pp. 179–214.

Cleaveland, L. C., Jensen, J., Goese, S., Bice, D. M., and Montanari, A., 2002, Cyclostratigraphic analysis of pelagic carbonates at Monte dei Corvi (Ancona, Italy) and astronomical correlation of the Serravallian-Tortonian boundary: Geology, 30, pp. 931–934.

Cloetingh, S., 1986, Intraplate stresses: a new tectonic mechanism for fluctuations of relative sea-level: Geology, 14, pp. 617–621.

Cloetingh, S., 1988, Intraplate stresses: a new element in basin analysis, in Kleinspehn K., and Paola, C., eds., New Perspectives in basin analysis: Springer-Verlag, New York, pp. 205–230.

Cloetingh, S., ed., 1991, Measurement, causes, and consequences of long-term sea-level changes: Journal of Geophysical Research, 96, No. B4, special issue, pp. 6584–6949.

Cloetingh, S., Gradstein, F. M., Kooi, H., Grant, A. C., and Kaminski, M., 1990, Plate reorganization: a cause of rapid late Neogene subsidence and sedimentation around the North Atlantic? Journal of the Geological Society, London, 147, pp. 495–506.

Cloetingh, S., and Kooi, H., 1990, Intraplate stresses: a new perspective on QDS and Vail's third-order cycles, in Cross, T. A., ed., Quantitative dynamic stratigraphy: Prentice-Hall, Englewood Cliffs, pp. 127–148.

Cloetingh, S., McQueen, H., and Lambeck, K., 1985, On a tectonic mechanism for regional sea-level variations: Earth and Planetary Science Letters, 75, pp. 157–166.

Coe, A. L., 2003, The sedimentary record of sea-level change: Cambridge University press, New York, 287p.

Cohen, S., 1986, Historical culture: on the recoding of an academic discipline: University of California Press, Berkley, 329p

Cole, S., 1992, Making science: Harvard University Press, Cambridge, MA, USA, 290p.

Cole, S. A., 1996, Which came first, the fossil or the fuel? Social Studies of Science, 26, pp. 733–766.

Collinson, J. D., ed., 1989, Correlation in hydrocarbon exploration: Graham and Trotman, London, 381p.

Conkin, B. M., and Conkin, J. E., eds., 1984, Stratigraphy: foundations and concepts: Benchmark Papers in Geology: Van Nostrand Reinhold, New York, 363p.

Cope, J. C. W., 1993, High resolution biostratigraphy, in Hailwood, E. A., and Kidd, R. B., eds., High resolution stratigraphy: Geological Society, London, Special Publication 70, pp. 257–265.

Cotillon, P., 1995, Constraints for using high-frequency sedimentary cycles in cyclostratigraphy, in House. M. R., and Gale, A. S., eds., Orbital forcing timescales and cyclostratigraphy: Geological Society, London, Special Publication 85, pp. 133–141.

Covault, J. A., Normark, W. R., Romans, B. W., and Graham, S. A., 2007, Highstand fans in the California borderland: the overlooked deep-water depositional systems: Geology, 35, pp. 783–786.

Cowie, J. W., 1986, Guidelines for boundary stratotypes: Episodes, 9, pp. 78–82.

Cox, B. M., 1990, A review of Jurassic chronostratigraphy and age indicators for the UK, in Hardman, R. F. P., and Brooks, J., eds., Tectonic events responsible for Britain's oil and gas reserves: Geological Society, London, Special Publication 55, pp. 169–190.

Cózar, P., Somerville, I. D., Rodríguez, S., Mas, R., and Medina-Varea, P., 2006, Development of a late Visean (Mississippian mixed carbonate/siliciclastic platform in the Guadalmellato Valley (southwestern Spain): Sedimentary Geology, 183, pp. 269–295.

Crampton, J. S., Schiøler, P., and Roncaglia, L., 2006, Detection of Late Cretaceous eustatic signatures using quantitative biostratigraphy: Geological Society of America Bulletin, 118, pp. 975–990.

Crane, D., 1972, Invisible colleges: diffusion of knowledge in scientific communities: University of Chicago Press, Chicago, Illinois.

Croll, J., 1864, On the physical cause of the change of climate during geological epochs: Philosophical Magazine, 28, pp. 435–436.

Cross, T. A., 1986, Tectonic controls of foreland basin subsidence and Laramide style deformation, western United States, in Allen, P. A., and Homewood, P., eds., Foreland basins: International Association of Sedimentologists Special Publication 8, pp. 15–39.

Cross, T. A., 1988, Controls on coal distribution in transgressive-regressive cycles, Upper Cretaceous, Western Interior, U.S.A., in Wilgus, C. K., Hastings, B. S., Kendall, C. G. St. C., Posamentier, H. W., Ross, C. A., and Van Wagoner, J. C., eds., Sea-level Changes: an integrated approach: Society of Economic Paleontologists and Mineralogists Special Publication 42, pp. 371–380.

Cross, T. A., ed., 1990, Quantitative dynamic stratigraphy: Prentice-Hall, Englewood Cliffs, New Jersey, 625p.

Cross, T. A., and Lessenger, M. A., 1988, Seismic stratigraphy: Annual Review of Earth and Planetary Sciences, 16, pp. 319–354.

Crough, S. T., and Jurdy, D. M., 1980, Subducted lithosphere, hotspots, and the geoid: Earth and Planetary Science Letters, 48, pp. 15–22.

Crowell, J. C., 1978, Gondwanan glaciation, cyclothems, continental positioning, and climate change: American Journal of Science, 278, pp. 1345–1372.

Crowley, K. D., 1984, Filtering of depositional events and the completeness of sedimentary sequences: Journal of Sedimentary Petrology, 54, pp. 127–136.

Crowley, T. J., and Baum, S. K., 1992, Modeling late Paleozoic glaciation: Geology, 20, pp. 507–510.

Curray, J. R., 1964, Transgressions and regressions, in Miller, R. L., ed., Papers in marine geology, Shepard Commemorative volume: MacMillan Press, New York, pp. 175–203.

Currie, B. S., 1997, Sequence stratigraphy of nonmarine Jurassic-Cretaceous rocks, central Cordilleran foreland-basin system: Geological Society of America Bulletin, 109, pp. 1206–1222.

Curtis, D. M., 1970, Miocene deltaic sedimentation, Louisiana Gulf Coast, in Morgan, J. P., ed., Deltaic sedimentation modern and ancient: Society of Economic Paleontologists and Mineralogists Special Publication 15, pp. 293–308.

Dahlen, F. A., 1984, Non cohesive critical Coulomb Wedges: an exact solution. Journal of Geophysical Research, 89, pp. 10125–10133.

Dahlen, F. A., and Suppe, J., 1984, Mechanics of fold-and-thrust belts and accretionary wedges: Cohesive Coulomb Theory. Journal of Geophysical Research, 89, pp. 10087–10101.

Dahlstrom, C. D. A., 1970, Structural geology in the eastern margin of the Canadian Rocky Mountains: Bulletin of Canadian Petroleum Geology, 18, pp. 332–406.

Dallmeyer, R. D., 1989, Late Paleozoic thermal evolution of crystalline terranes within portions of the U.S. Appalachian orogen; Chapter 9, in Hatcher, R. D. Jr., Thomas, W. A., and Viele, G. W. eds., The Appalachian–Ouachita orogen in the United States, The geology of North America, Geological Society of America, Boulder, CO, vol. F-2, pp. 417–444.

Dalrymple, R. W., Boyd, R., and Zaitlin, B. A., eds., 1994, Incised-valley systems: origin and sedimentary sequences: SEPM (Society for Sedimentary Geology) Special Publication 51, 391p.

Dalziel, I. W. D., 1991, Pacific margins of Laurentia and East Antarctica-Australia as a conjugate rift pair: evidence and implications for an Eocambrian supercontinent: Geology, 19, pp. 598–601.

Daradich, A., Pysklywec, R. N., and Mitrovica, J. X., 2002, Mantle flow modeling of the anomalous subsidence of the Silurian Baltic Basin: Geophysical Research Letters, 29, no. 6, pp. 20-1–20-4.

D'Argenio, B., Fischer, A. G., Silva, I. P., Weissert, H., and Ferreri, V., eds., 2004a, Cyclostratigraphy: approaches and case histories: Society for Sedimentary Geology (SEPM) Special Publication 81, 311p.

D'Argenio, B., Ferreri, V., Weissert, H., Amodio, A., Buonocunto, F. P., and Wissler, L., 2004b, A multidisciplinary approach to global correlation and geochronology. The Cretaceous shallow-water carbonates of southern Apennines, Italy, in D'Argenio, B., Fischer, A. G., Silva, I. P., Weissert, H., and Ferreri, V., eds., Cyclostratigraphy: approaches and case histories: Society for Sedimentary Geology (SEPM) Special Publication 81, pp. 103–122.

Davies, R. J., Posamentier, H. W., Wood, L. J., and Cartwright, J. A., eds., 2007, Seismic geomorphology: applications to hydrocarbon exploration and production: Geological Society, London, Special Publication 277.

Davis, D., Suppe, J. and Dahlen, F. A. 1983, Mechanics of fold-and-thrust belts and accretionary wedges. Journal of Geophysical Research, 88, pp. 1153–1172.

Dawson, L., and Prus, R., 1995, Postmodernism and linguistic reality versus symbolic interactionism and obdurate reality, in N. K. Denzin, ed., Studies in Symbolic Interaction, 17: JAI Press, Greenwich, Connecticut, pp. 105–124.

de Boer, P. L., and Smith, D. G., eds., 1994a, Orbital forcing and cyclic sequences: International Association of Sedimentologists Special Publication 19, 559p.

de Boer, P. L., and Smith, D. G., 1994b, Orbital forcing and cyclic sequences, in de Boer, P. L., and Smith, D. G., eds., Orbital forcing and cyclic sequences: International Association of Sedimentologists Special Publication 19, pp. 1–14.

DeCelles, P. G., 2004, Late Jurassic to Eocene evolution of the Cordilleran thrust belt and foreland basin system, western U.S.A.: American Journal of Science, 304, pp. 105–168.

DeCelles, P. G., and Giles, K. A., 1996, Foreland basin systems: Basin Research, 8, pp. 105–123.

DeCelles, P.G., Lawton, T.F., and Mitra, G., 1995, Thrust timing, growth of structural culminations, and synorogenic sedimentation in the type Sevier orogenic belt, western United States: Geology, 23, pp. 699–702.

DeConto, R. M., and Pollard, D., 2003, Rapid Cenozoic glaciation of Antarctica induced by declining atmospheric $CO_2$: Nature, 421, pp. 245–249.

DeLapparant, A. A. C., 1885, Traité de Géologie: Savy, Paris, 1504p.

Dennison, J. M., and Ettensohn, F. R., eds., 1994, Tectonic and eustatic controls on sedimentary cycles: Society for Sedimentary Geology (SEPM), Concepts in Sedimentology and Paleontology, 4, p. 264.

Dennison, J. M., and Head, J. W., 1975, Sea-level variations interpreted from the Appalachian basin Silurian and Devonian: American Journal of Science, 275, pp. 1089–1120.

Denzin, N. K., ed., 1970, Sociological methods: a sourcebook: Aldine, Chicago.

Denzin, N. K., 1992, Symbolic interactionism and cultural studies: the politics of interpretation: Blackwell, Oxford, 217p.

Deramond, J., Souquet, P., Fondecave-Wallez, M.-J., and Specht, M., 1993, Relationships between thrust tectonics and sequence stratigraphy surfaces in foredeeps: model and examples from the Pyrenees (Cretaceous-Eocene, France, Spain), in Williams, G. D., and Dobb, A., eds., Tectonics and seismic sequence stratigraphy: Geological Society, London, Special Publication 71, pp. 193–219.

Devine, P. E., 1991, Transgressive origin of channeled estuarine deposits in the Point Lookout Sandstone, northwestern New Mexico: a model for Upper Cretaceous, cyclic regressive parasequences of the U. S. Western Interior: American Association of Petroleum Geologists Bulletin, 75, pp. 1039–1063.

Dewey, J. F., 1980, Episodicity, sequence, and style at convergent plate boundaries, in Strangway, D. W., ed., The continental crust and its mineral deposits: Geological Association of Canada Special Paper 20, pp. 553–573.

Dewey, J. F., 1982, Plate tectonics and the evolution of the British Isles: Journal of the Geological Society, London, 139, pp. 371–412.

Dewey, J. F., and Pitman, W. C., 1998, Sea-level changes: mechanisms, magnitudes and rates, in Pindell, J. L., and Drake, C. L., eds., Paleogeographic evolution and non-glacial eustasy, northern South America: Society for Sedimentary Geology (SEPM) Special Publication 58, pp. 1–16.

Dickinson, W. R., 1974, Plate tectonics and sedimentation, in Dickinson, W. R., ed., Tectonics and sedimentation: Society of Economic Paleontologists and Mineralogists Special Publication 22, pp. 1–27.

Dickinson, W. R., 1980, Plate tectonics and key petrologic associations, in Strangway, D. W., ed., The continental crust and its mineral deposits: Geological Association of Canada Special Paper 20, pp. 341–360.

Dickinson, W. R., 1981, Plate tectonics and the continental margin of California, in Ernst, W. G., ed., The geotectonic development of California: Prentice-Hall Inc., Englewood Cliffs, New Jersey, pp. 1–28.

Dickinson, W. R., 1988, Provenance and sediment dispersal in relation to paleotectonics and paleogeography of sedimentary basins, in Kleinspehn, K. L., and Paola, C., eds., New perspectives in basin analysis: Springer Verlag, New York, pp. 1–25.

Dickinson, W. R., 1993, The Exxon global cycle chart: an event for every occasion? Discussion: Geology, 21, pp. 282–283.

Dickinson, W. R., 1995, Forearc basins, in Busby, C. J., and Ingersoll, R. V., eds., Tectonics of sedimentary basins: Blackwell Science, Oxford, pp. 221–261.

Dickinson, W. R., Armin, R. A., Beckvar, N., Goodlin, T. C., Janecke, S. U., Mark, R. A., Norris, R. D., Radel, G., and Wortman, A. A., 1987, Geohistory analysis of rates of sediment accumulation and subsidence for selected California basins, in Ingersoll, R. V., and Ernst, W. G., eds., Cenozoic basin development of coastal California: Rubey Volume VI; Prentice-Hall Inc., Englewood Cliffs, New Jersey, pp. 1–23.

Dickinson, W. R., and Gehrels, G. E., 2003, U-Pb ages of detrital zircons from Permian and Jurassic eolian sandstones from the Colorado Plateau, USA: paleogeographic implications: Sedimentary Geology, 163, pp. 29–66.

Dickinson, W. R., Soreghan, G. S., and Giles, K. A., 1994, Glacio-eustatic origin of Permo-Carboniferous stratigraphic cycles: evidence from the southern Cordilleran foreland region, in Dennison, J. M., and Ettensohn, F. R., eds., Tectonic and eustatic controls on sedimentary cycles: Society for Sedimentary Geology, Concepts in Sedimentology and Paleontology, 4, pp. 25–34.

Dixon, J., and Dietrich, J. R., 1988, The nature of depositional and seismic sequence boundaries in Cretaceous-Tertiary strata of the Beaufort-Mackenzie basin, in James, D. P., and Leckie, D. A., eds., Sequences, stratigraphy, sedimentology: surface and subsurface: Canadian Society of Petroleum Geologists Memoir 15, pp. 63–72.

Dixon, J., and Dietrich, J. R., 1990, Canadian Beaufort Sea and adjacent land areas, in Grantz, A., Johnson, L., and Sweeney, J. F., eds., The Arctic Ocean Region: Boulder, Colorado, Geological Society of America, The Geology of North America, L, pp. 239–256.

Dockal, J. A., and Worsley, T. R., 1991, Modeling sea level changes as the Atlantic replaces the Pacific: submergent versus emergent observers: Journal of Geophysical Research, 96B, pp. 6805–6810.

Dolan, J. F., 1989, Eustatic and tectonic controls on deposition of hybrid siliciclastic/carbonate basinal cycles: American Association of Petroleum Geologists Bulletin, 73, pp. 1233–1246.

Donovan, A. D., Baum, G. R., Blechschmidt, G. L., Loutit, T. S., Pflum, C. E., and Vail, P. R., 1988, Sequence stratigraphic setting of the Cretaceous-Tertiary boundary in central Alabama, in Wilgus, C. K., Hastings, B. S., Kendall, C. G. St. C., Posamentier, H. W., Ross, C. A., and Van Wagoner, J. C., eds., Sea-level Changes: an integrated approach: Society of Economic Paleontologists and Mineralogists Special Publication 42, pp. 299–307.

Donovan, D. T., and Jones, E. J. W., 1979, Causes of worldwide changes in sea level: Journal of the Geological Society, London, 136, pp. 187–192.

Donovan, R. N., 1975, Devonian lacustrine limestones at the margin of the Orcadian Basin, Scotland: Journal of the Geological Society, London, 131, pp. 489–510.

Donovan, R. N., 1978, The Middle Old Red Sandstone of the Orcadian Basin, in Friend, P. F., ed., The Devonian of Scotland: International Symposium on the Devonian System, Field Guide: Palaeontological Association, London, U.K., pp. 37–53.

Donovan, R. N., 1980, Lacustrine cycles, fish ecology, and stratigraphic zonation in the Middle Devonian of Caithness: Scottish Journal of Geology, 16, pp. 35–50 .

Dott, R. H., Jr., 1998, What is unique about geological reasoning? GSA Today, 10, pp. 15–18.

Dott, R. H., Jr., ed., 1992a, Eustasy: the ups and downs of a major geological concept: Geological Society of America Memoir, 180, 111p.

Dott, R. H., Jr., 1992b, An introduction to the ups and downs of eustasy, in Dott, R. H., Jr., ed., Eustasy: the historical ups and downs of a major geological concept: Geological Society of America Memoir 180, pp. 1–16.

Dott, R. H., Jr., 1992c, T. C. Chamberlin's hypothesis of diastrophic control of worldwide changes of sea level: a precursor of sequence stratigraphy, in Dott, R. H., Jr., ed., Eustasy: the ups and downs of a major geological concept: Geological Society of America Memoir 180, pp. 31–41.

Dott, R. H., Jr., and Bourgeois, J., 1982, Hummocky stratification: significance of its variable bedding sequences: Geological Society of America Bulletin, 93, pp. 663–680.

Doyle, J. A., 1977, Spores and pollen: the Potomac Group (Cretaceous) Angiosperm sequence, in Kauffman, E. G., and Hazel, J. E., eds., Concepts and methods of biostratigraphy: Dowden, Hutchinson and Ross Inc., Stroudsburg, Pennsylvania, pp. 339–364.

Doyle, P., and Bennett, M. R., eds., 1998, Unlocking the stratigraphical record: Advances in modern stratigraphy: John Wiley and Sons, Chichester, 532p.

Drummond, C. N., and Wilkinson, B. H., 1993a, On the use of cycle thickness diagrams as records of long-term sealevel change during accumulation of carbonate sequences: Journal of Geology, 101, pp. 687–702.

Drummond, C. N., and Wilkinson, B. H., 1993b, Carbonate cycle stacking patterns and hierarchies of orbitally forced eustatic sealevel change: Journal of Sedimentary Petrology, 63, pp. 369–377.

Drummond, C. N., and Wilkinson, B. H., 1996, Stratal thickness frequencies and the prevalence of orderedness in stratigraphic sequences: Journal of Geology, 104, pp. 1–18.

Duff, P. M. D., Hallam, A., and Walton, E. K., 1967, Cyclic sedimentation: Developments in Sedimentology, Elsevier, Amsterdam, 10, 280p.

Duk-Rodkin, A., and Hughes, O. L., 1994, Tertiary-Quaternary drainage of the pre-glacial Mackenzie Basin: Quaternary International, 22/23, pp. 221–241.

Dunbar, G. B., Dickens, G. R. and Carter, R. M., 2000, Sediment flux across the Great Barrier reef to the Queensland Trough over the last 300 ky: Sedimentary Geology, 133, pp. 49–92.

Eardley, A. J., 1951, Structural Geology of North America: Harper, New York, 624p.

Eberli, G., and Ginsburg, R. N., 1988, Aggrading and prograding infill of buried Cenozoic seaways, northwestern Great

Bahama Bank, in Bally, A. W., ed., Atlas of seismic stratigraphy: American Association of Petroleum Geologists Studies in Geology 27, 2, pp. 97–103.

Eberli, G., and Ginsburg, R. N., 1989, Cenozoic progradation of northwestern Great Bahama Bank, a record of lateral platform growth and sea-level fluctuations, in Crevello, P. D., Wilson, J. L., Sarg, J. F., and Read, J. F., eds., Controls on carbonate platform and basin development: Society of Economic Paleontologists and Mineralogists Special Publication 44, pp. 339–351.

Eberli, G. P., Swart, P. K., McNeill D. F., Kenter, J. A. M., Anselmetti, F. S., Melim, L. A., and Ginsburg, R. N., 1997, A synopsis of the Bahamas drilling project: results from two deep core borings drilled on the Great Bahama Bank: Proceedings of the Ocean Drilling Program, Initial Reports, vol. 166, pp. 23–41.

Edwards, L. E., 1984, Insights on why graphic correlation (Shaw's method) works: Journal of Geology, 92, pp. 583–597.

Edwards, L. E., 1985, Insights on why graphic correlation (Shaw's method) works: A reply [to discussion]: Journal of Geology, 93, pp. 507–509.

Edwards, L. E., 1989, Supplemented graphic correlation: a powerful tool for paleontologists and nonpaleontologists: Palaios, 4, pp. 127–143.

Ehrmann, W. U., and Mackensen, A., 1992, Sedimentological evidence for the formation of an East Antarctic ice sheet in Eocene/Oligocene time: Palaeogeography, Palaeoclimatology, Palaeoecology, 93, pp. 85–112.

Einsele, G., and Ricken, W., 1991, Larger cycles and sequences: Introductory remarks, in Einsele, G., Ricken, W., and Seilacher, A., eds., Cycles and events in stratigraphy: Springer-Verlag, Berlin, pp. 611–616.

Einsele, G., Ricken, W., and Seilacher, A., eds., 1991a, Cycles and events in stratigraphy: Springer-Verlag, Berlin, 955p.

Einsele, G., Ricken, W., and Seilacher, A., 1991b, Cycles and events in stratigraphy—basic concepts and terms, in Einsele, G., Ricken, W., and Seilacher, A., eds., Cycles and events in stratigraphy: Springer-Verlag, Berlin, pp. 1–19.

Einsele, G., and Seilacher, A., eds., 1982, Cyclic and event stratification: Springer-Verlag, Berlin, 536p.

Elder, W. P., Gustason, E. R., and Sageman, B. B., 1994, Correlation of basinal carbonate cycles to nearshore parasequences in the Late Cretaceous Greenhorn seaway, Western Interior, U.S.A.: Geological Society of America Bulletin, 106, pp. 892–902.

Elrick, M., and Read, J. F., 1991, Cyclic ramp-to-basin carbonate deposits, Lower Mississippian, Wyoming and Montana: a combined field and computer modeling study: Journal of Sedimentary Petrology, 61, pp. 1194–1224.

Embry, A. F., 1990, A tectonic origin for third-order depositional sequences in extensional basins – implications for basin modeling, in Cross, T. A., ed., Quantitative dynamic stratigraphy: Prentice-Hall, Englewood Cliffs, New Jersy, pp. 491–501.

Embry, A. F., 1993, Transgressive-regressive (T-R) sequence analysis of the Jurassic succession of the Sverdrup Basin, Canadian Arctic Archipelago: Canadian Journal of Earth Sciences, 30, pp. 301–320.

Embry, A. F., 1995, Sequence boundaries and sequence hierarchies: problems and proposals, in Steel, R. J.,

Felt, V. L., Johannessen, E. P., and Mathieu, C., eds., Sequence stratigraphy on the Northwest European margin: Norsk Petroleumsforening Special Publication 5, Elsevier, Amsterdam, pp. 1–11.

Embry, A. F., 1997, Global sequence boundaries of the Triassic and their identification in the Western Canadian Sedimentary Basin: Bulletin of Canadian Petroleum Geology, 45, pp. 415–433.

Embry, A. F., and Johannessen, E. P., 1992, T-R sequence stratigraphy, facies analysis and reservoir distribution in the uppermost Triassic-Lower Jurassic succession, western Sverdrup Basin, Arctic Canada, in Vorren, T. O., Bergsager, E., Dahl-Stamnes, O. A., Holter, E., Johansen, B., Lie, E., and Lund, T. B., eds., Arctic geology and petroleum potential: Norwegian Petroleum Society Special Publication 2, pp. 121–146.

Emery, D., and Myers, K. J., 1996, Sequence stratigraphy: Blackwell, Oxford, 297p.

Emiliani, C., 1955, Pleistocene temperatures: Journal of Geology, 63, pp. 538–578.

Engebretson, D. C., Cox, A., and Gordon, R. G., 1985, Relative motions between oceanic and continental plates in the Pacific basin: Geological Society of America Special paper 206, 59p.

Engel, A. E. G., and Engel, C. B., 1964, Continental accretion and the evolution of North America, in Subramaniam A. P., and Balakrishna, S., eds., Advancing frontiers in geology and geophysics: Indian Geophysical Union, Hyderabad, pp. 17–37.

Enos, P., and Samankassou, E., 1998, Lofer cyclothems revisited (Late Triassic, Northern Alps, Austria): Facies, 38, pp. 207–227.

Epting, M., 1989, Miocene carbonate buildups of central Luconia, offshore Sarawak, in Bally, A. W., ed., Atlas of seismic stratigraphy: American Association of Petroleum Geologists Studies in Geology 27, 3, pp. 168–173.

Erba, E., Castradori, D., Guasti, G., and Ripepe, M., 1992, Calcareous nannofossils and Milankovitch cycles: the example of the Albian Gault Clay Formation (southern England): Palaeogeography, Palaeoclimatology, Palaeoecology, 93, pp. 47–69.

Erikson, J. P., and Pindell, J. L., 1998, Sequence stratigraphy and relative sea-level history of the Cretaceous to Eocene passive margin of northeastern Venezuela and the possible tectonic and eustatic causes of stratigraphic development, in Pindell, J. L., and Drake, C. L., eds., Paleogeographic evolution and non-glacial eustasy, northern South America: Society for Sedimentary Geology (SEPM) Special Publication 58, pp. 261–281.

Erlich, R. N., Longo, A. P., Jr., and Hyare, S., 1993, Response of carbonate platform margins to drowning: evidence of environmental collapse, in Loucks, R. G., and Sarg, J. F., Jr., eds., Carbonate sequence stratigraphy – recent developments and applications: American Association of Petroleum Geologists, memoir 57, pp. 241–266.

Eschard, R. and Doligez, B., eds., 1993, Subsurface reservoir characterization from outcrop observations: Institut Français du Petrole. Éditions Technip, Paris, 189p.

Ettensohn, F. R., 1994, Tectonic control on formation and cyclicity of major Appalachian unconformities and associated stratigraphic sequences, in Dennison, J. M., and Ettensohn,

F. R., eds., Tectonic and eustatic controls on sedimentary cycles, SEPM Concepts in Sedimentology and Paelontology, 4, pp. 217–242.

Ettensohn, F. R., 2008, The Appalachian foreland basin in eastern United States, in Miall, A. D., ed., The Phanerozoic sedimentary basins of the United States and Canada: Sedimentary Basins of the World, Elsevier, Amsterdam, vol. 5, pp. 105–179.

Eugster, H. P., and Hardie, L. A., 1975, Sedimentation in an ancient playa-lake complex: The Wilkins Peak Member of the Green River Formation of Wyoming: Geological Society of America Bulletin, 86, pp. 319–334.

Eyles, N., 1993, Earth's glacial record and its tectonic setting: Earth Science Reviews, 35, pp. 1–248.

Eyles, N., 2008, Glacio-epochs and the supercontinent cycle after ~3.0 Ga: Tectonic boundary conditions for glaciation: Palaeogeography, Palaeoclimatology, Palaeoecology, 258, pp. 89–129.

Eyles, N., and Januszczak, N., 2004, 'Zipper-rift': a tectonic model for Neoproterozoic glaciations during the breakup of Rodinia after 750 Ma: Earth-Science Reviews, 65 pp. 1–73.

Fairbridge, R. W., 1961, Eustatic changes in sea level, in Ahrens, L. H., Press, F., Rankama, K., and Runcorn, S. K., eds., Physics and chemistry of the Earth: Pergamon Press, London, pp. 99–185.

Falvey, D. A., 1974, The development of continental margins in plate tectonic theory: Australian Petroleum Association Journal, 14, pp. 95–106.

Feeley, M. H., Moore, T. C., Jr., Loutit, T. S., and Bryant, W. R., 1990, Sequence stratigraphy of Mississippi fan related to oxygen isotope sea level index: American Association of Petroleum Geologists Bulletin, 74, pp. 407–424.

Ferm, J. C., 1975, Pennsylvanian cyclothems of the Appalachian Plateau, a retrospective view: U.S. Geological Survey Professional Paper 853, pp. 57–64.

Filer, J. K., 1994, High frequency eustatic and siliciclastic sedimentation cycles in a foreland basin, Upper Devonian, Appalachian Basin, in Dennison, J. M., and Ettensohn, F. R., eds., Tectonic and eustatic controls on sedimentary cycles: Society for Sedimentary Geology, Concepts in Sedimentology and Paleontology, 4, pp. 133–145.

Fischer, A. G., 1964, The Lofer cyclothems of the Alpine Triassic: Kansas Geological Survey Bulletin, 169, pp. 107–149.

Fischer, A. G., 1981, Climatic oscillations in the biosphere, in Nitecki, M. H., ed., Biotic crises in ecological and evolutionary time: Academic Press, New York, pp. 102–131.

Fischer, A. G., 1984, The two Phanerozoic supercycles, in Berggren, W. A., and Van Couvering, J., eds., Catastrophes and earth history: Princeton University Press, Princeton, New Jersey, pp. 129–150.

Fischer, A. G., 1986, Climatic rhythms recorded in strata: Annual Review of Earth and Planetary Sciences, 14, pp. 351–376.

Fischer, A. G., and Arthur, M. A., 1977, Secular variations in the pelagic realm, in Cook, H. E., and Enos, P., eds., Deep-water carbonate environments: Society of Economic Paleontologists and Mineralogists, Special Publication 25, pp. 19–50.

Fischer, A. G., and Bottjer, D. J., 1991, Orbital forcing and sedimentary sequences (introduction to Special Issue): Journal of Sedimentary Petrology, 61, pp. 1063–1069.

Fischer, A. G., D'Argenio, B., Silva, I. P., Weissert, H., and Ferreri, V., 2004, Cyclostratigraphic approach to Earth's history: an introduction, in D'Argenio, B., Fischer, A. G., Silva, I. P., Weissert, H., and Ferreri, V., eds., Cyclostratigraphy: approaches and case histories: Society for Sedimentary Geology (SEPM) Special Publication 81, pp. 5–16.

Fischer, A. G., Hilgen, F. J., and Garrison, R. E., 2009, Mediterranean contributions to cyclostratigraphy and astrochronology: Sedimentology, 56, pp. 63–94.

Fischer, A. G., and Roberts, L. T., 1991, Cyclicity in the Green River Formation (lacustrine Eocene) of Wyoming: Journal of Sedimentary Petrology, 61, pp. 1146–1154.

Fischer, A. G., and Schwarzacher, W., 1984, Cretaceous bedding rhythms under orbital control, in Berger, A., Imbrie, J., Hays, J., Kukla, G., and Saltzman, B., eds., Milankovitch and climate: NATO ASI Series, D. Reidel Publishing Company, Dordrecht, pp. 163–175.

Fisher, W. L., and McGowen, J. H., 1967, Depositional systems in the Wilcox Group of Texas and their relationship to occurrence of oil and gas: Transactions of the Gulf Coast Association of Geological Societies, 17, pp. 105–125.

Fisk, H. N., 1939, Depositional terrace slopes in Louisiana: Journal of Geomorphology, 2, pp. 181–200.

Fisk, H. N., 1944, Geological investigations of the alluvial valley of the lower Mississippi River: U. S. Army Corps of Engineers, Mississippi River Commission, Vicksburg, Mississippi, 78p.

Fjeldskaar, W., 1989, Rapid eustatic change—never globally uniform, in Collinson, J. D., ed., Correlation in hydrocarbon exploration: Graham and Trotman, London, pp. 13–19.

Flint, S. S., Turner, P., and Jolley, E. J., 1991, Depositional architecture of Quaternary fan-delta deposits of the Andean fore-arc: relative sea-level changes as a response to aseismic ridge subduction, in Macdonald, D. I. M., ed., 1991, Sedimentation, tectonics and eustasy: sea-level changes at active margins: International Association of Sedimentologists Special Publication 12, pp. 91–103.

Fluegeman, R. H., 2007, Unresolved issues in Cenozoic chronostratigraphy: Stratigraphy, 4, pp. 109–116.

Fortuin, A. R., and de Smet, M. E. M., 1991, Rates and magnitudes of late Cenozoic vertical movements in the Indonesian Banda Arc and the distinction of eustatic effects, in Macdonald, D. I. M., ed., 1991, Sedimentation, tectonics and eustasy: sea-level changes at active margins: International Association of Sedimentologists Special Publication 12, pp. 79–89.

Fouch, T. D., Lawton, T. F., Nichols, D. J., Cashion, W. B., and Cobban, W. A., 1983, Patterns and timing of synorogenic sedimentation in Upper Cretaceous rocks of central and northeast Utah, in Reynolds, M., and Dolly, E., eds., Mesozoic paleogeography of west-central United States: Society of Economic Paleontologists and Mineralogists, Rocky Mountain Section, Symposium, 2, pp. 305–334.

Frakes, L. A., 1979, Climates throughout geologic time: Elsevier, Amsterdam, 310p.

Frakes, L. A., 1986, Mesozoic-Cenozoic climatic history and causes of the glaciation, in Hsü, K. J., ed., Mesozoic and Cenozoic oceans: Geodynamics Series, American Geophysical Union, Washington, vol. 15, pp. 33–48.

Frakes, L. A., and Francis, J. E., 1988, A guide to Phanerozoic cold polar climates from high-latitude ice-rafting in the Cretaceous: Nature, 333, pp. 547–549.

Francis, J.E., and Frakes, L. A., 1993, Cretaceous climates: Sedimentology Review, 1, pp. 17–30.

Franseen, E. K., Goldstein, R. H., and Whitesell, T. E., 1993, Sequence stratigraphy of Miocene carbonate complexes, Las Negras area, southeastern Spain: implications for quantification of changes in relative sea level, in Loucks, R. G., and Sarg, J. F., eds., Carbonate sequence stratigraphy: American Association of Petroleum Geologists Memoir 57, pp. 409–434.

Franseen, E. K., Watney, W. L., and Kendall, C. G. St.C., eds., 1991, Sedimentary modeling: computer simulations and methods for improved parameter definition: Kansas Geological Survey Bulletin, 233, 524p.

Frazier, D. E., 1974, Depositional episodes: their relationship to the Quaternary stratigraphic framework in the northwestern portion of the Gulf Basin: Bureau of Economic Geology, University of Texas, Geological Circular, 74–1, 26p.

Fritz, W. H., Cecile, M. Norford, B. S., Morrow, D. W., and Geldsetzer, H. H. J., 1992, Cambrian to Middle Devonian assemblages, in Gabrielse, H., and Yorath, C. J., eds., Geology of the Cordilleran Orogen in Canada, Geological Survey of Canada, Geology of Canada, 4, pp. 153–218.

Frodeman, R., 1995, Geological reasoning: geology as an interpretive and historical science: Geological Society of America Bulletin, 107, pp. 960–968.

Fuchs, S., 1992,. The Professional quest for truth: a social theory of science and knowledge: State University of New York Press, Albany, NY.

Fulford, M. M., and Busby, C. J., 1993, Tectonic controls on non-marine sedimentation in a Cretaceous fore-arc basin, Baja California, Mexico, in Frostick, L. E., and Steel, R. J., eds., Tectonic controls and signatures in sedimentary successions: International Association of Sedimentologists Special Publication 20, pp. 301–333.

Fulthorpe, C. S., Camoin, G., Miller, K. G., and Droxler, A., eds., 2008, Pre-Quaternary sea-level changes: records and processes: Basin Research, 20, pp. 161–162.

Fulthorpe, C. S., 1991, Geological controls on seismic sequence resolution: Geology, 19, pp. 61–65.

Fulthorpe, C. S., and Carter, R. M., 1989, Test of seismic sequence methodology on a southern hemisphere passive margin: the Canterbury Basin, New Zealand: Marine and Petroleum Geology, 6, pp. 348–359.

Funnell, B. M., 1981, Mechanisms of autocorrelation: Journal of the Geological Society, London, 138, pp. 177–182.

Gale, A. S., Hardenbol, J., Hathaway, B., Kennedy, W. J., Young, J. R., and Phansalkar, V., 2002, Global correlation of Cenomanian (Upper Cretaceous) sequence: evidence for Milankovitch control on sea level: Geology, 30, pp. 291–294.

Gale, A. S., Huggett, J. M., Palike, H., Laurie, E., Hailwood, E. A., and Hardenbol, J., 2005, Correlation of Eocene-Oligocene marine and continental records: orbital cyclicity, magnetostratigraphy and sequence stratigraphy of the Solent Group, Isle of Wight, U.K.: Journal of the Geological Society, London, 162, pp. 1–15.

Gale, A. S., Jenkyns, H. C., Kennedy, W. J., and Corfield, R. M., 1993, Chemostratigraphy versus biostratigraphy: data from around the Ceonomanian/Turonian boundary: Journal of the Geological Society, London, 150, pp. 29–32.

Gale, A. S., Voigt, S., Sageman, B. B., and Kennedy, W. J., 2008, Eustatic sea-level record for the Cenomanian (Late Cretaceous)—Extension to the Western Interior Basin, USA: Geology, 36, pp. 859–862.

Galeotti, S., Rusciadelli, G., Sprovieri, M., Lanci, L., Gaudio, Al and Pekar, S., 2009, Sea-level control on facies architecture in the Cenomanian-Coniacian Apulian margin (Western Tethys): a record of glacio-eustatic fluctuations during the Cretaceous greenhouse? Palaeogeography, Paleoclimatology, Palaeoecology, 276, pp. 196–205.

Gallagher, K., Dumitru, T. A., and Gleadow, A. J. W., 1994, Constraints on the vertical motion of eastern Australia during the Mesozoic: Basin Research, 6, pp. 77–94.

Galloway, W. E., 1989a, Genetic stratigraphic sequences in basin analysis I: Architecture and genesis of flooding-surface bounded depositional units: American Association of Petroleum Geologists Bulletin, 73, p. 125–142.

Galloway, W. E., 1989b, Genetic stratigraphic sequences in basin analysis II: Application to northwest Gulf of Mexico Cenozoic basin: American Association of Petroleum Geologists Bulletin, 73, pp. 143–154.

Galloway, W. E., 2008, Depositional evolution of the Gulf of Mexico sedimentary basin, in Miall, A. D., ed., The Sedimentary Basins of the United States and Canada: Sedimentary basins of the World, vol. 5, K. J. Hsü, Series Editor, Elsevier Science, Amsterdam, pp. 505–549.

Galloway, W. E., and Brown, L. F., Jr., 1973, Depositional systems and shelf-slope relations on cratonic basin margin, uppermost Pennsylvanian of north-central Texas: American Association of Petroleum Geologists Bulletin, 57, pp. 1185–1218.

Galloway, W. E., 2008, Depositional evolution of the Gulf of Mexico sedimentary basin, in Miall, A. D., ed., Sedimentary Basins of the United States and Canada: Sedimentary basins of the World, vol. 5, K. J. Hsü, Series Editor, Elsevier Science, Amsterdam, pp. 505–549.

Geological Society of London, 1964, Geological Society Phanerozoic time-scale 1964, in The Phanerozoic time scale: Geological Society of London Quarterly Journal, 120, supplement, pp. 260–262.

George, T. N., 1978, Eustasy and tectonics: sedimentary rhythms and stratigraphical units in British Dinantian correlation: Proceedings of the Yorkshire Geological Society, 42, pp. 229–253.

Gibling, M. R., and Bird, D. J., 1994, Late Carboniferous cyclothems and alluvial paleovalleys in the Sydney Basin, Nova Scotia: Geological Society of America Bulletin, 106, pp. 105–117.

Gilbert, G. K., 1886, The inculcation of scientific method by example, with an illustration drawn from the Quaternary geology of Utah: American Journal of Science, 31, pp. 284–299.

Gilbert, G. K., 1890, Lake Bonneville, U. S. Geological Survey Monograph 1,438p.

Gilbert, G. K., 1895, Sedimentary measurement of geologic time: Journal of Geology, 3, pp. 121–127.

Ginsburg, R. N., and Beaudoin, B., eds., 1990, Cretaceous resources, events and rhythms: NATO ASI Series C, vol. 304, Kluwer Academic Publishers, Dordrecht, The Netherlands, 352p.

Glennie, K. W., ed., 1998, Petroleum geology of the North Sea, Fourth edition: Blackwell Science, Oxford, 636p.

Goldhammer, R. K., Dunn, P. A., and Hardie, L. A., 1987, High-frequency glacio-eustatic sea level oscillations with Milankovitch characteristics recorded in Middle Triassic platform carbonates in northern Italy: American Journal of Science, 287, pp. 853–892.

Goldhammer, R. K., Dunn, P. A., and Hardie, L. A., 1990, Depositional cycles, composite sea-level changes, cycle stacking patterns, and the hierarchy of stratigraphic forcing: examples from Alpine Triassic platform carbonates: Geological Society of America Bulletin, 102, pp. 535–562.

Goldhammer, R. K., and Harris, M. T., 1989, Eustatic controls on the stratigraphy and geometry of the Latemar buildup (Middle Triassic), the Dolomites of northern Italy, in Crevello, P. D., Wilson, J. L., Sarg, J. F., and Read, J. F., eds., Controls on carbonate platform and basin development: Society of Economic Paleontologists and Mineralogists Special Publication 44, pp. 323–338.

Goldhammer, R. K., Harris, M. T., Dunn, P. A., and Hardie, L. A., 1993, Sequence stratigraphy and systems tract development of the Latemar Platform, Middle Triassic of the Dolomites (northern Italy): outcrop calibration keyed by cycle stacking patterns, in Loucks, R. G., and Sarg, J. F., eds., Carbonate sequence stratigraphy: American Association of Petroleum Geologists Memoir 57, pp. 353–387.

Goldstein, R. H., and Franseen, E. K., 1995, Pinning points: a method providing quantitative constraints on relative sea-level history: Sedimentary Geology, 95, pp. 1–10.

Golinski, J., 1998,. Making natural knowledge: constructivism and the history of science: Cambridge University Press, Cambridge.

Grabau, A. W., 1906, Types of sedimentary overlap: Geological Society of America Bulletin, 17, pp. 567–636.

Grabau, A. W., 1913, Principles of stratigraphy: A. G. Seiler and Company, New York.

Grabau, A. W., 1936a, Revised classification of the Paleozoic System in the light of the pulsation theory: Geological Survey of China Bulletin, 15, pp. 23–51.

Grabau, A. W., 1936b, Oscillation or pulsation: International Geological Congress, Report of the 16th Session, USA, 1933, vol. 1, pp. 539–553.

Grabau, A. W., 1940, The rhythm of the ages: Henri Vetch, Peking, 561p.

Graciansky, P.-C. de, Hardenbol, J., Jacquin, T., and Vail, P. R., eds., 1998, Mesozoic and Cenozoic sequence stratigraphy of European basins, Society for Sedimentary Geology (SEPM) Special Publication 60, 786p.

Gradstein, F. M., Agterberg, F. P., and D'Iorio, M. A., 1990, Time in quantitative stratigraphy, in Cross, T. A., ed., Quantitative dynamic stratigraphy: Prentice Hall, Englewood Cliffs, New Jersy, pp. 519–542.

Gradstein, F. M., Agterberg, F. P., Ogg, J. G., Hardenbol, J., Van Veen, P., Thierry, J., and Zehui Zhang, 1995, A Triassic, Jurassic and Cretaceous time scale, in Berggren, W. A., Kent, D. V., Aubry, M.-P., and Hardenbol, J., eds., Geochronology, time scales and global stratigraphic correlation: Society for Sedimentary Geology Special Publication 54, pp. 95–126.

Gradstein, F. M., Ogg, J. G., and Smith, A. G., eds., 2004, A geologic time scale: Cambridge University Press, Cambridge, 610p.

Gradstein, F. M., Sandvik, K. O., and Milton, N. J., eds., 1998, Sequence stratigraphy: concepts and applications: Norwegian Petroleum Society Special Publication 8, Elsevier, Amsterdam, 437p.

Gräfe, K., and Wiedmann, J., 1998, Sequence stratigraphy on a carbonate ramp: the late Cretaceous Basco-Cantabrian Basin (northern Spain), in Graciansky, P.-C. de, Hardenbol, J., Jacquin, T., and Vail, P. R., eds., Mesozoic and Cenozoic sequence stratigraphy of European basins, Society for Sedimentary Geology (SEPM) Special Publication 60, pp. 333–341.

Gressly, A., 1838, Observations geologiques sur le Jura Soleurois: Nouv. Mem. Soc. Helv. Sci. Nat, 2, pp. 1–112.

Grippo, A., Fischer, A. G., Hinnov, L. A., Herbert, T. D., and Silva, I. P., 2004, Cyclostratigraphy and chronology of the Albian stage (Piobbico core, Italy), in D'Argenio, B., Fischer, A. G., Silva, I. P., Weissert, H., ,and Ferreri, V., eds., Cyclostratigraphy: approaches and case histories: Society for Sedimentary Geology (SEPM) Special Publication 81, pp. 57–81.

Guex, J., 1991, Biochronological correlations: Springer-Verlag, Berlin, 252p.

Gurnis, M., 1988, Large-scale mantle convection and the aggregation and dispersal of supercontinents: Nature, 332, pp. 695–699.

Gurnis, M., 1990, Bounds on global dynamic topography from Phanerozoic flooding of continental platforms: Nature, 344, pp. 754–756.

Gurnis, M., 1992, Long-term controls on eustatic and epeirogenic motions by mantle convection: GSA Today, 2, pp. 141–157.

Gurnis, M., and Torsvik, T. H., 1994, Rapid drift of large continents during the Late Precambrian and Paleozoic: Paleomagnetic constraints and dynamic models. Geology, 22, pp. 1023–1026.

Gutting, G., 1984, Paradigms, revolutions, and technology, in Laudan, R., ed., The nature of technological knowledge: are models of scientific change relevant? D. Reidel Publishing Company, Dordrecht, pp. 47–65.

Hailwood, E. A., and Kidd, R. B., eds., 1993, High resolution sequence stratigraphy: Geological Society, London, Special Publication 70, 357p.

Hall, J. R., 1990, Social interaction, culture, and historical studies, in Becker, H. S., and McCall, M. M., eds., Symbolic interaction and cultural studies: The University of Chicago Press, Chicago, pp. 2–45.

Hallam, A., 1963, Major epeirogenic and eustatic changes since the Cretaceous and their possible relationship to crustal structure: American Journal of Science, 261, pp. 397–423.

Hallam, A., 1975, Jurassic environments: Cambridge University Press, London and New York, 269p.

Hallam, A., 1978, Eustatic cycles in the Jurassic: Palaeogeography, Palaeoclimatology, Palaeoecology, 23, pp. 1–32.

Hallam, A., 1981, A revised sea-level curve for the early Jurassic: Journal of the Geological Society, London, 138, pp. 735–743.

Hallam, A., 1984, Pre-Quaternary sea-level changes: Annual Review of Earth and Planetary Sciences, 12, pp. 205–243.

Hallam, A., 1986, Origin of minor limestone-shale cycles: climatically induced or diagenetic? Geology, 14, pp. 609–612.

Hallam, A., 1988, A reevaluation of Jurassic eustasy in the light of new data and the revised Exxon curve, in Wilgus, C. K., Hastings, B. S., Kendall, C. G. St. C., Posamentier, H. W., Ross, C. A., and Van Wagoner, J. C., eds., Sea-level Changes: an integrated approach: Society of Economic Paleontologists and Mineralogists Special Publication 42, pp. 261–273.

Hallam, A., 1989, Great geological controversies, second edition: Oxford University Press, Oxford, 244p.

Hallam, A., 1991, Relative importance of regional tectonics and eustasy for the Mesozoic of the Andes, in Macdonald, D. I. M., ed., Sedimentation, tectonics and eustasy: sea-level changes at active margins: International Association of Sedimentologists Special Publication 12, pp. 189–200.

Hallam, A., 1992a, Phanerozoic sea-level changes: Columbia University Press, Irvington, New York, 224p.

Hallam, A., 1992b, Eduard Suess and European thought on Phanerozoic eustasy, in Dott, R. H., Jr., ed., Eustasy: the ups and downs of a major geological concept: Geological Society of America Memoir 180, pp. 25–29.

Hallam, A., 1998a, Lyells' views on organic progression, evolution and extinction, in Blundell, D. J., and Scott, A. C., eds., Lyell: The past is the key to the present: Geological Society of London Special Publication 143, pp. 133–136.

Hallam, A., 1998b, Interpreting sea level, in Doyle, P., and Bennett, M. R., eds., Unlocking the stratigraphical record: John Wiley and Sons, Chichester, pp. 421–439.

Hambrey, M. J., and Harland, W. B., eds., 1981, Earth's Pre-Pleistocene glacial record: Cambridge University Press, Cambridge, 1004p.

Hampson, G., Stollhofen, H., and Flint, S., 1999, A sequence stratigraphic model for the Lower Coal measures (Upper Carboniferous) of the Ruhr district, north-west Germany: Sedimentology, 46, pp. 1199–1231.

Hancock, J. M., 1977, The historic development of biostratigraphic correlation, in Kauffman, E. G. and Hazel, J. E., eds., Concepts and methods of biostratigraphy: Dowden, Hutchinson and Ross Inc., Stroudsburg, Pennsylvania, pp. 3–22.

Hancock, J. M., 1993a, Comments on the EXXON cycle chart for the Cretaceous system: Madrid, Cuadernos de Geologia Ib, Rica, No. 17, pp 3–24.

Hancock, J. M., 1993b, Transatlantic correlations in the Campanian-Maastrichtian stages by eustatic changes of sea-level, in Hailwood, E. A., and Kidd, R. B., eds., High resolution stratigraphy: Geological Society, London, Special Publication 70, pp. 241–256.

Hancock, J. M., and Kauffman, E. G., 1979, The great transgressions of the Late Cretaceous: Journal of the Geological Society, London, 136, pp. 175–186.

Haq, B. U., 1991, Sequence stratigraphy, sea-level change, and significance for the deep sea, in Macdonald, D. I. M., ed., 1991, Sedimentation, tectonics and eustasy: sea-level changes at active margins: International Association of Sedimentologists Special Publication 12, pp. 3–39.

Haq, B. U., Hardenbol, J., and Vail, P. R., 1987, Chronology of fluctuating sea levels since the Triassic (250 million years ago to present): Science, 235, pp. 1156–1167.

Haq, B. U., Hardenbol, J., and Vail, P. R., 1988a, Mesozoic and Cenozoic chronostratigraphy and cycles of sea-level change, in Wilgus, C. K., Hastings, B. S., Kendall, C. G. St. C., Posamentier, H. W., Ross, C. A., and Van Wagoner, J. C., eds., Sea-level Changes: an integrated approach: Society of Economic Paleontologists and Mineralogists Special Publication 42, pp. 71–108.

Haq, B. U., Vail, P. R., Hardenbol, J., and Van Wagoner, J. C., 1988b, Sea level history: Science, 241, pp. 596–602.

Haq, B. U., and Schutter, S. R., 2008, A chronology of Paleozoic sea-level changes: Science, 322, pp. 64–68.

Hardenbol, J., and Robaszynski, F., 1998, Introduction to the Upper Cretaceous, in Graciansky, P.-C. de, Hardenbol, J., Jacquin, T., and Vail, P. R., eds., Mesozoic and Cenozoic sequence stratigraphy of European basins, Society for Sedimentary Geology (SEPM) Special Publication 60, pp. 329–332.

Hardenbol, J., Thierry, J., Farley, M. B., Jacquin, T., Graciansky, P.-C., and Vailo, P. R., 1998, Mesozoic and Cenozoic sequence chronostratigraphic framework of European basins, in Graciansky, P.-C. de, Hardenbol, J., Jacquin, T., and Vail, P. R., eds., Mesozoic and Cenozoic sequence stratigraphy of European basins, Society for Sedimentary Geology (SEPM) Special Publication 60, pp. 3–13.

Hardenbol, J., Vail, P. R., and Ferrer, J., 1981, Interpreting paleoenvironments, subsidence history and sea-level changes of passive margins from seismic and biostratigraphy: Oceanologica Acta Supplement, 3, pp. 33–44.

Hardman, R. F. P., and Brooks, J., eds., 1990, Tectonic events responsible for Britain's oil and gas reserves: Geological Society, London, Special Publication 55, 404p.

Harland, W. B., 1978, Geochronologic scales, in Cohee, G. V., Glaessner, M. F., and Hedberg, H. D., eds., Contributions to the geologic time scale: American Association of Petroleum Geologists Studies in Geology 6, pp. 9–32.

Harland, W. B., Armstrong, R. L., Cox, A. V., Craig, L. E., Smith, A. G., and Smith, D. G., 1990, A geologic time scale, 1989: Cambridge Earth Science Series, Cambridge University Press, Cambridge, 263p.

Harper, C. W., Jr., and Crowley, K. D., 1985, Insights on why graphic correlation (Shaw's method) works: A discussion: Journal of Geology, 93, pp. 503–506.

Harries, P., ed., 2003, High resolution approaches in stratigraphic paleontology: Kluwer Academic Publications, Dordrecht, 474p.

Harrison, C. G. A., 1990, Long-term eustasy and epeirogeny in continents, in Revelle, R., ed., Sea-level change: National Research Council, Studies in Geophysics, National Academy Press: Washington, pp. 141–158.

Hart, B. S., and Plint, A. G., 1993, Tectonic influence on deposition in a ramp setting: Upper Cretaceous Cardium Formation, Alberta foreland basin: American Association of Petroleum Geologists Bulletin, 77, pp. 2092–2107.

Hartley, R. W., and Allen, P. A., 1994, Interior cratonic basins of Africa: relation to continental break-up and role of mantle convection: Basin Research, 6, pp. 95–113.

Hay, W. W., and Southam, J. R., 1978, Quantifying biostratigraphic correlation: Annual Review of Earth and Planetary Sciences, 6, pp. 353–375.

Hays, J. D., Imbrie, J., and Shackleton, N. J., 1976, Variations in the earth's orbit: pacemaker of the ice ages: Science, 194, pp. 1121–1132.

Hays, J. D., and Pitman, W. C., III, 1973, Lithospheric plate motion, sea level changes and climatic and ecological consequences: Nature, 246, pp. 18–22.

Hazel, J. E., Jr., 1994, Sedimentary response to intrabasinal salt tectonism in the Upper Triassic Chinle Formation, Paradox Basin, Utah. U.S. Geological Survey Bull., 2000-F, 34p.

Heckel, P. H., 1986, Sea-level curve for Pennsylvanian eustatic marine transgressive-regressive depositional cycles along midcontinent outcrop belt, North America: Geology, 14, pp. 330–334.

Heckel, P. H., 1990, Evidence for global (glacio-eustatic) control over Upper Carboniferous (Pennsylvanian) cyclothems in midcontinent North America, in Hardman, R. F. P., and Brooks, J., eds., Tectonic events responsible for Britains's oil and gas reserves: Geological Society, London, Special Publication 55, pp. 35–47.

Heckel, P. H., 1994, Evaluation of evidence for glacio-eustatic control over marine Pennsylvanian cyclothems in North America and consideration of possible tectonic effects, in Dennison, J. M., and Ettensohn, F. R., eds., Tectonic and eustatic controls on sedimentary cycles: Society for Sedimentary Geology, Concepts in Sedimentology and Paleontology, 4, pp. 65–87.

Hedberg, H. D., 1948, Time-stratigraphic classification: Geological Society of America Bulletin, 59, pp. 447–462.

Hedberg, H. D., ed., 1976, International stratigraphic guide: John Wiley and Sons, New York, 200p.

Heidegger, M., 1927, Sein und zeit: Neomarius Verlag, Tübingen, 488p.

Heidegger, M., 1962, Being and time: Macquarrie, J., and Robinson, E., translators, Harper and Row, New York, 589p.

Heinrich, H., 1988, Origin and consequences of cyclic ice-rafting in the Northeast Atlantic Ocean during the past 130,000 years: Quaternary Research, 29, pp. 142–152.

Heller, P. L., and Angevine, C. L., 1985, Sea-level cycles during the growth of Atlantic-type oceans: Earth and Planetary Science Letters, 75, pp. 417–426.

Heller, P. L., Angevine, C. L., Winslow, N. S., and Paola, C., 1988, Two-phase stratigraphic model of foreland-basin sequences: Geology, 16, pp. 501–504.

Heller, P. L., Beekman, F., Angevine, C. L., and Cloetingh, S. A. P. L., 1993, Cause of tectonic reactivation and subtle uplifts in the Rocky Mountain region and its effect on the stratigraphic record: Geology, 21, pp. 1003–1006.

Heller, P. L., and Paola, C., 1989, The paradox of Lower Cretaceous gravels and the initiation of thrusting in the Sevier orogenic belt, United States Western Interior: Geological Society of America Bulletin, 101, pp. 864–875.

Heller, P. L., and Paola, C., 1992, The large-scale dynamics of grain-size variation in alluvial basins, 2: application to syntectonic conglomerate: Basin Research, 4, pp. 91–102.

Herbert, T. D., Premoli Silva, I, Erba, E., and Fischer, A. G., 1995, Orbital chronology of Cretaceous-Paleocene marine sediments, in Berggren, W. A., Kent, D. V., Aubry, M.-P., and Hardenbol, J., eds., Geochronology, time scales and global stratigraphic correlation: Society for Sedimentary Geology Special Publication 54, pp. 81–93.

Herrle, J. O., Köβler, P., Friedrich, O., Erlenkeuser, H., and Hemleben, C., 2004, High-resolution carbon isotope records of the Aptian to Lower Albian from SE France and the Mazagan plateau (DSDP Site 545): a stratigraphic tool for paleoceanographic and paleobiologic reconstruction: Earth and Planetary Science Letters, 218, pp. 149–161.

Hewitt, J. P., 1997, Self and Society: a symbolic interactionist social psychology, Seventh edition: Allyn and Bacon, Boston, 263p.

Hilgen, F. J., 1991, Extension of the astronomically calibrated (polarity) time scale to the Miocene/Pliocene boundary: Earth and Planetary Sciences Letters, 107, pp. 349–368.

Hilgen, F., Brinkhuis, H., and Zachariasse, W.-J., 2006, Unit stratotypes for global stages: The Neogene perspective: Earth Science Reviews, 74, pp. 113–125.

Hilgen, F., Kuiper, K., Krijgsman, W., Snel, E., and van der Laan, E., 2007, Astronomical tuning as the basis for high resolution chronostratigraphy: the intricate history of the Messinian Salinity Crisis: Stratigraphy, 4, pp. 231–238.

Hinnov, L. A., 2000, New perspectives on orbitally forced stratigraphy: Annual Review of Earth and Planetary Sciences, 28, pp. 419–475.

Hinnov, L. A., and Goldhammer, R. K., 1991, Spectral analysis of the Triassic Latemar Limestone: Journal of Sedimentary Petrology, 61, pp. 1173–1193.

Hinnov, L. A., and Ogg, J. G., 2007, Cyclostratigraphy and the astronomical time scale: Stratigraphy, 4, pp. 239–251.

Hiroki, Y., 1994, Quaternary crustal movements from facies distribution in the Atsumi and Hamana areas, central Japan: Sedimentary Geology, 93, pp. 223–235.

Hiscott, R. N., Wilson, R. C. L., Gradstein, F. M., Pujalte, V., Garcia-Mondéjar, J., Boudreau, R. K., and Wishart, H. A., 1990, Comparative stratigraphy and subsidence history of Mesozoic rift basins of North Atlantic: American Association of Petroleum Geologists Bulletin, 74, pp. 60–76.

Hoffman, P. F., 1989, Speculations on Laurentia's first gigayear (2.0 to 1.0 Ga): Geology, 17, pp. 135–138.

Hoffman, P. F., 1991, Did the breakout of Laurentia turn Gondwanaland inside-out? Science, 252, pp. 1409–1412.

Hoffman, P. F., and Grotzinger, J. P., 1993, Orographic precipitation, erosional unloading, and tectonic style: Geology, 21, pp. 195–198.

Hoffman, P. F., and Schrag, D. P., 2000, Snowball Earth: Scientific American, 282, pp. 62–75.

Holbrook, J., Scott, R. W., and Oboh-Ikuenobe, F. E., 2006, Base-level buffers and buttresses: a model for upstream versus downstream control on fluvial geometry and architecture within sequences: Journal of Sedimentary Research, 76, pp. 162–174.

Holdsworth, B. K., and Collinson, J. D., 1988, Millstone Grit cyclicity revisited, in Besly, B. M., and Kelling, G., 1988, Sedimentation in a synorogenic basin complex: the Upper Carboniferous of northwest Europe: Blackie, Glasgow, pp. 132–152.

Holland, C. H., 1986, Does the golden spike still glitter? Journal of the Geological Society, London, 143, pp. 3–21.

Holland, C. H., 1998, Chronostratigraphy (global standard stratigraphy): A personal perspective, in Doyle, P., and Bennett, M. R., eds., Unlocking the stratigraphical record: John Wiley and Sons, Chichester, pp. 383–392.

Holmes, A., 1937, The age of the Earth: Harper, London.

House, M. R., 1985, A new approach to an absolute timescale from measurements of orbital cycles and sedimentary microrhythms: Nature, 315, pp. 721–725.

House, M. R., 1995, Orbital forcing timescales: an introduction, in House. M. R., and Gale, A. S., eds., Orbital forcing timescales and cyclostratigraphy: Geological Society, London, Special Publication 85, pp. 1–18.

House, M. R., and Gale, A. S., eds., 1995, Orbital forcing timescales and cyclostratigraphy: Geological Society, London, Special Publication 85, 210p.

Howell, P. D., and van der Pluijm, B. A., 1999, Structural sequences and styles of subsidence in the Michigan Basin: Geological Society of America Bulletin, 111, pp. 974–991.

Hubbard, R. J., 1988, Age and significance of sequence boundaries on Jurassic and Early Cretaceous rifted continental margins: American Association of Petroleum Geologists Bulletin, 72, pp. 49–72.

Huggett, R. J., 1991, Climate, earth processes and earth history: Springer-Verlag, New York, 281p.

Hunt, D., and Gawthorpe, R. L., eds., 2000, Sedimentary responses to forced regressions, Geological Society, London, Special Publication 172, 383p.

Hunt, D., and Tucker, M. E., 1992, Stranded parasequences an the forced regressive wedge systems tract: deposition during base-level fall: Sedimentary Geology, 81, pp. 1–9.

Imbrie, J., 1985, A theoretical framework for the Pleistocene ice ages: Journal of the Geological Society, London, 142, pp. 417–432.

Imbrie, J., and Imbrie, K. P., 1979, Ice ages: solving the mystery: Enslow, Hillside, New Jersey, 224p.

Imbrie, J., Hays, J. D., Martinson, D. G., McIntyre, A., Mix, A. C., Morley, J. J., Pisias, N. G., Prell, W. L., and Shackleton, N. J., 1984, The orbital theory of Pleistocene climate: support from a revised chronology of the marine ëO18 record, in Berger, A., Imbrie, J., Hays, J., Kukla, G., and Saltzman, B., eds., Milankovitch and climate: D. Reidel, Amsterdam, pp. 269–305.

Immenhauser, A., and Scott, R. W., 1999, Global correlation of middle Cretaceous sea-level events: Geology, 27, pp. 551–554.

Ingersoll, R. V., 1988, Tectonics of sedimentary basins: Geological Society of America Bulletin, 100, pp. 1704–1719.

International Subcommission on Stratigraphic Classification, 1987 Unconformity-bounded stratigraphic units: Geological Society of America Bulletin, 98, pp. 232–237.

Ito, M., 1992, High-frequency depositional sequences of the upper part of the Kazusa Group, a middle Pleistocene forearc basin fill on Boso Peninsula, Japan: Sedimentary Geology, 76, pp. 155–175.

Ito, M., 1995, Volcanic ash layers facilitate high-resolution sequence stratigraphy at convergent plate margins: an example from the Plio-Pleistocene forearc basin fill in the Boso Peninsula, Japan: Sedimentary Geology, 95, pp. 187–206.

Ito, M., and Masuda, F., 1988, Late Cenozoic deep-sea to fan-delta sedimentation in an arc-arc collision zone, central Honshu, Japan; sedimentary response to varying plate-tectonic regime, in Nemec, W., and Steel, R. J., eds., Fan deltas; sedimentology and tectonic settings: Blackie and Son, Glasgow, UK, pp. 400–418.

Ito, M., and O'Hara, S., 1994, Diachronous evolution of systems tracts in a depositional sequence from the middle Pleistocene paleo-Tokyo Bay, Japan: Sedimentology, 41, pp. 677–698.

Jaccarino, S. M., Lirer, F., Bonomo, S., Caruso, A., Stefano, A., Stefano, E., Foresi, L. M., Mazzel, R., Salvatorini, G., Sprovieri, M., Sprovieri, R., and Turco, E., 2004, Astrochronology of late Middle Miocene Mediterranean sections, in D'Argenio, B., Fischer, A. G., Silva, I. P., Weissert, H.,and Ferreri, V., eds., Cyclostratigraphy: approaches and case histories: Society for Sedimentary Geology (SEPM) Special Publication 81, pp. 27–44.

James, D. P., and Leckie, D. A., eds., 1988, Sequences, stratigraphy, sedimentology: surface and subsurface: Canadian Society of Petroleum Geologists Memoir 15, 586p.

James, N. P., 1983, Reef environment, in Scholle, P. A., Bebout, G., and Moore, C. H., eds., Carbonate depositional environments: American Association of Petroleum Geologists, memoir 33, pp. 345–444.

James, N. P., and Kendall, A. C., 1992, Introduction to carbonate and evaporite facies models, in Walker, R. G., and James, N. P., eds., Facies models: response to sea level change: Geological Association of Canada, Geotext 1, pp. 265–275.

Janssen, M. E., Stephenson, R. A., and Cloetingh, S., 1995, Temporal and spatial correlations between changes in plate motions and the evolution of rifted basins in Africa: Geological Society of America Bulletin, 107, pp. 1317–1332.

Jenkins, D. G., and Gamson, P., 1993, The late Cenozoic *Globorotalia truncatulinoides* datum plane in the Atlantic, Pacific and Indian Oceans, in Hailwood, E. A., and Kidd, R. B., eds., High resolution stratigraphy: Geological Society, London, Special Publication 70, pp. 127–130.

Jervey, M. T., 1988, Quantitative geological modeling of siliciclastic rock sequences and their seismic expression, in Wilgus, C. K., Hastings, B. S., Kendall, C. G. St. C., Posamentier, H. W., Ross, C. A., and Van Wagoner, J. C., eds., Sea level changes – an integrated approach: Society of Economic Paleontologists and Mineralogists Special Publication 42, pp. 47–69.

Johnson, D., 1933, Role of analysis in scientific investigation: Geological Society of America Bulletin, 44, pp. 461–493.

Johnson, G. D., Raynolds, R. G. H., and Burbank, D. W., 1986, Late Cenozoic tectonics and sedimentation in the north-western Himalayan foredeep: I. Thrust ramping and associated deformation in the Potwar region, in Allen, P. A., and Homewood, P., eds., Foreland basins: International Association of Sedimentologists Special Publication 8, pp. 273–291.

Johnson, J. G., 1971, Timing and coordination of orogenic, epeirogenic, and eustatic events: Geological Society of America Bulletin, 82, pp. 3263–3298.

Johnson, J. G., 1992a, Belief and reality in biostratigraphic zonation: Newsletters in Stratigraphy, 26, pp. 41–48.

Johnson, M. E., 1992b, A. W. Grabau's embryonic sequence stratigraphy and eustatic curve, in Dott, R. H., Jr., ed., Eustasy: the ups and downs of a major geological concept: Geological Society of America Memoir 180, pp. 43–54.

Joly, J., 1930, The surface-history of the earth, Second edition: Clarendon Press, Oxford, 211p.

Jones, B., and Desrochers, A., 1992, Shallow platform carbonates, in Walker, R. G., and James, N. P., eds., Facies models: response to sea level change: Geological Association of Canada, Geotext 1, pp. 277–301.

Jordan, T. E., 1981, Thrust loads and foreland basin evolution, Cretaceous, Western United States: American Association of Petroleum Geologists Bulletin, 65, pp. 2506–2520.

Jordan, T. E., 1995, Retroarc foreland and related basins, in Busby, C. J., and Ingersoll, R. V., eds., Tectonics of sedimentary basins: Blackwell Science, Oxford, pp. 331–362.

Jordan, T. E., and Flemings, P. B., 1991, Large-scale stratigraphic architecture, eustatic variation, and unsteady tectonism: a theoretical evaluation: Journal of Geophysical Research, 96B, pp. 6681–6699.

Jordan, T. E., Flemings, P. B., and Beer, J. A., 1988, Dating thrust-fault activity by use of foreland-basin strata, in Kleinspehn, K. L., and Paola, C., eds., New perspectives in basin analysis: Springer Verlag, New York, pp. 307–330.

Joy, M. P., Mitchell, C. E., and Adhya, S., 2000, Evidence of a tectonically driven sequence succession in the Middle Ordovician Taconic foredeep: Geology, 28, pp. 727–730.

Kamola, D. L., and Huntoon, J. E., 1995, Repetitive stratal patterns in a foreland basin sandstone and their possible tectonic significance: Geology, 23, pp. 177–180.

Kamp, P. J. J., and Turner, G. M., 1990, Pleistocene unconformity-bounded shelf sequences (Wanganui Basin, New Zealand) correlated with global isotope record: Sedimentary Geology, 68, pp. 155–161.

Karner, G. D., 1986, Effects of lithospheric in-plane stress on sedimentary basin stratigraphy: Tectonics, 5, pp. 573–588.

Katz, M. E., Wright J. D., Miller, K. G., Cramer, B. S., Fennel, K., and Falkowski, P. G., 2005, Biological overprint on the geological carbon cycle: Marine Geology, 217, pp. 323–338.

Kauffman, E. G., 1984, Paleobiogeography and evolutionary response dynamic in the Cretaceous Western Interior Seaway of North America, in Westerman, G. E., ed., Jurassic-Cretaceous biochronology and paleogeography of North America: Geological Association of Canada Special Paper 27, pp. 273–306.

Kauffman, E. G., and Caldwell, W. G. E., 1993, The Western Interior Basin in space and time, in Caldwell, W. G. E., and Kauffman, E. G., eds., Evolution of the Western Interior Basin: Geological Association of Canada Special Paper 39, pp. 1–30.

Kauffman, E. G., Elder, W. P., and Sageman, B. B., 1991, High-resolution correlation: a new tool in chronostratigraphy, in Einsele, G., Ricken, W., and Seilacher, A., eds., Cycles and events in stratigraphy: Springer-Verlag, Berlin, pp. 795–819.

Kauffman, E. G., and Hazel, J. E., eds., 1977a, Concepts and methods of biostratigraphy: Dowden, Hutchinson and Ross Inc., Stroudsburg, Pennsylvania, 658p.

Kauffman, E. G., and Hazel, J. E., 1977b, Preface, in Kauffman, E. G., and Hazel, J. E., eds., Concepts and methods of biostratigraphy: Dowden, Hutchinson and Ross Inc., Stroudsburg, Pennsylvania, pp. iii–v.A.

Kay, M., 1951, North American geosynclines: Geological Society of America, 48, p. 143.

Kendall, C. G. St.C., and Lerche, I., 1988, The rise and fall of eustasy, in Wilgus, C. K., Hastings, B. S., Kendall, C. G. St. C., Posamentier, H. W., Ross, C. A., and Van Wagoner, J. C., eds., Sea-level Changes: an integrated approach: Society of Economic Paleontologists and Mineralogists Special Publication 42, pp. 3–17.

Kennedy, W. J., and Cobban, W. A., 1977, The role of ammonites in biostratigraphy, in Kauffman, E. G., and Hazel, J. E., eds., Concepts and methods of biostratigraphy: Dowden, Hutchinson and Ross Inc., Stroudsburg, Pennsylvania, pp. 309–320.

Kennett, J. P., 1977, Cenozoic evolution of Antarctic glaciation, the circum-Antarctic Ocean, and their impact on global paleoceanography: Journal of Geophysical Research, 82, pp. 3843–3860.

Kerr, R. A., 1980, Changing global sea levels as a geologic index: Science, 209, pp. 483–486.

Kerr, R. A., 1984, Vail's sea-level curves aren't going away: Science, 226, pp. 677–678.

Kidd, R. B., and Hailwood, E. A., 1993, High resolution stratigraphy in modern and ancient marine sequences: ocean sediment cores to Paleozoic outcrop, in Hailwood, E. A., and Kidd, R. B., eds., High resolution stratigraphy: Geological Society, London, Special Publication 70, pp. 1–8.

Kidwell, S. M., 1988, Reciprocal sedimentation and noncorrelative hiatuses in marine-paralic siliciclastics: Miocene outcrop evidence: Geology, 16, pp. 609–612.

King, P. B., 1942, Permean of west Texas and southeaster New Mexico: American Association of Petroleum Geologists Bulletin, 26, pp. 535–763.

King, P. B., 1948, Geology of the southern Gudalupe Mountains, Texas, U. S. Geological Survey Professional Paper 215, 183p.

Klein, G. deV., 1992, Climatic and tectonic sea-level gauge for Midcontinent Pennsylvanian cyclothems: Geology, 20, pp. 363–366.

Klein, G. deV., 1994, Depth determination and quantitative distinction of the influence of tectonic subsidence and climate on changing sea level during deposition of midcontinent Pennsylvanian cyclothems, in Dennison, J. M., and Ettensohn, F. R., eds., Tectonic and eustatic controls on sedimentary cycles: Society for Sedimentary Geology, Concepts in Sedimentology and Paleontology, 4, pp. 35–50.

Klein, G. deV., and Willard, D. A., 1989, Origin of the Pennsylvanian coal-bearing cyclothems of North America: Geology, 17, pp. 152–155.

Klein, G. deV., and Kupperman, J. B., 1992, Pennsylvanian cyclothems: methods of distinguishing tectonically induced changes in sea level from climatically induced changes: Geological Society of America Bulletin, 104, pp. 166–175.

Klitgord, K. D., and Schouten, H., 1986, Plate kinematics of the central Atlantic, in Vogt, P. R., and Tucholke, B. E., eds., The Western North Atlantic region, The Geology of North America Volume M: Geological Society of America, pp. 351–378.

Klüpfel, W., 1917, Über die sedimente der flachsee im Lothringer Jura: Geologisch Rundschau, 7, pp. 98–109.

Kocurek, G., 1981, Significance of interdune deposits and bounding surfaces in aeolian dune sands: Sedimentology, 28, pp. 753–780.

Kocurek, G., ed., 1988a, Late Paleozoic and Mesozoic eolian deposits of the Western Interior of the United States: Sedimentary Geology, 56, 413p. (special issue).

Kocurek, G. A., 1988b, First-order and super bounding surfaces in eolian sequences – bounding surfaces revisited: Sedimentary Geology, 56, pp. 193–206.

Kocurek, G., and Havholm, K. G., 1993, Eolian sequence stratigraphy—a conceptual framework, in Weimer, P., and Posamentier, H. W., eds., Siliciclastic sequence stratigraphy: American Association of Petroleum Geologists Memoir 58, pp. 393–409.

Kolb, W., and Schmidt, H., 1991, Depositional sequences associated with equilibrium coastlines in the Neogene of southwestern Nicaragua, in Macdonald, D. I. M., ed., 1991, Sedimentation, tectonics and eustasy: sea-level changes at active margins: International Association of Sedimentologists Special Publication 12, pp. 259–272.

Kolla, V., Posamentier, H. W., and Eichenseer, H., 1995, Stranded parasequences and the forced regressive wedge systems tract: deposition during base-level fall—discussion: Sedimentary Geology, 95, pp. 139–145.

Kominz, M. A., 1984, Ocean ridge volumes and sea-level change – an error analysis, in Schlee, J. S., ed., Interregional unconformities and hydrocarbon accumulation: American Association of Petroleum Geologists Memoir 36, pp. 108–127.

Kominz, M. A., and Bond, G. C., 1990, A new method of testing periodicity in cyclic sediments: application to the Newark Supergroup: Earth and Planetary Science Letters, 98, pp. 233–244.

Kominz, M. A., and Bond, G. C., 1991, Unusually large subsidence and sea-level events during middle Paleozoic time: new evidence supporting mantle convection models for supercontinent assembly: Geology, 19, pp. 56–60.

Kominz, M. A., Browning, J. V., Miller, K. G., Sugarman, P. J., Mizintserva, S., and Scotese, C. R., 2008, late Cretaceous to Miocene sea-level estimates from the New Jersey and Delaware coastal plain coreholes: an error analysis: Basin Research, 20, pp. 211–226.

Kominz, M. A., and Pekar, S. F., 2001, Oligocene eustasy from two-dimensional sequence stratigraphic backstripping: Geological Society of America Bulletin, 113, pp. 291–304.

Kooi, H., and Cloetingh, S., 1992a, Lithospheric necking and regional isostasy at extensional basins 1. Subsidence and gravity modeling with an application to the Gulf of Lions margin (SE France): Journal of Geophysical Research, 97B, pp. 17553–17571.

Kooi, H., and Cloetingh, S., 1992b, Lithospheric necking and regional isostasy at extensional basins 2. Stress-induced vertical motions and relative sea level changes: Journal of Geophysical Research, 97B, pp. 17573–17591.

Korus, J. T., Kvale, E. P., Eriksson, K. A., and Joeckel, R. M., 2008, Compound paleovalleys fills in the Lower Pennyslvanian New River Formation, West Virginia, USA: Sedimentary Geology, 208, pp. 15–26.

Kuhn, T. S., 1962, The structure of scientific revolutions: Harvard University Press, Cambridge, 172p.

Kuhn, T. S., 1996, The structure of scientific revolutions, Third edition: The University of Chicago press, Chicago, 212p.

Laferriere, A. P., Hattin, D. E., and Archer, A. W., 1987, Effects of climate, tectonics, and sea-level changes on rhythmic bedding patterns in the Niobrara Formation (Upper Cretaceous), U.S. Western Interior: Geology, 15, pp. 233–236.

Lambeck, K., 1980, The earth's variable rotation: Cambridge University Press, Cambridge, 449p.

Lambeck, K., Cloetingh, S., and McQueen, H., 1987, Intraplate stresses and apparent changes in sea level: the basins of northwestern Europe, in Beaumont, C., and Tankard, A. J., eds., Sedimentary basins and basin-forming mechanisms: Canadian Society of Petroleum Geologists Memoir 12, pp. 259–268.

Langenheim, R. L., Jr., and Nelson, J. W., 1992, The cyclothemic concept in the Illinois Basin: A review, in Dott, R. H., Jr., ed., Eustasy: the ups and downs of a major geological concept: Geological Society of America Memoir 180, pp. 55–71.

Lapworth, C., 1879, On the tripartite classification of the Lower Palaeozoic rocks: Geological Magazine, new series, vol. 6, pp. 1–15.

Larson, R. L., 1991, Geological consequences of superplumes: Geology, 19, pp. 963–966.

Larson, R. L., and Pitman, W. C., III, 1972, World-wide correlation of Mesozoic magnetic anomalies and its implications: Geological Society of America Bulletin, 83, pp. 3645–3662.

Laskar, J., 1989, A numerical experiment on the chaotic behaviour of the Solar System: Nature, 338, pp. 237–238.

Laskar, J., 1999, The limits of Earth orbital calculations for geological time-scale use: Philosophical Transactions of the Royal Society, London, Series A, v. 357, pp. 1735–1759.

Latour, B., 1987, Science in Action: How to Follow Scientists and Engineers through Society: Open University Press, Milton Keynes, UK.

Law, J., 1980, Fragmentation and investment in sedimentology: Social Studies of Science, 10, pp. 1–22.

Law, J., and Lodge, P., 1984, Science for Social Scientists: MacMillan Press, London.

Lawton, T. F., 1986a, Compositional trends within a clastic wedge adjacent to a fold-thrust belt: Indianola Group, central Utah, U.S.A., in Allen, P. A., and Homewood, P., eds., Foreland basins: International Association of Sedimentologists Special Publication 8, pp. 411–423.

Lawton, T. F., 1986b, Fluvial systems of the Upper Cretaceous Mesaverde Group and Paleocene North Horn Formation, central Utah: a record of transition from thin-skinned to thick-skinned deformation in the foreland region, in Peterson, J. A., ed., Paleotectonics and sedimentation in the Rocky Mountain region, United States: American Association of Petroleum Geologists Memoir 41, pp. 423–442.

Lawver, L. A., Grantz, A., and Gahagan, L. M., 2002, Plate kinematic evolution of the present Arctic region since the Ordovician: in Miller, E. L., Grantz, A., and Klemperer, S. L., eds., Tectonic evolution of the Bering Shelf-Chukchi Sea-Arctic margin and adjacent landmasses: Geological Society of America Special Paper 360, pp. 333–358.

Leckie, D. A., 1986, Rates, controls, and sand-body geometries of transgressive-regressive cycles: Cretaceous Moosebar and

Gates Formations, British Columbia: American Association of Petroleum Geologists Bulletin, 70, pp. 516–535.

Leckie, D. A., 1994, Canterbury Plains, New Zealand—implications for sequence stratigraphic models: American Association of Petroleum Geologists Bulletin, 78, pp. 1240–1256.

Leckie, D. A., and Krystinik, L. F., 1993, Sequence stratigraphy: fact, fantasy, or work in progress (?): Canadian Society of Petroleum Geologists, Reservoir, 20, No. 8, pp. 2–3.

Leeder, M. R., 1988, Recent developments in Carboniferous geology: a critical review with implications for the British Isles and N.W. Europe: Proceedings of the Geologists' Association, 99, pp. 73–100.

Legarreta, L., and Uliana, M., 1991, Jurassic-Cretaceous marine oscillations and geometry of back-arc basin fill, central Argentine Andes, in Macdonald, D. I. M., ed., 1991, Sedimentation, tectonics and eustasy: sea-level changes at active margins: International Association of Sedimentologists Special Publication 12, pp. 429–450.

Leggett, J. K., McKerrow, W. S., Cocks, L. R. M., and Rickards, R. B., 1981, Periodicity in the Lower Paleozoic marine realm: Journal of the Geological Society, London, 138, pp. 167–176.

Leithold, E. L., 1994, Stratigraphical architecture at the muddy margin of the Cretaceous Western Interior Seaway, southern Utah: Sedimentology, 41, pp. 521–542.

Levorsen, A. I., 1943, Discovery thinking: American Association of Petroleum Geologists Bulletin, 27, pp. 887–928.

Lincoln, J. M., and Schlanger, S. O., 1991, Atoll stratigraphy as a record of sea level change: problems and prospects: Journal of Geophysical Research, 96B, pp. 6727–6752.

Linsley, P. N., Potter, H. C., McNab, G., and Racher, D., 1979, Beatrice field, Moray Firth, North Sea (abs.): American Association of Petroleum Geologists Bulletin, 63, p. 487.

Liu, S. F., Nummedal, D., Yin, P. G., and Luo, H. J., 2005, Linkage of Sevier thrusting episodes and Late Cretaceous foreland basin megasequences across southern Wyoming (USA): Basin Research, 17, pp. 487–506.

Loucks, R. G., and Sarg, J. F., eds., 1993, Carbonate sequence stratigraphy: American Association of Petroleum Geologists Memoir 57, 545p.

Loup, B., and Wildi, W., 1994, Subsidence analysis in the Paris Basin: a key to Northwest European intracontinental basins: Basin Research, 6, pp. 159–177.

Loutit, T. S., Hardenbol, J., Vail, P. R., and Baum, G. R., 1988, Condensed sections: the key to age dating and correlation of continental margin sequences, in Wilgus, C. K., Hastings, B. S., Kendall, C. G. St. C., Posamentier, H. W., Ross, C. A., and Van Wagoner, J. C., eds., Sea-level Changes: an integrated approach: Society of Economic Paleontologists and Mineralogists Special Publication 42, pp. 183–213.

Lumsden, D. N., 1985, Secular variations in dolomite abundance in deep marine sediments: Geology, 13, pp. 766–769.

Lyell, C., 1830–1833, Principles of Geology, 3 vols.: John Murray, London (reprinted by Johnson Reprint Corp., New York, 1969).

Macdonald, D. I. M., ed., 1991, Sedimentation, tectonics and eustasy: sea-level changes at active margins: International Association of Sedimentologists Special Publication 12, 518p.

Mackenzie, F. T., and Pigott, J. D., 1981, Tectonic controls of Phanerozoic sedimentary rock cycling: Journal of the Geological Society, London, 138, pp. 183–196.

MacLaren, C., 1842, Art. XVI—The glacial theory of Prof. Agassiz: Reprinted in American Journal of Science, 42, pp. 346–365.

MacLeod, N., and Keller, G., 1991, How complete are Cretaceous/Tertiary boundary sections? A chronostratigraphic estimate based on graphic correlation: Geological Society of America Bulletin, 103, pp. 1439–1457.

Mandelbrot, B. B., 1983, The fractal geometry of nature: Freeman, New York, 468p.

Mann, K. O., and Lane, H. R., eds., 1995, Graphic correlation: Society for Sedimentary Geology (SEPM), Special Publication 53, 263p.

Markwick, P. J., and Rowley, D. B., 1998, The geologic evidence for Triassic to Pleistocene glaciations: implications for eustasy, in Pindell, J. L., and Drake, C. L., eds., Paleogeographic evolution and non-glacial eustasy, Northern South America: Society for Sedimentary Geology (SEPM) Special Publication 58, pp. 17–43.

Martinsen, O. J., 1993, Namurian (Late Carboniferous) depositional systems of the Craven-Askrigg area, northern England: implications for sequence-stratigraphic models, in Posamentier, H. W., Summerhayes, C. P., Haq, B. U., and Allen, G. P., eds., Sequence stratigraphy and facies associations: International Association of Sedimentologists Special Publication 18, pp. 247–281.

Martinsen, O. J., and Helland-Hansen, W., 1994, Sequence stratigraphy and facies model of an incised valley fill: the Gironde Estuary, France—discussion: Journal of Sedimentary Research, B64, pp. 78–80.

Martinsen, O. J., Martinsen, R. S., and Steidtmann, J. R., 1993, Mesaverde Group (Upper Cretaceous), southeastern Wyoming: allostratigraphy versus sequence stratigraphy in a tectonically active area: American Association of Petroleum Geologists Bulletin, 77, pp. 1351–1373.

Marwick, P. J., and Rowley, D. B., 1998, The geological evidence for Triassic to Pleistocene glaciation: implications for eustasy, in Pindell, J. L., and Drake, C. L., eds., Paleogeographic evolution and non-glacial eustasy, northern South America: Society for Sedimentary Geology (SEPM) Special Publication 58, pp. 17–43.

Mascle, A., Puigdefàbregas, C., Luterbacher, H. P., and Fernàndez, M., eds., 1998, Cenozoic foreland basins of western Europe: Geological Society, London, Special Publication 134, 427p.

Masuda, F., 1994, Onlap and downlap patterns in Plio-Pleistocene forearc and backarc basin-fill successions, Japan: Sedimentary Geology, 93, pp. 237–246.

Matthews, R. K., 1984, Oxygen-isotope record of ice-volume history: 100 million years of glacio-isostatic sea-level fluctuation, in Schlee, J. S., ed., Interregional unconformities and hydrocarbon accumulation: American Association of Petroleum Geologists Memoir 36, pp. 97–107.

Matthews, R. K., 1988, Sea level history: Science, 241, pp. 597–599.

Matthews, R. K., and Frohlich, C., 1991, Orbital forcing of low-frequency glacioeustasy: Journal of Geophysical Research, 96 (B), pp. 6797–6803.

Matthews, R. K., and Frohlich, C., 2002, Maximum flooding surfaces and sequence boundaries: comparisons between observations and orbital forcing in the Cretaceous and Jurassic (65–190 Ma): GeoArabia, 7, pp. 503–538.

Matthews, R. K., and Poore, R. Z., 1980, Tertiary $\delta^{18}O$ record and glacioeustatic sea-level fluctuations: Geology, 8, pp. 501–504.

Matthews, S. C., and Cowie, J. W., 1979, Early Cambrian transgression: Journal of the Geological Society, London, 136, pp. 133–135.

Mattioli, E., Pittet, B., Palliani, R. B., Röhl, H-J., Schmid-Röhl, A., and Morettini, E., 2004, Phytoplankton evidence for the timing and correlation of palaeoceanographical changes during the early Toarcian oceanic anoxic events (Early Jurassic): Journal of the Geological Society, London, 161, pp. 685–693.

Maxwell, J. C., 1984, What is the lithosphere? Eos, 65, pp. 321–325.

May, J. A., and Warme, J. E., 1987, Synchronous depositional phases in west coast basins: eustasy or regional tectonics, in Ingersoll, R. V., and Ernst, W. G., eds., Cenozoic basin development of coastal California: Prentice-Hall Inc., Englewood Cliffs, New Jersey, pp. 25–46.

Mayer, L., 1987, Subsidence analysis of the Los Angeles Basin, in Ingersoll, R. V., and Ernst, W. G., eds., Cenozoic basin development of coastal California, Rubey Volume VI: Prentice-Hall Inc., Englewood Cliffs, New Jersey, pp. 299–320.

McArthur, J. M., and Howarth, R. J., 2004, Strontium isotope stratigraphy, in Gradstein, F. M., Ogg, J. G., and Smith, A. G., eds., A geologic time scale: Cambridge University Press, Cambridge, pp. 96–105.

McArthur, J. M., Thirlwall, M. F., Chen, M., Gale, A. S, and Kennedy, W. J., 1993, Strontium isotope stratigraphy in the Late Cretaceous: numerical calibration of the Sr isotope curve and intercontinental correlation for the Campanian: Paleoceanography, 8, pp. 859–873.

McCarthy, P. J., 2002, Micromorphology and development of interfluve paleosols: a case study from the Cenomanian Dunvegan Formation, NE British Columbia, Canada: Bulletin of Canadian Petroleum Geology, v. 50, pp. 158–177.

McCarthy, P. J., Faccini, U. F., and Plint, A. G., 1999, Evolution of an ancient coastal plain: palaeosols, interfluves and alluvial architecture in a sequence stratigraphic framework, Cenomanian Dunvegan Formation, NE British Columbia, Canada: Sedimentology, 46, pp. 861–891.

McCrossan, R. G., and Glaister, R. P., eds., 1964, Geological history of Western Canada: Alberta Society of Petroleum Geologists, 232p.

McDonough, K. J., and Cross, T. A., 1991, Late Cretaceous sea level from a paleoshoreline: Journal of Geophysical Research, 96B, pp. 6591–6607.

McGinnis, J. P., Driscoll, N. W., Karner, G. D., Brumbaugh, W. D., and Cameron, N., 1993, Flexural response of passive margins to deep-sea erosion and slope retreat: implications for relative sea-level change: Geology, 21, pp. 893–896.

McGowran, B., 2005, Biostratigraphy: Microfossils and geological time: Cambridge University Press, Cambridge, 459p.

McHugh, P., 1968, Defining the situation: Boobs-Merrill, Indianapolis.

McKenzie, D. P., 1978, Some remarks on the development of sedimentary basins: Earth and Planetary Science Letters, 40, pp. 25–32.

McKenzie, D. P., and Sclater, J. G., 1971, The evolution of the Indian Ocean since the Late Cretaceous: Geophysical Journal of the Royal Astronomical Society, 25, pp. 437–528.

McKerrow, W. S., 1979, Ordovician and Silurian changes in sea level: Journal of the Geological Society, London, 136, pp. 137–146.

McKinney, M. L., 1986, Biostratigraphic gap analysis: Geology, 14, pp. 36–38.

McLaren, D. J., 1970, Presidential address: time, life and boundaries: Journal of Paleontology, 44, pp. 801–813.

McLaren, D. J., 1973, The Silurian-Devonian boundary: Geological Magazine, 110, pp. 302–303.

McLaren, D. J., 1977, The Silurian-Devonian Boundary: Committee: a final report, in Martinsson, A., et al., The Silurian-Devonian boundary: International Union of Geological Sciences, pp. 1–34.

McLaren, D. J., 1978, Dating and correlation: A review, in Cohee, G. V., Glaessner, M. F., and Hedberg, H. D., eds., Contributions to the geologic time scale: American Association of Petroleum Geologists, Studies in Geology 6, pp. 1–7.

McLuhan, M., 1962, The Gutenberg Galaxy: University of Toronto Press, Toronto, Ontario, Canada.

McMillan, N. J., 1973, Shelves of Labrador Sea and Baffin Bay, Canada, in The Future Petroleum Provinces of Canada; Their Geology and Potential: Canadian Society of Petroleum Memoir 1, pp. 473–517.

McMillen, K. M., and Winn, R. D., Jr., 1991, Seismic facies of shelf, slope, and submarine fan environments of the Lewis Shale, Upper Cretaceous, Wyoming, in Weimer, P., and Link, M. H., eds., Seismic facies and sedimentary processes of submarine fans and turbidite systems: Springer-Verlag, New York, pp. 273–287.

McShea, D. W., and Raup, D. M., 1986, Completeness of the geological record: Journal of Geology, 94, pp. 569–574.

Melnyk, D. H., Smith, D. G., and Amiri-Garroussi, K., 1994, Filtering and frequency mapping as tools in subsurface cyclostratigraphy, with examples from the Wessex Basin, UK, in de Boer, P. L., and Smith, D. G., eds., Orbital forcing and cyclic sequences: International Association of Sedimentologists Special Publication 19, pp. 35–46.

Merriam, D. F., ed., 1964, Symposium on cyclic sedimentation: Kansas Geological Survey Bulletin 169, 636p.

Meyers, S. R., 2008, resolving Milakovitch controversies: the Latemar Limestone and the Eocene Green River Formation: Geology, 36, pp. 319–322.

Meyers, S. R., and Sageman, B. B., 2004, detection, quantification, and significance of hiatuses in pelagic and hemipelagic strata: Earth and Planetary Science Letters, 224, pp. 55–72.

Miall, A. D., 1978, Tectonic setting and syndepositional deformation of molasse and other nonmarine-paralic sedimentary basins: Canadian Journal of Earth Sciences, 15, pp. 1613–1632.

Miall, A. D., 1981, Alluvial sedimentary basins: tectonic setting and basin architecture, in Miall, A. D., ed., Sedimentation and tectonics in alluvial basins: Geological Association of Canada Special Paper 23, pp. 1–33.

Miall, A. D., 1984, Principles of sedimentary basin analysis: Springer-Verlag Inc., New York, 490p.

Miall, A. D., 1986, Eustatic sea-level change interpreted from seismic stratigraphy: a critique of the methodology with particular reference to the North Sea Jurassic record: American Association of Petroleum Geologists Bulletin, 70, pp. 131–137.

Miall, A. D., 1988, Reservoir heterogeneities in fluvial sandstones: lessons from outcrop studies: American Association of Petroleum Geologists Bulletin, 72, pp. 682–697.

Miall, A. D., 1991a, Stratigraphic sequences and their chronostratigraphic correlation: Journal of Sedimentary Petrology, 61, pp. 497–505.

Miall, A. D., 1991b, Hierarchies of architectural units in terrigenous clastic rocks, and their relationship to sedimentation rate, in Miall, A. D., and Tyler, N., eds., The three-dimensional facies architecture of terrigenous clastic sediments and its implications for hydrocarbon discovery and recovery: Society of Economic Paleontologists and Mineralogists, Concepts in Sedimentology and Paleontology, 3, pp. 6–12.

Miall, A. D., 1992, The Exxon global cycle chart: an event for every occasion? Geology, 20, pp. 787–790.

Miall, A. D., 1994, Sequence stratigraphy and chronostratigraphy: problems of definition and precision in correlation, and their implications for global eustasy: Geoscience Canada, 21, pp. 1–26.

Miall, A. D., 1995a, Whither stratigraphy? Sedimentary Geology, 100, pp. 5–20.

Miall, A. D., 1995b, Chapter 11: Collision-related foreland basins, in Busby, C. J., and Ingersoll, R. V., eds., Tectonics of sedimentary basins: Blackwell Science, Oxford, pp. 393–424.

Miall, A. D., 1996, The geology of fluvial deposits: sedimentary facies, basin analysis and petroleum geology: Springer-Verlag Inc., Heidelberg, 582p.

Miall, A. D., 1997, The geology of stratigraphic sequences, First edition: Springer-Verlag, Berlin, 433p.

Miall, A. D., 1999, Principles of sedimentary basin analysis, Third edition: Springer-Verlag Inc., New York, 616p.

Miall, A. D., 2004, Empiricism and model building in stratigraphy: the historical roots of present-day practices. Stratigraphy: American Museum of Natural History, 1, pp. 3–25.

Miall, A. D., 2008, Postscript: what have we learned, and where do we go from here?, in Miall, A. D., ed., The Sedimentary Basins of the United States and Canada: Sedimentary basins of the World, 5, K. J. Hsü, Series Editor, Elsevier Science, Amsterdam, pp. 573–591.

Miall, A. D., and Arush, M., 2001a, The Castlegate Sandstone of the Book Cliffs, Utah: sequence stratigraphy, paleogeography, and tectonic controls: Journal of Sedimentary Research, 71, pp. 536–547.

Miall, A. D, and Arush, M., 2001b, Cryptic sequence boundaries in braided fluvial successions: Sedimentology, 48, pp. 971–985.

Miall, A. D., and Blakey, R. C., 2008, The Phanerozoic tectonic and sedimentary evolution of North America, in Miall, A. D., ed., The Sedimentary Basins of the United States and Canada: Sedimentary basins of the World, vol. 5, K. J. Hsü, Series Editor, Elsevier Science, Amsterdam, pp. 1–29.

Miall, A. D., Catuneanu, O., Vakarelov, B., and Post, R., 2008, The Western Interior Basin, in Miall, A. D.,

ed., The Sedimentary Basins of the United States and Canada: Sedimentary basins of the World, vol. 5, K. J. Hsü, Series Editor, Elsevier Science, Amsterdam, pp. 329–362.

Miall, A. D., and Miall, C. E., 2001, Sequence stratigraphy as a scientific enterprise: the evolution and persistence of conflicting paradigms: Earth Science Reviews, 54, #4, pp. 321–348.

Miall, A. D., and Miall, C. E., 2004, Empiricism and Model-Building in stratigraphy: Around the Hermeneutic Circle in the Pursuit of Stratigraphic Correlation. Stratigraphy: American Museum of Natural History, 1, pp. 27–46.

Miall, C. E., and Miall, A. D., 2002, The Exxon factor: the roles of academic and corporate science in the emergence and legitimation of a new global model of sequence stratigraphy: Sociological Quarterly, 43, pp. 307–334.

Milankovitch, M., 1930, Mathematische klimalehre und astronomische theorie der klimaschwankungen, in Koppen, W., and Geiger, R., eds., Handbuch der klimatologie, I (A): Gebruder Borntraeger, Berlin.

Milankovitch, M., 1941, Kanon der Erdbestrahlung und seine Anwendung auf das Eiszeitenproblem: Akad. Royale Serbe, 133, 633p.

Millendorf, S. A., and Heffner, T., 1978, FORTRAN program for lateral tracing of time-stratigraphic units based on faunal assemblage zones: Computers and Geoscience, 4, pp. 313–318.

Miller, F. X., 1977, The graphic correlation method in biostratigraphy, in Kauffman, E. G., and Hazel, J. E., eds., Concepts and methods in biostratigraphy: Dowden, Hutchinson and Ross, Inc., Stroudsburg, Pennsylvania, pp. 165–186.

Miller, K. B., McCahon, T. J., and West, R. R., 1996, Lower Permian (Wolfcampian) paleosol-bearing cycles of the U.S. midcontinent: evidence of climatic cyclicity: Journal of Sedimentary Research, 66, pp. 71–84.

Miller, K. G., 1990, Recent advances in Cenozoic marine stratigraphic resolution: Palaios, 5, pp. 301–302.

Miller, K. G., 1994, The rise and fall of sea level studies: are we at a stillstand? Paleoceanography, 9, pp. 183–184.

Miller, K. G., and Kent, D. V., 1987, Testing Cenozoic eustatic changes: the critical role of stratigraphic resolution, in Ross, C. A., and Haman, D., eds., Timing and depositional history of eustatic sequences: constraints on seismic stratigraphy: Cushman Foundation for Foraminiferal Research, Special Publication 24, pp. 51–56.

Miller, K. G., Mountain, G. S., Browning, J. V., Kominz, M., Sugarman, P. J., Christi-Blick, N., Katz, M. E., and Wright, J. D., 1998, Cenozoic global sea level, sequences, and the New Jersey transect: results from coastal plain and continental slope drilling: Reviews of Geophysics, 36, pp. 569–601.

Miller, K. G., Barrera, E., Olsson, R. K., Sugarman, P. J., and Savin, S. S., 1999, Does ice drive early Maastrichtian eustasy? Geology, 27, pp. 783–786.

Miller, K. G., Sugarman, P. J., Browning, J. V., Kominz, M. A., Olsson, R. K., Feigenson, M. D., and Hernández, J. C., 2004, Upper Cretaceous sequences and sea-level history, New Jersey Coastal Plain: Geological Society of America Bulletin, 116, pp. 368–393.

Miller, K. G., Kominz, M. A., Browning, J. V., Wright, J. D., Mountain, G. S., Katz, M. E., Sugarman, P. J., Cramer, B. S., Christie-Blick, N., and Pekar, S. F., 2005a, The

Phanerozoic record of global sea-level change: Science, 310, pp. 1293–1298.

Miller, K. G., Wright, J. D., and Browning, J. V., 2005b, Visions of ice sheets in a greenhouse world: Marine Geology, 217, pp. 215–231.

Miller, K. G., Browning, J. V., Aubry, M.-P., Wade, B., Katz, M. E., Kulpecz, A. A., and Wright, J. D., 2008a, Eocene-Oligocene global climate and sea-level changes: St. Stephens Quarry, Alabama: Geological Society of America Bulletin, 120, pp. 34–53.

Miller, K. G., Wright, J. D., Katz, M. E., Browning, J. V., Cramer, B. S., Wade, B. S., and Mizintseva, S. F., 2008b, A view of Antarctic ice-sheet evolution from sea-level and deep-sea isotope changes during the Late Cretaceous-Cenozoic, in Cooper, A. K., Barrett, P. J., Stagg, H., Storey, B., Stump, E., Wise, W., and the 10th ISAES editorial team, eds., Antarctica: A keystone in a changing world: Proceedings of the 10th International Symposium on Antarctic Earth Sciences: The National Academies Press, Washington, DC, pp. 55–70.

Mitchum, R. M., Jr., and Uliana, M. A., 1988, Regional seismic stratigraphic analysis of Upper Jurassic-Lower Cretaceous carbonate depositional sequences, Neuquén Basin, Argentina, in Bally, A. W., ed., Atlas of seismic stratigraphy: American Association of Petroleum Geologists Studies in Geology 27, 2, pp. 206–212.

Mitchum, R. M., Jr., Vail, P. R., and Sangree, J. B., 1977a, Seismic stratigraphy and global changes of sea level, Part six: Stratigraphic interpretation of seismic reflection patterns in depositional sequences, in Payton, C. E., ed., Seismic stratigraphy—applications to hydrocarbon exploration; American Association of Petroleum Geologists Memoir 26, pp. 117–133.

Mitchum, R. M., Jr., Vail, P. R., and Thompson, S. III, 1977b, Seismic stratigraphy and global changes of sea level, Part 2, The depositional sequence as a basic unit for stratigraphic analysis, in Payton, C. E., ed., Seismic stratigraphy—applications to hydrocarbon exploration: American Association of Petroleum Geologists Memoir 26, pp. 53–62.

Mitchum, R. M., Jr., and Van Wagoner, J. C., 1991, High-frequency sequences and their stacking patterns: sequence-stratigraphic evidence of high-frequency eustatic cycles: Sedimentary Geology, 70, pp. 131–160.

Mitrovica, J. X., and Jarvis, G. T., 1985, Surface deflections due to transient subduction in a convecting mantle: Tectonophysics, 120, pp. 211–237.

Mitrovica, J. X., Beaumont, C., and Jarvis, G. T., 1989, Tilting of continental interiors by the dynamical effects of subduction: Tectonics, 8, pp. 1079–1094.

Molenaar, C. M., 1983, Principle reference section and correlation of Gallup Sandstone, northwestern New Mexico, in Hook, S. C., compiler, Contributions to mid-Cretaceous paleontology and stratigraphy of New Mexico – Part II: New Mexico Bureau of Mines and Mineral Resources Circular 185, pp. 29–40.

Molenaar, C. M., and Rice, D. D., 1988, Cretaceous rocks of the Western Interior Basin, in Sloss, L. L., ed., Sedimentary cover-North American Craton: U.S., The Geology of North America: Boulder, Colorado, Geological Society of America, D-2, pp. 77–82.

Molnar, P., and Tapponnier, P., 1975, Cenozoic tectonics of Asia: effects of a continental collision: Science, 189, pp. 419–426.

Montañez, I. P., and Read, J. F., 1992, Eustatic control on early dolomitization of cyclic peritidal carbonates: evidence from the early Ordovician Upper Knox Group, Appalachians: Geological Society of America Bulletin, 104, pp. 872–886.

Monty, C. L. V., 1968, D'Orbigny's concepts of stage and zone: Journal of Paleontology, 42, pp. 689–701.

Moore, R. C., 1936, Stratigraphic classification of the Pennsylvanian rocks of Kansas: Kansas Geological Survey Bulletin 22, p. 256.

Moore, R. C., 1964, Paleoecological aspects of Kansas Pennsylvanian and Permian cyclothems, in Merriam, D. F., ed., Symposium on cyclic sedimentation: Kansas Geological Survey Bulletin 169, pp. 287–380.

Moore, T. C., Jr., Pisias, N. G., and Dunn, D. A., 1982, Carbonate time series of the Quaternary and late Miocene sediments in the Pacific Ocean: a spectral comparison: Marine Geology, 46, pp. 217–233.

Moore, T. C., Jr., and Romine, K., 1981, In search of biostratigraphic resolution, in Warme, J. E., Douglas, R. G., and Winterer, E. L., eds., The Deep Sea Drilling Project: A decade of progress: Society of Economic Paleontologists and Mineralogists Special Publication 32, pp. 317–334.

Moran, K., Backman, J., Brinkhuis, H., Clemens, S., Cronin, T., Dickens, G., Eynaud, F., Gattacceca, J., Jakobsson, M., Jordan, R., Kaminski, M., King, J., Koc, N., Krylov, A., Martinez, N., Matthiessen, J., Moore, T., Onodera, J., O'Regan, M., Pälike, H., Rea, B., Rio, D., Sakamoto, T., Smith, D., Stein, R., St. John, K., Suto, I., Suzuki, N., Takahashi, K., Watanabe, M., Yamamoto, M., Frank, M., Kubik, P., Jokat, W., Kristoffersen, Y., McInroy, D., Farrell, J., 2006, The Cenozoic palaeoenvironment of the Arctic Ocean, Nature, 441, pp. 601–605.

Mörner, N.-A., 1994, Internal response to orbital forcing and external cyclic sedimentary sequences, in de Boer, P. L., and Smith, D. G., eds., Orbital forcing and cyclic sequences: International Association of Sedimentologists Special Publication 19, pp. 25–33.

Morton, R. A., and Price, W. A., 1987, late Quaternary sea-level fluctuations and sedimentary phases of the Texas coastal plain and shelf, in Nummedal, D., Pilkey, O. H., and Howard, J. D., eds., 1987, Sea-level fluctuation and coastal evolution: Society of Economic Paleontologists and Mineralogists Special Publication 41, pp. 181–198.

Mossop, G. D., and Shetsen, I., compilers, 1994, Geological Atlas of the Western Canada Sedimentary Basin: Canadian Society of Petroleum Geologists, 510p.

Moxon, I. W., and Graham, S. A., 1987, History and controls of subsidence in the Late Cretaceous-Tertiary Great Valley forearc basin, California: Geology, 15, pp. 626–629.

Mulkay, M., 1979, Science and the Sociology of Knowledge: George Allen and Unwin, London.

Müller, D., Sdrolias, M., Gaina, C., Steinberger, B. and Heine, C., 2008, Long-term sea level fluctuations driven by ocean basin dynamics. Science, 319, pp. 1357–1362.

Mundil, R., Zühlke, R., Bechstädt, T., Peterhänsel, A., Egenhoff, S. O., Oberli, F., Meier, M., Brack, P., and Reiber, H., 2003, Cyclicities in Triassic platform carbonates: synchronizing radio-isotopic and orbital clocks: Terra Nova, 15, pp. 81–87.

Murphy, M. A., 1977, On time-stratigraphic units: Journal of Paleontology, 51, pp. 213–219.

Murphy, M. A., 1988, Unconformity-bounded stratigraphic units: Discussion: Geological Society of America Bulletin, 100, pp. 155.

Murray, N., and Holman, M., 1999, The origin of chaos in the outer solar system: Science, 283, Issue 5409 (March 19), pp. 1877–1881.

Mutti, E., and Normark, W. R., 1987, Comparing examples of modern and ancient turbidite systems: problems and concepts; in Leggett, J. K., and Zuffa, G. G., eds., Marine clastic sedimentology: concepts and case studies: Graham and Trotman, London, pp. 1–38.

Naish, T. R., Field, B. D., Melhuish, A., Carter, R. M., Abbott, S. T., Edwards, S., Alloway, B. V., Wilson, G. S., Niessen, F., Barker, A., Browne, G. H., and Maslen, G., 2005, Integrated outcrop, drill core, borehole and seismic stratigraphic architecture of a cyclothemic, shallow-marine depositional system, Wanganui Basin, New Zealand, Journal of the Royal Society of New Zealand, 35, pp. 91–122.

Naish, T. R., and Kamp, P. J. J., 1995, Pliocene-Pleistocene marine cyclothems, Wanganui Basin, New Zealand: a lithostratigraphic framework: New Zealand Journal of Geology and Geophysics, 38, pp. 223–243.

Naish, T. R., and Kamp, P. J. J., 1997, Sequence stratigraphy of sixth-order (41 k.y.) Pliocene-Pleistocene cyclothems, Wanganui basin, new Zealand: a case for the regressive systems tract: Geological Society of America Bulletin, 109, pp. 978–999.

Naylor, M., and Sinclair, H. D., 2007, Punctuated thrust deformation in the context of doubly vergent thrust wedges: Implications for the localization of uplift and exhumation: Geology, 35, pp. 559–562.

Newell, N. D., 1962, Paleontological gaps and geochronology: Journal of Paleontology, 36, pp. 592–610.

Nichols, G., 1999, Sedimentology and Stratigraphy. Blackwell Science: Oxford, 355p.

Nielsen, S., Stephenson, R., and Thomsen, E., 2007, Dynamics of mid-Palaeocene north Atlantic rifting linked with European intra-plate deformations: Nature, 450, pp. 1071–1074.

Nijman, W., 1998, Cyclicity and basin axis shift in a piggyback basin: toward modeling of the Eocene Tremp-Ager basin, south Pyrenees, Spain, in Mascle, A., Puigdefàbregas, C., Luterbacher, H. P., and Fernàndez, M., eds., Cenozoic foreland basins of western Europe: Geological Society, London, Special Publication 134, pp. 135–162.

Nio, S. D., and Yang, C. S., 1991, Sea-level fluctuations and the geometric variability of tide-dominated sandbodies: Sedimentary Geology, 70, pp. 161–193.

North American Commission on Stratigraphic Nomenclature, 1983, North American Stratigraphic Code: American Association of Petroleum Geologists Bulletin, 67, pp. 841–875.

Nummedal, D., 1987, Preface, in Nummedal, D., Pilkey, O. H., and Howard, J. D., eds., Sea-level fluctuation and coastal evolution: Society of Economic Paleontologists and Mineralogists Special Publication 41, pp. iii–iv.

Nummedal, D., 1990, Sequence stratigraphic analysis of Upper Turonian and Coniacian strata in the San Juan Basin, New Mexico, USA, in Ginsburg, R. N., and Beaudoin, B., eds., Cretaceous resources, events and rhythms: Background and plans for research: Kluwer Academic Publishers, Dordrecht, p. 33–46.

Nummedal, D., and Molenaar, C. M., 1995, Sequence stratigraphy of ramp-type setting strand plain succession: the Gallup Sandstone, New Mexico, in Van Wagoner, J. C., and Bertram, G. T., eds., Sequence stratigraphy of foreland basins: American Association of Petroleum Geologists Memoir 64, pp. 277–310.

Nummedal, D., Pilkey, O. H., and Howard, J. D., eds., 1987, Sea-level fluctuation and coastal evolution: Society of Economic Paleontologists and Mineralogists Special Publication 41, 267p.

Nummedal, D., and Swift, D. J. P., 1987, Transgressive stratigraphy at sequence-bounding unconformities: some principles derived from Holocene and Cretaceous examples, in Nummedal, D., Pilkey, O. H., and Howard, J. D., eds., Sea-level fluctuation and coastal evolution; Society of Economic Paleontologists and Mineralogists Special Publication 41, pp. 241–260.

Nummedal, D., Wright, R., Swift, D. J. P., Tillman, R. W., and Wolter, N. R., 1989, Depositional systems architecture of shallow marine sequences, in Nummedal, D., and Wright, R., eds., Cretaceous shelf sandstones and shelf depositional sequences, Western Interior Basin, Utah, Colorado and New Mexico: Field Trip Guidebook T119, 28th International Geological Congress, American Geophysical Union, pp. 35–79.

Nystuen, J. P., 1998, History and development of sequence stratigraphy, in Gradstein, F. M., Sandvik, K. O., and Milton, N. J., eds., Sequence stratigraphy: concepts and applications: Norwegian Petroleum Society Special Publication 8, Elsevier, Amsterdam, pp. 31–116.

Oldale. H. S., and Munday, R. J., 1994, Devonian Beaverhill Lake Group of the Western Canada Sedimentary Basin, in Mossop, G. D., and Shetsen, I., compilers, Geological Atlas of the Western Canada Sedimentary Basin: Canadian Society of Petroleum Geologists, pp. 149–163.

Olsen, P. E., 1984, Periodicity of lake-level cycles in the Late Triassic Lockatong Formation of the Newark Basin (Newark Supergroup, New Jersey and Pennsylvania), in Berger, A., Imbrie, J., Hays, J., Kukla, G., and Saltzman, B., eds., Milankovitch and climate: NATO ASI Series, D. Reidel Publishing Company, Dordrecht, Part 1, pp. 129–146.

Olsen, P. E., 1986, A 40-million year lake record of Early Mesozoic orbital climatic forcing: Science, 234, pp. 842–848.

Olsen, P. E., 1990, Tectonic, climatic, and biotic modulation of lacustrine ecosystems—examples from Newark Supergroup of eastern North America, in Katz, B. J., ed., Lacustrine basin exploration: case studies and modern analogs: American Association of Petroleum Geologists Memoir 50, pp. 209–224.

Olsen, T., Steel, R. J., Høgseth, K., Skar, T., and Røe, S.-L., 1995, Sequential architecture in a fluvial succession: sequence stratigraphy in the Upper Cretaceous Mesaverde Group, Price Canyon, Utah: Journal of Sedimentary Research, B65, pp. 265–280.

Oppel, A., 1856–1858, Die Juraformation Englands, Frankreichs und des südwestlichen Deutschlands: Württemb. Naturwiss. Verein Jahresh., vol. xii–xiv (pp. 1–438, 1856; pp. 439–694, 1857; pp. 695–857, 1858), Stuttgart.

Orbigny, A. d', 1842–1843, Paléontologie Française. Description zoologique et géologique des tous les animaux mollusques et rayonnés fossiles de France. 2 (Gastropoda). Paris: Victor Masson, 456p.

Orbigny, A. d', 1849–1852, Cours élémentaire de paléontologie et de géologie stratigraphique: Victor Masson: Paris.

Oreskes, N., Shrader-Frechette, K., and Belitz, K., 1994, Verification, validation, and confirmation of numerical models in the earth sciences: Science, 263, pp. 641–646.

Osleger, D., and Read, J. F., 1991, Relation of eustasy to stacking patterns of meter-scale carbonate cycles, Late Cambrian, U.S.A.: Journal of Sedimentary Petrology, 61, pp. 1225–1252.

Osleger, D. A., and Read, J. F., 1993, Comparative analysis of methods used to define eustatic variations in outcrop: Late Cambrian interbasinal sequence development: American Journal of Science, 293, pp. 157–216.

Pang, M., and Nummedal, D., 1995, Flexural subsidence and basement tectonics of the Cretaceous Western Interior Basin, United States: Geology, 23, pp. 173–176.

Paola, C., 2000, Quantitative models of sedimentary basin filling: Sedimentology, 47 (supplement 1), pp. 121–178.

Parente, M., Frijia, G., Di Lucia, M., Jenkyns, H. C., Woodfine, R. G., and Baroncini, F., 2008, Stepwise extinction of larger foraminifers at the Cenomanian-Turonian boundary: a shallow-water perspective on nutrient fluctuations during Oceanic Anoxic Event 2 (Bonarelli Event): Geology, 2008, pp. 715–718.

Parkinson, N., and Summerhayes, C., 1985, Synchronous global sequence boundaries: American Association of Petroleum Geologists Bulletin, 69, pp. 685–687.

Pashin, J. C., 1994, Flexurally-influenced eustatic cycles in the Pottsville Formation (Lower Pennsylvanian), Black Warrior Basin, Alabama, in Dennison, J. M., and Ettensohn, F. R., eds., Tectonic and eustatic controls on sedimentary cycles: Society for Sedimentary Geology (SEPM), Concepts in Sedimentology and Paleontology, 4, 89–105.

Paul, C. R. C., Lamolda, M. A., Mitchell, S. F., Vaziri, M. R., Gorostidi, A. and Marshall, J. D., 1999, The Cenomanian/Turonian boundary at Eastbourne (Sussex, UK): a proposed European reference section: Palaeogeography, Paleoclimatology, Palaeoecology, 150, pp. 83–121.

Payton, C. E., ed., 1977, Seismic stratigraphy—applications to hydrocarbon exploration: American Association of Petroleum Geologists Memoir 26, 516p.

Peat, F. D., 1989, Cold fusion: the making of a scientific controversy: Contemporary Books: Chicago, 188p.

Pedley, M., and Grasso, M., 1991, Sea-level changes around the margins of the Catania-Gela Trough, and Hyblean Plateau, southeast Sicily (African-European plate convergence zone): a problem of Plio-Quaternary plate buoyancy? in Macdonald, D. I. M., ed., 1991, Sedimentation, tectonics and eustasy: sea-level changes at active margins: International Association of Sedimentologists Special Publication 12, pp. 451–464.

Peper, T., 1994, Tectonic and eustatic control on Late Albian shallowing (Viking and Paddy formations) in the Western Canada Foreland Basin: Geological Society of America Bulletin, 106, pp. 254–263.

Peper, T., Beekman, F., and Cloetingh, S., 1992, Consequences of thrusting and intraplate stress fluctuations for vertical motions in foreland basins and peripheral areas: Geophysical Journal International, 111, pp. 104–126.

Peper, T., and Cloetingh, S., 1995, Autocyclic perturbations of orbitally forced signals in the sedimentary record: Geology, 23, pp. 937–940.

Perlmutter, M. A., and Matthews, M. D., 1990, Global cyclostratigraphy—a model, in Cross, T. A., ed., Quantitative dynamic stratigraphy: Prentice Hall, Englewood Cliffs, New Jersey, pp. 233–260.

Perlmutter, M. A., and Plotnick, R. E., 2003, Hemispheric asymmetry of the marine stratigraphic record: conceptual proof of a unipolar ice cap, in: Cecil, C. B., and Edgar, N. T., eds., Climate controls on stratigraphy, Society of Economic Paleontologists and Mineralogists Special Publication 77, pp. 51–66.

Petrobras Exploration Department, 1988, Pará-Maranhâo Basin, Brazil, in Bally, A. W., ed., Atlas of seismic stratigraphy: American Association of Petroleum Geologists Studies in Geology 27, 2, pp. 179–183.

Phillips, J., 1860, Anniversary Address of the President: Geological Society of London, Quarterly Journal, 16, pp. xxxi–xlv.

Piller, W. E., Harzhauser, M., and Mandic, O., 2007, Miocene Central Paratethys stratigraphy – current status and future directions: Stratigraphy, 4, pp. 145–149.

Pindell, J. L., and Drake, C. L., eds., 1998, Paleogeographic evolution and non-glacial eustasy, Northern South America: Society for Sedimentary Geology (SEPM) Special Publication 58, 324p.

Pinous, O. V., Levchuk, M. A., and Sahagian, D. L., 2001, regional synthesis of the productive Neocomian complex of West Siberia: sequence stratigraphic framework: American Association of Petroleum Geologists Bulletin, 85, pp. 1713–1730.

Piper, D. J. W., and Normark, W. R., 2001, Sandy fans; from Amazon to Hueneme and beyond: American Association of Petroleum Geologists Bulletin, 85, pp. 1407–1438.

Pitman, W.C., III, 1978, Relationship between eustacy and stratigraphic sequences of passive margins: Geological Society of America Bulletin, 89, pp. 1389–1403.

Pitman, W C., III, 1979, The effect of eustatic sea level changes on stratigraphic sequences at Atlantic margins, in Watkins, J. S., Montadert, L., and Dickerson, P. W., eds., Geological and geophysical investigations of continental margins: American Association of Petroleum Geologists Memoir 29, pp. 453–460.

Pitman, W.C., III, and Golovchenko, X., 1988, Sea-level changes and their effect on the stratigraphy of Atlantic-type margins, in Sheridan, R. E., and Grow, J. A., eds., The Atlantic continental margin, The geology of North America, United States: Boulder, Colorado, Geological Society of America, vol. I-2, pp. 429–436.

Pitman, W. C., III, and Talwani, M., 1972, Sea-floor spreading in the North Atlantic, Geological Society of America Bulletin, 83, pp. 619–646.

Platt, J. P., 1988, The mechanics of frontal imbrication—A 1st-order analysis: Geologische Rundschau, 77, pp. 577–589.

Plint, A. G., 1988, Sharp-based shoreface sequences and "offshore bars" in the Cardium Formation of Alberta: their relationship to relative changes in sea level, in Wilgus, C. K., Hastings, B. S., Kendall, C. G. St. C., Posamentier, H. W., Ross, C. A., and Van Wagoner, J. C., eds., Sea-level Changes: an integrated approach: Society of Economic Paleontologists and Mineralogists Special Publication 42, pp. 357–370.

Plint, A. G., 1990, An allostratigraphic correlation of the Muskiki and Marshybank Formations (Coniacian-Santonian) in the foothills and subsurface of the Alberta Basin: Bulletin of Canadian Petroleum Geology, 38, pp. 288–306.

Plint, A. G., 1991, High-frequency relative sea-level oscillations in Upper Cretaceous shelf clastics of the Alberta foreland basin: possible evidence for a glacio-eustatic control? in Macdonald, D. I. M., ed., Sedimentation, tectonics and eustasy: sea-level changes at active margins: International Association of Sedimentologists Special Publication 12, pp. 409–428.

Plint, A. G., and Kreitner, M. A., 2007, Extensive thin sequences spanning Cretaceous foredeep suggest high-frequency eustatic control: Late Cenomanian Western Canada foreland basin: Geology, 35, pp. 735–738.

Plint, A. G., McCarthy, P. J., and Faccini, U. F., 2001, Nonmarine sequence stratigraphy: updip expression of sequence boundaries and systems tracts in a high-resolution framework: Cenomanian Dunvegan Formation, Alberta foreland basin, Canada: American Association of Petroleum Geologists Bulletin, 85, pp. 1967–2001.

Plint, A. G., Walker, R. G., and Bergman, K. M., 1986, Cardium Formation 6. Stratigraphic framework of the Cardium in subsurface: Bulletin of Canadian Petroleum Geology, 34, pp. 213–225.

Plint, A. G., Eyles, N., Eyles, C. H., and Walker, R. G., 1992, Control of sea level change, in Walker, R. G., and James, N. P., eds., Facies models: response to sea level change: Geological Association of Canada, pp. 15–25.

Plotnick, R. E., 1986, A fractal model for the distribution of stratigraphic hiatuses: Journal of Geology, 94, pp. 885–890.

Poag, C. W., and Sevon, W. D., 1989, A record of Appalachian denudation in postrift Mesozoic and Cenozoic sedimentary deposits of the U. S. Middle Atlantic continental margin: Geomorphology, 2, pp. 119–157.

Polletti, L., Premoli-Silva, I., Masetti, D., Pipan, M., and Claps, M., 2004, Orbitally driven fertility cycles in the Palaeocene pelagic sequences of the Southern Alps (Northern Italy): Sedimentary Geology, 164, pp. 35–54.

Pomar, L., 1993, High-resolution sequence stratigraphy in prograding Miocene carbonates: application to seismic interpretation, in Loucks, R. G., and Sarg, J. F., eds., Carbonate sequence stratigraphy: American Association of Petroleum Geologists Memoir 57, pp. 389–407.

Pomar, L., and Ward, W. C., 1999, reservoir-scale heterogeneity in depositional packages and diagenetic patterns on a reef-rimmed platform, Upper Miocene, Mallorca, Spain: American Association of Petroleum Geologists, 83, pp. 1759–1773.

Ponte, F. C., Fonseca, J. dR., and Carozzi, A. V., 1980, Petroleum habitats in the Mesozoic-Cenozoic of the continental margin of Brazil, in Miall, A. D., ed., Facts and principles of world petroleum occurrence: Canadian Society of Petroleum Geologists Memoir 6, pp. 857–886.

Popper, K. R., 1959, the logic of scientific discovery: Hutchinson, London, 479p.

Porebski, S. J., and Steel, R. J., 2003, Shelf-margin deltas: their stratigraphic significance and relation to deepwater sands: Earth-Science Reviews, 62, p. 283–326.

Porebski, S. J., and Steel, R. J., 2006, Deltas and sea-level change: Journal of Sedimentary Research, 76, pp. 390–403.

Posamentier, H. W., 2002, Ancient shelf ridges—a potentially significant component of the transgressive systems tract: case study from offshore northwest Java: American Association of Petroleum Geologists Bulletin, 86, pp. 75–106.

Posamentier, H. W., and Allen, G. P., 1993, Siliciclastic sequence stratigraphic patterns in foreland ramp-type basins: Geology, 21, pp. 455–458.

Posamentier, H. W., and Allen, G. P., 1999, Siliciclastic sequence stratigraphy—concepts and applications: Society for Sedimentary Geology (SEPM), Concepts in sedimentology and paleontology 7, 210p.

Posamentier, H. W., Allan, G. P., and James, D. P., 1992, High-resolution sequence stratigraphy – the East Coulee Delta, Alberta: Journal of Sedimentary Petrology, 62, pp. 310–317.

Posamentier, H. W., Jervey, M. T., and Vail, P. R., 1988, Eustatic controls on clastic deposition I—Conceptual framework, in Wilgus, C. K., Hastings, B. S., Kendall, C. G. St. C., Posamentier, H. W., Ross, C. A., and Van Wagoner, J. C., eds., Sea level Changes – an integrated approach: Society of Economic Paleontologists and Mineralogists Special Publication 42, pp. 109–124.

Posamentier, H. W., Summerhayes, C. P., Haq, B. U., and Allen, G. P., eds., 1993, Sequence stratigraphy and facies associations: International Association of Sedimentologists Special Publication 18, 644p.

Posamentier, H. W., and Vail, P. R., 1988, Eustatic controls on clastic deposition II—sequence and systems tract models, in Wilgus, C. K., Hastings, B. S., Kendall, C. G. St. C., Posamentier, H. W., Ross, C. A., and Van Wagoner, J. C., eds., Sea level Changes – an integrated approach: Society of Economic Paleontologists and Mineralogists Special Publication 42, pp. 125–154.

Potma, K., Wong, P. K., Weissenberger, J. A. W., and Gilhooly, M. G., 2001, Toward a sequence stratigraphic framework for the Frasnian of the Western Canada Basin, Bulletin of Canadian Petroleum Geology, 49, pp. 37–85.

Potter, P. E., 1978, Significance and origin of big rivers: Journal of Geology, 86, pp. 13–33.

Pratt, B. R., and James, N. P., 1986, The St. George Group (Lower Ordovician) of western Newfoundland: tidal flat island model for carbonate sedimentation in shallow epeiric seas: Sedimentology, 33, pp. 313–343.

Pratt, B. R., James, N. P., and Cowan, C. A., 1992, Peritidal carbonates, in Walker, R. G. and James, N. P., eds., Facies models: response to sea-level change: Geological Association of Canada, Geotext 1, pp. 303–322.

Preto, N., Hinnnov, L. A., Hardie, L. A., and De Zanche, V., 2001, Middle Triassic orbital signature recorded in the shallow-marine latemar carbonate buildup (Dolomites, Italy): Geology, 29, pp. 1123–1126.

Price, D. J. de S., 1961, Science Since Babylon: Yale University Press, New Haven, Conn.

Price, D. J. de S., 1963, Little Science, Big Science: Columbia University Press, New York.

Price, R. A., 1973, Large-scale gravitational flow of supracrustal rocks, southern Canadian Rockies, in DeJong, K. A., and

Scholten, R. A., eds., Gravity and tectonics: John Wiley, New York, pp. 491–502.

Price, R. A., Balkwill, H. R., Charlseworth, H. A. K., Cook, D. G., and Simony, P. S., 1972, The Canadian Rockies and tectonic evolution of the southeastern Canadian Cordillera, Excursion AC15, XXIV International Geological Congress, Montreal, 129p.

Prosser, S., 1993, Rift-related linked depositional systems and their seismic expression, in Williams, G. D., and Dobb, A., eds., Tectonics and seismic sequence stratigraphy: Geological Society, London, Special Publication 71, pp. 35–66.

Prothero, D. R., 2001, Magnetostratigraphic tests of sequence stratigraphic correlations from the southern California Paleogene: Journal of Sedimentary Research, 71, pp. 526–536.

Pysklywec, R. N., and Mitrovica, J. X., 1999, The Role of Subduction-Induced Subsidence in the Evolution of the Karoo Basin, Journal of Geology, 107, pp. 155–164.

Quinlan, G. M., 1987, Models of subsidence mechanisms in intracratonic basins, and their applicability to North American examples, in Beaumont, C., and Tankard, A. J., eds., Sedimentary basins and basin-forming mechanisms: Canadian Society of Petroleum Geologists Memoir 12, pp. 463–481.

Quinlan, G. M., and Beaumont, C., 1984, Appalachian thrusting, lithospheric flexure, and the Paleozoic stratigraphy of the eastern interior of North America: Canadian Journal of Earth Sciences, 21, pp. 973–996.

Pysklywec, R. N., and Mitrovica, J. X., 1998, Mantle flow mechanisms for the large scale subsidence of continental interiors. Geology, 26, pp. 687–690.

Rainbird, R. H., 1992, Anatomy of a large-scale braid-plain quartzarenite from the Neoproterozoic Shaler Group, Victoria Island, Northwest Territories, Canada: Canadian Journal of Earth Sciences, 29, pp. 2537–2550.

Rainbird, R. H., Heaman, L. M., and Young, G. M., 1992, Sampling Laurentia: detrital zircon geochronology offers evidence for an extensive Neoproterozoic river system originating from the Grenville orogen: Geology, 20, pp. 351–354.

Rainbird, R. H., McNicoll, V. J., Heaman, L. M., Abbott, J. G., Long, D. G. F., and Thorkelson, D. J., 1997, Pan-continental river system draining Grenville Orogen recorded by U-Pb and Sm-Nd geochronology of Neoproterozoic quartzarenites and mudrocks, northwestern Canada: Journal of Geology, 105, pp. 1–17.

Ramane, J., 2000, Should the golden spike glitter?—Comments to the paper of M.-P. Aubry et al.: Episodes, 23, pp. 211–213.

Ramsbottom, W. H. C., 1979, Rates of transgression and regression in the Carboniferous of NW Europe: Journal of the Geological Society, London, 136, pp. 147–153.

Rawson, P. F., and Riley, L. A., 1982, Latest Jurassic—Early Cretaceous events and the 'late Cimmerian unconformity' in the North Sea: American Association of Petroleum Geologists Bulletin, 66, pp. 2628–2648.

Ray, R. R., 1982, Seismic stratigraphic interpretation of the Fort Union Formation, western Wind River Basin: example of subtle trap exploration in a nonmarine sequence, in Halbouty, M. T., ed., The deliberate search for the subtle trap: American Association of Petroleum Geologists Memoir 32, pp. 169–180.

Read, J. F., and Goldhammer, R. K., 1988, Use of Fischer plots to define third-order sea-level curves in Ordovician peritidal cyclic carbonates, Appalachians: Geology, 16, pp. 895–899.

Reinson, G. E., 1992, Transgressive barrier island and estuarine systems, in Walker, R. G. and James, N. P., eds., Facies models: response to sea-level change: Geological Association of Canada, Geotext 1, pp. 179–194.

Revelle, R., ed., 1990, Sea-level change: National Research Council, Studies in Geophysics: National Academy Press, Washington, 234p.

Rial, J. A., 1999, Pacemaking the ice ages by frequency modulation of Earth's orbital eccentricity: Science, 285, pp. 564–568.

Rial, J. A., 2004, Earth's orbital eccentricity and the rhythm of the Pleistocene ice ages: the concealed pacemaker: Global and Planetary Change, 41, pp. 81–93.

Riba, O., 1976, Syntectonic unconformities of the Alto Cardener, Spanish Pyrenees, a genetic interpretation: Sedimentary Geology, 15, pp. 213–233.

Rich, J. L., 1951, Three critical environments of deposition and criteria for recognition of rocks deposited in each of them: Geological Society of America Bulletin, 62, pp. 1–20.

Richardson, R. M., 1992, Ridge forces, absolute plate motions, and the intraplate stress field: Journal of Geophysical Research, 97B, pp. 11739–11748.

Ricken, W., 1991, Time span assessment—an overview, in Einsele, G., Ricken, W., and Seilacher, A., eds., Cycles and events in stratigraphy: Springer-Verlag, Berlin, pp. 773–794.

Riedel, W. R., 1981, DSDP biostratigraphy in retrospect and prospect, in Warme, J. E., Douglas, R. G., and Winterer, E. L., eds., The Deep Sea Drilling Project: A decade of progress: Society of Economic Paleontologists and Mineralogists Special Publication 32, pp. 253–315.

Rine, J. M., Helmold, K. P., Bartlett, G. A., Hayes, B. J. R., Smith, D. G., Plint, A. G., Walker, R. G., and Bergman, K. M., 1987, Cardium Formation 6. Stratigraphic framework of the Cardium in subsurface: Discussions and reply: Bulletin of Canadian Petroleum Geology, 35, pp. 362–374.

Roberts, A. M., Yielding, G., and Badley, M. E., 1993, Tectonic and bathymetric controls on stratigraphic sequences within evolving half-grabens, in Williams, G. D., and Dobb, A., eds., Tectonics and seismic sequence stratigraphy: Geological Society, London, Special Publication 71, pp. 87–121.

Roberts, H. H., 1987, Modern carbonate-siliciclastic transitions: humid and arid tropical examples: Sedimentary Geology, 50, pp. 25–66.

Robertson, A. H. F., Eaton, S., Follows, E. J., and McCallum, J. E., 1991, The role of tectonics versus global sea-level change in the Neogene evolution of the Cyprus active margin, in Macdonald, D. I. M., ed., 1991, Sedimentation, tectonics and eustasy: sea-level changes at active margins: International Association of Sedimentologists Special Publication 12, pp. 331–369.

Rogers, J. J. W., 1996, A history of continents in the past three billion years: Journal of Geology, 104, pp. 91–107.

Rogers, J. J. W., and Santosh, M., 2004, Continents and supercontinents, Oxford University Press, Oxford, 304p.

Rona, P. A., 1973, Relations between rates of sediment accumulation on continental shelves, sea-floor spreading and eustacy inferred from the central North Atlantic: Geological Society of America Bulletin, 84, pp. 2851–2872.

Ronov, A. B., 1994, Phanerozoic transgressions and regressions on the continents: a quantitative approach based on areas flooded by the sea and areas of marine and continental deposition: American Journal of Sciences, 294, pp. 777–801.

Roof, S. R., Mullins, H. T., Gartner, S., Huang, T. C., Joyce, E., Prutzman, J., and Tjalmsa, L., 1991, Climatic forcing of cyclic carbonate sedimentation during the last 5.4 million years along the west Florida continental margin: Journal of Sedimentary Research, 61, pp. 1070–1088.

Ross, C. A. and Ross, J. R. P., 1988, Late Paleozoic transgressive-regressive deposition, in Wilgus, C. K., Hastings, B. S., Kendall, C. G. St. C., Posamentier, H. W., Ross, C. A., and Van Wagoner, J. C., eds., Sea-level Changes: an integrated approach: Society of Economic Paleontologists and Mineralogists Special Publication 42, pp. 227–247.

Ross, W. C., 1991, Cyclic stratigraphy, sequence stratigraphy, and stratigraphic modeling from 1964 to 1989: twenty-five years of progress?, in Franseen, E. K., Watney, W. L., and Kendall, C. G. St.C., Sedimentary modeling: computer simulations and methods for improved parameter definition: Kansas Geological Survey Bulletin 233, pp. 3–8.

Roth, P. H., and Bowdler, J. L., 1981, Middle Cretaceous calcareous nannoplankton biogeography and oceanography of the Atlantic Ocean, in Warme, J. E., Douglas, R. G., and Winterer, E. L., eds., The Deep Sea Drilling Project: A decade of progress: Society of Economic Paleontologists and Mineralogists Special Publication 32, pp. 517–546.

Rowley, D. B., and Markwick, P. J., 1992, Haq et al. eustatic sea level curve: implications for sequestered water volumes: Journal of Geology, 100, pp. 703–715.

Ruddiman, W. F., 2008, earth's climate: past and future, Second edition: W. H. Freeman and Company, New York, 388p.

Ruddiman, W. F., and Prell, W. L., 1997, Introduction to the uplift-climate connection, in Ruddiman, W. F., ed., Tectonic uplift and climate change, Plenum Press, New York, pp. 3–19.

Ruddiman, W. F., and Raymo, M. E., 1988, Northern hemisphere climatic regimes during the last 3 Ma: possible tectonic connections: Philosophical Transactions of the Royal Society, London, 318B, pp. 411–430.

Rudwick, M. J. S., 1982, Cognitive styles in geology, in Douglas, M., ed., Essays in the sociology of perception: Routledge and Kean Paul: London, pp. 219–241.

Rudwick, M., 1996, Geological travel and theoretical innovation: the role of 'Liminal' experience: Social Studies of Science, 26, pp. 143–159.

Rudwick, M., J. S., 1998, Lyell and the Principles of Geology, in Lyell: The past is the key to the present, edited by D. Blundell, and A. Scott. Geological Society of London Special Publication 143, pp. 3–15.

Runkel, A. C., Miller, J. F., McKay, R. M., Palmer, A. R., and Taylor, J. F., 2007, High-resolution sequence stratigraphy of lower Paleozoic sheet sandstones in central North America: the role of special conditions of cratonic interiors in development of stratal architecture: Geological Society of America Bulletin, 119, pp. 860–881.

Runkel, A. C., Miller, J. F., McKay, R. M., Palmer, A. R., and Taylor, J. F., 2008, The record of time in cratonic interior strata: does exceptionally slow subsidence necessarily result in exceptionally poor stratigraphic completeness? in Pratt, B. R., and Holmden, C., eds., Dynamics of epeiric seas: Geological Association of Canada Special Paper 48, pp. 341–362.

Russell, L. K., 1968, Oceanic ridges and eustatic changes in sea level: Nature, 218, pp. 861–862.

Russell, M., and Gurnis, M., 1994, The planform of epeirogeny: vertical motions of Australia during the Cretaceous: Basin Research, 6, pp. 63–76.

Rust, B. R., and Koster, E. H., 1984, Coarse alluvial deposits, in Walker, R. G., ed., Facies models, second edition: Geoscience Canada Reprint Series 1, pp. 53–69.

Ryer, T. A., 1977, Patterns of Cretaceous shallow-marine sedimentation, Coalville and Rockport areas, Utah: Geological Society of America Bulletin, 88, pp. 177–188

Ryer, T. A., 1983, Transgressive-regressive cycles and the occurrence of coal in some Upper Cretaceous strata of Utah: Geology, 11, pp. 201–210.

Ryer, T. A., 1984, Transgressive-regressive cycles and the occurrence of coal in some Upper Cretaceous strata of Utah, U. S. A., in Rahmani, R. A., and Flores, R. M., eds., Sedimentology of coal and coal-bearing sequences: International Association of Sedimentologists Special Publication 7, pp. 217–227.

Sadler, P. M., 1981, Sedimentation rates and the completeness of stratigraphic sections: Journal of Geology, 89, pp. 569–584.

Sadler, P. M., 2004, Quantitative biostratigraphy—Achieving finer resolution in global correlation: Annual Review of Earth and Planetary Sciences, 32, pp. 187–213.

Sadler, P. M., Osleger, D. A., and Montañez, I. P., 1993, On the labeling, length, and objective basis of Fischer plots: Journal of Sedimentary Petrology, 63, pp. 360–368.

Sageman, B. B., Myers, S. R., and Arthur, M. A., 2006, Orbital time scale and new C-isotope record for Cenomanian-Turonian boundary stratotype: Geology, 34, pp. 125–128.

Sageman, B. B., Rich, J., Arthur, M. A., Birchfield, G. E., and Dean, W. E., 1997, Evidence for Milankovitch periodicities in Cenomanian-Turonian lithologic and geochemical cycles, western interior, U.S.A.: Journal of Sedimentary Research, 67, pp. 286–302.

Sageman, B. Rich, J., Savrada, C. E., Bralower, T., Arthur, M. A., and Dean, W. E., 1998, Multiple Milankovitch cycles in the Bridge Creek Limestone (Cenomanian-Turonian), Western Interior Basin, in Arthur. M. A., and Dean, W. E., eds., Stratigraphy and paleoenvironments of the Cretaceous Western Interior Seaway, SEPM Concepts in Sedimentology and Paleontology, n. 6, pp. 153–171.

Sahagian, D. L., 1987, Epeirogeny and eustatic sea level changes as inferred from Cretaceous shoreline deposits: applications to the central and western United States: Journal of Geophysical Research, 92, pp. 4895–4904.

Sahagian, D., L., 1988, Epeirogenic movements of Africa as inferred from Cretaceous shoreline deposits: Tectonics, 7, pp. 125–138.

Sahagian, D. L., and Holland, S. M., 1991, Eustatic sea-level curve based on a stable frame of reference: preliminary results: Geology, 19, pp. 1209–1212.

Sahagian, D. L., Pinous, O., Olferiev, A., and Zakharov, V., 1996, Eustatic curve for the Middle Jurassic-Cretaceous based on Russian Platform and Siberian stratigraphy: zonal resolution: American Association of Petroleum Geologists Bulletin, 80, pp. 1433–1458.

Sahagian, D. L., and Watts, A. B., 1991, Introduction to the special section on measurement, causes, and consequences of long-term sea level changes: Journal of Geophysical Research, 96B, pp. 6585–6589.

Saller, A. H., Noah, J. T., Ruzuar, A. P., and Schneider, R., 2004, Linked lowstand delta to basin-floor fan deposition, off-shore Indonesia: an analog for deep-water reservoir systems: American Association of Petroleum Geologists Bulletin, 88, pp. 21–46.

Salvador, A., ed., 1994, International Stratigraphic Guide, Second edition: International Union of Geological Sciences, Trondheim, Norway, and Geological Society of America, Boulder, Colorado, 214p.

Sandberg, P. A., 1983, An oscillating trend in Phanerozoic non-skeletal carbonate mineralogy: Nature, 305, pp. 19–22.

Sanford, B. V., Thompson, F. J., and McFall, F. J., 1985, Plate tectonics – a possible controlling mechanism in the development of hydrocarbon traps in southwestern Ontario: Bulletin of Canadian Petroleum Geology, 33, pp. 52–71.

Sangree, J. B., and Widmier, J. M., 1977, Seismic stratigraphy and global changes of sea level, part 9: seismic interpretation of clastic depositional facies: American Association of Petroleum Geologists Memoir 26, pp. 165–184.

Sarg, J. F., 1988, Carbonate sequence stratigraphy, in Wilgus, C. K., Hastings, B. S., Kendall, C. G. St. C.,Posamentier, H. W., Ross, C. A., and Van Wagoner, J. C., eds., Sea level Changes – an integrated approach: Society of Economic Paleontologists and Mineralogists Special Publication 42, pp. 155–181.

Schenk, H. G., and Muller, S. W., 1941, Stratigraphic terminology: Geological Society of America Bulletin, 52, pp. 1419–1426.

Schlager, W., 1989, Drowning unconformities on carbonate platforms, in Crevello, P. D., Wilson, J. L., Sarg, J. F., and Read, J. F., eds., Controls on carbonate platforms and basin development: Society of Economic Paleontologists and Mineralogists Special Publication 44, pp. 15–25.

Schlager, W., 1991, Depositional bias and environmental change—important factors in sequence stratigraphy: Sedimentary Geology, 70, pp. 109–130.

Schlager, W., 1992, Sedimentology and sequence stratigraphy of reefs and carbonate platforms: American Association of Petroleum Geologists Continuing Education Course Notes Series 34, 71p.

Schlager, W., 1993, Accommodation and supply—a dual control on stratigraphic sequences: Sedimentary Geology, 86, pp. 111–136.

Schlager, W., 2004, Fractal nature of stratigraphic sequences, Geology, 32, pp. 185–188.

Schlager, W., 2005, Carbonate sedimentology and sequence stratigraphy: Society for Sedimentary Geology (SEPM) Concepts in Sedimentology and Paleontology #8, 200p.

Schlanger, S. O., and Jenkyns, H. C., 1976, Cretaceous oceanic anoxic events: causes and consequences: Geologie en Mijnbouw, 55, pp. 179–184.

Schlee, J. S., ed., 1984, Interregional unconformities and hydro-carbon accumulation: American Association of Petroleum Geologists Memoir 36, 184p.

Schmidt, H., and Seyfried, H., 1991, Depositional sequences and sequence boundaries in fore-arc coastal embayments: case histories from Central America, in Macdonald, D. I. M., ed., 1991, Sedimentation, tectonics and eustasy: sea-level changes at active margins: International Association of Sedimentologists Special Publication 12, pp. 241–258.

Scholle, P., 2006, An introduction and virtual field trip to the Permian reef complex, Guadalupe and Delaware Mountains, New Mexico-West Texas. http://geoinfo.nmt.edu/staff/scholle/guadalupe.html

Scholle, P., and Arthur, M. A., 1980, Carbon isotopic fluc-tuations in Cretaceous pelagic limestones: potential stratigraphic and petroleum exploration tool: American Association of Petroleum Geologists Bulletin, 64, pp. 67–87.

Schopf, T. J. M., 1974, Permo-Triassic extinctions: relation to sea-floor spreading: Journal of Geology, 82, pp. 129–143.

Schumm, S. A., 1993, River response to baselevel change: implications for sequence stratigraphy: Journal of Geology, 101, pp. 279–294.

Schwan, W., 1980, Geodynamic peaks in Alpinotype orogenies and changes in ocean-floor spreading during Late Jurassic—Late Tertiary time: American Association of Petroleum Geologists Bulletin, 64, pp. 359–373.

Sclater, J. G., and Christie, P. A. F., 1980, Continental stretching: an explanation of the post-mid-Cretaceous subsidence of the central North Sea Basin: Journal of Geophysical Research, 85, no. B7, pp. 3711–3739.

Schwarzacher, W., 1993, Cyclostratigraphy and the Milankovitch theory: Elsevier, Amsterdam, Developments in Sedimentology 52, 225p.

Schwarzacher, A., 2000, Repetition and cycles in stratigraphy: Earth Science Reviews, 50, pp. 51–75.

Sclater, J. G., Anderson, R. N., and Bell, M. L., 1971, The elevation of ridges and the evolution of the central eastern Pacific: Journal of Geophysical Research, 76, pp. 7888–7915.

Scopelli, G., Bellanca, A., Erba, E., Jenkyns, H. C., Neri, R., Tamagnini, P., Luciani, V., and Masetti, D., 2008, Cenomanian-Turonian carbonate and organiz-carbon isotope records, biostratigraphy and prevenance of a key section in NE Sicily, Italy: palaeoceanographic and palaeogeo-graphic implications: Palaeogeography, Paleoclimatology, Palaeoecology, 265, pp. 59–77.

Scott, R. W., Evetts, M. J., and Stein, J. A., 1993, Are seismic/depositional sequences time units? Testing by SHADS cores and graphic correlation: Offshore Technology Conference, Houston, Paper OTC 7110, pp. 269–276.

Scott, R. W., Frost, S. H., and Shaffer, B. L., 1988, Early Cretaceous sea-level curves, Gulf Coast and south-eastern Arabia, in Wilgus, C. K., Hastings, B. S., Kendall, C. G. St. C., Posamentier, H. W., Ross, C. A., and Van Wagoner, J. C., eds., Sea-level Changes: an inte-grated approach: Society of Economic Paleontologists and Mineralogists Special Publication 42, pp. 275–284.

Sengör, A. M. C., 1992, Unconformities in mountain belts: does the sequence stratigraphy make sense? 29th International Geological Congress, Kyoto, Japan, Abstracts, p. 294.

Seyfried, H., Astorga, A., Amann, H., Calvo, C., Kolb, W., Schmidt, H., and Winsemann, J., 1991, Anatomy of an evolving island arc: tectonic control in the south Central American fore-arc area, in Macdonald, D. I. M., ed., 1991, Sedimentation, tectonics and eustasy: sea-level changes at active margins: International Association of Sedimentologists Special Publication 12, pp. 217–240.

Shackleton, N. J., Crowhurst, S. J., Weedon G. P., and Laskar, J., 1999, Astronomical calibration of Oligocene-Miocene time, Philosophical Transactions of the Royal Society, London, Series A, 357, pp. 1907–1929.

Shanley, K. W., and McCabe, P. J., 1991, Predicting facies architecture through sequence stratigraphy—an example from the Kaiparowits Plateau, Utah: Geology, 19, pp. 742–745.

Shanley, K. W., and McCabe, P. J., 1994, Perspectives on the sequence stratigraphy of continental strata: American Association of Petroleum Geologists Bulletin, 78, pp. 544–568.

Shanmugam, G., and Moiola, R. J., 1982, Eustatic control of turbidites and winnowed turbidites: Geology, 10, pp. 231–235.

Shaub, E. J., Buffler, R. T., and Parsons, J. G., 1984, Seismic stratigraphic framework of deep central Gulf of Mexico Basin: American Association of Petroleum Geologists Bulletin, 68, pp. 1790–1802.

Shaw, A. B., 1964, Time in stratigraphy: McGraw-Hill, New York, 365p.

Shepard, F. P., and Wanless, H. R., 1935, Permo-Carboniferous coal series related to southern hemisphere glaciation: Science, 81, pp. 521–522.

Shurr, G. W., 1984, Geometry of shelf-sandstone bodies in the Shannon Sandstone of southeastern Montana, in Tillman, R. W., and Siemers, C. T., eds., Siliciclastic shelf sediments: Society of Economic Paleontologists and Mineralogists Special Publication 34, pp. 63–83.

Simmons, M. D., Berggren, W. A., Koshkarly, R. O., O'Neill, B. J. Scott, R. W., and Ziegler, W., 1997, Biostratigraphy and geochronology in the 21st century, in Paleontology in the 21st Century: International Senckenberg Conference, Frankfurt, 1977, Kleine Senckenbergreihe 25, pp. 87–97.

Simpson, G. H. D., 2006, Modelling interactions between fold-thrust belt deformation, foreland flexure and surface mass transport: Basin Research, 18, pp. 125–143.

Sinclair, H. D., Coakley, B. J., Allen, P. A., and Watts, A. B., 1991, Simulation of foreland basin stratigraphy using a diffusion model of mountain belt uplift and erosion: an example from the central Alps , Switzerland: Tectonics, 10, pp. 599–620.

Sinclair, I. K., Shannon, P. M., Williams, B. P. J., Harker, S. D., and Mooren, J. G., 1994, Tectonic control on sedimentary evolution of three North Atlantic borderland Mesozoic basins: Basin Research, 6, pp. 193–217.

Sleep, N. H., 1971, Thermal effects of the formation of Atlantic continental margins by continental breakup: Geophysical Journal of the Royal Astronomical Society, 24, pp. 325–350.

Sleep, N. H., 1976, Platform subsidence mechanisms and "eustatic" sea level changes: Tectonophysics, 36, pp. 45–56.

Sloan, R. J., and Williams, B. P. J., 1991, Volcano-tectonic control of offshore to tidal-flat regressive cycles from the Dunquin Group (Silurian) of southwest Ireland, in Macdonald, D. I. M., ed., 1991, Sedimentation, tectonics and eustasy: sea-level changes at active margins: International Association of Sedimentologists Special Publication 12, pp. 105–119.

Sloss, L. L., 1962, Stratigraphic models in exploration: American Association of Petroleum Geologists Bulletin, 46, pp. 1050–1057.

Sloss, L. L., 1963, Sequences in the cratonic interior of North America: Geological Society of America Bulletin, 74, pp. 93–113.

Sloss, L. L., 1972, Synchrony of Phanerozoic sedimentary-tectonic events of the North American craton and the Russian platform: 24th International Geological Congress, Montreal, Section 6, pp. 24–32.

Sloss, L. L., 1979, Global sea level changes: a view from the craton, in Watkins, J. S., Montadert, L., and Dickerson, P. W., eds., Geological and geophysical investigations of continental margins: American Association of Petroleum Geologists Memoir 29, pp. 461–468.

Sloss, L. L., 1982, The Midcontinent Province: United States, in Palmer, A. R., ed., Perspectives in regional geological syntheses: Geological Society of America, Decade of North American Geology Special Publication 1, pp. 27–39.

Sloss, L. L., 1984, Comparative anatomy of cratonic unconformities, in Schlee, J. S., ed., Interregional unconformities and hydrocarbon accumulation: American Association of Petroleum Geologists Memoir 36, pp. 1–6.

Sloss, L. L., 1988a, Forty years of sequence stratigraphy: Geological Society of America Bulletin, 100, pp. 1661–1665.

Sloss, L. L., 1988b, Tectonic evolution of the craton in Phanerozoic time, in Sloss, L. L., ed., Sedimentary cover—North American Craton: U.S.: The Geology of North America, Boulder, Colorado, Geological Society of America, D-2, pp. 25–51.

Sloss, L. L., 1991, The tectonic factor in sea level change: a countervailing view: Journal of Geophysical Research, 96B, pp. 6609–6617.

Sloss, L. L., Krumbein, W. C., and Dapples, E. C., 1949, Integrated facies analysis; in Longwell, C. R., ed., Sedimentary facies in geologic history: Geological Society of America Memoir 39, pp. 91–124.

Sloss, L. L., and Speed, R. C., 1974, Relationships of cratonic and continental margin episodes, in Dickinson, W. R., ed., Tectonics and sedimentation, Society of Economic Paleontologists and Mineralogists Special Publication 22, pp. 98–119.

Smith, A. G., 1993, Methods for improving the chronometric time scale, in Hailwood, E. A., and Kidd, R. B., eds., High resolution stratigraphy: Geological Society, London, Special Publication 70, pp. 9–25.

Smith, G. A., 1994, Climatic influences on continental deposition during late-stage filling of an extensional basin, southeastern Arizona: Geological Society of America Bulletin, 106, pp. 1212–1228.

Smith, W., 1815, A memoir to the map and delineation of the strata of England and Wales, with part of Scotland: John Carey, London, 51p.

Soares, P. C., Landim, P. M. B., and Fulfaro, V. J., 1978, Tectonic cycles and sedimentary sequences in the Brazilian intracratonic basins: Geological Society of America Bulletin, 89, pp. 181–191.

Sonnenfeld, M. D., and Cross, T. A., 1993, Volumetric partitioning and facies differentiation within the Permian Upper Sand Andres Formation of Last Change Canyon, Guadalupe Mountains, New Mexico, in Loucks, R. G., and Sarg, J. F., eds., Carbonate sequence stratigraphy, American Association of Petroleum Geologists memoir 57, pp. 435–474.

Southgate, P. N., Kennard, J. M., Jackson, M. J., O'Brien, P. E., and Sexton, M. J., 1993, Reciprocal lowstand clastic and highstand carbonate sedimentation, subsurface Devonian reef complex, Canning Basin, Western Australia, in Loucks, R. G., and Sarg, J. F., eds., Carbonate sequence stratigraphy: American Association of Petroleum Geologists Memoir 57, pp. 157–179.

Srinivasan, M. S., and Kennett, J. P., 1981, A review of planktonic foraminiferal biostratigraphy: applications in the equatorial and South Pacific, in Warme, J. E., Douglas, R. G., and Winterer, E. L., eds., The Deep Sea Drilling Project: A decade of progress: Society of Economic Paleontologists and Mineralogists Special Publication 32, pp. 395–432.

Srivastava, S. P., and Tapscott, C. R., 1986, Plate kinematics of the North Atlantic, in Vogt, P. R., and Tucholke, B. E., eds., The Western North Atlantic region, The Geology of North America: Boulder, Colorado, Geological Society of America, M, pp. 379–404.

Steckler, M. S., and Watts, A. B., 1978, Subsidence of the Atlantic-type continental margin off New York: Earth and Planetary Science Letters, 41, pp. 1–13.

Steel, R. J., Felt, V. L., Johannessen, E. P., and Mathieu, C., eds., 1995, Sequence stratigraphy on the Northwest European Margin, Norwegian Petroleum Society Special Publication 5, Elsevier, Amsterdam, 608p.

Stevenson, G. M., and Baars, D. L., 1986, The Paradox: a pull-apart basin of Pennsylvanian age, in Peterson, J. A., ed., Paleotectonics and sedimentation in the Rocky Mountain region, United States: American Association of Petroleum Geologists Memoir 41, pp. 513–539.

Stewart, J. A., 1986, Drifting continents and colliding interests: a quantitative application of the interests perspective: Social Studies of Science, 16, pp. 261–279.

Stewart, J. H., 1972, Initial deposits in the Cordilleran geosyncline: evidence of a Late Precambrian (<850 m.y.) continental separation: Geological Society of America Bulletin, 83, pp. 1345–1360.

Stewart, J. H., and Suczek, C. A., 1977, Cambrian and latest Precambrian paleogeography and tectonics in the western United States, in Stewart, J. H., Stevens, C. H., and Fritsche, A. E., eds., Paleozoic paleogeography of the western United States: Society of Economic Paleontologists and Mineralogists, Pacific Section, Pacific Coast Paleogeography Symposium 1, pp. 1–17.

Stille, H., 1924, Grudfragen der vergleichenden Tektonik: Bornstraeger, Berlin, 443p.

Stoakes, F. A., 1988, Evolution of the Upper Devonian of Western Canada, in, Bloy, G. R., and Charest, M., eds., Principles and Concepts for the Exploration of Reefs in the Western Canada Basin, Short Course Notes, Canadian Society of Petroleum Geologists, Calgary.

Stockmal, G. S., Cant, D. J., and Bell, J. S., 1992, Relationship of the stratigraphy of the Western Canada foreland basin to Cordilleran tectonics: insights from geodynamic models, in Macqueen, R. W., and Leckie, D. A., eds., Foreland basin and fold belts: American Association of Petroleum Geologists Memoir 55, pp. 107–124.

Stoll, H. M., and Schrag, D. P., 1996, Evidence for glacial control of rapid sea level changes in the Early Cretaceous: Science, 272, pp. 1771–1774.

Stoll, H. M., and Schrag, D. P., 2000, High-resolution stable isotope records from the Upper Cretaceous rocks of Italy and Spain: glacial episodes in a greenhouse planet? Geological Society of America Bulletin, 112, pp. 308–319.

Swart, P. K., and Eberli, G. P., 2005, The nature of the $\delta^{13}C$ of periplatform sediments: Implications for stratigraphy and the global carbon cycle: Sedimentary Geology, 175, pp. 115–129.

Switzer, S. B., Holland, W. G., Christie, D. S., Graf, G. C., Hedinger, A. S., McAuley, R. J., Wierzbicki, R. A., and Packard, J. J., 1994, Devonian Woodbend-Winterburn strata of the Western Canada Sedimentary Basin, in Mossop, G. D., and Shetsen, I., compilers, Geological Atlas of the Western Canada Sedimentary Basin: Canadian Society of Petroleum Geologists, pp. 165–202.

Suess, E., 1885–1909, Das antlitz der erde: Vienna: F. Tempsky.

Suess, E., 1906, The face of the Earth, translated by Sollas, W. J., Clarendon Press, Oxford, vol. 2, 556p.

Summerhayes, C. P., 1986, Sea level curves based on seismic stratigraphy: their chronostratigraphic significance: Palaeogeography, Palaeoclimatology, Palaeoecology, 57, pp. 27–42.

Surlyk, F., 1990, A Jurassic sea-level curve for East Greenland: Palaeogeography, Palaeoclimatology, Palaeoecology, 78, pp. 71–85.

Surlyk, F., 1991, Sequence stratigraphy of the Jurassic-lowermost Cretaceous of east Greenland: American Association of Petroleum Geologists Bulletin, 75, pp. 1468–1488.

Suter, J. R., Berryhill, H. L., Jr., and Penland, S., 1987, Late Quaternary sea-level fluctuations and depositional sequences, southwest Louisiana continental shelf, in Nummedal, D., Pilkey, O. H., and Howard, J. D., eds., 1987, Sea-level fluctuation and coastal evolution: Society of Economic Paleontologists and Mineralogists Special Publication 41, pp. 199–219.

Swart, P. K., and Eberli, G. P., 2005, The nature of the δ13C of periplatform sediments: Implications for stratigraphy and the global carbon cycle: Sedimentary Geology, 175, pp. 115–129.

Swift, D. J. P., Hudelson, P. M., Brenner, R. L., and Thompson, P., 1987, Shelf construction in a foreland basin: storm beds, shelf sandbodies, and shelf-slope depositional sequences in the Upper Cretaceous Mesaverde Group, Book Cliffs, Utah: Sedimentology, 34, pp. 423–457.

Swift, D. J. P., Oertel, G. F., Tillman, R. W., and Thorne, J. A., eds., 1991a, Shelf sand and sandstone bodies: International Association of Sedimentologists Special Publication 14, 532p.

Swift, D. J. P., and Thorne, J. A., 1991, Sedimentation on continental margins, I: a general model for shelf sedimentation, in Swift, D. J. P., Oertel, G. F., Tillman, R. W., and Thorne, J. A., eds., Shelf sand and sandstone bodies: geometry, facies and sequence stratigraphy: International Association of Sedimentologists Special Publication 14, pp. 3–31.

Tandon, S. K., and Gibling, M. R., 1994, Calcrete and coal in late Carboniferous cyclothems of Nova Scotia, Canada: climate and sea-level changes linked: Geology, 22, pp. 755–758.

Tankard, A. J., 1986, On the depositional response to thrusting and lithospheric flexure: examples from the Appalachian and Rocky Mountain basins, in Allen, P. A., and Homewood, P., eds., Foreland basins: International Association of Sedimentologists Special Publication 8, pp. 369–392.

Tankard, A. J., and Welsink, H. J., 1987, Extensional tectonics and stratigraphy of Hibernia oil field, Grand Banks, Newfoundland: American Association of Petroleum Geologists Bulletin, 71, pp. 1210–1232.

Tankard, A. J., Welsink, H. J., and Jenkins, W. A. M., 1989, Structural styles and stratigraphy of the Jeanne d'Arc Basin, Grand Banks of Newfoundland, in Tankard, A. J., and Balkwill, H. R., eds., Extensional tectonics and stratigraphy of the North Atlantic margins: American Association of Petroleum Geologists Memoir 46, pp. 265–282.

Tcherepanov, E. N., Droxler, A. W., Lapointe, P., and Mohn, K., 2008, Carbonate seismic stratigraphy of the Gulf of Papua mixed depositional system: Neogene stratigraphic signature and eustatic control: Basin Research, 20, pp. 185–209.

Teichert, C., 1958, Some biostratigraphical concepts: Geological Society of America Bulletin, 69, pp. 99–120.

Thierstein, H. R., 1981, Late Cretaceous nannoplankton and the change at the Cretaceous-Tertiary boundary, in Warme, J. E., Douglas, R. G., and Winterer, E. L., eds., The Deep Sea Drilling Project: A decade of progress: Society of Economic Paleontologists and Mineralogists Special Publication 32, pp. 355–394.

Thomas, M. A., and Anderson, J. B., 1994, Sea-level controls on the facies architecture of the Trinity/Sabine incised-valley system, Texas continental shelf, in Dalrymple, R. W., Boyd, R., and Zaitlin, B. A., eds., Incised-valley systems: origin and sedimentary sequences: SEPM (Society for Sedimentary Geology) Special Publication 51, pp. 63–82.

Thomas, W. I., 1931, The unadjusted girl: Little Brown and Company: Boston.

Thorne, J. A., 1992, An analysis of the implicit assumptions of the methodology of seismic sequence stratigraphy, in Watkins, J. S., Zhiqiang, F., and McMillen, K. J., eds., Geology and geophysics of continental margins: American Association of Petroleum Geologists Memoir 53, pp. 375–396.

Törnqvist, T. E., 1993, Holocene alternation of meandering and anastomosing fluvial systems in the Rhine-Meuse delta (central Netherlands) controlled by sea-level rise and subsoil erodibility: Journal of Sedimentary Petrology, 63, pp. 683–693.

Törnqvist, T. E., van Ree, M. H. M., and Faessen, E. L. J. H., 1993, Longitudinal facies architectural changes of a Middle Holocene anastomosing distributary system (Rhine-Meuse delta, central Netherlands): Sedimentary Geology, 85, pp. 203–219.

Torrens, H. S., 2001, Timeless order: William Smith (1769–1839) and the search for raw materials 1800–1820, in Lewis, C. L. E., and Knell, S. J., The age of the Earth: from 4004 BC to AD 2002: Geological Society of London Special Publication 190, pp. 61–83.

Toksöz, M. N., and Bird, P., 1977, Formation and evolution of marginal basins and continental plateaus, in Talwani, M., and Pitman, W. C., III, eds., Island arcs, deep-sea trenches and back-arc basins: Maurice Ewing Series 1, American Geophysical Union, pp. 379–393.

Torrens, H. S., 2002, Some personal thoughts on stratigraphic precision in the twentieth century, in Oldroyd, D. R., ed., The Earth inside and out: some major contributions to geology in the twentieth century: Geological Society of London Special Publication 192, pp. 251–272.

Trettin, H. P., ed., 1991, Geology of the Innuitian orogen and Arctic Platform of Canada and Greenland: Ottawa: Geological Survey of Canada, Geology of Canada, 3,569p.

Tsikos, H. Jenkyns, H. C., Walsworth-Bell, B., Petrizzo, M. R., Forster, A., Kolonic, S., Erba, E., Premoli-Silva, I., Baas, M., Wagner, T., and Sinninghe Damste, J. S., 2004, Carbon-isotope stratigraphy recorded by the Cenomanian-Turonian Oceanic Anoxic Event: correlation and implications based on three key localities: Journal of the Geological Society, London, 161, pp. 711–719.

Turner, G. M., Kamp, P. J. J., McIntyre, A. P., Hayton, S., McGuire, D. M., and Wilson, G. S., 2005, A coherent middle-Pliocene magnetostratigraphy, Wanganui Basin, New Zealand: Journal of the Royal Society of New Zealand, 35, pp. 197–228.

Uchupi, E., and Emery, K. O., 1991, Pangaean divergent margins: historical perspective: Marine Geology, 102, pp. 1–28.

Udden, J. A., 1912, Geology and mineral resources of the Peoria Quadrangle: U. S. Geological Survey Bulletin 506, 103p.

Ulrich, E. O., 1911, Revision of the Paleozoic systems: Geological Society of America Bulletin, 22, pp. 281–680.

Umbgrove, J. H. F., 1947, The pulse of the earth: Nijhoff: The Hague, 358p.

Umbgrove, J. H. F., 1950, Symphony of the earth: Nijhoff: The Hague, 220p,

Underhill, J. R., 1991, Controls on Late Jurassic seismic sequences, Inner Moray Firth, UK North Sea: A critical test of a key segment of Exxon's original global cycle chart: Basin Research, 3, p. 79–98.

Underhill, J R., and Partington, M. A., 1993a, Jurassic thermal doming and deflation in the North Sea: implications of the sequence stratigraphy evidence, in Parker, J. R., ed., Petroleum geology of northwest Europe: Proceedings of the 4th Conference, Bath, Geological Society, London, vol. 1, pp. 337–346.

Underhill, J R., and Partington, M. A., 1993b, Use of genetic sequence stratigraphy in defining and determining a regional tectonic control on the "Mid-Cimmerian unconformity"— implications for North Sea basin development and the global sea level chart, in Weimer, P., and Posamentier, H. W., eds., Siliciclastic sequence stratigraphy: American Association of Petroleum Geologists Memoir 58, pp. 449–484.

Vai, G. B., 2001, GSSP, IUGS and IGC: an endless story toward a common language in the Earth Sciences: Episodes, 24, pp. 29–31.

Vai, G. B., 2007, A history of chronostratigraphy: Stratigraphy, 4, pp. 83–97.

Vail, P. R., 1975, Eustatic cycles from seismic data for global stratigraphic analysis (abstract): American Association of Petroleum Geologists Bulletin, 59, pp. 2198–2199.

Vail, P. R., 1987, Seismic stratigraphy interpretation using sequence stratigraphy, Part 1: seismic stratigraphy interpretation procedure, in Bally, A. W., ed., Atlas of seismic stratigraphy: American Association of Petroleum Geologists Studies in Geology 27, 1, pp. 1–10.

Vail, P. R., 1992, The evolution of seismic stratigraphy and the global sea-level curve, in Dott, R. H., Jr., ed., Eustasy: the historical ups and downs of a major geological concept: Geological Society of America Memoir 180, pp. 83–91.

Vail, P. R., Audemard, F., Bowman, S. A., Eisner, P. N., and Perez-Crus, C., 1991, The stratigraphic signatures of tectonics, eustasy and sedimentology—an overview, in Einsele, G., Ricken, W., and Seilacher, A., eds., Cycles and events in stratigraphy: Springer-Verlag, Berlin, pp. 617–659.

Vail, P. R., Hardenbol, J., and Todd, R. G., 1984, Jurassic unconformities, chronostratigraphy and sea-level changes from seismic stratigraphy and biostratigraphy, in Schlee, J. S., ed., Interregional unconformities and hydrocarbon exploration: American Association of Petroleum Geologists Memoir 36, pp. 129–144.

Vail, P. R., Mitchum, R. M., Jr., Todd, R. G., Widmier, J. M., Thompson, S., III, Sangree, J. B., Bubb, J. N., and Hatlelid, W. G., 1977, Seismic stratigraphy and global changes of sea-level, in Payton, C. E., ed., Seismic stratigraphy – applications to hydrocarbon exploration: American Association of Petroleum Geologists Memoir 26, pp. 49–212.

Vail, P. R., and Todd, R. G., 1981, Northern North Sea Jurassic unconformities, chronostratigraphy and sea-level changes from seismic stratigraphy, in Illing, L. V., and Hobson, G. D., eds., Petroleum Geology of the continental shelf of northwest Europe: Institute of Petroleum, London, pp. 216–235.

Vail, P. F., and Wilbur, P. O., 1966, Onlap, key to worldwide unconformities and depositional cycles (abstract): American Association of Petroleum Geologists Bulletin, 50, p. 638.

Vakarcs, G., Hardenbol, J., Abreu, V. A., Várnai, P., and Tari, G., 1998, Oligocene-Miocene depositional sequences of the Central Paratethys and their correlation with regional stages, in Graciansky, P.-C. de, Hardenbol, J., Jacquin, T., and Vail, P. R., eds., Mesozoic and Cenozoic sequence stratigraphy of European basins, Society for Sedimentary Geology (SEPM) Special Publication 60, pp. 209–232.

Vakarelov, B. K., Bhattacharya, J. P., and Nebrigic, D. D., 2006, Importance of high-frequency tectonic sequences during greenhouse times of earth history: Geology, 34, pp. 797–800.

Valentine, J. W., and Moores, E., 1970, Plate tectonic regulation of faunal diversity and sea level: Nature, 228, pp. 657–669.

Valentine, J. W., and Moores, E., 1972, Global tectonics and the fossil record: Journal of Geology, 80, pp. 167–184.

Van Couvering, J. A., Castradori, D., Cita, M. B., Hilgen, F. J., and Rio, D., 2000, The base of the Zanclean Stage and of the Pliocene Series: Episodes, 23, pp. 179–187.

Van der Zwan, C. J., 2002, The impact of Milankovitch-scale climatic forcing on sediment supply: Sedimentary Geology, 147, pp. 271–294.

Vandenberghe, J., 1993, Changing fluvial processes under changing periglacial conditions: Z. Geomorph, N.F., 88, pp. 17–28.

Vandenberghe, J., Kasse, C., Bohnke, S., and Kozarski, S., 1994, Climate-related river activity at the Weichselian-Holocene transition: a comparative study of the Warta and Maas rivers: Terra Nova, 6, pp. 476–485.

Van Hinte, J. E., 1976a, A Jurassic time scale: American Association of Petroleum Geologists Bulletin, 60, pp. 489–497.

Van Hinte, J. E., 1976b, A Cretaceous time scale: American Association of Petroleum Geologists Bulletin, 60, pp. 498–516.

Van Houten, F. B., 1964, Cyclic lacustrine sedimentation, Upper Triassic Lockatong Formation, central New Jersey and adjacent Pennsylvania, in Merriam, D. F., ed., Symposium on cyclic sedimentation: Geological Survey of Kansas Bulletin 169, pp. 495–531.

Van Houten, F.B., 1981, The odyssey of molasse, in Miall, A. D., ed., Sedimentation and tectonics in alluvial basins: Geological Association of Canada Special Paper 23, pp. 35–48.

Van Siclen, D. C., 1958, Depositional topography—examples and theory: American Association of Petroleum Geologists Bulletin, 42, pp. 1897–1913.

Van Tassell, J., 1994, Evidence for orbitally-driven sedimentary cycles in the Devonian Catskill Delta complex, in Dennison, J. M., and Ettensohn, F. R., eds., Tectonic and eustatic controls on sedimentary cycles: Society for Sedimentary Geology, Concepts in Sedimentology and Paleontology, 4, pp. 121–131.1

van Veen, P. M., and Simonsen, B. T., 1991, Comment on "glacial-eustatic sea-level curve for early late Pennsylvanian sequence in north-central Texas and biostratigraphic correlation with curve for midcontinent North America": Geology, 19, pp. 91–92.

Van Wagoner, J. C., and Bertram, G. T., eds., 1995, Sequence stratigraphy of foreland basin deposits: American Association of Petroleum Geologists Memoir 64, 487 p.

Van Wagoner, J. C., Mitchum, R. M., Campion, K. M. and Rahmanian, V. D. 1990, Siliciclastic sequence stratigraphy in well logs, cores, and outcrops: American Association of Petroleum Geologists Methods in Exploration Series 7, 55p.

Van Wagoner, J. C., Nummedal, D., Jones, C. R., Taylor, D. R., Jennette, D. C., and Riley, G. W., 1991, Sequence stratigraphy applications to shelf sandstone reservoirs: American Association of Petroleum Geologists Field Conference Guidebook.

Van Wagoner, J. C., Mitchum, R. M., Jr., Posamentier, H. W., and Vail, P. R., 1987, Seismic stratigraphy interpretation using sequence stratigraphy, Part 2: key definitions of sequence stratigraphy, in Bally, A. W., ed., Atlas of seismic stratigraphy: American Association of Petroleum Geologists Studies in Geology 27, 1, pp. 11–14.

Varban, B. L., and Plint, A., 2008, Sequence stacking patterns in the Western Canada foredeep: influence of tectonics, sediment loading and eustasy on deposition of the Upper Cretaceous Kaskapau and Cardium formations: Sedimentology, 55, pp. 395–421.

Vecsei, A., Sanders, D. G. K., Bernoulli, D., Eberli, G. P., and Pignatti, J. S., 1998, Cretaceous to Miocene sequence stratigraphy and evolution of the Maiella carbonate platform margin, Italy, in Graciansky, P.-C. de, Hardenbol, J., Jacquin, T., and Vail, P. R., eds., Mesozoic and Cenozoic sequence stratigraphy of European basins, Society for Sedimentary Geology (SEPM) Special Publication 60, pp. 53–74.

Veeken, P. C. H., 2007, Seismic stratigraphy, basin analysis and reservoir characterization: Elsevier, Amsterdam, Seismic Exploration, vol. 37, 509p.

Veevers, J. J., 1990, Tectonic-climatic supercycle in the billion-year plate-tectonic eon: Permian Pangean icehouse alternates with Cretaceous dispersed-continents greenhouse: Sedimentary Geology, 68, pp. 1–16.

Veevers, J. J., and Powell, C. McA., 1987, Late Paleozoic glacial episodes in Gondwanaland reflected in transgressive-regressive depositional sequences in Euramerica: Geological Society of America Bulletin, 98, pp. 475–487.

Veizer, J., Ala, D., Azmy, K., Bruckschen, P., Buhl, D., Bruhn, F. et.al., 1999, $^{87}Sr/^{86}Sr$, $\delta^{13}C$, $\delta^{18}O$ evolution of Phanerozoic seawater Chemical Geology, 161, pp. 59–88.

Villien, A., and Kligfield, R. M., 1986, Thrusting and synorogenic sedimentation in central Utah, in Peterson, J. A., ed., Paleotectonics and sedimentation in the Rocky Mountain region, United States: American Association of Petroleum Geologists Memoir 41, pp. 281–307.

Vinogradov, A. P., and Nalivkin, V. D., eds., 1960, Atlas of lithopaleogeographical maps of the Russian Platform and its geosynclinal framing, Part 1—Late Precambrian and Paleozoic: Academy of Science, Moscow.

Vinogradov, A. P., Ronov, A. B., and Khain, V. E., eds., 1961, Atlas of lithopaleogeographical maps of the Russian Platform and its geosynclinal framing, Part II—Mesozoic and Cenozoic: Academy of Science, U.S.S.R, Moscow.

Vogt, W. P., 1999, Dictionary of statistics and methodology, Second edition: Sage Publications, Thousand Oaks, 318p.

Wagner, H. C., 1964, Pennsylvanian megacyclothems of Wilson County, Kansas, and speculations concerning their depositional environments: Kansas Geological Survey Bulletin 169, pp. 565–591.

Walker, J. C. G., and Zahnle, K. J., 1986, Lunar nodal tide and distance to the Moon during the Precambrian: Nature, 320, pp. 600–602.

Walker, R. G., 1973, Mopping up the turbidite mess, in Ginsburg, R. N., ed., Evolving concepts in sedimentology: Johns Hopkins University Press, Baltimore, pp. 1–37.

Walker, R. G., 1992, Facies, facies models and modern stratigraphic concepts, in Walker, R. G. and James, N. P., eds., Facies models: response to sea-level change: Geological Association of Canada, pp. 1–14.

Walker, R. G., and Eyles, C. H., 1988, Geometry and facies of stacked shallow-marine sandier upward sequences dissected by an erosion surface, Cardium Formation, Willesden Green, Alberta: American Association of Petroleum Geologists Bulletin, 72, pp. 1469–1494.

Walker, R. G. and James, N. P., eds., 1992, Facies models: response to sea-level change: Geological Association of Canada, 409p.

Wallace, W. L., 1969, Sociological theory: An introduction: Aldine de Gruyter: New York.

Wanless, H. R., 1950, Late Paleozoic cycles of sedimentation in the United States: 18th International Geological Congress, Algiers, pt. 4, pp. 17–28.

Wanless, H. R., 1964, Local and regional factors in Pennsylvanian cyclic sedimentation, in Merriam, D. F., ed., Symposium on cyclic sedimentation: Kansas Geological Survey Bulletin 169, pp. 593–606.

Wanless, H. R., 1972, Eustatic shifts in sea level during the deposition of Late Paleozoic sediments in the central United States, in Elam, J. G., and Chuber, S., eds., Cyclic sedimentation in the Permian Basin: West Texas Geological Society Symposium, pp. 41–54.

Wanless, H. R., 1991, Observational foundation for sequence modeling, in Franseen, E. K., Watney, W. L., and Kendall, C. G. St.C., eds., Sedimentary modeling: computer simulations and methods for improved parameter definition: Kansas Geological Survey Bulletin 233, pp. 43–62.

Wanless, H. R., and Shepard, E. P., 1936, Sea level and climatic changes related to Late Paleozoic cycles: Geological Society of America Bulletin, 47, pp. 1177–1206.

Wanless, H. R., and Weller, J. M., 1932, Correlation and extent of Pennsylvanian cyclothems: Geological Society of America Bulletin, 43, pp. 1003–1016.

Waschbusch, P. J., and Royden, L. H., 1992, Episodicity in foredeep basins: Geology, 20, pp. 915–918.

Washington, P. A., and Chisick, S. A., 1994, Foundering of the Cambro-Ordovician shelf margin: onset of Taconian orogenesis or eustatic drowning, in Dennison, J. M., and Ettensohn, F. R., eds., Tectonic and eustatic controls on sedimentary cycles: Society for Sedimentary Geology, Concepts in Sedimentology and Paleontology, 4, pp. 203–216.

Watson, R. A., 1983, A critique [of] chronostratigraphy: American Journal of Science, 283, pp. 173–177.

Watts, A. B., 1981, The U. S. Atlantic margin: subsidence history, crustal structure and thermal evolution: American Association of Petroleum Geologists, Education Course Notes Series #19, Chap. 2, 75p.

Watts, A. B., 1989, Lithospheric flexure due to prograding sediment loads: implications for the origin of offlap/onlap patterns in sedimentary basins: Basin Research, 2, pp. 133–144.

Watts, A. B., Karner, G. D., and Steckler, M. S., 1982, Lithospheric flexure and the evolution of sedimentary basins, in Kent, P., Bott, M. H. P., McKenzie, D. P., and Williams, C. A., eds., The evolution of sedimentary basins: Philosophical Transactions of the Royal Society, London, A305, pp. 249–281.

Watts, A. B., and Ryan, W. B. F., 1976, Flexure of the lithosphere and continental margin basins: Tectonophysics, 36, pp. 24–44.

Watts, A. B., and Thorne, J., 1984, Tectonics, global changes in sea level and their relationship to the stratigraphical sequences at the US Atlantic continental margin: Marine and Petroleum Geology, 1, pp. 319–321.

Weber, M. E., Wiedicke, M. H., Kudrass, H. R., Huebscher, C., and Erlenkeuser, H., 1997, Active growth of the Bengal Fan during sea-level rise and highstand: Geology, 25, pp. 315–318.

Weedon, G. P., 1986, Hemipelagic shelf sedimentation and climatic cycles: the basal Jurassic (Blue Lias) of South Britain: Earth and Planetary Science Letters, 76, pp. 321–335.

Weedon, G. P., 1993, The recognition and stratigraphic implications of orbital-forcing of climate and sedimentary cycles: Sedimentology Review, 1, pp. 31–50.

Weedon, G., 2003, Time series analysis and cyclostratigraphy, Cambridge University Press, Cambridge, 259p.

Weimer, P., 1990, Sequence stratigraphy, facies geometries, and depositional history of the Mississippi fan, Gulf of Mexico: American Association of Petroleum Geologists Bulletin, 74, pp. 425–453.

Weimer, P., and Posamentier, H. W., eds., 1993, Siliciclastic sequence stratigraphy: American Association of Petroleum Geologists Memoir 58, 492p.

Weimer, R. J., 1960, Upper Cretaceous stratigraphy, Rocky Mountain area: American Association of Petroleum Geologists Bulletin, 44, pp. 1–20.

Weimer, R. J., 1970, Rates of deltaic sedimentation and intra-basin deformation, Upper Cretaceous of Rocky Mountain region, in Morgan, J. P., ed., Deltaic sedimentation modern and ancient: Society of Economic Paleontologists and Mineralogists Special Publication 15, pp. 270–292.

Weimer, R. J., 1986, Relationship of unconformities, tectonics, and sea level change in the Cretaceous of the Western Interior, United States, in Peterson, J. A., ed., Paleotectonics and sedimentation in the Rocky Mountain region, United States: American Association of Petroleum Geologists Memoir 41, pp. 397–422.

Weller, J. M., 1930, Cyclical sedimentation of the Pennsylvanian Period and its significance: Journal of Geology, 38, pp. 97–135.

Welsink, H. J., Srivastava, S. P., and Tankard, A. J., 1989, Basin architecture of the Newfoundland continental margin and its relationship to ocean crust fabric during extension, in Tankard, A. J., and Balkwill, H. R., eds., Extensional tectonics and stratigraphy of the North Atlantic margins: American Association of Petroleum Geologists Memoir 46, pp. 197–213.

Welsink, H. J., and Tankard, A. J., 1988, Structural and stratigraphic framework of the Jeanne d'Arc Basin, Grand Banks, in Bally, A. W., ed., Atlas of seismic stratigraphy: American Association of Petroleum Geologists Studies in Geology 27, 2, pp. 14–21.

Weltje, G., and de Boer, P. L., 1993, Astronomically induced paleoclimatic oscillations reflected in Pliocene turbidite deposits on Corfu (Greece): implications for the interpretation of higher order cyclicity in ancient turbidite systems: Geology, 21, pp. 307–310.

Wescott, W. A., 1993, Geomorphic thresholds and complex response of fluvial systems—some implications for sequence stratigraphy: American Association of Petroleum Geologists Bulletin, 77, pp. 1208–1218.

Westphal, H., Munnecke, A., and Brandano, M., 2008, Effects of diagenesis on the astrochronological approach of defining stratigraphic boundaries in calcareous rhythmites: The Tortonian GSSP: Lethaia, 41, pp. 461–476.

Wheeler, H. E., 1958, Time-stratigraphy: American Association of Petroleum Geologists Bulletin, 42, pp. 1047–1063.

Wheeler, H. E., 1959a, Note 24—unconformity-bounded units in stratigraphy: American Association of Petroleum Geologists Bulletin, 43, pp. 1975–1977.

Wheeler, H. E., 1959b, Stratigraphic units in time and space: American Journal of Science, 257, pp. 692–706.

Wheeler, H. E., 1963, Post-Sauk and pre-Absaroka Paleozoic stratigraphic patterns in North America: American Association of Petroleum Geologists Bulletin, 47, pp. 1497–1526.

Whewell, W., 1872, The two antagonist doctrines of geology: Chapter 8, in History of the Inductive Sciences from the Earliest to the Present Time, vol. 2, Appleton-Century-Crofts, New York, pp. 586–598.

White, N., and Lovell, B., 1997, Measuring the pulse of a plume with the sedimentary record: Nature, 387, pp. 888–891.

Wignall, P. B., 1991, Ostracod and foraminifera micropaleontology and its bearing on biostratigraphy: a case study from the Kimmeridgian (Late Jurassic) of north west Europe: Palaios, 5, pp. 219–226.

Wilgus, C. K., Hastings, B. S., Kendall, C. G. St. C., Posamentier, H. W., Ross, C. A., and Van Wagoner, J. C., eds., 1988, Sea level changes – an integrated approach: Society of Economic Paleontologists and Mineralogists Special Publication 42, 407p.

Williams, D. F., 1988, Evidence for and against sea-level changes from the stable isotopic record of the Cenozoic, in Wilgus, C. K., Hastings, B. S., Kendall, C. G. St. C., Posamentier, H. W., Ross, C. A., and Van Wagoner, J. C., eds., Sea level Changes – an integrated approach: Society of Economic Paleontologists and Mineralogists Special Publication 42, pp. 31–36.

Williams, G. D., and Dobb, A., eds., 1993, Tectonics and seismic sequence stratigraphy: Geological Society, London, Special Publication 71, 226p.

Williams, G. E., 1981, Megacycles, Benchmark papers in Geology, vol. 57: Hutchinson Ross Publishing Company, Stroudsberg, Pennsylvania, 434p.

Williams, H. S., 1893, The elements of the geological time scale: Journal of Geology, 1, pp. 283–295.

Williams, H. S., 1894, Dual nomenclature in geological classification: Journal of Geology, 2, pp. 145–160.

Willis, A. J., 2000, Tectonic control of nested sequence architecture in the Sego Sandstone, Neslen Formation, and Upper Castlegate Sandstone (Upper Cretaceous), Sevier Foreland Basin, Utah, USA: Sedimentary Geology, 136, pp. 277–317.

Wilson, J. L., 1967, Cyclic and reciprocal sedimentation in Virgilian strata of southern New Mexico: Geological Society of America Bulletin, 78, pp. 805–818.

Wilson, J. L., 1975, Carbonate facies in geologic history: Springer-Verlag, New York, 471p.

Wilson, R. C. L., 1998, Sequence stratigraphy: a revolution without a cause? in Blundell, D. J., and Scott, A. C., eds., Lyell: the past is the key to the present: Geological Society of London, Special Publication 143, pp. 303–314.

Winchester, S., 2001, The map that changed the world: Williams Smith and the birth of modern geology: Harper Collins, New York, 329p.

Winsemann, J., and Seyfried, H., 1991, Response of deep-water fore-arc systems to sea-level changes, tectonic activity and volcaniclastic input in Central America, in Macdonald, D. I. M., ed., 1991, Sedimentation, tectonics and eustasy: sea-level changes at active margins: International Association of Sedimentologists Special Publication 12, pp. 273–292.

Wise, D. U., 1974, Continental margins, freeboard and the volumes of continents and oceans through time, in Burk, C. A., and Drake, C. L., eds., The geology of continental margins: Springer-Verlag, New York, pp. 45–58.

Wood, L. J., Ethridge, F. G., and Schumm, S. A., 1993, The effects of rate of base-level fluctuations on coastal plain, shelf and slope depositional systems: an experimental approach, in Posamentier, H. W., Summerhayes, C. P., Haq, B. U., and Allen, G. P., eds., Sequence stratigraphy and facies associations: International Association of Sedimentologists Special Publication 18, pp. 43–53.

Worrall, D. M., and Snelson, S., 1989, Evolution of the northern Gulf of Mexico, with emphasis on Cenozoic growth faulting and the role of salt, in Bally, A. W., and Palmer, A. R., eds., The geology of North America—an overview: Geological Society of America, The geology of North America, A, pp. 97–138.

Worsley, T. R., and Nance, R. D., 1989, Carbon redox and climate control through earth history: a speculative reconstruction: Palaeogeography, Palaeoclimatology, Palaeoecology, 75, pp. 259–282.

Worsley, T. R., Nance, D., and Moody, J. B., 1984, Global tectonics and eustasy for the past 2 billion years: Marine Geology, 58, pp. 373–400.

Worsley, T. R., Nance, D., and Moody, J. B., 1986, Tectonic cycles and the history of the earth's biogeochemical and paleoceanographic record: Paleoceanography, 1, pp. 233–263.

Worsley, T. R., Nance, R. D., and Moody, J. B., 1991, Tectonics, carbon, life, and climate for the last three billion years: a unified system? in Schneider, S. H., and Boston, P. J., eds., Scientists on Gaia: The MIT Press, Cambridge, Massachusetts, pp. 200–210.

Wortmann, U. G., Herrle, J. O., and Weissert, H., 2004, Altered carbon cycliign and couplped changes in Early Cretaceous wethering patterns: Evidence from integrated carbon isotope and sandstone records of the western Tethys: Earth and Planetary Science Letters, 220, pp. 69–82.

Wright, V. P., 1992, Speculations on the controls on cyclic peritidal carbonates: ice-house versus greenhouse eustatic controls: Sedimentary Geology, 76, pp. 1–5.

Wright, V. P., and Marriott, S. B., 1993, The sequence stratigraphy of fluvial depositional systems: the role of floodplain sediment storage: Sedimentary Geology, 86, pp. 203–210

Li, X., Jenkyns, H. C., Wang, C., Hu, X., Chen, X., Wei, Y., Huang, Y., and Cui, J., 2006, Upper Cretaceous carbon- and oxygen-isotope stratigraphy of hemipelagic carbonate facies from southern Tibet, China: Journal of the Geological Society, London, 163, pp. 375–382.

Yang, C.-S., and Nio, S.-D., 1993, Application of high-resolution sequence stratigraphy to the Upper Rotliegend in the Netherlands offshore, in Weimer, P., and Posamentier, H. W., eds., Siliciclastic sequence stratigraphy: Recent developments and applications: American Association of Petroleum Geologists Memoir 58, pp. 285–316.

Yang, C.-S., and Baumfalk, Y. A., 1994, Milankovitch cyclicity in the Upper Rotliegend Group of the Netherlands offshore, in de Boer, P. L., and Smith, D. G., eds., Orbital forcing and cyclic sequences: International Association of Sedimentologists, Special Publication 19, pp. 47–61.

Yongtai Y., and Miall, A. D., 2008, Marine transgression in the Mid-Cretaceous of the Cordilleran foreland basin reinterpreted as orogenic unloading deposits: Bulletin of Canadian Petroleum Geology, 56, pp. 179–198.

Yongtai Y., and Miall, A. D., 2009, Evolution of the northern Cordilleran foreland basin during the middle Cretaceous: Geological Society of America Bulletin, 121, pp. 483–501.

Yoshida, S., 2000, Sequence stratigraphy and facies architecture of the upper Blackhawk Formation and the Lower Castlegate Sandstone (Upper Cretaceous), Book Cliffs, Utah, USA: Sedimentary Geology, 136, pp. 239–276.

Yoshida, S., Willis, A., and Miall, A. D., 1996, Tectonic control of nested sequence architecture in the Castlegate Sandstone (Upper Cretaceous), Book Cliffs, Utah: Journal of Sedimentary Research, 66, pp. 737–748.

Yoshida, S., Steel, R. S, and Dalrymple, R. W., 2007, Changes in depositional processes – an ingredient in a new generation of sequence stratigraphic models: Journal of Sedimentary Research, 77, pp. 447–460.

Youle, J. C., Watney, W. L., and Lambert, L. L., 1994, Stratal hierarchy and sequence stratigraphy—Middle Pennsylvanian, southwestern Kansas, U.S.A., in Klein G. deV., ed., Pangea: Paleoclimate, tectonics, and sedimentation during accretion, zenith, and breakup of a supercontinent: Geological Society of America Special Paper 288, pp. 267–285.

Zalasiewicz, J., Smith, A., Brenchley, P., Evans, J., Knox, R., Riley, N., Gale, A., Gregory, F. J., Rushton, A., Gibbard, P., Hesselbo, S., Marshall, J., Oates, M., Rawson, P., and Trewin, N., 2004, Simplifying the stratigraphy of time: Geology, 32, pp. 1–4.

Zalasiewicz, J., Smith, A., Hounslow, M., Williams, M., Gale, A., Powell, J., Waters, C., Barry, T. L., Bown, P. R., Brenchley, P., Cantrill, D., Gibbard, P., Gregory, F. J., Knox, R., Marshall, J., oates, M., Rawson, P., Stone, P., and Trewin, N., 2007, The scale-dependence of strata-time relations: implications for stratigraphic classification: Stratigraphy, 4, pp. 139–144.

Zeng, H., and Hentz, T. F., 2004, High-frequency sequence stratigraphy from seismic sedimentology: applied to Micoene, Vermilion Block 50, Tiger Shoal area, offshore Louisiana: American Association of Petroleum Geologists Bulletin, 88, pp. 153–174.

Zhang, P-Z., Shen, A., Wang, M., Gan, W., Burgmann, R., Molnar, P., Wang, Q., Niu, Z., Sun, J., Wu, J., Hanrong, S., and Xinzhao, Y., 2004, Continuous deformation of the Tibetan Plateau from global positioning system data: Geology, 32, pp. 809–812.

Ziegler, A. M., 1965, Silurian marine communities and their environmental significance: Nature, 207, pp. 270–272.

Ziegler, A. M., Cocks, L. R. M., and McKerrow, W. S., 1968, The Llandovery transgression of the Welsh borderland: Paleontology, 11, pp. 736–782.

Ziegler, P. A., 1982, Geological atlas of western and central Europe: Shell Internationale Petroleum Maatschappij B.V., The Hague, The Netherlands.

Ziegler, P. A., 1988, Evolution of the Arctic-North Atlantic and the Western Tethys: American Association of Petroleum Geologists, Memoir 43, 198p.

Zeller, E. J., 1964, Cycles and psychology: Geological Survey of Kansas Bulletin, 169, pp. 631–636.

Zoback, M. L., 1992, First- and second-order patterns of stress in the lithosphere: the world stress map project: Journal of Geophysical Research, 97B, pp. 11703–11728.

Zühlke, R., Bechstädt, T., and Mundil, R., 2003, Sub-Milankovitch and Milankovitch forcing on a model Mesozoic carbonate platform—the Latemar (Middle Triassic, Italy): Terra Nova, 15, pp. 69–80.

# Author Index

# Subject Index

MIX
Papier aus verantwortungsvollen Quellen
Paper from responsible sources
FSC® C105338

If you have any concerns about our products,
you can contact us on
**ProductSafety@springernature.com**

In case Publisher is established outside the EU,
the EU authorized representative is:
**Springer Nature Customer Service Center GmbH
Europaplatz 3, 69115 Heidelberg, Germany**

Printed by Libri Plureos GmbH
in Hamburg, Germany